High-Dimensional Chaotic and Attractor Systems

International Series on
INTELLIGENT SYSTEMS, CONTROL AND AUTOMATION: SCIENCE AND ENGINEERING

VOLUME 32

Editor

Professor S. G. Tzafestas, National Technical University of Athens, Greece

High-Dimensional Chaotic and Attractor Systems

A Comprehensive Introduction

by

VLADIMIR G. IVANCEVIC
Defence Science and Technology Organisation, Adelaide, SA, Australia

and

TIJANA T. IVANCEVIC
The University of Adelaide, SA, Australia

 Springer

A C.I.P. Catalogue record for this book is available from the Library of Congress.

ISBN-10 1-4020-5455-6 (HB)
ISBN-13 978-1-4020-5455-6 (HB)
ISBN-10 1-4020-5456-4 (e-book)
ISBN-13 978-1-4020-5456-3 (e-book)

Published by Springer,
P.O. Box 17, 3300 AA Dordrecht, The Netherlands.

www.springer.com

Printed on acid-free paper

Dedicated to Nitya, Atma and Kali

Contents

Preface

If we try to describe real world in mathematical terms, we will see that *real life* is very often a *high–dimensional chaos*. Sometimes, by pushing hard', we manage to make order out of it; yet sometimes, we need simply to accept our life as it is. To be able to still live successfully, we need to *understand, predict,* and ultimately *control* this high–dimensional chaotic dynamics of life. This is the main theme of the present book. In our previous book, Geometrical Dynamics of Complex Systems, Vol. 31 in *Springer* book series *Microprocessor–Based and Intelligent Systems Engineering,* we developed the most powerful mathematical machinery to deal with high–dimensional nonlinear dynamics. In the present text, we consider the *extreme cases of nonlinear dynamics*, the high–dimensional chaotic and other attractor systems. Although they might look as examples of complete disorder – they still represent control systems, with their inputs, outputs, states, feedbacks, and stability.

Today, we can see a number of nice books devoted to nonlinear dynamics and chaos theory (see our reference list). However, *all* these books are only undergraduate, introductory texts, that are concerned *exclusively* with *oversimplified low–dimensional chaos*, thus providing only *an inspiration* for the readers to actually throw themselves into the real–life chaotic dynamics. As far as we know, there has not been published yet a single book on high–dimensional chaos. Without exception, all authors were happy to show that chaos can exist in simple dynamical systems, and that has its own value. But, at the same time, we can see chaos everywhere in our lives, only this chaos of life is not as simple as in undergraduate chaos texts. With the present text we intend to fill–up this gap.

High–Dimensional Chaotic and Attractor Systems: A Comprehensive Introduction is a graduate–level monographic textbook devoted to understanding, prediction and control of high–dimensional chaotic & attractor systems of real life. For all necessary mathematical background, we refer to our previous book, mentioned above.

The objective of the present book is to provide a serious reader with a serious scientific tool that will enable him to actually *perform* a competitive research in high–dimensional chaotic and attractor dynamics.

This book has nine Chapters. The first Chapter gives a textbook–like introduction into the low–dimensional attractors and chaos. This Chapter has an inspirational character, similar to other books on nonlinear dynamics and deterministic chaos. The second Chapter deals with Smale's topological transformations of stretching, squeezing and folding (of the system's phase–space), developed for the purpose of chaos theory. The third Chapter is devoted to the Poincaré's 3–body problem and basic techniques of chaos control, mostly of Ott–Grebogi–Yorke type. The fourth Chapter is a review of both Landau's and topological phase transition theory, as well as Haken's synergetics. The fifth Chapter deals with phase Synchronization in high–dimensional chaotic systems. The sixth Chapter presents high–tech Josephson junctions, the basic components for the future quantum computers. The seventh Chapter deals with fractals and fractional Hamiltonian dynamics. The 8th Chapter gives a review of modern techniques for dealing with turbulence, ranging from the parameter–space of the Lorenz attractor to the Lie symmetries. The last, 9th Chapter attempts to give a brief on the cutting edge techniques of the high–dimensional nonlinear dynamics (including geometries, gauges and solitons, and culminating into Cvitanovic's chaos field theory).

Note that Einstein's summation convention over repeated indices is used throughout the text (not in accordance with the chaos theory or neural networks, but in accordance with our Geometrical Dynamics book). Similarly, for the precise meaning of other frequently used symbols (only if it is really needed), see the *List of Symbols* in Geometrical Dynamics. The authors' intention has been to write the present book in a less formal style, to make it accessible to the wider audience. For its comprehensive reading the only necessary prerequisite is standard engineering mathematics (namely, calculus and linear algebra).

Adelaide,   *V. Ivancevic, Defence Science & Technology Organisation,*
August, 2006   *Australia, e-mail: Vladimir.Ivancevic@dsto.defence.gov.au*

T. Ivancevic, School of Mathematics, The University of Adelaide,
e-mail: Tijana.Ivancevic@adelaide.edu.au

Acknowledgments

The authors wish to thank Land Operations Division, Defence Science & Technology Organisation, Australia, for the support in developing the *Human Biodynamics Engine* (HBE) and all the HBE–related text in this monograph.

We also express our gratitude to *Springer* book series *Microprocessor–Based and Intelligent Systems Engineering*, and especially to the Editor, Professor Spyros Tzafestas.

1

Introduction to Attractors and Chaos

Recall that a *Newtonian deterministic system* is a system whose *present state is fully determined by its initial conditions* (at least, in principle), in contrast to a stochastic (or, random) system, for which the initial conditions determine the present state only partially, due to noise, or other external circumstances beyond our control. For a stochastic system, the present state reflects the past initial conditions plus the particular realization of the noise encountered along the way. So, in view of classical science, we have either deterministic or stochastic systems.

However, "Where chaos begins, classical science stops... After relativity and quantum mechanics, chaos has become the 20th century's third great revolution in physical sciences." [Gle87].

For a long time, scientists avoided the irregular side of nature, such as disorder in a turbulent sea, in the atmosphere, and in the fluctuation of wild–life populations [Sha06]. Later, the study of this unusual results revealed that irregularity, nonlinearity, or chaos was the organizing principle of nature [Gle87]. Thus nonlinearity, most likely in its extreme form of chaos, was found to be ubiquitous [Hil94, CD98]. For example, in theoretical physics, chaos is a type of moderated randomness that, unlike true randomness, contains complex

1

patterns that are mostly unknown [CD98]. Chaotic behavior appeared in the weather, the clustering of cars on an expressway, oil flowing in underground pipes [Gle87], convecting fluid, simple diode-circuits [Hil94], neural networks, digital filters, electronic devices, non-linear optics, lasers [CD98], and in complex systems like thrust-vectored fighter aircraft [Mos96]. No matter what the system, its behavior obeyed the same newly discovered law of nonlinearity and chaos [Gle87]. Thus, nonlinear dynamical system theory transcended the boundaries of different scientific disciplines, because it appeared to be a science of the global nature of systems [Sha06]. As a result, nonlinear dynamics found applications in physics, chemistry, meteorology, biology, medicine, physiology, psychology, fluid dynamics, engineering and various other disciplines. It has now become a common language used by scientists in various domains to study any system that obeys the same universal law [Gle87].

A modern scientific term *deterministic chaos* depicts an *irregular and unpredictable* time evolution of many (simple) deterministic dynamical systems, characterized by nonlinear coupling of its variables (see, e.g., [GOY87, YAS96, BG96, Str94]). Given an initial condition, the dynamic equation determines the dynamic process, i.e., every step in the evolution. However, the initial condition, when magnified, reveals a cluster of values within a certain error bound. For a regular dynamic system, processes issuing from the cluster are bundled together, and the bundle constitutes a predictable process with an error bound similar to that of the initial condition. In a chaotic dynamic system, processes issuing from the cluster diverge from each other exponentially, and after a while the error becomes so large that the dynamic equation losses its predictive power (see Figure 1.1).

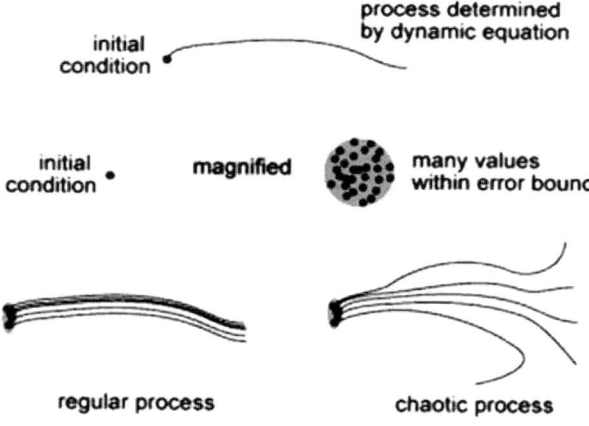

Fig. 1.1. Regular v.s. chaotic process.

For example, in a *pinball game*, any two trajectories that start out very close to each other separate exponentially with time, and in a finite (and in practice, a very small) number of bounces their separation $\delta x(t)$ attains the magnitude of L, the characteristic linear extent of the whole system. This property of sensitivity to initial conditions can be quantified as

$$|\delta x(t)| \approx \mathrm{e}^{\lambda t} |\delta x(0)|,$$

where λ, the *mean rate of separation of trajectories* of the system, is called the *Lyapunov exponent*. For any finite accuracy $|\delta x(0)| = \delta x$ of the initial data, the dynamics is predictable only up to a finite *Lyapunov time*

$$T_{Lyap} \approx -\frac{1}{\lambda} \ln |\delta x/L|,$$

despite the deterministic and infallible simple laws that rule the pinball motion.

However, a positive Lyapunov exponent does not in itself lead to chaos (see [CAM05]). One could try to play 1– or 2–disk pinball game, but it would not be much of a game; trajectories would only separate, never to meet again. What is also needed is mixing, the coming together again and again of trajectories. While locally the nearby trajectories separate, the interesting dynamics is confined to a globally finite region of the phase–space and thus the separated trajectories are necessarily folded back and can re-approach each other arbitrarily closely, infinitely many times. For the case at hand there are 2^n topologically distinct n bounce trajectories that originate from a given disk. More generally, the number of distinct trajectories with n bounces can be quantified as

$$N(n) \approx \mathrm{e}^{hn},$$

where the *topological entropy* h ($h = \ln 2$ in the case at hand) is the growth rate of the number of topologically distinct trajectories.

When a physicist says that a certain system "exhibits chaos", he means that the system obeys deterministic laws of evolution, but that the outcome is highly sensitive to small uncertainties in the specification of the initial state. The word "chaos" has in this context taken on a narrow technical meaning. If a deterministic system is locally unstable (positive Lyapunov exponent) and globally mixing (positive entropy), it is said to be *chaotic*.

While mathematically correct, the definition of chaos as "positive Lyapunov exponent + positive entropy" is useless in practice, as a measurement of these quantities is intrinsically asymptotic and beyond reach for systems observed in nature. More powerful is Poincaré's vision of chaos as the interplay of local instability (unstable periodic orbits) and global mixing (intertwining of their stable and unstable manifolds). In a chaotic system any open ball of initial conditions, no matter how small, will in finite time overlap with any other finite region and in this sense spread over the extent of the entire asymptotically accessible phase–space. Once this is grasped, the focus of theory shifts from attempting to predict individual trajectories (which is impossible) to a description of the geometry of the space of possible outcomes, and evaluation of averages over this space.

A definition of "turbulence" is even harder to come by. Intuitively, the word refers to *irregular behavior of an infinite–dimensional dynamical system described by deterministic equations of motion* – say, a bucket of boiling water – *described by the Navier–Stokes equations.* But in practice the word "turbulence" tends to refer to messy dynamics which we understand poorly. As soon as a phenomenon is understood better, it is reclaimed renamed as: "a route to chaos", or "spatio–temporal chaos", etc. (see [CAM05]).

Motivating Example: A Playground Swing

To gain the initial idea of the principles behind chaos and (strange) attractors, let us consider a common *playground swing*, which physically represents a *driven nonlinear pendulum*. Its dimensionless differential equation of motion (to be explained later in more detail) reads

$$\ddot{\theta} + \gamma\dot{\theta} + \sin\theta = F\cos(w_D t), \qquad (1.1)$$

where θ is the *pendulum angle* in radians, overdot represents the time derivative (as always), γ is the *damping parameter*, F is the *amplitude of the driving force*, while w_D is the *frequency* of that force.

Now, it is common to use the small angle approximation for this equation with $\sin\theta \sim \theta << 1$. This *linearization* allows the equation to be analytically integrated, but at the same time physically means that the pendulum will either undergo regular motion or, without a driving term, eventually stop swinging altogether. For the type of motion that we desire to study, this approximation is invalid and hence the equation can no longer be solved analytically.

Instead, in nonlinear dynamics, we would rewrite the original second–order ODE of the pendulum motion (1.1), either as a 2D *autonomous system* (suitable for numerical integration),

$$\dot{w} = -\gamma w - \sin\theta + F\cos(w_D t), \qquad \dot{\theta} = w, \qquad (1.2)$$

or, as a 3D autonomous system,

$$\dot{w} = -\gamma w - \sin\theta + F\cos\phi, \qquad \dot{\theta} = w, \qquad \dot{\phi} = w_D, \qquad (1.3)$$

where the new variable, ϕ, is called the *phase of the driver*.

The dynamical variables of the system, in this case the pendulum's angle θ and angular velocity w, are the coordinates defining the system's *phase–space*. In the 2D case (1.2), the variables can be plotted to display a *phase portrait* of the system's dynamical behavior. By varying the parameters of the equation for the nonlinear pendulum and then plotting the resulting phase portrait a wide range of behavior can be observed.

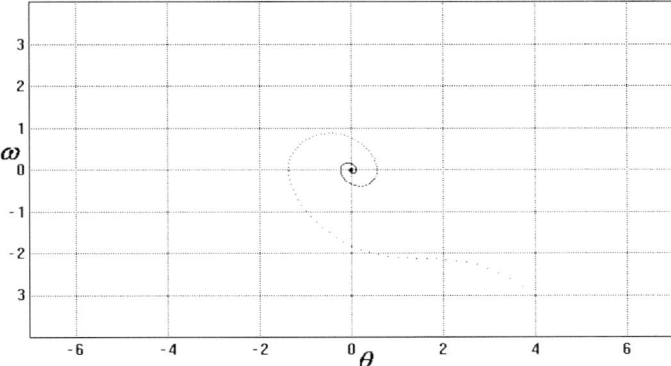

Fig. 1.2. Phase portrait of the driven nonlinear pendulum, showing the points spiralling into (0,0).

We start by setting the driver strength, F, to 0, the damping parameter γ to 0.5 and (θ, w) to $(4, -3)$. The phase portrait (Figure 1.2) shows the points spiralling into (0,0), which, in the pendulum's case, corresponds to the motionless and vertical pendulum, with the weight below the pivot. The phase portrait is then redrawn (Figure 1.3) with a different starting point at $(-3, 0)$. The points again spiral in, this time from the opposite direction. In fact, if we take any point, with one exception, the trajectory of the phase points will eventually end at (0,0). If the physics of the situation are considered then this makes perfect sense. The damping term, $\gamma\dot{\theta}$, means that the pendulum is losing energy (this would be through friction in a real system) and as such the swing is decreasing until the pendulum comes to rest. In other words, with a non–zero damping parameter γ and no driving force to replace the energy lose ($F = 0$), the pendulum is a *dissipative system*.

The point (0,0) is called an *attractor*. Roughly, *an attractor is a 'magnetic set' in the system's phase–space to which all neighboring trajectories converge*.[1] That is, an attractor is the subset of the phase–space with the following properties (see, e.g., [Str94]):

1. it is an invariant set (i.e., any trajectory that starts in it – stays in it for all time);
2. it attracts all trajectories that start sufficiently close to it;
3. it is minimal (it cannot contain one or more smaller attractors).

Ordinary (or, regular) attractors are *stable fixed–points* (which can exist both in linear and nonlinear dynamics) and *stable limit cycles* (which can exist only in nonlinear dynamics). Finally, in chaotic dynamics the most important geometrical objects are *strange attractors*, also called *chaotic attractors* or *fractal attractors* (that is, attractors with non–integer dimension), which are

---

[1] Opposite of an *attractor* is a *repeller*.

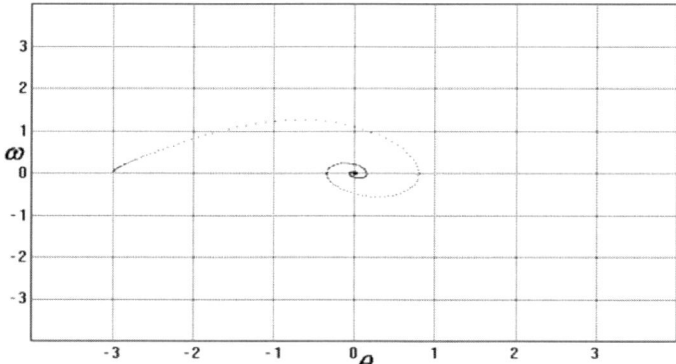

Fig. 1.3. Another phase portrait of the driven nonlinear pendulum with different initial conditions.

special attractors that exhibit sensitive dependence on initial conditions. A strange attractor typically arises when the phase–flow undergoes *stretching*, *squeezing* and *folding*. Trajectories on a strange attractor remain confined to a bounded region of phase–space, yet they separate from their neighbors exponentially fast.

Now, by approximating the nonlinear equations with linear forms and assuming that their behavior will not differ substantially around critical points it can be shown that the damped pendulum with initial conditions θ_0 and w_0 will spiral in towards the point $(2n\pi)$ closest to it, where n is an integer. A useful analogy could be to think of a series of bowls (imagine, somehow, that there are no gaps between them) with the points of $\theta = 2n\pi$ at the centers. These are *equilibrium points*, where the potential energy is the lowest. The peaks of the walls separating these 'bowls' or, *basins of attraction* are at the points $((2n+1)\pi, 0)$, representing unstable critical points called *saddle points*. At these saddle points the pendulum would be pointing vertically upwards and the merest push will send it back down again.

Suppose we now use a non–zero driving force F. We start off with a value for g that's not too strong, the intention is just to overcome the energy loss due to damping. If we plot the phase diagram we see that the attractor is no longer a single point at (0,0) but now a closed, almost elliptical, curve (Figure 1.4), called the *limit–cycle attractor*. A limit cycle is an *isolated closed trajectory*. If all neighboring trajectories approach the limit cycle, we say that it is *attracting* or *stable*. Otherwise, it is *repelling* or *unstable*. In our case, the pendulum is swinging back and forth tracing out the same path, undergoing regular motion.

If we increase the driving force a bit more, we now find that instead of just tracing out one loop, the pendulum must now swing through two loops until it reaches the same point again on the phase diagram. The pendulum's *period*

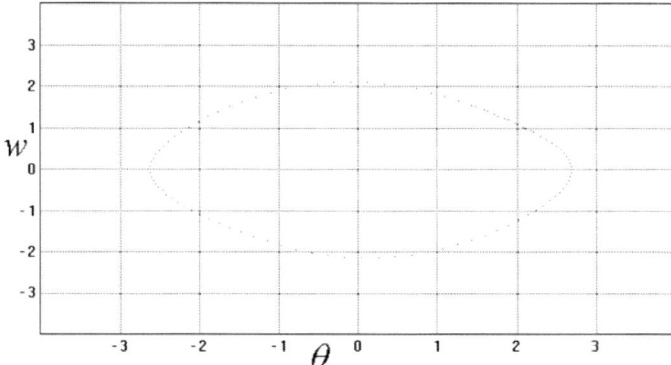

Fig. 1.4. The limit–cycle attractor in the phase portrait of the driven nonlinear pendulum.

has doubled. Increase a bit the driving force F and the period doubles once more: 4 loops appear (Figure 1.5). This doubling continues as F is increased until a point is reached where, in order to return to the same point, the pendulum must pass through an infinite number of swings. In other words, the motion of the pendulum ceases to be regular and becomes *chaotic.*

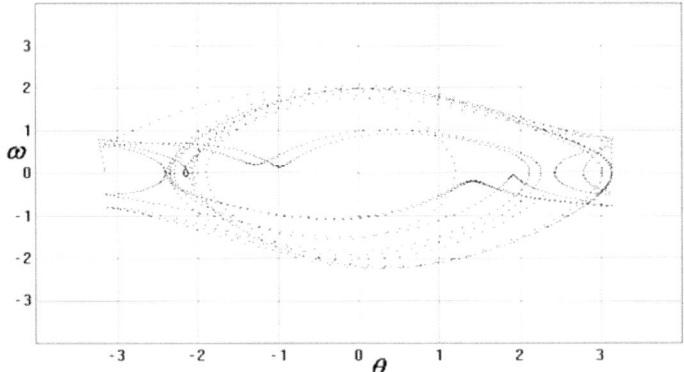

Fig. 1.5. Chaotic phase portrait of the driven nonlinear pendulum.

Using the so–called *Poincaré section* (taking 'snapshots' of the phase–space at time intervals equal to $t_n = \frac{2n\pi}{w+\phi}$, where n is an integer and ϕ is the above phase term), we find that regular motion with a single period means that only one point is plotted. A period doubling adds another point, and each extra period doubling adds two more points. When a chaotic state is reached then instead of points Poincaré section consists of long 'wavy' lines, composed themselves of bunches of lines. If we magnified one of these lines we would see that this was also composed of another bunch of lines. In fact we could continue the magnification indefinitely and we would still see much the same thing. Hence, the pendulum's attractor is now a fractal, just like the celebrated *Lorenz attractor* (see below).

Dynamics Tradition and Chaos

Therefore, traditionally, a dynamicist would believe that to write down a system's equations is to understand the system. How better to capture the essential features? For example, for a *playground swing*, the equation of motion ties together the pendulum's angle, its velocity, its friction, and the force driving it. But because of the little bits of nonlinearity in this equation, a dynamicist would find himself helpless to answer the easiest practical questions about the future of the system. A computer can simulate the problem numerically calculating each cycle (i.e., integrating the pendulum equation), but simulation brings its own problem: the tiny imprecision built into each calculation rapidly takes over, because this is a system with sensitive dependence on initial condition. Before long, the signal disappears and all that remains is noise [Gle87].

For example, in 1960s, Ed Lorenz from MIT created a simple weather model in which small changes in starting conditions led to a marked ('catastrophic') changes in outcome, called *sensitive dependence on initial conditions*, or popularly, the *butterfly effect* (i.e., "the notion that a butterfly stirring the air today in Peking can transform storm systems next month in New York, or, even worse, can cause a hurricane in Texas"). Thus long–range prediction of imprecisely measured systems becomes an impossibility.

At about the same time, Steve Smale from Berkeley studied an oscillating system (the Van der Pol oscillator) and found that his initial conjecture "that all systems tend to a steady state" – was not valid for certain nonlinear dynamical systems. He represented behavior of these systems with topological foldings called *Smale's horseshoe* in the system's phase–space. These foldings allowed graphical display of why points close together could lead to quite different outcomes, which is again sensitive dependence on initial conditions.

The unique character of chaotic dynamics may be seen most clearly by imagining the system to be started twice, but from slightly different initial conditions (in case of human motion, these are initial joint angles and angular velocities). We can think of this small initial difference as resulting from measurement error. For non–chaotic systems, this uncertainty leads only to an error in prediction that *grows linearly* with time. For chaotic systems, on the other hand, the error *grows exponentially* in time, so that the state of the system is essentially unknown after very short time. This phenomenon, firstly recognized by H. Poincaré, the father of topology, in 1913, which occurs only when the governing equations are nonlinear, with nonlinearly coupled variables, is known as *sensitivity to initial conditions*. Another type of sensitivity of chaotic systems is *sensitivity to parameters*: a small variation of system parameters (e.g., mass, length and moment of inertia of human body segments) results in great change of system output (dynamics of human movement).

If prediction becomes impossible, it is evident that a chaotic system can resemble a stochastic system, say a Brownian motion. However, the source of the irregularity is quite different. For chaos, the irregularity is part of the intrinsic dynamics of the system, not random external influences (for example, random muscular contractions in human motion). Usually, though, chaotic systems are predictable in the short–term. This *short–term predictability* is useful in various domains ranging from weather forecasting to economic forecasting.

Recall that some aspects of chaos have been known for over a hundred years. Isaac Newton was said to get headaches thinking about the 3−body problem (Sun, Moon, and Earth). In 1887, King Oscar II of Sweden announced a prize for anyone who could solve the $n-$body problem and hence demonstrate stability of the solar system. The prize was awarded to Henri Poincaré, who showed that even the 3−body problem has no analytical solution [Pet93, BG79]. He went on to deduce many of the properties of chaotic systems including the sensitive dependence on initial conditions. With the successes of linear models in the sciences and the lack of powerful computers, the work of these early nonlinear dynamists went largely unnoticed and undeveloped for many decades. In 1963, Ed Lorenz from MIT published a seminal paper [Lor63, Spa82] in which he showed that chaos can occur in systems of autonomous (no explicit time dependence) ordinary differential equations (ODEs) with as few as three variables and two quadratic nonlinearities. For continuous flows, the *Poincaré–Bendixson theorem* [HS74] implies the necessity of three variables, and chaos requires at least one nonlinearity. More explicitly, the theorem states that the long–time limit of any 'smooth' two–dimensional flow is either a fixed–point or a periodic solution. With the growing availability of powerful computers, many other examples of chaos were subsequently discovered in algebraically simple ODEs. Yet the sufficient conditions for chaos in a system of ODEs remain unknown [SL00].

So, *necessary condition* for *existence of chaos* satisfies any autonomous continuous–time dynamical system (a vector–field) of dimension three or higher, with at least two nonlinearly coupled variables (e.g., a single human swivel joint like a shoulder or hip, determined by three joint angles and three angular momenta). In case of non–autonomous continuous–time systems, chaos can happen in dimension two, while in case of discrete–time systems – even in dimension one. Now, whether the behavior (a flow), of any such system will actually be chaotic or not depends upon the values of its parameters and/or initial conditions. Usually, for some values of involved parameters, the system behavior is oscillating in a stable regime, while for another values of the parameters the behavior becomes chaotic, showing a *bifurcation*, or a *phase transition* – from one regime/phase to a totally different one. If a change in the system's behavior at the bifurcation point is really sharp, we could probably be able to recognize one of the celebrated polynomial *catastrophes* of R. Thom (see [Tho75, Arn92]). A series of such bifurcations usually depicts a *route to chaos*.

Chaos theory has developed special mathematical procedures to *understand* irregularity and unpredictability of low–dimensional nonlinear systems, including Poincaré sections, bifurcation diagrams, power spectra, Lyapunov exponents, period doubling, fractal dimension, stretching and folding, special identification and estimation techniques, etc. (see e.g., [Arn89, Arn78, Arn88, Arn93, YAS96, BG96]). Understanding these phenomena has enabled science to *control* the chaos (see, e.g., [OGY90, CD98]).

There are many practical reasons for *controlling* or *ordering chaos*. For example, in case of a distributed artificial intelligence system, which is usually characterized by a massive collection of decision–making agents, the fact that an agent's decision also depends on decisions made by other agents – leads to extreme complexity and nonlinearity of the overall system. More often than not, the information received by agents about the 'state' of the system may be 'tainted'. When the system contains imperfect information, its agents tend to make poor decisions concerning choosing an optimal problem–solving strategy or cooperating with other agents. This can result in certain chaotic behavior of the agents, thereby downgrading the performance of the entire system. Naturally, chaos should be reduced as much as possible, or totally suppressed, in these situations [CD98].

In contrast, recent research has shown that chaos may actually be useful under certain circumstances, and there is growing interest in utilizing the richness of chaos [Gle87, Mos96, DGY97]. Since a chaotic, or *strange attractor*[2] usually has embedded within it a dense set of unstable limit cycles, if any of these limit cycles can be stabilized, it may be desirable to stabilize one that characterizes certain maximal system performance [OGY90]. The key is, in a situation where a system is meant for multiple purposes, switching among different limit cycles may be sufficient for achieving these goals. If, on the other hand the attractor is not chaotic, then changing the original system configuration may be necessary to accommodate different purposes. Thus, when designing a system intended for multiple uses, purposely building chaotic dynamics into the system may allow for the desired flexibilities [OGY90].

Within the context of *brain dynamics*, there are suggestions that 'the controlled chaos of the brain is more than an accidental by–product of the brain complexity, including its myriad connections' and that 'it may be the chief property that makes the brain different from an artificial–intelligence machine [FS92]. The so–called *anti–control of chaos* has been proposed for solving the problem of driving the system trajectories of a human brain model away from the stable direction and, hence, away from the stable equilibrium (in the case

---

[2] *Strange attractor* is an attracting set that has zero measure in the embedding phase–space and has fractal dimension. Trajectories within a strange attractor appear to skip around randomly.

of a saddle type equilibrium), thereby preventing the periodic behavior of neuronal population bursting. Namely, in a spontaneously bursting neuronal network in vitro, chaos can be demonstrated by the presence of unstable fixed–point behavior. Chaos control techniques can increase the periodicity of such neuronal population bursting behavior. Periodic pacing is also effective in entraining such systems, although in a qualitatively different fashion. Using a strategy of anti–control such systems can be made less periodic. These techniques may be applicable to *in vivo* epileptic foci [SJD94].

Within the context of *heart dynamics*, traditionally in physiology, healthy dynamics has been regarded as regular and predictable, whereas disease, such as fatal arrythmias, aging, and drug toxicity, are commonly assumed to produce disorder and even chaos [Gol99, AGI98, IAG99, KFP91]. However, in the last two decades, laboratory studies produced evidence to show that:

1. The complex variability of healthy dynamics in a variety of physiological systems has features reminiscent of deterministic chaos; and
2. A wide class of disease processes (including drug toxicities and aging) may actually decrease, yet not completely eliminate, the amount of chaos or complexity in physiological systems (decomplexification).

These postulates have implications both for basic mechanisms in physiology as well as for clinical monitoring, including the problem of anticipating sudden cardiac death. In contrast to the prevalent belief of clinicians that healthy heart beats are regular, recent research on the inter–beat interval variations in healthy individuals shows that a normal heart rate apparently fluctuates in a highly erratic fashion. This turns out to be consistent with deterministic chaos [Gol99, AGI98, IAG99, KFP91].

Similar to the brain (and heart) dynamics, human biodynamics represents a highly nonlinear dynamics with several hundreds of degrees of freedom, many of which are naturally and nonlinearly coupled (see [II05, II06a. II06b]). Its hierarchical control system, neural motor controller, necessarily has to cope with the high–dimensional chaos.

Nevertheless, whether the purpose is to reduce 'bad' chaos or to induce 'good' ones, researchers felt strongly the necessity for chaos control [CD98].

Basic Terms of Nonlinear Dynamics

Recall that nonlinear dynamics is a language to talk about dynamical systems. Here, brief definitions are given for the basic terms of this language. All these terms will be illustrated at the pendulum example (see Introduction).

- *Dynamical system:* A part of the world which can be seen as a self–contained entity with some temporal behavior. In nonlinear dynamics, speaking about a dynamical system usually means to speak about an abstract mathematical system which is a model for such an entity. Mathematically, a dynamical system is defined by its *state* and by its *dynamics*. A pendulum is an example for a dynamical system.

- *State of a system:* A number or a vector (i.e., a list of numbers) defining the state of the dynamical system uniquely. For the free (un–driven) pendulum, the state is uniquely defined by the angle θ and the angular velocity $\dot{\theta} = d\theta/dt$. In the case of driving, the driving phase ϕ is also needed because the pendulum becomes a non–autonomous system. In spatially extended systems, the state is often a *field* (a scalar–field or a vector–field). Mathematically spoken, fields are functions with space coordinates as independent variables. The velocity vector–field of a fluid is a well–known example.
- *Phase space:* All possible states of the system. Each point in the phase–space corresponds to a unique state (see Figure 1.6). In the case of the free pendulum, the phase–space has 2D whereas for driven pendulum it has 3D. The dimension of the phase–space is infinite in cases where the system state is defined by a field.
- *Dynamics, or equation of motion:* The causal relation between the present state and the next state in the future. It is a deterministic rule which tells us what happens in the next time step. In the case of a continuous time, the time step is infinitesimally small. Thus, the equation of motion is an ordinary differential equation (ODE) (or a system of ODEs):

$$\dot{x} = f(x),$$

where x is the state and t is the time variable (overdot is the time derivative – as always). An example is the equation of motion of an un–driven and un–damped pendulum. In the case of a discrete time, the time steps are nonzero and the dynamics is a map:

$$x_{n+1} = f(x_n),$$

with the discrete time n. Note, that the corresponding physical time points t_n do not necessarily occur equidistantly. Only the order has to be the same. That is,

$$n < m \quad \Longrightarrow \quad t_n < t_m.$$

The dynamics is *linear* if the causal relation between the present state and the next state is linear. Otherwise it is *nonlinear*. If we have the case in which the next state is not uniquely defined by the present one, this is generally an indication that the *phase–space is not complete.* Thus, there are important variables determining the state which had been forgotten. This is a crucial point while modelling a real–life systems. Beside this, there are two important classes of systems where the phase–space is incomplete: the *non–autonomuous and stochastic systems.* A non–autonomous system has an equation of motion which depends explicitly on time. Thus, the dynamical rule governing the next state not only depends on the present state but also at the time it applies. A driven pendulum is a classical example of a *non–autonomuous system.* Fortunately, there is an easy way to make the phase–space complete: we simply include the time into the definition of the state. Mathematically, this is done by introducing a new state variable: t. Its dynamics reads

$$\dot{t} = 1, \qquad \text{or} \qquad t_{n+1} = t_n,$$

depending on whether time is continuous or discrete. For the periodically driven pendula, it is also natural to take the driving phase as the new state variable. Its equation of motion reads

$$\dot{\theta} = 2\pi w,$$

where w is the driving frequency (so that the angular driving frequency is $2\pi w$). On the other hand, in a *stochastic system*, the number and the nature of the variables necessary to complete the phase–space is usually unknown. Therefore, the next state can not be deduced from the present one. The deterministic rule is replaced by a stochastic one. Instead of the next state, it gives only the probabilities of all points in the phase–space to be the next state.

- *Orbit or trajectory:* A solution of the equation of motion. In the case of continuous time, it is a curve in phase–space parametrized by the time variable. For a discrete system it is an ordered set of points in the phase–space.
- *Phase Flow:* The mapping (or, map) of the whole phase–space of a continuous dynamical system onto itself for a given time step t. If t is an infinitesimal time step dt, the flow is just given by the right–hand side of the equation of motion (i.e., f). In general, the flow for a finite time step is not known analytically because this would be equivalent to have a solution of the equation of motion. For example, Figure 1.6 shows the *phase–flow* of a *damped pendulum* in the $(\theta, \dot{\theta})$–phase–plane.

Phase Plane: Nonlinear Dynamics without Chaos

The general form of a 2D vector–field on the phase plane (similar to one in Figure 1.6) is given by

$$\dot{x}_1 = f_1(x_1, x_2), \qquad \dot{x}_2 = f_2(x_1, x_2),$$

where f_i ($i = 1, 2$) are given function. By 'flowing along' the above vector–field, a *phase point* 'traces out' a solution $x_i(t)$, corresponding to a *trajectory* which is tangent to the vector–field. The entire phase plane is filled with trajectories (since each point can play the role of initial condition, depicting the so–called *phase portrait*. Every phase portrait has the following salient features (see [Str94]):

1. The fixed points, which satisfy: $f_i(x) = 0$, and correspond to the system's steady states or equilibria.

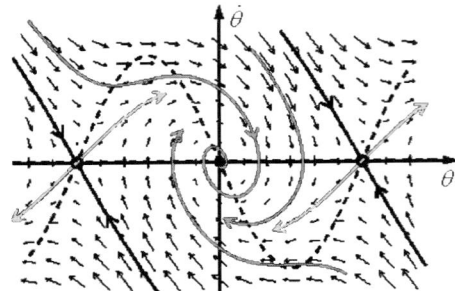

Fig. 1.6. Phase–portrait of a damped pendulum: Arrows denote the phase–flow, dashed line is a null–cline, filled dot is a stable fixed–point, open dot is an unstable fixed–point, dark gray curves are trajectories starting from sample initial points, dark lines with arrows are stable directions (manifolds), light lines with arrows are unstable directions (manifolds), the area between the stable manifolds is basin of attraction.

2. The closed orbits, corresponding to the *periodic solutions* (for which $x(t + T) = x(t)$, for all t, for some $T > 0$.

3. The specific *flow pattern*, i.e., the arrangement of trajectories near the fixed points and closed orbits.

4. The *stability* (attracting property) or *instability* (repelling property) of the fixed points and closed orbits.

Nothing more complicated than the fixed points and closed orbits can exist in the phase plane, according to the celebrated *Poincaré–Bendixson theorem*, which says that the dynamical possibilities in the phase plane are very limited. Specifically, *there cannot be chaotic behavior in the phase plane*. In other words, there is *no chaos in continuous 2D systems*.

However, there can exist chaotic behavior in *non–autonomous 2D continuous systems*, namely in the *forced nonlinear oscillators*, where explicit time–dependence actually represents the third dimension.

Free vs. Forced Nonlinear Oscillators

Here we give three examples of classical nonlinear oscillators, each in two modes: free (non–chaotic) and forced (possibly chaotic). For the simulation we use the technique called *time–phase plot*, combining an ordinary time plot with a phase–plane plot. We can see the considerable difference in complexity between unforced and forced oscillators (with all other parameters being the same). The reason for this is that *all forced 2D oscillators actually have dimension 3, although they are commonly written as a second–order ODE*. That is why for development of non–autonomous mechanics we use the *formalism of jet bundles*, see [II06b].

Spring

● Free (Rayleigh) spring (see Figure 1.7):

$$\dot{x} = y,$$
$$\dot{y} = -\frac{1}{m}(ax^3 + bx + cy),$$

where x is displacement, y is velocity, $m > 0$ is mass, $ax^3 + bx + cy$ is the restoring force of the spring, with $b > 0$; we have three possible cases: hard spring $(a > 0)$, linear (Hooke) spring $(a = 0)$, or soft spring $(a < 0)$.[3]

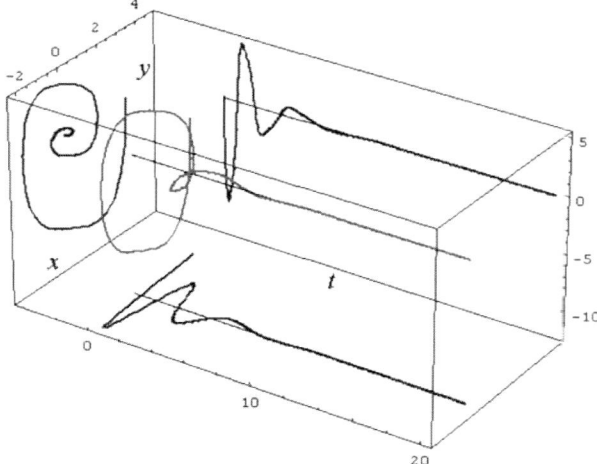

Fig. 1.7. Time–phase plot of the free hard spring with the following parameters: $m = 0.5\,kg$, $a = 1.3$, $b = 0.7$, $c = 0.5$, $x_0 = 3$, $y_0 = 0$, $t_{max} = 20\,s$. Simulated using $Mathematica^{TM}$.

● Forced (Duffing) spring (see Figure 1.8):

$$\dot{x} = y,$$
$$\dot{y} = -\frac{1}{m}(ax^3 + bx + cy) + F\cos(wt),$$
$$\dot{\theta} = w,$$

where F is the force amplitude, θ is the driving phase and w is the driving frequency; the rest is the same as above.

---

[3] In his book *The Theory of Sound*, Lord Rayleigh introduced a series of methods that would prove quite general, such as the notion of a *limit cycle* – a *periodic motion a system goes to regardless of the initial conditions.*

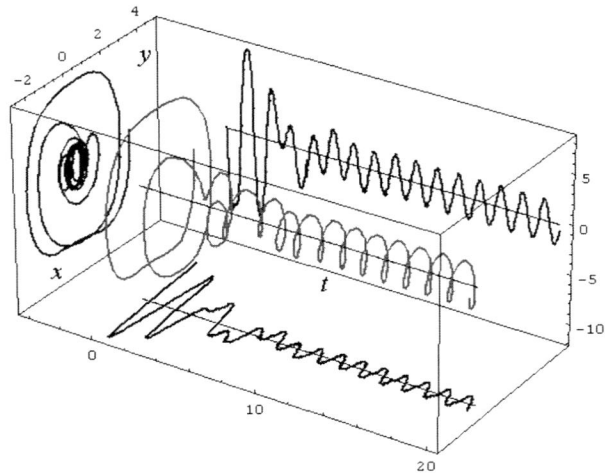

Fig. 1.8. Time–phase plot of the forced hard spring with the following parameters: $m = 0.5\,kg$, $a = 1.3$, $b = 0.7$, $c = 0.5$, $x_0 = 3$, $y_0 = 0$, $t_{max} = 20\,s$, $F = 10$, $w = 5$. Simulated using $Mathematica^{TM}$.

Self–Sustained Oscillator

- Free (Rayleigh) self–sustained oscillator (see Figure 1.9):

$$\dot{x} = y,$$
$$\dot{y} = -\frac{1}{CL}(x + By^3 - Ay),$$

  where x is current, y is voltage, $C > 0$ is capacitance and $L > 0$ is inductance; $By^3 - Ay$ (with $A, B > 0$) is the characteristic function of vacuum tube.

- Forced (Rayleigh) self–sustained oscillator (see Figure 1.10):

$$\dot{x} = y,$$
$$\dot{y} = -\frac{1}{CL}(x + By^3 - Ay) + F\cos(wt),$$
$$\dot{\theta} = w.$$

Van der Pol Oscillator

- Free Van der Pol oscillator (see Figure 1.11):

$$\dot{x} = y, \qquad\qquad (1.4)$$
$$\dot{y} = -\frac{1}{CL}[x + (Bx^2 - A)y].$$

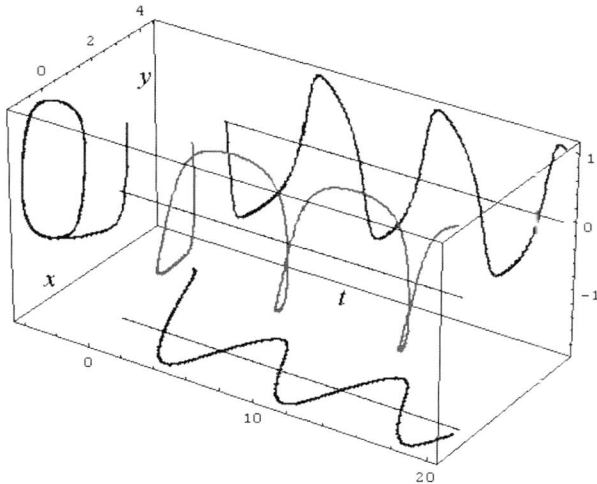

Fig. 1.9. Time–phase plot of the free Rayleigh's self–sustained oscillator with the following parameters: $A = 1.3$, $B = 1.5$, $C = 0.7$, $L = 1.5$, $x_0 = 3$, $y_0 = 0$, $t_{max} = 20\,s$. Simulated using $Mathematica^{TM}$.

- Forced Van der Pol oscillator (see Figure 1.12):

$$\dot{x} = y,$$
$$\dot{y} = -\frac{1}{CL}[x + (Bx^2 - A)y] + F\cos(wt).$$
$$\dot{\theta} = w.$$

1.1 Basics of Attractor and Chaotic Dynamics

Recall from [II06b] that the concept of *dynamical system* has its origins in *Newtonian mechanics*. There, as in other natural sciences and engineering disciplines, the evolution rule of dynamical systems is given implicitly by a relation that gives the state of the system only a short time into the future. This relation is either a differential equation or difference equation. To determine the state for all future times requires iterating the relation many times–each advancing time a small step. The iteration procedure is referred to as solving the system or integrating the system. Once the system can be solved, given an initial point it is possible to determine all its future points, a collection known as a *trajectory* or *orbit*. All possible system trajectories comprise its *flow* in the phase–space.

More precisely, recall from [II06b] that a *dynamical system* geometrically represents a *vector–field* (or, more generally, a *tensor–field*) in the system's

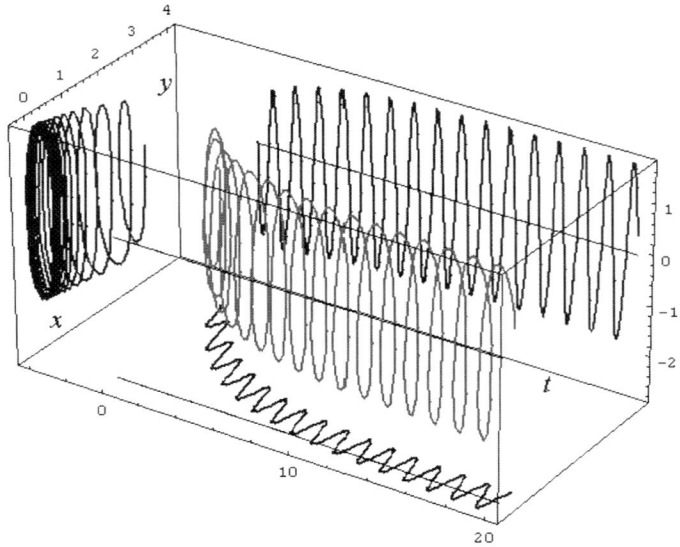

Fig. 1.10. Time–phase plot of the forced Rayleigh's self–sustained oscillator with the following parameters: $A = 1.3$, $B = 1.5$, $C = 0.7$, $L = 1.5$, $x_0 = 3$, $y_0 = 0$, $t_{max} = 20\,s$, $F = 10$, $w = 5$. Simulated using $Mathematica^{TM}$.

phase–space manifold M, which upon *integration* (governed by the celebrated *existence & uniqueness theorems for ODEs*) defines a *phase–flow* in M (see

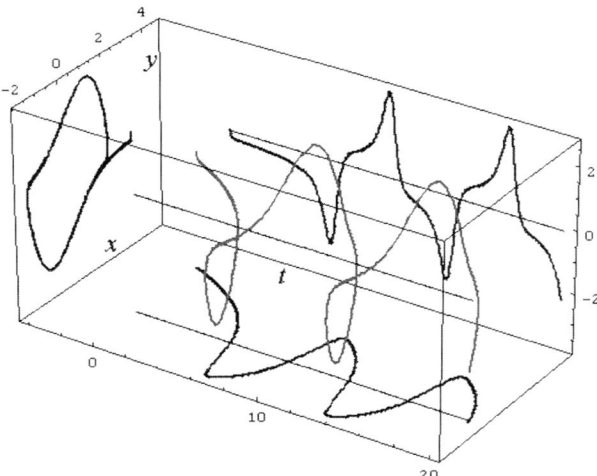

Fig. 1.11. Time–phase plot of the free Van der Pol oscillator with the following parameters: $A = 1.3$, $B = 1.5$, $C = 0.7$, $L = 1.5$, $x_0 = 3$, $y_0 = 0$, $t_{max} = 20\,s$. Simulated using $Mathematica^{TM}$.

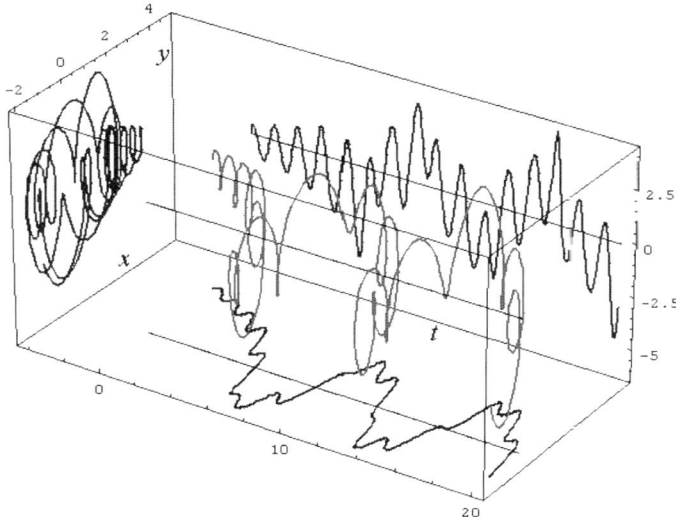

Fig. 1.12. Time–phase plot of the forced Van der Pol oscillator oscillator with the following parameters: $A = 1.3$, $B = 1.5$, $C = 0.7$, $L = 1.5$, $x_0 = 3$, $y_0 = 0$, $t_{max} = 20\,s$, $F = 10$, $w = 5$. Simulated using $Mathematica^{TM}$.

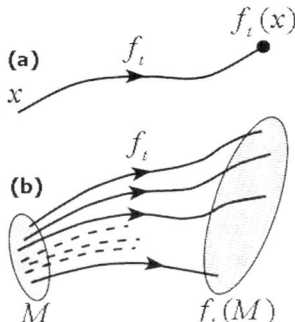

Fig. 1.13. Action of the *phase–flow* f_t in the phase–space manifold M: (a) Trajectory of a single initial point $x(t) \in M$, (b) Transporting the whole manifold M.

Figure 1.13). This phase–flow $f_t \in M$, describing the complete behavior of a dynamical system at every time instant, can be either linear, nonlinear or chaotic.

Before the advent of fast computers, solving a dynamical system required sophisticated mathematical techniques and could only be accomplished for a small class of linear dynamical systems. Numerical methods executed on computers have simplified the task of determining the orbits of a dynamical system.

For simple dynamical systems, knowing the trajectory is often sufficient, but most dynamical systems are too complicated to be understood in terms of individual trajectories. The difficulties arise because:

1. The systems studied may only be known approximately–the parameters of the system may not be known precisely or terms may be missing from the equations. The approximations used bring into question the validity or relevance of numerical solutions. To address these questions several notions of stability have been introduced in the study of dynamical systems, such as *Lyapunov stability* or *structural stability*. The stability of the dynamical system implies that there is a class of models or initial conditions for which the trajectories would be equivalent. The operation for comparing orbits to establish their equivalence changes with the different notions of stability.

2. The type of trajectory may be more important than one particular trajectory. Some trajectories may be periodic, whereas others may wander through many different states of the system. Applications often require enumerating these classes or maintaining the system within one class. Classifying all possible trajectories has led to the qualitative study of dynamical systems, that is, properties that do not change under coordinate changes. Linear dynamical systems and systems that have two numbers describing a state are examples of dynamical systems where the possible classes of orbits are understood.

3. The behavior of trajectories as a function of a parameter may be what is needed for an application. As a parameter is varied, the dynamical systems may have *bifurcation points* where the qualitative behavior of the dynamical system changes. For example, it may go from having only periodic motions to apparently erratic behavior, as in the transition to *turbulence* of a fluid.

4. The trajectories of the system may appear erratic, as if random. In these cases it may be necessary to compute averages using one very long trajectory or many different trajectories. The averages are well defined for ergodic systems and a more detailed understanding has been worked out for *hyperbolic systems*. Understanding the probabilistic aspects of dynamical systems has helped establish the foundations of statistical mechanics and of chaos.

Now, let us start 'gently' with chaotic dynamics. Recall that a dynamical system may be defined as a deterministic rule for the time evolution of state observables. Well known examples are *ODEs* in which time is continuous,

$$\dot{\mathbf{x}}(t) = \mathbf{f}(\mathbf{x}(t)), \qquad (\mathbf{x}, \mathbf{f} \in \mathbb{R}^n); \qquad (1.5)$$

and *iterative maps* in which time is discrete:

$$\mathbf{x}(t+1) = \mathbf{g}(\mathbf{x}(t)), \qquad (\mathbf{x}, \mathbf{g} \in \mathbb{R}^n). \qquad (1.6)$$

In the case of maps, the evolution law is straightforward: from $\mathbf{x}(0)$ one computes $\mathbf{x}(1)$, and then $\mathbf{x}(2)$ and so on. For ODE's, under rather general assumptions on $\mathbf{f}$, from an initial condition $\mathbf{x}(0)$ one has a unique trajectory $\mathbf{x}(t)$ for

$t > 0$ [Ott93]. Examples of regular behaviors (e.g., stable fixed–points, limit cycles) are well known, see Figure 1.14.

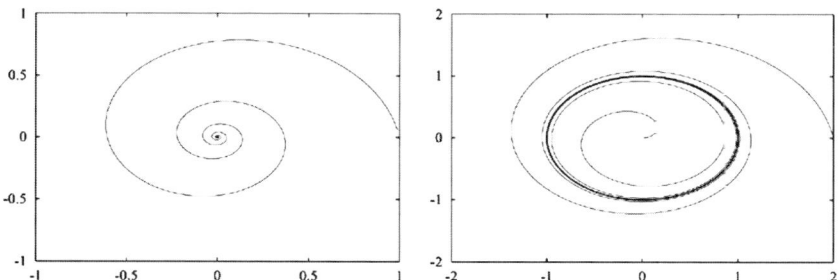

Fig. 1.14. Examples of regular attractors: fixed–point (left) and limit cycle (right). Note that limit cycles exist only in nonlinear dynamics.

A rather natural question is the possible existence of less regular behaviors i.e., different from stable fixed–points, periodic or quasi-periodic motion.

After the seminal works of Poincaré, Lorenz, Smale, May, and Hénon (to cite only the most eminent ones) it is now well established that the so called chaotic behavior is ubiquitous. As a relevant system, originated in the geophysical context, we mention the celebrated *Lorenz system* [Lor63, Spa82]

$$\dot{x} = -\sigma(x - y)$$
$$\dot{y} = -xz + rx - y \qquad (1.7)$$
$$\dot{z} = xy - bz$$

This system is related to the *Rayleigh–Bénard convection* under very crude approximations. The quantity x is proportional the circulatory fluid particle velocity; the quantities y and z are related to the temperature profile; σ, b and r are dimensionless parameters. Lorenz studied the case with $\sigma = 10$ and $b = 8/3$ at varying r (which is proportional to the Rayleigh number). It is easy to see by linear analysis that the fixed–point $(0,0,0)$ is stable for $r < 1$. For $r > 1$ it becomes unstable and two new fixed–points appear

$$C_{+,-} = (\pm\sqrt{b(r-1)}, \pm\sqrt{b(r-1)}, r-1), \qquad (1.8)$$

these are stable for $r < r_c = 24.74$. A nontrivial behavior, i.e., non periodic, is present for $r > r_c$, as is shown in Figure 1.15.

In this 'strange', chaotic regime one has the so called sensitive dependence on initial conditions. Consider two trajectories, $\mathbf{x}(t)$ and $\mathbf{x}'(t)$, initially very close and denote with $\Delta(t) = ||\mathbf{x}'(t) - \mathbf{x}(t)||$ their separation. Chaotic behavior means that if $\Delta(0) \to 0$, then as $t \to \infty$ one has $\Delta(t) \sim \Delta(0)\exp \lambda_1 t$, with $\lambda_1 > 0$ [BLV01].

Let us notice that, because of its chaotic behavior and its dissipative nature, i.e.,

$$\frac{\partial \dot{x}}{\partial x} + \frac{\partial \dot{y}}{\partial y} + \frac{\partial \dot{z}}{\partial z} < 0, \tag{1.9}$$

the attractor of the Lorenz system cannot be a smooth surface. Indeed the attractor has a self–similar structure with a fractal dimension between 2 and 3. The Lorenz model (which had an important historical relevance in the development of chaos theory) is now considered a paradigmatic example of a chaotic system.

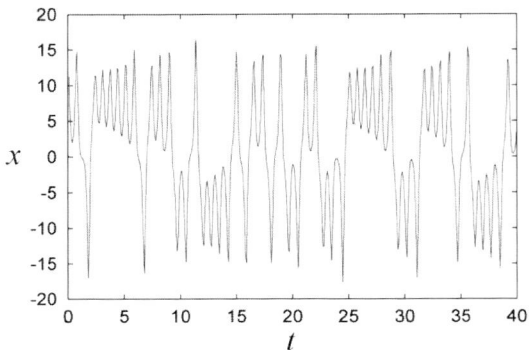

Fig. 1.15. Example of an aperiodic signal: the x variable of the Lorenz system (1.7) as function of time t, for $r = 28$.

Lyapunov Exponents

The sensitive dependence on the initial conditions can be formalized in order to give it a quantitative characterization. The main growth rate of trajectory separation is measured by the first (or maximum) *Lyapunov exponent*, defined as (see, e.g., [BLV01])

$$\lambda_1 = \lim_{t \to \infty} \lim_{\Delta(0) \to 0} \frac{1}{t} \ln \frac{\Delta(t)}{\Delta(0)}, \tag{1.10}$$

As long as $\Delta(t)$ remains sufficiently small (i.e., infinitesimal, strictly speaking), one can regard the separation as a tangent vector $\mathbf{z}(t)$ whose time evolution is

$$\dot{z}_i = \frac{\partial f_i}{\partial x_j}|_{\mathbf{x}(t)} \cdot z_j, \tag{1.11}$$

and, therefore,

$$\lambda_1 = \lim_{t \to \infty} \frac{1}{t} \ln \frac{||\mathbf{z}(t)||}{||\mathbf{z}(0)||}. \tag{1.12}$$

In principle, λ_1 may depend on the initial condition $\mathbf{x}(0)$, but this dependence disappears for ergodic systems. In general there exist as many Lyapunov exponents, conventionally written in decreasing order $\lambda_1 \geq \lambda_2 \geq \lambda_3 \geq ...$, as the independent coordinates of the phase–space [BGG80]. Without entering the details, one can define the sum of the first k Lyapunov exponents as the growth rate of an infinitesimal kD volume in the phase–space. In particular, λ_1 is the growth rate of material lines, $\lambda_1 + \lambda_2$ is the growth rate of $2D$ surfaces, and so on. A numerical widely used efficient method is due to Benettin et al. [BGG80].

It must be observed that, after a transient, the growth rate of any generic small perturbation (i.e., distance between two initially close trajectories) is measured by the first (maximum) Lyapunov exponent λ_1, and $\lambda_1 > 0$ means chaos. In such a case, the state of the system is unpredictable on long times. Indeed, if we want to predict the state with a certain tolerance Δ then our forecast cannot be pushed over a certain time interval T_P, called *predictability time*, given by [BLV01]:

$$T_P \sim \frac{1}{\lambda_1} \ln \frac{\Delta}{\Delta(0)}. \tag{1.13}$$

The above relation shows that T_P is basically determined by $1/\lambda_1$, seen its weak dependence on the ratio $\Delta/\Delta(0)$. To be precise one must state that, for a series of reasons, relation (1.13) is too simple to be of actual relevance [BCF02].

Kolmogorov–Sinai Entropy

Deterministic chaotic systems, because of their irregular behavior, have many aspects in common with stochastic processes. The idea of using stochastic processes to mimic chaotic behavior, therefore, is rather natural [Chi79, Ben84]. One of the most relevant and successful approaches is symbolic dynamics [BS93]. For the sake of simplicity let us consider a discrete time dynamical system. One can introduce a partition $\mathcal{A}$ of the phase–space formed by N disjoint sets $A_1, ..., A_N$. From any initial condition one has a trajectory

$$\mathbf{x}(0) \rightarrow \mathbf{x}(1), \mathbf{x}(2), ..., \mathbf{x}(n), ... \tag{1.14}$$

dependently on the partition element visited, the trajectory (1.14), is associated to a symbolic sequence

$$\mathbf{x}(0) \rightarrow i_1, i_2, ..., i_n, ... \tag{1.15}$$

where i_n $(n = 1, 2, ..., N)$ means that $\mathbf{x}(n) \in A_{i_n}$ at the step n, for $n = 1, 2, ...$. The coarse-grained properties of chaotic trajectories are therefore studied through the discrete time process (1.15).

An important characterization of symbolic dynamics is given by the *Kolmogorov–Sinai entropy* (KS), defined as follows. Let $C_r = (i_1, i_2, ..., i_n)$

be a generic 'word' of size n and $P(C_n)$ its occurrence probability, the quantity [BLV01]

$$H_n = \sup_A [-\sum_{C_n} P(C_n) \ln P(C_n)], \qquad (1.16)$$

is called *block entropy* of the $n-$sequences, and it is computed by taking the largest value over all possible partitions. In the limit of infinitely long sequences, the asymptotic entropy increment

$$h_{KS} = \lim_{n \to \infty} H_{n+1} - H_n, \qquad (1.17)$$

is the Kolmogorov–Sinai entropy. The difference $H_{n+1} - H_n$ has the intuitive meaning of average information gain supplied by the $(n+1)-$th symbol, provided that the previous n symbols are known. KS–entropy has an important connection with the positive Lyapunov exponents of the system [Ott93]:

$$h_{KS} = \sum_{\lambda_i > 0} \lambda_i. \qquad (1.18)$$

In particular, for low–dimensional chaotic systems for which only one Lyapunov exponent is positive, one has $h_{KS} = \lambda_1$.

We observe that in (1.16) there is a technical difficulty, i.e., taking the sup over all the possible partitions. However, sometimes there exits a special partition, called generating partition, for which one finds that H_n coincides with its superior bound. Unfortunately the generating partition is often hard to find, even admitting that it exist. Nevertheless, given a certain partition, chosen by physical intuition, the statistical properties of the related symbol sequences can give information on the dynamical system beneath. For example, if the probability of observing a symbol (state) depends only by the knowledge of the immediately preceding symbol, the symbolic process becomes a *Markov chain* (see [II06b]) and all the statistical properties are determined by the transition matrix elements W_{ij} giving the probability of observing a transition $i \to j$ in one time step. If the memory of the system extends far beyond the time step between two consecutive symbols, and the occurrence probability of a symbol depends on k preceding steps, the process is called *Markov process* of order k and, in principle, a k rank tensor would be required to describe the dynamical system with good accuracy. It is possible to demonstrate that if $H_{n+1} - H_n = h_{KS}$ for $n \geq k + 1$, k is the (minimum) order of the required Markov process [Khi57]. It has to be pointed out, however, that to know the order of the suitable Markov process we need is of no practical utility if $k \gg 1$.

Further Motivating Example: Pinball Game and Periodic Orbits

Confronted with a potentially chaotic dynamical system, we analyze it through a sequence of three distinct stages: (i) diagnose, (ii) count, (iii) measure. First we determine the intrinsic dimension of the system – the minimum number of coordinates necessary to capture its essential dynamics. If the system is very turbulent we are, at present, out of luck. We know only how to deal with the transitional regime between regular motions and chaotic dynamics in a few dimensions. That is still something; even an infinite–dimensional system such as a burning flame front can turn out to have a very few chaotic degrees of freedom. In this regime the chaotic dynamics is restricted to a space of low dimension, the number of relevant parameters is small, and we can proceed to step (ii); we count and classify all possible topologically distinct trajectories of the system into a hierarchy whose successive layers require increased precision and patience on the part of the observer. If successful, we can proceed with step (iii): investigate the weights of the different pieces of the system [CAM05].

With the game of pinball we are lucky: it is only a 2D system, free motion in a plane. The motion of a point particle is such that after a collision with one disk it either continues to another disk or it escapes. If we label the three disks by 1, 2 and 3, we can associate every trajectory with an itinerary, a sequence of labels indicating the order in which the disks are visited; for example, the two trajectories in figure 1.2 have itineraries 2313, 23132321 respectively. The itinerary is finite for a scattering trajectory, coming in from infinity and escaping after a finite number of collisions, infinite for a trapped trajectory, and infinitely repeating for a periodic orbit.[4] Such labelling is the simplest example of *symbolic dynamics*. As the particle cannot collide two times in succession with the same disk, any two consecutive symbols must differ. This is an example of *pruning*, a rule that forbids certain subsequences of symbols. Deriving pruning rules is in general a difficult problem, but with the game of pinball we are lucky, as there are no further pruning rules.[5]

Suppose you wanted to play a good game of pinball, that is, get the pinball to bounce as many times as you possibly can – what would be a winning strategy? The simplest thing would be to try to aim the pinball so it bounces many times between a pair of disks – if you managed to shoot it so it starts out in the periodic orbit bouncing along the line connecting two disk centers, it would stay there forever. Your game would be just as good if you managed to get it to keep bouncing between the three disks forever, or place it on any periodic orbit. The only rub is that any such orbit is unstable, so you have to aim very accurately in order to stay close to it for a while. So it is pretty clear that if one is interested in playing well, unstable periodic orbits are important – they form the skeleton onto which all trajectories trapped for long times cling.

---

[4] The words *orbit* and *trajectory* here are synonymous.

[5] The choice of symbols is in no sense unique. For example, as at each bounce we can either proceed to the next disk or return to the previous disk. the above 3–letter alphabet can be replaced by a binary $\{0, 1\}$ alphabet. A clever choice of an alphabet will incorporate important features of the dynamics, such as its symmetries.

Now, recall that a trajectory is *periodic* if it returns to its starting position and momentum. It is custom to refer to the set of periodic points that belong to a given periodic orbit as a *cycle*.

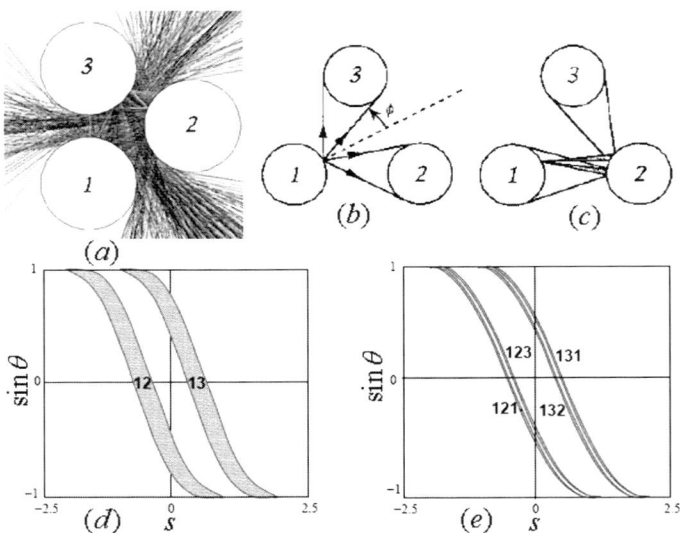

Fig. 1.16. A 3–disk pinball game. Up: (a) Elastic scattering around three hard disks (simulated in *Dynamics Solver*^{TM}); (b) A trajectory starting out from disk 1 can either hit another disk or escape; (c) Hitting two disks in a sequence requires a much sharper aim; the cones of initial conditions that hit more and more consecutive disks are nested within each other. Down: Poincaré section for the 3–disk pinball game, with trajectories emanating from the disk 1 with $x_0 = (arc-length, parallel\ momentum) = (s_0, p_0)$, disk radius: center separation ratio $a : R = 1 : 2.5$; (d) Strips of initial points M_{12}, M_{13} which reach disks 2, 3 in one bounce, respectively. (e) Strips of initial points M121, M131 M132 and M123 which reach disks 1, 2, 3 in two bounces, respectively; the Poincaré sections for trajectories originating on the other two disks are obtained by the appropriate relabelling of the strips (modified and adapted from [CAM05]).

Short periodic orbits are easily drawn and enumerated, but it is rather hard to perceive the systematics of orbits from their shapes. In mechanics a trajectory is fully and uniquely specified by its position and momentum at a given instant, and no two distinct phase–space trajectories can intersect. Their projections on arbitrary subspaces, however, can and do intersect, in rather unilluminating ways. In the pinball example, the problem is that we are looking at the projections of a 4D phase–space trajectories onto its 2D subspace, the configuration space. A clearer picture of the dynamics is obtained by constructing a phase–space Poincaré section.

The position of the ball is described by a pair of numbers (the spatial coordinates on the plane), and the angle of its velocity vector. As far as a classical dynamist is concerned, this is a complete description. Now, suppose that the pinball has just bounced off disk 1. Depending on its position and outgoing angle, it could proceed to either disk 2 or 3. Not much happens in between the bounces – the ball just travels at constant velocity along a straight line – so we can reduce the 4D flow to a 2D map f that takes the coordinates of the pinball from one disk edge to another disk edge. Let us state this more precisely: the trajectory just after the moment of impact is defined by marking s_n, the arc–length position of the nth bounce along the billiard wall, and $p_n = p \sin \phi_n$ is the momentum component parallel to the billiard wall at the point of impact (see Figure 1.16). Such a section of a flow is called a *Poincaré section*, and the particular choice of coordinates (due to Birkhoff) is particularly smart, as it conserves the phase–space volume. In terms of the Poincaré section, the dynamics is reduced to the *return map*

$$P : (s_n, p_n) \rightarrow (s_{n+1}, p_{n+1}),$$

from the boundary of a disk to the boundary of the next disk.

Next, we mark in the Poincaré section those initial conditions which do not escape in one bounce. There are two strips of survivors, as the trajectories originating from one disk can hit either of the other two disks, or escape without further ado. We label the two strips M_0, M_1. Embedded within them there are four strips, $M_{00}, M_{10}, M_{01}, M_{11}$ of initial conditions that survive for two bounces, and so forth (see Figure 1.16). Provided that the disks are sufficiently separated, after n bounces the survivors are divided into 2^n distinct strips: the M_ith strip consists of all points with itinerary $i = s_1 s_2 s_3 ... s_n$, $s = \{0, 1\}$. The unstable cycles as a skeleton of chaos are almost visible here: each such patch contains a periodic point $\overline{s_1 s_2 s_3 ... s_n}$ with the basic block infinitely repeated. Periodic points are skeletal in the sense that as we look further and further, the strips shrink but the periodic points stay put forever.

We see now why it pays to utilize a symbolic dynamics; it provides a navigation chart through chaotic phase–space. There exists a unique trajectory for every admissible infinite length itinerary, and a unique itinerary labels every trapped trajectory. For example, the only trajectory labelled by 12 is the 2–cycle bouncing along the line connecting the centers of disks 1 and 2; any other trajectory starting out as 12 . . . either eventually escapes or hits the 3rd disk [CAM05].

Now we can ask what is a good physical quantity to compute for the game of pinball? Such system, for which almost any trajectory eventually leaves a finite region (the pinball table) never to return, is said to be open, or a *repeller*. The repeller escape rate is an eminently measurable quantity. An example of such a measurement would be an unstable molecular or nuclear state which can be well approximated by a classical potential with the possibility of escape in certain directions. In an experiment many projectiles are injected into such a non–confining potential and their mean escape rate is measured. The numerical experiment might consist of injecting the pinball between the disks in some random direction and asking how many times the pinball bounces on the average before it escapes the region between the disks. On the other hand, for a theorist a good game of pinball consists in predicting accurately the asymptotic lifetime (or the escape rate) of the pinball.

Here we briefly show how Cvitanovic's *periodic orbit theory* [Cvi91] accomplishes this for us. Each step will be so simple that you can follow even at the cursory pace of this overview, and still the result is surprisingly elegant. Let us consider Figure 1.16 again. In each bounce, the initial conditions get thinned out, yielding twice as many thin strips as at the previous bounce. The total area that remains at a given time is the sum of the areas of the strips, so that the fraction of survivors after n bounces, or the *survival probability* is given by

$$\hat{\Gamma}_1 = \frac{|M_0|}{|M|} + \frac{|M_1|}{|M|}, \tag{1.19}$$

$$\hat{\Gamma}_2 = \frac{|M_{00}|}{|M|} + \frac{|M_{10}|}{|M|} + \frac{|M_{01}|}{|M|} + \frac{|M_{11}|}{|M|},$$

$$\dots$$

$$\hat{\Gamma}_n = \frac{1}{|M|} \sum_{i=1}^{(n)} |M_i|,$$

where $i = 01, 10, 11, \dots$ is a label of the ith strip (not a binary number), $|M|$ is the initial area, and $|M_i|$ is the area of the ith strip of survivors. Since at each bounce one routinely loses about the same fraction of trajectories, one expects the sum (1.19) to fall off exponentially with n and tend to the limit

$$\Gamma_{n+1}/\hat{\Gamma}_n = e^{-\gamma n} \rightarrow e^{-\gamma},$$

where the quantity γ is called the *escape rate* from the repeller. In [Cvi91] and subsequent papers, Cvitanovic has showed that the escape rate γ can be extracted from a highly convergent exact expansion by reformulating the sum (1.19) in terms of *unstable periodic orbits*.

1.2 A Brief History of Chaos Theory in 5 Steps

Now, without pretending to give a complete history of chaos theory, in this section we present only its most prominent milestones (in our view). For a number of other important contributors, see [Gle87]). Before we embark on the quick historical journey of chaos theory, note that classical mechanics has not stood still since the foundational work of its father, *Sir Isaac Newton*. The mechanical formalism that we use today was developed mostly by the three giants: *Leonhard Euler*, *Joseph Louis Lagrange* and *Sir William Rowan Hamilton*.

By the end of the 1800's the three problems that would lead to the notion of chaotic dynamics were already known: the *3–body problem* (see Figure 1.17), the *ergodic hypothesis*,[6] and *nonlinear oscillators* (see Figures 1.7–1.12).

---

[6] The second problem that played a key role in development of chaotic dynamics was the *ergodic hypothesis of Boltzmann*. Recall that *James Clerk Maxwell* and *Ludwig Boltzmann* had combined the mechanics of Newton with notions of probability in order to create statistical mechanics, deriving thermodynamics from the equations of mechanics. To evaluate the heat capacity of even a simple system, Boltzmann had to make a great simplifying assumption of ergodicity: that the dynamical system would visit every part of the phase–space allowed by conservations law equally often. This hypothesis was extended to other averages used in statistical mechanics and was called the ergodic hypothesis. It was reformulated by Poincaré to say that a trajectory comes as close as desired to any phase–space point.

Proving the ergodic hypothesis turned out to be very difficult. By the end of our own century it has only been shown true for a few systems and wrong for quite a few others. Early on, as a mathematical necessity, the proof of the hypothesis was broken down into two parts. First one would show that the mechanical system was ergodic (it would go near any point) and then one would show that it would go near each point equally often and regularly so that the computed averages made mathematical sense. Koopman took the first step in proving the ergodic hypothesis when he noticed that it was possible to reformulate it using the recently

1.2.1 *Henry Poincaré:* Qualitative Dynamics, Topology and Chaos

Chaos theory really started with *Henry Jules Poincaré*, the last mathematical universalist, the father of both *dynamical systems* and *topology* (which he

considered to be the two sides of the same coin). Together with the four dynamics giants mentioned above, Poincaré has been considered as one of the great scientific geniuses of all time.[7]

developed methods of *Hilbert spaces*. This was an important step that showed that it was possible to take a finite–dimensional nonlinear problem and reformulate it as a infinite–dimensional linear problem. This does not make the problem easier, but it does allow one to use a different set of mathematical tools on the problem. Shortly after Koopman started lecturing on his method, *John von Neumann* proved a version of the ergodic hypothesis, giving it the status of a theorem. He proved that if the mechanical system was ergodic, then the computed averages would make sense. Soon afterwards *George Birkhoff* published a much stronger version of the theorem (see [CAM05]).

[7] Recall that Henri Poincaré (April 29, 1854–July 17, 1912), was one of France's greatest mathematicians and theoretical physicists, and a philosopher of science. Poincaré is often described as the last 'universalist' (after Gauss), capable of understanding and contributing in virtually all parts of mathematics. As a mathematician and physicist, he made many original fundamental contributions to pure and applied mathematics, mathematical physics, and celestial mechanics. He was responsible for formulating the Poincaré conjecture, one of the most famous problems in mathematics. In his research on the three-body problem, Poincaré became the first person to discover a *deterministic chaotic system*. Besides, Poincaré introduced the modern principle of relativity and was the first to present the Lorentz transformations in their modern symmetrical form (Poincaré group). Poincaré discovered the remaining relativistic velocity transformations and recorded them in a letter to Lorentz in 1905. Thus he got perfect invariance of all of Maxwell's equations, the final step in the discovery of the theory of special relativity. As a mathematician and physicist, he made many original fundamental contributions to pure and applied mathematics, mathematical physics, and celestial mechanics. He was responsible for formulating the Poincaré conjecture, one of the most famous problems in mathematics. In his research on the three-body problem,

Poincaré conjectured and proved a number of theorems. Two of them related to chaotic dynamics are:

1. The *Poincaré–Bendixson theorem* says: Let F be a dynamical system on the real plane defined by

$$(\dot{x}, \dot{y}) = (f(x, y), g(x, y)),$$

where f and g are continuous differentiable functions of x and y. Let S be a closed bounded subset of the 2D phase–space of F that does not contain a stationary point of F and let C be a trajectory of F that never leaves S. Then C is either a limit–cycle or C converges to a limit–cycle. The Poincaré–Bendixson theorem limits the types of long term behavior that can be exhibited by continuous planar dynamical systems. One important implication is that a 2D continuous dynamical system cannot give rise to a *strange attractor*. If a strange attractor C did exist in such a system, then it could be enclosed in a closed and bounded subset of the phase–space. By making this subset small enough, any nearby stationary points could be excluded. But then the Poincaré–Bendixson theorem says that C is not a strange attractor at all – it is either a limit–cycle or it converges to a limit–cycle. The Poincaré–Bendixson theorem says that chaotic behavior can only arise in continuous dynamical systems whose phase–space has 3 or more dimensions. However, this restriction does not apply to discrete dynamical systems, where chaotic behavior can arise in two or even one–dimensional.

2. The *Poincaré–Hopf index theorem* says: Let M be a compact differentiable manifold and v be a vector–field on M with isolated zeroes. If M has boundary, then we insist that v be pointing in the outward normal direction along the boundary. Then we have the formula

$$\sum_i index_v = \chi(M),$$

where the sum is over all the isolated zeroes of v and $\chi(M)$ is the *Euler characteristic* of M. systems.

Poincaré became the first person to discover a *deterministic chaotic system*. Besides, Poincaré introduced the modern principle of relativity and was the first to present the Lorentz transformations in their modern symmetrical form (Poincaré group). Poincaré discovered the remaining relativistic velocity transformations and recorded them in a letter to Lorentz in 1905. Thus he got perfect invariance of all of Maxwell's equations, the final step in the discovery of the theory of special relativity.

Poincaré had the opposite philosophical views of Bertrand Russell and Gottlob Frege, who believed that mathematics were a branch of logic. Poincaré strongly disagreed, claiming that *intuition* was the *life of mathematics*. Poincaré gives an interesting point of view in his book 'Science and Hypothesis': "For a superficial observer, scientific truth is beyond the possibility of doubt; the logic of science is infallible, and if the scientists are sometimes mistaken, this is only from their mistaking its rule."

In 1887, in honor of his 60th birthday, Oscar II, King of Sweden offered a prize to the person who could answer the question "Is the Solar system stable?" Poincaré won the prize with his famous work on the *3–body problem*. He considered the Sun, Earth and Moon orbiting in a plane under their mutual gravitational attractions (see Figure 1.17). Like the pendulum, this system has some unstable solutions. Introducing a *Poincaré section*, he saw that *homoclinic tangles* must occur. These would then give rise to *chaos* and *unpredictability*.

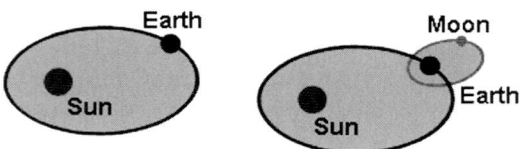

Fig. 1.17. The 2–body, problem solved by Newton (left), and the 3–body problem, first attacked by Poincaré, and still the point of active research (right).

Recall that trying to predict the motion of the Moon has preoccupied astronomers since antiquity. Accurate understanding of its motion was important for determining the longitude of ships while traversing open seas. The *Rudolphine Tables* of *Johannes Kepler* had been a great improvement over previous tables, and Kepler was justly proud of his achievements. Bernoulli used Newton's work on mechanics to derive the elliptic orbits of Kepler and set an example of how equations of motion could be solved by integrating. But the motion of the Moon is not well approximated by an ellipse with the Earth at a focus; at least the effects of the Sun have to be taken into account if one wants to reproduce the data the classical Greeks already possessed. To do that one has to consider the motion of three bodies: the Moon, the Earth, and the Sun. When the planets are replaced by point particles of arbitrary masses, the problem to be solved is known as the 3–body problem. The 3–body problem was also a model to another concern in astronomy. In the Newtonian model of the Solar system it is possible for one of the planets to go from an elliptic orbit around the Sun to an orbit that escaped its domain or that plunged right into it. Knowing if any of the planets would do so became the problem of the stability of the Solar system. A planet would not meet this terrible end if Solar system consisted of two celestial bodies, but whether such fate could befall in the 3–body case remained unclear.

After many failed attempts to solve the 3–body problem, natural philosophers started to suspect that it was impossible to integrate. The usual technique for integrating problems was to find the conserved quantities, quantities that do not change with time and allow one to relate the momenta and positions different times. The first sign on the impossibility of integrating the 3–body problem came from a result of Burns that showed that there were no conserved quantities that were polynomial in the momenta and positions. Burns' result did not preclude the possibility of more complicated conserved quantities. This problem was settled by Poincaré and Sundman in two very different ways.

In an attempt to promote the journal *Acta Mathematica*, *Gustaf Mittag-Leffler* got the permission of the King Oscar II of Sweden and Norway to establish a mathematical competition. Several questions were posed (although the king would have preferred only one), and the prize of 2500 kroner would go to the best submission. One of the questions was formulated by the 'father of modern analysis', *Karl Weierstrass*:

> "Given a system of arbitrary mass points that attract each other according to Newton's laws, under the assumption that no two points ever collide, try to find a representation of the coordinates of each point as a series in a variable that is some known function of time and for all of whose values the series converges uniformly.
> This problem, whose solution would *considerably extend our understanding of the Solar system...*"

Poincaré's submission won the prize. He showed that *conserved quantities that were analytic in the momenta and positions could not exist.* To show that he introduced methods that were very geometrical in spirit: the importance of phase flow, the role of *periodic orbits* and their cross sections, the *homoclinic points* (see [CAM05]).[8]

Poincaré pointed out that the problem was not correctly posed, and proved that a complete solution to it could not be found. His work was so impressive that in 1888 the jury recognized its value by awarding him the prize. He found that the evolution of such a system is often chaotic in the sense that a small perturbation in the initial state, such as a slight change in one body's initial position, might lead to a radically different later state. If the slight change

---

[8] The interesting thing about Poincaré's work was that it did not solve the problem posed. He did not find a function that would give the coordinates as a function of time for all times. He did not show that it was impossible either, but rather that it could not be done with the Bernoulli technique of finding a conserved quantity and trying to integrate. Integration would seem unlikely from Poincaré's prize–winning memoir, but it was accomplished by the Finnish–born Swedish mathematician Sundman, who showed that to integrate the 3–body problem one had to confront the 2–body collisions. He did that by making them go away through a trick known as regularization of the collision manifold. The trick is not to expand the coordinates as a function of time t, but rather as a function of 3 t. To solve the problem for all times he used a conformal map into a strip. This allowed Sundman to obtain a series expansion for the coordinates valid for all times, solving the problem that was proposed by Weirstrass in the King Oscar II's competition. Though Sundman's work deserves better credit than it gets, it did not live up to Weirstrass's expectations, and the series solution did not 'considerably extend our understanding of the Solar system.' The work that followed from Poincaré did.

is not detectable by our measuring instruments, then we will not be able to predict which final state will occur. One of the judges, the distinguished Karl Weierstrass, said, "This work cannot indeed be considered as furnishing the complete solution of the question proposed, but that it is nevertheless of such importance that its publication will inaugurate a new era in the history of celestial mechanics." Weierstrass did not know how accurate he was. In Poincaré's paper, he described new mathematical ideas such as homoclinic points. The memoir was about to be published in Acta Mathematica when an error was found by the editor. This error in fact led to further discoveries by Poincaré, which are now considered to be the beginning of *chaos theory*. The memoir was published later in 1890. Poincaré's research into orbits about Lagrange points and low-energy transfers was not utilized for more than a century afterwards.

In 1889 Poincaré proved that for the restricted three body problem no integrals exist apart from the Jacobian. In 1890 Poincaré proved his famous recurrence theorem, namely that in any small region of phase–space trajectories exist which pass through the region infinitely often. Poincaré published 3 volumes of 'Les méthods nouvelle de la mécanique celeste' between 1892 and 1899. He discussed convergence and uniform convergence of the series solutions discussed by earlier mathematicians and proved them not to be uniformly convergent. The stability proofs of Lagrange and Laplace became inconclusive after this result.

Poincaré introduced further topological methods in 1912 for the theory of stability of orbits in the 3–body problem. It fact Poincaré essentially invented topology in his attempt to answer stability questions in the three body problem. He conjectured that there are infinitely many periodic solutions of the restricted problem, the conjecture being later proved by George *Birkhoff*. The stability of the orbits in the three body problem was also investigated by Levi–Civita, Birkhoff and others (see [II06b] for technical details).

To examine chaos, Poincaré used the idea of a section, today called the *Poincaré section*, which cuts across the orbits in phase–space. While the original dynamical system always *flows in continuous time*, on the Poincaré section we can observe *discrete–time steps*. More precisely, the original *phase–space flow* (see [II06b]) is replaced by an *iterated map*, which reduces the dimension of the phase–space by one (see Figure 1.18). Later, to show what a Poincaré section would look like, Hénon devised a simple 2D–map, which is today called the *Hénon map*: $x_{new} = 1 - ax^2 + by$, $y_{new} = x$, with parameters $a = 1.4$, $b = 0.3$. Given any starting point, this map generates a sequence of points settling onto a chaotic attractor.

As an inheritance of Poincaré work, the chaos of the Solar system has been recently used for the SOHO project,[9] to minimize the fuel consumption need

---

[9] The SOHO project is being carried out jointly by ESA (European Space Agency) and NASA (US National Aeronautics and Space Administration), as a cooperative effort between the two agencies in the framework of the Solar Terrestrial

2D Poincaré section

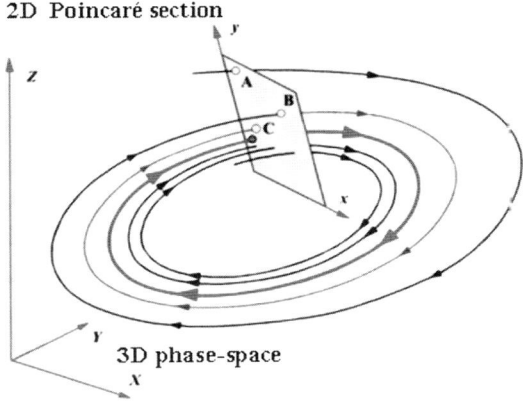

3D phase-space

Fig. 1.18. The 2D Poincaré section, reducing the 3D phase–space, using the iterated map: $x_{new} = F(x, y)$, $y_{new} = G(x, y)$.

for the space flights. Namely, in a rotating frame, a spacecraft can remain stationary at 5 *Lagrange's points* (see Figure 1.19).

Poincaré had two protegés in the development of chaos theory in the new world: *George D. Birkhoff* (see [Bir15, Bir27, Bir17] for the *Birkhoff curve shortening flow*), and *Stephen Smale*.

In some detail, the theorems of John von Neumann and George Birkhoff on the *ergodic hypothesis* (see footnote 6 above) were published in 1912 and 1913. This line of enquiry developed in two directions. One direction took an abstract approach and considered *dynamical systems as transformations of measurable spaces into themselves.* Could we classify these transformations in a meaningful way? This lead *Andrey N. Kolmogorov* to the introduction of the fundamental concept of *entropy* for dynamical systems. With entropy as a *dynamical invariant* it became possible to classify a set of abstract dynamical systems known as the *Bernoulli systems*.

Science Program (STSP) comprising SOHO and CLUSTER, and the International Solar–Terrestrial Physics Program (ISTP), with Geotail (ISAS–Japan), Wind, and Polar. SOHO was launched on December 2, 1995. The SOHO spacecraft was built in Europe by an industry team led by Matra and instruments were provided by European and American scientists. There are nine European Principal Investigators (PI's) and three American ones. Large engineering teams and more than 200 co–investigators from many institutions support the PI's in the development of the instruments and in the preparation of their operations and data analysis. NASA is responsible for the launch and mission operations. Large radio dishes around the world which form NASA's Deep Space Network are used to track the spacecraft beyond the Earth's orbit. Mission control is based at Goddard Space Flight Center in Maryland.

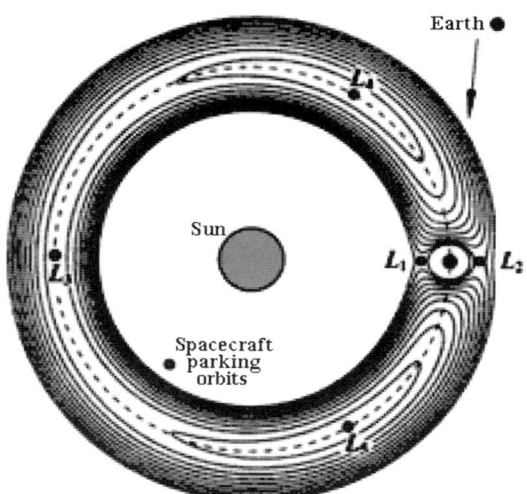

Fig. 1.19. The *Lagrange's points* ($L_1, ...L_5$) used for the space flights. Points L_1, L_2, L_3 on the Sun–Earth axis are unstable. The SOHO spacecraft used a *halo orbit* around L_1 to observe the Sun. The triangular points L_4 and L_5 are often stable. A Japanese rescue mission used a chaotic Earth–Moon trajectory.

The other line that developed from the ergodic hypothesis was in trying to find mechanical systems that are ergodic. *An ergodic system could not have stable orbits, as these would break ergodicity.* So, in 1898 *Jacques S. Hadamard* published a paper on *billiards*, where he showed that the *motion of balls on surfaces of constant negative curvature is everywhere unstable.* This dynamical system was to prove very useful and it was taken up by Birkhoff.

Marston Morse in 1923 showed that it was possible to enumerate the orbits of a ball on a surface of constant negative curvature.[10] He did this by introducing a symbolic code to each orbit and showed that the number of possible codes grew exponentially with the length of the code. With contributions by E. Artin, G. Hedlund, and *Heinz Hopf* it was eventually proven that the motion of a ball on a surface of constant negative curvature was ergodic. The

---

[10] Recall from [II06b] that in differential topology, the techniques of *Morse theory* give a very direct way of analyzing the topology of a manifold by studying differentiable functions on that manifold. According to the basic insights of Marston Morse, a differentiable function on a manifold will, in a typical case, reflect the topology quite directly. Morse theory allows one to find the so–called CW–structures and handle decompositions on manifolds and to obtain substantial information about their homology. Before Morse, Arthur Cayley and James Clerk Maxwell developed some of the ideas of Morse theory in the context of topography. Morse originally applied his theory to geodesics (critical points of the energy functional on paths). These techniques were later used by *Raoul Bott* in his proof of the celebrated Bott periodicity theorem.

importance of this result escaped most physicists, one exception being N.M. *Krylov*, who understood that a physical billiard was a dynamical system on a surface of negative curvature, but with the curvature concentrated along the lines of collision. Sinai, who was the first to show that a physical billiard can be ergodic, knew Krylov's work well.

On the other hand, the work of Lord Rayleigh also received vigorous development. It prompted many experiments and some theoretical development by B. *Van der Pol*, G. *Duffing*, and D. *Hayashi*. They found other systems in which the nonlinear oscillator played a role and classified the possible motions of these systems. This concreteness of experiments, and the possibility of analysis was too much of temptation for M. L. *Cartwright* and J.E. *Littlewood*, who set out to prove that many of the structures conjectured by the experimentalists and theoretical physicists did indeed follow from the equations of motion.

Also, G. Birkhoff had found a 'remarkable curve' in a 2D map; it appeared to be non–differentiable and it would be nice to see if a smooth flow could generate such a curve. The work of Cartwright and Littlewood lead to the work of N. Levinson, which in turn provided the basis for the horseshoe construction of Steve Smale.

In Russia, *Aleksandr M. Lyapunov* paralleled the methods of Poincaré and initiated the strong Russian dynamical systems school. A. *Andronov*[11] carried on with the study of nonlinear oscillators and in 1937 introduced together with *Lev S. Pontryagin*[12] the notion of *coarse systems*. They were formalizing the understanding garnered from the study of nonlinear oscillators, the understanding that many of the details on how these oscillators work do not affect the overall picture of the phase–space: there will still be limit cycles if one changes the dissipation or spring force function by a little bit. And changing the system a little bit has the great advantage of eliminating exceptional cases in the mathematical analysis. Coarse systems were the concept that caught Smale's attention and enticed him to study dynamical systems (see [CAM05]).

The path traversed from ergodicity to entropy is a little more confusing. The general character of entropy was understood by *Norbert Wiener*,[13] who seemed to have spoken to *Claude E. Shannon*.[14] In 1948 Shannon published his results on *information theory*, where he discusses the entropy of the shift transformation.

In Russia, *Andrey N. Kolmogorov* went far beyond and suggested a definition of the metric entropy of an area preserving transformation in order to classify Bernoulli shifts. The suggestion was taken by his student Ya.G.

---

[11] Recall that both the *Andronov–Hopf bifurcation* and a crater on the Moon are named after Aleksandr Andronov.

[12] the father of modern optimal control theory (see [II06b])

[13] the father of cybernetics

[14] the father of information theory

Sinai and the results published in 1959. In 1967 D.V. Anosov[15] and Sinai applied the notion of entropy to the study of dynamical systems. It was in the context of studying the entropy associated to a dynamical system that Sinai introduced *Markov partitions* (in 1968), which allow one *to relate dynamical systems and statistical mechanics*; this has been a very fruitful relationship. It adds measure notions to the topological framework laid down in Smale's dynamical systems paper. Markov partitions *divide the phase–space* of the dynamical system into nice little boxes that map into each other. Each box is labelled by a code and the dynamics on the phase–space maps the codes around, inducing a *symbolic dynamics*. From the number of boxes needed to cover all the space, Sinai was able to define the notion of entropy of a dynamical system. However, the relations with statistical mechanics became explicit in the work of *David Ruelle*.[16] Ruelle understood that the topology of the orbits could be specified by a symbolic code, and that one could associate an 'energy' to each orbit. The energies could be formally combined in a *partition function* (see [II06b]) to generate the invariant measure of the system.

1.2.2 *Stephen Smale:* Topological Horseshoe and Chaos of Stretching and Folding

The first deliberate, coordinated attempt to understand how global system's behavior might differ from its local behavior, came from topologist Steve Smale from the University of California at Berkeley. A young physicist, making a small talk, asked what Smale was working on. The answer stunned him: "Oscillators." It was absurd. Oscillators (pendulums, springs, or electric

---

[15] Recall that the *Anosov map* on a manifold M is a certain type of mapping, from M to itself, with rather clearly marked local directions of 'expansion' and 'contraction'. More precisely:

- If a differentiable map f on M has a hyperbolic structure on the tangent bundle, then it is called an *Anosov map*. Examples include the *Bernoulli map*, and *Arnold cat map*.
- If the Anosov map is a diffeomorphism, then it is called an *Anosov diffeomorphism*. Anosov proved that Anosov diffeomorphisms are *structurally stable*.
- If a flow on a manifold splits the tangent bundle into three invariant subbundles, with one subbundle that is exponentially contracting, and one that is exponentially expanding, and a third, non–expanding, non–contracting 1D sub–bundle, then the flow is called an *Anosov flow*.

[16] David Ruelle is a mathematical physicist working on statistical physics and dynamical systems. Together with *Floris Takens*, he coined the term *strange attractor*, and founded a modern *theory of turbulence*. Namely, in a seminal paper [RT71] they argued that, as a function of an external parameter, the *route to chaos in a fluid flow* is a transition sequence leading from stationary (S) to single periodic (P), double periodic (QP_2), triple periodic (QP_3) and, possibly, quadruply periodic (QP_4) motions, before the flow becomes chaotic (C).

circuits) where the sort of problem that a physicist finished off early in his training. They were easy. Why would a great mathematician be studying elementary physics? However, Smale was looking at nonlinear oscillators, chaotic oscillators – and seing things that physicists had learned no to see [Gle87].

Smale's 1966 Fields Medal honored a famous piece of work in high–dimensional topology, proving *Poincaré conjecture* for all dimensions greater than 4; he later generalized the ideas in a 107 page paper that established the *H–cobordism theorem* (this seminal result provides algebraic algebraic topological criteria for establishing that higher–dimensional manifolds are diffeomorphic).

After having made great strides in topology, Smale then turned to the study of nonlinear dynamical systems, where he made significant advances as well.[17] His first contribution is the famous *horseshoe map* [Sma67] that started–off significant research in dynamical systems and chaos theory.[18] Smale also outlined a mathematical research program carried out by many others. Smale is also known for injecting *Morse theory* into mathematical economics, as well as recent explorations of various theories of computation. In 1998 he compiled a list of 18 problems in mathematics to be solved in the 21st century. This list was compiled in the spirit of Hilbert's famous list of problems produced in 1900. In fact, Smale's list includes some of the original Hilbert problems. Smale's problems include the Jacobian conjecture and the Riemann hypothesis, both of which are still unsolved.

The *Smale horseshoe map* (see Figure 1.20) *is any member of a class of chaotic maps of the square into itself.* This topological transformation provided

---

[17] In the fall of 1961 Steven Smale was invited to Kiev where he met V.I. Arnol'd, (one of the fathers of modern geometrical mechanics [II06b]), D.V. Anosov, Sinai, and Novikov. He lectured there, and spent a lot of time with Anosov. He suggested a series of conjectures, most of which Anosov proved within a year. It was Anosov who showed that there are dynamical systems for which all points (as opposed to a nonwandering set) admit the hyperbolic structure, and it was in honor of this result that Smale named them *Axiom–A systems*. In Kiev Smale found a receptive audience that had been thinking about these problems. Smale's result catalyzed their thoughts and initiated a chain of developments that persisted into the 1970's.

[18] In his landmark 1967 Bulletin survey article entitled 'Differentiable dynamical systems' [Sma67], Smale presented his program for hyperbolic dynamical systems and stability, complete with a superb collection of problems. The major theorem of the paper was the $\Omega-$Stability Theorem: the global foliation of invariant sets of the map into disjoint stable and unstable parts, whose proof was a tour de force in the new dynamical methods. Some other important ideas of this paper are the existence of a horseshoe and enumeration and ordering of all its orbits, as well as the use of zeta functions to study dynamical systems. The emphasis of the paper is on the global properties of the dynamical system, on how to understand the topology of the orbits. Smale's account takes us from a local differential equation (in the form of vector fields) to the global topological description in terms of horseshoes.

a basis for understanding the chaotic properties of dynamical systems. Its basis are simple: A space is stretched in one direction, squeezed in another, and then folded. When the process is repeated, it produces something like a many–layered pastry dough, in which a pair of points that end up close together may have begun far apart, while two initially nearby points can end completely far apart.[19]

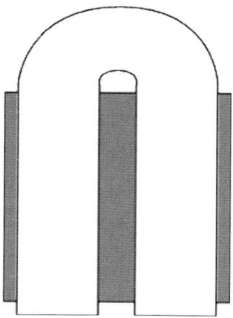

Fig. 1.20. The *Smale horseshoe map* consists of a sequence of operations on the unit square. First, stretch in the y–direction by more than a factor of two, then squeeze (compress) in the x–direction by more than a factor of two. Finally, fold the resulting rectangle and fit it back onto the square, overlapping at the top and bottom, and not quite reaching the ends to the left and right (and with a gap in the middle), as illustrated in the diagram. The shape of the stretched and folded map gives the horseshoe map its name. Note that it is vital to the construction process for the map to overlap and leave the middle and vertical edges of the initial unit square uncovered.

The horseshoe map was introduced by Smale while studying the behavior of the orbits of the *relaxation Van der Pol oscillator*. The action of the map is defined geometrically by squishing the square, then stretching the result into a long strip, and finally folding the strip into the shape of a horseshoe.

Most points eventually leave the square under the action of the map f. They go to the side caps where they will, under iteration, converge to a *fixed–point* in one of the caps. The points that remain in the square under repeated iteration form a *fractal set* and are part of the *invariant set* of the map f (see Figure 1.21).

The *stretching, folding* and *squeezing* of the horseshoe map are the essential elements that must be present in any chaotic system. In the horseshoe map the squeezing and stretching are uniform. They compensate each other so that

---

[19] Originally, Smale had hoped to explain all dynamical systems in terms of *stretching* and *squeezing* – with no folding, at least no folding that would drastically undermine a system's stability. But *folding* turned out to be necessary, and folding allowed sharp changes in dynamical behavior [Gle87].

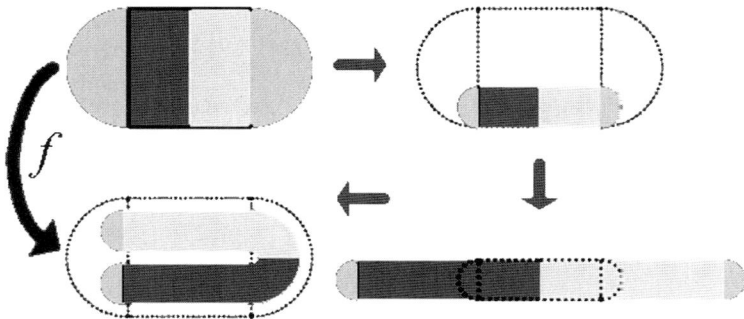

Fig. 1.21. The *Smale horseshoe map* f, defined by *stretching, folding* and *squeezing* of the system's phase–space.

the area of the square does not change. The folding is done neatly, so that the orbits that remain forever in the square can be simply described.

Repeating this generates the horseshoe attractor. If one looks at a cross section of the final structure, it is seen to correspond to a *Cantor set*.

The Smale horseshoe map is the set of basic topological operations for constructing an attractor consist of stretching (which gives sensitivity to initial conditions) and folding (which gives the attraction). Since *trajectories in phase–space cannot cross*, the repeated stretching and folding operations result in an object of great topological complexity. For any horseshoe map we have:

- There is an infinite number of periodic orbits;
- Periodic orbits of arbitrarily long period exist;
- The number or periodic orbits grows exponentially with the period; and
- Close to any point of the fractal invariant set there is a point of a periodic orbit.

More precisely, the horseshoe map f is a *diffeomorphism* defined from a region S of the plane into itself. The region S is a square capped by two semi–disks. The action of f is defined through the composition of three geometrically defined transformations. First the square is contracted along the vertical direction by a factor $a < 1/2$. The caps are contracted so as to remain semi-disks attached to the resulting rectangle. Contracting by a factor smaller than one half assures that there will be a gap between the branches of the horseshoe. Next the rectangle is stretched by a factor of $1/a$; the caps remain unchanged. Finally the resulting strip is folded into a horseshoe–shape and placed back into S.

The interesting part of the dynamics is the image of the square into itself. Once that part is defined, the map can be extended to a diffeomorphism by defining its action on the caps. The caps are made to contract and eventually map inside one of the caps (the left one in the figure). The extension of f to the caps adds a fixed–point to the *non–wandering set* of the map. To keep the class of horseshoe maps simple, the curved region of the horseshoe should not map back into the square.

The horseshoe map is one–to–one (1–1, or injection): any point in the domain has a unique image, even though not all points of the domain are the image of a point. The inverse of the horseshoe map, denoted by f^{-1}, cannot have as its domain the entire region S, instead it must be restricted to the image of S under f, that is, the domain of f^{-1} is $f(S)$.

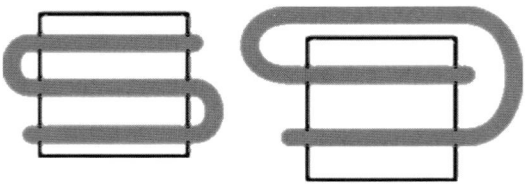

Fig. 1.22. Other types of horseshoe maps can be made by folding the contracted and stretched square in different ways.

By folding the contracted and stretched square in different ways, other types of horseshoe maps are possible (see Figure 1.22). The contracted square cannot overlap itself to assure that it remains 1–1. When the action on the square is extended to a diffeomorphism, the extension cannot always be done on the plane. For example, the map on the right needs to be extended to a diffeomorphism of the sphere by using a 'cap' that wraps around the equator.

The horseshoe map is an Axiom A diffeomorphism that serves as a model for the general behavior at a transverse *homoclinic point*, where the *stable and unstable manifolds* of a periodic point intersect.

The horseshoe map was designed by Smale to reproduce the chaotic dynamics of a *flow* in the neighborhood of a given periodic *orbit*. The neighborhood is chosen to be a small disk perpendicular to the orbit. As the system evolves, points in this disk remain close to the given periodic orbit, tracing out orbits that eventually intersect the disk once again. Other orbits diverge.

The behavior of all the orbits in the disk can be determined by considering what happens to the disk. The intersection of the disk with the given periodic orbit comes back to itself every period of the orbit and so do points in its neighborhood. When this neighborhood returns, its shape is transformed. Among the points back inside the disk are some points that will leave the disk neighborhood and others that will continue to return. The set of points that never leaves the neighborhood of the given periodic orbit form a fractal.

A symbolic name can be given to all the orbits that remain in the neighborhood. The initial neighborhood disk can be divided into a small number of regions. Knowing the sequence in which the orbit visits these regions allows the orbit to be pinpointed exactly. The visitation sequence of the orbits provide the so–called *symbolic dynamics*[20]

It is possible to describe the behavior of all initial conditions of the horseshoe map. An initial point $u_0 = x, y$ gets mapped into the point $u_1 = f(u_0)$. Its iterate is the point $u_2 = f(u_1) = f^2(u_0)$, and repeated iteration generates the orbit $u_0, u_1, u_2, ...$ Under repeated iteration of the horseshoe map, most orbits end up at the fixed–point in the left cap. This is because the horseshoe maps the left cap into itself by an *affine transformation*, which has exactly one fixed–point. Any orbit that lands on the left cap never leaves it and converges to the fixed–point in the left cap under iteration. Points in the right cap get mapped into the left cap on the next iteration, and most points in the square get mapped into the caps. Under iteration, most points will be part of orbits that converge to the fixed–point in the left cap, but some points of the square never leave.

Under forward iterations of the horseshoe map, the original square gets mapped into a series of horizontal strips. The points in these horizontal strips come from vertical strips in the original square. Let S_0 be the original square, map it forward n times, and consider only the points that fall back into the square S_0, which is a set of horizontal stripes $H_n = f^n(S_0) \cap S_0$. The points in the horizontal stripes came from the vertical stripes $V_n = f^{-n}(H_n)$, which are the horizontal strips H_n mapped backwards n times. That is, a point in V_n will, under n iterations of the horseshoe map, end up in the set H_n of vertical strips (see Figure 1.23).

Now, if a point is to remain indefinitely in the square, then it must belong to an *invariant set* Λ that maps to itself. Whether this set is empty or not has to be determined. The vertical strips V_1 map into the horizontal strips H_1, but not all points of V_1 map back into V_1. Only the points in the intersection of V_1 and H_1 may belong to Λ, as can be checked by following points outside the intersection for one more iteration. The intersection of the horizontal and vertical stripes, $H_n \cap V_n$, are squares that converge in the limit $n \to \infty$ to the invariant set Λ (see Figure 1.24).

The structure of invariant set Λ can be better understood by introducing a system of labels for all the intersections, namely a *symbolic dynamics*. The intersection $H_n \cap V_n$ is contained in V_1. So any point that is in Λ under iteration must land in the left vertical strip A of V_1, or on the right vertical

---

[20] Symbolic dynamics is the practice of modelling a dynamical system by a space consisting of infinite sequences of abstract symbols, each sequence corresponding to a state of the system, and a shift operator corresponding to the dynamics. Symbolic dynamics originated as a method to study general dynamical systems, now though, its techniques and ideas have found significant applications in data storage and transmission, linear algebra, the motions of the planets and many other areas. The distinct feature in symbolic dynamics is that time is measured in discrete intervals. So at each time interval the system is in a particular state. Each state is associated with a symbol and the evolution of the system is described by an infinite sequence of symbols (see text below).

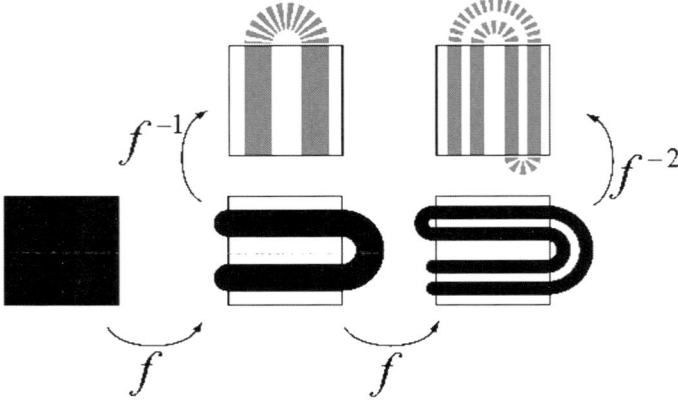

Fig. 1.23. Iterated horseshoe map: pre–images of the square region.

Fig. 1.24. Intersections that converge to the invariant set Λ.

strip B. The lower horizontal strip of H_1 is the image of A and the upper horizontal strip is the image of B, so $H_1 = f(A) \cap f(B)$. The strips A and B can be used to label the four squares in the intersection of V_1 and H_1 (see Figure 1.25) as:

$$\Lambda_{A \bullet A} = f(A) \cap A, \qquad \Lambda_{A \bullet B} = f(A) \cap B,$$
$$\Lambda_{B \bullet A} = f(B) \cap A, \qquad \Lambda_{B \bullet B} = f(B) \cap B.$$

The set $\Lambda_{B \bullet A}$ consist of points from strip A that were in strip B in the previous iteration. A dot is used to separate the region the point of an orbit is in from the region the point came from.

This notation can be extended to higher iterates of the horseshoe map. The vertical strips can be named according to the sequence of visits to strip A or strip B. For example, the set $ABB \subset V_3$ consists of the points from A that will all land in B in one iteration and remain in B in the iteration after that:

$$ABB = \{x \in A | f(x) \in B \ \text{ and } \ f^2(x) \in B\}.$$

Working backwards from that trajectory determines a small region, the set ABB, within V_3.

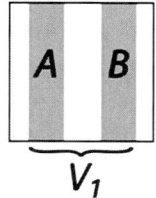

Fig. 1.25. The basic domains of the horseshoe map in symbolic dynamics.

The horizontal strips are named from their vertical strip pre–images. In this notation, the intersection of V_2 and H_2 consists of 16 squares, one of which is

$$\Lambda_{AB\bullet BB} = f^2(AB) \cap BB.$$

All the points in $\Lambda_{AB\bullet BB}$ are in B and will continue to be in B for at least one more iteration. Their previous trajectory before landing in BB was A followed by B.

Any one of the intersections $\Lambda_{P\bullet F}$ of a horizontal strip with a vertical strip, where P and F are sequences of As and Bs, is an affine transformation of a small region in V_1. If P has k symbols in it, and if $f^{-k}(\Lambda_{P\bullet F})$ and $\Lambda_{P\bullet F}$ intersect, then the region $\Lambda_{P\bullet F}$ will have a *fixed–point*. This happens when the sequence P is the same as F. For example, $\Lambda_{ABAB\bullet ABAB} \subset V_4 \cap H_4$ has at least one fixed–point. This point is also the same as the fixed–point in $\Lambda_{AB\bullet AB}$. By including more and more ABs in the P and F part of the label of intersection, the area of the intersection can be made as small as needed. It converges to a point that is part of a *periodic orbit of the horseshoe map*. The periodic orbit can be labelled by the simplest sequence of As and Bs that labels one of the regions the periodic orbit visits. For every sequence of As and Bs there is a periodic orbit.

The Smale horseshoe map is the same topological structure as the *homoclinic tangle*. To dynamically introduce homoclinic tangles, let us consider a classical engineering problem of *escape from a potential well*. Namely, if we have a motion, $x = x(t)$, of a damped particle in a well with potential energy $V = x^2/2 - x^3/3$ (see Figure 1.26) excited by a periodic driving force, $F \cos(wt)$ (with the period $T = 2\pi/w$), we are dealing with a nonlinear dynamical system given by [TS01]

$$\ddot{x} + a\dot{x} + x - x^2 = F \cos(wt). \tag{1.20}$$

Now, if the driving is switched off, i.e., $F = 0$, we have an autonomous 2D–system with the phase–portrait (and the safe basin of attraction) given in Figure 1.26 (below). The grey area of escape starts over the hilltop to infinity. Once we start driving, the system (1.20) becomes 3–dimensional, with its 3D phase–space. We need to see the basin in a *stroboscopic section* (see Figure 1.27). The hill–top solution still has an inset and and outset. As the driving

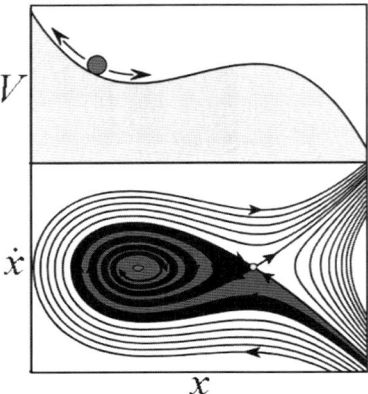

Fig. 1.26. Motion of a damped particle in a potential well, driven by a periodic force $F\cos(wt)$,. Up: potential $(x-V)$-plot, with $V = x^2/2 - x^3/3$; down: the corresponding phase $(x-\dot{x})$-portrait, showing the safe basin of attraction – if the driving is switched off $(F = 0)$.

increases, the inset and outset get tangled. They intersect one another an infinite number of times. The boundary of the safe basin becomes fractal. As the driving increases even more, the so–called fractal–fingers created by the homoclinic tangling, make a sudden incursion into the safe basin. At that point, the integrity of the in–well motions is lost [TS01].

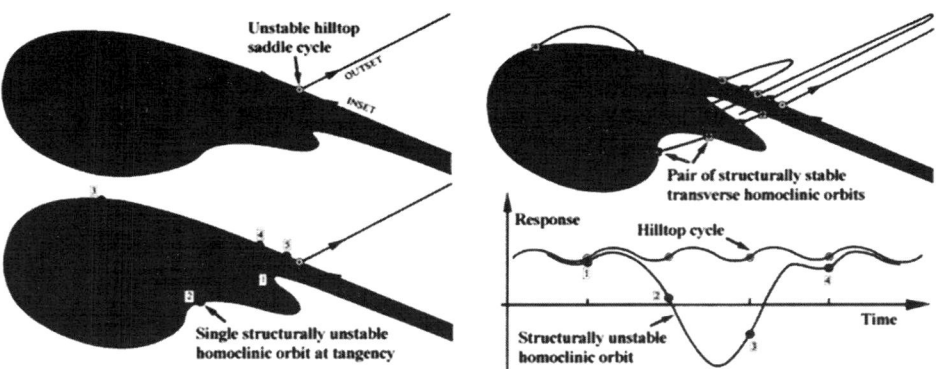

Fig. 1.27. Dynamics of a homoclinic tangle. The hill–top solution of a damped particle in a potential well driven by a periodic force. As the driving increases, the inset and outset get tangled.

Now, topologically speaking (referring to the Figure 1.28), let X be the point of intersection, with X' ahead of X on one manifold and ahead of X'' of the other. The map of each of these points TX' and TX'' must be ahead of the map of X, TX. The only way this can happen is if the manifold loops back and crosses itself at a new *homoclinic point*, i.e., a point where a stable and an unstable separatrix (invariant manifold) from the same fixed–point or same family intersect. Another loop must be formed, with T^2X another homoclinic point. Since T^2X is closer to the hyperbolic point than TX, the distance between T^2X and TX is less than that between $\bar{X}$ and TX. Area preservation requires the area to remain the same, so each new curve (which is closer than the previous one) must extend further. In effect, the loops become longer and thinner. The network of curves leading to a dense area of homoclinic points is known as a homoclinic tangle or tendril. Homoclinic points appear where chaotic regions touch in a *hyperbolic fixed–point*.

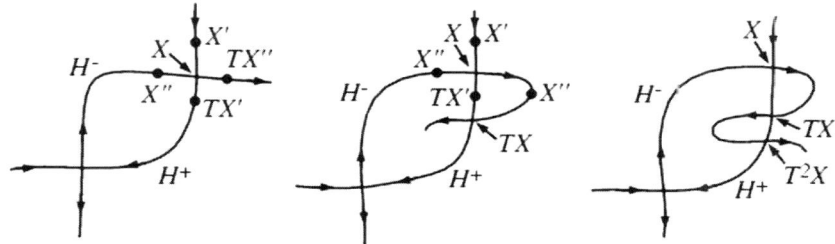

Fig. 1.28. More on homoclinic tangle (see text for explanation).

On the other hand, tangles are in general related to $n-$categories (see [II05, II06a, II06b]). Recall that in describing dynamical systems (processes) by means of $n-$categories, instead of classical starting with a *set of things:*

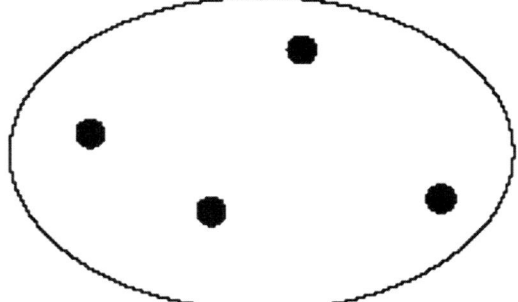

we can now start with a *category of things and processes between things:*

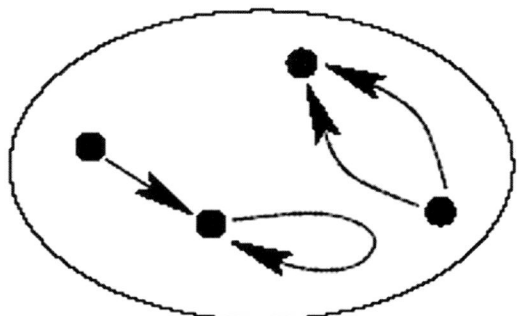

or, a 2−*category of things, processes, and processes between processes:*

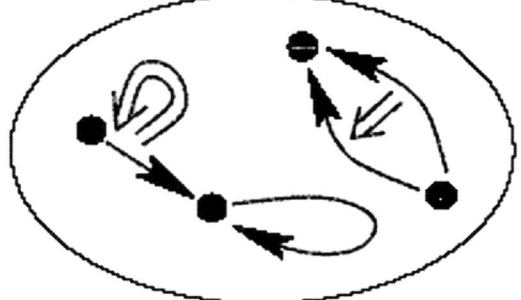

... and so on. In this way, topological $n-$categories form the natural framework for *high–dimensional chaos theory.*

1.2.3 *Ed Lorenz:* **Weather Prediction and Chaos**

Recall that an *attractor* is a set of system's states (i.e., points in the system's phase–space), invariant under the dynamics, towards which neighboring states in a given *basin of attraction* asymptotically approach in the course of dynamic evolution.[21] An attractor is defined as the smallest unit which cannot be itself decomposed into two or more attractors with distinct basins of attraction. This restriction is necessary since a dynamical system may have multiple attractors, each with its own basin of attraction.

Conservative systems do not have attractors, since the motion is periodic. For dissipative dynamical systems, however, volumes shrink exponentially, so attractors have 0 volume in nD phase–space.

In particular, a stable *fixed–point* surrounded by a dissipative region is an attractor known as a *map sink*.[22] Regular attractors (corresponding to 0 *Lyapunov exponents*) act as *limit cycles*, in which trajectories circle around

---

[21] A *basin of attraction* is a set of points in the system's phase–space, such that initial conditions chosen in this set dynamically evolve to a particular attractor.

[22] A *map sink* is a stable fixed–point of a map which, in a dissipative dynamical system, is an attractor.

a limiting trajectory which they asymptotically approach, but never reach. The so–called *strange attractors*[23] are bounded regions of phase–space (corresponding to positive Lyapunov characteristic exponents) having zero measure in the embedding phase–space and a *fractal dimension*. Trajectories within a strange attractor appear to skip around randomly.

In 1963, Ed Lorenz from MIT was trying to improve weather forecasting. Using a primitive computer of those days, he discovered the first *chaotic attractor*. Lorenz used three Cartesian variables, (x, y, z), to define *atmospheric convection*. Changing in time, these variables gave him a trajectory in a (Euclidean) 3D–space. From all starts, trajectories settle onto a chaotic, or *strange attractor*.[24]

---

[23] A strange attractor is an attracting set that has zero measure in the embedding phase–space and has fractal dimension. Trajectories within a strange attractor appear to skip around randomly.

[24] Edward Lorenz is a professor of meteorology at MIT who wrote the first clear paper on *deterministic chaos*. The paper was called 'Deterministic Nonperiodic Flow' and it was published in the Journal of Atmospheric Sciences in 1963. Before that, in 1960, Lorenz began a project to simulate weather patterns on a computer system called the Royal McBee. Lacking much memory, the computer was unable to create complex patterns, but it was able to show the interaction between major meteorological events such as tornados, hurricanes, easterlies and westerlies. A variety of factors was represented by a number, and Lorenz could use computer printouts to analyze the results. After watching his systems develop on the computer, Lorenz began to see patterns emerge, and was able to predict with some degree of accuracy what would happen next. While carrying out an experiment, Lorenz made an accidental discovery. He had completed a run, and wanted to recreate the pattern. Using a printout, Lorenz entered some variables into the computer and expected the simulation to proceed the same as it had before. To his surprise, the pattern began to diverge from the previous run, and after a few 'months' of simulated time, the pattern was completely different. Lorenz eventually discovered why seemingly identical variables could produce such different results. When Lorenz entered the numbers to recreate the scenario, the printout provided him with numbers to the thousandth position (such as 0.617). However, the computer's internal memory held numbers up to the millionth position (such as 0.617395); these numbers were used to create the scenario for the initial run. This small deviation resulted in a completely divergent weather pattern in just a few months. This discovery creates the groundwork of chaos theory: In a system, small deviations can result in large changes. This concept is now known as a *butterfly effect*.

Lorenz definition of chaos is: "The property that characterizes a dynamical system in which most orbits exhibit sensitive dependence." Dynamical systems (like the weather) are all around us. They have recurrent behavior (it is always hotter in summer than winter) but are very difficult to pin down and predict apart from the very short term. 'What will the weather be tomorrow?' – can be anticipated, but 'What will the weather be in a months time?' is an impossible question to answer.

More precisely, Lorenz reduced the *Navier–Stokes equations* for *convective Bénard fluid flow* (see section (8.77) below) into three first order coupled nonlinear differential equations, already introduced above as (1.7) and demonstrated with these the idea of sensitive dependence upon initial conditions and chaos (see [Lor63, Spa82]).

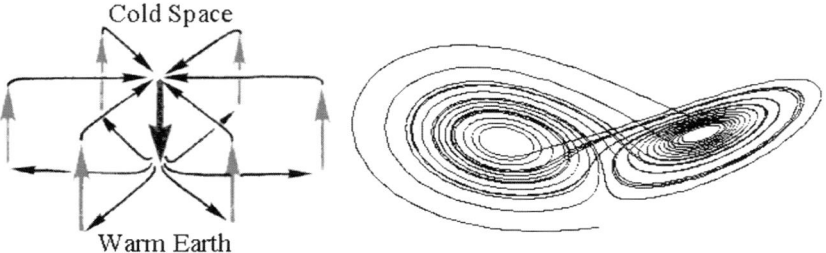

Fig. 1.29. Bénard cells, showing a typical vortex of a rolling air, with a warm air rising in a ring and a cool air descending in the center (left). A simple model of the Bénard cells provided by the celebrated 'Lorenz–butterfly' (or, 'Lorenz–mask') *strange attractor* (right).

We rewrite the celebrated Lorenz equations here as

$$\dot{x} = a(y - x), \qquad \dot{y} = bx - y - xz, \qquad \dot{z} = xy - cz, \qquad (1.21)$$

where x, y and z are dynamical variables, constituting the 3D *phase–space* of the *Lorenz system*; and a, b and c are the parameters of the system. Originally, Lorenz used this model to describe the unpredictable behavior of the weather, where x is the rate of convective overturning (convection is the process by which heat is transferred by a moving fluid), y is the horizontal temperature overturning, and z is the vertical temperature overturning; the parameters are: $a \equiv P-$proportional to the *Prandtl number* (ratio of the fluid viscosity of a substance to its thermal conductivity, usually set at 10), $b \equiv R-$proportional to the Rayleigh number (difference in temperature between the top and bottom of the system, usually set at 28), and $c \equiv K-$a number proportional to the physical proportions of the region under consideration (width to height ratio of the box which holds the system, usually set at 8/3). The Lorenz system (1.21) has the properties:

1. *Symmetry*: $(x, y, z) \to (-x, -y, z)$ for all values of the parameters, and

---

Lorenz showed that with a set of simple differential equations seemingly very complex turbulent behavior could be created that would previously have been considered as random. He further showed that accurate longer range forecasts in any chaotic system were impossible, thereby overturning the previous orthodoxy. It had been believed that the more equations you add to describe a system, the more accurate will be the eventual forecast.

2. The z–axis ($x = y = 0$) is *invariant* (i.e., all trajectories that start on it also end on it).

Nowadays it is well–known that the Lorenz model is a paradigm for low–dimensional chaos in dynamical systems in synergetics and this model or its modifications are widely investigated in connection with modelling purposes in meteorology, hydrodynamics, laser physics, superconductivity, electronics, oil industry, chemical and biological kinetics, etc.

The 3D *phase–portrait* of the Lorenz system (1.29) shows the celebrated *'Lorenz mask'*, a special type of *fractal attractor* (see Figure 1.29). It depicts the famous *'butterfly effect'*, (i.e., sensitive dependence on initial conditions) – the popular idea in meteorology that 'the flapping of a butterfly's wings in Brazil can set off a tornado in Texas' (i.e., a tiny difference is amplified until two outcomes are totally different), so that the long term behavior becomes impossible to predict (e.g., long term weather forecasting). The Lorenz mask has the following characteristics:

1. Trajectory does not intersect itself in three dimensions;
2. Trajectory is not periodic or transient;
3. General form of the shape does not depend on initial conditions; and
4. Exact sequence of loops is very sensitive to the initial conditions.

1.2.4 *Mitchell Feigenbaum:* A Constant and Universality

Mitchell Jay Feigenbaum (born December 19, 1944; Philadelphia, USA) is a mathematical physicist whose pioneering studies in chaos theory led to the discovery of the *Feigenbaum constant*.

In 1964 he began graduate studies at the MIT. Enrolling to study electrical engineering, he changed to physics and was awarded a doctorate in 1970 for a thesis on dispersion relations under Francis Low. After short positions at Cornell University and Virginia Polytechnic Institute, he was offered a longer–term post at Los Alamos National Laboratory to study turbulence. Although the group was ultimately unable to unravel the intractable theory of turbulent fluids, his research led him to study chaotic maps.

Many mathematical maps involving a single linear parameter exhibit apparently random behavior known as chaos when the parameter lies in a certain range. As the parameter is increased towards this region, the map undergoes bifurcations at precise values of the parameter. At first there is one stable point, then bifurcating to oscillate between two points, then bifurcating again to oscillate between four points and so on. In 1975 Feigenbaum, using the HP-65 computer he was given, discovered that the ratio of the difference between the values at which such successive *period–doubling bifurcations* (called the *Feigenbaum cascade*) occur tends to a constant of around 4.6692. He was then able to provide a mathematical proof of the fact, and showed that the same behavior and the same constant would occur in a wide class of mathematical functions prior to the onset of chaos. For the first time this universal result enabled mathematicians to take their first huge step to unravelling the apparently intractable 'random' behavior of chaotic systems. This 'ratio of convergence' is now known as the Feigenbaum constant.

More precisely, the Feigenbaum constant δ is a universal constant for functions approaching chaos via successive period doubling bifurcations. It was discovered by Feigenbaum in 1975, while studying the fixed–points of the iterated function $f(x) = 1 - \mu|x|^r$, and characterizes the geometric approach of the bifurcation parameter to its limiting value (see Figure 1.30) as the parameter μ is increased for fixed x [Fei79].

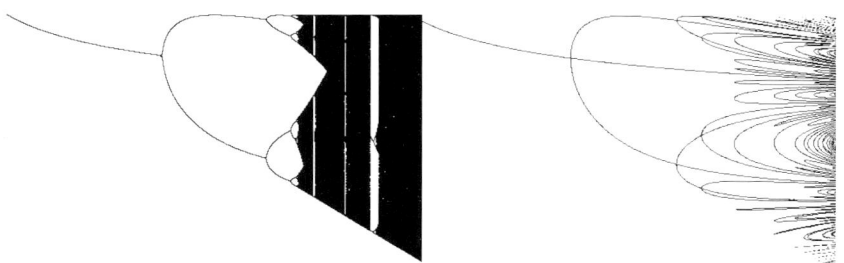

Fig. 1.30. Feigenbaum constant: approaching chaos via successive period doubling bifurcations. The plot on the left is made by iterating equation $f(x) = 1 - \mu|x|^r$ with $r = 2$ several hundred times for a series of discrete but closely spaced values of μ, discarding the first hundred or so points before the iteration has settled down to its fixed–points, and then plotting the points remaining. The plot on the right more directly shows the cycle may be constructed by plotting function $f^n(x) - x$ as a function of μ, showing the resulting curves for $n = 1, 2, 4$. Simulated in $Mathematica^{TM}$.

The Logistic map is a well known example of the maps that Feigenbaum studied in his famous Universality paper [Fei78].

In 1986 Feigenbaum was awarded the Wolf Prize in Physics. He has been Toyota Professor at Rockefeller University since 1986.

For details on Feigenbaum universality, see [Gle87].

1.2.5 *Lord Robert May:* Population Modelling and Chaos

Let $x(t)$ be the population of the species at time t; then the *conservation law* for the population is conceptually given by (see [Mur02])

$$\dot{x} = births - deaths + migration, \tag{1.22}$$

where $\dot{x} = dx/dt$. The above conceptual equation gave rise to a series of *population models*. The simplest continuous–time model, due to Thomas Malthus from 1798 [Mal798],[25] has no migration, while the birth and death terms are proportional to x,

$$\dot{x} = bx - dx \qquad \Longrightarrow \qquad x(t) = x_0 e^{(b-d)t}, \qquad (1.23)$$

where b, d are positive constants and $x_0 = x(0)$ is the initial population. Thus, according to the *Malthus model* (1.23), if $b > d$, the population grows exponentially, while if $b < d$, it dies out. Clearly, this approach is fairly over-simplified and apparently fairly unrealistic. (However, if we consider the past

---

[25] The Rev. Thomas Robert Malthus, FRS (February, 1766–December 23, 1834), was an English demographer and political economist best known for his pessimistic but highly influential views. Malthus's views were largely developed in reaction to the optimistic views of his father, Daniel Malthus and his associates, notably Jean-Jacques Rousseau and William Godwin. Malthus's essay was also in response to the views of the Marquis de Condorcet. In An Essay on the Principle of Population, first published in 1798, Malthus made the famous prediction that population would outrun food supply, leading to a decrease in food per person: "The power of population is so superior to the power of the earth to produce subsistence for man, that premature death must in some shape or other visit the human race. The vices of mankind are active and able ministers of depopulation. They are the precursors in the great army of destruction; and often finish the dreadful work themselves. But should they fail in this war of extermination, sickly seasons, epidemics, pestilence, and plague, advance in terrific array, and sweep off their thousands and tens of thousands. Should success be still incomplete, gigantic inevitable famine stalks in the rear, and with one mighty blow levels the population with the food of the world." This Principle of Population was based on the idea that population if unchecked increases at a geometric rate, whereas the food supply grows at an arithmetic rate. Only natural causes (eg. accidents and old age), misery (war, pestilence, and above all famine), moral restraint and vice (which for Malthus included infanticide, murder, contraception and homosexuality) could check excessive population growth. Thus, Malthus regarded his Principle of Population as an explanation of the past and the present situation of humanity, as well as a prediction of our future. The eight major points regarding evolution found in his 1798 *Essay* are: (i) Population level is severely limited by subsistence. (ii) When the means of subsistence increases, population increases. (iii) Population pressures stimulate increases in productivity. (iv) Increases in productivity stimulates further population growth. (v) Since this productivity can never keep up with the potential of population growth for long, there must be strong checks on population to keep it in line with carrying capacity. (vi) It is through individual cost/benefit decisions regarding sex, work, and children that population and production are expanded or contracted. (vii) Positive checks will come into operation as population exceeds subsistence level. (viii) The nature of these checks will have significant effect on the rest of the sociocultural system.

Evolutionists John Maynard Smith and Ronald Fisher were both critical of Malthus' theory, though it was Fisher who referred to the *growth rate r* (used in *logistic equation*) as the *Malthusian parameter*. Fisher referred to "...a relic of creationist philosophy..." in observing the fecundity of nature and deducing (as Darwin did) that this therefore drove natural selection. Smith doubted that famine was the great leveller that Malthus insisted it was.

and predicted growth estimates for the total world population from the 1900, we see that it has actually grown exponentially.)

This simple example shows that it is difficult to make long–term predictions (or, even relatively short–term ones), unless we know sufficient facts to incorporate in the model to make it a *reliable predictor*. In the long run, clearly, there must be some adjustment to such exponential growth. François Verhulst [Ver838, Ver845][26] proposed that a *self–limiting process* should operate when a population becomes too large. He proposed the so–called *logistic growth* population model,

$$\dot{x} = rx(1 - x/K), \tag{1.24}$$

where r, K are positive constants. In the Verhulst logistic model (4.7), the constant K is the *carrying capacity* of the environment (usually determined by the available sustaining resources), while the per capita birth rate $rx(1-x/K)$ is dependent on x. There are two steady states (where $\dot{x} = 0$) for (4.7): (i) $x = 0$ (unstable, since linearization about it gives $\dot{x} \approx rx$); and (ii) $x = K$ (stable, since linearization about it gives $\frac{d}{dt}(x - K) \approx -r(x - K)$, so $\lim_{t\to\infty} x = K$). The carrying capacity K determines the size of the stable steady state population, while r is a measure of the rate at which it is reached (i.e., the measure of the dynamics) – thus $1/r$ is a representative timescale of the response of the model to any change in the population. The solution of (4.7) is

$$x(t) = \frac{x_0 K e^{rt}}{[K + x_0(e^{rt} - 1)]} \qquad \Longrightarrow \qquad \lim_{t\to\infty} x(t) = K.$$

In general, if we consider a population to be governed by

$$\dot{x} = f(x), \tag{1.25}$$

where typically $f(x)$ is a nonlinear function of x, then the equilibrium solutions x^* are solutions of $f(x) = 0$, and are linearly stable to small perturbations if $\dot{f}(x^*) < 0$, and unstable if $\dot{f}(x^*) > 0$ [Mur02].

In the mid 20th century, ecologists realised that many species had no overlap between successive generations and so population growth happens in discrete–time steps x_t, rather than in continuous–time $x(t)$ as suggested by the conservative law (1.22) and its Maltus–Verhulst derivations. This leads to study *discrete–time models* given by *difference equations*, or, *maps*, of the form

$$x_{t+1} = f(x_t), \tag{1.26}$$

where $f(x_t)$ is some generic nonlinear function of x_t. Clearly, (1.26) is a discrete–time version of (1.25). However, instead of solving differential equations, if we know the particular form of $f(x_t)$, it is a straightforward matter to

---

[26] François Verhulst (October 28, 1804–February 15, 1849, Brussels, Belgium) was a mathematician and a doctor in number theory from the University of Ghent in 1825. Verhulst published in 1838 the logistic demographic model (4.7).

evaluate x_{t+1} and subsequent generations by simple recursion of (1.26). The skill in modelling a specific population's growth dynamics lies in determining the appropriate form of $f(x_t)$ to reflect known observations or facts about the species in question.

In 1970s, Robert May, a physicist by training, won the Crafoord Prize for 'pioneering ecological research in theoretical analysis of the dynamics of populations, communities and ecosystems', by proposing a simple *logistic map* model for the generic population growth (1.26).[27] May's model of population growth is the celebrated *logistic map* [May76, May73, May76],

$$x_{t+1} = r\, x_t\, (1 - x_t),\qquad\qquad(1.27)$$

where r is the *Malthusian parameter* that varies between 0 and 4, and the initial value of the population $x_0 = x(0)$ is restricted to be between 0 and 1. Therefore, in May's logistic map (1.27), the generic function $f(x_t)$ gets a specific quadratic form

$$f(x_t) = r\, x_t\, (1 - x_t).$$

For $r < 3$, the x_t have a single value. For $3 < r < 3.4$, the x_t oscillate between two values (see *bifurcation diagram*[28] on Figure 1.31). As r increases, bifurcations occur where the number of iterates doubles. These *period doubling bifurcations* continue to a limit point at $r_{lim} = 3.569944$ at which the period is 2^{∞} and the dynamics become chaotic. The r values for the first two bifurcations can be found analytically, they are $r_1 = 3$ and $r_2 = 1 + \sqrt{6}$. We can label the successive values of r at which bifurcations occur as r_1, r_2, ... The universal number associated with such period doubling sequences is called the *Feigenbaum number*,

$$\delta = \lim_{k \to \infty} \frac{r_k - r_{k-1}}{r_{k+1} - r_k} \approx 4.669.$$

This series of period–doubling bifurcations says that close enough to r_{lim} the distance between bifurcation points decreases by a factor of δ for each bifurcation. The complex *fractal pattern* got in this way shrinks indefinitely.

---

[27] Lord Robert May received his Ph.D. in theoretical physics from University of Sydney in 1959. He then worked at Harvard University and the University of Sydney before developing an interest in animal population dynamics and the relationship between complexity and stability in natural communities. He moved to Princeton University in 1973 and to Oxford and the Imperial College in 1988. May was able to make major advances in the field of population biology through the application of mathematics. His work played a key role in the development of *theoretical ecology* through the 1970s and 1980s. He also applied these tools to the study of disease and to the study of *bio–diversity*.

[28] A bifurcation diagram shows the possible long–term values a variable of a system can get in function of a parameter of the system.

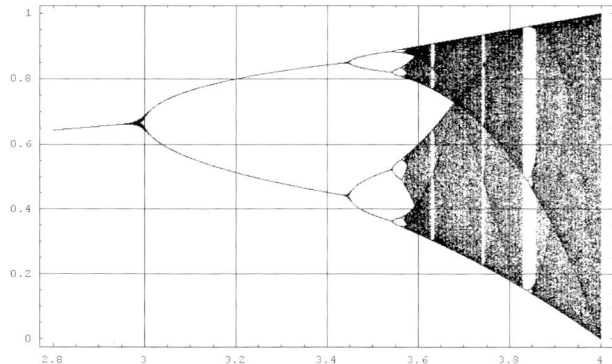

Fig. 1.31. Bifurcation diagram for the logistic map, simulated using $Mathematica^{TM}$.

1.2.6 *Michel Hénon:* A Special 2D Map and Its Strange Attractor

Michel Hénon (born 1931 in Paris, France) is a mathematician and astronomer. He is currently at the Nice Observatory. In astronomy, Hénon is well known for his contributions to stellar dynamics, most notably the problem of *globular cluster* (see [Gle87]). In late 1960s and early 1970s he was involved in dynamical evolution of star clusters, in particular the globular clusters. He developed a numerical technique using *Monte Carlo methods*, to follow the dynamical evolution of a spherical star cluster much faster than the so–called $n-$body methods. In mathematics, he is well known for the Hénon map, a simple discrete dynamical system that exhibits chaotic behavior. Lately he has been involved in the restricted $3-$body problem.

His celebrated *Hénon map* [Hen69] is a discrete–time dynamical system that is an extension of the *logistic map* (1.27) and exhibits a chaotic behavior. The map was introduced by Michel Hénon as a simplified model of the *Poincaré section* of the *Lorenz system* (1.21). This 2D–map takes a point (x, y) in the plane and maps it to a new point defined by equations

$$x_{n+1} = y_n + 1 - ax_n^2, \qquad y_{n+1} = bx_n,$$

The map depends on two parameters, a and b, which for the canonical Hénon map have values of $a = 1.4$ and $b = 0.3$ (see Figure 1.32). For the canonical values the Hénon map is chaotic. For other values of a and b the map may be chaotic, intermittent, or converge to a periodic orbit. An overview of the type of behavior of the map at different parameter values may be obtained from its orbit (or, bifurcation) diagram (see Figure 1.33). For the canonical map, an initial point of the plane will either approach a set of points known as the *Hénon strange attractor*, or diverge to infinity. The Hénon attractor is a fractal, smooth in one direction and a Cantor set in another. Numerical estimates yield a correlation dimension of 1.42 ± 0.02 (Grassberger, 1983) and

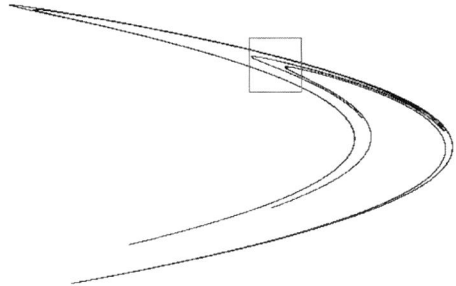

Fig. 1.32. *Hénon strange attractor* (see text for explanation), simulated using *Dynamics Solver*^{TM}.

a Hausdorff dimension of 1.261 ± 0.003 (Russel 1980) for the Hénon attractor. As a dynamical system, the canonical Hénon map is interesting because, unlike the logistic map, its orbits defy a simple description. The Hénon map maps

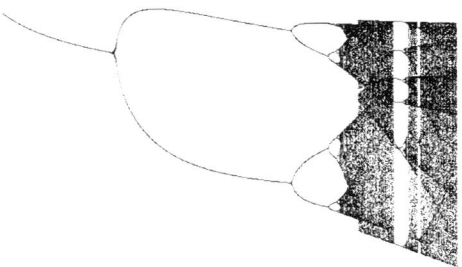

Fig. 1.33. Bifurcation diagram of the *Hénon strange attractor*, simulated using *Dynamics Solver*^{TM}.

two points into themselves: these are the invariant points. For the canonical values of a and b, one of these points is on the attractor: $x = 0.631354477...$ and $y = 0.189406343...$ This point is unstable. Points close to this fixed–point and along the slope 1.924 will approach the fixed–point and points along the slope -0.156 will move away from the fixed–point. These slopes arise from the linearizations of the *stable manifold* and *unstable manifold* of the fixed–point. The unstable manifold of the fixed–point in the attractor is contained in the strange attractor of the Hénon map. The Hénon map does not have a strange attractor for all values of the parameters a and b. For example, by keeping b fixed at 0.3 the bifurcation diagram shows that for a = 1.25 the Hénon map has a stable periodic orbit as an attractor. Cvitanovic *et al.* [CGP88] showed how the structure of the Hénon strange attractor could be understood in terms of unstable periodic orbits within the attractor.

For the (slightly modified) Hénon map: $x_{n+1} = ay_n + 1 - x_n^2$, $y_{n+1} = bx_n$, there are three *basins of attraction* (see Figure 1.34).

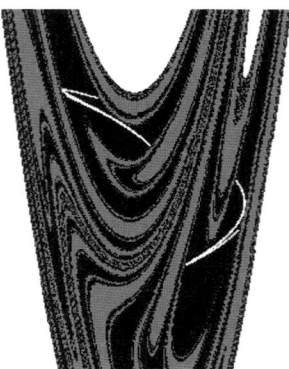

Fig. 1.34. Three basins of attraction for the Hénon map $x_{n+1} = ay_n + 1 - x_n^2$, $y_{n+1} = bx_n$, with $a = 0.475$.

The *generalized Hénon map* is a 3D–system (see Figure 1.35)

$$x_{n+1} = a\,x_n - z\,(y_n - x_n^2)), \qquad y_{n+1} = z\,x_n + a\,(y_n - x_n^2)), \qquad z_{n+1} = z_n,$$

where $a = 0.24$ is a parameter. It is an *area–preserving map*, and simulates the *Poincaré map* of period orbits in *Hamiltonian systems*. Repeated random initial conditions are used in the simulation and their gray–scale color is selected at random.

Other Famous 2D Chaotic Maps

1. The *standard map*:

$$x_{n+1} = x_n + y_{n+1}/2\pi, \qquad y_{n+1} = y_n + a\sin(2\pi x_n).$$

2. The *circle map*:

$$x_{n+1} = x_n + c + y_{n+1}/2\pi, \qquad y_{n+1} = by_n - a\sin(2\pi x_n).$$

3. The *Duffing map*:

$$x_{n+1} = y_n, \qquad y_{n+1} = -bx_n + ay_n - y_n^3.$$

4. The *Baker map*:

$$x_{n+1} = bx_n, \qquad y_{n+1} = y_n/a \qquad \text{if} \quad y_n \le a,$$
$$x_{n+1} = (1-c) + cx_n, \qquad y_{n+1} = (y_n - a)/(1-a) \qquad \text{if} \quad y_n > a.$$

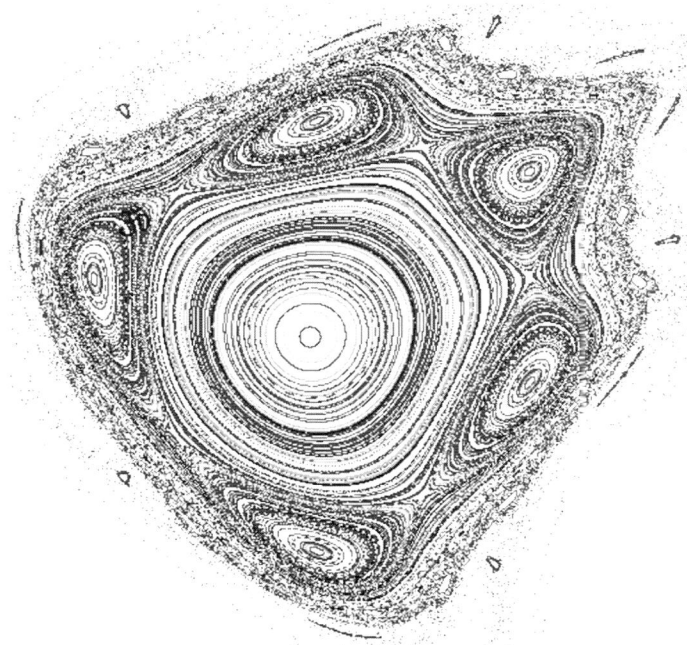

Fig. 1.35. Phase–plot of the *area–preserving generalized Hénon map*, simulated using $Dynamics\,Solver^{TM}$.

5. The *Kaplan–Yorke map*:

$$x_{n+1} = ax_n \bmod 1, \qquad y_{n+1} = -by_n + \cos(2\pi x_n).$$

6. The *Ott–Grebogi–Yorke map*:

$$x_{n+1} = x_n + w_1 + aP_1(x_n, y_n) \bmod 1,$$
$$y_{n+1} = y_n + w_2 + aP_2(x_n, y_n) \bmod 1,$$

where the nonlinear functions P_1, P_2 are sums of sinusoidal functions $A_{rs}^{(i)} \sin[2\pi(rx + sy + B_{rs}^{(i)})]$, with $(r, s) = (0, 1), (1, 0), (1, 1), (1, -1)$, while $A_{rs}^{(i)}, B_{rs}^{(i)}$ were selected randomly in the range $[0, 1]$.

1.3 Some Classical Attractor and Chaotic Systems

Here we present numerical simulations of several popular chaotic systems (see, e.g., [Wig90, BCB92, Ach97]). Generally, to observe chaos in continuous time system, it is known that the dimension of the equation must be three or higher. That is, *there is no chaos in any phase plane* (see [Str94]), we need the third dimension for chaos in continuous dynamics. However, note that *all forced oscillators have actually dimension 3, although they are commonly written as second–order ODEs.*[29] On the other hand, in discrete–time systems like logistic map or Hénon map, we can see chaos even if the dimension is one.

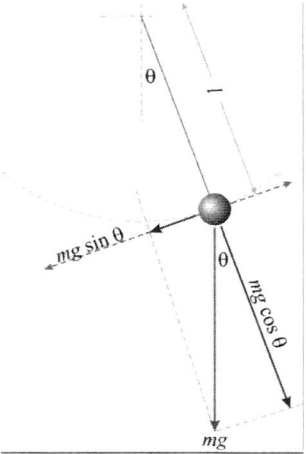

Fig. 1.36. Force diagram of a simple gravity pendulum.

Simple Pendulum

Recall (see [II05, II06a, II06b]) that a simple *un–damped pendulum* (see Figure 1.36), given by equation

$$\ddot{\theta} + \frac{g}{l}\sin\theta = 0, \qquad (1.28)$$

swings forever; it has closed orbits in a 2D phase–space (see Figure 1.37).

The conservative (un–damped) pendulum equation (1.28) does not take into account the effects of friction and dissipation. On the other hand, a simple *damped pendulum* (see Figure 1.36) is given by modified equation, including a damping term proportional to the velocity,

$$\ddot{\theta} + \gamma\dot{\theta} + \frac{g}{l}\sin\theta = 0,$$

with the positive constant damping γ. This pendulum settles to rest (see Figure 1.38). Its spiralling orbits lead to a point attractor (focus) in a 2D

---

[29] Both Newtonian equation of motion and RLC circuit can generate chaos, provided they have a forcing term. This forcing (driving) term in second–order ODEs is the motivational reason for development of the jet–bundle formalism for non–autonomous dynamics (see [II06b]).

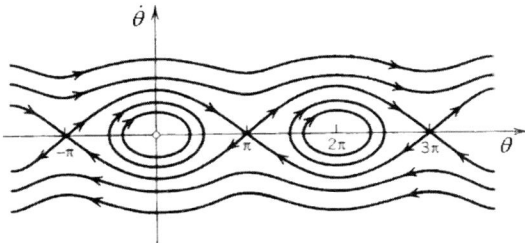

Fig. 1.37. Phase portrait of a simple gravity pendulum.

phase–space. All closed trajectories for periodic solutions are destroyed, and the trajectories spiral around one of the critical points, corresponding to the vertical equilibrium of the pendulum. On the phase plane, these critical points are stable spiral points for the underdamped pendulum, and they are stable nodes for the overdamped pendulum. The unstable equilibrium at the inverted vertical position remains an unstable saddle point. It is clear physically that damping means loss of energy. The dynamical motion of the pendulum decays due to the friction and the pendulum relaxes to the equilibrium state in the vertical position.

Finally, a *driven pendulum*, periodically forced by a force term $F \cos(w_D t)$, is given by equation (see our introductory example (1.1))

$$\ddot{\theta} + \gamma \dot{\theta} + \frac{g}{l} \sin \theta = F \cos(w_D t). \tag{1.29}$$

It has a 3D phase–space and can exhibit chaos (for certain values of its parameters, see Figure 1.39).

Van der Pol Oscillator

The unforced Van der Pol oscillator has the form of a second order ODE (compare with 1.4 above)

$$\ddot{x} = \alpha \left(1 - x^2\right) \dot{x} - \omega^2 x. \tag{1.30}$$

Its celebrated *limit cycle* is given in Figure 1.40. The simulation is performed with zero initial conditions and parameters $\alpha = \text{random}(0, 3)$, and $\omega = 1$. The Van der Pol oscillator was the first *relaxation oscillator*, used in 1928 as a model of human heartbeat (ω controls how much voltage is injected into the system, and α controls the way in which voltage flows through the system). The oscillator was also used as a model of an electronic circuit that appeared in very early radios in the days of vacuum tubes. The tube acts like a normal resistor when current is high, but acts like a negative resistor if the current is low. So this circuit pumps up small oscillations, but drags down large oscillations. α is a constant that affects how nonlinear the system is. For

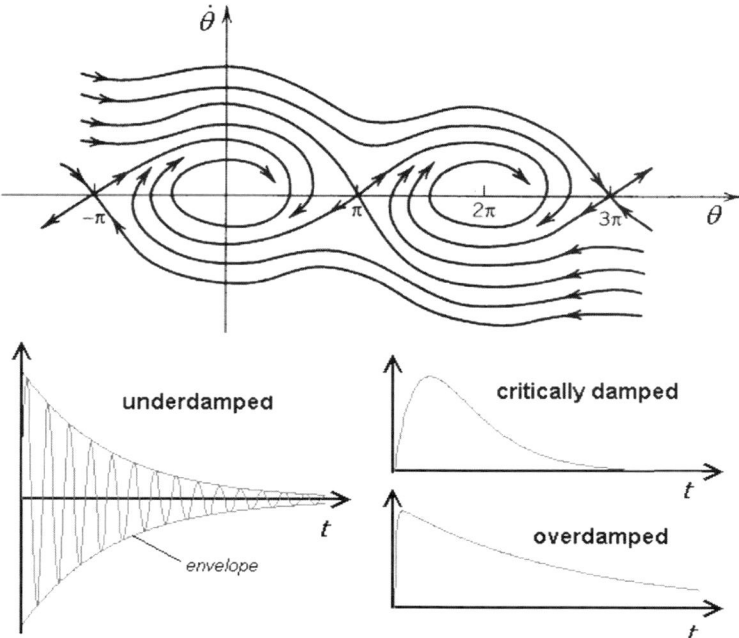

Fig. 1.38. A damped gravity pendulum settles to a rest: its phase portrait (up) shows spiralling orbits that lead to a focus attractor; its time plot (down) shows three common damping cases.

α equal to zero, the system is actually just a linear oscillator. As α grows the nonlinearity of the system becomes considerable.

The *sinusoidally–forced Van der Pol oscillator* is given by equation

$$\ddot{x} - \alpha \left(1 - x^2\right) \dot{x} + \omega^2 \, x = \gamma \cos(\phi t), \qquad (1.31)$$

where ϕ is the forcing frequency and γ is the amplitude of the forcing sinusoid.

Nerve Impulse Propagation

The nerve impulse propagation along the axon of a neuron can be studied by combining the equations for an excitable membrane with the differential equations for an electrical core conductor cable, assuming the axon to be an infinitely long cylinder. A well known approximation of FitzHugh [Fit61] and Nagumo [NAY60] to describe the propagation of voltage pulses $V(x,t)$ along the membranes of nerve cells is the set of coupled PDEs

$$V_{xx} - V_t = F(V) + R - I, \qquad R_t = c(V + a - bR), \qquad (1.32)$$

where $R(x,t)$ is the recovery variable, I the external stimulus and a, b, c are related to the membrane radius, specific resistivity of the fluid inside the membrane and temperature factor respectively.

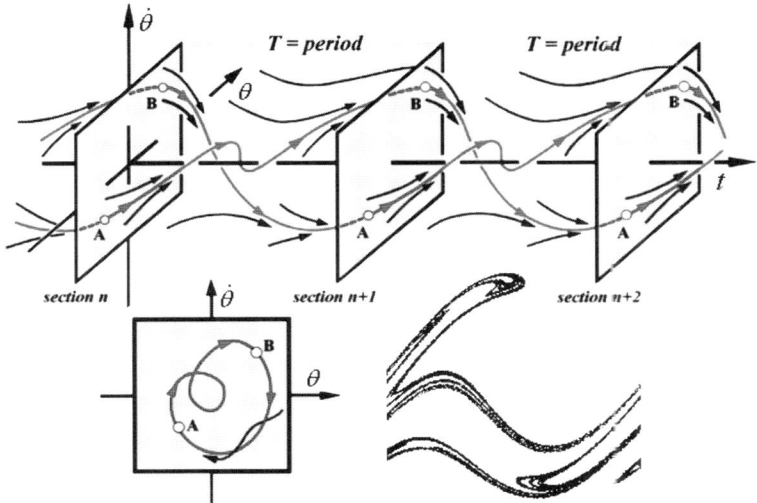

Fig. 1.39. A driven pendulum has a 3D phase–space with angle θ, angular velocity $\dot\theta$ and time t. Dashed lines denote steady states, while solid lines denote transients. Right–down we see a sample chaotic attractor (adapted and modified from [TS01]).

When the spatial variation of V, namely V_{xx}, is negligible, (1.32) reduces to the Van der Pol oscillator,

$$\dot V = V - \frac{V^3}{3} - R + I, \qquad \dot R = c(V + a - bR),$$

with $F(V) = -V + \frac{V^3}{3}$. Normally the constants in (1.32) satisfy the inequalities $b < 1$ and $3a + 2b > 3$, though from a purely mathematical point of view this need not be insisted upon. Then with a periodic (ac) applied membrane current $A_1 \cos\omega t$ and a (dc) bias A_0, the Van der Pol equation becomes

$$\dot V = V - \frac{V^3}{3} - R + A_0 + A_1 \cos\omega t, \qquad \dot R = c(V + a - bR). \qquad (1.33)$$

Further, (1.33) can be rewritten as a single second–order ODE by differentiating $\dot V$ with respect to time and using $\dot R$ for R,

$$\ddot V - (1 - bc)\left\{1 - \frac{V^2}{1 - bc}\right\}\dot V - c(b-1)V + \frac{bc}{3}V^3$$
$$= c(A_0 b - a) + A_1 \cos(\omega t + \phi), \qquad (1.34)$$

where $\phi = \tan^{-1}\frac{\omega}{bc}$. Using the transformation $x = (1 - bc)^{-(1/2)}V$, $t \longrightarrow t' = t + \frac{\phi}{\omega}$, (1.34) can be rewritten as

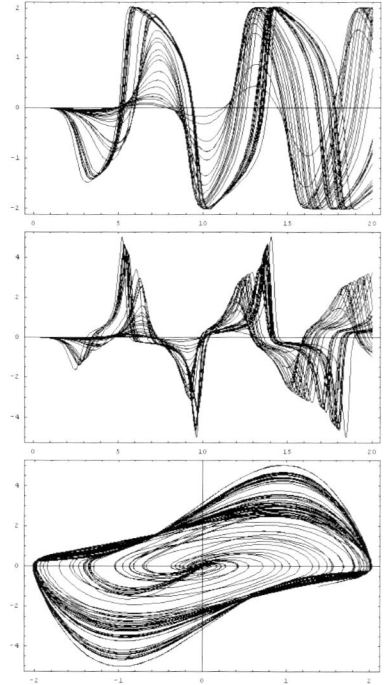

Fig. 1.40. Cascade of 30 unforced Van der Pol oscillators, simulated using $Mathematica^{TM}$; top–down: displacements, velocities and phase–plot (showing the celebrated limit cycle).

$$\ddot{x} + p(x^2 - 1)\dot{x} + \omega_0^2 x + \beta x^3 = f_0 + f_1 \cos \omega t, \qquad \text{where} \qquad (1.35)$$

$$p = (1 - bc), \qquad \omega_0^2 = c(1 - b), \qquad \beta = bc\frac{(1 - bc)}{3},$$

$$f_0 = c\frac{(A_0 b - a)}{\sqrt{1 - bc}}, \qquad f_1 = \frac{A_1}{\sqrt{1 - bc}}.$$

Note that (1.35), or its rescaled form

$$\ddot{x} + p(kx^2 + g)\dot{x} + \omega_0^2 x + \beta x^3 = f_0 + f_1 \cos \omega t, \qquad (1.36)$$

is the *Duffing–Van der Pol equation*. In the limit $k = 0$, we have the Duffing equation discussed below (with $f_0 = 0$), and in the case $\beta = 0$ ($g = -1$, $k = 1$) we have the forced van der Pol equation. Equation (1.36) exhibits a very rich variety of bifurcations and chaos phenomena, including quasi–periodicity, phase lockings and so on, depending on whether the potential $V = \frac{1}{2}\omega_0^2 x^2 + \frac{\beta x^4}{4}$ is i) a double well, ii) a single well or iii) a double hump [Lak97, Lak03].

Duffing Oscillator

The forced *Duffing oscillator* [Duf18] has the form similar to (1.31),

$$\ddot{x} + b\,\dot{x} - a\,x\,(1 - x^2) = \gamma \cos(\phi t). \qquad (1.37)$$

Stroboscopic *Poincaré sections* of a *strange attractor* can be seen (Figure 1.41), with the *stretch–and–fold* action at work. The simulation is performed with parameters: $a = 1$, $b = 0.2$, and $\gamma = 0.3$, $\phi = 1$. The Duffing equation is used to model a double well oscillator such as the magneto–elastic mechanical system. This system consists of a beam positioned vertically between two magnets, with the top end fixed, and the bottom end free to swing. The beam will be attracted to one of the two magnets, and given some velocity will oscillate about that magnet until friction stops it. Each of the magnets creates a fixed–point where the beam may come to rest above that magnet and remain there in equilibrium. However, when this whole system is shaken by a periodic forcing term, the beam may jump back and forth from one magnet to the other in a seemingly random manner. Depending on how big the shaking term is, there may be no stable fixed–points and no stable fixed cycles in the system.

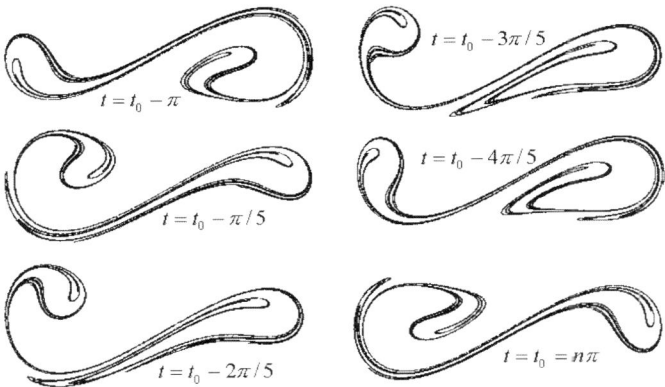

Fig. 1.41. Duffing strange attractor, showing stroboscopic Poincaré sections; simulated using *Dynamics Solver*^{TM}.

Rossler System

Classical *Rossler system* is given by equations

$$\dot{x} = -y - z, \qquad \dot{y} = x + b\,y, \qquad \dot{z} = b + z\,(x - a). \qquad (1.38)$$

Using the parameter values $a = 4$ and $b = 0.2$, the phase–portrait is produced (see Figure 1.42), showing the celebrated attractor. The system is credited to O. *Rossler* and arose from work in chemical kinetics.

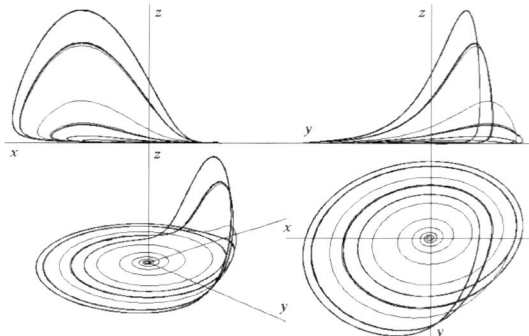

Fig. 1.42. The celebrated Rossler attractor, simulated using $Dynamics\,Solver^{TM}$.

Chua's Circuit

Chua's circuit is a simple electronic circuit that exhibits classic chaotic behavior. First introduced in 1983 by Leon O. Chua, its ease of construction has made it an ubiquitous real–world example of a chaotic system, leading some to declare it 'a paradigm for chaos'. It has been the subject of much study; hundreds of papers have been published on this topic (see [Chu94]).

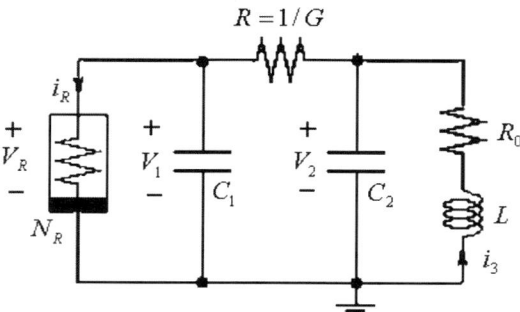

Fig. 1.43. Chua's circuit.

The *Chua's circuit* consists of two linear capacitors, two linear resistors, one linear inductor and a nonlinear resistor (see Figure 1.43). By varying the various circuit parameters, we can get complicated nonlinear and chaotic phenomena. Let us consider the case where we vary the conductance G of the resistor R and keep the other components fixed. In particular, we choose $L = 18\,mH, R_0 = 12.5\,Ohms, C_1 = 10\,nF, C_2 = 100\,nF$. The nonlinear resistor N_R (Chua's diode) is chosen to have a piecewise–linear $V - I$ characteristic of the form:

$$i = -\begin{cases} G_b v + G_a - G_b & \text{if} \quad v > 1, \\ G_a v & \text{if} \quad |v| < 1, \\ G_b v + G_b - G_a & \text{if} \quad v < -1 \end{cases}$$

with $G_a = -0.75757\,mS$, and $G_b = -0.40909\,mS$.

Starting from low $G-$values, the circuit is stable and all trajectories converge towards one of the two stable equilibrium points. As G is increased, a limit cycle appears due to a *Hopf–like bifurcation*. In order to observe the period–doubling route to chaos, we need to further increase G. At the end of the period–doubling bifurcations, we observe a chaotic attractor. Because of symmetry, there exists a twin attractor lying in symmetrical position with respect the the origin. As G is further increased, these two chaotic attractors collide and form a 'double scroll' chaotic attractor.

After normalization, the state equations for the Chua's circuit read:

$$\dot{x} = a(y - x - f(x)), \qquad \dot{y} = x - y + z, \qquad \dot{z} = -by - cz, \qquad (1.39)$$

where $f(x)$ is a nonlinear function to be manipulated to give various chaotic behaviors.

By using a specific form of the nonlinearity $f(x)$, a family of *multi–spiral strange attractors* have been generated in [Ala99] (see Figure 1.44).

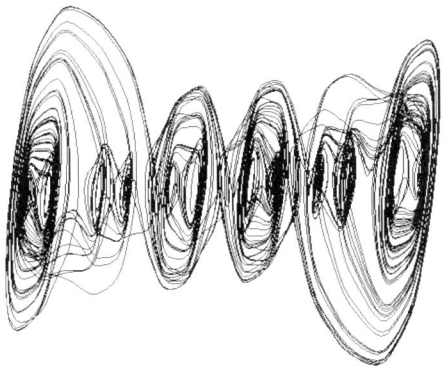

Fig. 1.44. A multi–spiral strange attractor of the Chua's circuit (modified from [Ala99]).

Inverted Pendulum

Stability of the *inverted driven pendulum* given by equation

$$\ddot{\theta} + k\,\dot{\theta} + (1 + a\,\sqrt{\phi}\cos(\phi t))\,\sin\theta = 0,$$

where θ is the angle, is simulated in Figure 1.45, using the parameter $a = 0.33$. It is possible to stabilize a mathematical pendulum around the upper vertical position by moving sinusoidally the suspension point in the vertical direction. Furthermore, the perturbed solution may be of two kinds: one goes to the vertical position while the other becomes periodic (see, e.g., [Ach97]).

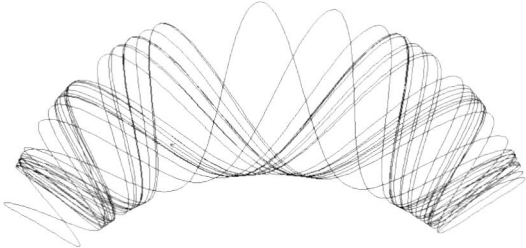

Fig. 1.45. Duffing strange attractor, showing stroboscopic Poincaré sections; simulated using $Dynamics\,Solver^{TM}$.

Elastic Pendulum

Elastic pendulum (Figure 1.46) of proper length l, mass m and elastic constant k is given by equation

$$\ddot{x} = x\,\sqrt{\dot{y}} + \cos y - a\,(x-1), \qquad \ddot{y} = -(2\,\dot{x}\,\dot{y} + \sin y)/x,$$

where the parameter $a = kl/mg = 0.4$. High values of a give raise to a simple pendulum.

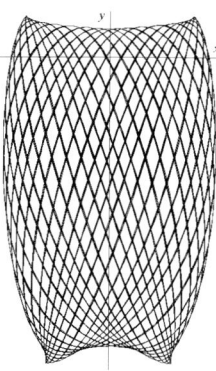

Fig. 1.46. Phase–portrait of an elastic pendulum showing *Lissajous curves*; simulated using $Dynamics\,Solver^{TM}$.

Lorenz–Maxwell–Haken System

In 1975, H. Haken showed [Hak83] that the *Lorenz equations* (1.29) were isomorphic to the *Maxwell–Haken laser equations*

$$\dot{E} = \sigma(P - E), \qquad \dot{P} = \beta(ED - P), \qquad \dot{D} = \gamma(\sigma - 1 - D - \sigma EP),$$

Here, the variables in the Lorenz equations, namely x, y and z correspond to the slowly varying amplitudes of the electric field E and polarization P and the inversion D respectively in the Maxwell–Haken equations. The parameters are related via $c = \frac{\gamma}{\beta}$, $a = \frac{\sigma}{\beta}$ and $b = \sigma + 1$, where γ is the relaxation rate of the inversion, β is the relaxation rate of the polarization, σ is the field relaxation rate, and σ represents the normalized pump power.

Autocatalator System

This 4D *autocatalator* system from *chemical kinetics* (see Figure 1.47) is defined as (see, e.g., [BCB92])

$$\dot{x}_1 = -a\,x_1, \quad \dot{x}_2 = a\,x_1 - b\,x_2 - x_2\,x_3^2, \quad \dot{x}_3 = b\,x_2 - x_3 + x_2\,x_3^2, \quad \dot{x}_4 = x_3.$$

The simulation is performed with parameters: $a = 0.002$, and $b = 0.08$.

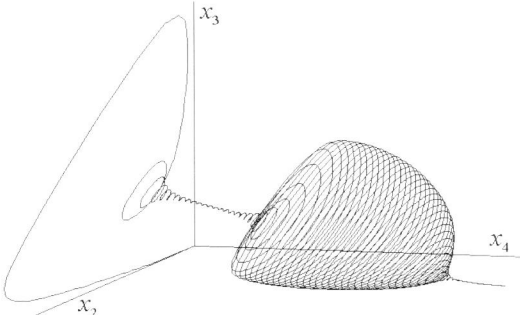

Fig. 1.47. 3D phase–portrait of the 4D autocatalator system, simulated using *Dynamics Solver*^{TM}.

Mandelbrot and Julia Sets

Recall that *Mandelbrot and Julia sets* (see Figure 1.48) are celebrated *fractals*. Recall that fractals are sets with *fractional dimension* (see Figure 1.49). The Mandelbrot and Julia sets are defined either by a quadratic *conformal $z-map$* [Man80a, Man80b]

$$z_{n+1} = z_n^2 + c,$$

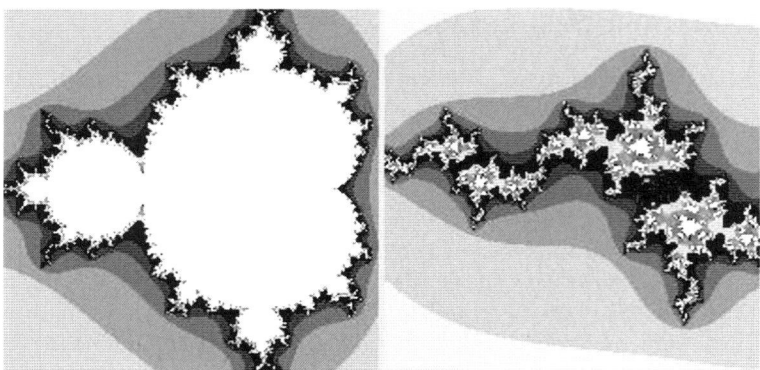

Fig. 1.48. The celebrated conformal Mandelbrot (left) and Julia (right) sets in the complex plane, simulated using *Dynamics Solver*^{TM}.

or by a real (x, y)−map

$$x_{n+1} = \sqrt{x_n} - \sqrt{y_n} + c_1, \qquad y_{n+1} = 2\, x_n\, y_n + c_2,$$

where c, c_1 and c_2 are parameters. For almost every c, this conformal transformation generates a fractal (probably, only for $c = -2$ it is not a fractal). Julia set J_c with $c \ll 1$, the *capacity dimension* is

$$d_{cap} = 1 + \frac{|c|^2}{4 \ln 2} + O(|c|^3).$$

The set of all points for which J_c is connected is the Mandelbrot set.[30]

Biomorphic Systems

Closely related to the Mandelbrot and Julia sets are *biomorphic systems*, which look like one–celled organisms. The term '*biomorph*' was proposed by C. Pickover from IBM [Pic86, Pic87]. Pickover's biomorphs inhabit the complex plane like the the Mandelbrot and Julia sets and exhibit a *protozoan morphology*. Biomorphs began for Pickover as a 'bug' in a program intended to probe the fractal properties of various formulas. He accidentally used an OR logical operator instead of an AND operator in the conditional test for the

---

[30] The Mandelbrot set has its place in complex–valued dynamics, a field first investigated by the French mathematicians Pierre Fatou [Fat19, Fat22] and Gaston Julia [Jul18] at the beginning of the 20th century. For general families of holomorphic functions, the boundary of the Mandelbrot set generalizes to the bifurcation locus, which is a natural object to study even when the connectedness locus is not useful. A related *Mandelbar set* was encountered by mathematician John Milnor in his study of parameter slices of real cubic polynomials; it is not locally connected; this property is inherited by the connectedness locus of real cubic polynomials.

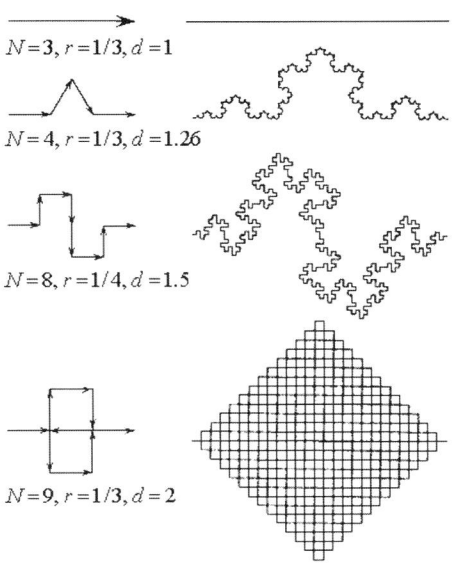

$N=3, r=1/3, d=1$

$N=4, r=1/3, d=1.26$

$N=8, r=1/4, d=1.5$

$N=9, r=1/3, d=2$

Fig. 1.49. Fractal dimension of curves in $\mathbb{R}^2$: $d = \frac{\log N}{\log 1/r}$.

size of z's real and imaginary parts. The cilia that project from the biomorphs are a consequence of this 'error'. Each biomorph is generated by multiple iterations of a particular conformal map,

$$z_{n+1} = f(z_n, c),$$

where c is a parameter. Each iteration takes the output of the previous operations as the input of the next iteration. To generate a biomorph, one first needs to lay out a grid of points on a rectangle in the complex plane [And01]. The coordinate of each point constitutes the real and imaginary parts of an initial value, z_0, for the iterative process. Each point is also assigned a pixel on the computer screen. Depending on the outcome of a simple test on the 'size' of the real and imaginary parts of the final value, the pixel is colored either black or white. The biomorphs presented in Figure 1.50 are generated using the following conformal functions:

1. $f(z, c) = z^3$,
2. $f(z, c) = z^3 + c,$ $c = 10$,
3. $f(z, c) = z^3 + c,$ $c = 10 - 10i$,
4. $f(z, c) = z^5 + c,$ $c = 0.77 - 0.77i$,
5. $f(z, c) = z^3 + \sin z + c,$ $c = 1 - i$,
6. $f(z, c) = z^6 + \sin z + c,$ $c = 0.5 - 0.5i$,
7. $f(z, c) = z^2 \sin z + c,$ $c = 0.78 - 0.78i$,
8. $f(z, c) = z^c,$ $c = 5 - i$,

9. $f(z,c) = |z|^c \sin z,$ $c = 4,$
10. $f(z,c) = |z|^c \cos z + c,$ $c = 3 + 3i,$
11. $f(z,c) = |z|^c(\cos z + z) + c,$ $c = 3 + 2i.$

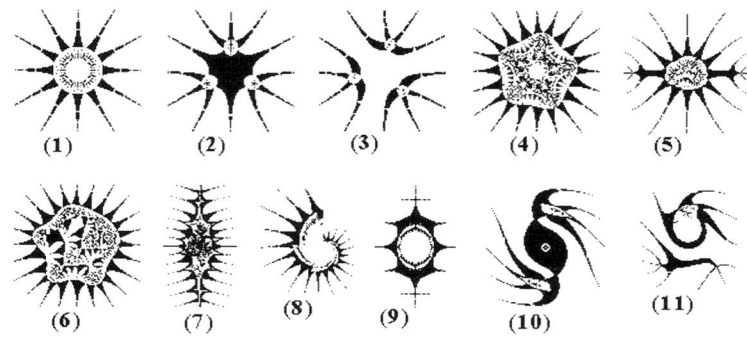

Fig. 1.50. *Pickover's biomorphs* (see text for details).

1.4 Basics of Continuous Dynamical Analysis

In this section we give basics of dynamical analysis as used in standard chaos theory (for more technical details, see [II06b]).

1.4.1 A Motivating Example

An example that illustrates many of the concepts of dynamical systems is the *ball in a rotating hoop* (see Figure 1.51). This system consists of a rigid hoop that hangs from the ceiling with a small ball resting in the bottom of the hoop. The hoop rotates with frequency ω about a vertical axis through its center (Figure 1.51–left). Now we consider varying ω, keeping the other parameters fixed. For small values of ω, the ball stays at the bottom of the hoop and that position is stable. Let us accept this in an intuitive sense for the moment, as we will have to define this concept carefully. However, when ω reaches the critical value ω_0, this point becomes unstable and the ball rolls up the side of the hoop to a new position $x(\omega)$, which is stable. The ball may roll to the left or to the right, depending, perhaps upon the side of the vertical axis to which it was initially leaning (Figure 1.51–right). The position at the bottom of the hoop is still a fixed–point, but it has become unstable. The solutions to the initial value problem governing the ball's motion are unique for all values of ω.

Using principles of mechanics one can show that the equations for this system are given by

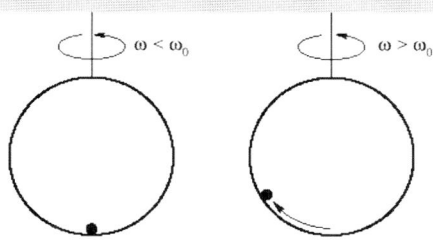

Fig. 1.51. The ball in the hoop system: left–stable, right–unstable (adapted from [Mar99]).

$$mR^2\ddot{\theta} = mR^2\omega^2 \sin\theta \cos\theta - mgR \sin\theta - \nu R\dot{\theta},$$

where R is the radius of the hoop, θ is the angle from the bottom vertical, m is the mass of the ball, g is the acceleration due to gravity, and ν is a coefficient of friction. To analyze this system, we use a *phase–plane analysis*; that is, we write the equation as a system:

$$\dot{x} = y, \qquad \dot{y} = \frac{g}{R}(\alpha \cos x - 1)\sin x - \beta y, \qquad (1.40)$$

where $\alpha = R\omega^2/g$ and $\beta = \nu/m$. This system of equations produces for each initial point in the xy–plane, a unique trajectory. That is, given a point (x_0, y_0) there is a unique solution $(x(t), y(t))$ of the equation that equals (x_0, y_0) at $t = 0$. This statement is proved by using general existence and uniqueness theory for ordinary differential equations. When we draw these curves in the plane, we get gures like those shown in Figure 1.52.

More generally, *equilibrium points* are points where the right hand side of the system (1.40) vanish. Technically, we say that the original stable fixed–point has become unstable and has split into two stable fixed–points. One can use some basic stability theory to show that $\omega_0 = \sqrt{g/R}$(see Figure 1.52). This is one of the simplest situations in which symmetric problems can have non–symmetric solutions and in which there can be multiple stable equilibria, so there is non–uniqueness of equilibria (even though the solution of the initial value problem is unique) [Mar99].

This example shows that in some systems the phase portrait can change as certain parameters are changed. Changes in the qualitative nature of phase portraits as parameters are varied are called *bifurcations*. Consequently, the corresponding parameters are often called *bifurcation parameters*. These changes can be simple such as the formation of new fixed–points, these are called *static bifurcations*, to *dynamic bifurcations* such as the formation of *periodic orbits*, that is, an orbit $x(t)$ with the property that $x(t + T) = x(t)$ for some T and all t, or more complex dynamical structures.

A very important bifurcation is called the *Hopf bifurcation* (more properly, the Poincaré–Andronov–Hopf bifurcation). This is a dynamic bifurcation in

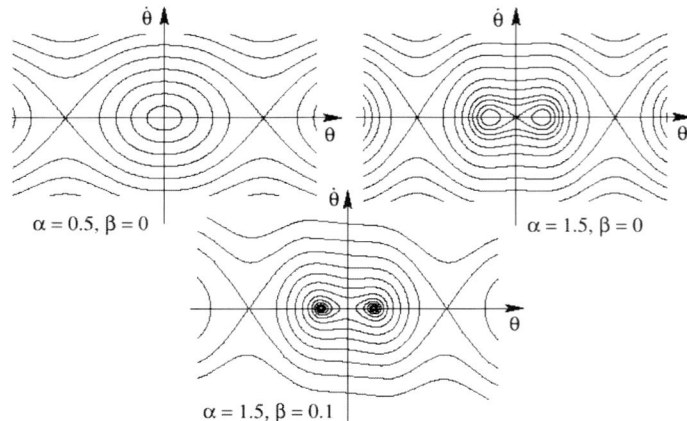

Fig. 1.52. The phase portrait for the ball in the hoop before and after the onset of instability for the case $g/R = 1$ (adapted from [Mar99]).

which, roughly speaking, a periodic orbit rather than another fixed–point is formed when an equilibrium looses stability. An everyday example of a Hopf bifurcation we all encounter is flutter. For example, when a television antenna 'sings' in the wind, there is probably a Hopf bifurcation occurring. An example that is physically easy to understand is flow through a tube. One considers a straight vertical rubber tube conveying fluid. The lower end is a nozzle from which the fluid escapes. This is called a *follower–load problem* since the water exerts a force on the free end of the tube which follows the movement of the tube. Those with any experience in a garden will not be surprised by the fact that the hose will begin to oscillate if the water velocity is high enough.

More precisely, *Hopf bifurcation* is a local bifurcation in which a fixed–point of a dynamical system loses stability as a pair of complex conjugate eigenvalues of the linearization around the fixed–point cross the imaginary axis of the complex plane. Under reasonably generic assumptions about the dynamical system, we can expect to see a small amplitude limit cycle branching from the fixed–point. The limit cycle is orbitally stable if its *Lyapunov coefficient* has a negative real part, and the bifurcation is supercritical. Otherwise it is unstable and the bifurcation is subcritical.

The normal form of a Hopf bifurcation is:

$$\dot{z} = z(\lambda + b|z|^2),$$

where z, b are both complex and λ is a parameter; b is the Lyapunov coefficient. Write $b = \alpha + i\beta$. If α is negative then there is a stable limit cycle for $\lambda > 0$, $z(t) = re^{i\omega t}$, where $r = \sqrt{-\lambda/\alpha}$ and $\omega = \beta r^2$. In this case we have the *supercritical bifurcation*. On the other hand, if α is positive then there is an unstable limit cycle for $\lambda < 0$. In this case we have the *subcritical bifurcation*.

Hopf bifurcations occur e.g., in the *Hodgkin–Huxley model for nerve membrane*, the *Belousov–Zhabotinsky reaction* and in the following simpler chemical system called the *Brusselator* as the parameter b changes:

$$\dot{x} = a + x^2 y - (b+1)x, \qquad \dot{y} = bx - x^2 y.$$

On the other hand, *symmetry* plays a very important role in many bifurcation problems. Already in the above ball in the hoop example, one sees that symmetry plays an important role. The 'perfect' system discussed so far has a symmetry in the sense that one can reflect the system in the vertical axis of the hoop and one gets an equivalent system; we say that the system has a $\mathbb{Z}_2$–symmetry in this case. This symmetry is manifested in the obvious symmetry in the phase portraits.

We say that a fixed–point has symmetry when it is fixed under the action of the symmetry. The straight down solution of the ball is thus symmetric, but the solutions that are to each side are not symmetric; we say that these solutions have undergone *solution symmetry breaking*.[31] This simple example already shows the important point that symmetric systems need not have symmetric solutions! In fact, the way solutions loose symmetry as bifurcations occur is a fundamental and important idea [Mar99].

Now, the solution symmetry breaking is distinct from the notion of system symmetry breaking, in which the whole system looses its symmetry. If this $\mathbb{Z}_2$–symmetry is broken, by setting the rotation axis a little off center, for example, then one side gets preferred, as in Figure 1.53.

Finally, if one modulates the frequency periodically by, say writing $\omega = \omega_0 + \epsilon \sin \Omega t$ (where Ω is another frequency and ϵ is a 'small parameter'), then the above equations have very complex solutions – this kind of complexity is the origin of the notion of *chaos*.

1.4.2 Systems of ODEs

More precisely, we will consider *systems of ordinary differential equations* (ODEs) of the form

$$\dot{x} = f(x), \tag{1.41}$$

where $x = (x_1, \ldots, x_n) \in \mathbb{R}^n$ and $f : \mathbb{R}^n \to \mathbb{R}^n$. Since the r.h.s of (1.41) does not depend on t explicitly, the ODE is called *autonomous ODE*. If f is a linear function, i.e.,

---

[31] Symmetry breaking usually occurs in *bifurcation sets*. For example, in a dynamical system given by the ODE

$$\ddot{x} + f(x; \mu) + \epsilon g(x) = 0,$$

which is structurally stable when $\mu \neq 0$, if a bifurcation diagram is plotted, treating μ as the *bifurcation parameter*, but for different values of ϵ, the case $\epsilon = 0$ is the symmetric *pitchfork bifurcation*. When $\epsilon \neq 0$, we say we have a pitchfork with broken symmetry.

Fig. 1.53. The ball in an off–center rotating hoop (adapted from [Mar99]).

$$f(x) = Ax,$$

where A is an $n \times n$ matrix of real numbers, we have a *linear ODE*. In general f is a *nonlinear function*. The vector $x \in \mathbb{R}^n$ is called the *state–vector* of the system, and $\mathbb{R}^n$ is called the *state–space*.

The function f can be interpreted as a *vector–field* on the state–space $\mathbb{R}^n$, since it associates with each $x \in \mathbb{R}^n$ an element $f(x)$ on $\mathbb{R}^n$, which can be interpreted as a vector

$$f(x) = (f_1(x), \dots, f_n(x))$$

situated at x.

A *solution* of the ODE (1.41) is a function $\psi : \mathbb{R}^n \to \mathbb{R}^n$ which satisfies

$$\dot{\psi}(t) = f(\psi(t)) \tag{1.42}$$

for all $t \in R$ (the domain of ψ may be a finite interval (α, β)).

The image of the solution function ψ in $\mathbb{R}^n$ is called an *orbit* of the ODE. Equation (1.42) implies that the vector–field f at x is tangent to the orbit through x. The state of the physical system that is being analyzed is represented by a point $x \in \mathbb{R}^n$. The evaluation of the system in time is described by the motion of this point along an orbit of the ODE in $\mathbb{R}^n$, with t as time. In this interpretation, the ODE implies that the *vector–field f is the velocity* of the moving point in the state–space (this should not be confused with the physical velocity of a physical particle).

One cannot hope to find exact solutions of a nonlinear ODE (1.41) for $n \geq 2$ (except in very special cases). One thus has to use either qualitative methods, perturbative methods, or numerical methods, in order to deduce the behavior of the physical system. The aim of *qualitative analysis* is to understand the qualitative behavior of typical solutions of the ODE, e.g., the *long–term behavior* as $t \to \infty$ of typical solutions. One is also interested in questions of *stability* and the possible existence of *bifurcations*.

The starting point in the qualitative analysis of an autonomous ODE (1.41) in $\mathbb{R}^n$ is to locate the zeros of the vector–field, i.e., to find all $c \in \mathbb{R}^n$ such that

$$f(a) = 0.$$

If $f(a) = 0$, then $\psi(t) = a$, for all $t \in \mathbb{R}$, and it is a solution of the ODE, since $\dot{\psi}(t) = f(\psi(t))$ is satisfied trivially for all $t \in \mathbb{R}$. A constant solution $\psi(t) = a$ describes an *equilibrium state* of the physical system, and hence the point $a \in \mathbb{R}^n$ is called an *equilibrium point* of the ODE. More precisely, given an ODE $\dot{x} = f(x)$ in $\mathbb{R}^n$, any point $a \in \mathbb{R}^n$ which satisfies $f(a) = 0$, is an equilibrium point of the ODE.

We are interested in the stability of equilibrium states. In order to address this question it is necessary to study the behavior of the orbits of the ODE close to the equilibrium points. The idea is to consider the linear approximation of the vector–field $f : \mathbb{R}^n \to \mathbb{R}^n$ at an equilibrium point. We thus assume that the function f is of class $C^1(\mathbb{R}^n)$ (i.e., that the partial derivatives of f exist and are C^0–functions on $\mathbb{R}^n$.)

The *derivative matrix* (or, *Jacobian matrix*) of $f : \mathbb{R}^n \to \mathbb{R}^n$ is the $n \times n$ matrix $Df(x)$ defined by

$$Df(x) = \left(\frac{\partial f_i}{\partial x_j} \right), \qquad (i, j = 1, \ldots, n),$$

where the f_i are the component functions of f.

The *linear approximation* of f is written in terms of the derivative matrix,

$$f(x) = f(a) + Df(a)(x - a) + R_1(x, a), \qquad (1.43)$$

where $Df(a)(x - a)$ denotes the $n \times n$ derivative matrix evaluated at a, acting on the vector $(x - a)$, and $R_1(x, a)$ is the *error term*. An important result from advanced calculus is that if f is of class C^0, then the magnitude of the error $\|R_1(x, a)\|$ tends to zero faster than the magnitude of the displacement $\|x - a\|$. Here $\|\cdot\|$ denotes the Euclidean norm on $\mathbb{R}^n$ (i.e., $\|x\| = \sqrt{x_1^2 + \cdots + x_n^2}$). This means that in general, $R_1(x, a)$ will be small compared to $Df(a)(x - a)$, for x sufficiently close to a.

If $a \in \mathbb{R}^n$ is an equilibrium point of the ODE $\dot{x} = f(x)$, we can use (1.43) to write the ODE in the form

$$\text{NL}: \qquad \dot{x} = Df(a)(x - a) + R_1(x, a),$$

assuming that f is of class C^0. We let $y = x - a$, and with the nonlinear ODE (NL) above, we associate the following linear ODE (L),

$$\text{L}: \qquad \dot{y} = Df(a)\, y,$$

which is called the *linearization* of the NL *at the equilibrium point* $a \in \mathbb{R}^n$. The question is when do solutions of L approximate the solutions of the NL near $x = a$? In general the approximation is valid, but in special situations, the approximation can fail.

1.4.3 Linear Autonomous Dynamics: Attractors & Repellors

Recall that *linear dynamical systems* can be solved in terms of simple functions and the behavior of all orbits can be classified (see Figure 1.54). In a linear system the phase–space is the nD Euclidean space, so any point in phase–space can be represented by a vector with n numbers. The analysis of linear systems is possible because they satisfy a *superposition principle*: if $u(t)$ and $w(t)$ satisfy the differential equation for the vector–field (but not necessarily the initial condition), then so will $u(t) + w(t)$.

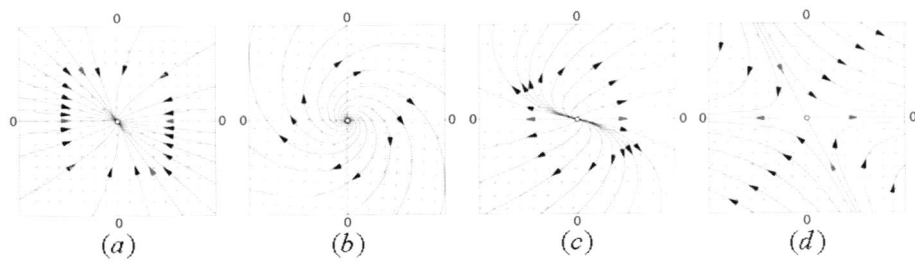

Fig. 1.54. Several examples of linear vector–fields, generated in Mathematica^{TM}: (a) stable node, (b) unstable focus, (c) unstable node, and (d) saddle (unstable).

Flow of a Linear ODE

Fundamental theorem for linear autonomous ODEs states that if A is an $n \times n$ real matrix then the initial value problem

$$\dot{x} = Ax, \qquad x(0) = a \in \mathbb{R}^n \tag{1.44}$$

has the unique solution

$$x(t) = e^{tA}a, \qquad \text{for all } t \in \mathbb{R}. \tag{1.45}$$

(Here a is the state at time $t = 0$ and e^{tA} is the state at time t). To prove the existence, let $x(t) = e^{tA}a$ then

$$\frac{dx}{dt} = \frac{d(e^{tA}a)}{dt} = Ae^{tA}a = Ax,$$
$$x(0) = e^0 a = Ia = a,$$

shows that $x(t)$ satisfies the initial value problem (1.44) (here I denotes the $n \times n$ identity matrix).

To prove the uniqueness, let $x(t)$ be any solution of (1.44). It follows that

$$\frac{d}{dt}\left[e^{-tA}x(t)\right] = 0.$$

Thus $e^{-tA}x(t) = C$, a constant. The initial condition implies that $C = a$ and hence $x(t) = e^{tA}a$.

The unique solution of the ODE (1.44) is given by (1.45) for all t. Thus, for each $t \in \mathbb{R}$, the matrix $e^{tA}a$ maps

$$a \mapsto e^{tA}a.$$

The set $\{e^{tA}\}_{t\in\mathbb{R}}$ is a $1-$parameter family of linear maps of $\mathbb{R}^n$ into $\mathbb{R}^n$, and is called the *linear flow* of the ODE (for comparison with the general flow notion, see [II06b]).

We write

$$g^t = e^{tA}$$

– to denote the flow. The flow describes the evolution in time of the physical system for all possible initial states. As the physical system evolves in time, one can think of the state vector x as a moving point in state space, its motion being determined by the flow $g^t = e^{tA}$. The linear flow satisfies two important properties, which also hold for nonlinear flows.

The linear flow $g^t = e^{tA}$ satisfies:

F1 : $g^0 = I$,   identity map,   and

F2 : $g^{t_1+t_2} = g^{t_1} \circ g^{t_2}$,   composition.

Note that properties F1 and F2 imply that the flow $\{g^t\}_{t\in\mathbb{R}}$ forms a *group* under composition of maps.

The flow g^t of the ODE (1.44) partitions the state–space $\mathbb{R}^n$ into subsets called *orbits*, defined by

$$\gamma(a) = \{g^t a : t \in \mathbb{R}\}.$$

The set $\gamma(a)$ is called the orbit of the ODE through a. It is the image in $\mathbb{R}^n$ of the solution curve $x(t) = e^{tA}a$. It follows that for $a, b \in \mathbb{R}^n$, either $\gamma(a) = \gamma(b)$ or $\gamma(a) \cap \gamma(b) = \emptyset$, since otherwise the uniqueness of solutions would be violated.

For example, consider

$$\dot{x} = Ax,   \text{for all} x \in \mathbb{R}^2;$$

with

$$A = \begin{pmatrix} 0 & 1 \\ -1 & 0 \end{pmatrix},$$

the linear flow is

$$e^{tA} = \begin{pmatrix} \cos t & \sin t \\ -\sin t & \cos t \end{pmatrix}.$$

The *action of the flow* on $\mathbb{R}^2$, $a \mapsto e^{tA}a$ corresponds to a *clockwise rotation about the origin*. Thus if $a \neq 0$, the orbit $\gamma(a)$ is a circle centered at the origin passing through a. The origin is a *fixed–point* of the flow, since $e^{tA}0 = 0$, for all $t \in \mathbb{R}$. The orbit $\gamma(0) = \{0\}$ is called a *point orbit*. All other orbits are called *periodic orbits* since $e^{2\pi A}a = a$, i.e., the flow maps onto itself after a time $t = 2\pi$ has elapsed.

Classification of Orbits of an ODE

1. If $g^t a = a$ for all $t \in \mathbb{R}$, then $\gamma(a) = \{a\}$ and it is called a *point orbit*. Point orbits correspond to equilibrium points.
2. If there exists a $T > 0$ such that $g^T a = a$, then $\gamma(a)$ is called a *periodic orbit*. Periodic orbits describe a system that evolves periodically in time.
3. If $g^t a \neq a$ for all $t \neq 0$, then $\gamma(a)$ is called a *non–periodic orbit*.

Note that:

1. Non–periodic orbits can be of great complexity even for linear ODEs if $n > 3$ (for nonlinear ODEs if $n > 2$).

2. A *solution curve* of an ODE is a parameterized curve and hence contains information about the flow of time t. The *orbits* are paths in state–space (or subsets of state space). Orbits which are not point orbits are *directed paths* with the direction defined by increasing time. The orbits thus do not provide detailed information about the flow of time.

For an autonomous ODE, the slope of the solution curves depend only on x and hence the tangent vectors to the solution curves define a vector–field $f(x)$ in $x-$space. Infinitely many solution curves may correspond to a single orbit. On the other hand, a non–autonomous ODE does not define a flow or a family of orbits.

Canonical Linear Flows in $\mathbb{R}^2$

Jordan Canonical Forms

For any 2×2 real matrix A, there exists a non–singular matrix P such that

$$J = P^{-1}AP,$$

and J is one of the following matrices:

$$\begin{pmatrix} \lambda_1 & 0 \\ 0 & \lambda_2 \end{pmatrix}, \quad \begin{pmatrix} \lambda & 1 \\ 0 & \lambda \end{pmatrix}, \quad \begin{pmatrix} \alpha & \beta \\ -\beta & \alpha \end{pmatrix}.$$

Two linear ODEs, $\dot{x} = Ax$ and $\dot{x} = Bx$, are linearly equivalent iff there exists a non–singular matrix P and a positive constant k such that

$$A = kP^{-1}BP.$$

In other words, the linear ODEs, $\dot{x} = Ax$ and $\dot{x} = Bx$ are *linearly equivalent* iff there exists an invertible matrix P and a positive constant k such that

$$Pe^{tA} = e^{ktB}P, \qquad \text{for all } t \in \mathbb{R}.$$

Case I: two eigen–directions

Jordan canonical form is

$$J = \begin{pmatrix} \lambda_1 & 0 \\ 0 & \lambda_2 \end{pmatrix}.$$

The flow is

$$e^{tJ} = \begin{pmatrix} e^{\lambda_1 t} & 0 \\ 0 & e^{\lambda_2 t} \end{pmatrix},$$

and the eigenvectors are $e_1 = \begin{pmatrix} 1 \\ 0 \end{pmatrix}$ and $e_2 = \begin{pmatrix} 0 \\ 1 \end{pmatrix}$. The solutions are $y(t) = e^{tJ}b$ for all $b \in \mathbb{R}^2$, i.e., $y_1 = e^{\lambda_1 t}b_1$ and $y_2 = e^{\lambda_2 t}b_2$.

Ia. $\lambda_1 = \lambda_2 < 0$:   *attracting focus;*
Ib. $\lambda_1 < \lambda_2 < 0$:   *attracting node;*
Ic. $\lambda_1 < \lambda_2 = 0$:   *attracting line;*
Id. $\lambda_1 < 0 < \lambda_2$:   *saddle;*
Ie. $\lambda_1 = 0 < \lambda_2$:   *repelling line;*
If. $0 < \lambda_1 < \lambda_2$:   *repelling node;*
Ig. $0 < \lambda_1 = \lambda_2$:   *repelling focus.*

Case II: one eigen–direction

Jordan canonical form is

$$J = \begin{pmatrix} \lambda & 1 \\ 0 & \lambda \end{pmatrix}.$$

The flow is

$$e^{tJ} = e^{\lambda_t} \begin{pmatrix} 1 & t \\ 0 & 1 \end{pmatrix},$$

and the single eigenvector is $e = \begin{pmatrix} 1 \\ 0 \end{pmatrix}$.

IIa. $\lambda < 0$:   *attracting Jordan node;*
IIb. $\lambda = 0$:   *neutral line;*
IIc. $\lambda > 0$:   *repelling Jordan node.*

Case III: no eigen–directions

Jordan canonical form is

$$J = \begin{pmatrix} \alpha & \beta \\ -\beta & \alpha \end{pmatrix}.$$

The given ODE is linearly equivalent to $\dot{y} = Jy$.

IIIa. $\alpha < 0$:   *attracting spiral;*
IIIb. $\alpha = 0$:   *center;*
IIIc. $\alpha > 0$:   *repelling spiral.*

In terms of the Jordan canonical form of two matrices A and B, the corresponding ODEs are linearly equivalent iff:

1. A and B have the same number of eigen–directions, and
2. The eigenvalues of A are multiple (k) of the eigenvalues of B.

Topological Equivalence

Now, cases **Ia**, **Ib**, **IIa**, and **IIIa** have common characteristic that all orbits approach the origin (an equilibrium point) as $t \to \infty$. We would like these flows to be 'equivalent' in some sense. In fact, it can be shown, that for all flows of these types, *the orbits of one flow can be mapped onto the orbits of the simplest flow* **Ia**, using a (nonlinear) map $h : \mathbb{R}^2 \to \mathbb{R}^2$, which is a *homeomorphism* on $\mathbb{R}^2$.

Recall that map $h : \mathbb{R}^n \to \mathbb{R}^n$ is a *homeomorphism* on $\mathbb{R}^n$ iff (i) h is one–to–one and onto, (ii) h is continuous, and (iii) h^{-1} is continuous. Two linear flows e^{tA} and e^{tB} on $\mathbb{R}^n$ are said to be *topologically equivalent* if there exists a homeomorphism h on $\mathbb{R}^n$ and a positive constant k such that

$$h(e^{tA}x) = e^{ktB}h(x), \qquad \text{for all } x \in \mathbb{R}^n, \text{ and for all } t \in \mathbb{R}.$$

A *hyperbolic* linear flow in $\mathbb{R}^2$ is one in which the real parts of the eigenvalues are all non–zero (i.e., $\mathcal{R}e(\lambda_i) \neq 0$, for $i = 1, 2$.)

Any hyperbolic linear flow in $\mathbb{R}^2$ is topologically equivalent to the linear flow e^{tA}, where A is one of the following matrices:

1. $A = \begin{pmatrix} -1 & 0 \\ 0 & -1 \end{pmatrix}$,   *standard sink.*

2. $A = \begin{pmatrix} 1 & 0 \\ 0 & 1 \end{pmatrix}$,   *standard source.*

3. $A = \begin{pmatrix} -1 & 0 \\ 0 & 1 \end{pmatrix}$,   *standard saddle.*

Any non–hyperbolic linear flow in $\mathbb{R}^2$ is linearly (and hence topologically) equivalent to the flow e^{tA}, where A is one of the following matrices:

$$\begin{pmatrix} 0 & 0 \\ 0 & 0 \end{pmatrix}, \quad \begin{pmatrix} 0 & -1 \\ 1 & 0 \end{pmatrix}, \quad \begin{pmatrix} 0 & 1 \\ 0 & 0 \end{pmatrix}, \quad \begin{pmatrix} -1 & 0 \\ 0 & 0 \end{pmatrix}, \quad \begin{pmatrix} 1 & 0 \\ 0 & 0 \end{pmatrix}.$$

These five flows are topologically equivalent.

1.4.4 Conservative versus Dissipative Dynamics

Recall (see [II05, II06a, II06b]) that *conservative–reversible systems* are in classical dynamics described by Hamilton's equations

$$\dot{q}^i = \partial_{p_i} H, \qquad \dot{p}_i = \partial_{q^i} H, \qquad (i = 1, ..., n), \tag{1.46}$$

with a *constant Hamiltonian energy function*

$$H = H(q, p) = E_{kin}(p) + E_{pot}(q) = E = \text{const.} \qquad (1.47)$$

Conservative dynamics visualizes the time evolution of a system in a *phase–space* P, in which the coordinates are q^i and p_i. The instantaneous state of the system is the *representative point* (q, p) in P. As time varies, the representative point (q, p) describes the *phase trajectory*. A particular case of a phase trajectory is the *position of equilibrium*, in which both $\dot{q}^i = 0$ and $\dot{p}_i = 0$.

Dissipative Systems

In addition to *conservative–reversible* systems, we must consider systems that give rise to *irreversible processes* and *dissipative structures* of *Nobel Laureate Ilya Prigogine* (see [NP77, II06a]).

A typical example is a chemical reaction in which a molecule of species A (say the hydroxyl radical OH) can combine with a molecule of species B say molecular hydrogen H_2) to produce one molecule of species C and one molecule of species D (respectively H_2O and atomic hydrogen H in our example). This process is symbolized

$$A + B \xrightarrow{k} C + D, \qquad (1.48)$$

in which k is the rate constant, generally a function of temperature and pressure. On the l.h.s of (1.48), the *reactants* A and B combine and disappear in the course of time, whereas on the r.h.s the *products* C and D are formed and appear as the reaction advances. The rate at which particles of species A are consumed is proportional to the frequency of encounters of molecules of A and B – which, if the system is dilute, is merely proportional to the product of their concentrations, c,

$$\dot{c}_A = -k \, c_A \, c_B. \qquad (1.49)$$

Clearly, if we reverse time, $t' = -t$, and denote by c'_A, c'_B the values of the concentrations as functions of t', (1.49) becomes

$$\dot{c}_A = k \, c_A \, c_B,$$

and describes a process in which c_A would be produced instead of being consumed. This is certainly not equivalent to the phenomenon described by (1.49).

Further examples are *heat conduction*, given by *Fourier equation*

$$\partial_t T = \kappa \, \nabla^2 T, \qquad \kappa > 0, \qquad \left(\partial_t \equiv \frac{\partial}{\partial_t} \right), \qquad (1.50)$$

and *diffusion*, described by *Fick equation*

$$\partial_t T = D \, \nabla^2 c, \qquad D > 0. \qquad (1.51)$$

Here T is the temperature, c is the concentration of a certain substance dissolved in the fluid, κ is the heat diffusivity coefficient and D is the mass diffusivity. Both experiments and these two equations show that when a slight temperature variation (respectively, inhomogeneity) is imposed in an isothermal (respectively, uniform) fluid, it will *spread out* and eventually disappear.

Again, if we reverse time, we get the completely different laws

$$\partial_t T = -\kappa \, \nabla^2 T, \qquad \partial_t c = -D \, \nabla^2 c,$$

describing a situation in which an initial temperature or concentration disturbance would be amplified rather than damped.

Both the *concentration* and the *temperature* variables are examples of so–called *even variables*, whose sign does not change upon time reversal. In contrast, the *momentum of a particle* and the *convection velocity of a fluid* are *odd variables*, since they are ultimately expressed as time derivatives of position–like variables and change their sign with time reversal.

This leads us to the following general property of the evolution equation of a dissipative system. Let $\{X_i\}$ denote a complete set of macroscopic variables of such a system. *Dissipative evolution laws* have the form

$$\partial_t X_i = F_i(\{X_j\}, \lambda), \tag{1.52}$$

where λ denote *control parameters*, and F_i are functions of $\{X_i\}$ and λ.

The basic feature of (1.52) is that, whatever the form of the functions F_i, in the absence of constraints they must reproduce the steady state of equilibrium

$$F_i(\{X_{j,eq}\}, \lambda_{eq}) = 0. \tag{1.53}$$

More generally, for a non–equilibrium steady state,

$$F_i(\{X_{j,s}\}, \lambda_s) = 0. \tag{1.54}$$

These relations impose certain restrictions. For instance, the evolution laws must ensure that positive values are attained for temperature or chemical concentrations that come up as solutions, or that detailed balance is attained. This is an important point, for it shows that the analysis of physical systems cannot be reduced to a mathematical game. In many respects physical systems may be regarded as highly atypical, specific, or nongeneric from the mathematical point of view. In these steady state relations, the *nonlinearity*, relating the control parameters λ to the steady state values $X_{j,s}$, begins to play the prominent role.

Thermodynamic Equilibrium

In mechanics, (static) equilibrium is a particular 'state of rest' in which both the velocities and the accelerations of all the material points of a system are equal to zero. By definition the net balance of forces acting on each point is zero at each moment. If this balance is disturbed, equilibrium will be broken. This is what happens when a piece of metal fractures under the effect of load (see [NP77, II06a]).

Now, the notion of *thermodynamic equilibrium* is sharply different. Contrary to mechanical equilibrium, the molecules constituting the system are subject to forces that are not balanced and move continuously in all possible directions unless the temperature becomes very low. 'Equilibrium' refers here to some collective properties $\{X_i\}$ characterizing the system as a whole, such as temperature, pressure, or the concentration of a chemical constituent.

Consider a system $\{X_i\}$ embedded in a certain environment $\{X_{ie}\}$. Dynamic role of the sets of properties $\{X_i\}$ and $\{X_{ie}\}$ resides primarily in their exchanges between the system and the environment. For instance, if the system is contained in a vessel whose walls are perfectly rigid, permeable to heat but impermeable to matter, one of these quantities will be identical to the temperature, T and will control the exchange of energy in the form of heat between the system and its environment.

We say that the system is in thermodynamic equilibrium if it is completely identified with its environment, that is, if the properties X_i and X_{ie} have identical values. In the previous example, thermodynamic equilibrium between the system and its surroundings is tantamount to $T = T_e$ at all times and at all points in space. But because the walls of the vessel are impermeable to matter, system and environment can remain highly differentiated in their chemical composition, c. If the walls become permeable to certain chemical substances i, thermodynamic equilibrium will prevail when the system and the environment become indistinguishable as far as those chemicals are concerned. In simple cases this means that the corresponding composition variables will satisfy the equality $c_i = c_{ie}$, but more generally equilibrium will be characterized by the equality for a quantity known as the chemical potential, $\mu_i = \mu_{ie}$. Similarly, if the walls of the vessel are not rigid, the system can exchange mechanical energy with its environment. Equilibrium will then also imply the equality of pressures, $p = p_e$.

According to the above definitions, equilibrium is automatically a stationary state, $\partial X_i / \partial t = 0$: the properties X_i do not vary with time. As they are identical in the properties X_i, the system and the environment have nothing to exchange. We express this by saying that there are no net *fluxes* across the system,

$$J_i^{eq} = 0. \tag{1.55}$$

Nonlinearity

Here is a simple example. Let X be the unique state variable, k a parameter, and let λ represent the applied constraint. We can easily imagine a mechanism such as $A \rightleftarrows X \rightleftarrows D$ in which X evolves according to

$$\dot{X} = \lambda - kX,$$

yielding a stationary state value given by $\lambda - kX_s = 0$, or $X_s = \lambda/k$. In the linear law linking the steady state value X_s to the control parameter λ the behavior is bound to be qualitatively similar to that in equilibrium, even in the presence of strongly correlated non–equilibrium constraints. In the nonlinear law linking the steady state value X_s to the control parameter λ there is an unlimited number of possible forms describing nonlinear dependencies. For the certain values of λ the system can present several distinct solutions.

Nonlinearity combined with *non–equilibrium constraints* allows for multiple solutions and hence for the diversification of the behaviors presented by a system (see [NP77, II06b]).

The Second Law of Thermodynamics

According to this law there exists a function of the state variables (usually chosen to be the *entropy, S*) of the system that varies monotonically during the approach to the unique final state of thermodynamic equilibrium:

$$\dot{S} \geq 0 \qquad \text{(for any isolated system)}. \tag{1.56}$$

It is usually interpreted as a *tendency to increased disorder*, i.e., an irreversible trend to maximum disorder.

The above interpretation of entropy and a second law is fairly obvious for systems of *weakly interacting particles*, to which the arguments developed by Boltzmann referred.

Let us now turn to non–isolated systems, which exchange energy or matter with the environment. The entropy variation will now be the sum of two terms. One, entropy flux, $d_e S$, is due to these exchanges; the other, entropy production, $d_i S$, is due to the phenomena going on within the system. Thus the entropy variation is

$$\dot{S} = \frac{d_i S}{dt} + \frac{d_e S}{dt}. \tag{1.57}$$

For an isolated system $d_e S = 0$, and (1.57) together with (1.56) reduces to $dS = d_i S \geq 0$, the usual statement of the second law. But even if the system is non–isolated, $d_i S$ will describe those (irreversible) processes that would still go on even in the absence of the flux term $d_e S$. We thus require the following extended form of the second law:

$$\dot{S} \geq 0 \qquad \text{(nonisolated system)}. \tag{1.58}$$

As long as $d_i S$ is strictly positive, irreversible processes will go on continuously within the system. Thus, $d_i S > 0$ is equivalent to the condition of dissipativity as time irreversibility. If, on the other hand, $d_i S$ reduces to zero, the process will be reversible and will merely join neighboring states of equilibrium through a slow variation of the flux term $d_e S$.

Among the most common irreversible processes contributing to d_iS are chemical reactions, heat conduction, diffusion, viscous dissipation, and relaxation phenomena in electrically or magnetically polarized systems. For each of these phenomena two factors can be defined: an appropriate internal *flux*, J_i, denoting essentially its rate, and a driving *force*, X_i, related to the maintenance of the non–equilibrium constraint. A most remarkable feature is that d_iS becomes a *bilinear form* of J_i and X_i. The following table summarizes the fluxes and forces associated with some commonly observed irreversible phenomena (see [NP77, II06a])

$$
\begin{bmatrix}
\textbf{Phenomenon} & \textbf{Flux} & \textbf{Force} & \textbf{Rank} \\
\text{Heat conduction} & \text{Heat flux, } \mathbf{J}_{th} & grad(1/T) & \text{Vector} \\
\text{Diffusion} & \text{Mass flux, } \mathbf{J}_d & \text{-}[grad(\mu/T) - \mathbf{F}] & \text{Vector} \\
\text{Viscous flow} & \text{Pressure tensor, } \mathbf{P} & (1/T)\, grad\, \mathbf{v} & \text{Tensor} \\
\text{Chemical reaction} & \text{Rate of reaction, } \omega & \text{Affinity of reaction} & \text{Scalar}
\end{bmatrix}
$$

In general, the fluxes J_k are very complicated functions of the forces X_i. A particularly simple situation arises when their relation is linear, then we have the celebrated *Onsager relations* (named after *Nobel Laureate Lars Onsager*),

$$J_i = L_{ik}X_k, \qquad (1.59)$$

in which L_{ik} denote the set of *phenomenological coefficients*. This is what happens near equilibrium where they are also symmetric, $L_{ik} = L_{ki}$. Note, however, that certain states far from equilibrium can still be characterized by a linear dependence of the form of (1.59) that occurs either accidentally or because of the presence of special types of regulatory processes.

Geometry of Phase Space

Now, we reduce (1.52) to the *temporal* systems, in which there is no space dependence in the operator F_i, so that $\partial \to d$, and we have

$$\dot{X}_i = F_i(\{X_j\}, \lambda), \qquad i = 1, ..., n. \qquad (1.60)$$

Moreover, we restrict ourselves to autonomous systems, for which F_i does not depend explicitly on time, a consequence being that the trajectories in phase–space are invariant. Note that in a Hamiltonian system n must be even and F_i must reduce to the characteristic structure imposed by (1.46).

A first kind of phase–space trajectory compatible with (1.60) is given by

$$\dot{X}_i = 0. \qquad (1.61)$$

It includes as particular cases the states of mechanical equilibrium encountered in conservative systems and the steady states encountered in dissipative systems. In phase–space such trajectories are quite degenerate, since they are given by the solutions of the n algebraic equations for n unknowns, $F_i = 0$. They are represented by *fixed–points*.

If (1.61) is not satisfied, the representative point will not be fixed but will move along a phase–space trajectory defining a curve. The line element along this trajectory for a displacement corresponding to $(dX_1, ..., dX_n)$ along the individual axes is given by Euclidean metrics

$$ds^2 = dX_i dX_j = F_i F_j \, dt, \qquad (i, j = 1, ..., n). \qquad (1.62)$$

Thus, the projections of the tangent of the curve along the axes are given by

$$\frac{dX_\alpha}{ds} = \frac{F_\alpha}{\sqrt{F_i F_j}}, \qquad (1.63)$$

and are well defined everywhere. The points belonging to such curves are called *regular points*. In contrast, the tangent on the fixed–points is *ill–defined* because of the simultaneous vanishing of all F_i's. Therefore, the fixed–points could be also referred to as the singular points of the flow generated by (1.60). The set of fixed–points and phase–space trajectories constitutes the *phase portrait* of a dynamical system.

One property that plays a decisive role in the structure of the phase portrait relates to the *existence–uniqueness theorem* of the solutions of ordinary differential equations. This important result of A. Cauchy asserts that *under quite mild conditions on the functions F_i, the solution corresponding to an initial condition not on a fixed–point exists and is unique for all times in a certain interval $(0, \tau)$, whose upper bound τ depends on the specific structure of the functions F_i.* In the phase–space representation, the theorem automatically rules out the intersection of two trajectories in any regular point (see [NP77, II06a]).

A second structure of great importance is the *existence and structure of invariant sets of the flow*. By this we mean objects embedded in the phase–space that are bounded and are mapped onto themselves during the evolution generated by (1.60). An obvious example of an invariant set is the ensemble of fixed–points. Another is a closed curve in phase–space representing a periodic motion.

The *impossibility of self–intersection of the trajectories* and the *existence of invariant sets* of a certain form (fixed–points, limit circles,...) determine, to a large extent, the structure of the phase portrait in 2D phase–spaces, and through it the type of behavior that may arise. *In three or more dimensions*, however, the constraints imposed by these properties are much less severe, since the trajectories have many more possibilities to avoid each other by 'gliding' within the 'gaps' left between invariant sets, thus implying the *possibility for chaos*.

In principle, the solution of (1.60) constitutes a *well–posed problem*, in the sense that *a complete specification of the state $(X_1, ..., X_n)$ at any one time allows prediction of the state at all other times.* But in many cases such a complete specification may be operationally meaningless. For example, in a Hamiltonian system composed of particles whose number is of the order of

Avogadro's number, or in a chaotic regime, it is no longer meaningful to argue in terms of individual trajectories. New modes of approach are needed, and one of the most important is a description of the system in terms of *probability concepts*. For this purpose we consider not the rather special case of a single system, but instead focus attention on the *Gibbs ensemble* of a very large number of identical systems, which are in general in different states, but all subject to exactly the same constraints. They can therefore be regarded as emanating from an initial ensemble of systems whose representative phase points were contained in a certain phase–space volume V_0 (see [NP77, II06a]).

1.4.5 Basics of Nonlinear Dynamics

In this subsection we give the basics of nonlinear dynamics, so to make the following text as self–contained as possible (for more details, see [II06b]).

Real 1 − −DOF Hamiltonian Dynamics

The basic structural unit of the biomechanics is a *uniaxial rotational joint*, geometrically representing a *constrained SO(2) group of plane rotations*. In other words, this is a one–DOF dynamical system, represented in a differential formulation as a *vector–field*, or in an integral form as a *phase–flow*, on a 2D phase–space manifold M, *cylinder* $\mathbb{R}\times S^1$, being itself a *cotangent bundle* T^*M of the joint $SO(2)-configuration$ manifold, *circle* S^1.

A vector–field $X(t)$ on the momentum phase–space manifold M can be given by a system of *canonical equations of motion*

$$\dot{q} = f(q, p, t, \mu), \qquad \dot{p} = g(q, p, t, \mu), \qquad (1.64)$$

where t is time, μ is a parameter, $q \in S^1$, $p \in \mathbb{R}\times S^1$ are *coordinates* and *momenta*, respectively, while f and g are smooth functions on the phase–space $\mathbb{R}\times S^1$.

If time t does not explicitly appear in the functions f and g, the vector–field X is called *autonomous*. In this case equation (1.64) simplifies as

$$\dot{q} = f(q, p, \mu), \qquad \dot{p} = g(q, p, \mu). \qquad (1.65)$$

By a *solution curve* of the vector–field X we mean a map $x = (q, p)$, from some interval $I \subset \mathbb{R}$ into the phase–space manifold M, such that $t \mapsto x(t)$. The map $x(t) = (q(t), p(t))$ geometrically represents a curve in M, and equations (1.64) or (1.65) give the tangent vector at each point of the curve.

To specify an *initial condition* on the vector–field X, by

$$x(t, t_0, x_0) = (q(t, t_0, q_0), p(t, t_0, p_0)),$$

geometrically means to distinguish a solution curve by a particular point $x(t_0) = x_0$ in the phase–space manifold M. Similarly, it may be useful to explicitly display the parametric dependence of solution curves, as

$x(t, t_0, x_0, \mu) = (q(t, t_0, q_0, \mu_q), p(t, t_0, p_0, \mu_p))$, where μ_q, μ_p denote $q-$dependent and $p-$dependent parameters, respectively.

The solution curve $x(t, t_0, x_0)$ of the vector–field X, may be also referred as the *phase trajectory* through the point x_0 at $t = t_0$. Its graph over t is refereed to as an *integral curve*; more precisely, *graph*

$x(t, t_0, x_0) \equiv \{(x, t) \in M \times \mathbb{R} : x = x(t, t_0, x_0), t \in I \subset \mathbb{R}\}$.

Let $x_0 = (q_0, p_0)$ be a point on M. By the *orbit through* x_0, denoted $O(x_0)$, we mean the set of points in M that lie on a trajectory passing through x_0; more precisely, for $x_0 \in U$, U open in M, the orbit through x_0 is given by $O(x_0) = \{x \in \mathbb{R} \times S^1 : x = x(t, t_0, x_0), t \in I \subset \mathbb{R}\}$.

Consider a general autonomous vector–field X on the phase–space manifold M, given by equation $\dot{x} = f(x)$, $x = (q, p) \in M$. An *equilibrium solution*, *singularity*, or *fixed–point* of X is a point $\bar{x} \in M$ such that $f(\bar{x}) = 0$, i.e., a solution which does not change in time.

Any solution $\bar{x}(t)$ of an autonomous vector–field X on M is *stable vector–field* if solutions starting 'close' to $\bar{x}(t)$ at a given time remain close to $\bar{x}(t)$ for all later times. It is *asymptotically stable* if nearby solutions actually converge to $\bar{x}(t)$ as $t \longrightarrow \infty$. In order to determine the stability of $\bar{x}(t)$ we must understand the nature of solutions near $\bar{x}(t)$, which is done by *linearization* of the vector–field X. The solution of the linearized vector–field Y is asymptotically stable if all eigenvalues have negative real parts. In that case the fixed–point $x = \bar{x}$ of associated nonlinear vector–field X is also asymptotically stable. A fixed–point $\bar{x}$ is called *hyperbolic point* if none of the eigenvalues of Y have zero real part; in that case the orbit structure near $\bar{x}$ is essentially the same for X and Y.

In the case of autonomous vector–fields on M we have also an important property of *Hamiltonian flow*. If $x(t) = (q(t), p(t))$ is a solution of $\dot{x} = f(x)$, $x \in M$, then so is $x(t + \tau)$ for any $\tau \in \mathbb{R}$. Also, for any $x_0 \in M$ there exists only one solution of an autonomous vector–field passing through this point. The autonomous vector–field

$$\dot{x} = f(x)$$

has the following properties:

1. $x(t, x_0)$ is C^k;
2. $x(0, x_0) = x_0$; and
3. $x(t + s, x_0) = x(t, x(s, x_0))$.

These properties show that the solutions of an autonomous vector–field form a *1–parameter family of diffeomorphisms* of the phase–space manifold M. This is refereed to as a *phase–flow* and denoted by $\phi_t(x)$ or $\phi(t, x)$.

Consider a flow $\phi(t, x)$ generated by vector–field $\dot{x} = f(x)$. A point $x_0 = (q_0, p_0)$ on M is called an $\omega-$*limit point* of $x, = (q, p) \in M$, denoted $\omega(x)$, if there exists a sequence $\{t_i\}$, $t_i \mapsto \infty$, such that $\phi(t_i, x) \mapsto x_0$. Similarly, $\alpha-$*limit points* are defined by taking a sequence $\{t_i\}$, $t_i \mapsto -\infty$. The set of all

$\omega-$limit points of a flow is called the $\omega-limit\ set$. The $\alpha-$limit set is similarly defined.

A point $x_0 = (q_0, p_0)$ on M is called $non-wandering$ if for any open neighborhood $U \subset M$ of x_0, there exists some $t \neq 0$ such that $\phi(t, U) \cap U \neq 0$. The set of all non–wandering points of a flow is called the $non-wandering\ set$ of that particular map or flow.

A closed invariant subset $A \subset M$ is called an $attracting\ set$ if there is some open neighborhood $U \subset M$ of A such that $\phi(t, x) \in U$ and $\phi(t, x) \mapsto \infty$ for any $x \in U$ and $t \geq 0$. The $domain$ or $basin\ of\ attraction$ of A is given by $\cup_{t \leq 0} \phi(t, U)$. In practice, a way of locating attracting sets is to first find a $trapping\ region$, i.e., a closed, connected subset $V \subset M$ such that for any $t \geq 0$ $\phi(t, V) \subset V$. Then $\cap_{t \, 0} \phi(t, V) = A$ is an $attracting\ set$.

As a first example of one–DOF dynamical systems, let us consider a vector–field $x = (q, p) \in \mathbb{R} \times \mathbb{R}$ of a simple harmonic oscillator, given by equations

$$\dot{q} = p, \qquad \dot{p} = -q. \tag{1.66}$$

Here, the $solution$ passing through the point $(q, p) = (1, 0)$ at $t = 0$ is given by $(q(t), p(t)) = (\cos t, -\sin t)$; the $integral\ curve$ passing through $(q, p) = (1, 0)$ at $t = 0$ is given by $\{(q, p, t) \in \mathbb{R} \times \mathbb{R} \times \mathbb{R} : (q(t), p(t)) = (\cos t, -\sin t)\}$, for all $t \in \mathbb{R}$; the $orbit$ passing through $(q, p) = (1, 0)$ is given by the circle $q^2 + p^2 = 1$.

A one–DOF dynamical system is called $Hamiltonian\ system$ if there exists a $first\ integral$ or a function of the dependent variables (q, p) whose level curves give the orbits of the vector–field $X = X_H$, i.e., a total–energy $Hamiltonian$ $function$ $H = H(q, p) : U \to \mathbb{R}$, ($U$ open set on the phase–space manifold M), such that the vector–field X_H is given by $Hamilton's\ canonical\ equations$ (1.64) or (1.65). Here the first, $\dot{q}-$equation, is called the $velocity\ equation$ and serves as a definition of the $momentum$, while the second, $\dot{p}-$equation is called the $force\ equation$, and represents the $Newtonian\ second\ law\ of\ motion$.

The simple harmonic oscillator (1.66) is a Hamiltonian system with a Hamiltonian function $H = \frac{p^2}{2} + \frac{q^2}{2}$. It has a $fixed-point$ – $center$ (having purely imaginary eigenvalues) at $(q, p) = (0, 0)$ and is surrounded by a 1–parameter family of periodic orbits given by the Hamiltonian H.

A nice example of one–DOF dynamical system with a Hamiltonian structure is a $damped\ Duffing\ oscillator$ (see, e.g., [Wig90]). This is a $plane$ Hamiltonian vector–field $x = (q, p) \in \mathbb{R}^2$, given by Hamilton's equations

$$\dot{q} = p \equiv f(q, p), \qquad \dot{p} = q - q^3 - \delta p \equiv g(q, p, \delta), \qquad \delta \geq 0. \tag{1.67}$$

For the special parameter value $\delta = 0$, we have an $un-damped$ Duffing oscillator with a $first\ integral$ represented by Hamiltonian function $H = \frac{p^2}{2} - \frac{q^2}{2} + \frac{q^4}{4}$, where $\frac{p^2}{2}$ corresponds to the $kinetic\ energy$ (with a mass scaled to unity), and $-\frac{q^2}{2} + \frac{q^4}{4} \equiv V(x)$ corresponds to the $potential\ energy$ of the oscillator.

In general, if the first integral, i.e., a Hamiltonian function H, is defined by $H = \frac{p^2}{2} + V(x)$, then the momentum is given by $p = \pm\sqrt{2}\sqrt{H - V(x)}$. All one–DOF Hamiltonian systems are *integrable* and all the solutions lie on *level curves* of the Hamiltonian function, which are topologically equivalent with the circle S^1. This is actually a general characteristic of all $n-$DOF integrable Hamiltonian systems: their bounded motions lie on nD *invariant tori* $T^n = S^1 \times \cdots \times S^1$, or *homoclinic orbits*. The homoclinic orbit is sometimes called a *separatrix* because it is the boundary between two distinctly different types of motion.

For example, in case of a damped Duffing oscillator (1.67) with $\delta \neq 0$, we have

$$\partial_q f + \partial_p g = -\delta,$$

and according to the *Bendixon's criterion* for $\delta > 0$ it has no closed orbits.

The vector–field X given by equations (1.67) has three fixed–points given by $(q, p) = (0, 0), (\pm 1, 0)$. The *eigenvalues* $\lambda_{1,2}$ of the associated linearized vector–field are given by $\lambda_{1,2} = -\delta/2 \pm \frac{1}{2}\sqrt{\delta^2 + 4}$, for the fixed–point $(0, 0)$, and by $\lambda_{1,2} = -\delta/2 \pm \frac{1}{2}\sqrt{\delta^2 - 8}$, for the fixed–point $(\pm 1, 0)$. Hence, for $\delta > 0$, $(0, 0)$ is *unstable* and $(\pm 1, 0)$ are *asymptotically stable*; for $\delta = 0$, $(\pm 1, 0)$ are *stable in the linear approximation* (see, e.g., [Wig90]).

Another example of one–DOF Hamiltonian systems, representing the actual basis of the human $SO(2)-$joint dynamics, is a *simple pendulum* (again, all physical constants are scaled to unity), given by Hamiltonian function $H = \frac{p^2}{2} - \cos q$. This is the first integral of the *cylindrical* Hamiltonian vector–field $(q, p) \in S^1 \times \mathbb{R}$, defined by canonical equations

$$\dot{q} = p, \qquad \dot{p} = -\sin q.$$

This vector–field has fixed–points at $(0, 0)$, which is a *center* (i.e., the eigenvalues are purely imaginary), and at $(\pm\pi, 0)$, which are *saddles*, but since the phase–space manifold is the cylinder, these are really the same point.

The basis of human arm and leg dynamics represents the *coupling* of two uniaxial, $SO(2)-$joints. The study of two DOF Hamiltonian dynamics we shall start with the most simple case of two linearly coupled linear un–damped oscillators with parameters scaled to unity. Under general conditions we can perform a change of variables to canonical coordinates (the 'normal modes') (q^i, p_i), $i = 1, 2$, so that the vector–field X_H is given by

$$\dot{q}^1 = p_1, \qquad \dot{q}^2 = p_2, \qquad \dot{p}_1 = -\omega_1^2 q^1, \qquad \dot{p}_2 = -\omega_2^2 q^2.$$

This system is integrable, since we have two independent functions of (q^i, p_i), i.e., Hamiltonians

$$H_1 = \frac{p_1^2}{2} + \frac{\omega_1^2(q^1)^2}{2}, \qquad H_2 = \frac{p_2^2}{2} + \frac{\omega_2^2(q^2)^2}{2}.$$

The level curves of these functions are compact sets (topological circles); therefore, the orbits in the 4D phase–space $\mathbb{R}^4$ actually lie on the two–torus T^2. By making the appropriate change of variables, it can be shown (see, e.g., [Wig90]) that the whole dynamics of the two linearly coupled linear un–damped oscillators is actually contained in the equations

$$\dot{\theta}_1 = \omega_1, \qquad \dot{\theta}_2 = \omega_2, \qquad (\theta_1, \theta_2) \in S^1 \times S^2 \equiv T^2. \qquad (1.68)$$

The flow on the two–torus T^2, generated by (1.68), is simple to compute and is given by

$$\theta_1(t) = \omega_1 t + \theta_{1_0}, \qquad \theta_1(t) = \omega_1 t + \theta_{1_0}, \qquad (\mathrm{mod}\ 2\pi),$$

and θ_1 and θ_2 are called the longitude and latitude. However, orbits under this flow will depend on how ω_1 and ω_2 are related. If ω_1 and ω_2 are *commensurate* (i.e., the equation $m\omega_1 + n\omega_2 = 0$, $(n, m) \in Z$ has solutions), then every phase curve of (1.68) is closed. However, if ω_1 and ω_2 are *incommensurate* i.e., upper equation has no solutions), then every phase curve of (1.68) is everywhere dense on T^2.

Somewhat deeper understanding of Hamiltonian dynamics is related to the method of *action–angle variables*. The easiest way to introduce this idea is to consider again a simple harmonic oscillator (1.66). If we transform equations (1.66) into polar coordinates using $q = r \sin \theta$, $p = r \cos \theta$, then the equations of the vector–field become $\dot{r} = 0$, $\dot{\theta} = 1$, having the obvious solution $r = $ const, $\theta = t + \theta_0$. For this example polar coordinates work nicely because the system (1.66) is linear and, therefore, all of the periodic orbits have the same period.

For the general, nonlinear one–DOF Hamiltonian system (1.64) or (1.65) we will seek a coordinate transformation that has the same effect. Namely, we will seek a coordinate transformation $(q, p) \mapsto (\theta(q, p), I(q, p))$ with inverse transformation $(\theta, I) \mapsto (q(I, \theta), p(I, \theta))$ such that the vector–field X_H in the action–angle (θ, I) coordinates satisfies the following conditions: (i) $\dot{I} = 0$; (ii) θ changes linearly in time on the closed orbits with $\dot{\theta} = \Omega(I)$. We might even think of I and θ heuristically as 'nonlinear polar coordinates'. In such a coordinate system Hamiltonian function takes the form $H = H(I)$, and also, $\Omega(I) = \partial_I H$, i.e., specifying I specifies a periodic orbit.

The action variable $I(q, p)$ geometrically represents an area enclosed by any closed curve, which is constant in time. It is defined as an integral $I = \frac{1}{2\pi} \int_H p\, dq$, where H denotes the periodic orbit defined by $H(q, p) = H = $ const. If the period of each periodic orbit defined by $H(q, p) = H = $ const is denoted by $T(H)$, the angle variable $\theta(q, p)$ is defined by

$$\theta(q, p) = \frac{2\pi}{T(H)} t(q, p),$$

where $t = t(q, p)$ represents the time taken for the solution starting from (q_0, p_0) to reach (q, p).

For the system with Hamiltonian $H = \frac{p^2}{2} + V(x)$ and momentum $p = \pm\sqrt{2}\sqrt{H - V(x)}$ the action is given by $I = \frac{\sqrt{2}}{\pi} \int_{q_{min}}^{q_{max}} \sqrt{H - V(q)}\, dq$, and the angle is given by $\theta(q, p) = \frac{2\pi}{T(H)} \int_{q_{min}}^{q_{max}} \frac{dq}{\sqrt{2}\sqrt{H - V(q)}}$.

Closely related to the action–angle variables is the *perturbation theory* (see [Nay73]). To explain the main idea of this theory, let us consider an $\epsilon-perturbed$ vector–field periodic in t which can be in component form given as (with $(q, p) \in \mathbb{R}^2$)

$$\dot{q} = f_1(q, p) + \epsilon g_1(q, p, t, \epsilon), \qquad \dot{p} = f_2(q, p) + \epsilon g_2(q, p, t, \epsilon). \qquad (1.69)$$

Setting $\epsilon = 0$ we get the *un–perturbed* Hamiltonian system with a smooth scalar–valued function $H(q, p)$ for which holds

$$f_1(q, p) = \frac{\partial H(q, p)}{\partial p}, \qquad f_2(q, p) = -\frac{\partial H(q, p)}{\partial q},$$

so, the perturbed system (1.69) gets the symmetric canonical form

$$\dot{q} = \frac{\partial H(q, p)}{\partial p} + \epsilon g_1(q, p, t, \epsilon), \qquad \dot{p} = -\frac{\partial H(q, p)}{\partial q} + \epsilon g_2(q, p, t, \epsilon).$$

The perturbation (g_1, g_2) need not be Hamiltonian, although in the case where perturbation is Hamiltonian versus the case where it is not, the dynamics are very different.

Now, if we transform the coordinates of the perturbed vector–field using the action–angle transformation for the un–perturbed Hamiltonian vector–field, we get

$$\dot{I} = \epsilon \left(\frac{\partial I}{\partial q} g_1 + \frac{\partial I}{\partial p} g_2 \right) \equiv \epsilon F(I, \theta, t, \epsilon), \qquad (1.70)$$

$$\dot{\theta} = \Omega(I) + \epsilon \left(\frac{\partial \theta}{\partial q} g_1 + \frac{\partial \theta}{\partial p} g_2 \right) \equiv \Omega(I) + \epsilon G(I, \theta, t, \epsilon),$$

where

$$F(I, \theta, t, \epsilon) = \frac{\partial I}{\partial q}(q(I, \theta), p(I, \theta))\, g_1((q(I, \theta), p(I, \theta), t, \epsilon)$$
$$+ \frac{\partial I}{\partial p}(q(I, \theta), p(I, \theta))\, g_2((q(I, \theta), p(I, \theta), t, \epsilon),$$

$$G(I, \theta, t, \epsilon) = \frac{\partial \theta}{\partial q}(q(I, \theta), p(I, \theta))\, g_1((q(I, \theta), p(I, \theta), t, \epsilon)$$
$$+ \frac{\partial \theta}{\partial p}(q(I, \theta), p(I, \theta))\, g_2((q(I, \theta), p(I, \theta), t, \epsilon).$$

Here, F and G are 2π periodic in θ and $T = 2\pi/\omega$ periodic in t.

Finally, we shall explain in brief the most important idea in the dynamical systems theory, the idea of *Poincaré maps*. The idea of reducing the study of continuous time systems (flows) to the study of an *associated discrete–time system (map)* is due to Poincaré who first utilized it in the end of the last century in his studies of the three body problem in celestial mechanics. Nowadays, virtually any discrete–time system that is associated with an ordinary differential equation is refereed to as a *Poincaré map* (see, e.g., [Wig90]). This technique offers several advantages in the study of dynamical systems, including dimensional reduction, global dynamics and conceptual clarity. However, construction of a Poincaré map requires some knowledge of the phase–space of a dynamical system. One of the techniques which can be used for construction of Poincaré maps is the perturbation method.

To construct the Poincaré map for the system (1.70), we have to rewrite it as an autonomous system

$$\dot{I} = \epsilon F(I, \theta, \phi, \epsilon), \qquad \dot{\theta} = \Omega(I) + \epsilon G(I, \theta, \phi, \epsilon), \qquad \dot{\phi} = \omega, \qquad (1.71)$$

(where $(I, \theta, \phi) \in \mathbb{R}^+ \times S^1 \times S^1$. We construct a global *cross–section* Σ to this vector–field defined as $\Sigma^{\phi_0} = \{(I, \theta, \phi) | \phi = \phi_0\}$. If we denote the (I, θ) components of solutions of (1.71) by $(I_\epsilon(t), \theta_\epsilon(t))$ and the (I, θ) components of solutions of (1.71) for $\epsilon = 0$ by $(I_0, \Omega(I_0)t + \theta_0)$, then the perturbed Poincaré map is given by

$$P_\epsilon : \Sigma^{\phi_0} \to \Sigma^{\phi_0}, \qquad (I_\epsilon(0), \theta_\epsilon(0)) \mapsto (I_\epsilon(T), \theta_\epsilon(T)),$$

and the *m*th *iterate* of the Poincaré map is given by

$$P_\epsilon^m : \Sigma^{\phi_0} \to \Sigma^{\phi_0}, \qquad (I_\epsilon(0), \theta_\epsilon(0)) \mapsto (I_\epsilon(mT), \theta_\epsilon(mT)).$$

Now we can approximate the solutions to the perturbed problem as linear, constant–coefficient approximation

$$I_\epsilon(t) = I_0 + \epsilon I_1(t) + O(\epsilon^2), \qquad \theta_\epsilon(t) = \theta_0 + \Omega(I_0)t + \epsilon \theta_1(t) + O(\epsilon^2),$$

where we have chosen $I_\epsilon(0) = I_0$, $\theta_\epsilon(0) = \theta_0$.

As a last example of one–DOF Hamiltonian dynamics we shall analyze a *damped, forced Duffing oscillator*, given by canonical equations [Wig90]

$$\dot{q} = p, \qquad \dot{p} = q - q^3 - \delta p + \gamma \cos \omega t, \qquad \delta, \gamma, \omega \geq 0, \ (q, p) \in \mathbb{R}^2. \ (1.72)$$

where δ, γ, and ω are real parameters physically meaning *dissipation, amplitude of forcing* and *frequency*, respectively.

The *perturbed* system (1.72) is given by

$$\dot{q} = p, \qquad \dot{p} = q - q^3 + \epsilon(\gamma \cos \omega t - \delta p), \qquad (1.73)$$

where ϵ−perturbation is assumed small. Then the *un–perturbed* system is given by

$$\dot{q} = p, \qquad \dot{p} = q - q^3,$$

and is conservative with Hamiltonian function

$$H(q, p) = \frac{p^2}{2} - \frac{q^2}{2} + \frac{q^4}{4}. \tag{1.74}$$

In the un–perturbed phase–space all orbits are given by the level sets of the Hamiltonian (1.74). There are three equilibrium points at the following coordinates: $(q, p) = (\pm 1, 0)$ – centers, and $(q, p) = (0, 0)$ – saddle. The saddle point is connected to itself by two homoclinic orbits given by

$$q_+^0(t) = (\sqrt{2}(\cosh t)^{-1}, -\sqrt{2}(\cosh t)^{-1} \tanh t), \qquad q_-^0(t) = -q_+^0(t).$$

There are two families of *periodic orbits* $q_\pm^k(t)$, where k represents the *elliptic modulus* related to the Hamiltonian by $H(q_\pm^k(t)) \equiv H(k) = \frac{k^2 - 1}{(2 - k^2)^2}$, inside the corresponding homoclinic orbits $q_\pm^0(t)$, with the period $T(k) = 2K(k)\sqrt{2 - k^2}$ ($K(k)$ is the complete elliptic integral of the first kind.

Also, there exists a family of periodic orbits outside the homoclinic orbits with the period $T(k) = 4K(k)\sqrt{k^2 - 1}$.

The perturbed system (1.73) can be rewritten as a third–order autonomous system

$$\dot{q} = p, \qquad \dot{p} = q - q^3 + \epsilon(\gamma \cos \phi - \delta p), \qquad \dot{\phi} = \omega,$$

where $(q, p, \phi) \in \mathbb{R}^2 \times S^1$, S^1 is the circle of length $2\pi/\omega$ and $\phi(t) = \omega t + \phi_0$. We form the global cross–section to the flow

$$\Sigma^{\phi_0} = \{(q, p, \phi) | \phi = \phi_0 \in [0, 2\pi/\omega]\}$$

and the associated Poincaré map is given by

$$P : \Sigma^{\phi_0} \to \Sigma^{\phi_0}, \qquad (q(0), p(0)) \mapsto (q(2\pi/\omega), p(2\pi/\omega)).$$

A detailed analysis of the perturbed Poincaré map for the damped, forced Duffing oscillator is related to the *Melnikov function* and can be found in [Wig90].

Complex One–DOF Hamiltonian Dynamics

Global complex analysis represents another powerful tool for analyzing the uni-axial joint dynamics, which can be easily generalized to $n-$DOF human–like musculo–skeletal chains. Setting $z = q + ip$, $z \in \mathbb{C}$, $i = \sqrt{-1}$, Hamilton's equations $\dot{q} = \partial H/\partial p$, $\dot{p} = -\partial H/\partial q$ may be written in *complex notation* as [AM78, MR99, Wig90]

$$\dot{z} = -2i \frac{\partial H}{\partial \bar{z}}. \tag{1.75}$$

Let U be an open set in the *complex phase–space manifold* M_C (i.e., manifold M modelled on $\mathbb{C}$). A C^0 function $\gamma : [a,b] \to U \subset M_C,\quad t \mapsto \gamma(t)$ represents a *solution curve* $\gamma(t) = q(t) + ip(t)$ of a complex Hamiltonian system (1.75). For instance, the curve $\gamma(\theta) = \cos\theta + i\sin\theta,\quad 0 \le \theta \le 2\pi$ is the unit circle. $\gamma(t)$ is a *parameterized curve*. We call $\gamma(a)$ the *beginning point*, and $\gamma(b)$ the *end point* of the curve. By a *point on the curve* we mean a point w such that $w = \gamma(t)$ for some $t \in [a,b]$.

The derivative $\dot{\gamma}(t)$ is defined in the usual way, namely

$$\dot{\gamma}(t) = \dot{q}(t) + i\dot{p}(t),$$

so that the usual rules for the derivative of a sum, product, quotient, and chain rule are valid. The *speed* is defined as usual to be $|\dot{\gamma}(t)|$. Also, if $f : U \to M_C$ represents a *holomorphic*, or *analytic* function, then the composite $f \circ \gamma$ is differentiable (as a function of the real variable t) and $(f \circ \gamma)'(t) = f'(\gamma(t))\,\dot{\gamma}(t)$.

A *path* represents a sequence of C^1-curves,

$$\gamma = \{\gamma_1, \gamma_2, \ldots, \gamma_n\},$$

such that the end point of γ_j, $(j = 1, \ldots, n)$ is equal to the beginning point of γ_{j+1}. If γ_j is defined on the interval $[a_j, b_j]$, this means that

$$\gamma_j(b_j) = \gamma_{j+1}(a_{j+1}).$$

We call $\gamma_1(a_1)$ the *beginning point* of γ_j, and $\gamma_n(b_n)$ the *end point* of γ_j. The path is said to *lie in an open set* $U \subset M_C$ if each curve γ_j lies in U, i.e., for each t, the point $\gamma_j(t)$ lies in U.

An open set U is *connected* if given two points α and β in U, there exists a path $\gamma = \gamma_1, \gamma_2, \ldots, \gamma_n$ in U such that α is the beginning point of γ_1 and β is the end point of γ_n; in other words, if there is a path γ in U which joins α to β. If U is a connected open set and f a holomorphic function on U such that $f' = 0$, then f is a constant. If g is a function on U such that $f' = g$, then f is called a *primitive* of g on U. Primitives can be either find out by integration or written down directly.

Let f be a C^0-function on an open set U, and suppose that γ is a curve in U, meaning that all values $\gamma(t)$ lie in U for $a \le t \le b$. The *integral of f along γ* is defined as

$$\int_\gamma f = \int_\gamma f(z) = \int_a^b f(\gamma(t))\,\dot{\gamma}(t)\,dt.$$

For example, let $f(z) = 1/z$, and $\gamma(\theta) = e^{i\theta}$. Then $\dot{\gamma}(\theta) = ie^{i\theta}$. We want to find the value of the integral of f over the circle, $\int_\gamma dz/z$, so $0 \le \theta \le 2\pi$. By definition, this integral is equal to $\int_0^{2\pi} ie^{i\theta}/e^{i\theta}\,d\theta = i\int_0^{2\pi} d\theta = 2\pi i$.

The *length* $L(\gamma)$ is defined to be the integral of the speed, $L(\gamma) = \int_a^b |\dot{\gamma}(t)|\,dt$.

If $\gamma = \gamma_1, \gamma_2, \ldots, \gamma_n$ is a path, then the integral of a C^0−function f on an open set U is defined as $\int_\gamma f = \sum_{i=1}^n \int_{\gamma_i} f$, i.e., the sum of the integrals of f over each curve γ_i ($i = 1, \ldots, n$ of the path γ. The *length of a path* is defined as $L(\gamma) = \sum_{i=1}^n L(\gamma_i)$.

Let f be continuous on an open set $U \subset M_C$, and suppose that f has a primitive g, that is, g is holomorphic and $g' = f$. Let α, β be two points in U, and let γ be a path in U joining α to β. Then $\int_\gamma f = g(\beta) - g(\alpha)$; this integral is independent of the path and depends only on the beginning and end point of the path.

A *closed path* is a path whose beginning point is equal to its end point. If f is a C^0−function on an open set $U \subset M_C$ admitting a holomorphic primitive g, and γ is any closed path in U, then $\int_\gamma f = 0$.

Let γ, η be two paths defined over the same interval $[a, b]$ in an open set $U \subset M_C$. Recall (see Introduction) that γ is *homotopic* to η if there exists a C^0−function $\psi : [a, b] \times [c, d] \to U$ defined on a rectangle $[a, b] \times [c, d] \subset U$, such that $\psi(t, c) = \gamma(t)$ and $\psi(t, d) = \eta(t)$ for all $t \in [a, b]$. For each number $s \in [c, d]$ we may view the function $|psi_s(t) = \psi(t, s)$ as a continuous curve defined on $[a, b]$, and we may view the family of continuous curves ψ_s as a *deformation* of the path γ to the path η. It is said that the homotopy ψ *leaves the end points fixed* if we have $\psi(a, s) = \gamma(a)$ and $\psi(b, s) = \gamma(b)$ for all values of $s \in [c, d]$. Similarly, when we speak of a homotopy of closed paths, we assume that each path ψ_s is a closed path.

Let γ, η be paths in an open set $U \subset M_C$ having the same beginning and end points. Assume that they are homotopic in U. Let f be holomorphic on U. Then $\int_\gamma f = \int_\eta f$. The same holds for closed homotopic paths in U. In particular, if γ is homotopic to a point in U, then $\int_\gamma f = 0$. Also, it is said that an open set $U \subset M_C$ is *simply connected* if it is connected and if every closed path in U is homotopic to a point.

In the previous example we found that

$$\frac{1}{2\pi I} \int_\gamma \frac{1}{z}\, dz = 1,$$

if γ is a circle around the origin, oriented counterclockwise. Now we define for any closed path γ its *winding number* with respect to a point α to be

$$W(\gamma, \alpha) = \frac{1}{2\pi i} \int_\gamma \frac{1}{z - \alpha}\, dz,$$

provided the path does not pass through α. If γ is a closed path, then $W(\gamma, \alpha)$ is an integer.

A closed path $\gamma \in U \subset M_C$ is *homologous to 0 in U* if

$$\int_\gamma \frac{1}{z - \alpha}\, dz = 0,$$

for every point α not in U, or in other words, $W(\gamma, \alpha) = 0$ for every such point.

Similarly, let γ, η be closed paths in an open set $U \subset M_C$. We say that they are *homologous in* U, and write $\gamma \sim \eta$, if $W(\gamma, \alpha) = W(\eta, \alpha)$ for every point α in the complement of U. We say that γ is *homologous to* 0 *in* U, and write $\gamma \sim 0$, if $W(\gamma, \alpha) = 0$ for every point α in the complement of U.

If γ and η are closed paths in U and are homotopic, then they are homologous. If γ and η are closed paths in U and are close together, then they are homologous.

Let $\gamma_1, \ldots, \gamma_n$ be curves in an open set $U \subset M_C$, and let $m_1, \ldots, m_n$ be integers. A formal sum $\gamma = m_1\gamma_1 + \cdots + m_n\gamma_n = m_i\gamma_i$ is called a *chain in* U. The chain is called *closed* if it is a finite sum of closed paths. If γ is the chain as above, then $\int_\gamma f = m_i \int_{\gamma_i} f$. If γ and η are closed chains in U, then $W(\gamma + \eta, \alpha) = W(\gamma, \alpha) + W(\eta, \alpha)$. We say that γ and η are *homologous in* U, and write $\gamma \sim \eta$, if $W(\gamma, \alpha) = W(\eta, \alpha)$ for every point α in the complement of U. We say that γ is *homologous to* 0 *in* U, and write $\gamma \sim 0$, if $W(\gamma, \alpha) = 0$ for every point α in the complement of U.

Cauchy's theorem states that if γ is a closed chain in an open set $U \subset M_C$, and γ is homologous to 0 in U, then $\int_\gamma f = 0$. If γ and η are closed chains in U, and $\gamma \sim \eta$ in U, then $\int_\gamma f = \int_\eta f$.

It follows from Cauchy's theorem that if γ and η are homologous, then $\int_\gamma f = \int_\eta f$ for all holomorphic functions f on U [AM78, Wig90].

Library of Basic Hamiltonian Systems

In this subsection we present some basic Hamiltonian systems [Put93] (commonly used in biodynamics and robotics see [II05, II06a, II06b]).

1D Harmonic Oscillator

In this case we have $\{p, q\}$ as canonical coordinates on $\mathbb{R}^2$

$$M = T^*\mathbb{R} \simeq \mathbb{R}^2, \qquad \omega = dp \wedge dq,$$
$$H = \frac{1}{2}\left(p^2 + y\right), \qquad X_H = p\frac{\partial}{\partial q} - q\frac{\partial}{\partial p},$$

and Hamilton's equations read

$$\dot{q} = p, \qquad \dot{p} = -q.$$

For each $f, g \in C^k(\mathbb{R}^2, \mathbb{R})$ the Poisson bracket is given by

$$\{f, g\}_\omega = \frac{\partial f}{\partial q}\frac{\partial g}{\partial p} - \frac{\partial f}{\partial p}\frac{\partial g}{\partial q}.$$

Complex Plane

Let $T^*\mathbb{R} \simeq \mathbb{R}^2$ have the canonical symplectic structure $\omega = dp \wedge dq$. Writing $z = q + ip$, we have

$$\omega = \frac{1}{2i}\, dz \wedge d\bar{z}, \qquad X_H = i\left(\frac{\partial H}{\partial z}\frac{\partial}{\partial z} - \frac{\partial H}{\partial \bar{z}}\frac{\partial}{\partial \bar{z}}\right),$$

$$\{f,g\}_\omega = \frac{i}{2}\left(\frac{\partial f}{\partial z}\frac{\partial g}{\partial \bar{z}} - \frac{\partial f}{\partial \bar{z}}\frac{\partial g}{\partial z}\right),$$

so, the Hamilton's equations, $\dot{q} = \partial_p H$, $\dot{p} = -\partial_q H$, become

$$\dot{z} = -2i\frac{\partial H}{\partial \bar{z}}.$$

2D Harmonic Oscillator

In this case we have $\{q^1, y, p_1, p_2\}$ as canonical coordinates on $\mathbb{R}^4$

$$M = T^*\mathbb{R}^2 \simeq \mathbb{R}^4, \qquad \omega = dp_1 \wedge dq^1 + dp_2 \wedge dq^2,$$

$$H = \frac{1}{2}\left[p_1^2 + p_2^2 + (q^1)^2 + (y)^2\right].$$

The functions $f = p_i p_j + q^i q^j$ and $g = p_i q^j + p_j q^i$, (for $i,j = 1,2$), are constants of motion.

nD Harmonic Oscillator

In this case we have $(i = 1, ..., n)$

$$M = T^*\mathbb{R}^n \simeq \mathbb{R}^{2n}, \qquad \omega = dp_i \wedge dq^i,$$

$$H = \frac{1}{2}\sum_{i=1}^n \left[p_i^2 + (q^i)^2\right].$$

The system is integrable in an open set of $T^*\mathbb{R}^n$ with:

$$K_1 = H, \qquad K_2 = p_2^2 + (y)^2, \qquad ..., \qquad K_n = p_n^2 + (q^n)^2.$$

Toda Molecule

Consider three mass–points on the line with coordinates q^i, $(i = 1,2,3)$, and satisfying the ODEs:

$$\ddot{q}^i = -\partial_{q^i} U, \qquad \text{where} \qquad U = e^{q^1 - q^2} + e^{q^2 - q^3} - e^{q^3 - q^1}.$$

This is a Hamiltonian system with $\{q^i, p_i\}$ as canonical coordinates on $\mathbb{R}^6$,

$$M = T^*\mathbb{R}^3 \simeq \mathbb{R}^6, \qquad \omega = dp_i \wedge dq^i,$$

$$H = \frac{1}{2}\left(p_1^2 + p_2^2 + p_3^2\right) + U.$$

The Toda molecule (1.4.5) is an integrable Hamiltonian system in an open set of $T^*\mathbb{R}^3$ with:

$$K_1 = H, \qquad K_2 = p_1 + p_2 + p_3,$$
$$K_3 = \frac{1}{9}\left(p_1 + p_2 + p_3\right)\left(p_2 + p_3 - 2p_1\right)\left(p_3 + p_1 - 2p_2\right) - \left(p_1 + p_2 - 2p_3\right)e^{q^1 - q^2}$$
$$- \left(p_2 + p_3 - 2p_1\right)e^{q^2 - q^3} - \left(p_3 + p_1 - 2p_2\right)e^{q^3 - q^1}.$$

3–Point Vortex Problem

The motion of three–point vortices for an ideal incompressible fluid in the plane is given by the equations:

$$\dot{q}^j = -\frac{1}{2\pi}\sum_{i \neq j}\Gamma_i\left(p_j - p_i\right)/r_{ij}^2,$$
$$\dot{p}_j = \frac{1}{2\pi}\sum_{i \neq j}\Gamma_i\left(q^i - q^j\right)/r_{ij}^2,$$
$$r_{ij}^2 = \left(q^i - q^j\right)^2 + \left(p_j - p_i\right)^2,$$

where $i, j = 1, 2, 3$, and Γ_i are three nonzero constants. This mechanical system is Hamiltonian if we take:

$$M = T^*\mathbb{R}^3 \simeq \mathbb{R}^6, \qquad \omega = dp_i \wedge dq^i, \qquad (i = 1, ..., 3),$$
$$H = -\frac{1}{4\pi}\Gamma_i\Gamma_j \ln\left(r_{ij}\right).$$

Moreover, it is integrable in an open set of $T^*\mathbb{R}^3$ with:

$$K_1 = H, \qquad K_2 = \Gamma_i\left[\left(q^i\right)^2 + p_i^2\right],$$
$$K_3 = \left(\Gamma_i q^i\right)^2 + K_2^2.$$

The Newton's Second Law as a Hamiltonian System

In the case of conservative forces, Newton's law of motion can be written on $\mathbb{R}^{3n}$ as

$$m_i\ddot{q}^i = -\partial_{q^i}U, \qquad (i = 1, 2, ..., 3n).$$

Its symplectic formulation reads:

$$M = T^*\mathbb{R}^3 \simeq \mathbb{R}^6, \qquad \omega = dp_i \wedge dq^i,$$
$$H = \frac{p_i^2}{2m_i} + U.$$

The Hamiltonian vector–field X_H is

$$X_H = \left(\frac{p_i}{m_i} \, \partial_{q^i} - \partial_{q^i} U \, \partial_{p_i} \right),$$

giving the Hamilton's equations

$$\dot{q}^i = \frac{p_i}{m_i}, \qquad \dot{p}_i = -\partial_{q^i} U.$$

Rigid Body

The configuration space of the *rigid body* is $SO(3)$, the group of proper orthogonal transformations of $\mathbb{R}^3$ to itself, while the corresponding phase–space is its cotangent bundle, $T^*SO(3)$. The motion of a rigid body is a *geodesic* with respect to a *left–invariant Riemannian metric* (the inertia tensor) on $SO(3)$. The *momentum map* $J : P \to \mathbb{R}^3$ for the *left* $SO(3)-$action is *right* translation to the identity. We identify $\mathfrak{so}(3)^*$ with $\mathfrak{so}(3)$ via the *Killing form* and identify $\mathbb{R}^3$ with $\mathfrak{so}(3)$ via the map $v \mapsto \hat{v}$, where $\hat{v}(w) = v \times w$ ($\times$ being the standard cross product). Points in $\mathfrak{so}(3)^*$ are regarded as the left reduction of $T^*SO(3)$ by $G = SO(3)$ and are the angular momenta as seen from a *body–fixed frame*.

A Segment of a Human–Like Body

A *rigid body with a fixed–point* is a basic model of a single segment of the human (or robot) body. This is a left–invariant Hamiltonian mechanical system on the phase–space $T^*SO(3)$. The differentiable structure on $SO(3)$ is defined using the traditional Euler angles $\{\varphi, \psi, \theta\}$.

More precisely, a local chart is given by [Put93]

$$(\varphi, \psi, \theta) \in \mathbb{R}^3 \longmapsto A \in SO(3), \qquad 0 < \varphi, \psi < 2\pi; \qquad 0 < \theta < \pi,$$

where

$$A = \begin{bmatrix} \cos\psi\cos\varphi - \cos\theta\sin\varphi\sin\psi & \cos\psi\cos\varphi + \cos\theta\cos\varphi\sin\psi & \sin\theta\sin\psi \\ -\sin\psi\cos\varphi - \cos\theta\sin\varphi\sin\psi & -\sin\psi\sin\varphi + \cos\theta\cos\varphi\cos\psi & \sin\theta\cos\psi \\ \sin\theta\sin\varphi & -\sin\theta\cos\varphi & \cos\theta \end{bmatrix}$$

The corresponding conjugate momenta are denoted by $p_\varphi, p_\psi, p_\theta$, so $\{\varphi, \psi, \theta, p_\varphi, p_\psi, p_\theta\}$ is the phase–space $T^*SO(3)$. Thus, we have

$$M = T^*SO(3), \qquad \omega = dp_\varphi \wedge d\varphi + dp_\psi \wedge d\psi + dp_\theta \wedge d\theta, \qquad H = \frac{1}{2}K,$$

$$K = \frac{[(p_\varphi - p_\psi \cos\theta)\sin\psi + p_\theta \sin\theta\cos\psi]^2}{I_1 \sin^2\theta}$$

$$+ \frac{[(p_\varphi - p_\psi \cos\theta)\cos\psi - p_\theta \sin\theta\sin\psi]^2}{I_2 \sin^2\theta} + \frac{p_\psi^2}{I_3},$$

where I_1, I_2, I_3 are the moments of inertia, diagonalizing the inertia tensor of the body.

The Hamilton's equations are

$$\dot{\varphi} = \frac{\partial H}{\partial p_\varphi}, \qquad \dot{\psi} = \frac{\partial H}{\partial p_\psi}, \qquad \dot{\theta} = \frac{\partial H}{\partial p_\theta},$$

$$\dot{p}_\varphi = -\frac{\partial H}{\partial \varphi}, \qquad \dot{p}_\psi = -\frac{\partial H}{\partial \psi}, \qquad \dot{p}_\theta = -\frac{\partial H}{\partial \theta}.$$

For each $f, g \in C^k(T^*SO(3), \mathbb{R})$ the Poisson bracket is given by

$$\{f, g\}_\omega = \frac{\partial f}{\partial \varphi}\frac{\partial g}{\partial p_\varphi} - \frac{\partial f}{\partial p_\varphi}\frac{\partial g}{\partial \varphi} + \frac{\partial f}{\partial \psi}\frac{\partial g}{\partial p_\psi} - \frac{\partial f}{\partial p_\psi}\frac{\partial g}{\partial \psi}$$

$$+ \frac{\partial f}{\partial \theta}\frac{\partial g}{\partial p_\theta} - \frac{\partial f}{\partial p_\theta}\frac{\partial g}{\partial \theta}.$$

The Heavy Top

The heavy top is by definition a rigid body moving about a fixed–point in a 3D space [Put93]. The rigidity of the top means that the distances between points of the body are fixed as the body moves. In this case we have

$$M = T^*SO(3),$$
$$\omega = dp_\varphi \wedge d\varphi + dp_\psi \wedge d\psi + dp_\theta \wedge d\theta,$$
$$H = \frac{1}{2}K + mgl\cos\theta,$$
$$K = \frac{[(p_\varphi - p_\psi\cos\theta)\sin\psi + p_\theta\sin\theta\cos\psi]^2}{I_1\sin^2\theta}$$
$$+ \frac{[(p_\varphi - p_\psi\cos\theta)\cos\psi - p_\theta\sin\theta\sin\psi]^2}{I_2\sin^2\theta} + \frac{p_\psi^2}{I_3},$$

where I_1, I_2, I_3 are the moments of inertia, m is the total mass, g is the gravitational acceleration and l is the length of the vector determining the center of mass at $t = 0$.

The Hamilton's equations are

$$\dot{\varphi} = \frac{\partial H}{\partial p_\varphi}, \qquad \dot{\psi} = \frac{\partial H}{\partial p_\psi}, \qquad \dot{\theta} = \frac{\partial H}{\partial p_\theta},$$

$$\dot{p}_\varphi = -\frac{\partial H}{\partial \varphi}, \qquad \dot{p}_\psi = -\frac{\partial H}{\partial \psi}, \qquad \dot{p}_\theta = -\frac{\delta H}{\partial \theta}.$$

For each $f, g \in C^k(T^*SO(3), \mathbb{R})$ the Poisson bracket is given by

$$\{f, g\}_\omega = \frac{\partial f}{\partial \varphi}\frac{\partial g}{\partial p_\varphi} - \frac{\partial f}{\partial p_\varphi}\frac{\partial g}{\partial \varphi} + \frac{\partial f}{\partial \psi}\frac{\partial g}{\partial p_\psi} - \frac{\partial f}{\partial p_\psi}\frac{\partial g}{\partial \psi}$$

$$+ \frac{\partial f}{\partial \theta}\frac{\partial g}{\partial p_\theta} - \frac{\partial f}{\partial p_\theta}\frac{\partial g}{\partial \theta}.$$

and $J = \begin{pmatrix} 0 & 1 & 0 \\ -1 & 0 & 0 \\ 0 & 0 & 0 \end{pmatrix}$. In these coordinates u_0 is the center of mass, v_0 is total linear momentum, and total angular momentum is

$$A = u_0 \times v_0 + u_1 \times v_1 + u_2 \times v_2.$$

See [MH92] for further details.

$n-$DOF Hamiltonian Dynamics

Classically, $n-$DOF Hamiltonian dynamics combines the ideas of differential equations and variational principles (see [AM78, Arn89, MR99, Wig90]). As Hamilton first realized, many of the systems of mechanics and optics can be put into the special form

$$\dot{q}^i = \frac{\partial H}{\partial p_i}(q^i, p_i, t), \qquad \dot{p}_i = -\frac{\partial H}{\partial q^i}(q^i, p_i, t), \qquad (i = 1, \ldots, n),$$

or an associated variational form (summing upon the repeated index is used in the following text)

$$\delta \int (p_i dq^i - H) \, dt = 0.$$

Here the state of the system is given as a point $(q^1, \ldots, q^n, p_1, \ldots, p_n)$ in *phase-space*, the q's are the configuration coordinates, the p's are the momenta, t is time, and $H = H(q^i, p_i, t)$ is a total–energy function called Hamiltonian. The variables (q^i, p_i) are called *canonical* coordinates.

If $H = H(q^i, p_i)$ does not depend explicitly on time, the system is said to be *autonomous*. In this case, it is easy to verify that H is conserved. The search for other conserved quantities led to a new notion of solving Hamiltonian systems. Instead of finding formulae for the coordinates as a function of time, one searches for constants of the motion (*integrals*). If one can find n integrals $I_i(q^i, p_i)$ which are in *involution*:

$$[I_i, I_j] = \frac{\partial I_i}{\partial q^k} \frac{\partial I_j}{\partial p_k} - \frac{\partial I_i}{\partial p_k} \frac{\partial I_j}{\partial q^k} = 0, \qquad (i \neq j),$$

(here the square brackets denote the Poisson bracket) and *independent* (the vectors ∇I_i are independent 'almost everywhere'), then associated variables ϕ_i can be derived which evolve linearly in time:

$$\dot{\phi}^i = \frac{\partial H}{\partial I_i}(I_i).$$

Such a system is *integrable* in the sense of Liouville [Arn89]. If the sets $I = \text{const}$ are bounded, then they are nD *tori* T^n in phase–space. Choosing irreducible cycles, γ_i, on the tori, one can define a preferred set of integrals

$J_i = \int_{\gamma_i} p_i dq^i$, called *action variables*, for which the corresponding ϕ_i are *angle variables* mod 1 on T^n. The quantities $\omega^i(J) = \frac{\partial H}{\partial J_i}(J_i)$ are called the *frequencies* on T^n.

Another feature of Hamiltonian systems noticed by Liouville is the preservation of phase–space volume $\int (dq)^n (dp)^n$. A more general result is that Poincaré's integral $\int p_i dq^i$ is conserved around any loop following the flow [Arn89]. This is the property that really distinguishes Hamiltonian differential equations from general ones.

The major problem with the notion of integrability is that *most systems are not integrable*. This was first appreciated when Poincaré proved that the *circular restricted 3–body problem has no integral analytic in the mass ratio*. The perturbation expansions which gave excellent predictions of motion of the planets do not converge. The basic reason is that among the invariant tori of integrable systems is a dense subset on which the frequencies ω^i are *commensurate*, i.e, $m_i \omega^i = 0$ for some non–zero integer vector m_i; however, most systems have no commensurate tori, because they can be destroyed by arbitrarily small perturbation.

Poincaré went on to examine what really does happen. The key technique he used was geometric analysis: instead of manipulating formulae for canonical transformations as Jacobi and others did, he pictured the *orbits* in phase–space. An important step in this *qualitative theory* of differential equations was the idea of *surface of section*. If Σ is a *codimension–one* surface (i.e., of dimension one less than that of the phase–space) transverse to a *flow*, then the sequence $\{x_j\}$ of successive intersections of an orbit with Σ provides a lot of information about that orbit. For example, if $\{x_j\}$ is periodic then it corresponds to a periodic orbit. If $\{x_j\}$ is confined to a subset of codimension m on Σ then so is the orbit of the flow, etc.. The flow induces a map of Σ to itself; the map takes a point in Σ to the point at which it first returns to Σ (assuming there is on). Since the surface of section has one dimension less than the phase–space it is easier to picture the dynamics of the return map than the flow. In fact, for Hamiltonian systems one can do even better; since H is conserved, Σ decomposes into a 1–parameter family of codimension two surfaces parameterized by the value of the energy, a reduction of two dimensions.

This led Poincaré to the ideas of *stable* and *unstable manifolds* for *hyperbolic* periodic orbits, which are extensions of the stable and unstable *eigen–spaces* for associated linear systems, and their intersections, known as *hetero–* and *homo–clinic points*, whose orbits converge to one periodic orbit in the past and to another (or the same) in the future. He showed that having intersected once, the invariant manifolds must intersect infinitely often. Moreover the existence of one heteroclinic orbit implies the existence of an infinity of others.

The distance between the stable and unstable manifolds can be quantified by Melnikov's integral. This leads to a technique for proving the non–existence of integrals for a slightly perturbed, integrable Hamiltonian.

For integrable systems, nearby orbits separate linearly in time; however, dynamical systems can have exponentially separating orbits. Let δx be a tangent vector at the phase–space point x and δx_t be the evolved vector following the orbit of x. Then the average rate of exponentiation of δx_t is the *Lyapunov exponent* λ,

$$\lambda(x, \delta x) = \lim_{t \to \infty} 1/t \ln |\delta x_t|.$$

If λ is nonzero, then the predictions one can make will be valid for a time only logarithmic in the precision. Therefore, although deterministic in principle, a system need not be predictable in practice.

A concrete example of the complexity of behavior of typical Hamiltonian systems is provided by the *Smale horseshoe*, a type of invariant set found near homoclinic orbits. Its points can be labelled by doubly infinite sequences of 0's and 1's corresponding to which half of a horseshoe shaped set the orbit is in at successive times. For every sequence, no matter how complicated, there is an orbit which has that symbol sequence. This implies, e.g., that a simple pendulum in a sufficiently strongly modulated time–periodic gravitational field has an initial condition such that the pendulum will turn over once each period when there is 1 in the sequence and not if there is a 0 for any sequence of 0's and 1's.

1.4.6 Ergodic Systems

In many dynamical systems it is possible to choose the coordinates of the system so that volume (really an nD volume) in phase–space is invariant. This happens for mechanical systems derived from Newton's laws as long as the coordinates are the position and the momentum and the volume is measured in units of *position × momentum*. The flow f_t takes points of a subset A into the points $f_t(A)$ and invariance of the phase–space means that $\text{vol}(A) = \text{vol}(f_t(A))$.

In the Hamiltonian formalism (see [II06b]), given a coordinate it is possible to derive the appropriate (generalized) momentum such that the associated volume is preserved by the flow. The volume is said to be computed by the *Liouville measure*.

In a Hamiltonian system not all possible configurations of position and momentum can be reached from an initial condition. Because of energy conservation, only the states with the same energy as the initial condition are accessible. The states with same energy form an energy shell Ω, a sub–manifold of the phase–space. The energy shell has its Liouville measure that is preserved.

For systems where the volume is preserved by the flow, Poincaré discovered the *recurrence theorem*: Assume the phase–space has a finite Liouville volume and let F be a phase–space volume–preserving map and A a subset of the phase–space. Then almost every point of A returns to A infinitely often. The *Poincaré recurrence theorem* was used by Zermelo to object to Boltzmann's derivation of the increase in entropy in a dynamical system of colliding atoms.

One of the questions raised by Boltzmann's work was the *possible equality between time averages and space averages*, what he called the *ergodic hypothesis*. The hypothesis states that the length of time a typical trajectory spends in a region A equals $\mathrm{vol}(A)/\mathrm{vol}(\Omega)$.

The ergodic hypothesis turned out not to be the essential property needed for the development of *statistical mechanics* and a series of other ergodic–like properties were introduced to capture the relevant aspects of physical systems. The invariance of the Liouville measure on the energy surface Ω is essential for the *Boltzmann factor* $\exp(-\beta H)$ used in the statistical mechanics of Hamiltonian systems. This idea has been generalized by Sinai, Bowen, and Ruelle to a larger class of dynamical systems that includes dissipative systems. SRB measures replace the Boltzmann factor and they are defined on *attractors of chaotic systems*.

1.5 Continuous Chaotic Dynamics

The prediction of the behavior of a system when its evolution law is known, is a problem with an obvious interest in many fields of scientific research. Roughly speaking, within this problem two main areas of investigation may be identified [CFL97]:

A) The definition of the 'predictability time'. If one knows the initial state of a system, with a precision $\delta_o = |\delta\mathbf{x}(0)|$, what is the maximum time T_p within which one is able to know the system future state with a given tolerance $\delta_{\max}$?

B) The understanding of the relaxation properties. What is the relation between the mean response of a system to an external perturbation and the features of its un–perturbed state [Lei78]? Using the terminology of statistical mechanics, one wants to reduce 'non–equilibrium' properties, such as relaxation and responses, to 'equilibrium' ones, such as correlation functions [KT85].

A remarkable example of type–A problem is the weather forecasting, where one has to estimate the maximum time for which the prediction is enough accurate. As an example of type–B problem, of geophysical interest, one can mention a volcanic eruption which induces a practically instantaneous change in the temperature. In this case it is relevant to understand how the difference between the atmospheric state after the eruption and the hypothetical un–perturbed state without the eruption evolves in time. In practice one wants to understand how a system absorbs, on average, the perturbation $\delta f(\tau)$ of a certain quantity f (e.g., the temperature), just looking at the statistical features, as correlations, of f in the un–perturbed regime. This is the so called fluctuation/relaxation problem [Lei75].

As far as problem–A is concerned, in the presence of deterministic chaos, a rather common situation, the distance between two initially close trajectories diverges exponentially:

$$|\delta \mathbf{x}(t)| \sim \delta_o \exp(\lambda t), \tag{1.76}$$

where λ is the *maximum Lyapunov exponent* of the system [BGS76]. From (4.13) it follows:

$$T_p \sim \frac{1}{\lambda} \ln \left(\frac{\delta_{max}}{\delta_o} \right). \tag{1.77}$$

Since the dependence on δ_{max} and δ_o is very weak, T_p appears to be proportional to the inverse of the Lyapunov exponent. We stress however that (4.14) is just a naive answer to the predictability problem, since it does not take into account the following relevant features of the chaotic systems:

i) The Lyapunov exponent is a global quantity, i.e., it measures the average exponential rate of divergence of nearby trajectories. In general there are finite-time fluctuations of this rate, described by means of the so called effective Lyapunov exponent $\gamma_t(\tau)$. This quantity depends on both the time delay τ and the time t at which the perturbation acted [PSV87]. Therefore, the predictability time T_p fluctuates, following the $\gamma-$variations [CJP93].
ii) In systems with many degrees of freedom one has to understand how a perturbation grows and propagates trough the different degrees of freedom. In fact, one can be interested in the prediction on certain variables, e.g., those associated with large scales in weather forecasting, while the perturbations act on a different set of variables, e.g., those associated to small scales.
iii) If one is interested into non–infinitesimal perturbations, and the system possess many characteristic times, such as the eddy turn–over times in fully developed turbulence, then T_p is determined by the detailed mechanism of propagation of the perturbations through different degrees of freedom, due to nonlinear effects. In particular, T_p may have no relation with λ [ABC96].

In addition to these points, when the evolution of a system is ruled by a set of ODEs,

$$\dot{\mathbf{x}} = \mathbf{f}(\mathbf{x}, t), \tag{1.78}$$

which depend periodically on time,

$$\mathbf{f}(\mathbf{x}, t + T) = \mathbf{f}(\mathbf{x}, t), \tag{1.79}$$

one can have a kind of 'seasonal' effect, for which the system shows an alternation, roughly periodic, of low and high predictability. This happens, for example, in the recently studied case of stochastic resonance in a chaotic deterministic system, where one observes a roughly periodic sequence of chaotic and regular evolution intervals [CFP94].

As far as problem–B is concerned, it is possible to show that in a chaotic system with an invariant measure $P(\mathbf{x})$, there exists a relation between the mean response $\langle \delta x_j(\tau) \rangle_P$ after a time τ from a perturbation $\delta x_i(0)$, and a suitable correlation function. Namely, one has the following equation [CFL97]

$$R_{ij}(\tau) \equiv \frac{\langle \delta x_j(\tau) \rangle_P}{\delta x_i(0)} = \left\langle x_j(\tau) \frac{\partial S(\mathbf{x}(0))}{\partial x_i} \right\rangle_P, \qquad (1.80)$$

where $S(\mathbf{x}) = -\ln P(\mathbf{x})$. Equation (1.80) ensures that the mean relaxation of the perturbed system is equal to some correlation of the un–perturbed system. As, in general, one does not know $P(\mathbf{x})$, (1.80) provides only a qualitative information.

1.5.1 Dynamics and Non–Equilibrium Statistical Mechanics

Recall that statistical mechanics, which was created at the end of the 19-th century by such people as Maxwell, Boltzmann, and Gibbs, consists of two rather different parts: equilibrium and non–equilibrium statistical mechanics. The success of equilibrium statistical mechanics has been spectacular. It has been developed to a high degree of mathematical sophistication, and applied with success to subtle physical problems like the study of critical phenomena. Equilibrium statistical mechanics also has highly nontrivial connections with the mathematical theory of smooth dynamical systems and the physical theory of quantum fields. By contrast, the progress of non–equilibrium statistical mechanics has been much slower. We still depend on the insights of Boltzmann for our basic understanding of irreversibility, and this understanding remains rather qualitative. Further progress has been mostly on dissipative phenomena close to equilibrium: Onsager reciprocity relations [Ons35], Green–Kubo formula [Gre51, Kub57], and related results. Yet, there is currently a strong revival of non–equilibrium statistical mechanics, based on taking seriously the complications of the underlying microscopic dynamics.

In this subsection, following [Rue78, Rue98, Rue99], we provide a general discussion of *irreversibility*. It is a fact of common observation that the behavior of bulk matter is often irreversible: if a warm and a cold body are put in contact, they will equalize their temperatures, but once they are at the same temperature they will not spontaneously go back to the warm and cold situation. Such facts have been organized into a body of knowledge named *thermodynamics*. According to thermodynamics, a number called *entropy* can be associated with macroscopic systems which are (roughly speaking) locally in equilibrium. The definition is such that, when the system is isolated, its entropy can only increase with time or stay constant (the celebrated *Second Law of thermodynamics*). A strict increases in entropy corresponds to an irreversible process. Such processes are also called *dissipative structures*, because they dissipate noble forms of energy (like mechanical energy) into heat. The flow of a viscous fluid, the passage of an electric current through a conductor, or a chemical reaction like a combustion are typical dissipative phenomena. The purpose of non–equilibrium statistical mechanics is to explain irreversibility on the basis of microscopic dynamics, and to give quantitative predictions for dissipative phenomena. In what follows we shall assume that the microscopic dynamics is classical. Unfortunately we shall have little to say about the quantum case.

Dynamics and Entropy

We shall begin with a naive discussion, and then see how we have to modify it to avoid the difficulties that will arise. The microscopic time evolution (f^t) which we want to consider is determined by an evolution equation

$$\dot{x} = F(x), \tag{1.81}$$

in a phase–space M. More precisely, for an isolated system with Hamiltonian $H = H(q, p)$, we may rewrite (1.81) as

$$\frac{d}{dt} \begin{pmatrix} p \\ q \end{pmatrix} = \begin{pmatrix} -\partial_q H \\ \partial_p H \end{pmatrix}, \tag{1.82}$$

where p, and q are ND for a system with N DOF, so that the phase–space M is $2ND$ manifold. Note that we are interested in a macroscopic description of a macroscopic system where N is large, and microscopic details cannot be observed. In fact many different points (p, q) in phase–space have the same macroscopic appearance. It appears therefore reasonable to describe a macroscopic state of our system by a probability measure $m(dx)$ on M. At equilibrium, $m(dx)$ should be an (f^t)-invariant measure. Remember now that the Hamiltonian time evolution determined by (1.82) preserves the energy and the volume element $dp\, dq$ in phase–space. This leads to the choice of the *micro–canonical ensemble* [Rue99]

$$m_K(dq\, dp) = \frac{1}{\Omega_K} \delta(H(q, p) - K) dq\, dp, \tag{1.83}$$

to describe the equilibrium state of energy K (Ω_K is a normalization constant, and the *ergodic hypothesis* asserts that m_K is ergodic). Note that the support of m_K is the energy shell

$$M_K = \{(q, p) : H(q, p) = K\}.$$

If the probability measure $m(dq\, dp)$ has density $\underline{m}(q, p)$ with respect to $dq\, dp$, we associate with the state described by this measure the entropy

$$S(\underline{m}) = -\int dq\, dp\, \underline{m}(q, p) \log \underline{m}(q, p). \tag{1.84}$$

This is what Lebowitz [Leb93] calls the *Gibbs entropy*, it is the accepted expression for the entropy of equilibrium states. There is a minor technical complication here. While (1.84) gives the right result for the canonical ensemble one should, for the micro–canonical ensemble, replace the reference measure $dp\,dq$ in (1.83) by $\delta(H(q,p) - K)dq\,dp$. This point is best appreciated in the light of the theory of equivalence of ensembles in equilibrium statistical mechanics. For our purposes, the easiest is to replace δ in (1.83) by the characteristic function of $[0, \epsilon]$ for small ϵ. The measure m_K is then still invariant, but no longer ergodic. Note also that the traditional definition of the entropy has a factor k (*Bolzmann constant*) in the right–hand side of (1.84), and it remains reasonable outside of equilibrium, as information theory for instance would indicate. Writing x instead of (q, p), we know that the density of $\widehat{m} = f^{t*}m$ is given by

$$\widehat{m}(x) = \frac{m(f^{-t}x)}{J_t(f^{-t}x)},$$

where J_t is the Jacobian determinant of f^t. We have thus [Rue99]

$$S(\underline{m}) = -\int dx\,\underline{m}(x) \log \underline{m}(x), \tag{1.85}$$

$$S(\widehat{\underline{m}}) = -\int dx\,\widehat{\underline{m}}(x) \log \widehat{\underline{m}}(x) = -\int dx\,\frac{m(f^{-t}x)}{J_t(f^{-t}x)} \log \frac{m(f^{-t}x)}{J_t(f^{-t}x)},$$

$$= -\int dx\,\underline{m}(x) \log \frac{\underline{m}(x)}{J_t(x)} = S(\underline{m}) + \int dx\,\underline{m}(x) \log J_t(x) \tag{1.86}$$

In the Hamiltonian situation that we are considering for the moment, since the volume element is preserved by f^t, $J_t = 1$, and therefore $S(\widehat{m}) = S(m)$. So, we have a problem: the entropy seems to remain constant in time. In fact, we could have expected that there would be a problem because the Hamiltonian evolution (1.82) has *time–reversal invariance* (to be discussed later), while we want to prove an increase of entropy which does not respect time reversal. We shall now present Boltzmann's way out of this difficulty.

Classical Boltzmann Theory

Let us cut M into cells c_i so that the coordinates p, q have roughly fixed values in a cell. In particular all points in c_i are macroscopically equivalent (but different c_i may be macroscopically indistinguishable). Instead of describing a macroscopic state by a probability measure m, we may thus give the weights $m(c_i)$. Now, time evolution will usually distort the cells, so that each $f^t c_i$ will now intersect a large number of cells. If the initial state m occupies N cells with weights $1/N$ (taken to be equal for simplicity), the state $f^{*t}m$ will occupy (thinly) N^t cells with weights $1/N^t$, where N^t may be much larger than N. If the c_i have side h, we have [Rue99]

$$\log N = -m(c_i) \log m(c_i) \approx -\int dx\,\underline{m}(x) \log(\underline{m}(x)h^{2N}) = S(\underline{m}) - 2N \log h,$$

i.e., the entropy associated with m is roughly $\log N + 2N \log h$. The apparent entropy associated with $f^{t*}m$ is similarly $\log N^t + 2N \log h$; it differs from $S(\widehat{m})$ because the density $\widehat{m}$, which may fluctuate rapidly in a cell, has been replaced by an average for the computation of the apparent entropy. The entropy increase $\log N^t - \log N$ is due to the fact that the initial state m is concentrated in a small region of phase–space, and becomes by time evolution spread (thinly) over a much larger region. The time evolved state, after a little smoothing (*coarse graining*), has a strictly larger entropy than the initial state. This gives a microscopic interpretation of the second law of thermodynamics. In specific physical examples (like that of two bodies in contact with initially different temperatures) one sees that the time evolved state has correlations (say between the microscopic states of the two bodies, after their temperatures have equalized) which are macroscopically unobservable. In the case of a macroscopic system locally close to equilibrium (small regions of space have a definite temperature, pressure,...) the above classical ideas of Boltzmann have been expressed in particularly clear and compelling manner by [Leb93]. He defines a local *Boltzmann entropy* to be the equilibrium entropy corresponding to the temperature, pressure,... which approximately describe locally a non–equilibrium state. The integral over space of the Boltzmann entropy is then what we have called apparent entropy, and is different from the Gibbs entropy defined by (1.84). These ideas would deserve a fuller discussion, but here we shall be content to refer to [Leb93], and to Boltzmann's original works. There is still some opposition to Boltzmann's ideas (notably by I. Prigogine and his school, see [Pri62]), but most workers in the area accept them, and so shall we. We shall however develop a formalism which is both rather different from and completely compatible with the ideas of Boltzmann and Lebowitz just discussed.

Description of States by Probability Measures

In the above discussion, we have chosen to describe states by probability measures. One may object that the state of a (classical) system is represented by a point in phase–space rather than by a probability measure. But for a many-particle system like those of interest in statistical mechanics it is practically impossible to know the position of all the particles, which changes rapidly with time anyway. The information that we have is macroscopic or statistical and it is convenient to take as description of our system a probability distribution ρ on phase–space compatible with the information available to us. Trying to define this ρ more precisely at this stage leads to the usual kind of difficulties that arise when one wants to get from a definition what should really come as a result of theory. For our later purposes it is technically important to work with states described by probability measures. Eventually, we shall get results about individual points of phase–space (true almost everywhere with respect to some measure). From a physical point of view, it would be desirable to make use here of points of phase–space which are typical for a macroscopic state of our system. Unfortunately, a serious discussion of this point seems out of reach in the present state of the theory.

Beyond Boltzmann

There are two reasons why one would want to go beyond the ideas presented above. One concerns explicit calculations, like that of a rate of entropy production; the other concerns situations far from equilibrium. We discuss these two points successively. If one follows Boltzmann and uses a decomposition of phase–space into cells to compute entropy changes, the result need not be monotone in time, and will in general depend on the particular decomposition used. Only after taking a limit $t \to \infty$ can one let the size of cells tend to 0, and get a result independent of the choice of coarse graining. This leaves open the problem of computing the entropy production per unit time. The idea of using local Boltzmann entropies works only for a macroscopic system locally close to equilibrium, and one may wonder what happens far from equilibrium. In fact one finds statements in the literature that biological processes are far from equilibrium (which is true), and that they may violate the second law of thermodynamics (which is not true). To see that life processes or other processes far from equilibrium cannot violate the second law, we can imagine a power plant fueled by sea water: it would produce electric current and reject colder water in the sea. Inside the plant there would be some life form or other physico-chemical system functioning far from equilibrium and violating the second law of thermodynamics. The evidence is that this is impossible, even though we cannot follow everything that happens inside the plant in terms of Boltzmann entropies of systems close to equilibrium. In fact, Boltzmann's explanation of irreversibility reproduced above applies also here, and the only unsatisfactory feature is that we do not have an effective definition of entropy far from equilibrium. The new formalism which we shall introduce below is quite in agreement with the ideas of Boltzmann which we have described. We shall however define the physical entropy by (1.84) (Gibbs entropy). To avoid the conclusion that this entropy does not change in time for a Hamiltonian time evolution, we shall idealize our physical setup differently, and in particular introduce a thermostat [Rue99].

Thermostats

Instead of investigating the approach to equilibrium as we have done above following Boltzmann, we can try to produce and study non–equilibrium steady states. To keep a finite system outside of equilibrium we subject it to non-Hamiltonian forces. We consider thus an evolution of the form (1.81), but not (1.82). Since we no longer have conservation of energy, $x(t)$ cannot be expected to stay in a bounded region of phase–space. This means that the system will heat up. Indeed, this is what is observed experimentally: dissipative systems produce heat. An experimentalist will eliminate excess heat by use of a thermostat, and if we want to study non–equilibrium steady states we have to introduce the mathematical equivalent of a thermostat. In the lab, the system in which we are interested (called *small system*) would be coupled with a *large system* constituting the thermostat. The obvious role of the large system is to take up the heat produced in the small system. At the same time, the thermostat allows entropy to be produced in the small system by a mechanism discussed above: microscopic correlations which exist between the small and the large system are rendered unobservable by the time evolution. An exact study of the pair small+large system would lead to the same problems that we have met above with Boltzmann's approach. For such an approach, see Jakšić and Pillet [JP98], where the large system has infinitely many noninteracting degrees of freedom. Studying the *diffusion and loss of correlations* in an infinite interacting system (say a system of particles with Hamiltonian interactions) appears to be very difficult in general, because the same particles may interact again and again, and it is hard to keep track of the correlations resulting from earlier interactions. This difficulty was bypassed by Lanford when he studied the Boltzmann equation in the Grad limit [Lan75], because in that limit, two particles that collide once will never see each other again.. Note however that the thermostats used in the lab are such that their state changes as little as possible under the influence of the small system. For instance the small system will consist of a small amount of fluid, surrounded by a big chunk of copper constituting the large system: because of the high thermal conductivity of copper, and the bulk of the large system, the temperature at the fluid-copper interface will remain constant to a good precision. In conclusion, the experimentalist tries to build an *ideal thermostat*, and we might as well do the same. Following [Rue99], we replace thus (1.81) by

$$\dot{x} = F(x) + \Theta(\omega(t), x),$$

where the effect $\Theta(\omega(t), x)$ of the thermostat depends on the state x of the small system, and on the state $\omega(t)$ of the thermostat, but the time evolution $t \rightarrow \omega(t)$ does not depend on the state x of the small system. We may think of $\omega(t)$ as random (corresponding to a *random thermostat*), but the simplest choice is to take ω constant, and use $\Theta(x) = \Theta(\omega, x)$ to keep $x(t)$ on a compact manifold M. For instance if M is the manifold $\{x : h(x) = K\}$, we may take

$$\Theta(x) = -\frac{(F(x) \cdot h(x))}{(\mathrm{grad}h(x) \cdot h(x))} \, \mathrm{grad}h(x).$$

This is the so-called Gaussian thermostat [Hoo86, EM90]. We shall be particularly interested later in the isokinetic thermostat, which is the special case where $x = (q, p)$ and $h(x) = p/2m$ (kinetic energy).

Non–Equilibrium Steady States

We assume for the moment that the phase–space of our system is reduced by the action of a thermostat to be a compact manifold M. The time evolution equation on M has the same form as (1.81):

$$\dot{x} = F(x), \tag{1.87}$$

where the vector–field F on M now describes both the effect of nonhamiltonian forces and of the thermostat. Note that (1.87) describes a general smooth evolution, and one may wonder if anything of physical interest is preserved at this level of generality. Perhaps surprisingly, the answer is yes, as we see when we ask what are the physical stationary states for (1.87). We start with a probability measure m on M such that $m(dx) = \underline{m}(x)\, dx$, where dx is the volume element for some Riemann metric on M (for simplicity, we shall say that m is absolutely continuous). At time t, m becomes $(f^{t})^{*}m$, which still is *absolutely continuous*. If $(f^{t})^{*}m$ has a limit ρ when $t \to \infty$, then ρ is invariant under time evolution, and in general singular with respect to the Riemann volume element dx (a time evolution of the form (1.87) does not in general have an absolutely continuous invariant measure). The probability measures

$$\rho = \lim_{t \to \infty} (f^{t})^{*}m,$$

or more generally [Rue99]

$$\rho = \lim_{t \to \infty} \frac{1}{t} \int_{0}^{t} d\tau\, (f^{\tau})^{*}m, \tag{1.88}$$

(with m absolutely continuous) are natural candidates to describe non–equilibrium stationary states, or non–equilibrium steady states. Examples of such measures are the SRB states discussed later.

Entropy Production

We return now to our calculation (1.86) of the entropy production:

$$S(\widehat{m}) - S(\underline{m}) = \int dx\, \underline{m}(x) \log J_{t}(x),$$

where $\widehat{m}$ is the density of $\widehat{m} = f^{t*}m$. This is the amount of entropy gained by the system under the action of the external forces and thermostat in time t. The amount of entropy produced by the system and given to the external world in one unit of time is thus (with $J = J_{1}$)

$$e_{f}(m) = - \int m(dx) \log J(x).$$

Notice that this expression makes sense also when m is a singular measure. The average entropy production in t units of time is [Rue99]

$$\frac{1}{t} \sum_{k=0}^{t-1} e_{f}(f^{k}m) = \frac{1}{t}[S(\underline{m}) - S(\widehat{m})]. \tag{1.89}$$

When $t \to \infty$, this tends according to (1.88) towards

$$e_f(\rho) = - \int \rho(dx) \log J(x), \tag{1.90}$$

which is thus the entropy production in the state ρ (more precisely, the entropy production per unit time in the non–equilibrium steady state ρ). Using (1.87) we can also write

$$e_f(\rho) = - \int \rho(dx) \operatorname{div} F(x). \tag{1.91}$$

Notice that the entropy S in (1.89) is bounded above (see the above definition (1.85)), so that $e_f(\rho) \geq 0$, and in many cases $e_f(\rho) > 0$ as we shall see later. Notice that for an arbitrary probability measure μ (invariant or not), $e_f(\mu)$ may be positive or negative, but the definition (1.88) of ρ makes a choice of the direction of time, and results in positive entropy production. It may appear paradoxical that the state ρ, which does not change in time, constantly gives entropy to the outside world. The solution of the paradox is that ρ is (normally) a singular measure and therefore has entropy $-\infty$: the non–equilibrium steady state ρ is thus a bottomless source of entropy.

Recent Idealization of Non–Equilibrium Processes

We have now reached a new framework idealizing non–equilibrium processes. Instead of following Boltzmann in his study of *approach to equilibrium*, we try to understand the *non–equilibrium steady states* as given by (1.88). For the definition of the entropy production $e_f(\rho)$ we use the rate of phase–space contraction (1.90) or (1.91). Later we shall discuss the SRB states, which provide a mathematically precise *definition of the non–equilibrium steady states*. After that, the idea is to use SRB states to make interesting physical predictions, making hyperbolicity assumptions (this will be explained later) as strong as needed to get results. Such a program was advocated early by Ruelle, but only recently were interesting results actually obtained, the first one being the *fluctuation theorem* of Gallavotti and Cohen [GC95a, GC95b].

Diversity of Non–Equilibrium Regimes

Many important non–equilibrium systems are locally close to equilibrium, and the classical non–equilibrium studies have concentrated on that case, yielding such results as the Onsager reciprocity relations and the Green-Kubo formula. Note however that chemical reactions are often far from equilibrium. More exotic non–equilibrium systems of interest are provided by *metastable states*. Since quantum measurements (and the associated *collapse of wave packets*) typically involve metastable states, one would like to have a reasonable fundamental understanding of those states. Another class of exotic systems are *spin glasses*, which are almost part of equilibrium statistical mechanics, but evolve slowly, with extremely long relaxation times.

Further Discussion of Non–equilibrium steady states

Recall that above we have proposed a definition (1.88) for non–equilibrium steady states ρ. We now make this definition more precise and analyze it further. Write [Rue99]

$$\rho = \text{w.lim}_{t\to\infty} \frac{1}{t} \int_0^t d\tau \, (f^\tau)^* m, \qquad (1.92)$$

$$m \qquad \text{a.c. probability measure,} \qquad (1.93)$$

$$\rho \qquad \text{ergodic.} \qquad (1.94)$$

In (1.92), w.lim is the weak or vague limit defined by $(\text{w.lim}\, m_t)(\Phi) = \lim(m_t(\Phi))$ for all continuous $\Phi : M \to \mathbf{C}$. The set of probability measures on M is compact and metrizable for the vague topology. There are thus always sequences (t_k) tending to ∞ such that

$$\rho = \text{w.lim}_{k\to\infty} \frac{1}{t_k} \int_0^{t_k} d\tau \, (f^\tau)^* m$$

exists; ρ is automatically $(f^\tau)^*$ invariant. By (1.93), we ask that the probability measure m be a.c. (absolutely continuous) with respect to the Riemann volume element dx (with respect to any metric) on M. The condition (1.94) is discussed below. Physically, we consider a system which in the distant past was in equilibrium with respect to a Hamiltonian time evolution

$$\dot{x} = F_0(x), \qquad (1.95)$$

and described by the probability measure m on the energy surface M. According to conventional wisdom, m is the *micro–canonical ensemble*, which satisfies (1.93), and is ergodic with respect to the time evolution (1.95) when the *ergodic hypothesis* is satisfied. Even if the ergodic hypothesis is accepted, the physical justification of the micro–canonical ensemble remains a delicate problem, which we shall not further discuss here.. For our purposes, we might also suppose that m is an ergodic component of the micro–canonical ensemble or an integral over such ergodic components, provided (1.93) is satisfied. We assume now that, at some point in the distant past, (1.95) was replaced by the time evolution

$$\dot{x} = F(x), \qquad (1.96)$$

representing nonhamiltonian forces plus a thermostat keeping x on M; we write the general solution of (1.96) as $x \mapsto f^t x$. We are interested in time averages of $f^t x$ for m–almost all x. Suppose therefore that

$$\rho_x = \text{w.lim}_{t\to\infty} \frac{1}{t} \int_0^t d\tau \, \delta_{f^\tau x} \qquad (1.97)$$

exists for m–almost all x. In particular, with ρ defined by (1.92), we have

$$\rho = \int m(dx)\,\rho_x. \tag{1.98}$$

If (1.94) holds. Suppose that ρ is not ergodic but that $\rho = \alpha\rho' + (1 - \alpha)\rho''$ with ρ' ergodic and $\alpha \neq 0$. Writing $S = \{x : \rho_x = \rho'\}$, we have $m(S) = \alpha$ and $\rho' = \int m'(dx)\,\rho_x$ with $m' = \alpha^{-1}\chi_S.m$. Therefore, (1.92–1.94) hold with ρ, m replaced by ρ', m'., (1.98) is equivalent to

$$\rho_x = \rho \qquad \text{for } m-\text{almost all } x.$$

($\Leftarrow$ is obvious; if $\Rightarrow$ did not hold (1.98) would give a non-trivial decomposition $\rho = \alpha\rho' + (1-\alpha)\rho''$ in contradiction with ergodicity). As we have just seen, the ergodic assumption (1.94) allows us to replace the study of (1.97) by the study of (1.92), with the condition (1.93). This has interesting consequences, as we shall see, but note that (1.92–1.94) are not always satisfyable simultaneously (consider for instance the case $F = 0$). To study non–equilibrium steady states we shall modify or strengthen the conditions (1.92–1.94) in various ways. We may for simplicity replace the continuous time $t \in \mathbb{R}$ by a discrete time $t \in \mathbb{Z}$.

Uniform Hyperbolicity: Anosov Diffeomorphisms and Flows

Let M be a compact connected manifold. In what follows we shall be concerned with a time evolution (f^t) which either has discrete time $t \in \mathbb{Z}$, and is given by the iterates of a *diffeomorphism* f of M, or has continuous time $t \in \mathbb{R}$, and is the *flow* generated by some vector–field F on M. We assume that f or F are of class $C^{1+\alpha}$ (Hölder continuous first derivatives). The Anosov property is an assumption of strong chaoticity (in physical terms) or uniform hyperbolicity (in mathematical terms. For background see for instance Smale [Sma67] or Bowen [Bow75]). Choose a Riemann metric on M. The diffeomorphism f is Anosov if there are a continuous Tf–invariant splitting of the tangent bundle: $TM = E^s \oplus E^u$ and constants $C > 0$, $\theta > 1$ such that [Rue99]

$$\text{for all } t \geq 0 : \quad \begin{array}{ll} \|Tf^t\xi\| \leq C\theta^{-t} & \text{if } \xi \in E^s \\ \|Tf^{-t}\xi\| \leq C\theta^{-t} & \text{if } \xi \in E^u. \end{array} \tag{1.99}$$

One can show that $x \mapsto E^s_x$, E^u_x are Hölder continuous, but not C^1 in general. The flow (f^t) is Anosov if there are a continuous (Tf^t)–invariant splitting. Remember that F is the vector–field generating the flow: $F(x) = df^t x/dt|_{t=0}$. Therefore $\mathbb{R}.F$ is the 1D subbundle in the direction of the flow. $TM = \mathbb{R}.F \oplus E^s \oplus E^u$ and constants $C > 0$, $\theta > 1$ such that (1.99) again holds. In what follows we shall assume that the periodic orbits are dense in M. This is conjectured to hold automatically for Anosov diffeomorphisms, but there is a counterexample for flows (see [fW80]). Since M is connected, we are thus dealing with what is called technically a *mixing Anosov diffeomorphism* f, or a *transitive Anosov flow* (f^t). There is a powerful method, called *symbolic dynamics*, for the study of Anosov diffeomorphisms and flows. Recall that symbolic dynamics (see [Sin68a]) is based on the existence of *Markov partitions* [Sin68b, Rat69].

Uniform Hyperbolicity: Axiom A Diffeomorphisms and Flows

Smale [[Sma67]] has achieved an important generalization of Anosov dynamical systems by imposing hyperbolicity only on a subset Ω (the *non–wandering set*) of the manifold M. A point $x \in M$ is *wandering point* if there is an open set $O \ni x$ such that $O \cap f^t O \neq \emptyset$ for $|t| > 1$. The points of M which are not wandering constitute the non–wandering set Ω. A diffeomorphism or flow satisfies *Axiom A* if the following conditions hold (Aa) there is a continuous (Tf^t)–invariant splitting of $T_\Omega M$ (the tangent bundle restricted to the non–wandering set) verifying the above hyperbolicity conditions. (Ab) the periodic orbits are dense in Ω. Under these conditions, Ω is a finite union of disjoint compact (f^t)-invariant sets B (called *basic sets*) on which (f^t) is topologically transitive. (f^t) is topologically transitive on B if there is $x \in B$ such that the orbit $(f^t x)$ is dense in B.. This result is known as Smale's *spectral decomposition* theorem. If there is an open set $U \supset B$ such that $\cap_{t \geq 0} f^t U = B$, the basic set B is called an *attractor*. The set $\{x \in M : \lim_{t \to +\infty} d(f^t x, B) = 0\} = \cup_{t \geq 0} f^{-t} U$ is the *basin of attraction* of the attractor B. Let B be a basic set. Given $x \in B$ and $\epsilon > 0$, write [Rue99]

$$W^s_{x,\epsilon} = \{y \in M : d(f^t y, f^t x) < \epsilon \text{ for } t \geq 0, \text{ and } \lim_{t \to +\infty} d(f^t y, f^t x) = 0\},$$
$$W^u_{x,\epsilon} = \{y \in M : d(f^t y, f^t x) < \epsilon \text{ for } t \leq 0, \text{ and } \lim_{t \to -\infty} d(f^t y, f^t x) = 0\}$$

Then for sufficienty small ϵ, $W^s_{x,\epsilon}$ and $W^u_{x,\epsilon}$ are pieces of smooth manifolds, called *locally stable* and *locally unstable* manifold respectively. There are also *globally* stable and unstable manifolds defined by

$$W^s_x = \{y \in M : \lim_{t \to +\infty} d(f^t y, f^t x) = 0\},$$
$$W^u_x = \{y \in M : \lim_{t \to -\infty} d(f^t y, f^t x) = 0\},$$

and tangent to E^s_x, E^u_x respectively. A basic set B is an attractor if and only if the stable manifolds $W^s_{x,\epsilon}$ for $x \in B$ cover a neighborhood of B. Also, a basic set B is an attractor if and only if the unstable manifolds $W^u_{x,\epsilon}$ for $x \in B$ are contained in B (see [BR75]). Markov partitions, symbolic dynamics (see [Bow70, Bow73]), and shadowing (see [Bow75]) are available on Axiom A basic sets as they were for Anosov dynamical systems.

SRB States on Axiom A Attractors

Let us cut an attractor B into a finite number of small cells such that the unstable manifolds are roughly parallel within a cell. Each cell is partitionned into a continuous family of pieces of local unstable manifolds, and we get thus a partition (Σ_α) of B into small pieces of unstable manifolds. If ρ is an invariant probability measure on B (for the dynamical system (f^t)), and if its conditional measures σ_α with respect to the partition (Σ_α) are a.c. with respect to the Riemann volume $d\sigma$ on the unstable manifolds, ρ is called an

SRB measure. The study of SRB measures is transformed by use of symbolic dynamics into a problem of statistical mechanics: one can characterize SRB states as Gibbs states with respect to a suitable interaction. Such Gibbs states can in turn be characterized by a variational principle. In the end one has a variational principle for SRB measures, which we shall now describe. It is convenient to consider a general basic set B (not necessarily an attractor) and to define generalized SRB measures. First we need the concept of the *time entropy* $h_f(\mu)$, where $f = f^1$ is the time 1 map for our dynamical system, and μ an $f-$invariant probability measure on B. This entropy (or Kolmogorov–Sinai invariant) has the physical meaning of mean information production per unit time by the dynamical system (f^t) in the state μ (see for instance [6] for an exact definition). The time entropy $h_f(\mu)$ should not be confused with the *space entropy* S and the entropy production rate e_f which we have discussed above. We also need the concept of *expanding Jacobian* J^u. Since $T_x f$ maps E^u_x linearly to E^u_{fx}, and volume elements are defined in E^u_x, E^u_{fx} by a Riemann metric, we can define a volume expansion rate $J^u(x) > 0$. It can be shown that the function $\log J^u : B \to \mathbb{R}$ is Hölder continuous. We say that the $(f^t)-$invariant probability measure ρ is a *generalized SRB measure* if it makes maximum the function

$$\mu \quad \mapsto \quad h_f(\mu) - \mu(\log J^u). \tag{1.100}$$

One can show that there is precisely one generalized SRB measure on each basic set B; it is ergodic and reduces to the unique SRB measure when B is an attractor. The value of the maximum of (9) is 0 precisely when B is an attractor, it is < 0 otherwise. If m is a measure absolutely continuous with respect to the Riemann volume element on M, and if its density $\underline{m}$ vanishes outside of the basin of attraction of an attractor B, then [Rue99]

$$\rho = \text{w.lim}_{t \to \infty} \frac{1}{t} \int_0^t d\tau \, (f^\tau)^* m \tag{1.101}$$

defines the unique SRB measure on B. We also have

$$\rho = \text{w.lim}_{t \to \infty} \frac{1}{t} \int_0^t d\tau \, \delta_{f^\tau x},$$

when x is in the domain of attraction of B and outside of a set of measure 0 for the Riemann volume. The conditions (1.92–1.94) and also (1.97) are thus satisfied, and the SRB state ρ is a natural non–equilibrium steady state. Note that if B is a *mixing*, i.e., for any two nonempty open sets $O_1, O_2 \subset B$, there is T such that $O_1 \cap f^t O_2 \neq \emptyset$ for $|t| \geq T$. attractor, (10) can be replaced by the stronger result

$$\rho = \text{w.lim}_{t \to \infty} (f^t)^* m.$$

See [Sin72, Rue76, BR75] for details.

SRB States Without Uniform Hyperbolicity

Remarkably, the theory of SRB states on Axiom A attractors extends to much more general situations. Consider a smooth dynamical system (f^t) on the compact manifold M, without any hyperbolicity condition, and let ρ be an ergodic measure for this system. Recall that the *Oseledec theorem* [Ose68, Rue79] permits the definition of *Lyapunov exponents* $\lambda_1 \leq \ldots \leq \lambda_{\dim M}$ which are the rates of expansion, ρ−almost everywhere, of vectors in TM. The λ_i are real numbers, positive (expansion), negative (contraction), or zero (neutral case). Pesin theory [Pes76, Pes77] allows the definition of stable and unstable manifolds ρ−almost everywhere. These are smooth manifolds; the dimension of the stable manifolds is the number of negative Lyapunov exponents while the dimension of the unstable manifold is the number of positive Lyapunov exponents. Consider now a family (Σ_α) constituted of pieces of (local) unstable manifolds, and forming, up to a set of ρ−measure 0, a partition of M. As in Section 4 above we define the conditional measures σ_α of ρ with respect to (Σ_α). If the measures σ_α are absolutely continuous with respect to the Riemann volumes of the corresponding unstable manifolds, we say that ρ is an SRB measure. For a C^2 diffeomorphism, the above definition of SRB measures is equivalent to the following condition (known as *Pesin formula*) for an ergodic measure ρ:

$$h(\rho) = \sum \text{positive Lyapunov exponents for } \rho,$$

(see [LS82, LY85] for the nontrivial proof). This is an extension of the result of Section 4, where the sum of the positive Lyapunov exponents is equal to $\rho(\log J^u)$. Note that in general $h(\mu)$ is $\leq$ the sum of the positive exponents for the ergodic measure μ (see [Rue78]). Suppose that the time t is discrete (diffeomorphism case), and that the Lyapunov exponents of the SRB state ρ are all different from zero: this is a weak (nonuniform) hyperbolicity condition. In this situation, there is a measurable set $S \subset M$ with positive Riemann volume such that [PS89]

$$\lim_{n \to \infty} \frac{1}{n} \sum_{k=0}^{n-1} \delta_{f^k x} = \rho,$$

for all $x \in S$. This result shows that ρ has the properties required of a non−equilibrium steady state. One expects that for continuous time (flow case), if supp ρ is not reduced to a point and if the Lyapunov exponents of the SRB state ρ except one. There is one zero exponent corresponding to the flow direction. are different from 0, there is a measurable set $S \subset M$ with positive Riemann volume such that [Rue99]

$$\rho = \text{w.lim}_{t \to \infty} \frac{1}{t} \int_0^t d\tau \, \delta_{f^\tau x}, \qquad \text{when } x \in S.$$

See [LS82, LY85, Via97] for details.

1.5.2 Statistical Mechanics of Nonlinear Oscillator Chains

Now, consider a model of a finite nonlinear chain of n $d-$dimensional oscillators, coupled to two Hamiltonian heat reservoirs initially at different temperatures T_L, T_R, each of which is described by a dD wave equation. A natural goal is to get a usable expression for the invariant (marginal) state of the chain analogous to the Boltzmann–Gibbs prescription $\mu = Z^{-1} \exp(-H/T)$ which one has in equilibrium statistical mechanics [BT00]. We assume that the Hamiltonian $H(p,q)$ of the isolated chain has the form

$$ H(p,q) = \sum_{i=1}^{n} \frac{p_i^2}{2} + \sum_{i=1}^{n} U^{(1)}(q_i) + \sum_{i=1}^{n-1} U^{(2)}(q_i - q_{i+1}) \equiv \sum_{i=1}^{n} \frac{p_i^2}{2} + V(q), \quad (1.102) $$

where q_i and p_i are the coordinate and momentum of the ith particle, and where $U^{(1)}$ and $U^{(2)}$ are C^k confining potentials, i.e., $\lim_{|q| \to \infty} V(q) = +\infty$.

The coupling between the reservoirs and the chain is assumed to be of dipole approximation type and it occurs at the boundary only: the first particle of the chain is coupled to one reservoir and the nth particle to the other heat reservoir. At time $t = 0$ each reservoir is assumed to be in thermal equilibrium, i.e., the initial conditions of the reservoirs are distributed according to (Gaussian) Gibbs measure with temperature $T_1 = T_L$ and $T_n = T_R$ respectively. Projecting the dynamics onto the phase–space of the chain results in a set of integro–differential equations which differ from the Hamiltonian equations of motion by additional force terms in the equations for p_1 and p_n. Each of these terms consists of a deterministic integral part independent of temperature and a Gaussian random part with covariance proportional to the temperature. Due to the integral (memory) terms, the study of the long–time limit is a difficult mathematical problem. But by a further appropriate choice of couplings, the integral parts can be treated as auxiliary variables r_1 and r_n, the random parts become Markovian. Thus we get (see [EPR99] for details) the following system of Markovian stochastic differential equations (SDEs) on the extended phase–space $\mathbb{R}^{2dn+2d}$: For $x = (p, q, r)$ we have

$$ \begin{aligned} \dot{q}_1 &= p_1, & \dot{p}_1 &= -\nabla_{q_1} V(q) + r_1, \\ \dot{q}_j &= p_j, & \dot{q}_j &= -\nabla_{q_j} V(q), & (j = 2, \ldots, n-1) \\ \dot{q}_n &= p_n, & \dot{q}_n &= -\nabla_{q_n} V(q) + r_n, & (1.103) \\ dr_1 &= -\gamma(r_1 - \lambda^2 q_1)dt + (2\gamma\lambda^2 T_1)^{1/2} dw_1, \\ dr_n &= -\gamma(r_n - \lambda^2 q_1)dt + (2\gamma\lambda^2 T_n)^{1/2} dw_n. \end{aligned} $$

In equation (1.103), $w_1(t)$ and $w_n(t)$ are independent dD Wiener processes, and λ^2 and γ are constants describing the couplings.

Now introduce a generalized Hamiltonian $G(p, q, r)$ on the extended phase–space, given by

$$ G(p,q,r) = \sum_{i=1}^{n} \left(\frac{r_i^2}{2\lambda^2} - r_i q_i \right) + H(p,q), \quad (1.104) $$

where $H(p,q)$ is the Hamiltonian of the isolated systems of oscillators given by (1.102). We also introduce the parameters ε (the mean temperature of the reservoirs) and η (the relative temperature difference):

$$\varepsilon = \frac{T_1 + T_n}{2}, \qquad \eta = \frac{T_1 - T_n}{T_1 + T_n}. \tag{1.105}$$

Using (1.104), the equation (1.103) takes the form [BT00]

$$\dot{q} = \nabla_p G, \qquad \dot{p} = -\nabla_q G, \tag{1.106}$$
$$dr = -\gamma\lambda^2\nabla_r G dt + \varepsilon^{1/2}(2\gamma\lambda^2 D)^{1/2}dw,$$

where $p = (p_1,\ldots,p_n)$, $q = (q_1,\ldots,q_n)$, $r = (r_1, r_n)$ and where D is the $2d \times 2d$ matrix given by

$$D = \begin{pmatrix} 1+\eta & 0 \\ 0 & 1-\eta \end{pmatrix}.$$

The function G is a Lyapunov function, non–increasing in time, for the deterministic part of the flow (1.106). If the system is in equilibrium, i.e, if $T_1 = T_n = \varepsilon$ and $\eta = 0$, it is not difficult to check that the generalized Gibbs measure

$$\mu_\varepsilon = Z^{-1}\exp\left(-G(p,q,r)/\varepsilon\right),$$

is an invariant measure for the Markov process solving (1.106).

1.5.3 Geometrical Modelling of Continuous Dynamics

Here we give a paradigm of geometrical modelling and analysis of complex continuous dynamical systems (see [II06b] for technical details). This is essentially a *recipe* how to develop a *covariant formalism on smooth manifolds*, given a certain physical, or bio–physical, or psycho–physical, or socio–physical system, here labelled by a generic name: 'physical situation'. We present this recipe in the form of the following five–step algorithm.

(I) So let's start: given a certain physical situation, the first step in its predictive modelling and analysis, that is, in applying a powerful differential–geometric machinery to it, is to associate with this situation *two* independent coordinate systems, constituting two independent smooth Riemannian manifolds. Let us denote these two coordinate systems and their respective manifolds as:

- *Internal coordinates:* $x^i = x^i(t)$, $(i = 1,\ldots,m)$, constituting the mD *internal configuration manifold*: $M^m \equiv \{x^i\}$; and
- *External coordinates:* $y^e = y^e(t)$, $(e = 1,\ldots,n)$, constituting the nD *external configuration manifold*: $N^n \equiv \{y^e\}$.

The main example that we have in mind is a standard robotic or bio-dynamic (loco)motion system, in which x^i denote internal joint coordinates, while y^e denote external Cartesian coordinates of segmental centers of mass. However, we believe that such developed methodology can fit a generic physical situation.

Therefore, in this first, engineering step (I) of our differential–geometric modelling, we associate to the given natural system, not one but two different and independent smooth configuration manifolds, somewhat like viewing from two different satellites a certain place on Earth with a football game playing in it.

(II) Once that we have precisely defined two smooth manifolds, as two independent views on the given physical situation, we can apply our differential–geometric modelling to it and give it a natural physical interpretation. More precisely, once we have two smooth Riemannian manifolds, $M^m \equiv \{x^i\}$ and $N^n \equiv \{y^e\}$, we can formulate two smooth maps between them:[32]

$$f : N \to M, \text{ given by coordinate transformation: } x^i = f^i(y^e), \quad (1.107)$$
and
$$g : M \to N, \text{ given by coordinate transformation: } y^e = g^e(x^i). \quad (1.108)$$

If the Jacobian matrices of these two maps are nonsingular (regular), that is if their Jacobian determinants are nonzero, then these two maps are mutually inverse, $f = g^{-1}$, and they represent standard *forward and inverse kinematics*.

(III) Although, maps f and g define some completely general nonlinear coordinate (functional) transformations, which are even unknown at the moment, there is something linear and simple that we know about them (from calculus). Namely, the corresponding infinitesimal transformations are linear and homogenous: from (1.107) we have (applying everywhere Einstein's summation convention over repeated indices)

$$dx^i = \frac{\partial f^i}{\partial y^e}\, dy^e, \quad (1.109)$$

while from (1.108) we have

$$dy^e = \frac{\partial g^e}{\partial x^i}\, dx^i. \quad (1.110)$$

Furthermore, (1.109) implies the linear and homogenous transformation of *internal velocities*,

$$v^i \equiv \dot{x}^i = \frac{\partial f^i}{\partial y^e}\, \dot{y}^e, \quad (1.111)$$

while (1.110) implies the linear and homogenous transformation of *external velocities*,

---

[32] This obviously means that we are working in the category of smooth manifolds.

$$u^e \equiv \dot{y}^e = \frac{\partial g^e}{\partial x^i}\, \dot{x}^i. \tag{1.112}$$

In this way, we have defined *two velocity vector–fields*, the internal one: $v^i = v^i(x^i, t)$ and the external one: $u^e = u^e(y^e, t)$, given respectively by the two nonlinear systems of ODEs, (1.111) and (1.112).[33]

(IV) The next step in our differential–geometrical modelling/analysis is to define second derivatives of the manifold maps f and g, that is the two *acceleration vector–fields*, which we will denote by $a^i = a^i(x^i, \dot{x}^i, t)$ and $w^e = w^e(y^e, \dot{y}^e, t)$, respectively. However, unlike simple physics in linear Euclidean spaces, these two acceleration vector–fields on manifolds M and N are not the simple time derivatives of the corresponding velocity vector–fields ($a^i \neq \dot{v}^i$ and $w^e \neq \dot{u}^e$), due to the existence of the *Levi–Civita connections* ∇_M and ∇_N on both M and N. Properly defined, these two acceleration vector–fields respectively read:

$$a^i = \dot{v}^i + \Gamma^i_{jk} v^j v^k = \ddot{x}^i + \Gamma^i_{jk}\dot{x}^j \dot{x}^k, \quad \text{and} \tag{1.113}$$

$$w^e = \dot{u}^e + \Gamma^e_{hl} u^h u^l = \ddot{y}^e + \Gamma^e_{hl}\dot{y}^h \dot{y}^l, \tag{1.114}$$

where Γ^i_{jk} and Γ^e_{hl} denote the (second–order) *Christoffel symbols* of the connections ∇_M and ∇_N.

Therefore, in the step (III) we gave the first–level model of our physical situation in the form of two ordinary vector–fields, the first–order vector–fields (1.111) and (1.112). For some simple situations (e.g., modelling ecological systems), we could stop at this modelling level. Using physical terminology we call them velocity vector–fields. Following this, in the step (IV) we have defined the two second–order vector–fields (1.113) and (1.114), as a connection–base derivations of the previously defined first–order vector–fields. Using physical terminology, we call them 'acceleration vector–fields'.

(V) Finally, following our generic physical terminology, as a natural next step we would expect to define some kind of generic Newton–Maxwell force–fields. And we can actually do this, with a little surprise that individual forces involved in the two force–fields will not be vectors, but rather the dual objects called 1–forms (or, 1D differential forms). Formally, we define the two *covariant force–fields* as

$$F_i = \mathfrak{m} g_{ij} a^j = \mathfrak{m} g_{ij}(\dot{v}^j + \Gamma^j_{ik} v^i v^k) = \mathfrak{m} g_{ij}(\ddot{x}^j + \Gamma^j_{ik}\dot{x}^i \dot{x}^k), \text{ and} \tag{1.115}$$

$$G_e = \mathfrak{m} g_{eh} w^h = \mathfrak{m} g_{eh}(\dot{u}^h + \Gamma^h_{el} u^e u^l) = \mathfrak{m} g_{eh}(\ddot{y}^h + \Gamma^h_{el}\dot{y}^e \dot{y}^l), \tag{1.116}$$

where $\mathfrak{m}$ is the mass of each single segment (unique, for simplicity), while $g_{ij} = g^M_{ij}$ and $g_{eh} = g^N_{eh}$ are the two *Riemannian metric tensors* corresponding to the manifolds M and N. The two force–fields, F_i defined by (1.115) and G_e defined by (1.116), are generic force–fields corresponding to the manifolds M and N,

---

[33] Although transformations of differentials and associated velocities are linear and homogeneous, the systems of ODE's define nonlinear vector–fields, as they include Jacobian (functional) matrices.

which represent the *material cause* for the given physical situation. Recall that they can be physical, bio–physical, psycho–physical or socio–physical force–fields. Physically speaking, they are the *generators* of the corresponding dynamics and kinematics.

Main geometrical relations behind this fundamental paradigm, forming the *covariant force functor* [II06b], are depicted in Figure 1.55.

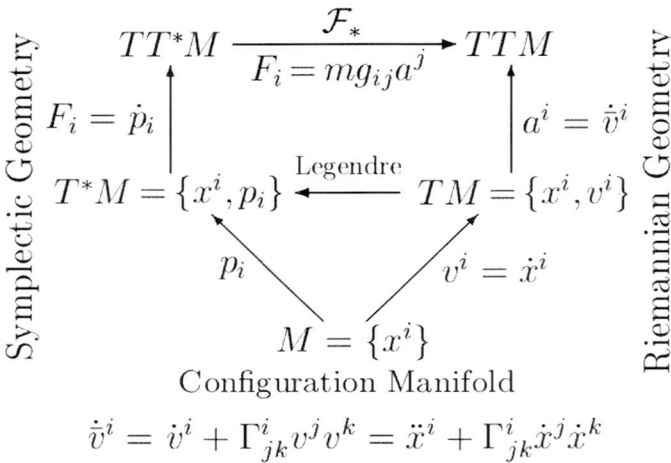

Fig. 1.55. The covariant force functor, including the main relations used by differential–geometric modelling of complex continuous systems.

1.5.4 Lagrangian Chaos

A problem of great interest concerns the study of the spatial and temporal structure of the so–called passive fields, indicating by this term passively quantities driven by the flow, such as the temperature under certain conditions [Mof83]. The equation for the evolution of a passive scalar field $\theta(\mathbf{x}, t)$, advected by a velocity field $\mathbf{v}(\mathbf{x}, t)$, is

$$\partial_t \theta + \nabla \cdot (\mathbf{v}\,\theta) = \chi\,\nabla^2 \theta, \tag{1.117}$$

where $\mathbf{v}(\mathbf{x}, t)$ is a given velocity field and χ is the molecular diffusion coefficient.

The problem (1.117) can be studied through two different approaches. Either one deals at any time with the field θ in the space domain covered by the fluid, or one deals with the trajectory of each fluid particle. The two approaches are usually designed as 'Eulerian' and 'Lagrangian', although both of them are due to Euler [Lam45]. The two points of view are in principle equivalent.

The motion of a fluid particle is determined by the differential equation

$$\dot{\mathbf{x}} = \mathbf{v}(\mathbf{x}, t), \tag{1.118}$$

which also describes the motion of test particles, for example a powder embedded in the fluid, provided that the particles are neutral and small enough not to perturb the velocity field, although large enough not to perform a Brownian motion. Particles of this type are commonly used for flow visualization in fluid mechanics experiments, see [Tri88]. Let us note that the true equation for the motion of a material particle in a fluid can be rather complicated [CFP90].

It is now well established that even in regular velocity field the motion of fluid particles can be very irregular [Hen66, Are84]. In this case initially nearby trajectories diverge exponentially and one speaks of *Lagrangian chaos*. In general, chaotic behaviors can arise in 2D flow only for time dependent velocity fields in 2D, while it can be present even for stationary velocity fields in 3D.

If $\chi = 0$, it is easy to realize that (1.117) is equivalent to (1.118). In fact, we can write

$$\theta(\mathbf{x}, t) = \theta_o(T^{-t}\mathbf{x}), \tag{1.119}$$

where $\theta_o(\mathbf{x}) = \theta(\mathbf{x}, t = 0)$ and T is the formal evolution operator of (1.118) ,

$$\mathbf{x}(t) = T^t \mathbf{x}(0). \tag{1.120}$$

Taking into account the molecular diffusion χ, (1.117) is the Fokker–Planck equation of the Langevin equation [Cha43]

$$\dot{\mathbf{x}} = \mathbf{v}(\mathbf{x}, t) + \eta(t), \tag{1.121}$$

where η is a Gaussian process with zero mean and variance

$$\langle \eta_i(t)\, \eta_j(t') \rangle = 2\chi \delta_{ij}\, \delta(t - t'). \tag{1.122}$$

In the following we will consider only incompressible flow

$$\nabla \cdot \mathbf{v} = 0, \tag{1.123}$$

for which the dynamical system (1.118) is conservative. In 2D, the constraint (8.44) is automatically satisfied assuming

$$v_1 = \frac{\partial \psi}{\partial x_2}, \qquad v_2 = -\frac{\partial \psi}{\partial x_1}, \tag{1.124}$$

where $\psi(\mathbf{x}, t)$ is the *stream function*. Inserting (8.45) into (1.118) the evolution equations become

$$\dot{\mathbf{x}}_1 = \frac{\partial \psi}{\partial x_2}, \qquad \dot{\mathbf{x}}_2 = -\frac{\partial \psi}{\partial x_1}. \tag{1.125}$$

Formally (8.46) is a Hamiltonian system with the Hamiltonian given by the stream function ψ.

- a) λ_E is the mean exponential rate of the increasing of the uncertainty in the knowledge of the velocity field (which is, by definition, independent on the Lagrangian motion);
- b) λ_L estimates the rate at which the distance $\delta x(t)$ between two fluid particles initially close increases with time, when the velocity field is given, i.e., a particle pair in the same Eulerian realization;
- c) λ_T is the rate of growth of the distance between initially close particle pairs, when the velocity field is not known with infinite precision.

There is no general relation between λ_E and λ_L. One could expect that in presence of a chaotic velocity field the particle motion has to be chaotic. However, the inequality $\lambda_L \geq \lambda_E$ – even if generic – sometimes does not hold, e.g., in some systems like the Lorenz model [FPV88] and in generic $2n$ flows when the Lagrangian motion happens around well defined vortex structures [BBP94] as discussed in the following. On the contrary, one has [CFP91]

$$\lambda_T = \max(\lambda_E, \lambda_L). \tag{1.137}$$

Lagrangian Chaos in 2D–Flows

Let us now consider the 2D *Navier–Stokes equations* with periodic boundary conditions at low Reynolds numbers, for which we can expand the stream function ψ in Fourier series and takes into account only the first F terms [BF79],

$$\psi = -i \sum_{j=1}^{F} k_j^{-1} Q_j e^{i\mathbf{k}_j \mathbf{x}} + \text{c.c.}, \tag{1.138}$$

where c.c. indicates the complex conjugate term and $\mathbf{Q} = (Q_1, \ldots, Q_F)$ are the Fourier coefficients. Inserting (1.138) into the Navier–Stokes equations and by an appropriate time rescaling, we get the system of F ordinary differential equations

$$\dot{Q}_j = -k_j^2 Q_j + \sum_{l,m} A_{jlm} Q_l Q_m + f_j, \tag{1.139}$$

in which f_j represents an external forcing.

Franceschini and coworkers have studied this truncated model with $F = 5$ and $F = 7$ [BF79]. The forcing were restricted to the 3th mode $f_j = \text{Re}\,\delta_{j,3}$. For $F = 5$ and $\text{Re} < \text{Re}_1 = 22.85\ldots$, there are four stable stationary solutions, say $\hat{\mathbf{Q}}$, and $\lambda_E < 0$. At $\text{Re} = \text{Re}_1$, these solutions become unstable, via a Hopf bifurcation [MM75], and four stable periodic orbits appear, still implying $\lambda_E = 0$. For $\text{Re}_1 < \text{Re} < \text{Re}_2 = 28.41\ldots$, one thus finds the stable limit cycles:

$$\mathbf{Q}(t) = \hat{\mathbf{Q}} + (\text{Re} - \text{Re}_1)^{1/2} \delta\mathbf{Q}(t) + O(\text{Re} - \text{Re}_1), \tag{1.140}$$

where $\delta\mathbf{Q}(t)$ is periodic with period

$$T(\text{Re}) = T_0 + O(\text{Re} - \text{Re}_1) \qquad T_0 = 0.7328\ldots \tag{1.141}$$

At Re = Re_2, these limit cycles lose stability and there is a period doubling cascade toward Eulerian chaos.

Let us now discuss the Lagrangian behavior of a fluid particle. For Re < Re_1, the stream function is asymptotically stationary, $\psi(\mathbf{x}, t) \longrightarrow \widehat{\psi}(\mathbf{x})$, and the corresponding 1D Hamiltonian is time-independent, therefore Lagrangian trajectories are regular. For Re = $Re_1 + \epsilon$ the stream function becomes time dependent [BLV01]

$$\psi(\mathbf{x}, t) = \widehat{\psi}(\mathbf{x}) + \sqrt{\epsilon}\, \delta\psi(\mathbf{x}, t) + O(\epsilon), \qquad (1.142)$$

where $\widehat{\psi}(\mathbf{x})$ is given by $\widehat{\mathbf{Q}}$ and $\delta\psi$ is periodic in $\mathbf{x}$ and in t with period T. The region of phase–space, here the real 2D space, adjacent to a separatrix is very sensitive to perturbations, even of very weak intensity. Figure 1.56 shows the structure of the separatrices, i.e., the orbits of infinite periods at Re = $Re_1 - 0.05$.

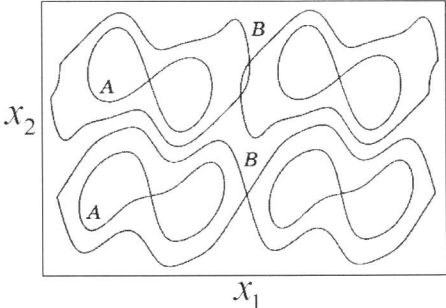

Fig. 1.56. Structure of the separatrices in the 5-mode model (1.138) with Re = $Re_1 - 0.05$ (adapted from [BLV01]).

Indeed, generically in 1D Hamiltonian systems, a periodic perturbation gives origin to stochastic layers around the separatrices where the motion is chaotic, as consequence of unfolding and crossing of the stable and unstable manifolds in domains centered at the hyperbolic fixed–points [Chi79, Ott93].

Chaotic and regular motion for small $\epsilon = Re_1 - Re$ can be studied by the Poincaré map

$$\mathbf{x}(nT) \longrightarrow \mathbf{x}(nT + T). \qquad (1.143)$$

The period $T(\epsilon)$ is computed numerically. The size of the stochastic layers rapidly increase with ϵ. At $\epsilon = \epsilon_c \approx 0.7$ they overlap and it is practically impossible to distinguish between regular and chaotic zones. At $\epsilon > \epsilon_c$ there is always diffusive motion.

We stress that this scenario for the onset of Lagrangian chaos in 2D fluids is generic and does not depend on the particular truncated model. In fact, it is only related to the appearance of stochastic layers under the effects of small time–dependent perturbations in 1D integrable Hamiltonian systems. As consequence of a general features of 1D Hamiltonian systems we expect that a stationary stream function becomes time periodic through a Hopf bifurcation as occurs for all known truncated models of Navier–Stokes equations.

We have seen that there is no simple relation between Eulerian and Lagrangian behaviors. In the following, we shall discuss two important points [BLV01]:

- (i) what are the effects on the Lagrangian chaos of the transition to Eulerian chaos, i.e., from $\lambda_E = 0$ to $\lambda_E > 0$.
- (ii) whether a chaotic velocity field ($\lambda_E > 0$) always implies an erratic motion of fluid particles.

The first point can be studied again within the $F = 5$ modes model (1.139). Increasing Re, the limit cycles bifurcate to new double period orbits followed by a period doubling transition to chaos and a strange attractor appears at $\mathrm{Re}_c \approx 28.73$, where λ_E becomes positive. These transitions have no signature on Lagrangian behavior, i.e., the onset of Eulerian chaos has no influence on Lagrangian properties.

This feature should be valid in most situations, since it is natural to expect that in generic cases there is a strong separation of the characteristic times for Eulerian and Lagrangian behaviors.

The second point – the conjecture that a chaotic velocity field always implies chaotic motion of particles – looks very reasonable. Indeed, it appears to hold in many systems [CFP91]. Nevertheless, one can find a class of systems where it is false, e.g., the equations (1.131), (1.132) may exhibit Eulerian chaoticity $\lambda_E > 0$, even if $\lambda_L = 0$ [BBP94].

Consider for example the motion of N point vortices in the plane with circulations Γ_i and positions $(x_i(t), y_i(t))$ $(i = 1, ..N)$ [Are83]:

$$\Gamma_i \dot{x}_i = \frac{\partial H}{\partial y_i}, \qquad \Gamma_i \dot{y}_i = -\frac{\partial H}{\partial x_i}, \qquad \text{where} \qquad (1.144)$$

$$H = -\frac{1}{4\pi} \Gamma_i \Gamma_j \ln r_{ij}, \qquad \text{and} \qquad r_{ij}^2 = (x_i - x_j)^2 + (y_i - y_j)^2. \quad (1.145)$$

The motion of N point vortices is described in an Eulerian phase–space with $2ND$. Because of the presence of global conserved quantities, a system of three vortices is integrable and there is no exponential divergence of nearby trajectories in phase–space. For $N \geq 4$, apart from non generic initial conditions and/or values of the parameters Γ_i, the system is chaotic [Are83].

The motion of a passively advected particle located in $(x(t), y(t))$ in the velocity field defined by (4.31) is given

$$\dot{x} = -\frac{\Gamma_i}{2\pi} \frac{y - y_i}{R_i^2}, \qquad \dot{y} = \frac{\Gamma_i}{2\pi} \frac{x - x_i}{R_i^2}, \qquad (1.146)$$

where $\qquad R_i^2 = (x - x_i)^2 + (y - y_i)^2.$

Let us first consider the motion of advected particles in a three-vortices (integrable) system in which $\lambda_E = 0$. In this case, the stream function for the advected particle is periodic in time and the expectation is that the advected particles may display chaotic behavior. The typical trajectories of passive particles which have initially been placed respectively in close proximity of a vortex center or in the background field between the vortices display a very different behavior. The particle seeded close to the vortex center displays a regular motion around the vortex and thus $\lambda_L = 0$; by contrast, the particle in the background field undergoes an irregular and aperiodic trajectory, and λ_L is positive.

We now discuss a case where the Eulerian flow is chaotic i.e., $N = 4$ point vortices [BLV01]. Let us consider again the trajectory of a passive particle deployed in proximity of a vortex center. As before, the particle rotates around the moving vortex. The vortex motion is chaotic; consequently, the particle position is unpredictable on large times as is the vortex position. Nevertheless, the Lagrangian Lyapunov exponent for this trajectory is zero (i.e., two initially close particles around the vortex remain close), even if the Eulerian Lyapunov exponent is positive, see Figure 1.57.

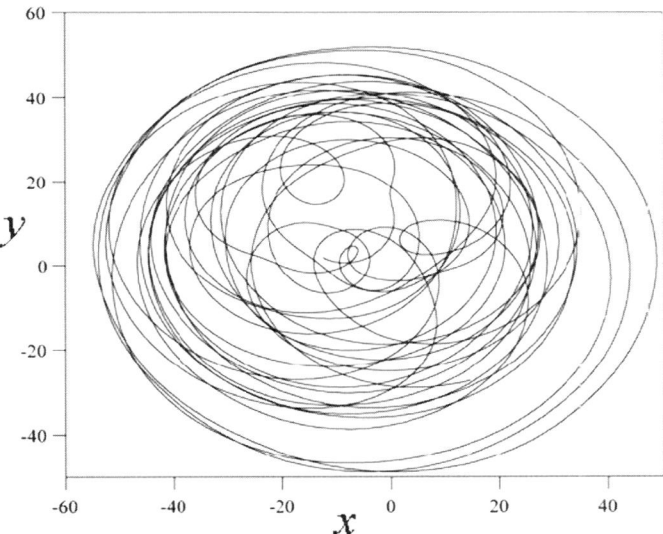

Fig. 1.57. Particle trajectories in the 4–vortex system. Eulerian dynamics in this case is chaotic. The left panel shows a regular Lagrangian trajectory while the right panel shows a chaotic Lagrangian trajectory. The different behavior of the two particles is due to different initial conditions (adapted from [BLV01]).

This result indicates once more that there is no strict link between Eulerian and Lagrangian chaoticity.

One may wonder whether a much more complex Eulerian flow, such as $2n$ turbulence, may give the same scenario for particle advection: i.e., regular trajectories close to the vortices and chaotic behavior between the vortices. It has been shown that this is indeed the case [BBP94] and that the chaotic nature of the trajectories of advected particles is not strictly determined by the complex time evolution of the turbulent flow.

We have seen that there is no general relation between Lagrangian and Eulerian chaos. In the typical situation Lagrangian chaos may appear also for regular velocity fields. However, it is also possible to have the opposite situation, with $\lambda_L = 0$ in presence of Eulerian chaos, as in the example of Lagrangian motion inside vortex structures. As an important consequence of this discussion we remark that it is not possible to separate Lagrangian and Eulerian properties in a measured trajectory. Indeed, using the standard methods for data analysis [GP83a], from Lagrangian trajectories one extracts the total Lyapunov exponent λ_T and not λ_L or λ_E.

1.6 Standard Map and Hamiltonian Chaos

Chaos as defined in the theory of dynamical systems has been recognized in recent years as a common structural feature of phase–space flows. Its characterization involves geometric, dynamic, and symbolic aspects. Concepts developed in the framework of Poincaré–Birkhoff and Kolmogorov–Arnold–Moser theories describe the transition from regular to chaotic behavior in simple cases, or locally in phase–space. Their global application to real systems tends to interfere with non–trivial bifurcation schemes, and computer experimentation becomes an essential tool for comprehensive insight [Ric01, RSW90].

The prototypic example for *Hamiltonian chaos* is the standard map M, introduced by the plasma physicist B.V. Chirikov [Chi79]. It is an area–preserving map of a cylinder (or annulus) with coordinates (φ, r) onto itself,

$$
M : \quad \begin{pmatrix} \varphi \\ r \end{pmatrix} \mapsto \begin{pmatrix} \varphi' \\ r' \end{pmatrix} = \begin{pmatrix} \varphi + 2\pi(1 - r') \\ r + \mu \sin \varphi \end{pmatrix},
$$

where φ is understood to be taken modulo 2π. The map has been celebrated for catching the essence of chaos in conservative systems, at least locally in the neighborhood of resonances. The number μ plays the role of a perturbation parameter. For $\mu = 0$, the map is integrable: all lines $r = const$ are invariant sets, with angular increment $\Delta\varphi = 2\pi(1 - r) = 2\pi W$. For obvious reasons, W is called *winding number*. For integer r, it is 0 mod 1; each point of the line is a fixed–point (see Figure 1.58). For rational $r = p/q$ with p, q coprime, the line carries a continuum of periodic orbits with period q, winding around the cylinder p/q times per period. However, when r is irrational, the orbit of an initial point never returns to it but fills the line densely. This makes for an important qualitative difference of rational and irrational invariant lines.

The dependence of W on r is called the *twist* r of the (un–perturbed) map. Here, $\tau = W/r = -1$. This has an intuitive interpretation: between any two integer values of r the map is a twist of the cylinder by one full turn. The behavior of such *twist maps* under small perturbations μ was the central issue of the stability discussion of which Poincaré's 'Méthodes Nouvelles' in 1892 and the discovery of the Kolmogorov–Arnold–Moser (KAM) theorem (see [AA68]) were the major highlights.

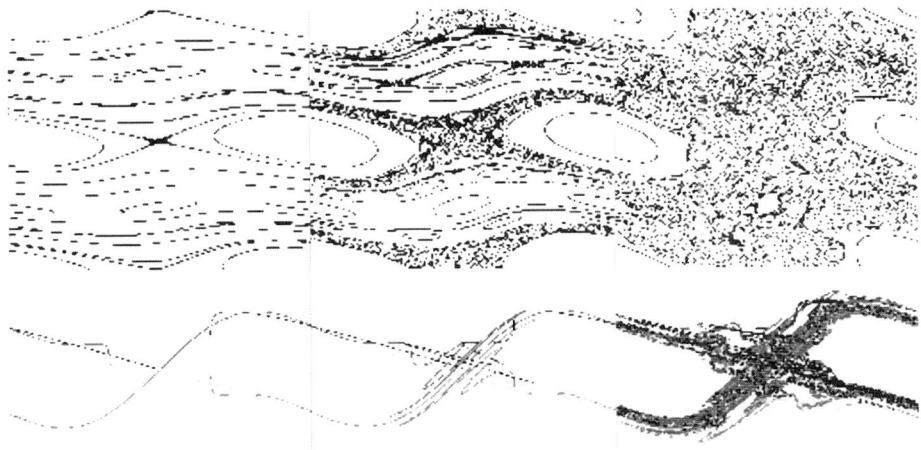

Fig. 1.58. Standard Map: Upper left: 20 trajectories with $\mu = 0.08$; the angle $\varphi \in [0, 2\pi]$ is plotted along the horizontal, the variable $r \in [0, 2]$ – along the vertical axis. Upper middle: the same for $\mu = 0.15$. Upper right: the same for $u = 0.25$. Lower left: beginning of the stable manifold (going from lower–left to upper–right) and unstable manifold (going from upper–left to lower–right) of the hyperbolic point $(\varphi, r) = 9\pi, 1)$ for $\mu = 0.25$. Lower middle: the same continued for two more iterations. Lower right: the same for additional 9 iterations (adapted from [Ric01]).

Consider the upper part of Figure 1.58. For perturbations $\mu = 0.08, 0.15, 0.25$, it shows 20 orbits generated from (almost) randomly chosen initial points. Three kinds of structures can be discerned: (i) islands, or chains of islands, surrounding elliptic periodic orbits; (ii) clouds of points centered at hyperbolic periodic orbits and surrounding the islands; (iii) the two left pictures contain invariant lines extending from left to right, i e., surrounding the cylinder, whereas the right picture exhibits 'global chaos' in the sense that a single trajectory can wander along the vertical cylinder axis. These features illustrate three fundamental theorems that govern the *behavior of invariant sets under small perturbations*:

• *Poincaré–Birkhoff Theorem*: Poincaré's last theorem, proved by Birkhoff in 1913, states that of the uncountable infinity of periodic orbits on a given rational invariant set, only a finite number survives, half of them elliptic, the other half hyperbolic. More precisely, lines with rational winding number p/q decay into chains of q islands (sometimes, for reasons of symmetry, a finite multiple of q). Their centers are elliptic resonances E^k, ($k = 1, ..., q$), i.e., fixed–points of the qth iterate M^q of the map. The Jacobian of M^q at E^k has eigenvalues of the form $\cos \rho \pm \mathrm{i} \sin \rho$, where $\cos \rho = \frac{1}{2} \mathrm{trace}\, M^q$ and $|\mathrm{trace}\, M^q| < 2$. The number ρ is a winding number of higher order; when points on small islands are parametrized by an angle, ρ is the angular increment between a point and its qth iterate. The islands of a given chain are separated by hyperbolic fixed–points H^k of M^q, ($k = 1, ..., q$); the Jacobian at H^k has eigenvalues λ_u and λ_s with $|\lambda_u| > 1$ and $\lambda_s = 1/\lambda_u$. The corresponding eigenvectors $\mathbf{e}_u$ and $\mathbf{e}_s$ are, respectively, the *unstable and stable directions* of the hyperbolic orbits. Poincaré and Birkhoff found that for arbitrarily small perturbations, this breakup of rational lines is generic.

- *Kolmogorov–Arnold–Moser (KAM) Theorem*: It took some 70 years after Poincaré before the survival of invariant lines under small perturbations could be established with mathematical certainty [AA68].[34] For systems with n degrees of freedom and sufficiently small and smooth (but otherwise

---

[34] KAM–theorem is a result in dynamical systems about the persistence of quasi-periodic motions under small perturbations. The theorem partly resolves the small-divisor problem that arises in the perturbation theory of classical mechanics. The problem is whether or not a small perturbation of a conservative dynamical system results in a lasting quasi–periodic orbit. The original breakthrough to this problem was given by Kolmogorov in 1954. This was rigorously proved and extended by Arnold (in 1963 for analytic Hamiltonian systems) and Moser (in 1962 for smooth twist maps), and the general result is known as the KAM–theorem. The KAM–theorem, as it was originally stated, could not be applied to the motions of the solar system, although Arnold used the methods of KAM to prove the stability of elliptical orbits in the planar 3–body problem. The KAM–theorem is usually stated in terms of trajectories in phase–space of an integrable Hamiltonian system. The motion of an integrable system is confined to a doughnut–shaped surface, an *invariant torus*. Different initial conditions of the integrable Hamiltonian system will trace different invariant tori in phase–space. Plotting any of the coordinates of an integrable system would show that they are quasi–periodic. The KAM–theorem states that if the system is subjected to a weak nonlinear perturbation, some of the invariant tori are deformed and others are destroyed. The ones that survive are those that have 'sufficiently irrational' frequencies (this is known as the non–resonance condition). This implies that the motion continues to be quasi–periodic, with the independent periods changed (as a consequence of the non–degeneracy condition). The KAM–theorem specifies quantitatively what level of perturbation can be applied for this to be true. An important consequence of the KAM–theorem is that for a large set of initial conditions the motion remains perpetually quasi–periodic. The methods introduced by Kolmogorov, Arnold, and Moser have developed into a large body of results related to quasi–periodic motions. Notably, it has been extended to non–Hamiltonian systems (starting with Moser), to non–perturbative situations (as in the work of Michael Herman) and to

arbitrary) perturbations, the KAM–theorem guarantees the existence of fD *Liouville tori* in the $(2f - 1)$D *energy surface*. For $f = 2$, this is an extremely important stability result because orbits between two invariant tori are confined: if there is chaos, it cannot be global. For larger f, the situation is more delicate and still not fully explored.

The KAM–theorem also gives a hint as to which irrational tori are the most robust. The point is that one may distinguish various degrees of irrationality. A number W is called 'more irrational' than a number W' if rational approximations p/q with q smaller than a given $q_{\max}$ tend to be worse for W than for W':

$$\min_{p,q}\left|W - \frac{p}{q}\right| > \min_{p,q}\left|W' - \frac{p}{q}\right|.$$

The numbers W which qualify in the KAM–theorem for being sufficiently irrational, are characterized by constants $c > 0$ and $\nu \geq 2$ in an estimate $\left|W - \frac{p}{q}\right| \geq \frac{c}{q^{\nu}}$, for arbitrary integers p and q. The set of W for which c and ν exist has positive measure. ν cannot be smaller than 2 because the *Liouville theorem* asserts that if p_k/q_k is a continued fraction approximation to W, then there exists a number C (independent of k) such that $|W - p_k/q_k| < C/q_k^2$. On the other hand, it is known that quadratic irrationals, i.e., solutions of quadratic equations with integer coefficients, have periodic continued fraction expansions which implies that the corresponding approximations behave asymptotically as [RSW90]

$$\left|W - \frac{p_k}{q_k}\right| = \frac{c_1}{q_k^2} + (-1)^k \frac{c_2}{q_k^4} + O\left(q_k^{-6}\right).$$

It is also known that there are no better rational approximations to a given W than its continued fraction approximations. In this sense, the quadratic irrationals are the 'most irrational' numbers because they have the smallest possible $\nu = 2$. Furthermore, among the quadratic irrationals, the so–called *noble numbers* have the largest possible $c_1 = 1/\sqrt{5}$; they are defined by continued fractions

$$W = w_0 + \cfrac{1}{w_1 + \cfrac{1}{w_2 + \cfrac{1}{w_3 + \ldots}}} = [w_0, w_1, w_2, w_3, \ldots],$$

where $w_k = 1$ for k greater than some K. Hence, invariant lines (or tori) with noble winding numbers tend to be the most robust. And finally, within the class of noble numbers, the so–called *golden number*

---

systems with fast and slow frequencies. The non–resonance and non–degeneracy conditions of the KAM–theorem become increasingly difficult to satisfy for systems with more degrees of freedom. As the number of dimensions of the system increase the volume occupied by the tori decreases. The *KAM–tori* that are not destroyed by perturbation become invariant Cantor sets, termed *Cantori*.

$g = [0, 1, 1, 1, \ldots] = \frac{1}{2}(\sqrt{5} - 1) = 0.618\,034$, as well as $1 - g = g^2$ and $1 + g = 1/g$ are the 'champions of irrationality' because they have the smallest possible $c_2 = 1/(5\sqrt{5})$. More generally, the most irrational noble number between neighboring resonances p_1/q_1 and p_2/q_2 (for example, $2/5$ and $1/2$) is obtained with the so–called *Farey construction* [Sch91] $p_{n+1}/q_{n+1} = (p_{n-1} + p_n)/(q_{n-1} + q_n)$.

It has been found in numerous studies based on the discovery of the KAM–theorem, that when a Hamiltonian system can in some way be described as a perturbed twist map, the following scenario holds with growing perturbation. First, many small resonances (chains of q islands) and many irrational tori (invariant lines) coexist. Then the islands with small q grow; the chaos bands associated with their hyperbolic orbits swallow islands with large q and the less irrational tori. At one point, only one noble *KAM–torus* and few conspicuous resonances survive, before with further increasing perturbation the last torus decays and gives way to global chaos. More detailed analysis shows that a decaying KAM–torus leaves traces in the chaotic sea: invariant Cantor sets which act as 'permeable barriers' for the phase–space flow [Ric01, RSW90].

- *Smale–Zehnder Theorem*: E. Zehnder (1973) proved that each hyperbolic point H^k of the Poincaré–Birkhoff scenario is the center of a chaotic band, in the sense that it contains a *Smale horseshoe* (see Figure 1.59). To see this, consider a point P close to H^k on its *unstable eigen–direction* $\mathbf{e}_u$ and let P' be its image under the standard map M^q. Now iterate the line PP' with M^q. Step by step, this generates the *unstable manifold* of H^k, shown in the lower part of Figure 1.59: starting at the center, the two parts evolve towards the lower right and the upper left. It is seen from the succession of the three pictures that the manifold develops a folded structure of increasing complexity as the iteration goes on.[35] Eventually it densely fills a region of full measure. In the last picture shown here, the barriers associated with the former golden KAM–torus have not yet been penetrated, but it is believed that the closure of the unstable manifold is the entire chaos band connected to H^k.

Similarly, the *stable manifold* can be generated by backward iteration along the *stable eigen–direction* $\mathbf{e}_s$. The decisive point in Zehnder's theorem is the demonstration that the stable and unstable manifolds have transverse intersections, called homoclinic points (or heteroclinic if the two manifolds belong to different hyperbolic orbits, but this difference is not important). Once this is established, it may be concluded that part of the map has the character of a horseshoe map, hence it contains an invariant set on which the dynamics is chaotic. Two strips in the neighborhood of the central hyperbolic point, one red the other green, each stretched along a folded piece of the stable manifold and crossing the unstable manifold at a homoclinic point, are mapped a number of times and returned near the hyperbolic

---

[35] The manifold may not intersect itself because the map is injective.

point in a transverse orientation. In that process they are contracted towards and expanded along the unstable manifold. The intersection pattern of the strips contains a thin invariant Cantor set on which an iterate of the standard map is conjugate to the shift map on bi–infinite sequences of two symbols. This implies the existence of chaos. Zehnder showed that

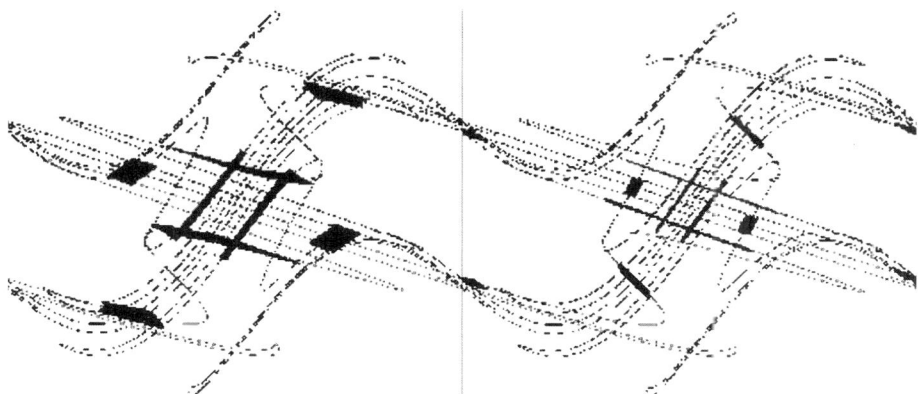

Fig. 1.59. Smale horseshoe in the standard map for $\mu = 0.4$; the angle $\varphi \in [0, 2\pi]$ is plotted along the horizontal, the variable $r \in [0, 2]$ – along the vertical axis. Left: the intersection of the two pairs of strips near the center contains an invariant set for M^3. Right: the same for M^4 (adapted from [Ric01]).

this situation is generic even for arbitrarily small perturbations μ, where it is impossible to give a graphical illustration because the homoclinic tangle is exponentially thin, and the number of iterations needed to return to the hyperbolic point very large. The message for our subsequent discussion of the three body problem is that when rational invariant sets of a twist map are perturbed, they break up into elliptic centers of order and hyperbolic centers of chaos. The invariant sets constructed on the basis of Figure 1.59 are of course not the whole story for the chaos band connected with the central hyperbolic point. We have shown the two perhaps most obvious horseshoes for this case, but an infinity of other horseshoes may be identified in the same chaotic region, using the same strategy with other homoclinic intersections. Each of them has a fractal dimension not much larger than 1, whereas the entire chaos band – the closure of the stable or unstable manifolds – is believed to have dimension 2. This has not yet been proven. Moreover, the computation of Lyapunov exponents is easy for a single horseshoe, but averaging them over the chaotic region is numerically delicate, analytically hopeless. So there remain important open questions [Ric01, RSW90].

If F is a continuous function in Ω, and Λ is a chaotic attractor then it is a *Cantor set*.

1.7.1 Examples of Dynamical Maps

Here we consider some examples of dynamical maps $f : \Omega \longrightarrow \Omega$. The first one is the one–side shift map σ of symbolic dynamics which we introduce to familiarize the reader with the notation.

One–Sided Shift Map σ

The continuous map σ defined by

$$\sigma\left(\mathbf{S}\left(0\right), \mathbf{S}\left(1\right), ...\right) = \left(\mathbf{S}\left(1\right), \mathbf{S}\left(2\right), ...\right),$$

is chaotic in Ω (see [Dev89, KH95, Rob95]).

Note that σ is non–invertible and its action loses the information carried by the binary state $\mathbf{S}\left(0\right)$. The meaning and usefulness of this map is quite clear in the context of symbolic dynamics when the Conley–Moser conditions are satisfied [Mos73]. There one studies, in general, a non–invertible function $f : \Xi \longrightarrow \Xi$ where Ξ is a Cantor set embedded in $\mathbb{R}^N$. The set Ξ is divided in 2^N sectors I_α $\alpha = 0, 1, ..., 2^N$. Then it is possible to establish a topological conjugation between f and σ through a homeomorphism ψ, so that the following diagram commutes [Wig90]:

$$
\begin{array}{ccc}
\Xi & \overset{f}{\longrightarrow} & \Xi \\
\psi \downarrow & & \downarrow \psi \\
\Omega & \overset{\sigma}{\longrightarrow} & \Omega
\end{array}
$$

Moreover, let $S = \psi\left(x\right)$, then $\mathbf{S}\left(n\right)$ is the binary decomposition of the label α, such that $f^n\left(x\right) \in I_\alpha$.

Chaotic Maps with Non–Trivial Attractors in Ω

The shift map can be modified to create maps which are homeomorphic to the shift map on an asymptotically stable transitive subset of the space of symbols. In the following, we introduce two very simple examples. Firstly, take the space of symbols Ω with $N = 2$, homeomorphic to $\Xi \times \Xi$ where Ξ is the space of symbols with $N = 1$, that is the space of semi–infinite sequences $S = \left(S_0, S_1, S_2, ...\right)$. Then consider the function $f_c : \Xi \times \Xi \to \Xi \times \Xi$ given by $f_c = \sigma \times \zeta$. Where σ is the usual shift function and ζ is a right inverse of the shift function defined as follows [WZ98]:

$$\zeta\left(S_0, S_1, S_2, ...\right) = \left(0, S_0, S_1, S_2, ...\right).$$

It is easy to check that ζ is a continuous function, and of course so is the shift: so f_c is continuous. The set $\Xi \times \{0\}$ is an asymptotically stable transitive set, on which the restriction of f_c is the shift map σ.

As another example, consider the space Ω with $N = 1$. It can be split into the disjoint union of two Cantor sets $\Omega = \Lambda_0 \cup \Lambda_1$. Where Λ_0 is the set of sequences such that $S_0 = 0$ and an analogous fashion for Λ_1. Take the continuous function $f_\pi = \pi \circ \sigma$, where σ is the shift map and π projects Ω in Λ_0 such that:

$$\pi \left(S_0, S_1, S_2, ... \right) = \left(0, S_1, S_2, ... \right).$$

Then the action of f_π is given by,

$$f_\pi \left(S_0, S_1, S_2, ... \right) = \left(0, S_2, S_3, ... \right).$$

It is easy to check that Λ_0 is a chaotic attractor of f_π.

Chaotic Maps in Ω Induced Through Chaotic Maps in Cantor Subsets of $\mathbb{R}^N$

Here, we consider a homeomorphism which relates a Cantor set $\chi \subset \mathbb{R}^N$ to the space Ω and allows one to construct chaotic maps in Ω from chaotic maps in χ through topological conjugation. Let $\chi \subset \mathbb{R}^N$ be the Cantor set that results from taking the Cartesian product of N Cantor sets χ_i;

$$\chi = \bigotimes_{i=1}^{N} \chi_i,$$

where the ith component χ_i is constructed by suppressing from the interval $[0, 1]$ the open middle $1/a_i$ part, $i = 1, \ldots, N$, $a_i > 1$, and repeating this procedure iteratively with the sub-intervals. Now, we define a map $\phi : \Omega \longrightarrow \chi$ by [WZ98]:

$$\phi_i \left(S \right) = \sum_{n=1}^{\infty} \left(l_{n-1} - l_n \right) S_i \left(n - 1 \right), \qquad \text{where}$$

$$l_n = \frac{1}{2^n} \left(1 - \frac{1}{a_i} \right)^n,$$

is the length of each of the remaining 2^n intervals at the nth step of the construction of χ_i. If Ω is endowed with the metric (1.148) and $\chi \subset \mathbb{R}^N$ with the standard Euclidean metric, is easy to show that ϕ is a homeomorphism.

Now, if we have a map $f : \mathbb{R}^N \longrightarrow \mathbb{R}^N$ which is chaotic in χ we can construct a map $F : \Omega \longrightarrow \Omega$ which is chaotic in Ω, and is defined through the commutation of the following diagram:

$$\begin{array}{ccc} \chi & \xrightarrow{f} & \chi \\ \phi \uparrow & & \uparrow \phi \\ \Omega & \xrightarrow{F} & \Omega \end{array}$$

This leads to an interesting practical application of the homeomorphism ϕ, to realize computer simulations of chaotic systems on Cantor sets. If, for example, one iterates the *logistic map*

$$f(x) = \mu x (1 - x), \qquad \text{for} \;\; \mu \geq 4,$$

with a floating-point variable, the truncation errors nudge the trajectory away from the Cantor set and eventually $x \to -\infty$. The homeomorphism ϕ suggests a natural solution to this, which is to iterate the truncated binary states rather than the floating–point variable. To iterate the dynamics, one computes $x_i = \phi_i(S)$ for all $i = 1, \dots, N$ by assuming that the truncated bits are all equal to zero, then applies f to get $x' = f(x)$. Since x' generally does *not* belong to the Cantor set (because of truncation errors), in the process of constructing $S' = \phi^{-1}(x')$, at some n one will find that this point does not belong to either the interval corresponding to $S_i(n) = 0$ or to $S_i(n) = 1$. This truncation error can be corrected by moving to the extremity of the interval which lies closest to x_i'. In this way, truncation errors are not allowed to draw the trajectory away from the Cantor set $\chi \subset \mathbb{R}^N$.

Binary Systems with Memory

Now we will define a map $\Gamma : \Omega \longrightarrow \Omega$ which is very useful to analyze binary systems with causal deterministic dynamics on N bits, such as neural networks, cellular automata, and neural networks with memory (see, e.g., [SK86]). Let

$$\gamma_i : \Omega \longrightarrow \{0, 1\}, \qquad (i = 1, ..., N), \tag{1.150}$$

be a set of continuous or discontinuous functions. The map $\Gamma : \Omega \longrightarrow \Omega$ is then defined by:

$$\Gamma_i(S) = (\gamma_i(S), S_i(0), S_i(1), \dots),$$
$$\text{or, short–hand,} \qquad \Gamma(S) = (\gamma(S), S). \tag{1.151}$$

Such maps have the following properties.

The shift map (1.150) is a left inverse of Γ since from (1.151) $\sigma \circ \Gamma(S) = S$. If Ω has an attracting set $\Lambda \subset \Omega$, then σ is also a right inverse in the restriction of Γ to Λ, so that, $\Gamma|_\Lambda^{-1} = \sigma$. For all $S \in \Lambda$ there is $S' \in \Lambda$ such that $\Gamma(S') = S$. Since

$$\Gamma(S') = (\gamma(S'), S') = S, \qquad \text{and} \qquad S = (\mathbf{S}(0), S_1),$$

where $S_1 \equiv (\mathbf{S}(1), \mathbf{S}(2), \dots)$, one sees that $S' = S_1$. Thus,

$$\Gamma \circ \sigma(S) = \Gamma(S_1) = \Gamma(S') = S.$$

Γ has an attracting set Λ contained properly in Ω. Given S there are 2^N states $S' = (\mathbf{S}'(0), S)$ of which only one, $\Gamma(S) = (\gamma(S), S)$, belongs to $\Gamma(\Omega)$. Therefore the set

$$\Lambda \equiv \bigcap_{n \geq 0} \Gamma^n(\Omega)$$

is a proper subset of Ω [WZ98].

If Γ is continuous, then it is not sensitive to initial conditions. Γ is a continuous map on a compact set, so it is uniformly continuous. Therefore there exists a $\delta > 0$ such that for any $S \in \Omega$,

$$d(S', S) < \delta \quad \Rightarrow \quad \gamma(S) = \gamma(S') \quad \Rightarrow \quad d(\Gamma(S), \Gamma(S')) < \delta/2,$$

where the distance function is given by (1.148). Applying the same argument to each iterate $\Gamma^k(S)$ shows that $d(\Gamma^k(S), \Gamma^k(S')) < \delta/2^k$, which contradicts sensitivity to initial conditions.

If Γ is continuous, then the attractor Λ is finite. We know that Λ exists. The property then follows from the previous one. Indeed, if Γ is not sensitive to initial conditions, then there is a $n > 0$ such that for all $S \in \Omega$, $S' \in \mathcal{N}_n(S)$,

$$\lim_{k \to \infty} d\left(\Gamma^k(S) - \Gamma^k(S')\right) = 0.$$

The set $A \subset \Omega$ defined by $S \in A$ iff for all $m > n, \mathbf{S}(m) = 0$, has a finite number of elements, namely $2^{N \times n}$. The whole space Ω is the union of the n−neighborhoods of each element of A, and as we just showed the map Γ is contracting in each such neighborhood, so the number of points in the attractor cannot be greater than the number of elements of A, namely $2^{N \times n}$.

Discrete–time neural networks and cellular automata are binary dynamical systems in which the values of the state variables S_i, $i = 1, \ldots, N$, at time t depend on the state variables at time $t - 1$. These systems are described by a function Γ such that the functions γ_i depend only on the components $\mathbf{S}(0)$. Therefore, all points $S' \in \mathcal{N}_n(S)$ for $n > 0$ have the same evolution so that these systems are not sensitive to initial conditions. One can recover a very rough approximation of sensitive dependence on initial conditions by considering the growth of *Hamming distance* with time, rather than the metric (1.148) of symbolic dynamics. However, one cannot describe the behavior of these systems to be approximately chaotic: They are well known to have attractors that consist of a collection of periodic limit–cycles, and the points of these limit–cycles are scattered over configuration space without any effective lower–dimensional structure. In particular, given any one point on the attractor there is usually no other point 'nearby', even in the weak sense of the Hamming distance, that also belongs to the attractor. This fact makes most practical uses of chaos theory in prediction and control inapplicable.

Compact Topology for Neural Networks and Cellular Automata

Since discrete neural networks and cellular automata in general are systems in which all the variables have the same type of interactions, it is natural to consider the Hamming distance as the metric (it is in fact the most widely used metric in the literature, see e.g., [CKS05] and the references therein). We have already seen that the topological structure which the Hamming distance confers to the phase–space does not conduce to chaotic behavior in the sense that we understand it even if we extend the phase–space to Ω. However, not all the neural network and cellular automata models confer the same type of interactions to neurons, so the use the Hamming distance for the metric is not so compelling. The use of a different metric can lead to a completely different topology. The resulting system will in general display a very different dynamical behavior. For example, the map

$$x_{n+1} = \alpha x_n$$

produces quite different dynamical behaviors for $x_n \in \mathbb{R}$ and $x_n \in S^1$.

So, let us consider systems which evolve according to the rule

$$R_i(t+1) = f_i(R(t)), \tag{1.152}$$

$R_i = 0, 1;$ $(i = 1, ..., M)$ and take for the metric

$$d(S, S') = \sum_{n=0}^{M} \frac{1}{2^n} \, d_n(S - S'). \tag{1.153}$$

These systems include neural networks and cellular automata as particular examples, but where the weight of the different neurons drops off as 2^{-n}. The metric (1.153) remains well defined in the limit $M \to \infty$ and once again we get the space Ω. In fact (1.152) and (1.153) with $M \to \infty$ are equivalent to (1.148) and (1.149) with $N = 1$ and $S_1(n) = R_n$. As we will see in the next subsection, these systems can have a correlation dimension which is less than or equal to one.

1.7.2 Correlation Dimension of an Attractor

Recall that in the theory of dynamical systems in $\mathbb{R}^N$ one is interested in calculating the fractal dimension of the attractor in which the system evolves. To do so, following the method of [GP83a, GP83b] one defines the correlation function $\mathcal{C}(\rho)$ as the average of the number of neighbors S_t, $S_{t'}$, with $S_t = F^t(S)$, which have a distance smaller than ρ. Since in $\mathbb{R}^N$ the volume of a sphere of radius ρ grows like ρ^N, one identifies the correlation dimension D_a of the attractor with the growth rate in $\mathcal{C}(\rho) \sim \rho^{D_a}$. This leads to the definition of the correlation dimension as [WZ98]

$$D_a = \lim_{\rho, \rho' \to 0} \left(\frac{\log(\mathcal{C}(\rho)) - \log(\mathcal{C}(\rho'))}{\log(\rho) - \log(\rho')} \right). \tag{1.154}$$

In order to have an analogous methodology to compute correlation dimensions in Ω, it is necessary to know how many states S' are within a distance less than ρ from a given point S. Since Ω is homogeneous we can take $S = 0$. To do the calculation we make Ω into a finite space by truncating the semi–infinite sequence to only T slices, and take the limit $T \to \infty$ in the end, that is:

$$\mathcal{C}\left(\rho\right) = \lim_{T\to\infty} \frac{1}{2^{NT}} \sum_{\{S\}} \Theta\left(\rho - d\left(S,0\right)\right),$$

where the distance is given by (1.148). Expressing $\Theta(x)$ in terms of its *Fourier transform*

$$\omega\left(k\right) = \pi\delta\left(k\right) - \frac{i}{k}, \qquad \text{we have}$$

$$\mathcal{C}\left(\rho\right) = \lim_{T\to\infty} \frac{1}{2^{NT}} \frac{1}{2\pi} \int_{-\infty}^{+\infty} dk\ \omega\left(k\right) e^{ik\rho} \sum_{\{S\}} e^{-ikd(S,0)}.$$

The sum over $\{S\}$ can be evaluated easily obtaining

$$\sum_{\{S\}} e^{-ikd(S,0)} = 2^{NT} e^{-ikN} \left(\prod_{n=0}^{T} \cos\frac{k}{2^{n+1}}\right)^{N}.$$

Using the identity $\sin k/k = \prod_{n=0}^{\infty} \cos\frac{k}{2^{n+1}}$ we get the integral

$$\mathcal{C}\left(\rho\right) = \frac{1}{2\pi} \int_{-\infty}^{+\infty} dk\ \omega\left(k\right) \left(\frac{\sin k}{k}\right)^{N} e^{ik(\rho-N)},$$

which may be evaluated by standard complex–variable methods, to get the final result for the correlation function in Ω,

$$\mathcal{C}\left(\rho\right) = \frac{1}{2^{N} N!} \sum_{k=0}^{[\rho/2]} (-1)^{k} \binom{N}{k} (\rho - 2k)^{N}. \tag{1.155}$$

So we see that the scaling in Ω is not a power law as in $\mathbb{R}^{N}$. However in the definition of the attractor dimension one is interested in calculating $\mathcal{C}\left(\rho\right)$ for $\rho \to 0$. For $\rho \leq 2$ equation (1.155) has the form

$$\mathcal{C}\left(\rho\right) = \frac{1}{2^{N} N!} \rho^{N}.$$

Therefore, the same techniques applied in $\mathbb{R}^{N}$ can be used in Ω, in particular an effective 'attractor dimension' is given by (1.154).

1.8 Basic Hamiltonian Model of Biodynamics

Recall that to describe the biodynamics of human–like movement, namely our *covariant force law*, $F_i = mg_{ij}a^j$, we can also start from generalized Hamiltonian vector–field X_H describing the behavior of the *human–like locomotor system* (see [II05, II06a, II06b])

$$\dot{q}^i = \frac{\partial H}{\partial p_i} + \frac{\partial R}{\partial p_i}, \qquad (1.156)$$

$$\dot{p}_i = F_i - \frac{\partial H}{\partial q^i} + \frac{\partial R}{\partial q^i}, \qquad (1.157)$$

where the vector–field X_H is generating time evolution, or *phase–flow*, of $2n$ *system variables*: n generalized coordinates (joint angles q^i) and n generalized momenta (joint angular momenta p_i), $H = H(q, p)$ represents the system's conservative energy: kinetic energy + various mechano–chemical potentials, $R = R(q, p)$ denotes the nonlinear dissipation of energy, and $F_i = F_i(t, q, p, \sigma)$ are external control forces (biochemical energy inputs). The *system parameters* include inertia tensor with mass distribution of all body segments, stiffness and damping tensors for all joints (labelled by index i, which is, for geometric reasons, written as a subscript on angle variables, and as a superscript on momentum variables), as well as amplitudes, frequencies and time characteristics of all active muscular forces.

The equation (1.156) is called the *velocity equation*, representing the *flow* of the system (analogous to current in electrodynamics), while the equation (1.157) is a Newton–like *force equation*, representing the *effort* of the system (analogous to voltage). Together, these two functions represent Hamiltonian formulation of the *biomechanical force–velocity relation* of A.V. Hill [Hil38]. From engineering perspective, their (inner) product, *flow · effort*, represents the total system's *power*, equal to the time–rate–of–change of the total system's energy (included in H, R and F_i functions). And energy itself is transformed into the *work* done by the system.

Now, the reasonably accurate musculo–skeletal biodynamics would include say a hundred DOF, which means a hundred of joint angles and a hundred of joint momenta, which further means a hundred of coupled equations of the form of (1.156–1.157). And the *full coupling* means that each angle (and momentum) includes the information of all the other angles (and momenta), the *chain coupling* means that each angle (and momentum) includes the information of all the previous (i.e., children) angles (and momenta), the *nearest neighbor coupling* includes the information of the nearest neighbors, etc.

No matter which coupling we use for modelling the dynamics of human motion, one thing is certain: the *coupling is nonlinear*. And we obviously have to fight chaos within several hundreds of variables.

Wouldn't it be better if we could somehow be able to get a synthetic information about the whole musculo–skeletal dynamics, synthesizing the hundreds of equations of motion of type (1.156–1.157) into a small number of equations describing the time evolution of the so–called *order parameters*? If we could do something similar to principal component analysis in multivariate statistics and neural networks, to get something like 'nonlinear factor dynamics'?

Starting from the basic system (1.156–1.157), on the lowest, *microscopic level of human movement organization*, the *order parameter equations of macroscopic synergetics* can be (at least theoretically), either exactly derived along the lines of *mezoscopic synergetics*, or phenomenologically stated by the use of the certain biophysical analogies and nonlinear identification and control techniques (a highly complex nonlinear system like human locomotor apparatus could be neither identified nor controlled by means of standard linear engineering techniques).

For the self–organization problems in biodynamics, see section 4.4.4 below.

2

Smale Horseshoes and Homoclinic Dynamics

This Chapter is a continuation of the section 1.2.2 from Introduction. Here we give a review of current development of the Smale Horseshoe techniques of topological stretching, folding and squeezing of the phase–spaces of the chaotic systems (e.g., Duffing oscillator, see Figure 2.1), with the associated symbolic dynamics.

2.1 Smale Horseshoe Orbits and Symbolic Dynamics

Orbits of the Smale horseshoe map are described combinatorially in terms of their natural symbolic coding. Homoclinic orbits have a naturally associated trellis, which can be considered to be a subset of a homoclinic tangle, and provides a powerful framework for studying problems of forcing and braid equivalence. Each trellis has a natural 1D representative map, closely related to the train tracks for periodic orbits [CB88], and these give another convenient combinatorial representation and computational tool. Finally, by thickening graphs to get thick graphs, we get a link back to homoclinic and periodic orbits as described by braid types.

In this section, following [Col02, Col05], in which the orbits of the Smale horseshoe map were described combinatorially in terms of their natural symbolic coding, we present the model of the Smale horseshoe map $F \colon S^2 \to S^2$ (see Figure 2.2). The stadium–shaped domain shown, consisting of two hemispheres and a rectangle R, is mapped into itself as in an orientation-preserving way as indicated by the dotted lines, with a stable fixed–point a in the left hemisphere. The map is then extended to a homeomorphism of S^2 with a repelling fixed–point at ∞ whose basin includes the complement of the stadium domain. The saddle fixed–point of negative index (i.e., positive eigenvalues) is denoted p.

The non–wandering set $\Omega(F)$ consists of the fixed–points a and ∞, together with a Cantor set: $\Lambda = \{x \in S^2 : F^n(x) \in R \text{ for all } n \in \mathbb{Z}\}$.

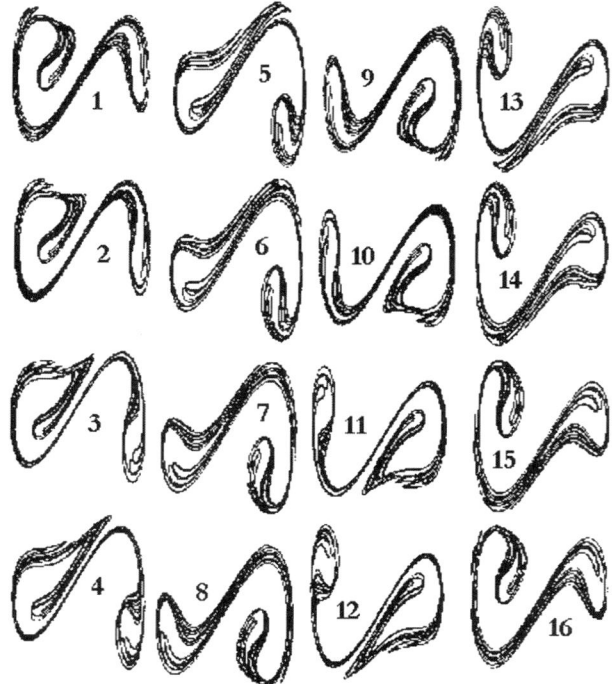

Fig. 2.1. Stroboscopic Poincaré sections (in order from 1 to 16) of a Duffing strange attractor. We can see phase–space stretching, folding and squeezing of the Duffing oscillator 1.37; simulated using *Dynamics Solver*^{TM}.

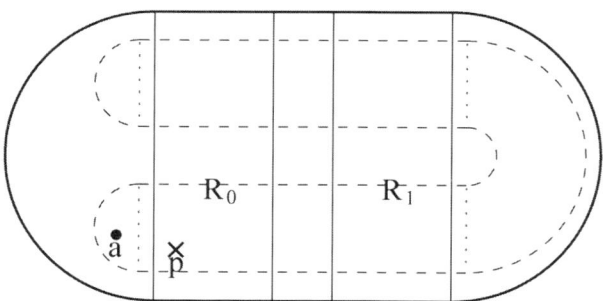

Fig. 2.2. The Smale horseshoe map.

Since Λ is contained in the union of the rectangles R_0 and R_1, *symbolic dynamics* can be introduced in the usual way, providing a so–called *itinerary homeomorphism*: $k\colon \Lambda \to \Sigma_2 = \{0,1\}^{\mathbb{Z}}$, with the property that $\sigma(k(x)) = k(F(x))$ for all $x \in \Lambda$ (where $\sigma\colon \Sigma_2 \to \Sigma_2$ is the shift map).

The itinerary $k(x)$ of a point x is periodic of period n iff x is a period n point of F. The following definition makes it possible to describe *periodic orbits* uniquely: The *code* $c_P \in \{0,1\}^n$ of a period n orbit P of F is given by the first n symbols of the itinerary of the rightmost point of P.

Since the uni–modal order on $\Sigma_2^+ = \{0,1\}^N$ reflects the normal horizontal ordering of points, the elements of $\{0,1\}^n$ which are codes of period n orbits are those which are *maximal* in the sense of the following definition: The *uni–modal order* $\prec$ on Σ_2^+ is defined as follows: if $s = s_0 s_1 \ldots, t = t_0 t_1 \ldots \in \Sigma_2^+$ have $s_n \neq t_n$, but agree on all earlier symbols, then $s \prec t$ iff $\sum_{i=0}^n s_i$ is even. A word $w \in \{0,1\}^n$ is *maximal* if $\sigma^j(\overline{w}) \prec \overline{w}$ for $1 \leq j < n$.

The author of [Col02, Col05] has also been concerned with orbits which are homoclinic to p, and the term *homoclinic orbit* will be used exclusively to mean such orbits. The points of a homoclinic orbit therefore have itineraries containing only finitely many 1s, and can thus be described as follows: Let H be a homoclinic orbit of the horseshoe. The *core* of H is the longest word in the itinerary of a point of H which begins and ends with 1

The *signature* of H is equal to the length of the core minus one. Thus, for example, the homoclinic orbit containing the point of itinerary $\overline{0}110 \cdot 0101\overline{0}$ has core 1100101 and signature 6. The *primary* homoclinic orbits are those with cores 1 and 11; these two orbits have the same homoclinic braid type, are forced by every other homoclinic orbit, but do not force any other periodic or homoclinic orbit. By contrast, the orbits with cores 111 and 101 will be shown to force all periodic and homoclinic orbits of the horseshoe [Han99].

Given a rational number $q = m/n \in (0,1/2]$ (with $(m,n) = 1$), define the *rational word* $c_q \in \{0,1\}^{n+1}$, by

$$c_{q,i} = \begin{cases} 1 \text{ if } \left(\frac{m(i-1)}{n}, \frac{m(i+1)}{n} \right) \text{ contains an integer,} \\ 0 \text{ otherwise,} \end{cases} \qquad (\text{for } 0 \leq i \leq n).$$

The word $c_{m/n}$ can be determined by drawing a line from $(0,0)$ to (n,m) in the plane. There is a 1 in the ith position if the line crosses an integer value in $i - 1 < x < i + 1$. Note in particular that, with the exception of the initial and final symbols, 1s always occur in blocks of even length. Using these rational words, it is possible to define the *height* of a horseshoe periodic orbit: this is a braid type invariant taking rational values in $(0, 1/2]$. The next result [Hal94] motivates the definition: let q and r be rationals in $(0, 1/2)$ with $q < r$. Then $c_q 0, c_q 1, c_r 0$, and $c_r 1$ are maximal, and

$$\overline{c_r 0} \prec \overline{c_r 1} \prec \overline{c_q 0} \prec \overline{c_q 1}.$$

Let P be a horseshoe periodic orbit with code c_P. Then the *height* $q(P) \in [0, 1/2]$ of P is given by

$$q(P) = \inf\{q \in \mathbb{Q} \cap (0, 1/2] : q = 1/2 \text{ or } \overline{c_q 0} \prec \overline{c_P}\}.$$

That is, the itineraries $\overline{c_q 0}$ are uni–modally ordered inversely to q: the height of P describes the position in this chain of the itinerary of the rightmost point of P. Although it is not obvious from the definition, $q(P)$ is always a strictly positive rational, which can be computed algorithmically (see [Hal94] for details).

There are several important classes of orbits which can be defined in terms of the rational words:

- The *rotation-compatible* periodic orbits of rotation number m/n are period n orbits whose codes agree with $c_{m/n}$ up to the final symbol. Thus, for example, the rotation-compatible orbits of rotation number $3/10$ have codes 100110110_1^0. The two rotation-compatible orbits of a given rotation number have the same braid type (this is generally true for orbits whose codes differ only in their final symbol), and force nothing but a fixed–point.

- The *no bogus transition (NBT)* periodic orbits of height m/n are the period $n+2$ orbits whose codes start with $c_{m/n}$. Thus, for example, the NBT orbits of height $3/10$ have codes 100110110011_1^0. These orbits force all periodic orbits dictated by the uni–modal order, that is, all periodic orbits P with $\overline{c_P} \prec \overline{c_{m/n}^0 1}$. In particular, by the definition of height, they force all periodic orbits of height greater than m/n.

- The *star* homoclinic orbits are those whose core is equal to $c_{m/n}$ for some m/n.

A conjectural description [CH01] of the structure of the forcing relation on the set of homoclinic and periodic orbits of the horseshoe can be given in terms of the *decorations* of the orbits. The definition depends on the following result [Hal94]: Let P be a horseshoe periodic orbit of height m/n which is not rotation–compatible. Then P has period at least $n + 2$, and c_P has $c_{m/n}$ as an initial word.

Let P be a period k horseshoe orbit of height $q = m/n$ which is not rotation–compatible or NBT. Then the *decoration* of P is the word $w \in \{0,1\}^{k-n-3}$ such that $c_P = c_{q1}^0 w_1^0$. (The empty decoration is denoted $\cdot$). The decoration of a rotation–compatible orbit is defined to be $\circlearrowleft$, and that of an NBT orbit to be $*$.

Let H be a homoclinic horseshoe orbit whose core c_H has length $k \geq 4$. Then the *decoration* of H is the word $w \in \{0,1\}^{k-4}$ such that $c_H = 1_1^0 w_1^0 1$. The decoration of the primary homoclinic orbits (with cores 1 and 11) is defined to be $\circlearrowleft$, and that of the homoclinic orbits with cores 111 and 101 is defined to be $*$.

A periodic orbit of height q and decoration w is denoted P_q^w: thus $c_{P_q^w} = c_{q_1^0} w_1^0$, provided that $w \neq \circlearrowleft, *$. The notation is justified by the result of [CH03], that all of the (four or fewer) periodic orbits of height q and decoration w have the same braid type. A homoclinic orbit of decoration w is denoted P_0^w: again, all the homoclinic orbits with the same decoration have the same homoclinic braid type [CH02].

Note that a decoration w may not be compatible with all heights, since $c_{q_1^0} w_1^0$ may not be a maximal word when q is large. For example, the word $c_q 0 w 0$ with $q = 2/5$, $c_q = 101101$ and $w = 1001$ is not maximal, since the periodic sequence $\overline{101101010010}$ has maximal word 100101011010 with $q = 1/3$, $c_q = 1001$ and $w = 101101$.

The *scope* q_w of a decoration w is defined to be the supremum of the values of q for which a periodic orbit of height q and decoration w exists.

The following result is from [CH02]: Let w be a decoration. If $w = \circlearrowleft$ or $w = *$, then $q_w = 1/2$. Otherwise, q_w is the height of the periodic orbit containing a point of itinerary $\overline{10w0}$. If $0 < q < q_w$, then each of the four words $c_{q_1^0} w_1^0$ (or each of the two appropriate words when $w = \circlearrowleft$ or $w = *$) is the code of a height q periodic orbit, while if $1/2 \geq q > q_w$, then none of the words is the code of a height q periodic orbit.

Thus, the set $\mathcal{D}_w$ of periodic and homoclinic orbits of the horseshoe of decoration w is given by

$$\mathcal{D}_w = \{P_q^w : 0 \leq q <= q_w\},$$

where the notation $<=$ indicates that $q = q_w$ is possible for some decorations but not for others. Moreover, the union of the sets $\mathcal{D}_w$ is the set of all periodic and homoclinic orbits of the horseshoe.

Given decorations w and w', write $w \succeq w'$ if the homoclinic orbit P_0^w forces the homoclinic orbit $P_0^{w'}$; and write $w \sim w'$ if the two homoclinic orbits have the same homoclinic braid type. Then we have:

1. If $q < q'$ and $w \succeq w'$, then P_q^w forces $P_{q'}^{w'}$.
2. P_q^w and $P_{q'}^{w'}$ have the same braid type iff $q = q'$ and $w \sim w'$.

In particular, 1. implies that each family $\mathcal{D}_w$ of orbits with a given decoration is linearly ordered by forcing, the order being the reverse of the usual order on heights. If this conjecture is true, then determining whether or not one horseshoe periodic orbit forces another, and whether or not two horseshoe periodic orbits have the same braid type, depends on being able to carry out the corresponding computations for homoclinic orbits.

2.1.1 Horseshoe Trellis

Let $f : S^2 \longrightarrow S^2$ be a diffeomorphism, and p be a hyperbolic saddle fixed–point of f. Then a *trellis for f (at p)* is a pair $T = (T^U, T^S)$, where T^U and T^S are intervals in $W^U(f; p)$ and $W^S(f; p)$ respectively containing p. (Here, $W^U(f; P)$ and $W^S(f; P)$ denote the unstable and stable manifolds, respectively, of f at p.) Given a trellis $T = (T^U, T^S)$, denote by T^V the set of intersections of T^U and T^S. The trellis is *transverse* if all of its intersection points are transverse. All trellises considered in this section are transverse, hence the word *trellis* will be understood to mean *transverse trellis*.

Let $T = (T^U, T^S)$ be a trellis. A *segment* of T is a closed subinterval of either T^U or T^S with endpoints in T^V but interior disjoint from T^V. The segment is called *unstable* or *stable* according as it is a subinterval of T^U or of T^S. Let $T = (T^U, T^S)$ be a trellis. Then a *region* of T is the closure of a component of $S^2 \setminus (T^U \cup T^S)$.

A *bigon* of T is a region bounded by two segments (one unstable and one stable). Let p be the fixed–point of the horseshoe map F with code 0, and let $W^U(F;p)$ and $W^S(F;p)$ be the unstable and stable manifolds of p. Let $Q = (\ldots, q_{-2}, q_{-1}, q_0, q_1, q_2, \ldots)$ be the homoclinic orbit of F with code $\overline{0}1\overline{0}$, with the points labelled in such a way that the itinerary $k(q_i)$ of q_i is $\sigma^i(\overline{0} \cdot 1\overline{0})$. Then $W^U(F;p)$ passes successively through the points $\ldots, q_{-2}, q_{-1}, q_0, q_1, q_2, \ldots$, while $W^S(F;p)$ passes successively through the points $\ldots, q_2, q_1, q_0, q_{-1}, q_{-2}, \ldots$.

Given $i \geq 0$ and $j \leq 0$, denote by $\overline{T}_i^U$ an interval in $W^U(F;p)$ with end intersections p and q_i, and by T_j^S the interval in $W^S(F;p)$ with endpoints p and q_j. For $n \geq 0$, a *full horseshoe trellis of signature n* is a trellis $T = (T^U, T^S)$ for F, where $T^S = T_j^S$ for some j, and T^U is a closed neighborhood of T_i^U in $W^U(F;p)$ for $i = j + n$ such that all intersections of T^U with T^S lie T_i^U. It is clear that all full horseshoe trellises as defined above are differentiably–conjugate; for definiteness, we will usually choose either $i = 1$ or $i = \lfloor n/2 \rfloor$. We see that the endpoints of T^S are intersection points, but the endpoints of T^U are not.

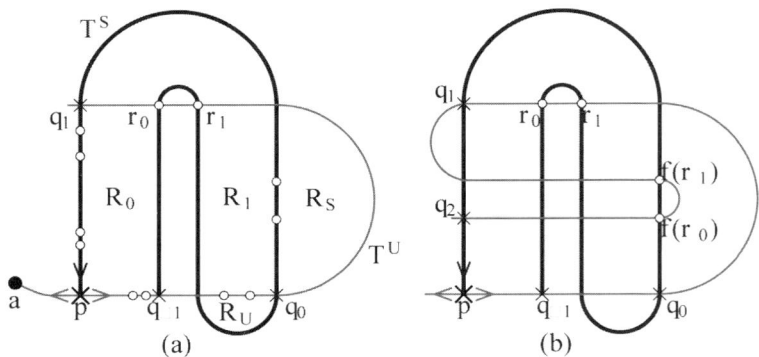

Fig. 2.3. The full horseshoe trellis: (a) with signature 2; (b) with signature 3 (adapted from [Col02, Col05]).

For example, the full horseshoe trellis of signature 2 is depicted in Figure 2.3(a). The chaotics dynamics is supported in the regions labelled R_0 and R_1. All points in R_U are in the basin of the attracting fixed–point a, and all points in the interior of R_U are in the basin of the repelling point at infinity. The point r_0 has itinerary $\overline{01} \cdot 01\overline{0}$, and the point r_1 has itinerary $\overline{01} \cdot 11\overline{0}$. The full horseshoe trellis of signatures 3 is depicted in Figure 2.3(b).

The regions of a trellis can be used to introduce symbolic dynamics: Let f be a diffeomorphism with trellis T. Then a bi–infinite sequence

$$\ldots R_{-2} R_{-1} R_0 R_1 R_2 \ldots$$

of regions of T is an *itinerary* for an orbit (x_i) of f if $x_i \in R_i$ (for all i).

Given a trellis T for a diffeomorphism f, a *pruning isotopy* is an isotopy which removes the intersections on the boundary of one or more bigons of F. To be more precise, it is an isotopy from f to a diffeomorphism f' which has a trellis T' got from T by removing such intersections. There are two possibilities; we can either remove both intersections of a single bigon, or remove intersections from two neighboring bigons, changing the orientation of the crossing at the remaining intersection. However, an isotopy of the diffeomorphism f supported in some open set U will also change the trellis outside of U. If we are trying to reduce the number of intersections of T, we need to ensure that no other intersections are created when we remove intersections locally. This gives rise to the notion of an *inner bigon*.

A bigon B is *inner* if $B \cap \bigcup_{n \in \mathbb{Z}} f^n(T^V) = B \cap T^V$. i.e., A bigon is B inner if the only intersections of B with the orbits of the intersection points of the trellis are the vertices of B.

Let T be a trellis of a diffeomorphism f. Suppose either that B is an inner bigon with vertices v_0 and v_1, or that B_0 and B_1 are inner bigons with a common vertex v and other vertices v_0 and v_1 on different orbits. Then there is a diffeomorphism $h : S^2 \to S^2$, which we can take to be supported on a neighborhood U of B or $B_0 \cup B_1$, such that $f' = f \circ h$ has a trellis T' with the same intersections apart from those on the orbits of v_0 and v_1 under f.[1]

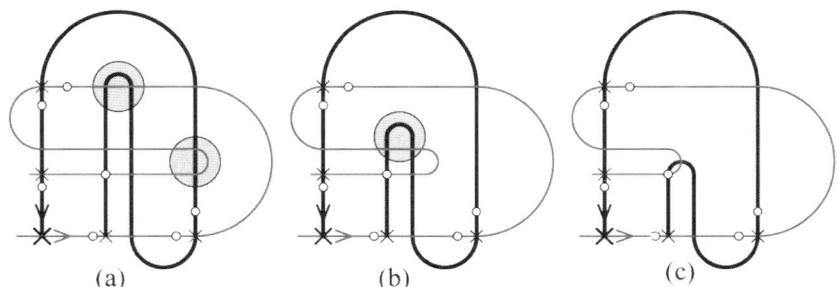

Fig. 2.4. Pruning away bigons in the horseshoe trellis (adapted from [Col02, Col05]).

For example, Figure 2.4 (a) depicts the full horseshoe trellis of signature 3, and a shaded neighborhood U of a bigon B, together with its image. Pruning

---

[1] Note that f' is isotopic to f, and has a trellis T' obtained by removing all the intersections of T contained in the orbit of U.

away the bigon B yields a diffeomorphism f' with trellis T' as shown in (b). A further pruning isotopy yields a diffeomorphism with the trellis depicted in (c).

A trellis got by pruning the full horseshoe trellis as in this example is called a horseshoe trellis. More precisely, a *horseshoe trellis* is a trellis T got from the full horseshoe trellis by pruning away a sequence of bigons. A horseshoe trellis can be associated to each homoclinic orbit of the horseshoe, by pruning away as many bigons as possible without touching the homoclinic orbit. It is trivial that the signature of a horseshoe homoclinic orbit H is equal to the least integer n such that H is an intersection of the full horseshoe trellis of signature n.

Let H be a horseshoe homoclinic orbit of signature n. The *trellis forced by H* is the trellis T got from the full horseshoe trellis of signature n by pruning away as many bigons as possible which do not contain a point of H.

For example, the white circles in Figure 2.4 represent points of the homoclinic orbit H with code $\bar{0}10010\bar{0}$ (which thus has signature 3). The trellis of Figure 2.4c) is thus the trellis forced by this homoclinic orbit. Note that every bigon has a point of H on its boundary.

This method makes it possible to determine whether or not two horseshoe homoclinic orbits have the same homoclinic braid type: Let T and T' be horseshoe trellises for diffeomorphism f and f' respectively. We say that $(f;T)$ and $(f';T')$ have the same *trellis type* if there is a diffeomorphism g isotopic to f relative to T, and a homeomorphism $h : S^2 \to S^2$ such that $h(T) = T'$ and $h^{-1} \circ f' \circ h = g$.

Following [Col02, Col05], we denote the trellis type containing $(f;T)$ by $[f;T]$. For horseshoe trellises, the trellis type is determined by the geometry of the trellis: Let T and T' be horseshoe trellises for diffeomorphism f and f' respectively. Then $(f;T)$ and $f';T')$ have the same trellis type iff T and T' are diffeomorphic. Since a horseshoe trellis type is fully determined by the geometry of the trellis, we define the type of a horseshoe trellis T to be the type of $(f;T)$ for any diffeomorphism f with trellis T which can be obtained by pruning away bigons.

Horseshoe trellises $(f;T)$ and $(f';T')$ have the same type iff the trellises T and T' are homeomorphic, and this occurs iff the orderings of the intersections on the stable and unstable manifolds are the same. Let $(f;T)$ be a horseshoe trellis, with intersections $T^V = \{v_i : i = 0 \ldots n-1\}$ such that $v_i <_u v_{i+1}$ (i.e., v_i is closer to p along the unstable manifold). Then the *relative ordering* of the unstable and stable manifolds is the permutation π_T such that $v_{\pi_T(i)} <_s v_{\pi_T(j)}$ iff $\pi_T(i) < \pi_T(j)$.

The following result gives a computable criterion for the equivalence of horseshoe trellises. Let T and T' be horseshoe trellises. Then T and T' have the same trellis type iff $\pi_T = \pi_{T'}$.

In particular, homoclinic orbits can only have the same braid type if they have the same signature (i.e., their cores have the same length). More precisely, let H and H' be horseshoe homoclinic orbits of signatures n and n', and let

T and T' be the trellises of signature $m \geq \max\{n, n'\}$ forced by them. Then H and H' have the same homoclinic braid type iff T and T" have the same trellis type.

2.1.2 Trellis–Forced Dynamics

The reason for the terminology 'trellis *forced* by a homoclinic orbit H' is that any diffeomorphism having a homoclinic orbit of the homoclinic braid type of H has a fixed–point and an associated trellis of the given trellis type. Moreover, it is straightforward to compute the dynamics forced by a trellis map, using techniques similar to those of Bestvina–Handel [BH95] to represent this forced dynamics by a graph map. It follows that the dynamics of this graph map is forced by the homoclinic orbit.

Let T be a horseshoe trellis, and G be a tree embedded in S^2. Then G is *compatible* with T if

1. G is disjoint from T^U.
2. G intersects each segment of T^S exactly once. The vertices of G are disjoint from T^S, and each edge of G intersects T^S at most once.

The edges containing points of W are called *control edges*. Let $(f; T)$ be a horseshoe trellis map, and let G be a tree compatible with T. Let $W = G \cap T^S$. A tree map $g \colon (G, W) \longrightarrow (G, W)$ is *compatible with* $(f; T)$ if

1. $g(w_i) = w_j$ whenever $w_i, w_j \in W$ are the intersections of G with stable segments S_i, S_j satisfying $f(S_i) \subseteq S_j$.
2. g maps each control edge to a control edge.

Let $(f; T)$ be a horseshoe trellis map, and $g \colon (G, W) \to (G, W)$ be a tree map compatible with $(f; T)$. Then g is the *tree representative* of $(f; T)$ if

1. Every valence 1 or 2 vertex of G is the endpoint of a control edge.
2. g is locally injective away from control edges (i.e., every $x \in G$ has a neighborhood U in G such that if y_1, y_2 are distinct points of U with $g(y_1) = g(y_2)$, then at least one of y_1 and y_2 lies in a control edge).
3. g is piecewise linear on each edge of G.

An algorithm for computing the tree representative of a horseshoe trellis map (and, more generally, the tree representative of an arbitrary trellis map) can be found in [Col02]. The reason for the use of control edges is technical, and primarily concerned with the details of the algorithm. Since the set of control edges is invariant under the tree representative, and since the concern here is with non-wandering dynamics, the tree representative can be simplified by collapsing control edges to points.

Let $g \colon (G, W) \to (G, W)$ be the tree representative of a horseshoe trellis map $(f; T)$. Then the *restricted tree representative* of $(f; T)$ is $g \colon (\tilde{G}, \tilde{W}) \to (\tilde{G}, \tilde{W})$, where $\tilde{G} = \bigcap_{n=0}^{\infty} g^n(G)$, and $\tilde{W} = \tilde{G} \cap W$. The *topological tree representative* is got by collapsing all control edges to points.

The restricted tree representative contains all of the non-wandering points of the tree representative, and so carries all of its topological entropy. The topological tree representative is essentially unique: Let H and H' be horseshoe homoclinic orbits, and let g and g' be the associated topological tree representatives. Then g and g' are conjugate iff H and H' have the same homoclinic braid type.

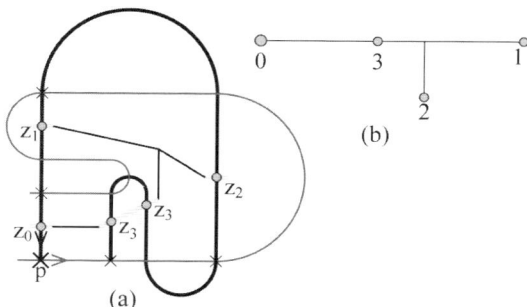

Fig. 2.5. (a) The restricted tree representative of the trellis of signature 3 forced by $\overline{0}1111\overline{0}$. (b) The topological tree representative of the orbit (adapted from [Col02, Col05]).

For example, Figure 2.5(a) depicts the restricted tree representative of the trellis of signature 3 forced by the homoclinic orbit $\overline{0}1111\overline{0}$. The control edges are labelled z_0, z_1, z_2, z_3 and z_3', and map under g as

$$z_3, z_3' \mapsto z_2 \mapsto z_1 \mapsto z_0 \mapsto z_0.$$

The topological tree representative of the orbit is shown in Figure 2.5(b).

The following conventions are used in labelling topological tree representatives. The fixed–point corresponding to the fixed–point p of the horseshoe is labelled 0, and a preimage x of p is labelled with the least integer n satisfying $g^n(x) = p$. Thus, each point labelled n is mapped to a point labelled $n - 1$. Where necessary, we use primes to distinguish nth preimages. We do not necessarily label all nth preimages of v_0, but will always label points of the topological tree representative which are valence 1 vertices or at the *fold vertices* at which the tree map is not locally injective. In many cases, this labelling alone is enough information to determine the tree map g.

For example, Figure 2.6 shows the trellis forced by $\overline{0}1_1^0 11_1^0 1\overline{0}$ and its topological tree representative. The fold points and valence 1 vertices are marked with black dots.

Let f be a diffeomorphism with a horseshoe trellis T which has restricted tree representative g. Then we have [Col02, Col05]:

1. For every orbit of g there is an orbit of f with the same itinerary.

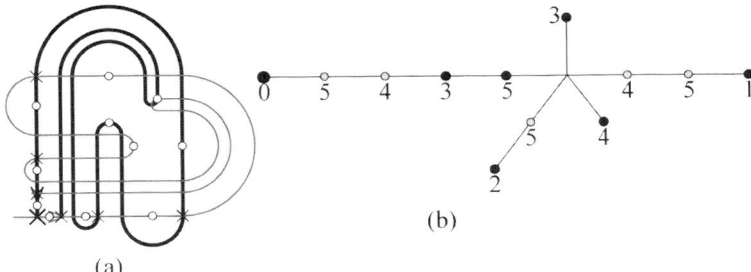

Fig. 2.6. Trellis forced by $\overline{0}1111110\overline{0}$ and its topological tree representative (adapted from [Col02, Col05]).

2. For each periodic orbit of g, there exists a periodic orbit of f with the same period and itinerary.
3. If Y is an orbit of g which is homoclinic to the fixed control edge z_0 then there is a homoclinic orbit X of f with the same itinerary as Y.
4. $h_{\text{top}}(f) \geq h_{\text{top}}(g)$.

These results also hold for the topological tree representative, since the orbits of the restricted tree representative project to the topological tree representative.

The previous theorem shows that the topological tree representative of a trellis type give a good description of the orbits up to itinerary, but we often also want information about the braid type of the orbits, which is not directly given by the tree representative. Following Franks and Misiurewicz [FM93] we 'thicken' the topological tree representative to get the *thick tree representative*. The *thick graph* $\widehat{G} \subset \mathbb{R}^2$ is a set consisting of *thick vertices* $\widehat{V}$ and *thick edges* $\widehat{E}$. There is a bijection $\widehat{\cdot}$ between vertices and edges of G and thick vertices and thick edges of $\widehat{G}$, a projection $\pi : \widehat{G} \to G$ taking $\widehat{v}$ to v and $\widehat{e}$ to e which preserves the stable leaves. There is also an embedding $i : G \hookrightarrow \widehat{G}$ such that $i(v)$ lies in the interior of $\widehat{v}$, and $i(e)$ intersects $\widehat{e}$ in an unstable leaf, and intersects no other thick edges of $\widehat{G}$.

A *thick tree representative* $\widehat{g}$ of g is an embedding $g : \mathbb{R}^2 \to \mathbb{R}^2$ such that $\widehat{G}$ is a global attractor of $\widehat{g}$. $\widehat{g}$ is a contraction on the set of thick vertices, uniformly contracts leaves of the stable foliation of the set of thick edges $\widehat{E}$ and uniformly expands the unstable foliation of $\widehat{E}$. Further, $\widehat{g}(\widehat{v}) \subset \widehat{g(v)}$ for each vertex v of G, and $\widehat{g}(\widehat{e})$ crosses the same thick edges of $\widehat{G}$ in the same order as $g(e)$ crosses edges of g. The thick tree representative is unique up to topological conjugacy.

Not all maps defined on trees embedded in S^2 give rise to thick tree maps, as edges may cross each other, resulting in a lack of embedding. However, it is always possible to thicken the topological tree representative of a trellis type, and hence get the *thick tree representative*. It can be shown, though we do not do so here, that a horseshoe trellis type T with a single homoclinic forcing

the nonwandering set. In particular, all the mentioned properties hold for a dense set of C^1 vector–fields on M. It is interesting to observe that neither structural stability nor Axiom A is dense in the space of C^1 vector–fields on M, for all $n \geq 3$.

To state our results in a precise way we use the following notation. M is a compact boundaryless n–manifold, $\mathcal{X}^1(M)$ is the space of C^1 vector–fields endowed with the C^1 topology. Given $X \in \mathcal{X}^1(M)$, X_t denotes the flow induced by X. The ω–limit set of p is the set $\omega_X(p)$ of accumulation points of the positive orbit of p. The α–limit set of p is $\alpha_X(p) = \omega_{-X}(p)$, where $-X$ denotes the time–reversed flow of X. The nonwandering set $\Omega(X)$ of X is the set of p such that for every neighborhood U of p and $T > 0$ there is $t > T$ such that $X_t(U) \cap U \neq \emptyset$. Clearly $\Omega(X)$ is closed, nonempty and contains any ω–limit (α–limit) set. A compact invariant set B of X is Ω–*isolated* if $\Omega(X) \setminus B$ is closed. B is *isolated* if $B = \cap_{t \in \mathbb{R}} X_t(U)$ for some compact neighborhood U of B (in this case U is called *isolating block*). We denote by $\mathrm{Per}(X)$ the union of the periodic orbits of X and $\mathrm{Crit}(X)$ the set formed by the union of $\mathrm{Per}(X)$ and the singularities of X.

A set is *transitive* for X if it is the ω–limit set of one of its orbits. A transitive set Λ of X is *maximal transitive* if it contains every transitive set T of X satisfying $\Lambda \cap T \neq \emptyset$. Note that a maximal transitive set is maximal with respect to the inclusion order. In [BD99, BDP99] it was asked whether every homoclinic class $H_f(p)$ of a generic diffeomorphism f satisfies the property that if T is a transitive set of f and $p \in T$, then $T \subset H_f(p)$. M. Arnaud [Arn99] also considered homoclinic classes for C^1 diffeomorphisms on M, and in particular she gives a positive answer to this question [Arn99]. On the other hand, item (i) of the theorem below states that generically any transitive set of a C^1 vector–field intersecting the homoclinic class is included in it, and thus the diffeomorphism version of it extends this result of Arnaud.

If Λ is a compact invariant set of X, we denote

$$W^s_X(\Lambda) = \{q \in M : \mathrm{dist}(X_t(q), \Lambda) \longrightarrow 0, t \longrightarrow \infty\}, \qquad \text{and}$$

$$W^u_X(\Lambda) = \{q \in M : \mathrm{dist}(X_t(q), \Lambda) \longrightarrow 0, t \longrightarrow -\infty\},$$

where dist is the metric on M. These sets are called respectively the stable and unstable set of Λ. We shall denote $W^s_X(p) = W^s_X(\mathcal{O}_X(p))$ and $W^u_X(p) = W^u_X(\mathcal{O}_X(p))$ where $\mathcal{O}_X(p)$ is the orbit of p. We say that Λ is *saturated* if $W^s_X(\Lambda) \cap W^u_X(\Lambda) = \Lambda$.

A *cycle* of X is a finite set of compact invariant sets $\Lambda_0, \Lambda_1, \ldots, \Lambda_n$ such that $\Lambda_n = \Lambda_0$, and $\Lambda_0, \Lambda_1, \ldots, \Lambda_{n-1}$ are disjoint, and

$$(W^u_X(\Lambda_i) \setminus \Lambda_i) \cap (W^s_X(\Lambda_{i+1}) \setminus \Lambda_{i+1}) \neq \emptyset$$

for all $i = 0, \ldots, n - 1$.

A compact invariant set Λ of X is *hyperbolic* if there is a continuous tangent bundle decomposition $E^s \oplus E^X \oplus E^u$ over Λ such that E^s is contracting, E^u

is expanding and E^X denotes the direction of X. We say that $p \in \mathrm{Crit}(X)$ is hyperbolic if $\mathcal{O}_X(p)$ is a hyperbolic set of X.

Recall that the *stable manifold theorem* [HPS70] asserts that $W^s_X(p)$ is an immersed manifold tangent to $E^s \oplus E^X$ for every p in a hyperbolic set Λ of X. Similarly for $W^u_X(p)$. This remark applies when $\Lambda = \mathcal{O}_X(p)$ for some $p \in \mathrm{Crit}(X)$ hyperbolic. As already defined, the homoclinic class associated to a hyperbolic periodic orbit p of X, $H_X(p)$, is the closure of the transversal intersection orbits in $W^s_X(p) \cap W^u_X(p)$.

We say that X is *Axiom A* if $\Omega(X)$ is both hyperbolic and the closure of $\mathrm{Crit}(X)$. The non wandering set of a nonsingular Axiom A flow splits in a finite disjoint union of homoclinic classes [PT93].

Another interesting property of homoclinic classes for Axiom A vector–fields is their continuous dependence on the periodic orbit data, that is, the map $p \in \mathrm{Per}(X) \longrightarrow H_X(p)$ is upper–semi–continuous.

In general, we say that a compact set sequence Λ_n *accumulates* on a compact set Λ if for every neighborhood U of Λ there is $n_0 > 0$ such that $\Lambda_n \subset U$ for all $n \geq n_0$. Note that this kind of accumulation is weaker than the usual Hausdorff metric accumulation.

If $\mathcal{Y}$ denotes a metric space, then $\mathcal{R} \subset \mathcal{Y}$ is *residual* in $\mathcal{Y}$ if $\mathcal{R}$ contains a countable intersection of open–dense subsets of $\mathcal{Y}$. Clearly a countable intersection of residual subsets of $\mathcal{Y}$ is a residual subset of $\mathcal{Y}$. For example, the set of *Kupka–Smale vector–fields* $\mathcal{KS}^1(M)$ on M is a residual subset of $\mathcal{X}^1(M)$ [MP92]. Recall that a vector–field is Kupka–Smale if all its periodic orbits and singularities are hyperbolic and the invariant manifolds of such elements intersect transversally.

The following properties hold for a residual subset of vector–fields X in $\mathcal{X}^1(M)$ [CMP00]:

(i) The homoclinic classes of X are maximal transitive sets of X. In particular, different homoclinic classes of X are disjoint.

(ii) The homoclinic classes of X are saturated.

(iii) The homoclinic classes of X depends continuously on the periodic orbit data, that is, the map $p \in \mathrm{Per}(X) \rightarrow H_X(p)$ is upper–semi–continuous.

(iv) A homoclinic class of X is isolated if and only if it is Ω-isolated.

(v) The hyperbolic homoclinic classes of X are isolated.

(vi) There is no cycle of X formed by homoclinic classes of X.

(vii) X has finitely many homoclinic classes if and only if the union of the homoclinic classes of X is closed and every homoclinic class of X is isolated.

When M has dimension three we obtain the following corollaries using the above theorem, [Lia83, Man82]. Recall that an isolated set Λ of a C^r vector–field X is C^r robust transitive $(r \geq 1)$ if it exhibits an isolating block U such that, for every vector–field Y C^r close to X, $\cap_{t \in \mathbb{R}} Y_t(U)$ is both transitive and nontrivial for Y.

The following properties are equivalent for a residual set of nonsingular 3-dimensional C^1 vector–fields X and every nontrivial homoclinic class $H_X(p)$ of X:

(i) $H_X(p)$ is hyperbolic.

(ii) $H_X(p)$ is isolated.

(iii) $H_X(p)$ is C^1 robust transitive for X.

The following properties are equivalent for a residual set of nonsingular 3-dimensional C^1 vector–fields X:

(i) X is Axiom A.

(ii) X has finitely many homoclinic classes.

(iii) The union of the homoclinic classes of X is closed and every homoclinic class of X is isolated.

The equivalence between the Items (i) and (ii) of the above corollary follows from [Lia83, Man82]. It shows how difficult is to prove the genericity of vector–fields exhibiting finitely many homoclinic classes.

2.2.1 Lyapunov Stability

Here we establish some useful properties of *Lyapunov stable sets*. A reference for Lyapunov stability theory is [BS70].

Recall we have denoted by X_t, $t \in I\!\!R$ the flow generated by $X \in \mathcal{X}^1(M)$. Given $A \subset M$ and $R \subset I\!\!R$ we set $X_R(A) = \{X_t(q) : (q, t) \in A \times R\}$. We denote $\mathrm{Cl}(A)$ the closure of A, and $\mathrm{int}(A)$ the interior of A. If $\epsilon > 0$ and $q \in M$ we set $B_\epsilon(q)$ the ϵ–ball centered at q.

A compact subset $A \subseteq M$ is *Lyapunov stable* for X if for every open set U containing A there exists an open set V containing A such that $X_t(V) \subset U$ for every $t \geq 0$. Clearly a Lyapunov stable set is forward invariant.

The following lemma summarizes some classical properties of Lyapunov stable sets [BS70]:

Let Λ^+ be a Lyapunov stable set of X. Then,

(i) If $x_n \in M$ and $t_n \geq 0$ satisfy $x_n \to x \in \Lambda^+$ and $X_{t_n}(x_n) \to y$, then $y \in \Lambda^+$;

(ii) $W_X^u(\Lambda^+) \subset \Lambda^+$;

(iii) if Γ is a transitive set of X and $\Gamma \cap \Lambda^+ \neq \emptyset$, then $\Gamma \subset \Lambda^+$.

We are interested in invariant compact sets $\Lambda = \Lambda^+ \cap \Lambda^-$ of X, where Λ^+ is Lyapunov stable set for X and Λ^- is Lyapunov stable set for the reversed flow $-X$. We shall call such sets *neutral* for the sake of simplicity. As we shall see in the next section, homoclinic classes are neutral sets for generic C^1 vector–fields on closed n–manifolds.

Elementary properties of neutral sets are:

Let Λ be a neutral set of X. Then,

(i) Λ is saturated;

(ii) Λ is transitive for X if and only if Λ is maximal transitive for X. In particular, different transitive neutral sets of X are disjoint.

Let $\Lambda = \Lambda^+ \cap \Lambda^-$ with $\Lambda^\pm$ being Lyapunov stable for $\pm X$. Clearly $W_X^u(\Lambda) \subset \Lambda^+$. Similarly, $W_X^s(\Lambda) \subset \Lambda^-$. Hence

$$W_X^u(\Lambda) \cap W_X^s(\Lambda) \subset \Lambda^+ \cap \Lambda^- = \Lambda.$$

Conversely, $\Lambda \subset W^u_X(\Lambda) \cap W^s_X(\Lambda)$ since Λ is invariant. Now, if Γ is a transitive set intersecting Λ, then $\Gamma \subset \Lambda^+$ and $\Gamma \subset \Lambda^-$. Thus, $\Gamma \subset \Lambda^+ \cap \Lambda^- = \Lambda$, and so, Λ is maximal transitive. The converse is obvious. Different transitive neutral sets of X are maximal transitive, and so, they are necessarily disjoint.

Note that a *Smale horseshoe* with a first tangency is an example of a maximal transitive set which is not neutral. This example also provides a hyperbolic homoclinic class which is not neutral.

There is no cycle of X formed by transitive neutral sets [CMP00].

By contradiction, suppose that there exists a cycle $\Lambda_0, \ldots, \Lambda_n$ of X such that every Λ_i is a transitive neutral set of X. Recall $\Lambda_n = \Lambda_0$.

Set $\Lambda_i = \Lambda_i^+ \cap \Lambda_i^-$ where each $\Lambda_i^\pm$ is Lyapunov stable for $\pm X$. Choose

$$x_i \in (W^u_X(\Lambda_i) \setminus \Lambda_i) \cap (W^s_X(\Lambda_{i+1}) \setminus \Lambda_{i+1})$$

according to the definition.

We claim that $x_i \in \Lambda_0^-$ for every i. Indeed, as $W^s_X(\Lambda_0) \subset \Lambda_0^-$ one has $x_{n-1} \in \Lambda_0^-$. Assume by induction that $x_i \in \Lambda_0^-$ for some i. As $x_i \in W^u_X(\Lambda_i)$, the backward invariance of Λ_0^- implies

$$\Lambda_0^- \cap \Lambda_i \supset \alpha_X(x_i) \neq \emptyset.$$

If Λ is neutral for X, then for every neighborhood U of Λ there exists a neighborhood $V \subset U$ of Λ such that

$$\Omega(X) \cap V \subset \cap_{t \in \mathbb{R}} X_t(U).$$

Let U be a neighborhood of a neutral set Λ of X. Choose $U' \subset \mathrm{Cl}(U') \subset U$ with U' being another neighborhood of Λ. Then there is a neighborhood $V \subset U'$ of Λ so that [CMP00]:
  (1) $t \geq 0$ and $p \in V \cap X_{-t}(V) \implies X_{[0,t]}(p) \subseteq U'$.
  (2) $t \leq 0$ and $p \in V \cap X_t(V) \implies X_{[-t,0]}(p) \subseteq U'$.

Indeed, it were not true then there would exist a neighborhood U of Λ and sequences $p_n \to \Lambda$, $t_n > 0$ such that $X_{t_n}(p_n) \to \Lambda$ but $X_{[0,t_n]}(p_n) \not\subseteq U'$. Choose $q_n \in X_{[0,t_n]}(p_n) \setminus U'$. Write $q_n = X_{t'_n}(p_n)$ for some $t'_n \in [0, t_n]$ and assume that $q_n \to q$ for some $q \notin U'$. Let $\Lambda = \Lambda^+ \cap \Lambda^-$ with $\Lambda^\pm$ Lyapunov stable for $\pm X$. Since Λ^+ is Lyapunov stable for X and $t'_n > 0$, we have that $q \in \Lambda^+$. On the other hand, as we can write $q_n = X_{t'_n - t_n}(X_{t_n}(p_n))$ where $t'_n - t_n > 0$ and $X_{t_n}(p_n) \to \Lambda$, we have that $q \in \Lambda^-$. This proves that $q \in \Lambda$, a contradiction since $q \notin U'$.

Next we prove that $\Omega(X) \cap V \subseteq \cap_{t \in \mathbb{R}} X_t(U)$. Indeed, choose $q \in \Omega(X) \cap V$. By contradiction, we assume that there is $t_0 > 0$ (say) such that $X_{t_0}(q) \notin U$. Then, there is a ball $B_\epsilon(q) \subseteq V$ such that $X_{t_0}(B_\epsilon(q)) \cap \mathrm{Cl}(U') = \emptyset$. As $q \in \Omega(X)$ there exists $t > t_0$ such that $B_\epsilon(q) \cap X_{-t}(B_\epsilon(q)) \neq \emptyset$. Pick $p \in B_\epsilon(q) \cap X_{-t}(B_\epsilon(q))$. By (1) above one has $X_{[0,t]}(p) \subseteq U$ since $B_\epsilon(q) \subset V$. This contradicts $X_{t_0}(p) \in X_{t_0}(B_\epsilon(q))$ and $X_{t_0}(B_\epsilon(q)) \cap \mathrm{Cl}(U') = \emptyset$.

A first consequence of the above lemma is the following corollary. Given compact subsets $A, B \subset M$ we denote $\mathrm{dist}(A, B) = \inf\{\mathrm{dist}(a, b); a \in A, b \in$

B}. If Λ is a neutral set of X and Λ_n is a sequence of transitive sets of X such that dist$(\Lambda_n, \Lambda) \to 0$ as $n \to \infty$, then Λ_n accumulates on Λ.

Let Λ_n and Λ as in the statement. Fix a neighborhood U of Λ and let $V \subset U$ be the neighborhood of Λ obtained by the previous lemma. As dist$(\Lambda_n, \Lambda) \to 0$ as $n \to \infty$ we have that $\Lambda_n \cap V \neq \emptyset$ for every n large. Let q_n the dense orbit of Λ_n. Clearly $q_n \in \Omega(X)$. We can assume that $q_n \in V$ for n large, and so, $q_n \in \Omega(X) \cap V$. Then, $X_t(q_n) \in U$ for every t. In particular, $\Lambda_n = \omega_X(q_n) \subset$ Cl(U).

A neutral set is isolated if and only if it is Ω−isolated [CMP00]. Any saturated Ω−isolated set Λ of X is isolated. Indeed, since Λ is Ω−isolated, there is $U \supset \Lambda$ open such that Cl$(U) \cap \Omega(X) = \Lambda$. This U is an isolating block for Λ. For if $x \in \cap_{t \in \mathbb{R}} X_t(U)$, then $\omega_X(x) \cup \alpha_X(x) \subset$ Cl$(U) \cap \Omega(X) \subset \Lambda$. So, $x \in W^s_X(\Lambda) \cap W^u_X(\Lambda) = \Lambda$. This proves that $\cap_{t \in \mathbb{R}} X_t(U) \subset \Lambda$. The opposite inclusion follows since Λ is invariant.

Transitive hyperbolic neutral sets are isolated. It suffices to show that transitive neutral hyperbolic sets Λ are Ω−isolated. Suppose by contradiction that Λ is not Ω−isolated. Then, there is a sequence $p_n \in \Omega(X) \setminus \Lambda$ converging to $p \in \Lambda$. Fix U a neighborhood of Λ and let V be given ias above for U. We can assume that $p_n \in V$ for every n. As p_n is non wandering for X, for every n there are sequences $q_i \in V \to p_n$ and $t_i > 0$ such that $X_{t_i}(q_i) \to p_n$ as $i \to \infty$. We have $X_{[0,t_i]}(q_i) \subset U$ for every i. So, we can construct a periodic pseudo orbit of X arbitrarily close to p_n. By the *shadowing lemma for flows* ([KH95]) applied to the hyperbolic set Λ, such a periodic pseudo orbit can be shadowed by a periodic orbit. This proves that $p_n \in$ Cl(Per(X)). As the neighborhood U is arbitrary, we can assume that $p_n \in$ Per(X) for every n. Note that $\mathcal{O}_X(p_n)$ converges to Λ.

As Λ is transitive we have that if $E^s \oplus E^X \oplus E^u$ denotes the corresponding hyperbolic splitting, then dim$(E^s) = s$ and dim$(E^u) = u$ are constant in Λ. Clearly neither $s = 0$ nor $u = 0$ since Λ is not Ω−isolated. As $\mathcal{O}_X(p_n)$ converges to Λ both the local stable and unstable manifolds of p_n have dimension s, u respectively. Moreover, both invariant manifolds have uniform size as well. This implies that $W^u_X(p_n) \cap W^s_X(p) \neq \emptyset$ and $W^s_X(p_n) \cap W^u_X(p) \neq \emptyset$ for n large. As $p_n \in Per(X)$ and $p \in \Lambda$, we conclude by the *Inclination lemma* [MP92] that $p_n \in$ Cl$(W^s_X(p) \cap W^u_X(p))$. As $p \in \Lambda$, $W^{s,u}_X(p) \subset W^{s,u}_X(\Lambda)$. So, $p_n \in$ Cl$(W^s_X(\Lambda) \cap W^u_X(\Lambda))$. As Λ is saturated, $W^s_X(\Lambda) \cap W^u_X(\Lambda) = \Lambda$ and hence $p_n \in$ Cl$(\Lambda) = \Lambda$. But this is impossible since $p_n \in \Omega(X) \setminus \Lambda$ by assumption. This concludes the proof.

Now, denote by $\mathcal{F}$ the collection of all isolated transitive neutral sets of X. A sub collection $\mathcal{F}'$ of $\mathcal{F}$ is finite if and only if $\cup_{\Lambda \in \mathcal{F}'} \Lambda$ is closed. Obviously $\cup_{\Lambda \in \mathcal{F}'} \Lambda$ is closed if $\mathcal{F}'$ is finite. Conversely, suppose that $\cup_{\Lambda \in \mathcal{F}'} \Lambda$ is closed. If $\mathcal{F}'$ were infinite then, it would exist sequence $\Lambda_n \in \mathcal{F}'$ of (different) sets accumulating some $\Lambda \in \mathcal{F}'$. We have $\Lambda_n \subseteq U$ for some isolating block U of Λ and n large. And then, we would have that $\Lambda_n = \Lambda$ for n large, a contradiction.

2.2.2 Homoclinic Classes

The main result of this subsection is the following *equivalence theorem* [CMP00]: There is a residual subset $\mathcal{R}$ of $\mathcal{X}^1(M)$ such that every homoclinic class of every vector–field in $\mathcal{R}$ is neutral. As a corollary, the following properties are equivalent for $X \in \mathcal{R}$ and every compact invariant set Λ of X :

(i) Λ is a transitive neutral set with periodic orbits of X.

(ii) Λ is a homoclinic class of X.

(iii) Λ is a maximal transitive set with periodic orbits of X.

Also, for $X \in \mathcal{R}$, a non singular compact isolated set of X is neutral and transitive if and only if it is a homoclinic class.

There exists a residual set $\mathcal{R}$ of $\mathcal{X}^1(M)$ such that, for every $Y \in \mathcal{R}$ and $\sigma \in \mathrm{Crit}(Y)$, $\mathrm{Cl}(W_X^u(\sigma))$ is Lyapunov stable for X and $\mathrm{Cl}(W_X^s(\sigma))$ is Lyapunov stable for $-X$.

There exists a residual set $\mathcal{R}$ in $\mathcal{X}^1(M)$ such that every $X \in \mathcal{R}$ satisfies

$$H_X(p) = \mathrm{Cl}(W_X^u(p)) \cap \mathrm{Cl}(W_X^s(p)), \qquad \text{for all } p \in \mathrm{Per}(X).$$

This lemma was proved in [MP01] when σ is a singularity and the same proof works when σ is a periodic orbit.

Now, recall that M is a closed $n-$manifold, $n \geq 3$. We denote 2_c^M the space of all compact subsets of M endowed with the Hausdorff topology. Recall that $\mathcal{KS}^1(M) \subset \mathcal{X}^1(M)$ denotes the set of Kupka–Smale C^1 vector–fields on M.

Given $X \in \mathcal{X}^1(M)$ and $p \in \mathrm{Per}(X)$ we denote $\Pi_X(p)$ the period of p. We set $\Pi_X(p) = 0$ if p is a singularity of X.

If $T > 0$ we denote

$$\mathrm{Crit}_T(X) = \{p \in \mathrm{Crit}(X) : \Pi_X(p) < T\}.$$

If $p \in \mathrm{Crit}(X)$ is hyperbolic, then there is a continuation $p(Y)$ of p for Y close enough to X so that $p(X) = p$.

Note that if $X \in \mathcal{KS}^1(M)$ and $T > 0$, then

$$\mathrm{Crit}_T(X) = \{p_1(X), \cdots, p_k(X)\}$$

is a finite set. Moreover,

$$\mathrm{Crit}_T(Y) = \{p_1(Y), \cdots, p_k(Y)\}$$

for every Y close enough to X.

Let $\mathcal{Y}$ be a metric space. A set-valued map

$$\Phi : \mathcal{Y} \longrightarrow 2_c^M$$

is *lower semi-continuous* at $Y_0 \in \mathcal{Y}$ if for every open set $U \subset M$ one has $\Phi(Y_0) \cap U \neq \emptyset$ implies $\Phi(Y) \cap U \neq \emptyset$ for every Y close to Y_0. Similarly, we say that Φ is *upper semi-continuous* at $Y_1 \in \mathcal{Y}$ if for every compact set $K \subset M$

one has $\Phi(Y_1) \cap K = \emptyset$ implies $\Phi(Y) \cap K = \emptyset$ for every Y close to Y_1. We say that Φ is *lower semi-continuous* if it is lower semi-continuous at every $Y_0 \in \mathcal{Y}$. A well known result [Kur68] asserts that if $\Phi : \mathcal{X}^1(M) \longrightarrow 2_c^M$ is a lower semi-continuous map, then it is upper semi-continuous at every Y in a residual subset of $\mathcal{X}^1(M)$.

The following lemma is the flow version of [Arn99, Hay00, Hay97, Hay98]:

Let $Y \in \mathcal{X}^1(M)$ and $x \notin \mathrm{Crit}(Y)$. For any C^1 neighborhood $\mathcal{U}$ of Y there are $\rho > 1$, $L > 0$ and $\epsilon_0 > 0$ such that for any $0 < \epsilon \le \epsilon_0$ and any two points $p, q \in M$ satisfying

(a) $p, q \notin B_\epsilon(Y_{[-L,0]}(x))$,
(b) $\mathcal{O}_Y^+(p) \cap B_{\epsilon/\rho}(x) \ne \emptyset$, and
(c) $\mathcal{O}_Y^-(q) \cap B_{\epsilon/\rho}(x) \ne \emptyset$,

There is $Z \in \mathcal{U}$ such that $Z = Y$ off $B_\epsilon(Y_{[-L,0]}(x))$ and that $q \in \mathcal{O}_Z^+(p)$.

Given $X \in \mathcal{X}^1(M)$ we denote by $\mathrm{Per}_T(X)$ the set of periodic orbits of X with period $< T$.

If $X \in \mathcal{KS}^1(M)$ and $T > 0$ then there are a neighborhood $\mathcal{V}_{X,T} \ni X$ and a residual subset $\mathcal{P}_{X,T}$ of $\mathcal{V}_{X,T}$ such that if $Y \in \mathcal{P}_{X,T}$ and $p \in \mathrm{Per}_T(Y)$ then

$$H_Y(p) = \mathrm{Cl}(W_Y^u(p)) \cap \mathrm{Cl}(W_Y^s(p)).$$

There is a neighborhood $\mathcal{V}_{X,T} \ni X$ such that [CMP00]

$$\mathrm{Per}_T(Y) = \{\sigma_1(Y), \ldots, \sigma_m(Y)\} \quad \text{for all} \quad Y \in \mathcal{V}_{X,T}.$$

For each $1 \le i \le m$, let $\Psi_i : \mathcal{V}_{X,T} \ni Y \mapsto H_Y(\sigma_i(Y)) \in 2_c^M$. Note that Ψ_i, for all i, is lower semi-continuous by the persistence of transverse homoclinic orbits. So, there is a residual subset $\mathcal{P}_{X,T}^i$ of $\mathcal{V}_{X,T}$ such that Ψ_i is upper semi–continuous in $\mathcal{P}_{X,T}^i$. Set $\mathcal{P}_{X,T} = \mathcal{KS}^1(M) \cap (\cap \mathcal{P}_{X,T}^i) \cap \mathcal{R}$, where $\mathcal{R}$ is the residual set given above. Then $\mathcal{P}_{X,T}$ is residual in $\mathcal{V}_{X,T}$.

Let us prove that $\mathcal{P}_{X,T}$ satisfies the conclusion of the lemma. For this, let $\sigma \in \mathrm{Per}_T(Y)$ for some $Y \in \mathcal{P}_{X,T}$. Then $\sigma = \sigma_i(Y)$ for some i, and so $\Psi_i(Y) = H_Y(\sigma)$.

Suppose, by contradiction, that $H_Y(\sigma) \ne \mathrm{Cl}(W_Y^u(\sigma)) \cap \mathrm{Cl}(W_Y^s(\sigma))$. Then there is $x \in \mathrm{Cl}(W_Y^u(\sigma)) \cap \mathrm{Cl}(W_Y^s(\sigma)) \setminus H_Y(\sigma)$.

We have either

(a) $x \notin \mathrm{Crit}(Y)$, or
(b) $x \in \mathrm{Crit}(Y)$.

It is enough to prove the lemma in case (a). Indeed, suppose that case (b) holds. As Y is Kupka–Smale we have that $\mathcal{O}_Y(x)$ is hyperbolic. Clearly $\mathcal{O}_Y(x)$ is neither a sink or a source and so $W_Y^s(x) \setminus \mathcal{O}_Y(x) \ne \emptyset$ and $W_Y^u(x) \setminus \mathcal{O}_Y(x) \ne \emptyset$. Note that $\mathrm{Cl}(W_Y^u(\sigma))$ is Lyapunov stable since $Y \in \mathcal{R}$. As $x \in \mathrm{Cl}(W_Y^u(\sigma))$ we conclude that $W_Y^u(x) \subseteq \mathrm{Cl}(W_Y^u(\sigma))$. As $x \in \mathrm{Cl}(W_Y^s(\sigma))$, there is $x' \in \mathrm{Cl}(W_Y^s(\sigma)) \cap (W_Y^u(x) \setminus \mathcal{O}_Y(x))$ arbitrarily close to x (for this use the Grobman–Hartman Theorem). Obviously $x' \notin \mathrm{Crit}(Y)$. If $x' \in H_Y(\sigma)$ we would have that $x \in H_Y(\sigma)$ since $\alpha_Y(x') = \mathcal{O}_Y(x)$ contradicting $x \notin H_Y(\sigma)$. Henceforth $x' \in \mathrm{Cl}(W_Y^u(\sigma)) \cap \mathrm{Cl}(W_Y^s(\sigma)) \setminus H_Y(\sigma)$ and $x' \notin \mathrm{Crit}(Y)$.

Now, as $x \notin H_Y(\sigma)$, there is a compact neighborhood K of x such that $K \cap H_Y(\sigma) = \emptyset$. As Ψ_i is upper semi-continuous at Y, there is a neighborhood $\mathcal{U}$ of Y such that

$$K \cap H_Z(\sigma(Z)) = \emptyset, \qquad \text{for all } Z \in \mathcal{U}. \tag{2.1}$$

Let ρ, L, ϵ_0 be the constants for $Y \in \mathcal{X}^1(M)$, x, and $\mathcal{U}$ as above. As $x \notin \mathrm{Crit}(Y)$, $Y_{[-L,0]}(x) \cap \mathcal{O}_Y(\sigma) = \emptyset$. Then, there is $0 < \epsilon < \epsilon_0$ such that $\mathcal{O}_Y(\sigma) \cap B_\epsilon(Y_{[-L,0]}(x)) = \emptyset$ and $B_\epsilon(x) \subseteq K$.

Now, choose an open set V containing $\mathcal{O}_Y(\sigma)$ such that $V \cap B_\epsilon(Y_{-L,0]}(x)) = \emptyset$.

As $x \in \mathrm{Cl}(W_Y^u(\sigma))$, one can choose $p \in W_Y^u(\sigma) \setminus \{\sigma\} \cap V$ such that

$$\mathcal{O}_Y^+(p) \cap B_{\epsilon/\rho}(x) \neq \emptyset.$$

Similarly, as $x \in \mathrm{Cl}(W_Y^s(\sigma))$, one can choose $q \in W_Y^s(\sigma) \setminus \{\sigma\} \cap V$ such that

$$\mathcal{O}_Y^-(q) \cap B_{\epsilon/\rho}(x) \neq \emptyset.$$

We can assume that $\mathcal{O}_Y^-(p) \subset V$ and $\mathcal{O}_Y^+(q) \subset V$. Henceforth

$$(\mathcal{O}_Y^-(p) \cup \mathcal{O}_Y^+(q)) \cap B_\epsilon(Y_{[-L,0]}(x)) = \emptyset. \tag{2.2}$$

Observe that

$$q \notin \mathcal{O}_Y^+(p)$$

for, otherwise, p would be a homoclinic orbit of Y passing through K contradicting (2.1).

By construction ϵ, p, q satisfy (b) and (c) in the previous lemma above. As $p, q \in V$ and $V \cap B_\epsilon(Y_{[-L,0]}(x)) = \emptyset$ we have that that ϵ, p, q also satisfy (a). Then, there is $Z \in \mathcal{U}$ such that $Z = Y$ off $B_\epsilon(Y_{[-L,0]}(x))$ and $q \in \mathcal{O}_Z^+(p)$. Clearly $\sigma(Z) = \sigma$ and by (2.2) we have $p \in W_Z^u(\sigma)$ and $q \in W_Z^s(\sigma)$ since $Z = Y$ off $B_\epsilon(Y_{[-L,0]}(x))$. Hence $\mathcal{O} = \mathcal{O}_Z(p) = \mathcal{O}_Z(q)$ is a homoclinic orbit of σ.

As $q \notin \mathcal{O}_Y^+(p)$, we have that $\mathcal{O} \cap B_\epsilon(x) \neq \emptyset$. Perturbing Z we can assume that $\mathcal{O}$ is transverse, i.e. $\mathcal{O} \subseteq H_Z(\sigma)$. As $\mathcal{O} \cap B_\epsilon(x) \neq \emptyset$ and $B_\epsilon(x) \subset K$ we would obtain $K \cap H_Z(\sigma(Z)) \neq \emptyset$ contradicting (2.1).

Now, fix $T > 0$. For any $X \in \mathcal{KS}^1(M)$ consider $\mathcal{V}_{X,T}$ and $\mathcal{P}_{X,T}$ as above. Choose a sequence $X^n \in \mathcal{KS}^1(M)$ such that $\{X^n : n \in \mathbb{N}\}$ is dense in $\mathcal{X}^1(M)$ (recall that $\mathcal{X}^1(M)$ is a separable metric space). Denote $\mathcal{V}_{n,T} = \mathcal{V}_{X^n,T}$ and $\mathcal{P}_{n,T} = \mathcal{P}_{X^n,T}$. Define

$$\mathcal{O}^T = \cup_n \mathcal{V}_{n,T} \quad \text{and} \quad \mathcal{P}^T = \cup_n \mathcal{P}_{n,T}.$$

Clearly $\mathcal{O}^T$ is open and dense in $\mathcal{X}^1(M)$.

$\mathcal{P}^T$ is residual in $\mathcal{O}^T$ [CMP00]. Indeed, for any n there is a sequence $D_{k,n,T}$, $k \in \mathbb{N}$, such that

$$\mathcal{P}_{n,T} = \cap_k D_{k,n,T},$$

and $D_{k,n,T}$ is open and dense in $\mathcal{V}_{n,T}$ for any k. As

$$\mathcal{P}^T = \cup_n \mathcal{P}_{n,T} = \cup_n (\cap_k D_{k,n,T}) = \cap_k (\cup_n D_{k,n,T})$$

and $\cup_n D_{k,n,T}$ is open and dense in $\cup_n \mathcal{V}_{n,T} = \mathcal{O}^T$ we conclude that $\mathcal{P}^T$ is residual in $\mathcal{O}^T$.

In particular, $\mathcal{P}^T$ is residual in $\mathcal{X}^1(M)$ for every T. Set $\mathcal{P} = \cap_{N \in \mathbb{N}} \mathcal{P}^N$. It follows that $\mathcal{P}$ is residual in $\mathcal{X}^1(M)$. Choose $X \in \mathcal{P}$, $p \in \mathrm{Per}(X)$ and $N_0 \in \mathbb{N}$ bigger than $\Pi_X(p) + 1$. By definition $X \in \mathcal{P}^{N_0}$, and so, $X \in \mathcal{P}_{X^n, N_0}$ for some n. As $N_0 > \Pi_X(p)$ we have $p \in \mathrm{Per}_{N_0}(X)$. Then we have

$$H_X(p) = \mathrm{Cl}(W_X^u(p)) \cap \mathrm{Cl}(W_X^s(p)),$$

as claimed above. For more details, see [CMP00].

2.3 Complex–Valued Hénon Maps and Horseshoes

A theory analogous to both the theories of polynomial–like maps and Smale's real horseshoes has been developed in [Obe87, Obe00] for the study of the dynamics of maps of two complex variables. In partial analogy with polynomials in a single variable there are the Hénon maps in two variables as well as higher dimensional analogues. From *polynomial–like maps*, *Hénon–like maps* and *quasi–Hénon–like maps* are defined following this analogy. A special form of the latter is the complex horseshoe. The major results about the real horseshoes of Smale remain true in the complex setting. In particular: (i) trapping fields of cones(which are sectors in the real case) in the tangent spaces can be defined and used to find horseshoes, (ii) the dynamics of a horseshoe is that of the two–sided shift on the symbol space on some number of symbols which depends on the type of the horseshoe, and (iii) transverse intersections of the stable and unstable manifolds of a hyperbolic periodic point guarantee the existence of horseshoes.

Recall that the study of the subject of the dynamics of complex analytic functions of one variable goes back to the early 1900's with the publication of several papers by Pierre Fatou and Gaston Julia around 1920, including the long memoirs [Fat19, Jul18]. Compared with the theory of complex analytic functions of a single variable, the theory of complex analytic maps of several variables is quite different. In particular, the theory of omitted values, including normal families, is not paralleled in the several complex variables case. This difference was really exposed by Fatou himself and L. Bieberbach in the 1920's in [Fat22, Bla84]. They showed the existence of the *Fatou–Bieberbach domains*: open subsets of $\mathbb{C}^n$ whose complements have nonempty interior and yet are the images of $\mathbb{C}^n$ under an injective analytic map. This is contrary to the one variable case where the image of every non-constant analytic function on $\mathbb{C}$ omits at most a single point. [Obe87, Obe00] attempted to understand these Fatou–Bieberbach domains, which arose naturally as the basins of attractive fixed–points of analytic automorphisms of $\mathbb{C}^n$; the basins were the image of the map conjugating the given automorphism to its linear part at the given fixed–point. For example, consider the map

$$F : \begin{pmatrix} x \\ y \end{pmatrix} \mapsto \begin{pmatrix} x^2 + 9/32 - y/8 \\ x \end{pmatrix},$$

which has two fixed–points, of which $(3/8, 3/8)$ is attractive with its linear part having resonant eigenvalues $1/4$ and $1/2$ (that is, $1/4 = (1/2)^2$). Moreover, none of the points in the region $\left\{ (x, y) \mid |y| < 4|x|^2/3, \quad |x| > 4 \right\}$, remain bounded under iteration of F. So the basin of $(3/8, 3/8)$ is not all of $\mathbb{C}^2$.

2.3.1 Complex Henon–Like Maps

In the time between Fatou and the present, most of the attention of those studying dynamical systems has been limited to maps in the real. This is somewhat surprising for two reasons. First, small perturbations of the coefficients of polynomial terms of real maps are liable to have large effects. For example, the number and periods of the periodic cycles may change. In the complex, the behavior is more uniform. Second, the major tools of complex analysis do not apply. These include the theory of normal families and the naturally contracting Poincaré metric together with the contracting map fixed–point theorem. Recently, the maps $F_{a,c} : \mathbb{R}^2 \to \mathbb{R}^2$, of the type

$$F_{a,c} : \begin{pmatrix} x \\ y \end{pmatrix} \mapsto \begin{pmatrix} x^2 + c - ay \\ x \end{pmatrix}, \qquad \text{with} \qquad a \neq 0,$$

have received much attention. Hénon first studied these maps numerically and they have become known as the *Hénon maps* [Hen69]. This family contains, up to conjugation, most of the most interesting of the simplest nonlinear polynomial maps of two variables. However, Hénon maps are still rather poorly understood and indeed the original question concerning the existence of a strange attractor for any values of the parameters is still unresolved today [Obe87, Obe00].

Despite the differences between the real and complex theories and the one variable and several variable theories, much of the development of the subject of complex analytic dynamics in several variables has been conceived through analogy. Recently, Hénon maps have started to be examined in the complex, that is, with both the variables and the parameters being complex. Also, there exist analogous maps of higher degree, called the *generalized Hénon maps*, of the form

$$\begin{pmatrix} x \\ y \end{pmatrix} \mapsto \begin{pmatrix} p(x) - ay \\ x \end{pmatrix},$$

where p is a polynomial of degree at least two and $a \neq 0$. Note that these are always invertible with inverses given by

$$\begin{pmatrix} x \\ y \end{pmatrix} \mapsto \begin{pmatrix} y \\ (p(y) - x)/a \end{pmatrix}.$$

For polynomials p of degree d, these maps are called *Hénon maps of degree d*.

Inspired by the definition of the *polynomial–like maps* [DH85], which was designed to capture the topological essence of polynomials on some disc or, more generally, on some open subset of $\mathbb{C}$ isomorphic to a disc, the *Hénon–like maps* have been defined in [Obe87, Obe00]. A polynomial–like map of degree d is a triple (U, U', f), where U and U' are open subsets of $\mathbb{C}$ isomorphic to discs, with U' relatively compact in U, and $f : U' \to U$ analytic and proper of degree d. Note that it is convenient to think of polynomial–like of degree d as meaning an analytic map $f : U \to \mathbb{C}$ such that $f(\partial U) \subset \mathbb{C} \setminus \overline{U}$ and $f|_{\partial U}$ of degree d (see Figure 2.9, which gives examples of the behavior, pictured in $\mathbb{R}^2$, that should be captured by the definition of Hénon–like maps of degree 2; in each case, the crescent-shaped region is the image of the square with A' the image of A, etc). It seems clear that the behaviors described by

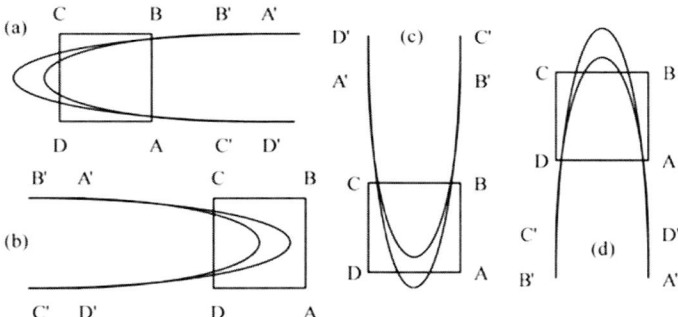

Fig. 2.9. The Hénon–like map of degree 2 (adapted from [Obe87, Obe00]).

(a) and (b) versus (c) and (d) in Figure 2.9 must be described differently, albeit analogously, trading 'horizontal' for 'vertical' [Obe87, Obe00]. In the following, d will always be an arbitrary fixed integer greater than one. Let $\pi_1, \pi_2 : \mathbb{C}^2 \to \mathbb{C}$ be the projections onto the first and second coordinates, respectively. We will consider a bidisc $B = D_1 \times D_2 \subset \mathbb{C}^2$, where $D_1, D_2 \subset \mathbb{C}$ are discs. Vertical and horizontal 'slices' of B are denoted respectively by

$$V_x = \{x\} \times D_2 \qquad \text{and} \qquad H_y = D_1 \times \{y\}, \qquad \text{for all} \quad x \in D_1, y \in D_2.$$

We will be considering maps of the bi–disc, $F : \overline{B} \to \mathbb{C}^2$, together with a map denoted by $F^{-1} : \overline{B} \to \mathbb{C}^2$, which is the inverse of F, where that makes sense. Now, for each $(x, y) \in B$, we can define

$$F_{1,y} = \pi_1 \circ F \circ (\text{Id} \times y) : D_1 \to \mathbb{C}, \qquad F_{2,x}^{-1} = \pi_2 \circ F^{-1} \circ (x \times \text{Id}) : D_2 \to \mathbb{C},$$
$$F_{2,x} = \pi_2 \circ F \circ (x \times \text{Id}) : D_2 \to \mathbb{C}, \qquad F_{1,y}^{-1} = \pi_1 \circ F^{-1} \circ (\text{Id} \times y) : D_1 \to \mathbb{C}.$$

The map $F : \overline{B} \to \mathbb{C}^2$ is a *Hénon-like map of degree* d if there exists a map $G : \overline{B} \to \mathbb{C}^2$ such that: (i) both F and G are injective and continuous on $\overline{B}$ and analytic on B, (ii) $F \circ G = \text{Id}$ and $G \circ F = \text{Id}$, where each makes sense; hence, we can rename G as F^{-1}, (iii) For all $x \in D_1$ and $y \in D_2$, either (i) $F_{1,y}$ and $F_{2,x}^{-1}$ are polynomial–like of degree d, or (ii) $F_{2,x}$ and $F_{1,y}^{-1}$ are polynomial-like of degree d. Depending on whether F satisfies condition (i) or (ii), we call it *horizontal* or *vertical* [Obe87, Obe00].

Now, let $\partial B_V = \partial\overline{D}_1 \times \overline{D}_2$ and $\partial B_H = \overline{D}_1 \times \partial\overline{D}_2$ be the 'vertical and horizontal boundaries'. If $F : \overline{B} \to \mathbb{C}^2$ is a Hénon–like map then either

$$F(\partial B_V) \subset \mathbb{C}^2 \setminus \overline{B} \quad \text{and} \quad F^{-1}(\partial B_H) \subset \mathbb{C}^2 \setminus \overline{B} \quad \text{or}$$
$$F^{-1}(\partial B_V) \subset \mathbb{C}^2 \setminus \overline{B} \quad \text{and} \quad F(\partial B_H) \subset \mathbb{C}^2 \setminus \overline{B}.$$

This follows from the fact that the boundary of a polynomial-like map is mapped outside of the closure of its domain. Note that these are equivalent to

$$F(\partial B_V) \cap \overline{B} = \emptyset \quad \text{and} \quad F^{-1}(\partial B_H) \cap \overline{B} = \emptyset \quad \text{or}$$
$$F^{-1}(\partial B_V) \cap \overline{B} = \emptyset \quad \text{and} \quad F(\partial B_H) \cap \overline{B} = \emptyset.$$

The class of polynomial–like maps is stable under small perturbations [DH85] and the same is true for Hénon–like maps. More precisely, suppose $F : \overline{B} \to \mathbb{C}^2$ is Hénon–like of degree d. Let $H : \overline{B} \to \mathbb{C}^2$ be injective, continuous on $\overline{B}$, and analytic on B. If $\|H\|$ is sufficiently small, then $F - H$ is also Hénon–like of degree d.

For example, a simple computation shows that if F is the Hénon map (of degree 2) with parameters a and c, and D_R is the disc of radius R, with

$$R > (1/2)\left(1 + |a| + \sqrt{(1 + |a|)^2 + 4|c|}\right),$$

then $F : \overline{D}_R^2 \to \mathbb{C}^2$ is a horizontal Hénon–like map of degree 2. This R is exactly what is required so that $F(\partial\overline{D}_R \times \overline{D}_R) \cap \overline{D}_R^2 = \emptyset$ and $F^{-1}(\overline{D}_R \times \partial\overline{D}_R) \cap \overline{D}_R^2 = \emptyset$. Of course, their inverses are vertical Hénon–like maps.

More generally, consider the maps $G : \mathbb{C}^2 \to \mathbb{C}^2$ of the form

$$G : \begin{pmatrix} x \\ y \end{pmatrix} \mapsto \begin{pmatrix} x^d + c - ay \\ x \end{pmatrix},$$

with $a \neq 0$ and $d \geq 2$. When $d = 2$, we are back to the previous example. The lower bound on R came from solving the inequality

$$R^d - (1 + |a|)R - 4|c| > 0$$

for R. Note that when $d = 2$ we already had $R > 1$. Therefore, the same lower bound will work here as well. Of course, better lower bounds can be found.

Analogous to the invariant sets defined for Hénon maps, we can define the following sets for Hénon–like maps:

$$K_+ = \{\, z \in B \,|\, F^{\circ n}(z) \in B \quad \text{for all } n > 0 \,\},$$
$$K_- = \{\, z \in B \,|\, F^{\circ -n}(z) \in B \quad \text{for all } n > 0 \,\},$$
$$J_\pm = \partial K_\pm, \qquad K = K_+ \cap K_-, \qquad J = J_+ \cap J_-.$$

For every d, all Hénon–like maps of degree d have the same number of periodic cycles, counted with multiplicity, as a polynomial of degree d [Obe87, Obe00].

2.3.2 Complex Horseshoes

Here, following [Obe87, Obe00], we define and analyze complex analogs of Smale horseshoes. Using a criterion analogous to the one given by Moser [Mos73] in the real case, we will show that many Hénon maps are complex horseshoes. In particular, actual Hénon maps (of degree 2) are complex horseshoes when $|c|$ is sufficiently large.

Now, recall that in Figure 2.9 above, only (a) and (c) appear to be horseshoes. Basically, we would like to say that a horizontal Hénon–like map F of degree d is a complex horseshoe of degree d if the projections

$$\pi_1 : \bigcap_{0 \le m \le n} F^{\circ m}(B) \to \mathbb{C} \qquad \text{and} \qquad \pi_2 : \bigcap_{0 \le m \le n} F^{\circ -m}(B) \to \mathbb{C}$$

are *trivial fibrations* with fibers disjoint unions of d^n discs (see [II06b]). However, this is not general enough for our purpose here, so we give a definition with weaker conditions, which encompasses the Hénon-like maps defined above. Instead of requiring B to be an actual bi–disc, B may be an embedded bi–disc. More precisely, letting $D \subset \mathbb{C}$ be the open unit disc, assume that there is an embedding, $\varphi : \overline{D}^2 \to \mathbb{C}^2$, which is analytic on D^2 and such that $B = \varphi(D^2)$ and, naturally, $\overline{B} = \varphi(\overline{D}^2)$. By $\partial B_H = \varphi(\overline{D} \times \partial \overline{D})$ and $\partial B_V = \varphi(\partial \overline{D} \times \overline{D})$ we denote the horizontal and vertical boundaries of B. Also, define horizontal and vertical slices $H_y = \varphi(D \times \{y\})$ and $V_x = \varphi(\{x\} \times D)$ for all $x, y \in D$.

Consider the maps $F : \overline{B} \to \mathbb{C}^2$ which are injective and continuous on $\overline{B}$ and analytic on B, and such that either

$$F(\overline{B}) \cap \partial B_H = \emptyset \qquad \text{and} \qquad \overline{B} \cap F(\partial B_V) = \emptyset, \qquad \text{or}$$
$$\overline{B} \cap F(\partial B_H) = \emptyset \qquad \text{and} \qquad F(\overline{B}) \cap \partial B_V = \emptyset.$$

Under these conditions, for all $y \in D$,

$$\pi_1 \circ \varphi^{-1} : F(H_y) \cap B \to D$$

is a proper map [Obe87, Obe00]. Such a proper map has a degree and since the degree is integer–valued and continuous in y, this defines a constant, the degree of such a map F. Now a class of maps generalizing the Hénon–like maps can be defined as follows.

$$F : \overline{B} \to \mathbb{C}^2$$

is a *quasi–Hénon–like map of degree d* if there exists a map $G : \overline{B} \to \mathbb{C}^2$ such that: (i) both F and G are injective and continuous on $\overline{B}$ and analytic on B, (ii) $F \circ G = \mathrm{Id}$ and $G \circ F = \mathrm{Id}$, where each makes sense; therefore, we can rename G as F^{-1}, and (iii) We have either

$$F(\overline{B}) \cap \partial B_H = \emptyset \quad \text{and} \quad \overline{B} \cap F(\partial B_V) = \emptyset, \qquad \text{or}$$
$$\cap F(\partial B_H) = \emptyset \quad \text{and} \quad F(\overline{B}) \cap \partial B_V = \emptyset.$$

The degree is $d \geq 2$. Moreover, call F either *horizontal* or *vertical* according to whether it satisfies (a) or (b), respectively. Also, Hénon–like maps of degree d are quasi–Hénon–like of degree d.

Now, using the notion of quasi–Hénon–like maps, complex horseshoes may be defined as follows [Obe87, Obe00]. A *complex horseshoe of degree d* is a quasi–Hénon–like map of degree d, $F : \overline{B} \to \mathbb{C}^2$, such that, for all integers $n > 0$, depending on if F is horizontal or vertical, then either the projections

$$\pi_1 \circ \varphi^{-1} : \bigcap_{0 \leq m \leq n} F^{\circ m}(B) \to \mathbb{C} \quad \text{and} \quad \pi_2 \circ \varphi^{-1} : \bigcap_{0 \leq m \leq n} F^{\circ -m}(B) \to \mathbb{C}$$

or

$$\pi_2 \circ \varphi^{-1} : \bigcap_{0 \leq m \leq n} F^{\circ m}(B) \to \mathbb{C} \quad \text{and} \quad \pi_1 \circ \varphi^{-1} : \bigcap_{0 \leq m \leq n} F^{\circ -m}(B) \to \mathbb{C},$$

respectively, are trivial fibrations with fibers disjoint unions of d^n discs.

In the context of Hénon–like maps, the following results will show the close relation between complex horseshoe maps of degree d and polynomial-like maps of degree d whose critical points escape immediately. The following definition, borrowed from Moser [Mos73], is the key tool in this study of complex horseshoes [Ale68]. Let M be a differentiable manifold, $U \subset M$ an open subset, and $f : U \to M$ a differentiable map. A field of cones $\mathcal{C} = (C_x \subset T_x M)_{x \in U}$ on U is an f–*trapping* field if: (i) $C_\mathbf{x}$ depends continuously on x, and (ii) whenever $x \in U$ and $f(x) \in U$, then $d_x f(C_x) \subset C_{f(x)}$.

Now we consider the connection between trapping fields of cones and complex horseshoes. Let $F : \overline{B} \to \mathbb{C}^2$ be a quasi–Hénon–like map of degree d. The following are equivalent: (i) $F : \overline{B} \to \mathbb{C}^2$ is a complex horseshoe of degree d, (ii) there exist continuous, positive functions $\alpha(\mathbf{z})$ and $\beta(\mathbf{z})$ on $\overline{B}$ such that the field of cones $C_z = \{ (\xi_1, \xi_2) : |\xi_2| < \alpha(z)|\xi_1| \}$ is F–trapping and the field of cones $C'_z = \{ (\xi_1, \xi_2) : |\xi_1| < \beta(z)|\xi_2| \}$ is F^{-1}–trapping, and (iii) $F(\overline{B}) \cap \overline{B}$ and $F^{-1}(\overline{B}) \cap \overline{B}$ both have d connected components. Note that (ii) $\Rightarrow$ (i) is borrowed from Moser and that the implication (i) $\Rightarrow$ (ii) is what the contractive nature of complex analytic maps gives us for free. (iii) arises naturally in the proof of (ii) $\Rightarrow$ (i). When considering a map which is actually Hénon–like, consideration of the critical points of $F_{1,y}$ and $F_{2,x}^{-1}$ (or $F_{2,x}$ and $F_{1,y}^{-1}$) in light of the equivalences above yields the following. Let $F : \overline{D_1} \times \overline{D_2} \to \mathbb{C}^2$ be a Hénon–like map of degree d. The following are equivalent: (i) $F : \overline{D_1} \times \overline{D_2} \to \mathbb{C}^2$ is a complex horseshoe of degree d, and (ii) for all $(x, y) \in \overline{D_1} \times \overline{D_2}$, the critical values of the polynomial–like maps $F_{1,y}$ and $F_{2,x}^{-1}$ (or $F_{2,x}$ and $F_{1,y}^{-1}$) lie outside of $\overline{D_1}$ and $\overline{D_2}$, respectively. The diameters of the discs in the fibers above tend to 0 with n [Obe87, Obe00].

This criterion can be used to show that for each a there exists $r(a)$ such that if $|c| > r(a)$, then the Hénon map $F_{a,c}$ is a complex horseshoe. Of course, in the real locus this was known [DN79, New80], except that then c has to be taken very negative. When c is large and positive, all the 'horseshoe behavior' is complex. More precisely, For each $a \neq 0$ and each c such that $|c| > \left(5/4 + \sqrt{5}/2\right)(1 + |a|)^2$, there exists an R such that $F_{a,c} : \overline{D_R}^2 \to \mathbb{C}^2$ is a complex horseshoe. The key to showing that the former field of cones is F−trapping is the observation that $F(x, y) \in \overline{D_R}^2$ implies that $|x^2 + c - ay| \leq R$ which implies that

$$|x|^2 \geq |c| - R(1 + |a|).$$

This result says essentially everything about Hénon maps in the parameter range to which it applies [Obe87, Obe00].

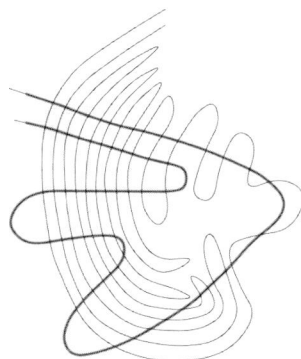

Fig. 2.10. Smale horseshoes from transverse homoclinic points.

Now, suppose that $F : \mathbb{C}^2 \to \mathbb{C}^2$ is an analytic map. Recall that a point q is a *homoclinic point* of F if there exists a positive integer k such that

$$\lim_{n \to \infty} F^{\circ kn}(q) = \lim_{n \to \infty} F^{\circ -kn}(q) = p, \tag{2.3}$$

where the limits exist. Note that the limit point, p, in (2.3) is a hyperbolic periodic point of F of period least k satisfying (2.3). To rephrase this definition,

q is in both the *stable and unstable manifolds* of F at p, which are denoted by W^s and W^u, respectively. We call a homoclinic point *transversal*, if these invariant manifolds intersect transversally. Also, note that the invariant manifolds W^s and W^u tend to intersect in lots of points in $\mathbb{C}^2$ unless F is linear or affine. This is quite different from the case of these invariant manifolds in $\mathbb{R}^2$. These intersections are almost always transversal. In the real domain, Smale showed that in a neighborhood of a transversal homoclinic point there exist horseshoes (see Figure 2.10). Here we give an analogous result in the case of complex horseshoes: for every positive integer $d \geq 2$, there exists an embedded bi–disc, B_d, centered at p and a positive integer $N = N(d)$ such that $F^{\circ kN} : B_d \to \mathbb{C}^2$ is a complex horseshoe of degree d. There exist $D_s \subset W^s$ and $D_u \subset W^u$ isomorphic to discs with $D_s, D_u \subset U$ and a positive integer n such that $F^{\circ n}(D_u)$ intersects D_s in exactly d points and the following conditions hold

$$F^{\circ n}(D_u) \cap \partial \overline{D_s} = \emptyset \qquad \text{and} \qquad F^{\circ n}(\partial \overline{D_u}) \cap D_s = \emptyset.$$

Note that the set $F^{\circ n}(D_u) \cap D_s$ must consist of finitely many points for all positive integers n. If $F^{\circ n}(D_u) \cap \partial \overline{D_s} \neq \emptyset$, then take D_s to be slightly smaller. If $F^{\circ n}(\partial \overline{D_u}) \cap D_s \neq \emptyset$, then take D_u to be slightly smaller. In either case, make the adjustments to keep the same number of points in $F^{\circ n}(D_u) \cap D_s$. If there are more than d points in $F^{\circ n}(D_u) \cap D_s$, then deform D_u slightly, making it smaller, to exclude some points from $F^{\circ n}(D_u) \cap D_s$. This can be done in an orderly way. In particular, there is a natural metric on D_s — —-the Poincaré metric of a disc and consider the point x of $F^{\circ n}(D_u) \cap D_s$ which is furthest from p in this metric (or one such point if more than one has this property). Now deform D_u by taking out $F^{\circ -n}(x)$ and staying clear of the other preimages of points in $F^{\circ n}(D_u) \cap D_s$ [Obe87, Obe00].

There exists a nonnegative integer m such that, if $D_{u,m} = F^{\circ -m}(D_u)$ and $U_m = D_{u,m} \times D_s$, then $F^{\circ n+m}|_{U_m}$ is a Hénon–like map of degree d.

2.4 Chaos in Functional Delay Equations

In this section, following [Lan99, ILW92], we will try to find and describe complicated solution behavior in a *functional differential equation* (FDE) as simple as possible. Let E, F be normed spaces, and $M \subset E$ be an open subset. For $k \in \mathbb{N}_0$, the space of bounded C^k–maps $f : M \to F$ with bounded derivatives up to order k is denoted by $BC^k(M, F)$. The *norm* on this space is given by

$$|f|_{C^k} = \max_{j=0,\dots,k} \sup_{x \in M} |D^j f(x)|,$$

with the usual norm on $L_c(E, F)$.[2] For compact intervals $I \subset \mathbb{R}$, the space $C^k(I, F)$ is defined analogously. When convenient, we write $(BC^k(M, F), |\ |_{C^k})$ and $(C^k(I, F), |\ |_{C^k})$.

---

[2] The space $BC^k(M, F)$ is a Banach space if F is a Banach space.

Let $C = C^0([-1,0],\mathbb{R})$. If $I \subset \mathbb{R}$ is an interval, $t \in \mathbb{R}$, $[t-1,t] \subset I$ and $x : I \longrightarrow \mathbb{R}$ is continuous, the segment $x_t \in C$ is defined by

$$x_t(s) = x(t+s), \qquad (s \in [-1,0]).$$

If $G \in BC^1(C,\mathbb{R})$ and $\varphi \in C$, the initial value problem

$$\dot{x}(t) = G(x_t), \qquad x_0 = \varphi,$$

has a unique solution $x^{\varphi,G} : [-1,\infty) \longrightarrow \mathbb{R}$ (see [HV93, DGV95]). The map

$$\Phi^G : \mathbb{R}_0^+ \times C \longrightarrow C, \qquad (t,\varphi) \mapsto x_t^{\varphi,G}$$

is a *continuous semi–flow*. The restriction $\Phi^G_{|(1,\infty)\times C}$ is C^1, and $D_2\Phi^G$ exists on $\mathbb{R}_0^+ \times C$ [Lan95a].

If $g : \mathbb{R} \longrightarrow \mathbb{R}$ and $G(\psi) = g(\psi(-1))$ for all $\psi \in C$, then the FDE

$$(G) : \dot{x}(t) = G(x_t) \qquad \text{is equivalent to}$$
$$(g) : \dot{x}(t) = g(x(t-1)).$$

We also write Φ^g for the semi–flow Φ^G in this case. It is easy to see that if $g \in BC^1(\mathbb{R},\mathbb{R})$, then $G \in BC^1(C,\mathbb{R})$, and

$$DG(\psi)\chi = g'(\psi(-1))\chi(-1) \qquad \text{for} \quad \psi, \chi \in C.$$

Note that, if $G(\psi) = -\mu\psi(0) + g(\psi(-1)),$
then (G) reads as $\dot{x}(t) = -\mu x(t) + g(x(t-1)).$

FDEs of type (G) are used as models in engineering and biology (see, e.g., [Dri77]).

Now, numerical solutions that appear to be *chaotic* have been observed in a number of experiments; among the earliest are the works of Mackey and Glass [MG77] and Lasota [Las77, LW76] on models of type (μ, g) for physiological control processes with delayed negative feedback (see [GP83] for a computation of dimension–like quantities from numerical trajectories of the *Mackey–Glass equation*).

If 'chaos' is to be described in mathematical terms, the first question is how to render an analytical notion of *chaotic behavior*. One way of doing this is to demand the existence of a subset of the state space on which the dynamics is equivalent to the index shift on a space of symbol sequences indexed with $\mathbb{Z}$. Since the work of Morse [Mor21, MH38], the symbol shift is a standard model for irregular, *quasi–random motion*. Typically one extracts a *Poincaré map Π* from the semi–flow and proves that trajectories $(x_n)_{n\in\mathbb{Z}}$ of Π in some open set W can be described by symbol sequences $(a_n)_{n\in\mathbb{Z}}$. The value of a_n indicates in which of several disjoint subsets of W the point x_n is contained. The existence of such a shift embedding follows if Π has a hyperbolic fixed–point and a

trajectory transversally homoclinic to that fixed–point. For diffeomorphisms
in finite dimension, this result is due to Smale [Sma67]. A shadowing lemma
approach that yields a description of trajectories in some neighborhood of the
homoclinic orbit was presented by [Pal88, KS90]. For non–invertible maps in
infinite dimension, corresponding results which even give a complete symbolic
coding of all trajectories in some neighborhood were proved by [EL86a], and,
using a generalized shadowing lemma, by [SW90].

The *transversally homoclinic* framework applies to all presently known examples of delay equations for which complicated motion is analytically proven.
For the smoothed step functions from the earliest examples by [Wal81, HW83],
complicated orbits were first constructed by embedding an interval map into a
Poincaré map for the semi–flow; the interval map had properties as in [LY77].
Later, [HL86b] remarked that these examples fit into the setting of transversally homoclinic points for a Poincaré map. A detailed proof of this statement
and a robustness result were given in [Lan95a]. In the more recent examples from [Wal89, LW96], the approach was from the beginning to construct
transversally *homoclinic points*.

Hale and Sternberg [HS88] obtained numerical evidence for the occurrence
of *homoclinic orbits* in the Mackey–Glass equation.

Let us briefly comment on the scope and the limitations of shift embedding
results in the 'transversally homoclinic' setting. On the one hand, the trajectories described by symbol sequences are robust under C^1-perturbations of the
equation (see [Lan95a, LW96]). In this sense, the presence of *shift dynamics*
is an essential feature of the dynamical system.

On the other hand, the points on the *chaotic trajectories* comprise only
a 'thin' *Cantor set*, and these trajectories are dynamically unstable. It may
well be that a typical solution does not spend much time in the vicinity of
the special solutions captured by the symbolic coding. Results that capture
complicated behavior on 'large' subsets of phase–space, for dynamical systems given by simple analytical expressions, are not easy to get even in low
finite dimensions (compare with [BC85, BC91, Laz94]). The observations from
[DW95] suggest that, numerically, erratic motion can take place for every initial value.

Now, if $g(0) = 0$, then the following

$$\text{Monotonicity Condition} \qquad (M): \qquad g' < 0$$

excludes erratic solutions, according to the *Poincaré–Bendixson theorem* for
delay equations proved by [MS96]. It was shown by [Wal95] that, under condition (M), equation (μ, g) with $\mu \geq 0$ has a global two–dimensional attractor
homeomorphic to a disc, and the planar dynamics on this attractor cannot be
complicated. Let us define the set S of data with at most one change of sign
by

$$S = \{ \varphi \in C \setminus \{0\} : \text{ There exist } a \in [-1, 0] \text{ and } j \in \{0, 1\} \text{ with}$$
$$0 \leq (-1)^j \varphi(t) \text{ for } t \leq a \qquad \text{and } (-1)^j \varphi(t) \leq 0 \text{ for } a \leq t.$$

Functions defined on an interval unbounded to the right are called *eventually slowly oscillating* if $x_t \in S$holds for all sufficiently large t. Further, $x : \mathbb{R} \longrightarrow \mathbb{R}$is called *slowly oscillating* if $x_t \in S$ for all $t \in \mathbb{R}$. The set $S$is of dominant importance for the semi–flow induced by equation (g), if condition (M)holds: $S$is invariant, and it was shown in [MW94] that the domain of attraction into $S$was open and dense in C [Lan99, ILW92].

Now consider the

Negative Feedback Condition (NF) : $x \cdot g(x) < 0$ for $x \neq 0$,

which is weaker than (M). The set S is also invariant if only condition (NF) holds, and the domain of attraction into S is open. Density of this domain in C is not yet proved, but likely. Condition (NF) still has a restrictive effect on the possible complexity of the semi–flow: It was shown by [Mal88] that under (NF), the set of all bounded solutions which are defined on all of $\mathbb{R}$ admits a *Morse decomposition*. This is a partition into N invariant subsets $S_1, \ldots, S_N$, and into solutions that are asymptotic to some S_j for $t \longrightarrow -\infty$ and to some S_k with $k < j$ for $t \longrightarrow \infty$. The index $i \in \{1, \ldots, N\}$ describes the number of zeroes per time unit, and the solutions in the 'lowest' *Morse set S_1* are the slowly oscillating solutions. The Morse decomposition result means that, if the behavior of the solutions inside each S_j is ignored, the orbit structure looks like the attractor of a gradient flow – equilibria and connecting orbits. However, it was shown in [LW96] that (NF) does not exclude complicated motion, and that erratic solutions of (g) can be found within the slowly oscillating class. The example function g from [LW96] has a rather complicated shape with at least two extrema.

2.4.1 Poincaré Maps and Homoclinic Solutions

Here, following [Lan99, ILW92], we set up a framework for the analytic description of erratic solutions of FDEs. Symbolic coding of the solutions, and their robustness under perturbation of the equation, is expressed in terms of associated Poincaré maps.

A sequence $(\chi_n)_{n \in \mathbb{Z}}$ of points from the domain of a map P is called a *trajectory* of P if $P(\chi_n) = \chi_{n+1}$ for $n \in \mathbb{Z}$.

Let X be a *Banach space*, $H \subset X$ a closed hyperplane, and $\Phi : \mathbb{R}_0^+ \times X \longrightarrow X$ a semi–flow. Let $U \subset C$ and $\varphi \in U$. A map $P : U \longrightarrow H$ is called a *Poincaré map* for Φ if the following conditions hold:

(i) There exist an open interval I with $\inf I > 0$ and a map $\tau : U \longrightarrow I$ with the property

$$\text{for all } \ t \in I, \psi \in U : \quad \Phi(t, \psi) \in H \quad \Longleftrightarrow \quad t = \tau(\psi).$$

(ii) $P(\psi) = \Phi(\tau(\psi), \psi))$ for all $\psi \in U$.

If $n \in \mathbb{N}$ and $H_i \subset X$ are hyper–planes, and $U_i \subset X$ are pairwise disjoint, and

$$P_i : U_i \longrightarrow H_i, \qquad \psi \mapsto \Phi(\tau_i(\psi), \psi), \qquad (i = 1, \ldots, n)$$

are Poincaré maps as above, then the map Π, defined as

$$\Pi : \bigcup_{i=1}^{n} U_i \longrightarrow \bigcup_{i=1}^{n} H_i, \qquad \Pi(\psi) = P_i(\psi) \qquad \text{if} \ \ \psi \in U_i$$

is also called a Poincaré map.

If Π is a Poincaré map, then every trajectory $(\chi_k)_{k \in \mathbb{Z}}$ of Π defines a function $x : \mathbb{R} \longrightarrow X$ with the properties [Lan99]:

(i)
$$x(t + s) = \Phi(s, x(t)) \qquad \text{for} \ \ t \in \mathbb{R}, s \geq 0;$$

(ii) There is a sequence $(t_k)_{k \in \mathbb{Z}}$ with $x(t_k) = \chi_k \ (k \in \mathbb{Z})$.

For every $k \in \mathbb{Z}$ there exists a unique $i(k) \in \{1, \ldots, n\}$ with $\chi_k \in U_{i(k)}$ and $\Pi(\chi_k) = \Phi(\tau_{i(k)}(\chi_k), \chi_k)$. Let us define

$$t_0 = 0, \qquad t_{k+1} = t_k + \tau_{i(k)}(\chi_k) \qquad \text{for} \ \ k \geq 0, \qquad \text{and}$$
$$t_{k-1} = t_k - \tau_{i(k-1)}(\chi_{k-1}) \qquad \text{for} \ \ k \leq 0.$$

Since τ_i maps into an interval I_i with $\inf I_i > 0 \ (i = 1, \ldots, n)$, we have $\sum_{k=0}^{n} t_{\pm k} \to \pm \infty$ for $n \to \infty$. Let us define $x : \mathbb{R} \to \mathbb{R}$ by $x(t) = \Phi(t - t_{k-1}, \chi_{k-1})$ if $t \in [t_{k-1}, t_k)$, $k \in \mathbb{Z}$. Then property (ii) is satisfied. Let $t \in \mathbb{R}, s \geq 0$. There exist uniquely determined $k, l \in \mathbb{Z}$ with $k \leq l$, with $t \in [t_{k-1}, t_k)$ and $t + s \in [t_{l-1}, t_l)$. Then we have

$$x(t + s) = \Phi(t + s - t_{l-1}, \chi_{l-1}) = \Phi(t + s - t_{l-1}, \Pi^{l-1-(k-1)}\chi_{k-1})$$
$$= \Phi(t + s - t_{l-1}, \Phi(\sum_{j=k-1}^{l-2} \tau_{i(j)}(\chi_j), \chi_{k-1}))$$
$$= \Phi(t + s - t_{l-1}, \Phi(\sum_{j=k-1}^{l-2} t_{j+1} - t_j, \chi_{k-1}))$$
$$= \Phi(t + s - t_{l-1}, \Phi(t_{l-1} - t_{k-1}, \chi_{k-1}))$$
$$= \Phi(s, \Phi(t - t_{k-1}, \chi_{k-1})) = \Phi(s, x(t)).$$

Let $H \subset X$ be a hyperplane. There exists a continuous linear functional $h \in X^*$ and $c \in \mathbb{R}$ such that $H = h^{-1}(\{c\})$. With $H_0 = \ker h$, one has $H = \varphi + H_0$ for every $\varphi \in H$.[3] If $I \subset \mathbb{R}^+$ is open and Φ is C^1 on $I \times X$, then smooth Poincaré maps are typically obtained from the *Implicit Function Theorem*, if the conditions $\Phi(t^*, \varphi) \in H$ and $D_1\Phi(t^*, \varphi)1 \notin H_0$ are satisfied (see, e.g., [Lan95a, DGV95]).

---

[3] Note that so far we have not required any smoothness properties of τ and P.

Poincaré maps associated with functional differential equations of type (G) can be described as functions not only of the initial value $\varphi \in C$ but also of the nonlinearity G. The following theorem is from [Lan95a]. For $F \in BC^1(C, \mathbb{R})$, we denote by $\Phi(\cdot, \cdot, F) : \mathbb{R}_0^+ \times C \longrightarrow C$ the semi–flow generated by equation (F). It follows from [Lan95a] that the map $\Phi : \mathbb{R}_0^+ \times C \times (BC^1(C, \mathbb{R}), |\ |_{C^1}) \longrightarrow C$ is continuous, and its restriction to $(1, \infty) \times C \times BC^1(C, \mathbb{R})$ is C^1.

Let $h \in C^*$ be a continuous linear functional, $c \in \mathbb{R}$, and set $H = h^{-1}(\{c\}) \subset C$. Assume $t_0 > 1$, that $\varphi \in C$, $G \in BC^1(C, \mathbb{R})$ and $x_{t_0}^{\varphi, G} \in H$, $h(\dot{x}_{t_0}^{\varphi, G}) \neq 0$. Let $\mathcal{U} \subset C$ be an open neighborhood of the set $\{x_s^{\varphi, G} : 0 \leq s \leq t_0\}$.

Then there exist bounded open neighborhoods $U \subset C$ of φ and $\mathcal{B} \subset BC^1(C, \mathbb{R})$ of G and a bounded C^1 map $\tau : U \times \mathcal{B} \longrightarrow \mathbb{R}$ with the following properties [Lan99]:

(i) For $F \in \mathcal{B}$, the map $P_F : U \longrightarrow H \subset C$, $\psi \mapsto \Phi(\tau(\psi, F), \psi, F)$ is a Poincaré map for $\Phi(\cdot, \cdot, F)$, and $P_F \in BC^1(U, C)$. Further, $\tau(\varphi, G) = t_0$, so that $P_G(\varphi) = \Phi(t_0, \varphi, G)$.

(ii) For $\psi \in U, F \in \mathcal{B}$ and $s \in [0, \tau(\psi, F)]$ one has $\Phi(s, \psi, F) \in \mathcal{U}$.

(iii) If $\tilde{G} \in \mathcal{B}$ and $D\tilde{G}_{|\mathcal{U}}$ is uniformly continuous on $\mathcal{U}$, then $DP_{\tilde{G}}$ is uniformly continuous, and the map $\mathcal{B} \ni F \mapsto P_F \in (BC^1(U, C), |\ |_{C^1})$ is continuous at $\tilde{G}$.

We want to describe erratic solutions of delay equations by 'chaotic' trajectories of a Poincaré map Π. Such trajectories exist in a neighborhood of a transversally homoclinic orbit of Π. We recall this notion, and we assume the reader to be familiar with local invariant manifolds at a hyperbolic fixed–point.

Let U be an open subset of the Banach space X, and let $\Pi : U \longrightarrow X$ be a C^1 map. Let $z \in U$ be a hyperbolic fixed–point of Π and let W^u, W^s denote local *unstable and stable manifolds* of Π at z, respectively. A trajectory $(x_n)_{n \in \mathbb{Z}}$ of Π is called a *transversally homoclinic trajectory* (or orbit) if the following condition is satisfied: There exists $n_0 \in \mathbb{N}$ such that for $m, n \in \mathbb{Z}$, $n \geq n_0, m \leq -n_0$, one has $x_m \in W^u, x_n \in W^s$, and $Df^{n-m}(x_m)$ maps the tangent space $T_{x_m} W^u$ isomorphically to a direct summand of the tangent space $T_{x_n} W^s$.

For $J, M \in \mathbb{N}$, define Σ_{JM} as the set of sequences $(a_n)_{n \in \mathbb{Z}} \in \{0, \dots, J\}^{\mathbb{Z}}$ composed of blocks $12 \dots J$ and of blocks of at least M zeroes. Σ_{JM} is a *metric space* with the metric

$$d((a_n), (b_n)) = \sup_{n \in \mathbb{Z}} 2^{-|n|} |a_n - b_n|.$$

Let us define the *shift operator*

$$\sigma_{JM} : \Sigma_{JM} \longrightarrow \Sigma_{JM}, \qquad (a_n) \mapsto (a_{n+1}).$$

If a C^1 map Π has a transversally homoclinic orbit, all trajectories in a neighborhood of this orbit can be described by the symbol sequences from some space Σ_{JM} ([SW90]). The first result of this type for non–invertible maps in infinite dimension is due to [HL86a].

Our next aim is to establish a *link between solutions of a delay equation homoclinic to a periodic solution and homoclinic orbits of associated Poincaré maps*. We do this for general equations of type (G) first, and then specialize to equations where the transversality criterion from [LW95] applies. Assume $G \in BC^1(C, \mathbb{R})$. Let $y : \mathbb{R} \longrightarrow \mathbb{R}$ be a periodic solution of (G) with minimal period $\eta > 0$ and $H = h^{-1}(\{c\}) \subset C$ a hyperplane as above. Assume $y_0 \in H$, $h(\dot{y}_\eta) \neq 0$. Let $\mathcal{U} \subset C$ be an open bounded neighborhood of the set $\{y_s : s \in \mathbb{R}\}$. Then there exist bounded open neighborhoods $U \subset C$ of y_0 and $\mathcal{B} \subset BC^1(C, \mathbb{R})$ of G and Poincaré maps $P_F : U \longrightarrow H$ ($F \in \mathcal{B}$) with the properties described above.

Let us define $U_H = U \cap H$ and set $\pi = P_{G|U_H} : U_H \longrightarrow H$. Now we set $H_0 = \ker h$. The hyperplane $H = y_0 + H_0$ is, in general, not a vector space, but we can identify H with the Banach space H_0 via the map $H_0 \ni v \mapsto y_0 + v$. Thus π is a C^1 map on an open subset of a Banach space, and $D\pi(\psi) \in L_c(H_0, H_0)$ for $\psi \in U_H$. Obviously $\pi(y_0) = y_0$.

Assume that the fixed–point y_0 of π is hyperbolic and unstable; let $W^u, W^s \subset U_H$ denote local unstable and stable manifolds of π at y_0. Then there exists an open neighborhood $\Lambda \subset U_H$ of y_0 such that π maps $\Lambda \cap W^u$ diffeomorphically onto W^u. Assume that $z : \mathbb{R} \longrightarrow \mathbb{R}$ is a solution of (G) different from y and that $(t_n)_{n \in \mathbb{Z}}$ is a strictly increasing sequence in $\mathbb{R}$ with the following properties:

(i) $z_{t_n} = \pi(z_{t_{n-1}}) \in W^u \cap \Lambda$ for $n < 0$;
(ii) $z_{t_0} = \pi(z_{t_{-1}}) \in W^u$;
(iii) $z_{t_1} \in W^s$, $\dot{z}_{t_1} \notin H_0$;
(iv) $z_{t_{n+1}} = \pi(z_{t_n}) \in W^s$ for $n \geq 1$.

Then there exist a bounded set $\mathcal{V} \subset C$, open bounded neighborhoods $U_0 \subset U_H$ of y_0 in H, U_1 of z_{t_0} in H and $\mathcal{G}$ of G in $BC^1(C, \mathbb{R})$, and C^1 Poincaré maps $\Pi_F : U_0 \cup U_1 \longrightarrow H \subset C$, with $F \in \mathcal{G}$, satisfying the following properties [Lan99, ILW92]:

a) $\operatorname{dist}(U_0, U_1) > 0$;
b) $\{z_{t_n} : n \neq 0\} \subset U_0$;
c) $\Pi_{G|U_0} = \pi_{|U_0}$;
d) $\Pi_G(z_{t_0}) = z_{t_1}$;
e) if $\tilde{G} \in \mathcal{G}$ and $D\tilde{G}_{|\mathcal{V}}$ is uniformly continuous, then $D\Pi_{\tilde{G}}$ is uniformly continuous, and the map $\mathcal{G} \ni F \mapsto \Pi_F \in BC^1(U_0 \cup U_1, H)$ is continuous at $\tilde{G}$;
f) with $\chi_n = z_{t_n}$ ($n \in \mathbb{Z}$), the sequence (χ_n) is a homoclinic orbit of Π_G.

Choose a bounded open neighborhood $\mathcal{U}_1$ of $z_t t \in \mathbb{R}$ in C. With $\mathcal{U}$ from the passage preceding the lemma, we set $\mathcal{V} = \mathcal{U} \cup \mathcal{U}_1$. Note that, since (G) is autonomous, we have with $\varphi = z_{t_0}$ that $x^{\varphi, G}(t) = z(t_0 - t)$ for $t \geq 0$. The conditions $\dot{z}_{t_1} \notin H_0, z_{t_1} \in W^s \subset H$ imply the existence of bounded open neighborhoods $U_1' \subset C$ of z_{t_0} and $\mathcal{B}_1$ of G in $BC^1(C,\mathbb{R})$ and of C^1 Poincaré maps $P_F^{(1)} : U_1' \to H \subset C$ as above.

With U from the passage before the statement of the lemma, we have $z_{t_n} \in U$ for all $n \in \mathbb{Z}$. Since W^u and W^s are local unstable and stable manifolds of π, properties (i) and (iv) imply that $z_{t_n} \to y_0$ for $n \to \pm\infty$, so the set $Z = \{y_0\} \cup z_{t_n} n \in \mathbb{Z} \backslash \{0\}$ is compact. Note that $z_{t_0} \notin Z$, since the z_{t_n} $(n \leq 0)$ are pairwise different and different from y_0, and since $W^u \cap W^s = \{y_0\}$. We can choose disjoint bounded open neighborhoods U_0 and U_1 of Z resp. z_{t_0} in H such that $Z \subset U_0 \subset U_H$, $U_1 \subset U_1' \cap U_H$, and $\mathrm{dist}(U_0, U_1) > 0$. Assertions a) and b) are then true. Recall $\mathcal{B}$ from the passage preceding the lemma; set $\mathcal{G} = \mathcal{B}_1 \cap \mathcal{B}$. For $F \in \mathcal{G}$, we define Π_F by

$$\Pi_F(\psi) = \begin{cases} P_F^{(1)}(\psi), & \psi \in U_1, \\ P_F(\psi), & \psi \in U_0. \end{cases}$$

The maps Π_F are Poincaré maps.

We get robust 'chaotic' trajectories of Poincaré maps if the homoclinic orbit (χ_n) above has the transversality property. For arbitrary sets $A \subset B$ and maps $f : A \longrightarrow B$, we write $\mathrm{traj}(f, A)$ for the set of all trajectories of $f \in A$. We denote the maximal invariant subset of $f \in A$ by $\Omega(f, A)$; then

$$x \in \Omega(f, A) \qquad \Longleftrightarrow \qquad \text{there is } (x_n) \in \mathrm{traj}(f, A), \qquad x_0 = x.$$

If A is a bounded subset of a normed space with norm $|\ |$, we give $\mathrm{traj}(f, A)$ the topology induced by the metric

$$d((x_n), (y_n)) = \sup_{n \in \mathbb{Z}} 2^{-|n|} |x_n - y_n|.$$

The map f induces a shift operator

$$\hat{f} : \mathrm{traj}(f, A) \longrightarrow \mathrm{traj}(f, A), \qquad (x_n) \mapsto (x_{n+1}) = (f(x_n)).$$

We do not repeat the definition of the term *hyperbolic set* for non–invertible maps, which is used in the theorem below (see [SW90, LW96]). We just mention that the main difference in comparison to diffeomorphisms is that one gives up the invariance of the unstable directions under the derivative of the map.

In the above situation, we assume in addition that

a) the homoclinic orbit (χ_n) of Π_G is transversal in the above sense;
b) $D\tilde{G}_{|\mathcal{V}}$ is uniformly continuous.

Then there exists $\bar{\epsilon}_0 > 0$ such that for $\epsilon_0 \in (0, \bar{\epsilon}_0]$ there exist $J, M \in \mathbb{N}$ open subsets $V_0, \ldots, V_J$ of $U_0 \subset U_1$ and a neighborhood $\mathcal{F} \subset \mathcal{G}$ of G in $BC^1(C, \mathbb{R})$ such that the following properties hold [Lan99]:

(i) $\mathrm{dist}(V_i, V_j) > 0$ if $i, j \in \{0, \ldots, J\}, i \neq j$;

(ii) $\operatorname{diam}(V_i) \leq \epsilon_0, \qquad i \in \{0, \dots, J\}$.

(iii) Now we set $W = V_0 \cup \dots \cup V_J$ and define the 'position' map $p : W \longrightarrow \{0, \dots, J\}$,

$p(x) = i$ if $x \in V_i$. For every $F \in \mathcal{F}$, the map

$$\operatorname{traj}(\Pi_F, W) \longrightarrow \{0, \dots, J\}, \qquad (x_n) \mapsto (p(x_n))$$

maps into Σ_{JM} and induces a homeomorphism

$$\hat{p} : \ \operatorname{traj}(\Pi_F, W) \longrightarrow \Sigma_{JM}, \qquad \text{with} \qquad \hat{p} \circ \hat{\Pi}_F = \sigma_{JM} \circ \hat{p};$$

(iv) for every $F \in \mathcal{F}, \qquad \Omega(\Pi_F, W)$ is a hyperbolic set for Π_F.

Conditions b) and e) above imply that $D\Pi_G$ is uniformly continuous. This fact, together with condition a), allows us to apply results from [Lan95b]. We get $\bar{\epsilon}_0 > 0$ such that for $\epsilon_0 \in (0, \bar{\epsilon}_0]$ there exist $J, M \in \mathbb{N}$ and open subsets $V_0, \dots, V_J$ of $U_0 \cup U_1$ with properties (i) and (ii); further, setting $W = V_0 \cup \dots \cup V_J$, the position map p induces a homeomorphism

$$\tilde{p} : \operatorname{traj}(\Pi_G, W) \to \Sigma_{JM}.$$

Let $\epsilon_0 \in (0, \bar{\epsilon}_0]$. From [Lan95b], we get $\tilde{\epsilon} > 0$ such that the $\tilde{\epsilon}$−neighborhood of $\Omega(\Pi_G, W)$ is contained in W, and with the following property: For every $\epsilon \in (0, \tilde{\epsilon}]$ there is a neighborhood Γ_ϵ of Π_G in $BC^1(U_0 \cup U_1, H)$ such that the following two statements hold for $\phi \in \Gamma_\epsilon$:

1) For every trajectory $(x_n) \in \operatorname{traj}(\Pi_G, W)$ there exists a unique trajectory $(y_n) \in \operatorname{traj}(\phi, W)$ with $|y_n - x_n| \leq \epsilon \ (n \in \mathbb{Z})$, and the corresponding map

$$h : \operatorname{traj}(\Pi_G, W) \to \operatorname{traj}(\phi, W)$$

is a homeomorphism.

2) The set $\Omega(\phi, W)$ is a *hyperbolic set* for ϕ.

Choose $\epsilon \in (0, \tilde{\epsilon}]$ such that $\epsilon < \min_{i \neq j, i, j \in \{1, \dots, n\}} \operatorname{dist}(V_i, V_j)$. Then, for $x \in \Omega(\Pi_G, W)$, one has $x \in V_j$ for some $j \in \{0, \dots, J\}$, and for $y \in W$ with $|x - y| \leq \epsilon$, one also has $y \in V_j$, so that $p(x) = p(y)$.

For $\phi \in \Gamma_\epsilon$, the map

$$\tilde{p} \circ h^{-1} : \operatorname{traj}(\phi, W) \to \Sigma_{JM}$$

is a *homeomorphism*. If $(y_n) = h((x_n)) \in \operatorname{traj}(\phi, W)$, we have $|x_n - y_n| \leq \epsilon \ (n \in \mathbb{Z})$, and

$$(\tilde{p} \circ h^{-1})((y_n)_{n \in \mathbb{Z}}) = \tilde{p}((x_n)_{n \in \mathbb{Z}}) = (p(x_n)_{n \in \mathbb{Z}}) = (p(y_n))_{n \in \mathbb{Z}}.$$

Thus p induces the homeomorphism $\hat{p} = \tilde{p} \circ h^{-1}$. For $(y_n) \in \operatorname{traj}(\phi, W)$, one has

$$(\hat{p} \circ \hat{\phi})((y_n)) = \hat{p}(\phi(y_n)) = \hat{p}((y_{n+1})) = (p(y_{n+1}))$$
$$= \sigma_{JM}[(p(y_n))] = (\sigma_{JM} \circ \hat{p})((y_n)).$$

Condition b) and e) above show that there exists a neighborhood $\mathcal{F} \subset \mathcal{G}$ of G in $BC^1(C, \mathbb{R})$ such that for all $F \in \mathcal{F}$, one has $\Pi_F \in \Gamma_\epsilon$. It is now obvious that properties (iii) and (iv) hold for $F \in \mathcal{F}$.

In the above result, we had assumed the transversality property of the homoclinic orbit (χ_n). It has to be verified when we construct our example, and we employ the criterion from [LW95] that describes transversality in terms of oscillation properties. We now specialize to the case $G(\psi) = g(\psi(-1))$, that is, to delay equations of type (g). If x is a solution of (g), the *variational equation* along x is [Lan99]

$$\dot{v}(t) = g'(x(t - 1))v(t - 1).$$

Let $g \in BC^1(\mathbb{R}, \mathbb{R})$. Assume that $y : \mathbb{R} \longrightarrow \mathbb{R}$ is a periodic solution of (g) with minimal period $\eta > 0$. Assume further that the following conditions hold:

(i) y is slowly oscillating;
(ii) g satisfies (NF);
(iii) $g' < 0$ on the range $y(\mathbb{R})$ of y.
(iv) The spectrum of the *monodromy operator* $D_2\Phi^g(\eta, y_0)$ contains a point $\lambda^u \in (1, \infty)$.

As above, let π be a Poincaré map associated with y and ω. Then y_0 is a hyperbolic fixed–point of π with 1D unstable space. If W^u, W^s are local unstable and stable manifolds of π at y_0 then

$$\dim W^u = \operatorname{codim} W^s = 1.$$

The assertions are proved in [LW95]. In that reference, it was assumed that $g' < 0$ on a neighborhood of $y(\mathbb{R})$, which follows from compactness of $y(\mathbb{R})$ and condition (iii).

In the above situation (with the homoclinic solution z), let the underlying equation be (g) with $g \in BC^1(C, \mathbb{R})$, and set $G(\psi) = g(\psi(-1))$ $(\psi \in C)$.

a) Then $DG_{|V}$ is uniformly continuous.

b) Assume in addition that y and g satisfy the above conditions. Now we set $\varphi = z_{t_0} \in W^u$ and let $\omega \in H$ be a vector that spans the 1D tangent space $T_\varphi W^u$. Consider the solution $w : [t_0 - 1, \infty) \longrightarrow \mathbb{R}$ of the variational equation (g, z) with $w_{t_0} = \omega$. Assume that the following condition is satisfied:
Transversality Condition (T) :   For all $(a, b) \in \mathbb{R}^2 \setminus \{(0, 0)\}$, the function $a \cdot \dot{z} + b \cdot w$ is eventually slowly oscillating. Then the homoclinic orbit (χ_n) of Π_G above is transversal, and the above conclusions hold.

c) Under the conditions of b), the domain $U_0 \cup U_1$ of Π_G can be chosen such that, for all trajectories of Π_G, the corresponding solutions of (g) are slowly oscillating [Lan99].

Since $\mathcal{V}$ is bounded, the set $V = \psi(-1)\psi \in \mathcal{V} \subset \mathbb{R}$ is bounded, and g' is uniformly continuous on V; in particular, $g'_{|V}$ is uniformly continuous. For $\psi, \tilde{\psi} \in \mathcal{V}, \chi \in C$ we have

$$|DG(\psi)\chi - DG(\tilde{\psi})\chi| \leq |g'(\psi(-1)) - g'(\tilde{\psi}(-1))||\chi|,$$

and hence uniform continuity of $DG_{|V}$ follows from uniform continuity of $g'_{|V}$.

We have to show that above transversality property holds. Now we set $n_0 = 1$. We have

$$\chi_n = z_{t_n} \in W^u \cap \Lambda \text{ for } n \leq -1 \text{ and } \chi_n \in W^s \text{ for } n \geq 1.$$

Let $m \leq -1, n \geq 1$. Let ω_m be a vector that spans $T_{\chi_m} W^u$. From the fact that π induces a diffeomorphism on $W^u \cap \Lambda$ we see that there exists $\lambda \neq 0$ such that

$$D\pi^{|m|}(\chi_m)\omega_m = \lambda\omega.$$

It follows from assertions b) and c) above that also

$$D\Pi_G^{|m|}(\chi_m)\omega_m = \lambda\omega. \tag{2.4}$$

For $j \geq 1$, let $pr_j \in L_c(H_0, H_0)$ denote the projection onto H_0 parallel to $\dot{z}_{t_j}$. Recall that the derivative $D\Pi^G$ is given by the derivative of the semi-flow $D_2\Phi^G$, followed by projection onto H_0 (see [DGV95]), and that $D_2\Phi^G$ is given by solutions of the variational equation (g, z). In formulas, we have

$$D\Pi_G(z_{t_0})\omega = pr_1(D_2\Phi^G(t_1 - t_0, z_{t_0})\omega) = pr_1(w_{t_1}).$$

There exists $\alpha \in \mathbb{R}$ with

$$D\Pi_G(z_{t_0})\omega = w_{t_1} - \alpha\dot{z}_{t_1}. \tag{2.5}$$

Next, we have

$$D\Pi_G^{n-1}(z_{t_1})(w_{t_1} - \alpha\dot{z}_{t_1}) = [pr_n \circ D_2\Phi^G(t_n - t_1, z_{t_1})](w_{t_1} - \alpha\dot{z}_{t_1})$$
$$= pr_n D_2\Phi^G(t_n - t_1, z_{t_1})w_{t_1} - \underbrace{\alpha \cdot pr_n \dot{z}_{t_n}}_{=0}$$
$$= pr_n w_{t_n},$$

and there exists $\beta \in \mathbb{R}$ with

$$D\Pi_G^{n-1}(z_{t_1})(w_{t_1} - \alpha\dot{z}_{t_1}) = w_{t_n} - \beta\dot{z}_{t_n}. \tag{2.6}$$

Formulas (2.4), (2.5) and (2.6) together give

$$D\Pi_G^{n-m}(\chi_m)\omega_m = D\Pi_G^{n-1}(z_{t_1})D\Pi_G(z_{t_0})D\Pi_G^{|m|}(\chi_m)$$
$$= \lambda w_{t_n} - \lambda\beta\dot{z}_{t_n}.$$

Now we set $\omega_n = D\Pi_G^{n-m}(\chi_m)\omega_m$. Since codim $T_{\chi_n}W^s = 1$, we have to show $\omega_n \notin T_{\chi_n}W^s$. Let ζ denote the solution of (g) with initial value $\zeta_0 = \chi_n$, and let ν denote the solution of the variational (g, ζ) with $\nu_0 = \omega_n$. We then have

$$\zeta(\cdot) = z(t_n + \cdot), \nu(\cdot) = \lambda w(t_n + \cdot) - \lambda\beta\dot{z}(t_n + \cdot). \tag{2.7}$$

The transversality criterion from [LW95] shows that $\omega_n \notin T_{\chi_n}W^s$ if, for all $(a, b) \in \mathbb{R}^2 \setminus \{(0,0)\}$, the function $a\zeta + b\nu$ is eventually slowly oscillating (here we use that $g' < 0$ on $y(\mathbb{R})$). In view of (2.7), the latter condition is equivalent to the property that for all $(a, b) \in \mathbb{R}^2 \setminus \{(0,0)\}$ the function $a\dot{z} + bw$ is eventually slowly oscillating, which is just condition (T). Consequently, the homoclinic orbit (χ_n) is transversal. In view of assertion a), we see that the above hypotheses.

Recall that U_0 and U_1 are neighborhoods of y_0 and z_{t_0}, respectively. Since y is slowly oscillating, there exists $T > 0$ with $y_T > 0$ ([LW95]). Hence there exists a neighborhood $U_{0,+}$ of y_0 in C such that $x_T^\psi > 0$ for all $\psi \in U_{0,+}$. As a first consequence, it follows from this property and from convergence of z_t to the orbit of y for $t \to -\infty$ that there exist arbitrarily large $t > 0$ with $z_{-t} > 0$. Hence, since the set S is invariant under the semi–flow induced by the negative feedback equation (g), the solution z is slowly oscillating. Repeating the above argument for z_{t_0} instead of y_0, we see that there exists a neighborhood $U_{1,+}$ of z_{t_0} in C and a $T_1 > 0$ such that $x_{T_1}^\psi > 0$ for all $\psi \in U_{1,+}$. Assertion c) now follows if we choose U_0 and U_1 as subsets of $U_{0,+}$ and $U_{1,+}$, respectively.

2.4.2 Starting Value and Targets

As in [LW96], we start with a function $g \in BC^1(\mathbb{R}, \mathbb{R})$ with the property: $g' < 0$, g' is decreasing on an interval $[0, A]$ $(A > 0)$. Further, $g(-x) = -g(x)$ $(x \in \mathbb{R})$, (g) has a slowly oscillating unstable periodic solution $y : \mathbb{R} \longrightarrow \mathbb{R}$ with $y(\mathbb{R}) \subset [-A, A]$. Also, y has the symmetry

$$y(t - 2) = -y(t), \qquad \text{for all } t \in \mathbb{R},$$

and is called a *Kaplan–Yorke solution*, since such symmetric solutions were first obtained by [KY75]. The phase of y is chosen such that

$$y(-1) = 0, \qquad y > 0 \qquad \text{on } (-1, 1).$$

In the notation of [LW96], we have $(y, g) \in YG$. On the space $C = C^0([-1, 0], \mathbb{R})$, we write $|\ |$ for $|\ |_{C^0}$, unless the more explicit notation seems favorable. For a normed space $(E, |\ |)$ and $r > 0$, we use the notation $E(r)$ for the open ball with radius r in E.

Now we set $H = \{\psi \in C : \psi(-1) = 0\}$ and let $\mathcal{O} = \{y_t : t \in \mathbb{R}\}$ denote the orbit of y in C. We have $y_0 = y_4 \in H$, $\dot{y}_4 \notin H$. Let us recall some facts from [LW96]: There exists an open neighborhood U_P of y_0 in H and a Poincaré map $P : U_P \longrightarrow H \subset C$. We have $P \in BC^1(U_P, H)$, $P(y_0) = y_0$, and

$DP(y_0)$ is hyperbolic with unstable space $U \subset H$, stable space $Z \subset H$, and $\dim U = 1$.

There exists a norm $\| \ \|$ on H equivalent to $| \ |_{C^0}$ such that $DP(y_0)$ is expanding on U and contracting on Z w.r. to $\| \ \|$, and such that

$$\|v + w\| = \max\{\|v\|, \|w\|\} \qquad \text{for } v \in Z, \ w \in U.$$

For $\delta > 0$ and a subspace $W \subset H$, we set

$$W_{\| \ \|}(\delta) = \{\psi \in W \ : \ \|\psi\| < \delta\} \quad \text{and} \quad W(\delta) = \{\psi \in W \ : \ |\psi|_{C^0} < \delta\}.$$

Now we set

$$N_\delta = \{\psi \in H \ : \ \|\psi - y_0\| < \delta\} = y_0 + Z_{\| \ \|}(\delta) + U_{\| \ \|}(\delta).$$

For $\psi \in C$, let $x^\psi : [-1, \infty) \longrightarrow \mathbb{R}$ denote the solution of (g) with $x_0 = \psi$.

We can choose $\bar{\delta} \in (0, 1/4]$ with the following properties: For $\delta \in (0, \bar{\delta}]$, the local unstable and stable sets $W^u(y_0, P, N_\delta)$, $W^s(y_0, P, N_\delta)$ coincide with the graphs

$$y_0 + \{\psi + w^u(\psi) : \ \psi \in U_{\| \ \|}(\delta)\}, \qquad y_0 + \{\psi + w^s(\psi) : \ \psi \in Z_{\| \ \|}(\delta)\}$$

of C^1-maps

$$w^u : U_{\| \ \|}(\delta) \longrightarrow Z_{\| \ \|}(\delta) \subset Z, \qquad w^s : Z_{\| \ \|}(\delta) \longrightarrow U_{\| \ \|}(\delta) \subset U,$$

respectively. One has

$$w^u(0) = 0, \qquad w^s(0) = 0, \qquad Dw^u(0) = 0, \qquad Dw^s(0) = 0.$$

For $f : \mathbb{R} \longrightarrow \mathbb{R}$ and $M \subset \mathbb{R}$, by a modification of f on $\mathbb{R} \setminus M$ we mean a function $\tilde{f} : \mathbb{R} \longrightarrow \mathbb{R}$ with $\tilde{f}(x) = f(x)$ for $x \in M$. We will construct modifications of g outside the range of y. The aim is that solutions of the modified equation which start in some unstable set $W^u(y_0, P, N_\delta)$ have their segments at a time approximately equal to 3 in a small neighborhood of y_0.

We first prove some preparatory statements that hold for *all* sufficiently small $\delta > 0$, where δ is essentially a measure for the distance of the starting value in $W^u(y_0, P, N_\delta)$ to y_0. Later on, we pick an appropriate fixed $\delta > 0$.

There exist $k_1 \in (0, 1]$, $k_2 > 0$, $k_3 \in (0, 1]$ such that the following properties hold for all $\delta \in (0, \bar{\delta}]$ [Lan99, ILW92]:

a) $\psi \in N_{k_1\delta} \Longrightarrow |\psi - y_0| < \delta$;
b) for all $\psi \in W^s(y_0, P, N_{k_1\delta})$ and for all $t \geq 0 : \qquad x_t^\psi \in \mathcal{O} + C(\delta)$;
c)
$$\psi \in H, \ |\psi - y_0| \leq k_3\delta \Longrightarrow x^\psi(1 - \delta) \geq k_2\delta.$$

There exist $K > 0$ and $\lambda > 0$ with the property

for all $\psi \in W^s(y_0, P, N_{\bar{\delta}})$: for all $t \geq 0$: $\mathrm{dist}(x_t^\psi, \mathcal{O}) \leq K \exp(-\lambda t)|\psi - y_0|$,

where 'dist' refers to $|\ | = |\ |_{C^0}$. Choose $T > 1$ such that $K \exp(-\lambda T) < 1$. There exists $c > 0$ such that

for all $\psi, \chi \in C$ for all $t \in [0, T]$: $|x^\psi(t) - x^\chi(t)| \leq c|\chi - \psi|$.

There exist constants $c_1, c_2 > 0$ such that the estimates $\|\ \| \leq c_1|\ |$ and $|\ | \leq c_2\|\ \|$ hold. Now we set $k_1 = \frac{1}{2(cc_2 + c_2 + 1)}$. For $\psi \in N_{k_1\delta}$, one has $|\psi - y_0| \leq c_2 k_1 \delta < \delta$, hence property a) holds.

If $\psi \in W^s(y_0, P, N_{k_1\delta})$ then $|x^\psi(t) - y(t)| \leq cc_2 k_1 \delta$ for $t \in [0, T]$, so $\mathrm{dist}(x_t^\psi, \mathcal{O}) \leq \max\{c_2 k_1 \delta; cc_2 k_1 \delta\} < \delta$ for $t \in [0, T]$. For $t \geq T$, we have

$$\mathrm{dist}(x_t^\psi, \mathcal{O}) \leq K \exp(-\lambda T)|\psi - y_0| \leq 1 \cdot c_2 k_1 \delta < \delta,$$

and property b) is proved.[4]

For $t \in (1/2, 1], \psi \in C$ with $|\psi - y_0| \leq 1$, one has

$$|\dot{x}^\psi(t) - \dot{y}(t)| = |g(x^\psi(t-1)) - g(y(t-1))| \leq |g|_{C^1}|\psi - y_0|.$$

Now we set

$$\mu = \min |g(y(s))| | s \in [-1/2, 0], \qquad k_3 = \min\{\frac{\mu}{2(c + |g|_{C^1})}, 1\}.$$

On $[1/2, 1]$ we have $|\dot{y}(t)| = |g(y(t-1))| \geq \mu > 0$, which implies $y(1-\delta) \geq \mu \cdot \delta$ for $\delta \in [0, 1/2]$. For $\delta \in (0, \bar{\delta}]$ and $\psi \in H$ with $|\psi - y_0| \leq k_3 \delta$, one has

$$x^\psi(1-\delta) = x^\psi(1) + \int_1^{1-\delta} \dot{x}^\psi(s)ds$$

$$\geq \underbrace{y(1)}_{=0} + \int_1^{1-\delta} \dot{y}(s)ds - |x^\psi(1) - y(1)| - \int_1^{1-\delta} |\dot{y}(s) - \dot{x}^\psi(s)|ds$$

$$\geq y(1-\delta) - ck_3\delta - \delta|g|_{C^1}k_3\delta$$

$$\geq (\mu - (c + |g|_{C^1})k_3)\delta \geq \frac{\mu}{2}\delta.$$

Property c) follows if we put $k_2 = \mu/2$.

It was shown in [LW96] that there exists an eigenvector $\chi^u \in H$ of $DP(y_0)$ with $U = \mathbb{R} \cdot \chi^u$, $\chi^u(0) > 0$, $\dot{\chi}^u > 0$ on $[-1, 0)$. We can choose χ^u such that $\|\chi^u\| = 1$. Let π^u, π^s denote the projections onto U, respectively Z, according to $H = U \oplus Z$. Recall the set $S \subset C$ of segments with at most one sign change.

---

[4] Note that $|x^\psi(1) - y(1)| = |x^\psi(1)| \leq c|\psi - y_0|$ for $\psi \in C$.

It is known that $S \cap Z = \emptyset$ (see [LW95]). Since $y_0 \in S$, we have $\pi^u y_0 \neq 0$, and so there exists a unique real number $\eta_0 > 0$ with $\pi^u y_0 = \eta_0 \cdot \chi^u$. Let us define $i : \mathbb{R} \longrightarrow U, \ r \mapsto r \cdot \eta_0 \cdot \chi^u$. The map i is an isomorphism, and $i^{-1}(\pi^u(y_0)) = 1$. Now we set

$$\mathcal{H} : N_{\bar{\delta}} \longrightarrow \mathbb{R}, \qquad \mathcal{H}(\psi) = i^{-1}[\pi^u(\psi - y_0) - w^s(\pi^s(\psi - y_0))].$$

Then, for all $\delta \in (0, \bar{\delta}]$ and $\psi \in N_\delta$, we have the equivalence

$$\psi \in W^s(y_0, P, N_\delta) \iff \mathcal{H}(\psi) = 0. \tag{2.8}$$

We provide a Lipschitz estimate for $\mathcal{H}$: For $\chi \in U$, $\chi = \lambda \cdot \chi^u$, we have $||\chi^u|| = \lambda$ and

$$|i^{-1}\chi| = |i^{-1}(\frac{\lambda}{\eta_0} \cdot \eta_0 \chi^u)| = \frac{|\lambda|}{\eta_0} = \frac{1}{\eta_0}||\chi||.$$

Since w^s is C^1 and $Dw^s(0) = 0$, there exists $\tilde{\delta} \in (0, \bar{\delta}]$ such that w^s has Lipschitz constant 1 on $Z_{||\ ||}(\tilde{\delta})$. Since the operator norms $||\pi^u||$ and $||\pi^s||$ are equal to one, we get for $\chi, \psi \in N_{\tilde{\delta}}$:

$$|\mathcal{H}(\psi) - \mathcal{H}(\chi)| \leq \frac{1}{\eta_0}[||\pi^u|| \cdot ||\psi - y_0|| + ||\pi^s|| \cdot ||\psi - y_0||] = \frac{2}{\eta_0}||\psi - y_0||.$$

We provide 'target' segments. Later, we want to steer the phase curves of solutions starting in some unstable manifold of y_0 into a neighborhood of these targets.

Now we set $\varphi_\lambda = (1 + \lambda)y_0$ for $\lambda \in \mathbb{R}$. There exists $\lambda_0 \in (0, 1)$ such that, for $\lambda \in [-\lambda_0, \lambda_0]$, one has [Lan99]

$$\mathcal{H}(\varphi_\lambda) \geq \frac{1}{2}\lambda \text{ if } \lambda > 0, \qquad \mathcal{H}(\varphi_\lambda) \leq \frac{1}{2}\lambda \text{ if } \lambda < 0.$$

Since $w^s(0) = 0$, $Dw^s(0) = 0$, there exists $r \in (0, 1)$ such that for $\chi \in Z$ with $||\chi|| \leq r$,

$$||w^s(\chi)|| \leq \frac{\eta_0}{2(||\pi^s y_0|| + 1)}||\chi||.$$

Now we set $\lambda_0 = \frac{r}{(||\pi^s y_0|| + 1)}$. For $\lambda \in [-\lambda_0, \lambda_0]$, we have

$$||\pi^s(\varphi_\lambda - y_0)|| = |\lambda| \cdot ||\pi^s y_0|| \leq \lambda_0 ||\pi^s y_0|| \leq r, \qquad \text{and}$$
$$w^s(\pi^s(\varphi_\lambda - y_0))|| \leq \frac{\eta_0}{2(||\pi^s y_0|| + 1)}|\lambda| \cdot ||\pi^s y_0|| \leq \frac{\eta_0}{2}|\lambda|.$$

It follows that, for $\lambda \in [-\lambda_0, \lambda_0]$, one has

$$|i^{-1}(w^s(\varphi_\lambda - y_0))| \leq \frac{1}{\eta_0}\frac{\eta_0}{2}|\lambda| = \frac{|\lambda|}{2}.$$

We get

$$\mathcal{H}(\varphi_\lambda) = i^{-1}[\pi^u(\lambda y_0) - w^s(\pi^s(\varphi_\lambda - y_0))]$$
$$= \lambda \underbrace{i^{-1}(\pi^u y_0)}_{=1} - i^{-1} w^s(\pi^s(\varphi_\lambda - y_0))$$
$$\in [\lambda - \tfrac{|\lambda|}{2}, \lambda + \tfrac{|\lambda|}{2}],$$

and the assertion follows.

There exist $\delta' \in (0, \tilde{\delta}]$ and constants $M_2 > 0$, $k_4 \in (0,1]$, $d_2 > 0$ such that for all $\iota \in (0, \delta']$ there exist functions $\psi_+, \psi_- \in N_\iota$ with the following properties [Lan99]:

(i) ψ_+, ψ_- are C^2, $\quad \dot{\psi}_\pm > 0$ on $[-1,0)$, $\dot{\psi}_\pm(0) = 0$, and $\ddot{\psi}_\pm < 0$ on $[-1,0]$, and $\ddot{\psi}_\pm(t) \le -d_2 < 0$ for $t \in [-1/2, 0]$.
(ii) $|\psi_\pm|_{C^2} \le M_2$.
(iii)For all $\psi \in \psi_+ + \overline{H(k_4\iota)}$ and for all $\chi \in \psi_- + \overline{H(k_4\iota)}$, one has $\psi, \chi \in N_\iota$, and

$$\mathcal{H}(\psi) > 0 > \mathcal{H}(\chi).$$

Choose λ_0 as above and choose $\delta' \in (0, \tilde{\delta}]$ such that $\delta' < 2||y_0|| \cdot \lambda_0$. Now we set

$$M_2 = (\lambda_0 + 1)|y_0|_{C^2} + 2 \quad \text{and} \quad k_4 = \min\{\frac{\eta_0}{32(||y_0|| + 1)}, \frac{1}{5}\}.$$

It follows from (g) that there exists $D_2 > 0$ such that $y_0''(t) \le -D_2 < 0$ for $t \in [-1/2, 0]$. Now we set $d_2 = (1 - \lambda_0)D_2$. Let $\iota \in (0, \delta']$ be given. Now we set $\lambda = \frac{\iota}{2(||y_0||+1)}$; then $\lambda \le \lambda_0$. With φ_λ as above, we have

$$||\varphi_\lambda - y_0|| = ||\lambda y_0|| \le \frac{\iota}{2} < \tilde{\delta}, \text{ and } \mathcal{H}(\varphi_\lambda) \ge \frac{\lambda}{2}.$$

If $\psi \in \varphi_\lambda + H_{||\,||}(2k_4\iota)$ then

$$||\psi - y_0|| \le 2k_4\iota + ||\varphi_\lambda - y_0|| = (2k_4 + 1/2)\iota < \iota \le \tilde{\delta},$$

so we have $\psi, \varphi_\lambda \in N_{\tilde{\delta}}$. Using the *Lipschitz estimate* for $\mathcal{H}$, we get for all $\psi \in \varphi_\lambda + H_{||\,||}(2k_4\iota)$:

$$\mathcal{H}(\psi) \ge \mathcal{H}(\varphi_\lambda) - \frac{2}{\eta_0} \cdot 2k_4\iota \ge \frac{\lambda}{2} - \frac{\iota}{8(||y_0|| + 1)} = \frac{\iota}{8(||y_0|| + 1)}. \tag{2.9}$$

Now we set $p(x) = (x-1)(x+1)$ for $x \in [-1, 0]$. Then

$$p'(x) = 2x < 0 \quad \text{for} \quad x \in [-1, 0),$$
$$p'(0) = 0, \qquad p''(x) = 2 \quad \text{for } x \in [-1, 0], \qquad \text{and}$$
$$|p|_{C^2} = \max\{|p|_{C^0}, |p'|_{C^0}, |p''|_{C^0}\} \le \max\{1, 2, 2\} = 2.$$

Consider the functions $\varphi_{\lambda,\epsilon} = \varphi_\lambda - \epsilon \cdot p$ for $\epsilon \in (0, 1)$. These functions are C^2 and have the following properties:

$$\varphi'_{\lambda,\epsilon} = \underbrace{(1+\lambda)y'_0}_{\geq 0} - \epsilon p' > 0 \quad \text{on } [-1,0), \varphi'_{\lambda,\epsilon}(0) = 0,$$

$$\varphi''_{\lambda,\epsilon} = \underbrace{(1+\lambda)y''_0}_{\leq 0} - 2\epsilon < 0 \text{ on } [-1,0], \quad \text{and}$$

$$\varphi''_{\lambda,\epsilon}(x) \leq \varphi''_\lambda(x) = (1+\lambda)y''_0(x) \leq -(1-\lambda_0)D_2 = -d_2 < 0$$

for $x \in [-1/2, 0]$. Further, we have

$$|\varphi_{\lambda,\epsilon}|_{C^2} \leq |\varphi_\lambda|_{C^2} + \epsilon|p|_{C^2} \leq (\lambda_0 + 1)|y_0|_{C^2} + 2 = M_2.$$

Choose $\epsilon \in (0,1]$ such that $||\varphi_{\lambda,\epsilon} - \varphi_\lambda|| < k_4\iota$ and set $\psi_+ = \varphi_{\lambda,\epsilon}$. Then ψ_+ has the properties asserted in (i) and (ii). Let $\psi \in \psi_+ - H_{||}(k_4\iota)$; then $\psi \in \varphi_\lambda + H_{||}(2k_4\iota)$, so (2.9) shows that $\mathcal{H}(\psi) > 0$. Further, since $||\varphi_\lambda - y_0|| = ||\lambda y_0|| \leq \iota/2$, and $\iota/2 + 2k_4\iota < (1/2 + 2/5)\iota < \iota$, we have $\psi_\pm \in N_\iota$. Consequently, ψ_+ also has the properties stated in item (iii). The construction of ψ_- satisfying conditions (i)–(iii) is analogous, setting $\lambda = -\frac{\iota}{2(||y_0||+1)}$.

The next proposition is the main step towards the choice of an initial segment through which there is a backward solution of (g), the phase curve of which converges to the periodic orbit $\mathcal{O}$. Later, we will find a modified equation such that the backward phase curve is preserved and the forward phase curve also converges to $\mathcal{O}$.

For every $\hat{\delta} \in (0, \bar{\delta}]$, there exist $\varphi \in W^u(y_0, P, N_{\bar{\delta}})$ and $\delta_1 \in (0, A - y(0))$ with the subsequent properties [Lan99]. With k_3 we can set $\delta = \frac{2|\varphi - y_0|}{k_3}$, so that $|\varphi - y_0| = k_3\delta/2$.

a) $\delta \leq \hat{\delta}$, $\varphi(0) < A$, $\varphi(0) - y(0) > \delta_1 \geq k_3\delta/8$, $\delta_1 \leq \bar{\delta}$. Further, $\varphi(-1/2) \geq y(-1/2)/2$.
b) There exists a solution $X : \mathbb{R} \longrightarrow \mathbb{R}$ of (g) with $X_0 = \varphi$.
c) $\dot{X} > 0$ on $[-1,0)$, $\dot{X} < 0$ on $(0,1]$, $X(-1) = 0$.
d) There exist $\theta_-, \theta_+ \in (0, 1/4)$ such that

$$X(t) > y(0) + \delta_1 \text{ for } t \in (-\theta_-, \theta_+);$$
$$X(-\theta_-) = X(\theta_+) = y(0) - \delta_1;$$
$$-y(0) - 1/2 < X(t) < y(0) + \delta_1 \text{ for } t \in (-\infty, -\theta_-);$$
$$-y(0) - 1/2 \leq X(t) \text{ for } t \in (-\infty, 1].$$

e) $\theta_- \leq \delta$, $\theta_+ \leq \delta$.
f)

$$|g(y(t)) - g(X(t))| \leq y(0)/4 \quad \text{for } t \in [0,1].$$

g) The tangent space $T_\varphi W^u(y_0, P, N_{\bar{\delta}})$ is spanned by a segment $\omega \in H$ with $\omega(0) > 0$.
h) The solution $w : [-1, \infty) \longrightarrow \mathbb{R}$ of the variational equation (g, X) with $w_0 = \omega$ satisfies $w(t) > 0$ for $t \in [-\theta_-, \theta_+]$.

1. Assume that properties b), c) and d) hold for some φ with $\varphi(0) > y(0)$ and some $\delta_1 \in (0, \varphi(0) - y(0))$, and for corresponding numbers θ_-, θ_+. Then the analogous properties hold for every $\tilde{\delta}_1 \in [\delta_1, \varphi(0) - y(0))$, with the corresponding numbers $\tilde{\theta}_-, \tilde{\theta}_+$. These $\tilde{\theta}_\pm$ are uniquely determined, in view of c), and we have

$$\lim_{\tilde{\delta}_1 \to \varphi(0) - y(0)} \tilde{\theta}_\pm = 0.$$

2. It was shown in [LW96] that there exists $\varphi \in W^u(y_0, P, N_{\bar{\delta}})$ such that properties b),c),d), f) and g) hold. t is clear from the construction that these properties can be achieved with φ arbitrarily close to y_0.

3. All $\varphi \in W^u(y_0, P, N_{\bar{\delta}})$ with $\varphi(0) > y(0)$ and sufficiently close to y_0 are of the form

$$\varphi = \varphi_r = y_0 + r \cdot \chi^u + w^u(r \cdot \chi^u)$$

for some uniquely determined $r > 0$. If r is so small that

$$|w^u(r\chi^u)| \leq \frac{1}{3} r |\chi^u|, \tag{2.10}$$

then, using $|\chi^u| = \chi^u(0)$, we get

$$|\varphi_r(0) - y(0)| \geq r\chi^u(0) - \frac{1}{3} r\chi^u(0) = \frac{2}{3} r \cdot \chi^u(0) \tag{2.11}$$

$$= \frac{1}{2} \cdot \frac{4}{3} r |\chi^u| \geq \frac{1}{2} |r\chi^u + w^u(r \cdot \chi^u)| = \frac{1}{2} |\varphi_r - y_0|.$$

4. Let $\hat{\delta} \in (0, \bar{\delta}]$. In view of steps 2 and 3, we can choose $r > 0$ so small that $\varphi = \varphi_r$ satisfies

$$\varphi(0) < A, \qquad |\varphi - y_0| \leq \frac{k_3 \hat{\delta}}{2}, \qquad \varphi(0) - y(0) \leq \bar{\delta}, \qquad \varphi(-1/2) \geq y(-1/2)/2, \tag{2.12}$$

and that properties b), c), d), f) and g) hold with some $\tilde{\delta}_1 \in (0, \varphi(0) - y(0))$ and corresponding numbers $\tilde{\theta}_\pm$. Then, setting $\delta = 2|\varphi - y_0|/k_3$, we have $\delta \leq \hat{\delta}$. Using step 1), we can choose $\delta_1 \in [\tilde{\delta}_1, \varphi(0) - y(0))$ so close to $\varphi(0) - y(0)$ that properties b), c), d), f) and g) also hold with δ_1 and the corresponding numbers $\theta_\pm$ instead of $\tilde{\delta}_1$ and $\tilde{\theta}_\pm$, and that, in addition, the following conditions are satisfied:

(i) $\theta_-, \theta_+ \leq \delta$, so that property e) holds;
(ii) property h) holds (here we use the continuity of w);
(iii) $\delta_1 \geq \frac{\varphi(0) - y(0)}{2}$.

It remains to check that a) holds. In view of (2.12) and the choice of δ_1 and δ, we only have to prove the inequalities $\delta_1 \leq \bar{\delta}$ and $\delta_1 \geq k_3\delta/8$. First, $\delta_1 \leq \varphi(0) - y(0) \leq \bar{\delta}$. Second, from (ii) above and from (2.11), we get

$$\delta_1 \geq \frac{\varphi(0) - y(0)}{2} \geq \frac{1}{4} |\varphi - y_0| = k_3\delta/8.$$

We are now ready to choose an initial value and targets. Now we set

$$\gamma_2 = |g'(0)||g(y_0(-1/2)/2)| > 0 \quad \text{and} \quad c_g = \sqrt{\frac{|g|_{C^0}}{\gamma_2}}.$$

Recall that $\delta' \leq \tilde{\delta} \leq \bar{\delta} \leq 1/4$. Choose $\hat{\delta} \in (0, \delta']$ such that the following inequalities hold for all $\delta \in (0, \hat{\delta}]$:

(i)
$$y(2 - \delta) - y(1 + \delta) \leq -y(0)/2; \tag{2.13}$$

(ii) $2\delta|g|_{C^0} \leq y(0)/8$;
(iii) $12\delta|g|_{C^1} \leq 1/2$;
(iv) $2c_g\sqrt{\delta} \leq 1/2 - \delta, \quad \delta \leq c_g \cdot \sqrt{\delta}$;
(v) $10M_2c_g\delta^{3/2} \leq \frac{k_4k_1k_3\delta}{64}$;
(vi) $\frac{9M_2}{2}\delta^2 \leq \frac{k_4k_1k_3\delta}{32}$.

From now on, let φ, $\delta \in (0, \hat{\delta}]$, X, $\theta_\pm$, $\delta_1 \in [k_3\delta/4, k_3\tilde{c}/2]$, and ω be as above. We set $\iota = k_1\delta_1$. Then $\iota \leq \delta_1 < \delta \leq \hat{\delta} \leq \delta'$, since $k_1, k_3 \leq 1$. With this number ι we get $\psi_+, \psi_- \in N_\iota$ with the properties listed above. Now we set

$$\psi_s = (1 - s)\psi_- + s\psi_+ \in N_\iota \quad \text{for } s \in [0, 1];$$

these functions ψ_s will be the targets.

2.4.3 Successive Modifications of g

We get our example by successive deformations of the original nonlinearity g, which satisfies $g' < 0$. The modifications are such that the unstable periodic solution y of (g), together with local unstable and stable manifolds, is preserved. The initial value φ from the unstable manifold leads to a homoclinic solution for the finally obtained equation.

The methods are partially similar to the ones used in the previous example from [LW96]. However, obtaining the simpler shapes of both the nonlinearity and the homoclinic solution requires finer techniques for estimates on solution behavior, compared to [LW96]. In addition, the necessity to avoid zeroes of the derivative of the nonlinearity (except for one) leads to some technical complications.

If $h : \mathbb{R} \longrightarrow \mathbb{R}$ is a continuous modification of g on $\mathbb{R} \setminus [-y(0) - 1, y(0) + \delta_1]$ then X satisfies (h) on $(-\infty, 1 - \theta_-]$. It follows that there exists a unique solution of (h) defined on $\mathbb{R}$ which coincides with X on $(-\infty, 1 - \theta_-]$. We denote this solution by $z(\cdot, h)$. Obviously,

$$\text{dist}(z(\cdot, h)_t, \mathcal{O}) \longrightarrow 0 \quad \text{as} \quad t \longrightarrow -\infty, \quad \text{and} \quad z(\cdot, h)_0 = \varphi.$$

Similarly, if h is C^1, we use the notation $w(\cdot, h) : [-1, \infty) \longrightarrow \mathbb{R}$ for the solution of the variational equation $(h, z(\cdot, h))$ with initial value $w(\cdot, h)_0 = \omega$, and we have $w(\cdot, h) = w(\cdot, g)$ on $[-1, 1 - \theta_-]$.

We want to find a modification g^* of g such that the phase curve of $z(\cdot, g^*)$ is homoclinic to $\mathcal{O}$. Further, we want to achieve that the solutions $\dot{z}(\cdot, g^*)$ and $w(\cdot, g^*)$ of the variational equation $(g^*, z(\cdot, g^*))$ satisfy the above assumptions.

Let $\tilde{g} \in C^1(\mathbb{R}, \mathbb{R})$ be a modification of g on $\mathbb{R} \setminus [-y(0) - 1, y(0) + \delta_1]$. If there exists $t^* > 0$ with $z(\cdot, \tilde{g})_{t^*} \in W^s(y_0, P, N_\iota)$ then the solution $z(\cdot, \tilde{g})$ of $(\tilde{g})$ has a phase curve homoclinic to the orbit of y [Lan99]. Now we set

$$\psi = z(\cdot, \tilde{g})_{t^*} \in W^s(y_0, P, N_\iota) = W^s(y_0, P, N_{k_1 \delta_1}).$$

We know that the solution $x^\psi : [-1, \infty) \to \mathbb{R}$ of (g) with $x_0^\psi = \psi$ satisfies $x_t^\psi \in \mathcal{O} + C(\delta_1)$ for $t \geq 0$, so x^ψ is also a solution of $(\tilde{g})$. It follows that $z(t_* + t, \tilde{g}) = x^\psi(t)$ for $t \geq 0$, so $\operatorname{dist}(z(\cdot, \tilde{g})_t, \mathcal{O}) \to 0$ for $t \to \infty$.

Let $a, b \in \mathbb{R}$, $a < b$ and $\zeta \in C^0([a, b], \mathbb{R})$. Now we set $m = \min \zeta$, and $M = \max \zeta$. Assume $m < M$ and that

$$\mu(\{s \in [a, b] : \zeta(s) \in \{m, M\}\}) = 0, \tag{2.14}$$

where μ denotes the *Lebesgue measure*. Let $K > 0$, $B > 0$, $M_0 \in (0, B)$, $\alpha_0 > 0$, $\alpha_1 > 0$, $M_1 < 0$ and $\eta > 0$ be given such that

$$M_0(b - a) < K < B(b - a) \qquad \text{and} \qquad \alpha_0(b - a) < K. \tag{2.15}$$

Then there exists $r \in (0, \frac{M - m}{2})$ and a continuous map

$$(0, \alpha_0] \times (0, \alpha_1] \longrightarrow (C^0([m, M], \mathbb{R}), |\,|_\infty),$$
$$(m_0, m_1) \qquad \mapsto h(\cdot, m_0, m_1)$$

such that, for every $(m_0, m_1) \in (0, \alpha_0] \times (0, \alpha_1]$, the function $h = h(\cdot, m_0, m_1)$ has the following properties:

(i) $h \in C^1([m, M], \mathbb{R})$;
(ii)

$$h(m) = m_0, \qquad h'(m) = m_1, \qquad h(M) = M_0, \qquad h'(M) = M_1;$$

(iii)

$$0 < h(x) < B \qquad \text{for} \quad x \in [m, M], h'((m + M)/2) = 0,$$
$$h'(x) > 0 \qquad \text{for } x \in [m, (m + M)/2),$$
$$h'(x) < 0 \qquad \text{for } x \in ((m + M)/2, M].$$

(iv)

$$\int_a^b h(\zeta(s))ds = K;$$

(v)

$$h(x) \geq \frac{K}{2(b-a)} \quad \text{for } x \in [m+r, M-r];$$

(vi)

$$\int_{\substack{s \in [a,b] \\ \zeta(s) \in [m, m+r]}} h(\zeta(s))ds \leq \eta, \qquad \int_{\substack{s \in [a,b] \\ \zeta(s) \in [M-r, M]}} h(\zeta(s))ds \leq \eta.$$

Take a function $\chi \in BC^1(\mathbb{R}, \mathbb{R})$ with $\chi(x) = 0$ for $x \leq 0$, $\dot{\chi} > 0$ on $(0,1)$ and $\chi(x) = 1$ for $x \geq 1$. It follows from (2.15) that there exist $\lambda_1, \lambda_2 > 0$ with

$$\max\{\alpha_0, M_0\} < \lambda_1 < \lambda_2 < B, \text{ and}$$

$$\frac{K}{2} < \lambda_1(b-a) < K < \lambda_2(b-a). \tag{2.16}$$

Choose a function $\beta \in C^1([m, M], \mathbb{R})$ with $\beta(m) = \beta'(m) = \beta'(M) = \beta(M) = 0$ and with $\beta'(x) > 0$ for $x \in (m, (m+M)/2)$, $\beta'(x) < 0$ for $x \in ((m+M)/2, M)$. Choose $\rho_0 \in (0, (M-m)/2]$ such that

$$\lambda_2 + \rho_0 |\beta|_{C^0} < B. \tag{2.17}$$

For $m_0 \in (0, \alpha_0]$, $m_1 \in (0, \alpha_1]$, $\rho \in (0, \rho_0]$ and $\lambda \in \mathbb{R}$, we define a function $f_{\rho,\lambda} = f_{\rho,\lambda,m_0,m_1} : [m, M] \to \mathbb{R}$, as

$$f_{\rho,\lambda}(x) = \begin{cases} [m_0 + m_1(x-m)](1 - \chi((x-m)/\rho)) + \lambda \cdot \chi((x-m)/\rho) \\ \qquad \text{for } x \in [m, m+\rho], \\ \lambda \qquad \text{for } x \in (m+\rho, M-\rho), \\ [M_0 + M_1(x-M)](1 - \chi((M-x)/\rho)) + \lambda \cdot \chi((M-x)/\rho) \\ \qquad \text{for } x \in [M-\rho, M]. \end{cases}$$

Now we set $g_{\rho,\lambda} = g_{\rho,\lambda,m_0,m_1} = f_{\rho,\lambda,m_0,m_1} + \rho\beta$ for ρ, λ, m_0, m_1 as above. Note that both $f_{\rho,\lambda}$ and $g_{\rho,\lambda}$ are in $C^1([m, M], \mathbb{R})$ and satisfy the boundary conditions that are imposed on the function h in assertion (ii).
For $\lambda > 0$ and $\rho \in (0, \rho_0]$ with

$$m_0 + m_1\rho < \lambda \quad \text{and} \quad M_0 + |M_1|\rho < \lambda, \tag{2.18}$$

the function $g = f_{\rho,\lambda,m_0,m_1}$ satisfies

$$g'(x) > 0 \quad \text{for} \quad x \in [m, (m+M)/2),$$
$$g'(x) < 0 \quad \text{for} \quad x \in ((m+M)/2), M].$$

Let $\lambda > 0$ and $\rho \in (0, \rho_0]$ satisfy the above assumptions. First, $g'(m) = f'_{\rho,\lambda}(m) = m_1 > 0$. For $x \in (m, m+\rho)$, one has with $\xi = (x-m)/\rho$ that
[Lan99]

$$g'(x) = m_0(-\chi'(\xi)/\rho) + \underbrace{m_1(1 - \chi(\xi))}_{\geq 0}$$

$$+m_1(x - m)(-\chi'(\xi)/\rho) + \lambda\chi'(\xi)/\rho + \rho\beta'(x)$$

$$\geq (\lambda - m_0)\chi'(\xi)/\rho - m_1(x - m)\chi'(\xi)/\rho + \rho\beta'(x)$$

$$\geq \frac{\lambda - m_0 - m_1\rho}{\rho}\chi'(\xi) \text{ (note that } \beta' > 0 \text{ on } (m, m + \rho) \subset (m, (m + M)/2))$$

$$> 0.$$

For $x \in [m + \rho, (m + M)/2)$ respectively $x \in ((m + M)/2, M - \rho]$, we have $g'(x) = \rho\beta'(x) > 0$ respectively $\ldots < 0$

For $x \in (M - \rho, M)$ and $\xi = (M - x)/\rho$,

$$g'(x) = M_0\chi'(\xi)/\rho + \underbrace{M_1(1 - \chi(\xi))}_{\leq 0}$$

$$+ M_1(x - M)\chi'(\xi)/\rho - \lambda\chi'(\xi)/\rho + \underbrace{\rho\beta'(x)}_{\leq 0}$$

$$\leq \frac{M_0 + |M_1|\rho - \lambda}{\rho}\chi'(\xi) < 0.$$

Finally, $g'(M) = M_1 < 0$.

According to (2.16), we can choose $\rho_1 \in (0, \rho_0]$ such that $\max\{\alpha_0 + \alpha_1\rho_1, M_0 + |M_1|\rho_1\} < \lambda_1$. Then, for $m_0 \in (0, \alpha_0]$, $m_1 \in (0, \alpha_1]$, $\lambda \in [\lambda_1, \lambda_2]$ and $\rho \in (0, \rho_1]$, condition (2.18) is satisfied. Therefore, we have

$$|g_{\rho,\lambda,m_0,m_1}|_{C^o} \leq g_{\rho,\lambda,m_0,m_1}((m + M)/2) \leq \lambda + \rho|\beta|_{C^o} \leq \lambda_2 + \rho_0|\beta|_{C^o} < B. \tag{2.19}$$

Now we set $\epsilon_1 = K - \lambda_1(b - a)$, $\epsilon_2 = \lambda_2(b - a) - K$. For $\rho \in (0, \rho_1]$, we set $E_\rho = s \in [a, b]\zeta(s) \in [m, m + \rho] \cup [M - \rho, M]$. It follows from condition (2.14) and the $\sigma-$continuity of the Lebesgue measure that there exists $\rho_2 \in (0, \rho_1]$ such that for $\rho \in (0, \rho_2]$ one has

$$\mu(E_\rho) < \min\left\{\eta/B, \frac{\min\{\epsilon_1, \epsilon_2\}}{3(B + \lambda_2(b - a) + \lambda_2)}\right\}. \tag{2.20}$$

Choose $\rho_3 \in (0, \rho_2]$ such that

$$\rho_3(b - a)|\beta|_{C^o} < \frac{\min\{\epsilon_1, \epsilon_2\}}{3}.$$

Let now $\rho \in (0, \rho_3]$, $\lambda \in [\lambda_1, \lambda_2]$, $m_0 \in (0, \alpha_0]$, $m_1 \in (0, \alpha_1]$. Using (2.19) and (2.20), we get

$$\left| \int_a^b g_{\rho,\lambda,m_0,m_1}(\zeta(s))ds - \lambda(b-a) \right|$$

$$\leq \mu(E_\rho)(B + \lambda(b-a))$$

$$+ \left| \int_{s \in [a,b] \setminus E_\rho} g_{\rho,\lambda,m_0,m_1}(\zeta(s))ds - \lambda(b-a) \right|$$

$$\leq \frac{\min\{\epsilon_1, \epsilon_2\}}{3} + \left| \int_{s \in [a,b] \setminus E_\rho} [\lambda + \rho\beta(\zeta(s))]ds - \lambda(b-a) \right|$$

$$= \frac{\min\{\epsilon_1, \epsilon_2\}}{3} + \left| \lambda(-\mu(E_\rho)) + \int_{s \in [a,b] \setminus E_\rho} \rho\beta(\zeta(s))ds \right|$$

$$\leq \frac{\min\{\epsilon_1, \epsilon_2\}}{3} + \lambda_2\mu(E_\rho) + \rho(b-a)|\beta|_{C^0}$$

$$< \frac{\min\{\epsilon_1, \epsilon_2\}}{3} + \frac{\min\{\epsilon_1, \epsilon_2\}}{3} + \frac{\min\{\epsilon_1, \epsilon_2\}}{3} = \min\{\epsilon_1, \epsilon_2\}.$$

Consequently, we have for $m_0 \in (0, \alpha_0]$, $m_1 \in (0, \alpha_1]$ that

$$\int_a^b g_{\rho_3,\lambda_1,m_0,m_1}(\zeta(s))ds < K < \int_a^b g_{\rho_3,\lambda_2,m_0,m_1}(\zeta(s))ds. \qquad (2.21)$$

The definition of the functions $g_{\rho_3,\lambda,m_0,m_1}$ shows that the map

$$\gamma : [\lambda_1, \lambda_2] \times [0, \alpha_0] \times [0, \alpha_1] \ni (\lambda, m_0, m_1) \mapsto g_{\rho_3,\lambda,m_0,m_1} \in (C^0([m, M], \mathbb{R}), |\ |_{C^0})$$

is continuous. It follows that the map

$$j \ : \ [\lambda_1, \lambda_2] \times [0, \alpha_0] \times [0, \alpha_1] \to \mathbb{R}$$

$$(\lambda, m_0, m_1) \mapsto \int_a^b g_{\rho_3,\lambda,m_0,m_1}(\zeta(s))ds$$

is continuous. j is differentiable w.r. to the first argument, and one has

$$\partial_1 j(\lambda, m_0, m_1) = \int_{s \in [a,b], \zeta(s) \in [m, m+\rho_3]} \chi((\zeta(s) - m)/\rho_3)ds$$

$$+ \int_{s \in [a,b], \zeta(s) \in [M-\rho_3, M]} \chi((M - \zeta(s))/\rho_3)ds + \int_{s \in [a,b] \setminus E_\rho} 1\, ds$$

$$\geq \mu([a,b] \setminus E_\rho) > 0. \qquad (2.22)$$

It follows from (2.21) and (2.22) that for all $(m_0, m_1) \in (0, \alpha_0] \times (0, \alpha_1]$ there exists a unique $\lambda(m_0, m_1)$ with $j(\lambda(m_0, m_1), m_0, m_1) = K$. Further,

(2.22) and the Implicit Function Theorem imply that the map $\lambda : (m_0, m_1) \mapsto \lambda(m_0, m_1)$ is continuous. Now we set

$$h(\cdot, m_0, m_1) = g_{\rho_3, \lambda(m_0, m_1), m_0, m_1} \quad \text{for} \quad (m_0, m_1) \in (0, \alpha_0] \times (0, \alpha_1],$$

and set $r = \rho_3$. Continuity of $\lambda(\cdot, \cdot)$ and of the map γ above imply that the map $(m_0, m_1) \mapsto h(\cdot, m_0, m_1) \in (C^0([m, M], \mathbb{R}), |\ |_{C^0})$ is continuous.

We check the asserted properties (i)–(vi) now: Properties (i) and (ii) are clear from the definition of the functions $g_{\rho, \lambda}$. Property (iii) follows from (2.19). Property (iv) holds, since

$$\int_a^b h(\zeta(s), m_0, m_1) ds = j(\lambda(m_0, m_1), m_0, m_1) = K.$$

Property (v) follows from the definition of the $g_{\rho, \lambda}$ and from the first estimate in the second line of (2.16). Property (vi) follows from property (iii) together with the estimate $B\mu(E_\rho) \leq \eta$ (see (2.20)).

We now construct the first modification g_1 of g, which will steer the solution $z(\cdot, g_1)$ to negative values away from the range of the periodic solution y. Parallel to controlling the behavior of $z(\cdot, g_1)$, we have to keep track of properties of $w(\cdot, g_1)$, which will finally make the proof of transversality possible.

There exists a modification g_1 of g on $[y(0) + \delta_1, \infty)$ with the following properties [Lan99, ILW92]:

a) $g_1 \in BC^1(\mathbb{R}, \mathbb{R})$, and g_1 satisfies (NF);
b) $g_1' < 0$ on $[0, \infty)$;
c) $z(1, g_1) < 0$, and $z(\cdot, g_1)$ has a unique zero t_0 in $(1 - \delta, 1)$.
   Setting $t_{\min} = t_0 + 1$, we have
d) $\dot{z}(t, g_1) < 0$ for $t \in (0, t_{\min})$ and $\dot{z}(t, g_1) > 0$ for $t \in (t_{\min}, 2 + \theta_+]$;
e) $z(1 + \theta_+, g_1) \leq -y(0) - 1$;
f) $w(t_{\min}, g_1) < 0$.

Choose $\eta \in BC^1(\mathbb{R}, \mathbb{R})$ with $\eta(r) = 0$ for $r \leq y(0) + \delta_1$ and with $\dot{\eta}(r) < 0$ for $r > y(0) + \delta_1$. Now we set $g_\lambda = g + \lambda \cdot \eta$ for $\lambda > 0$. The functions g_λ satisfy (NF), and $g_\lambda' < 0$ on $[0, \infty)$. The solutions $z(\cdot, g_\lambda)$ and $w(\cdot, g_\lambda)$ are defined for $\lambda > 0$, and $z(t, g_\lambda) = X(t)$ for $t \in (-\infty, 1 - \theta_-], w(t, g_\lambda) = w(t, g)$ for $t \in [-1, 1 - \theta_-]$. Since $\theta_- \leq \delta$, we have $z(1 - \delta, g_\lambda) = X(1 - \delta)$. Since $|\varphi - y_0| = k_3 \delta / 2$ and $\delta \leq \hat{\delta} \leq \bar{\delta}$, we see that $X(1 - \delta) \geq k_2 \delta > 0$. Hence we have

$$z(1 - \delta, g_\lambda) > 0. \tag{2.23}$$

Now, $z(1, g_\lambda) = z(1 - \delta, g_\lambda) + \displaystyle\int_{-\delta}^{0} (g + \lambda\eta)(X(s))ds$

$$\leq X(1 - \delta) + \int_{-\delta}^{0} g(X(s))ds + \lambda \int_{-\theta_-}^{0} \eta(X(s))ds$$

$$= X(1) + \lambda \int_{-\theta_-}^{0} \eta(X(s))ds.$$

The last integral is strictly negative, because the interval $X([-\theta_-, 0]) = [y(0)+\delta_1, \varphi(0)]$ does not consist of one point only. Hence we get $\lim_{\lambda\to\infty} z(1, g_\lambda) = -\infty$. For $t \in (1, 1 + \theta_+)$, we have

$$\dot{z}(t, g_\lambda) \in (g + \lambda\eta)(X[(0, \theta_+)]) = (g + \lambda\eta)[((y(0) + \delta_1, \varphi(0))] \subset (-\infty, 0],$$

so that also

$$\lim_{\lambda\to\infty} z(1 + \theta_+, g_\lambda) = -\infty.$$

There exists $\lambda_0 > 0$ such that for $\lambda > \lambda_0$ one has

$$z(1, g_\lambda) < 0, \qquad z(1 + \theta_+, g_\lambda) < -y(0) - 1. \qquad (2.24)$$

Similarly, we get $\dot{w}(t, g_\lambda) < 0$ on $[1 - \theta_-, 1 + \theta_+]$, and

$$\lim_{\lambda\to\infty} w(1, g_\lambda) = -\infty, \qquad \lim_{\lambda\to\infty} w(1 + \theta_+, g_\lambda) = -\infty.$$

Note that with $W = \sup |w(s, g)| \ -1 \leq s \leq 1 - \theta_-$, we have $V \geq \omega(0) > 0$, and

$$\sup |w(s, g_\lambda)| \ -1 \leq s \leq 1 - \theta_- = W \qquad \text{for all } \lambda > 0.$$

Choose $\lambda_1 \geq \lambda_0$ such that, for all $\lambda \geq \lambda_1$,

$$|w(1 + \theta_+, g_\lambda)| = \max |w(s, g_\lambda)| -1 \leq s \leq 1 + \theta_+, \qquad \text{and}$$
$$w(1 + \theta_+, g_\lambda) \leq -3|g|_{C^1} W. \qquad (2.25)$$

For $\lambda > 0$, property (NF) shows that $z(\cdot, g_\lambda)$ is strictly decreasing on $[0, 1]$, and therefore

$$z(t, g_\lambda) \leq z(\theta_+, g_\lambda) = y(0) + \delta_1 \qquad \text{for } t \in [\theta_+, 1].$$

It follows that

$$g_\lambda(z(t - 1, g_\lambda)) = g(z(t - 1, g_\lambda)) \qquad \text{for } t \in [1 + \theta_+, 2].$$

For $t \in [1 + \theta_+, 2 - \theta_-]$, we get the following estimate from the definition of W:

Now we set $B = (K + 1)/\tilde{\theta}$. Finally, define $z_3 = z(1 + \theta_+ + \tilde{\theta}, g_1)$. Note that the choice of α_1 and the inequality $\tilde{\theta} < 1/4$ imply that

$$\text{for all } s \in [0, 1], \qquad 0 < \frac{\ddot{\psi}_s(-1)}{\dot{z}(1 + \theta_+ + \tilde{\theta}, g_1)} \leq \alpha_1. \qquad (2.34)$$

There exist a continuous map

$$[0, 1] \ni s \mapsto g_s \in (BC^0(\mathbb{R}, \mathbb{R}), |\ |_{C^0})$$

and numbers $\tau_1, \tau_2 \in (1 + \theta_+, 1 + \theta_+ + \tilde{\theta})$ and $z_{\max} \in (z_3, z_0)$ satisfying the subsequent properties for all $s \in [0, 1]$ (we set $z_i = z(\tau_i, g_1)$ $(i = 1, 2)$).

a) $g_s \in BC^1(\mathbb{R}, \mathbb{R})$, and g_s satisfies (NF). $g_s = g_1$ on $[z_0, \infty) \supset [-y(0) - 1, \infty)$, and

$$g_s' > 0 \text{ on } (-\infty, z_{\max}), \qquad g_s' < 0 \text{ on } (z_{\max}, \infty).$$

b) $g_s(z_{\max}) \leq \frac{K+1}{\tilde{\theta}}$, $\qquad g_s \geq \frac{K}{2\theta}$ on $[z_2, z_1]$.
c)

$$\int_{1+\theta_+}^{\tau_1} g_s(z(t, g_1))dt \leq y(0)/16,$$

$$\int_{\tau_2}^{1+\theta_+ + \tilde{\theta}} g_s(z(t, g_1))dt \leq y(0)/16,$$

$$\int_{1+\theta_+}^{1+\theta_+ + \tilde{\theta}} g_s(z(t, g_1))dt = -z(2 + \theta_+, g_1).$$

d) $g_s(z(t, g_1)) = \dot{\psi}_s(t - (2 + \theta_+ + \tilde{\theta}))$ for $t \in [1 + \theta_+ + \tilde{\theta}, 2 - \delta]$.
e) $g_s(x) \leq M_2$ for $x \leq z_3$ [Lan99].

Now, if we set

$$a = 1 + \theta_+, \qquad b = 1 + \theta_+ + \tilde{\theta}, \qquad \zeta = z(\cdot, g_1)[a, b],$$

condition (2.14) is satisfied, since ζ is strictly decreasing, and we have $m = \min \zeta = z_3$, $M = \max \zeta = z_0$. The inequalities in (2.15) are satisfied, in view of (2.33) and the choice of B. We get a number $r \in (0, (z_0 - z_3)/2)$ and a continuous map $h : (0, \alpha_0] \times (0, \alpha_1] \to (C^0([z_3, z_0], \mathbb{R}), |\ |_{C^0})$ with the above properties. We now define g_s for $s \in [0, 1]$.

$$g_s(x) = \begin{cases} g_1(x) & \text{for } x \geq z_0; \\[2mm] h(x, \dot{\psi}_s(-1), \dfrac{\ddot{\psi}_s(-1)}{\dot{z}(1+\theta_+ +\tilde{\theta}, g_1)}) & \text{for } x \in [z_3, z_0]; \\[3mm] \dot{\psi}_s(\tau(x) - (2+\theta_+ +\tilde{\theta})) & \text{for } x \in (z_4, z_3): \\[3mm] \dot{\psi}_s(-\delta-\theta_+ -\tilde{\theta}) \cdot \exp[(x-z_4) \cdot \dfrac{\ddot{\psi}_s(-\delta-\theta_+ -\tilde{\theta})}{\dot{\psi}_s(-\delta-\theta_+ -\tilde{\theta}) \cdot \dot{z}(2-\delta, g_1)}] \\[2mm] \hspace{3cm} \text{for } x \in (-\infty, z_4]. \end{cases}$$

Now we set $\tau_1 = \tau(z_0 - r)$ and $\tau_2 = \tau(z_3 + r)$, so that $1 + \theta_+ < \tau_1 < \tau_2 < 1 + \theta_+ + \tilde{\theta}$. Note that g_s is well-defined on $[z_3, z_0]$ since $\dot{\psi}_s(-1) \in (0, \alpha_0]$ and $\dfrac{\ddot{\psi}_s(-1)}{\dot{z}(1+\theta_+ +\tilde{\theta}, g_1)} \in (0, \alpha_1]$. Note also that for $x \in (z_4, z_3)$, we have $\tau(x) \in (1+\theta_+ +\tilde{\theta}, 2-\delta)$, so $\tau(x) - (2+\theta_+ +\tilde{\theta}) \in (-1, -\delta-\theta_+ -\tilde{\theta})$, which shows that g_s is well-defined on (z_4, z_3). We now prove the asserted properties of the functions g_s. For $s \in [0,1]$, let $g_{s,1}, g_{s,2}, g_{s,3}$ and $g_{s,4}$ denote the restrictions of g_s to $I_1 = [z_0, \infty)$, $I_2 = [z_3, z_0]$, $I_3 = (z_4, z_3)$, $I_4 = (-\infty, z_4]$, respectively. It is clear that $g_{s,i} \in BC^1(I_i, \mathbb{R})$, $(i = 1, ..., 4)$. The map $[0,1] \ni s \mapsto \psi_s \in (C^2([-1,0], \mathbb{R}), |\ |_{C^2})$ is continuous, and the map

$$E : \mathbb{R}^+ \times \mathbb{R}^- \to (BC^0((-\infty, z_4], \mathbb{R}), |\ |_{C^0}),$$

$$E(\alpha, \beta)(x) = \exp((x-z_4) \frac{\beta}{\alpha \dot{z}(2-\delta, g_1)})$$

is continuous. In view of continuity of the map h, it is now obvious that the maps $[0,1] \ni s \mapsto g_{s,i} \in (BC^0(I_i, \mathbb{R}), |\ |_{C^0})$ are continuous for $i = 1, ..., 4$. $g_s \in BC^1(\mathbb{R}, \mathbb{R})$ for $s \in [0,1]$, and the following map is continuous,

$$[0,1] \ni s \mapsto g_s \in (BC^0(\mathbb{R}, \mathbb{R}), |\ |_{C^0})$$

It suffices to show that the functions g_s are continuously differentiable at the points z_0, z_3, z_4. Using $g_1 = g$ on $(-\infty, y(0) + \delta_1)$, and recalling $M = z_0$, we see that

$$\lim_{x \to z_0+} g_s(x) = g(z_0) = M_0 = \lim_{x \to z_0-} h(x, ...) = \lim_{x \to z_0-} g_s(x), \qquad \text{and}$$

$$\lim_{x \to z_0+} g_s'(x) = g_1'(z_0) = g'(z_0) = M_1 = \lim_{x \to z_0-} \partial_1 h(x, ...) = \lim_{x \to z_0-} g_s'(x).$$

Similarly,

$$\lim_{x \to z_3+} g_{s,2}(x) = h(z_3, ...) = h(m, ...) = \dot{\psi}_s(-1)$$

$$= \dot{\psi}_s(\tau(z_3) - (2+\theta_+ +\tilde{\theta})) = \lim_{x \to z_3-} g_{s,3}(x), \qquad \text{and}$$

$$\lim_{x \to z_3+} g_{s,2}'(x) = \partial_1 h(z_3, ...) = \frac{\ddot{\psi}_s(-1)}{\dot{z}(1+\theta_+ +\tilde{\theta}, g_1)}$$

$$= \ddot{\psi}_s(-1) \cdot \tau'(z_3) = \lim_{x \to z_3-} g_{s,3}'(x),$$

where the last equality comes from the chain rule. Finally,

$$\lim_{x \to z_4+} g_{s,3}(x) = \dot{\psi}_s(\tau(z_4) - (2 + \theta_+ + \tilde{\theta})) = \dot{\psi}_s(-\delta - \theta_+ - \tilde{\theta})$$

$$= \lim_{x \to z_4-} g_{s,4}(x), \qquad \text{and}$$

$$\lim_{x \to z_4+} g'_{s,3}(x) = \ddot{\psi}_s(-\delta - \theta_+ - \tilde{\theta})\tau'(z_4)$$

$$= \frac{\ddot{\psi}_s(-\delta - \theta_+ - \tilde{\theta})}{\dot{z}(2 - \delta, g_1)} = \lim_{x \to z_4-} g'_{s,4}(x).$$

It follows from $0 < h$ and from $\dot{\psi}_s > 0$ on $[-1, 0)$ that $g_s(x) > 0$ for $x \le z_0$. Since g_1 satisfies (NF), it follows that g_s also satisfies (NF).

Regarding the remaining assertions of a): $g_s(x) = g_{s,1}(x) = g_1(x)$ for $x \ge z_0$, and $g'_{s,4}(x) > 0$ for $x \le z_4$, $g'_{s,3}(x) = \ddot{\psi}_s(...)\tau'(x) > 0$ for $x \in (z_4, z_3)$. Setting $z_{\max} = (z_0 + z_3)/2$, it follows that $g'_{s,2} > 0$ on $[z_3, z_{\max})$, $g'_{s,2} < 0$ on $(z_{\max}, z_0]$. Since $g'_1 < 0$, we have $g'_s < 0$ on $[z_0, \infty)$. Together, we get the asserted properties of g'_s.

Regarding b): We have $g_s(z_{\max}) = h(z_{\max}, ...) < B = \frac{K+1}{\tilde{\theta}}$. For $x \in [z_2, z_1] = [z_3 + r, z_0 - r] = [m + r, M - r]$, we have $g_s(x) = h(x, ...) \ge K/2\tilde{\theta}$.

Regarding c): $\zeta < 0$ implies that $s \in [a, b]\zeta(s) \in [m, m + r] = [\tau_2, 1 + \theta_+ + \tilde{\theta}]$ and $s \in [a, b]\zeta(s) \in [M - r, M] = [1 + \theta_+, \tau_1]$. The first two estimates of assertion c) now follow from the definition of τ_1 and τ_2, and the choice of η.

Regarding d): For $t \in [1 + \theta_+ + \tilde{\theta}, 2 - \delta]$, one has $z(t, g_1) \in [z_4, z_3]$, and

$$g_s(z(t, g_1)) = g_{s,3}(z(t, g_1)) = \dot{\psi}_s(t - (2 + \theta_+ + \tilde{\theta})).$$

Regarding e): Recall the definition of ψ_s. For $x \le z_3$, we have

$$g_s(x) \le g_s(z_3) = \dot{\psi}_s(-1) = s\dot{\psi}_+(-1) + (1 - s)\dot{\psi}_-(-1)$$
$$\le \max\{|\psi_+|_{C^2}, |\psi_-|_{C^2}\} \le M_2.$$

Together with the estimate on the solutions $z(\cdot, g_1)$, the remark below is a preparation for estimating the error in the approximation of the targets ψ_s. For $s \in [0, 1]$ and $x \in [z_4, z_4 + 2\delta|g|_{C^0}]$, one has $0 < g_s(x) \le 5M_2 c_g \sqrt{\delta}$.

Now we set $z_4^+ = z_4 + 2\delta|g|_{C^0}$. Let $t \in [3/2, 2 - \delta]$. Since $\theta_- \le \delta$, we have $z(\cdot, g_1) = X(\cdot)$ on $(-\infty, t - 1]$. It follows that

$$X(t - 1) \in [X(1 - \delta), X(1/2)] \subset [0, X(\theta_+)] = [0, y(0) + \delta_1],$$

and $g = g_1$ on this interval. Hence,

$$|g'_1(z(t - 1, g_1))| = |g'(X(t - 1))| \ge |g'(0)|,$$

since $y(0) + \delta_1 < A$ and g' is decreasing on $[0, A]$. We get

$$|g_1(z(t-2,g_1))| = |g(X(t-2))| \geq |g(X(-1/2))| = |g(\varphi(-1/2))|$$
$$\geq |g(y(-1/2)/2)|.$$

Recall the definitions of γ_2 and c_g which were given before (2.13). From (g_1), we now get

$$\ddot{z}(t,g_1) \geq |g'(0)||g(y(-1/2)/2)| = \gamma_2 \text{ for } t \in [3/2, 2-\delta]. \tag{2.35}$$

Taylor expansion of $z(\cdot, g_1)$ at $2-\delta$ gives, for $r \in [0, 1/2 - \delta]$,

$$z(2-\delta-r,g_1) = z(2-\delta,g_1) + \dot{z}(2-\delta,g_1)(-r) + \ddot{z}(2-\delta-\sigma) \cdot r^2/2,$$

for some $\sigma \in [0,r]$. Since $\dot{z}(2-\delta,g_1) < 0$, we see from (2.35) that

$$z(2-\delta-r,g_1) \geq z(2-\delta,g_1) + \frac{\gamma_2}{2}r^2 = z_4 + \frac{\gamma_2}{2}r^2.$$

From the choice of δ (see (2.13),(iv)), we have that

$$r_+ = 2c_g\sqrt{\delta} \in [0, 1/2 - \delta],$$

and hence we get

$$z(2-\delta-r_+,g_1) \geq z_4 + \frac{\gamma_2}{2}4c_g^2\delta = z_4 + 2|g|_{C^0}\delta. \tag{2.36}$$

Recall the inverse function τ of $z(\cdot, g_1)[1+\theta_+, 2-\delta]$. Let now $x \in [z_4, z_4 + 2\delta|g|_{C^0}]$. Then the choice of $\tilde{\theta}$ (see (2.31)) shows that

$$x \in [z_4, z_0 - y(0)/8] \subset [z_4, z_0 - y(0)/16] \subset [z_4, z_3].$$

Moreover, we see from (2.36) that $\tau(x) \geq 2 - \delta - r_+$. It follows from property (NF) of g_s that, for $s \in [0,1]$, we have

$$0 < g_s(x) = g_s(z(\tau(x),g_1)) = \dot{\psi}_s(\tau(x)) - (2 + \theta_+ + \tilde{\theta}))$$
$$\leq \dot{\psi}_s(0) + |\ddot{\psi}_s|_{C^0} \cdot (2 + \theta_+ + \tilde{\theta} - \tau(x)).$$

Now we have $\dot{\psi}_s(0) = 0$. With estimate (2.30) and the choice of δ (see (2.13),(iv)), one gets

$$g_s(x) \leq M_2(2 + \theta_+ + \tilde{\theta} - (2 - \delta - r_+)) = M_2(\theta_+ + \tilde{e} + \delta + r_+)$$
$$\leq M_2(3\delta + 2c_g\sqrt{\delta}) \leq 5M_2c_g\sqrt{\delta}.$$

We now show that the nonlinearities g_s already lead to solutions that closely approximate the targets ψ_s. This would be sufficient in order to get a homoclinic solution from a shooting argument, but we still have to arrange the transversality.

Now we set $t^* = 3 + \theta_+ + \tilde{\theta}$. For $s \in [0,1]$, we have the following properties [Lan99]:

a) $z(t, g_s) = z(t, g_1)$ for $t \in (-\infty, 2 + \theta_+]$;
b) $\dot{z}(t, g_s) > 0$ for $t \in (t_{\min}, 3 + \theta_+ + \tilde{\theta})$;
c) With $\chi_s = z(\cdot, g_s)_{t^*} \in C$, we have $\chi_s \in H$ and $|\chi_s - \psi_s| \leq \frac{k_4 \iota}{2}$.

Let $s \in [0, 1]$. Regarding a): We have

$$z_0 = z(1 + \theta_+, g_1) \leq -y(0) - 1,$$

and we know that $g_s = g_1$ on $[z_0, \infty)$.
$z(t, g_1) \geq z_0$ for $t \in (-\infty, 1 + \theta_+]$.
For $t \in (-\infty, 1 - \theta_-]$, we have that

$$z(t, g_1) = X(t) \geq -y(0) - 1/2 > z_0.$$

For $t \in [1 - \theta_-, 1 + \theta_+] \subset [0, t_0 + 1]$, we have that $z(t, g_1) \geq z(1 + \theta_+, g_1) = z_0$.
It follows from (g_s) and from $g_s = g_1$ on $[z_0, \infty)$ that $z(\cdot, g_s) = z(\cdot, g_1)$ on $(-\infty, 2 + \theta_+]$.
Regarding b): Using assertion a), the property $z(\cdot, g_1) < 0$ on $[1 + \theta_+, 1 + \theta_+ + \tilde{\theta}]$, and property (NF) of g_s, we conclude

$$\dot{z}(t, g_s) > 0 \quad \text{on} \quad (t_{\min}, 2 + \theta_+ + \tilde{\theta}]. \tag{2.37}$$

It follows that

$$z(2 + \theta_+ + \tilde{\theta}, g_s) = z(2 + \theta_+, g_1) + \int_{1+\theta_+}^{1+\theta_+ +\tilde{\theta}} g_s(z(t, g_1))dt = 0. \tag{2.38}$$

Regarding c): From (2.38), we have $\chi_s(-1) = 0$, which means that $\chi_s \in H$.
For $t \in [2 + \theta_+ + \tilde{\theta}, 3 - \delta] = [t^* - 1, t^* - (\delta + \theta_+ + \tilde{\theta})]$, we have that

$$\dot{z}(t, g_s) = g_s(z(t - 1, g_1)) = \dot{\psi}_s(t - (3 + \theta_+ + \tilde{\theta})). \tag{2.39}$$

It follows from (2.38) and (2.39) that

$$z(t^* + t, g_s) = \psi_s(t) \text{ for } t \in [-1, -\delta - \theta_+ - \tilde{\theta}]. \tag{2.40}$$

In order to prove c), it therefore suffices to prove

$$|z(t, g_s) - \psi_s(t - t^*)| \leq \frac{k_4 \iota}{2} \text{ for } t \in [3 - \delta, 3 + \theta_+ + \tilde{\theta}]. \tag{2.41}$$

Regarding (2.41): For $t \in [3 - \delta, 3 + \theta_+ + \tilde{\theta}]$, we have, using (2.40), that

$$|z(t, g_s) - \psi_s(t - t^*)| \leq |z(t, g_s) - z(3 - \delta, g_s)| + |z(3 - \delta, g_s) - \psi_s(t - t^*)|$$
$$= |z(t, g_s) - z(3 - \delta, g_s)| + |\psi_s(-\delta - \theta_+ - \tilde{\theta}) - \psi_s(t - t^*)| \tag{2.42}$$

For $t \in [2 - \delta, 2 + \theta_+ + \tilde{\theta}]$, one has $z(t, g_s) \in [z(t_{\min}, g_1), 0]$ and therefore $\dot{z}(t, g_s) \geq 0$ for $t \in [3 - \delta, 3 + \theta_+ + \tilde{\theta}]$. Since also $\dot{\psi}_s \geq 0$, we can estimate the last two terms in (2.42) by $|z(3 + \theta_+ + \tilde{\theta}, g_s) - z(3 - \delta, g_s)|$ and by $|\psi_s(0) - \psi_s(-\delta - \theta_+ - \tilde{\theta})|$, respectively. In order to prove (2.41), it therefore suffices to prove the following two estimates:

$$|z(3 + \theta_+ + \tilde{\theta}, g_s) - z(3 - \delta, g_s)| \leq \frac{k_4 \iota}{4}, \qquad (2.43)$$

$$|\psi_s(0) - \psi_s(-\delta - \theta_+ - \tilde{\theta})| \leq \frac{k_4 \iota}{4}. \qquad (2.44)$$

Regarding (2.43): For $t \in [3 - \delta, 3 + \theta_+]$, we have $t - 1 \in [2 - \delta, 2 + \theta_+]$ and $z(t - 1, g_s) = z(t - 1, g_1)$. One gets

$$0 < \dot{z}(t, g_s) = g_s(z(t - 1, g_1)) \leq 5 M_2 c_g \sqrt{\delta}.$$

With the choice of δ (see (2.13)(v)), the estimate on δ_1 and the definition of $\iota = k_1 \delta_1$ we get

$$|z(3 + \theta_+, g_s) - z(3 - \delta, g_s)| \leq 5 M_2 c_g \sqrt{\delta}(\theta_+ + \delta)$$
$$\leq 10 M_2 c_g \delta^{3/2} (\text{ use } \theta_+ \leq \delta) \qquad (2.45)$$
$$\leq \frac{k_4 k_3 k_1 \delta}{64} \leq \frac{k_4 k_1 \delta_1}{8} = \frac{k_4 \iota}{8}.$$

Recall the numbers τ_1 and τ_2. For $t \in [2 + \theta_+, \tau_1 + 1]$, one has $t - 1 \leq \tau_1 < 2 + \theta_+$, and in view of assertion a) and the definition of z_1 and z_0 one gets

$$z(t - 1, g_s) = z(t - 1, g_1) \in [z_1, z_0] \subset (-\infty, 0).$$

Hence, $g_s(z(t - 1, g_s)) > 0$, and we have that

$$z(t, g_s) \quad = z(2 + \theta_+, g_s) + \int_{1+\theta_+}^{t-1} g_s(z(\tau, g_1)) d\tau$$
$$\leq z(2 + \theta_+, g_s) + \int_{1+\theta_+}^{\tau_1} g_s(z(\tau, g_1)) d\tau$$
$$\leq z(2 + \theta_+, g_s) + y(0)/16$$
$$= z(2 + \theta_+, g_1) + y(0)/16, \qquad \text{which implies}$$
$$z(t, g_s) \quad \leq z_0 - y(0)/16 \leq z(1 + \theta_+ + \tilde{\theta}, g_1) = z_\varepsilon.$$

Now we have

$$g_s(z(t, g_s)) \leq M_2 \qquad \text{for} \quad t \in [2 + \theta_+, \tau_1 + 1]. \qquad (2.46)$$

For $t \in [\tau_2 + 1, 2 + \theta_+ + \tilde{\theta}] \subset [2 + \theta_+, 2 + \theta_+ + \tilde{\theta}] \subset (t_{\min}, 3 + \theta_+ + \tilde{\theta})$, we have from assertion b) and from (2.38) that

$$0 = z(2 + \theta_+ + \tilde{\theta}, g_s) \geq z(t, g_s) \geq z(\tau_2 + 1, g_s), \qquad \text{so one gets}$$

$$z(\tau_2 + 1, g_s) = - \int_{\tau_2}^{1+\theta_+ + \tilde{\theta}} g_s(z(t, g_1)) ds \geq -y(0)/16 > -y(0).$$

Hence, we have

$$|g_s(z(t, g_s))| = |g(z(t, g_s))| \le |g|_{C^0},$$ (2.47)

for $t \in [\tau_2 + 1, 2 + \theta_+ + \tilde{\theta}]$. For $x \in (-\infty, 0] \setminus [z_3, z_0]$, we have that

$$|g_s(x)| \le \max\{|g|_{C^0}, M_2\}.$$

Further,

$$z_0 - z_3 \le \int_{\theta_+}^{\theta_+ + \tilde{\theta}} |g(X(t))| dt \le |g|_{C^0} \tilde{\theta}.$$

For $t \in [\tau_1 + 1, \tau_2 + 1]$, we have $\dot{z}(t, g_s) \in g_s([z_2, z_1])$, so that we have $\dot{z}(t, g_s) \ge \frac{K}{2\tilde{\theta}}$. Also, we get the following estimate

$$
\begin{aligned}
|\int_{\tau_1 + 1}^{\tau_2 + 1} g_s(z(t, g_s)) dt| &\le | \int_{\substack{\{t \in [\tau_1 + 1, \tau_2 + 1]: \\ z(t, g_s) \in [z_3, z_0]\}}} \dots | + | \int_{\substack{\{t \in [\tau_1 + 1, \tau_2 + 1]: \\ z(t, g_s) \notin [z_3, z_0]\}}} \dots | \\
&\le \int_{\substack{\{t \in [\tau_1 + 1, \tau_2 + 1]: \\ z(t, g_s) \in [z_3, z_0]\}}} |g_s(z(t, g_s))| \frac{\dot{z}(t, g_s)}{K/2\tilde{\theta}} dt + (\tau_2 - \tau_1) \max\{|g|_{C^0}, M_2\} \\
&\le \frac{2\tilde{\theta}}{K} g_s(z_{\max}) \int_{\substack{\{t \in [\tau_1 + 1, \tau_2 + 1]: \\ z(t, g_s) \in [z_3, z_0]\}}} \dot{z}(t, g_s) dt + \tilde{\theta} \cdot \max\{\dots\} \\
&\le \frac{2\tilde{\theta}}{K} \frac{K+1}{\tilde{\theta}} (z_0 - z_3) + \tilde{\theta} \cdot \max\{\dots\} \\
&\le [\frac{2(K+1)}{K} |g|_{C^0} + \max\{|g|_{C^0}, M_2\}] \tilde{\theta}.
\end{aligned}
$$ (2.48)

Putting together the estimates (2.46), (2.47) and (2.48), we get

$$
\begin{aligned}
&|z(3 + \theta_+ + \tilde{\theta}, g_s) - z(3 + \theta_+, g_s)| \\
&\le |\int_{2+\theta_+}^{\tau_1+1} g_s(z(t, g_s))| + |\int_{\tau_1+1}^{\tau_2+1} \dots| + |\int_{\tau_2+1}^{2+\theta_++\tilde{\theta}} \dots| \\
&\le \tilde{\theta} M_2 + \tilde{\theta}[\frac{2(K+1)}{K} |g|_{C^0} + \max\{|g|_{C^0}, M_2\}] + \tilde{\theta}|g|_{C^0}.
\end{aligned}
$$

Using (2.32) and the definition of ι, we can estimate the last term by

$$\frac{k_4 k_1 k_3 \delta}{64} \le \frac{k_4 k_1 \delta_1}{8} = \frac{k_4 \iota}{8}.$$

Together with (2.45), we get (2.43).

Using Taylor expansion of ψ_s at 0, and the properties $\theta_+ \le \delta, \tilde{\theta} \le \delta$, one gets

$$|\psi_s(0) - \psi_s(-\theta_+ - \tilde{\theta} - \delta)| \le |\underbrace{\dot{\psi}_s(0)}_{=0}| + M_2 \frac{(\theta_+ + \tilde{\theta} + \delta)^2}{2} \le \frac{9 M_2}{2} \delta^2.$$

With (2.13)(vi) and, as before, the relation between δ and δ_1, we can estimate the last term by

$$\frac{k_4 k_1 k_3 \delta}{32} \leq \frac{k_4 k_1 \delta_1}{4} = \frac{k_4 \iota}{4}.$$

2.4.4 Transversality

We have seen that the solutions $z(\cdot, g_s)$ reach the targets ψ_s ($s \in [0, 1]$), up to an error of at most $k_4 \iota / 2$. We could already use these nonlinearities to get a homoclinic solution, but we need to add a perturbation that makes sure that the homoclinic solution will be transversal. The perturbation must be such that it essentially keeps the accuracy by which the targets ψ_s are approximated. The method to achieve this is as in [LW96]; we add a perturbation to the g_s which has small C^0- norm, but 'large' C^1-norm, so that it has significant influence on the solutions of the variational equations $(g_s, z(\cdot, g_s))$, but little influence on the solutions $z(\cdot, g_s)$.

Recall condition (T) above. We will actually prove a stronger property, namely, that segments of solutions $a \cdot \dot{z} + b \cdot w$ $\quad ((a, b) \in \mathbb{R}^2 \setminus \{(0, 0)\})$ are all contained in the set S at a *fixed* time which is independent of a and b (in fact, at time $t^* + 1$). Proving this stronger property requires a perturbation that acts, so to speak, violently on the solutions of the variational equations $(g_s, z(\cdot, g_s))$.

Let us set: $z_{\min} = z(t_{\min}, g_1)$. We want to apply above results to $z(\cdot, g_1)$ and $w(\cdot, g_1)$ at $t_{\min}$. We know that $z(\cdot, g_1)$ is C^2 (twice differentiable) since it is a solution of (g_1) on all of $\mathbb{R}$. For $s \in [0, 1]$, we have

$$\dot{z}(t_{\min}, g_s) = \dot{z}(t_{\min}, g_1) = 0, \qquad \ddot{z}(t_{\min}, g_s) = \ddot{z}(t_{\min}, g_1) = g'(0)\dot{z}(t_0, g_1) > 0.$$

Now, from above results, we have numbers $s_1, s_2 \in [2 - \delta, 2]$ and $d > 0$, such that for every $\rho \in (0; d]$ there exist unique numbers $\sigma_1(\rho), \sigma_2(\rho)$ with the properties [Lan99]:

$$s_1 \leq \sigma_1(\rho) < t_{\min} < \sigma_2(\rho) \leq s_2,$$
$$z(t) \leq z_{\min} + \rho \qquad \text{for } t \in [\sigma_1(\rho), \sigma_2(\rho)],$$
$$z(t) > z_{\min} + \rho \qquad \text{for } t \in [s_1, \sigma_1(\rho)) \cup (\sigma_2(\rho), s_2].$$

So, we get a number $\rho \in (0; d]$ such that $\gamma(\sigma_2(\rho) - \sigma_1(\rho)) \leq \delta$, and a function $h \in BC^1(\mathbb{R}, \mathbb{R})$ with the following properties :

(i)
$$h(x) = 0 \qquad \text{for} \qquad x \geq z_{\min} + \rho; \qquad (2.49)$$

(ii) $h(x) = h(z_{\min})$ for $x \leq z_{\min}$;
(iii) $|h(x)| \leq \gamma$ for $x \in \mathbb{R}$;
(iv)

$$\text{sign } h(x) = -\text{ sign}(w(t_{\min}), g_1) \cdot \text{ sign}(-3W_1) = -1 \quad \text{for } x \in (-\infty, z_{\min} + \rho);$$

(v)

$$\text{sign } h'(z(t,g_1))w(t,g_1) = \text{sign}(-3W_1) = -1 \quad \text{for } t \in (\sigma_1(\rho), \sigma_2(\rho)) \setminus \{t_{\min}\},$$
$$\text{so} \quad h'(z(t,g_1)) > 0 \quad \text{and} \quad w(t,g_1) < 0 \quad \text{for the set;}$$

(vi)

$$\int_{\sigma_1(\rho)}^{\sigma_2(\rho)} h'(z(t,g_1))w(t,g_1)dt = -3W_1.$$

Note that $z(\cdot, g_s) = z(\cdot, g_1)$, $w(\cdot, g_s) = w(\cdot, g_1)$on $[2 - \delta, 2]$, and that the function $h$from above does not depend on s. Note also that $g_s$and $g_s + h$are both modifications of $g_1$on $(-\infty, z(1 + \theta_+, g_1)] \subset (-\infty, -y(0) - 1]$, and that $z(t, g_1) \geq z(1 + \theta_+, g_1)$ for $t \in (-\infty, 1 + \theta_+]$. It follows that

$$z(\cdot, g_s) = z(\cdot, g_s + h) = z(\cdot, g_1) \quad \text{on } (-\infty, 2 + \theta_+], \quad \text{and}$$
$$w(\cdot, g_s) = w(\cdot, g_s + h) = w(\cdot, g_1) \quad \text{on } [-1, 2 + \theta_+]. \tag{2.50}$$

We get the following properties for the nonlinearities $g_s + h$ and the solutions $z(\cdot, g_s + h)$, respectively $w(\cdot, g_s + h)$.

We have

(i)

$$z(t, g_s + h) = z(t, g_s) \quad \text{for } t \in (-\infty, \sigma_1(\rho) + 1), \quad \text{and}$$
$$w(t, g_s + h) \quad = w(t, g_s) \quad \text{for } t \in [-1, \sigma_1(\rho) + 1];$$

(ii)

$$|z(t, g_s + h) - z(t, g_s)| \leq \Delta \quad \text{for } t \in [\sigma_1(\rho) + 1, \sigma_2(\rho) + 1];$$

(iii)

$$w(\sigma_2(\rho) + 1, g_s + h) = w(\sigma_2(\rho) + 1, g_s) - 3W_1;$$

(iv)

$$\dot{z}(t, g_s + h) = \dot{z}(t, g_s) \quad \text{and} \quad \dot{w}(t, g_s + h) = \dot{w}(t, g_s)$$
$$\text{for } t \in [\sigma_2(\rho) + 1, \sigma_1(\rho) + 2].$$

Therefore we have

$$|z(t, g_s + h) - z(t, g_s)| \leq \Delta \text{ for } t \in [\sigma_1(\rho) + 1, \sigma_1(\rho) + 2]. \tag{2.51}$$

The functions $g_s + h$ do *not* satisfy (NF), and their derivatives may have more than one zero. We modify these functions to new nonlinearities f_s such that the above properties are preserved, and that f_s satisfies (NF) and f_s' has only one zero. Note that (2.49),(iii) and the choice of γ imply $(g_s + h)(z_{\min} - 1) > 0$ for $s \in [0, 1]$.

For $s \in [0,1]$, set $\lambda_s = \frac{(g_s + h)'(z_{\min} - 1)}{(g_s + h)(z_{\min} - 1)}$, and define f_s by

$$f_s(x) = \begin{cases} (g_s + h)(x) \text{ for } x > z_{\min} - 1, \\[2mm] (g_s + h)(z_{\min} - 1) \cdot \dfrac{e^{\lambda_s(x - (z_{\min} - 1))}}{1 + [x - (z_{\min} - 1)]^2} \text{ for } x \le z_{\min} - 1. \end{cases}$$

For $s \in [0,1]$, the function f_s is in $BC^1(\mathbb{R}, \mathbb{R})$, satisfies (NF), and $f_s' > 0$ on $(-\infty, z_{\max})$, $f_s' < 0$ on $(z_{\max}, \infty)$.

The map $[0,1] \ni s \mapsto f_s \in (BC^0(\mathbb{R}, \mathbb{R}), |\ |_{C^0})$ is continuous. Further we have [Lan99],

$$\begin{aligned} z(t, f_s) &= z(t, g_s + h) \text{ for } t \in (-\infty, t^* + 1], \\ w(t, f_s) &= w(t, g_s + h) \text{ for } t \in [-1, t^* + 1]. \end{aligned} \tag{2.52}$$

Let $s \in [0,1]$. First,

$$\lim_{x \to z_{\min} - 1-} f_s(x) = (g_s + h)(z_{\min} - 1) \qquad \text{and}$$

$$\lim_{x \to z_{\min} - 1-} f_s'(x) = \lambda_s f_s(z_{\min} - 1) = (g_s + h)'(z_{\min} - 1)$$
$$= \lim_{x \to z_{\min} - 1+} f_s'(x).$$

The property $f_s \in BC^1(\mathbb{R}, \mathbb{R})$ now follows from $g_s + h \in BC^1(\mathbb{R}, \mathbb{R})$ and the definition of f_s. Note that $f_s'(x) > 0$ for $x \le z_{\min} - 1$. From (2.49), (i),(ii) and (v), we have $h' \ge 0$. Since $g_s' > 0$ on $(-\infty, z_{\max})$, we have $f_s' > 0$ on $(-\infty, z_{\max})$. We have $f_s(x) = g_s(x)$ for $x \ge z_{\min} + d$ since $h(x) = 0$ for these x. Hence, $f_s' = g_s' < 0$ on $(z_{\max}, \infty)$. In order to prove (NF) for f_s, it suffices to show that $f_s > 0$ on $(-\infty, z_{\min} + d]$. It is obvious that $f_s(x) > 0$ for $x \in (-\infty, z_{\min} - 1]$. For $x \in (z_{\min} - 1, z_{\min} + d]$, we have $g_s(x) > g_s(z_{\min} - 1)$, and, recalling the definition of γ,

$$\begin{aligned} f_s(x) = (g_s + h)(x) &> g_s(z_{\min} - 1) + h(x) \\ &\ge 2\gamma - \gamma = \gamma > 0. \end{aligned} \tag{2.53}$$

It follows that the map $[0,1] \ni s \mapsto g_s + h \in (BC^0(\mathbb{R}, \mathbb{R}), |\ |_{C^0})$ is continuous. The definition of the functions g_s on $(-\infty, z_4]$, together with the property $z_{\min} + d \le z_4$ and with continuity of the map

$$[0,1] \ni s \mapsto \psi_s \in (C^2([-1,0], \mathbb{R}), |\ |_{C^2})$$

implies that the maps $[0,1] \ni s \mapsto \lambda_s \in \mathbb{R}$ and

$$[0,1] \ni s \mapsto f_s(-\infty, z_{\min} - 1] \in BC^0((-\infty, z_{\min} - 1], \mathbb{R}), |\ |_{C^0})$$

are continuous.

Using $\sigma_1(\rho) > 2 - \delta$, we see that $z(t, g_s + h) = z(t, g_s)$ for $t \in (-\infty, 3 - \delta]$. We get

$$z(t, g_s + h) = z(t, g_s) \geq z_{\min} \text{ for } t \in (-\infty, 3 - \delta], \qquad \text{and}$$
$$z(t, g_s + h) = z(t, g_s) < 0 \text{ for } t \in [2 - \delta, 2 + \theta_+ + \tilde{\theta}).$$

It follows from (2.53) that $(g_s + h)(x) > 0$ for $x \in (z_{\min}, 0)$ (although $g_s + h$ does not satisfy (NF) for all x). Hence we get that $\dot{z}(t, g_s + h) > 0$ for $t \in [3 - \delta, 3 + \theta_+ + \tilde{\theta}]$, so that together we get $z(t, g_s + h) \geq z_{\min}$ for $t \in (-\infty, t^*]$. Property (2.52) is now a consequence of $f_s = g_s + h$ on $[z_{\min}, \infty)$.

In the next lemma, we prove that the segments $z(\cdot, f_s)_{t^*}$ still approximate the targets ψ_s ($s \in [0, 1]$), and that, in addition, a transversality property holds uniformly with respect to $s \in [0, 1]$. This uniformity is necessary for our method of proof: We will find an $s^* \in (0, 1)$ with the property that the orbit of $z(\cdot, f_{s^*})$ is homoclinic to $\mathcal{O}$ by an intermediate value argument – therefore we have no information about where in $[0, 1]$ this s^* lies.

Recall the set S of segments with at most one sign change. Let $s \in [0, 1]$. The functions $z(\cdot, f_s)$ and $w(\cdot, f_s)$ have the following properties [Lan99]:

a)
$$z(\cdot, f_s)_{t^*} \in H, \ |z(\cdot, f_s)_{t^*} - \psi_s| \leq k_4 \iota.$$

b) If the condition

$$|z(t, f_s)| < y(0) + \delta_1 \quad \text{for all } t \in [t^* - 1, t^* + 1] \tag{2.54}$$

holds, then we have

$$\text{for all } (a, b) \in \mathbb{R}^2 \setminus \{(0, 0)\} : [a\dot{z}(\cdot, f_s) + b \cdot w(\cdot, f_s)]_{t^*+1} \in S.$$

Let $s \in [0, 1]$. Regarding a): Recall property (2.52). We get

$$z(t, f_s) = z(t, g_s) \qquad \text{for } t \in (-\infty, \sigma_1(\rho) + 1]. \tag{2.55}$$

In particular we have that $z(t^* - 1, f_s) = 0$, so $z(\cdot, f_s)_{t^*} \in H$.

Combining (2.52) and (2.51), and observing $\sigma_1(\rho) + 2 \leq 2 + 2 \leq t^* + 1$, we get

$$|z(t, f_s) - z(t, g_s)| \leq \Delta \qquad \text{for } t \in [\sigma_1(\rho) + 1, \sigma_1(\rho) + 2]. \tag{2.56}$$

Since $\sigma_1(\rho) + 2 > 2 - \delta + 2 > 3 + \theta_+ + \tilde{\theta} = t^*$, we conclude that

$$|z(\cdot, f_s)_{t^*} - \psi_s| \leq |z(\cdot, f_s)_{t^*} - z(\cdot, g_s)_{t^*}| + |z(\cdot, g_s)_{t^*} - \psi_s|$$
$$\leq \Delta + k_4 \iota / 2 = k_4 \iota.$$

Regarding b): Assume condition (2.54). It follows that we have $w(\sigma_2(\rho) + 1, g_s + h) \leq W_0 - 3W_1 \leq -2W_1$. We set

$$\bar{t}(s) = \min t \in [\sigma_1(\rho) + 1, \sigma_2(\rho) + 1] w(t, g_s + h) \le 0;$$

then $\bar{t}(s) < \sigma_2(\rho) + 1$. Note also that

$$3 - \delta < \sigma_1(\rho) + 1 \le \bar{t}(s) < \sigma_2(\rho) + 1 < 3. \tag{2.57}$$

$w(\cdot, f_s)$ has the following properties [Lan99].

In view of (2.52), it suffices to prove the corresponding assertions for the function $w(\cdot, g_s + h)$.

For $t \in [\sigma_1(\rho) + 1, \sigma_2(\rho) + 1]$, we have $z(t - 1, g_1) \le z_{\min} + d \le z_4 < z_3 < z_{\max}$. Hence we get from (2.50), from (2.49),(v), and from $g_s' > 0$ on $(-\infty, z_{\max})$ that

$$\begin{aligned} \dot{w}(t, g_s + h) &= (g_s + h)'(z(t - 1, g_s + h))w(t - 1, g_s + h) \\ &= (g_s + h)'(z(t - 1, g_1))w(t - 1, g_1) < 0. \end{aligned}$$

For $t \in [t^* - 1, \sigma_1(\rho) + 1]$, we have from the definition of W_0 that

$$|w(t, g_s + h)| = |w(t, g_s)| \le W_0.$$

Let $t \in [\sigma_1(\rho) + 1, \bar{t}(s)]$. If $\bar{t}(s) = \sigma_1(\rho) + 1$ then

$$|w(t, g_s + h)| = |w(\sigma_1(\rho) + 1, g_s)| \le W_0,$$

and otherwise one has

$$w(t, g_s + h) \in [0, w(\sigma_1(\rho) + 1, g_s)] \subset [0, W_0].$$

We have $g_s + h = g$ on $[-y(0) - \delta_1, y(0) + \delta_1]$. So, we get that for $t \in [t^*, \bar{t}(s) + 1]$

$$|\dot{w}(t, g_s + h)| = |g'(z(t - 1, f_s))w(t - 1, g_s + h)| \le |g|_{C^1} W_0.$$

For $t \in [\sigma_1(\rho) + 1, \sigma_2(\rho) + 1]$, we have

$$\begin{aligned} w(t, g_s + h) &\in [w(\sigma_2(\rho) + 1, g_s + h), w(\sigma_1(\rho) + 1, g_s + h)] \\ &= [w(\sigma_2(\rho) + 1, g_s) - 3W_1, w(\sigma_1(\rho) + 1, g_s)]. \end{aligned}$$

Using the definition of W_0 and W_1, one sees that

$$w(t, g_s + h) \in [-W_0 - 3W_1, W_0] \subset [-4W_1, 4W_1]. \tag{2.58}$$

For $t \in [\sigma_2(\rho) + 1, t^*]$ we get from $\sigma_1(\rho) + 2 > 4 - \delta > t^*$ that

$$\begin{aligned} w(t, g_s + h) = w(t, g_s) - 3W_1 &\in [-W_0 - 3W_1, W_0 - 3W_1] \\ &\subset [-4W_1, -2W_1]. \end{aligned} \tag{2.59}$$

Now we have $[\bar{t}(s) + 1, t^* + 1] \subset [\sigma_1(\rho) + 2, t^* + 1]$. Using condition (2.54) one gets that for $t \in [\bar{t}(s) + 1, t^* + 1]$,

$$|\dot{w}(t, g_s + h)| = |g'(z(t-1, f_s))w(t-1, g_s + h)| \leq |g|_{C^1} 4W_1.$$

For $t \in [t^*, t^* + 1]$, we get from properties (iv), (iii) and (v) that

$$w(t, g_s + h) \leq -2W_1 + (\bar{t}(s) + 1 - t^*)|g|_{C^1} W_0 + (t^* + 1 - (\bar{t}(s) + 1))4|g|_{C^1} W_1.$$

We see that

$$w(t, g_s + h) \leq -2W_1 + |g|_{C^1} W_0 + (\theta_+ + \tilde{\theta} + \delta)4|g|_{C^1} W_1.$$

With the choice of $\theta_+, \tilde{\theta}, \delta$ (see (2.13),(iii)) and of W_1, it follows that

$$w(t, g_s + h) \leq -2W_1 + \frac{1}{2}W_1 + 12\delta|g|_{C^1} W_1$$

$$\leq -2W_1 + \frac{1}{2}W_1 + \frac{1}{2}W_1 = -W_1.$$

Combining (2.57), the definition of $\bar{t}(s)$, and (i) with the second property from (iv), we get that $w(t, g_s + h) < 0$ for $t \in (\bar{t}(s), t^*]$. For these t, condition (2.54) together with (2.52) shows that $(g_s + h)'(z(t, g_s + h)) = g'(z(t, f_s)) < 0$. The assertion now follows from the variational equation $(g_s + h, z(\cdot, g_s + h))$.

So, we have [Lan99]:

(i) $z(\cdot, f_s)$ is C^2 on $[t^*, t^* + 1]$;
(ii) $\dot{z}(t^*, f_s) = 0$, $\dot{z}(t, f_s) \in [-|g|_{C^1}, 0)$ for $t \in (t^*, t^* + 1]$;
(iii)$\ddot{z}(t, f_s) \leq 0$ for $t \in [t^*, t^* + 1]$;
(iv)$\ddot{z}(t, f_s) \leq -\zeta_0$ for $t \in [t^*, \bar{t}(s) + 1]$.

Property (i) is a consequence of (f_s) and of $t^* > 1$.
From assertion a) we know $z(t^* - 1, f_s) = 0$, so $\dot{z}(t^*, f_s) = 0$. One sees that

$$z(t, f_s) = z(t, g_s + h) = z(t, g_s) < 0 \text{ for } t \in (t_0, t^* - 1),$$

and this interval contains $[t^* - 2, t^* - 1)$. Now (NF) implies that

$$\dot{z}(\cdot, f_s) > 0 \text{ on } [t^* - 1, t^*), \text{ and } z(\cdot, f_s) > 0 \text{ on } (t^* - 1, t^*]. \tag{2.60}$$

Using (NF) again, we infer $\dot{z}(\cdot, f_s) < 0$ on $(t^*, t^* + 1]$. Condition (2.54) and the property $f_s = g$ on $[-y(0) - \delta_1, y(0) + \delta_1]$ imply that $\dot{z}(t, f_s) \geq -|g|_{C^1}$ on this interval.

For $t \in [t^*, t^* + 1]$, (2.54) implies

$$\ddot{z}(t, f_s) = \underbrace{g'(z(t-1, f_s))}_{<0} \cdot \underbrace{\dot{z}(t-1, f_s)}_{\geq 0} \leq 0. \tag{2.61}$$

Let $t \in [t^*, \bar{t}(s) + 1] \subset [t^*, 4] \subset [t^*, t^* + 1]$. From (2.54), we have

$$g'(z(t-1, f_s)) \leq \max g'(x) \ x \in [-y(0) - \delta_1, y(0) + \delta_1] < 0.$$

Further, $t - 2 \in [t^* - 2, \bar{t}(s) - 1] \subset [t^* - 2, 2] = [1 + \theta_+ + \tilde{\theta}, 2]$. We have

$$z(t - 2, f_s) = z(t - 2, g_1) \in [z_{\min}, z_3].$$

Using the definition of f_s, the monotonicity properties of f_s, and (2.53), one gets

$$\dot{z}(t - 1, f_s) = f_s(z(t - 2, f_s)) \geq f_s(z_{\min}) \geq \gamma > 0.$$

Now (2.61) shows that

$$\ddot{z}(t, f_s) \leq \max g'(x) \ x \in [-y(0) - \delta_1, y(0) + \delta_1] \cdot \gamma = -\zeta_0.$$

In case $b = 0$, we have $a \neq 0$, and assertion b) is equivalent to $[\dot{z}(\cdot, f_s)]_{t^* + 1} \in S$.

In case $b \neq 0$, we set $c = a/b$, and the assertion is equivalent to

$$[c\dot{z}(\cdot, f_s) + w(\cdot, f_s)]_{t^* + 1} \in S. \tag{2.62}$$

First case: $c \geq \frac{-|g|_{C^1} W_0}{\zeta_0}$. Then we have from the choice of W_1 that for $t \in [t^*, t^* + 1]$,

$$c\dot{z}(t, f_s) + w(t, f_s) \leq \frac{|g|_{C^1} W_0}{\zeta_0} |g|_{C^1} - W_1 \leq -W_1/2 < 0.$$

from which (2.62) follows.

Second case: $c < \frac{-|g|_{C^1} W_0}{\zeta_0}$. Then, for $t \in [t^*, \bar{t}(s) + 1]$, we have that

$$c\ddot{z}(t, f_s) + \dot{w}(t, f_s) > \frac{-|g|_{C^1} W_0}{\zeta_0} (-\zeta_0) - |g|_{C^1} W_0 = 0.$$

Further, for $t \in (\bar{t}(s) + 1, t^* + 1]$, we have

$$c\ddot{z}(t, f_s) + \dot{w}(t, f_s) \geq \dot{w}(t, f_s) > 0.$$

Together, we get that the segment $[c\dot{z}(\cdot, f_s) + w(\cdot, f_s)]_{t^* + 1}$ is strictly increasing, and therefore (2.62) holds [Lan99].

2.4.5 Transversally Homoclinic Solutions

We are now ready to conclude our construction of a nonlinearity with only one extremum that generates a semi–flow with transversally homoclinic orbits. The missing step is an intermediate value argument.

a) There exists $s^* \in [0, 1]$ such that the solution $z(\cdot, f_{s^*})$ has a phase curve homoclinic to the orbit $\mathcal{O}$ of the periodic solution y of $\langle f_{s^*} \rangle$.

b) The nonlinearity f_{s^*} satisfies (NF), and

$$f'_{s^*} > 0 \text{ on } (-\infty, z_{\max}), \qquad f'_{s^*} < 0 \text{ on } (z_{\max}, \infty).$$

c) The above assumptions are satisfied by (f_{s^*}), the periodic solution y, by $\varphi = z(\cdot, f_{s^*})_0$ and by $w(\cdot, f_{s^*})$. From above analysis we get a Poincaré map with the property that the maximal invariant set of P is hyperbolic. The action of P on the trajectories in this set is conjugate to a symbol shift. The solutions of (f_{s^*}) corresponding to these trajectories are slowly oscillating [Lan99].

Now we set $\Omega_s = [z(\cdot, f_s)]_{t^*}$ for $s \in [0, 1]$. Recall that $\psi_0 = \psi_-$, $\psi_1 = \psi_+$. We have $\Omega_s \in N_\iota \subset H$, and $\mathcal{H}(\Omega_1) > 0 > \mathcal{H}(\Omega_0)$. Continuous dependence on initial data, together with continuity of the map $[0, 1] \ni s \mapsto f_s \in (BC^0(\mathbb{R}, \mathbb{R}), |\ |_{C^0})$ and continuity of $\mathcal{H}$ implies that the map $[0, 1] \ni s \mapsto \mathcal{H}(\Omega_s) \in C$ is continuous. It follows from the *Intermediate Value Theorem* that there exists $s^* \in [0, 1]$ with $\mathcal{H}(\Omega_{s^*}) = 0$. Now property (2.8), applied to $\iota \in (0, \bar{\delta}]$, shows that $\Omega_{s^*} \in W^s(y_0, P, N_\iota)$. We have that $z(\cdot, f_{s^*})$ is a solution of (f_{s^*}) with phase curve homoclinic to the orbit of y. Assertion a) is proved.

Regarding assertion c): Now we set $\Omega^* = \Omega_{s^*} \in W^s(y_0, P, N_\iota)$. Recall that $\iota = k_1 \delta_1$ and $\delta_1 \le \bar{\delta}$. We see that the solution x^{Ω^*} of the unmodified (g) with $x_0^{\Omega^*} = \Omega^*$ satisfies for all $t \ge 0$: $x_t^{\Omega^*} \in \mathcal{O} + C(\delta_1)$, so that we have

$$x^{\Omega^*}(t) \in (-y(0) - \delta_1, y(0) + \delta_1) \qquad \text{for all } t \in [-1, \infty).$$

Since $f_s^* = g$ on $(-y(0) - \delta_1, y(0) + \delta_1)$, we get

$$z(t^* + t, f_{s^*}) = x^{\Omega^*}(t) \in (-y(0) - \delta_1, y(0) + \delta_1) \qquad \text{for all } t \in [-1, \infty).$$

Moreover,

$$f_{s^*}'(z(t, f_{s^*})) = g'(z(t, f_{s^*})) < 0 \text{ for } t \ge t^* - 1,$$

so that the variational equation $(f_{s^*}, z(\cdot, f_{s^*}))$ has a negative coefficient for $t \ge t^* - 1$. It follows from this property that the function $a \cdot \dot{z}(\cdot, f_{s^*}) + b \cdot w(\cdot, f_{s^*})$ is eventually slowly oscillating for all $(a, b) \in \mathbb{R}^2 \setminus \{(0, 0)\}$. Hence condition (T), and the conclusions follow.

3

3–Body Problem and Chaos Control

This Chapter addresses fundamental mechanical *sources of chaos* (namely classical mechanics of chaos in the Solar system), as well as the basic techniques for *chaos control* (mainly, the so–called OGY–control).

3.1 Mechanical Origin of Chaos

Recall that Poincaré 3–body problem (Figure 1.17) was the first legitimate mechanical source of chaos. Also, recall from [II05, II06b] that the *restricted 3–body problem* describes the motion of an infinitesimal mass moving under the gravitational effect of the two finite masses, called primaries, which move in circular orbits around their center of mass on account of their mutual attraction and the infinitesimal mass not influencing the motion of the primaries.

There is a vast literature on the restricted 3–body problem. Among other things, there are investigations of the equilibriums points and their stability, investigations of the existence, stability and bifurcation of periodic orbits, and investigations of collisions and ejection orbits. The restricted problem is said to be a limit of the 3–body problem as one of the masses tends to zero, and so to each result for the restricted problem there should be a corresponding result for the full 3–body problem. Indeed, there are many such results for the 3–body problem, but usually the proofs for the 3–body problem are similar to, but independent of, the proofs for the restricted problem.

By the *restricted problem* we shall mean the circular, planar or spatial, restricted 3–body problem, and by the *full problem* we shall mean the planar or spatial 3–body problem. The restricted problem is a Hamiltonian system of differential equations which describes the motion of an infinitesimal particle (the *satellite*) moving under the gravitational influence of two particles of finite mass (the *primaries*) which are moving on a circular orbit of the Kepler problem (see [MS00]).

Since the motion of the primaries is given, the restricted problem has two degrees of freedom for the planar problem and three degrees of freedom for the spatial problem. However, the full problem has six degrees of freedom in the planar case and nine degrees of freedom in the spatial case. Thus, at first the restricted problem seems too small to reflect the full complexity of the full problem; but when the symmetries of the full problem are taken into account the dimension gap narrows considerably.

The Hamiltonian of the full problem is invariant under Euclidean motions, i.e., translations and rotations, which begets the integrals of linear and angular momentum. Translations and rotations give rise to ignorable coordinates. Holding the integrals fixed and dropping the ignorable coordinates reduces the full problem from six to three degrees of freedom in the planar case and from nine to four degrees of freedom in the spatial case. Thus the full problem on the reduced space is only one degree of freedom larger than the restricted problem in either the planar or the spatial case. We shall call the full 3–body problem on the reduced space the *reduced 3–body problem* or simply the *reduced problem.*

The question addressed in this secton is the relation between the reduced problem with one small mass and the restricted 3–body problem. The goal is to illustrate a set of methods and results that assure that much of what is known about the restricted problem can be carried over to the reduced problem. We will not attempt to delineate every such result, since this would be too tedious. Once a few examples have been seen the general pattern can be gleaned.

3.1.1 Restricted 3–Body Problem

The classical restricted 3–body problem can be generalized to include the force of radiation pressure, the *Poynting–Robertson effect* (see Figure 3.1) and *oblateness effect* (see [KI04, KSI06]).

John Poynting [Poy03] considered the effect of the absorption and subsequent re–emission of sunlight by small isolated particles in the solar system. His work was later modified by H. Robertson [Rob37] who used a precise relativistic treatments of the first order in the ratio of he velocity of the particle to that of light.[1] The PR–force is equal to

---

[1] The *Poynting–Robertson (PR) effect*, also known as PR–drag, named after John Henry Poynting and Howard Percy Robertson, is a process by which solar radiation causes dust particles in a solar system to slowly spiral inward. The effect is due to the orbital motion of the dust grains causing the radial push of the solar radiation to be offset slightly, slowing its orbit. The effect can be interpreted in two ways, depending on the reference frame chosen. From the perspective of the grain of dust, the Sun's radiation appears to be coming from slightly forward of the direct line to the center of its orbit because the dust is moving perpendicular to the radiation's movement. This angle of aberration is extremely small since the radiation is moving at the speed of light and the dust grain is moving many

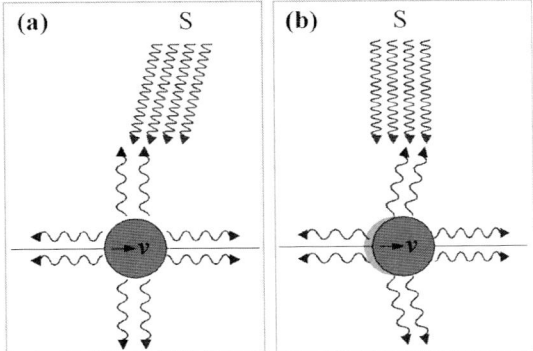

Fig. 3.1. Poynting–Robertson (PR) effect: radiation from the Sun (S) and thermal radiation from a particle seen (a) from an observer moving with the particle; and (b) from an observer at rest with respect to the Sun.

$$F_{PR} = \frac{Wv}{c^2} = \frac{r^2}{4c^2}\sqrt{\frac{GM_sL_s^2}{R^5}},$$

where W is the power radiated from the particle (equal to the incoming radiation), v is the particle's velocity, c is the speed of light, r is the object's radius, G is the universal gravitational constant, M_s is the Sun's mass, L_s is the solar luminosity and R is the object's orbital radius.[2]

_____

orders of magnitude slower than that. From the perspective of the solar system as a whole, the dust grain absorbs sunlight entirely in a radial direction. However, the dust grain's motion relative to the Sun causes it to re–emit that energy unevenly (more forward than aft), causing an equivalent change in angular momentum (a sort of recoil).

[2] Since the gravitational force goes as the cube of the object's radius (being a function of its volume) whilst the power it receives and radiates goes as the square of that same radius (being a function of its surface), the PR–effect is more pronounced for smaller objects. Also, since the Sun's gravity varies as one over R^2 whereas the PR–force varies as one over $R^{2.5}$, the latter gets relatively stronger as the object approaches the Sun, which tends to reduce the eccentricity of the object's orbit in addition to dragging it in. Dust particles sized a few micrometers need a few thousand years to get from 1 AU distance to distances where they evaporate. There is a critical size at which small objects are so affected by radiation pressure that the latter actually cancels the Sun's gravitation altogether. For rocky particles, this size is about $0.1\mu m$ in diameter. If the particles are already in motion at their creation, radiation pressure does not need to cancel gravity completely to move the particles out of the solar system, so the critical size gets a bit larger. The PR–effect still affects these small particles, but they will be blown out of the solar system by the Sun's light before the PR–force works any significant change in their motion. Poynting conceived of the force in the then-dominant 'luminiferous aether' view of the propagation of light; Robertson is the one who redid the demonstration from a relativistic standpoint and confirmed the result.

The effect of radiation pressure and PR–drag in the restricted 3–body problem has been studied by [CLS96, Che70, Sch80], who discussed the position as well as the stability of the Lagrangian equilibrium points when radiation pressure, PR–drag force are included. Later [Mur94] systematically discussed the dynamical effect of general drag in the planar circular restricted three body problem; [LZJ95] examined the effect of radiation pressure, PR–drag and solar–wind drag in the restricted 3–body problem.

Moser conditions [Mos62], *Arnold theorem* [Arn61] and *Liapunov theorem* [Lya56] played a significant role in deciding the nonlinear stability of an equilibrium point. Applying Arnold theorem, [Leo62] examined the nonlinear stability of triangular points, while Moser gave same modifications in Arnold theorem. Then [DD67] investigated the nonlinear stability of triangular points by applying Moser's modified version of the Arnold theorem.

Also, the effect of perturbations on the nonlinear stability of triangular points was studied in [BH83], while [Nie94] investigated the nonlinear stability of the libration points in the photo–gravitational restricted three body problem. [MI95] studied second order normalization in the generalized restricted problem of three bodies, smaller primary being an oblate spheroid, while [Ish97] studied nonlinear stability in the generalised restricted three body problem.

In [KI04, KSI06] the authors have performed *Birkhoff normalization* of the Hamiltonian. For this they have utilized the so–called *Henrard method* and expanded the coordinates of the third body in a *double d'Alembert series*. They have found the values of first and second order components. The second order components were obtained as solutions of the two PDEs. They have employed the first condition of KAM theorem in solving those equations. The first and second order components were affected by radiation pressure, oblateness and PR–drag.

Equations of Motion for the Restricted 3–Body Problem

The equations of motions are

$$\ddot{x} - 2n\dot{y} = U_x, \quad \text{where} \quad U_x = \frac{\partial U_1}{\partial x} - \frac{W_1 N_1}{r_1^2}$$

$$\ddot{y} + 2n\dot{x} = U_y, \quad \text{where} \quad U_y = \frac{\partial U_1}{\partial y} - \frac{W_1 N_2}{r_1^2}, \qquad \text{with}$$

$$U_1 = \frac{n^2(x^2 + y^2)}{2} + \frac{(1-\mu)q_1}{r_1} + \frac{\mu}{r_2} + \frac{\mu A_2}{2r_2^3}, \qquad \text{and}$$

$$r_1^2 = (x + \mu)^2 + y^2, \qquad r_2^2 = (x + \mu - 1)^2 + y^2, \qquad n^2 = 1 + \frac{3}{2}A_2,$$

$$N_1 = \frac{(x + \mu)[(x + \mu)\dot{x} + y\dot{y}]}{r_1^2} + \dot{x} - ny, \quad N_2 = \frac{y[(x + \mu)\dot{x} + y\dot{y}]}{r_1^2} + \dot{y} + n(x + \mu),$$

$$\text{with} \quad W_1 = \frac{(1 - \mu)(1 - q_1)}{c_d}, \qquad \mu = \frac{m_2}{m_1 + m_2} \leq \frac{1}{2},$$

where m_1, m_2 are masses of the primaries, $A_2 = \frac{r_e^2 - r_p^2}{5r^2}$ are the oblateness coefficient, r_e and r_p are the equatorial and polar radii respectively, r is the distance between primaries; $q = (1 - \frac{F_p}{F_g})$ is the mass reduction factor expressed in terms of the particle's radius a, density ρ and radiation pressure efficiency factor χ, i.e., $q = 1 - \frac{5.6 \times 10^{-5}\chi}{a\rho}$. Assumption $q = const$ is equivalent to neglecting fluctuation in the beam of solar radiation and the effect of solar radiation, the effect of the planet's shadow, obviously $q \leq 1$. If triangular equilibrium points are given by $U_x = 0$, $U_y = 0$, $z = 0$, $y \neq 0$, then we have [KI04, KSI06]

$$x_* = x_0 \left[1 - \frac{nW_1[(1 - \mu)(1 + \frac{5}{2}A_2) + \mu(1 - \frac{A_2}{2})\frac{\delta^2}{2}]}{3\mu(1 - \mu)y_0 x_0} - \frac{\delta^2}{2}\frac{A_2}{x_0} \right],$$

$$y_* = y_0 \left[1 - \frac{nW_1\delta^2[2\mu - 1 - \mu(1 - \frac{3A_2}{2})\frac{\delta^2}{2} + 7(1 - \mu)\frac{A_2}{2}]}{3\mu(1 - \mu)y_0^3} - \frac{\delta^2(1 - \frac{\delta^2}{2})A_2}{y_0^2} \right]^{1/2},$$

where $\quad x_0 = \frac{\delta^2}{2} - \mu, \qquad y_0 = \pm\delta(1 - \frac{\delta^2}{4})^{1/2} \qquad$ and $\qquad \delta = q_1^{1/3}.$

Birkhoff Normalization

The Lagrangian function of the problem can be written as [KI04, KSI06]

$$L = \frac{1}{2}(\dot{x}^2 + \dot{y}^2) + n(x\dot{y} - \dot{x}y) + \frac{n^2}{2}(x^2 + y^2) + \frac{(1 - \mu)q_1}{r_1} + \frac{\mu}{r_2} + \frac{\mu A_2}{2r_2^3}$$

$$+ W_1 \left[\frac{(x + \mu)\dot{x} + y\dot{y}}{r_1^2} - n \arctan \frac{y}{(x + \mu)} \right],$$

and the Hamiltonian is $H = -L + P_x\dot{x} + P_y\dot{y}$, where P_x, P_y are the momenta coordinates given by

$$P_x = \frac{\partial L}{\partial \dot{x}} = \dot{x} - ny + \frac{W_1}{2r_1^2}(x + \mu), \qquad P_y = \frac{\partial L}{\partial \dot{y}} = \dot{y} + nx + \frac{W_1}{2r_1^2}y.$$

For simplicity, we suppose $q_1 = 1 - \epsilon$, with $|\epsilon| \ll 1$ then coordinates of triangular equilibrium points L_4 can be written as the form

$$x = \frac{\gamma}{2} - \frac{\epsilon}{3} - \frac{A_2}{2} + \frac{A_2\epsilon}{3} - \frac{(9 + \gamma)}{6\sqrt{3}}nW_1 - \frac{4\gamma\epsilon}{27\sqrt{3}}nW_1,$$

$$y = \frac{\sqrt{3}}{2} \left[1 - \frac{2\epsilon}{3}\frac{A_2}{3} + \frac{2A_2\epsilon}{9} - \frac{(1 + \gamma)}{9\sqrt{3}}nW_1 - \frac{4\gamma\epsilon}{27\sqrt{3}}nW_1 \right],$$

where $\gamma = 1 - 2\mu$. We shift the origin to L_4. For that, we can apply the change

$$x \to x_* + x, \qquad y \to y_* + y, \qquad \text{and let} \qquad a = x_* + \mu, \qquad b = y_*,$$

so that

$$a = \frac{1}{2}\left(-\frac{2\epsilon}{3} - A_2 + \frac{2A_2\epsilon}{3} - \frac{(9+\gamma)}{3\sqrt{3}}nW_1 - \frac{8\gamma\epsilon}{27\sqrt{3}}nW_1\right),$$

$$b = \frac{\sqrt{3}}{2}\left(1 - \frac{2\epsilon}{-}\frac{A_2}{3} + \frac{2A_2\epsilon}{9} - \frac{(1+\gamma)}{9\sqrt{3}}nW_1 - \frac{4\gamma\epsilon}{27\sqrt{3}}nW_1\right),$$

Expanding L in power series of x and y, we get

$$L = L_0 + L_1 + L_2 + L_3 + \cdots \qquad \text{and}$$
$$H = H_0 + H_1 + H_2 + H_3 + \cdots = -L + P_x\dot{x} + P_y\dot{y},$$

where $L_0, L_1, L_2, L_3 \ldots$ are constants, first order term, second order term $\ldots$ respectively. Third order term H_3 of Hamiltonian can be written as

$$H_3 = -L_3 = -\frac{1}{3!}\left\{x^3T_1 + 3x^2yT_2 + 3xy^2T_3 + y^3T_4 + 6T_5\right\}, \qquad \text{where}$$

$$T_1 = \frac{3}{16}\left[\frac{16}{3}\epsilon + 6A_2 - \frac{979}{18}A_2\epsilon + \frac{(143+9\gamma)}{6\sqrt{3}}nW_1 + \frac{(459+376\gamma)}{27\sqrt{3}}nW_1\epsilon + \right.$$
$$\left.\gamma\{14 + \frac{4\epsilon}{3} + 25A_2 - \frac{1507}{18}A_2\epsilon - \frac{(215+29\gamma)}{6\sqrt{3}}nW_1 - \frac{2(1174+169\gamma)}{27\sqrt{3}}nW_1\epsilon\}\right],$$

$$T_2 = \frac{3\sqrt{3}}{16}\left[14 - \frac{16}{3}\epsilon + \frac{A_2}{3} - \frac{367}{18}A_2\epsilon + \frac{115(1+\gamma)}{18\sqrt{3}}nW_1 - \frac{(959-136\gamma)}{27\sqrt{3}}nW_1\epsilon + \right.$$
$$\left.\gamma\left\{\frac{32\epsilon}{3} + 40A_2 - \frac{382}{9}A_2\epsilon + \frac{(511+53\gamma)}{6\sqrt{3}}nW_1 - \frac{(2519-24\gamma)}{27\sqrt{3}}nW_1\epsilon\right\}\right],$$

$$T_3 = \frac{-9}{16}\left[\frac{8}{3}\epsilon + \frac{203A_2}{6} - \frac{625}{54}A_2\epsilon - \frac{(105+15\gamma)}{18\sqrt{3}}nW_1 - \frac{(403-114\gamma)}{81\sqrt{3}}nW_1\epsilon + \right.$$
$$\left.\gamma\left\{2 - \frac{4\epsilon}{9} + \frac{55A_2}{2} - \frac{797}{54}A_2\epsilon + \frac{(197+23\gamma)}{18\sqrt{3}}nW_1 - \frac{(211-32\gamma)}{81\sqrt{3}}nW_1\epsilon\right\}\right],$$

$$T_4 = \frac{-9\sqrt{3}}{16}\left[2 - \frac{8}{3}\epsilon + \frac{23A_2}{3} - 44A_2\epsilon - \frac{(37+\gamma)}{18\sqrt{3}}nW_1 - \frac{(219+253\gamma)}{81\sqrt{3}}nW_1\epsilon + \right.$$
$$\left.\gamma\left\{4\epsilon + \frac{88}{27}A_2\epsilon + \frac{(241+45\gamma)}{18\sqrt{3}}nW_1 - \frac{(1558-126\gamma)}{81\sqrt{3}}nW_1\epsilon\right\}\right],$$

$$T_5 = \frac{W_1}{2(a^2+b^2)^3}\left[(a\dot{x}+b\dot{y})\left\{3(ax+by) - (bx-ay)^2\right\}\right.$$
$$\left. -2(x\dot{x}+y\dot{y})(ax+by)(a^2+b^2)\right].$$

In order to perform *Birkhoff normalization*, following [KI04, KSI06], we use Henrard's method for which the coordinates (x, y) of infinitesimal body, to be expanded in double d'Alembert series $x = \sum_{n\geq 1} B_n^{1,0}$, $\quad y = \sum_{j\geq n} B_j^{0,1}$ where the homogeneous components $B_n^{1,0}$ and $B_n^{0,1}$ of degree n are of the form

$$\sum_{0\leq m\leq n} I_1^{\frac{n-m}{2}} I_2^{\frac{m}{2}} \sum_{(p,q)} C_{n-m,m,p,q} \cos\left(p\phi_1 + q\phi_2\right) + S_{n-m,m,p,q} \sin\left(p\phi_1 + q\phi_2\right).$$

The condition in double summation are (i) p runs over those integers in the interval $0 \leq p \leq n - m$ that have the same parity as $n - m$ (ii) q runs over those integers in the interval $-m \leq q \leq m$ that have the same parity as m. Here I_1, I_2 are the action momenta coordinates which are to be taken as constant of integer, ϕ_1, ϕ_2 are angle coordinates to be determined as linear functions of time in such a way that $\dot\phi_1 = \omega_1 + \sum_{n\geq 1} f_{2n}(I_1,I_2)$, $\dot\phi_2 = \omega_2 + \sum_{n\geq 1} g_{2n}(I_1,I_2)$ where ω_1, ω_2 are the basic frequencies, f_{2n} and g_{2n} are of the form

$$f_{2n} = \sum_{0\leq m\leq n} f'_{2(n-m),2m} I_1^{n-m} I_2^m, \qquad g_{2n} = \sum_{0\leq m\leq n} g'_{2(n-m),2m} I_1^{n-m} I_2^m,$$

The first order components $B_1^{1,0}$, and $B_1^{0,1}$ in I_1, I_2 are the values of x and y given by

$$X = JT, \qquad \text{where} \qquad X = \begin{bmatrix} x \\ y \\ p_x \\ p_y \end{bmatrix}, \qquad J = [J_{ij}]_{1\leq i\leq j\leq 4}, \qquad T = \begin{bmatrix} Q_1 \\ Q_2 \\ p_1 \\ p_2 \end{bmatrix},$$

$$P_i = (2I_i\omega_i)^{1/2} \cos\phi_i, \quad Q_i = \left(\frac{2I_i}{\omega_i}\right)^{1/2} \sin\phi_i, \quad (i = 1,2),$$

$$B_1^{1,0} = J_{13}\sqrt{2\omega_1 I_1}\cos\phi_1 + J_{14}\sqrt{2\omega_2 I_2}\cos\phi_2, \tag{3.1}$$

$$B_1^{0,1} = J_{21}\sqrt{\frac{2I_1}{\omega_1}}\sin\phi_1 + J_{22}\sqrt{\frac{2I_2}{\omega_2}}\sin\phi_2 + J_{23}\sqrt{2I_1\omega_1}\cos\phi_1 + J_{24}\sqrt{2I_2\omega_2}\sin\phi_2,$$

where

$$
\begin{aligned}
J_{13} = {} & \frac{l_1}{2\omega_1 k_1}\left\{1 - \frac{1}{2l_1^2}\left[\epsilon + \frac{45A_2}{2} - \frac{717A_2\epsilon}{36} + \frac{(67+19\gamma)}{12\sqrt{3}}nW_1 - \frac{(431-3\gamma)}{27\sqrt{3}}nW_1\epsilon\right]\right.\\
& + \frac{\gamma}{2l_1^2}\left[3\epsilon - \frac{29A_2}{36} - \frac{(187+27\gamma)}{12\sqrt{3}}nW_1 - \frac{2(247+3\gamma)}{27\sqrt{3}}nW_1\epsilon\right]\\
& - \frac{1}{2k_1^2}\left[\frac{\epsilon}{2} - 3A_2 - \frac{73A_2\epsilon}{24} + \frac{(1-9\gamma)}{24\sqrt{3}}nW_1 + \frac{(53-39\gamma)}{54\sqrt{3}}nW_1\epsilon\right]\\
& - \frac{\gamma}{4k_1^2}\left[\epsilon - 3A_2 - \frac{299A_2\epsilon}{72} - \frac{(6-5\gamma)}{12\sqrt{3}}nW_1 - \frac{(266-93\gamma)}{54\sqrt{3}}nW_1\epsilon\right]\\
& + \frac{\epsilon}{4l_1^2 k_1^2}\left[\frac{3A_2}{4} + \frac{(33+14\gamma)}{12\sqrt{3}}nW_1\right] + \frac{\gamma\epsilon}{8l_1^2 k_1^2}\left.\left[\frac{347A_2}{36} - \frac{(43-8\gamma)}{4\sqrt{3}}nW_1\right]\right\}
\end{aligned}
$$

$$J_{14} = \frac{l_2}{2\omega_2 k_2} \left\{ 1 - \frac{1}{2l_2^2} \left[\epsilon + \frac{45A_2}{2} - \frac{717A_2\epsilon}{36} + \frac{(67+19\gamma)}{12\sqrt{3}} nW_1 - \frac{(431-3\gamma)}{27\sqrt{3}} nW_1\epsilon \right] \right.$$
$$- \frac{\gamma}{2l_2^2} \left[3\epsilon - \frac{293A_2}{36} + \frac{(187+27\gamma)}{12\sqrt{3}} nW_1 - \frac{2(247+3\gamma)}{27\sqrt{3}} nW_1\epsilon \right]$$
$$- \frac{1}{2k_2^2} \left[\frac{\epsilon}{2} - 3A_2 - \frac{73A_2\epsilon}{24} + \frac{(1-9\gamma)}{24\sqrt{3}} nW_1 + \frac{(53-39\gamma)}{54\sqrt{3}} nW_1\epsilon \right]$$
$$+ \frac{\gamma}{2k_2^2} \left[\epsilon - 3A_2 - \frac{299A_2\epsilon}{72} - \frac{(6-5\gamma)}{12\sqrt{3}} nW_1 - \frac{(268-9\gamma)}{54\sqrt{3}} nW_1\epsilon \right]$$
$$\left. - \frac{\epsilon}{4l_2^2 k_2^2} \left[\frac{33A_2}{4} + \frac{(1643-93\gamma)}{216\sqrt{3}} nW_1 \right] + \frac{\gamma\epsilon}{4l_2^2 k_2^2} \left[\frac{737A_2}{72} - \frac{(13+2\gamma)}{\sqrt{3}} nW_1 \right] \right\}$$

$$J_{21} = -\frac{4n\omega_1}{l_1 k_1} \left\{ 1 + \frac{1}{2l_1^2} \left[\epsilon + \frac{45A_2}{2} - \frac{717A_2\epsilon}{36} + \frac{(67+19\gamma)}{12\sqrt{3}} nW_1 - \frac{(413-3\gamma)}{27\sqrt{3}} nW_1\epsilon \right] \right.$$
$$- \frac{\gamma}{2l_1^2} \left[3\epsilon - \frac{293A_2}{36} + \frac{(187+27\gamma)}{12\sqrt{3}} nW_1 - \frac{2(247+3\gamma)}{27\sqrt{3}} nW_1\epsilon \right]$$
$$- \frac{1}{2k_1^2} \left[\frac{\epsilon}{2} - 3A_2 - \frac{73A_2\epsilon}{24} + \frac{(1-9\gamma)}{24\sqrt{3}} nW_1 + \frac{(53-39\gamma)}{54\sqrt{3}} nW_1\epsilon \right]$$
$$- \frac{\gamma}{4k_1^2} \left[\epsilon - 3A_2 - \frac{299A_2\epsilon}{72} - \frac{(6-5\gamma)}{12\sqrt{3}} nW_1 - \frac{(268-93\gamma)}{54\sqrt{3}} nW_1\epsilon \right]$$
$$\left. + \frac{\epsilon}{8l_1^2 k_1^2} \left[\frac{33A_2}{4} + \frac{(68-10\gamma)}{24\sqrt{3}} nW_1 \right] + \frac{\gamma\epsilon}{8l_1^2 k_1^2} \left[\frac{242A_2}{9} + \frac{(43-8\gamma)}{4\sqrt{3}} nW_1 \right] \right\}$$

$$J_{22} = \frac{4n\omega_2}{l_2 k_2} \left\{ 1 + \frac{1}{2l_2^2} \left[\epsilon + \frac{45A_2}{2} - \frac{717A_2\epsilon}{36} + \frac{(67+19\gamma)}{12\sqrt{3}} nW_1 - \frac{(413-3\gamma)}{27\sqrt{3}} nW_1\epsilon \right] \right.$$
$$- \frac{\gamma}{2l_2^2} \left[3\epsilon - \frac{293A_2}{36} + \frac{(187+27\gamma)}{12\sqrt{3}} nW_1 - \frac{2(247+3\gamma)}{27\sqrt{3}} nW_1\epsilon \right]$$
$$+ \frac{1}{2k_2^2} \left[\frac{\epsilon}{2} - 3A_2 - \frac{73A_2\epsilon}{24} + \frac{(1-9\gamma)}{24\sqrt{3}} nW_1 + \frac{(53-39\gamma)}{54\sqrt{3}} nW_1\epsilon \right]$$
$$- \frac{\gamma}{4k_2^2} \left[\epsilon - 3A_2 - \frac{299A_2\epsilon}{72} - \frac{(6-5\gamma)}{12\sqrt{3}} nW_1 - \frac{(268-93\gamma)}{54\sqrt{3}} nW_1\epsilon \right]$$
$$\left. + \frac{\epsilon}{4l_2^2 k_2^2} \left[\frac{33A_2}{4} + \frac{(34+5\gamma)}{12\sqrt{3}} nW_1 \right] + \frac{\gamma\epsilon}{8l_2^2 k_2^2} \left[\frac{75A_2}{2} + \frac{(43-8\gamma)}{4\sqrt{3}} nW_1 \right] \right\}$$

$$J_{23} = \frac{\sqrt{3}}{4\omega_1 l_1 k_1} \left\{ 2\epsilon + 6A_2 + \frac{37A_2\epsilon}{2} - \frac{(13+\gamma)}{2\sqrt{3}} nW_1 + \frac{2(79-7\gamma)}{9\sqrt{3}} nW_1\epsilon \right.$$

$$-\gamma \left[6 + \frac{2\epsilon}{3} + 13A_2 - \frac{33A_2\epsilon}{2} + \frac{(11-\gamma)}{2\sqrt{3}} nW_1 - \frac{(186-\gamma)}{9\sqrt{3}} nW_1\epsilon \right]$$

$$+ \frac{1}{2l_1^2} \left[51A_2 + \frac{(14+8\gamma)}{3\sqrt{3}} nW_1 \right] - \frac{\epsilon}{k_1^2} \left[3A_2 + \frac{(19+6\gamma)}{6\sqrt{3}} nW_1 \right]$$

$$- \frac{\gamma}{2l_1^2} \left[6\epsilon + 135A_2 - \frac{808A_2\epsilon}{9} - \frac{(67+19\gamma)}{2\sqrt{3}} nW_1 - \frac{(755+19\gamma)}{9\sqrt{3}} nW_1\epsilon \right]$$

$$- \frac{\gamma}{2k_1^2} \left[3\epsilon - 18A_2 - \frac{55A_2\epsilon}{4} - \frac{(1-9\gamma)}{4\sqrt{3}} nW_1 + \frac{(923-60\gamma)}{12\sqrt{3}} nW_1\epsilon \right]$$

$$\left. + \frac{\gamma\epsilon}{8l_1^2 k_1^2} \left[\frac{9A_2}{2} + \frac{(34-5\gamma)}{2\sqrt{3}} nW_1 \right] \right\}$$

$$J_{24} = \frac{\sqrt{3}}{4\omega_2 l_2 k_2} \left\{ 2\epsilon + 6A_2 + \frac{37A_2\epsilon}{2} - \frac{(13+\gamma)}{2\sqrt{3}} nW_1 + \frac{2(79-7\gamma)}{9\sqrt{3}} nW_1\epsilon \right.$$

$$-\gamma \left[6 + \frac{2\epsilon}{3} + 13A_2 - \frac{33A_2\epsilon}{2} + \frac{(11-\gamma)}{2\sqrt{3}} nW_1 - \frac{(186-\gamma)}{9\sqrt{3}} nW_1\epsilon \right]$$

$$- \frac{1}{2l_2^2} \left[51A_2 + \frac{(14+8\gamma)}{3\sqrt{3}} nW_1 \right] - \frac{\epsilon}{k_2^2} \left[3A_2 + \frac{(19+6\gamma)}{6\sqrt{3}} nW_1 \right]$$

$$- \frac{\gamma}{2l_2^2} \left[6\epsilon + 135A_2 - \frac{808A_2\epsilon}{9} - \frac{(67+19\gamma)}{2\sqrt{3}} nW_1 - \frac{(755+19\gamma)}{9\sqrt{3}} nW_1\epsilon \right]$$

$$- \frac{\gamma}{2k_1^2} \left[3\epsilon - 18A_2 - \frac{55A_2\epsilon}{4} - \frac{(1-9\gamma)}{4\sqrt{3}} nW_1 + \frac{(923-60\gamma)}{12\sqrt{3}} nW_1\epsilon \right]$$

$$\left. - \frac{\gamma\epsilon}{4l_1^2 k_1^2} \left[\frac{99A_2}{2} + \frac{(34-5\gamma)}{2\sqrt{3}} nW_1 \right] \right\},$$

with $\quad l_j^2 = 4\omega_j^2 + 9, (j = 1, 2) \quad$ and $\quad k_j^2 = 2\omega_1^2 - 1, \quad k_2^2 = -2\omega_2^2 + 1.$

In order to find out the second order components $B_2^{1,0}, B_2^{0,1}$ we consider Lagrange's equations of motion

$$\frac{d}{dt}\left(\frac{\partial L}{\partial \dot{x}}\right) - \frac{\partial L}{\partial x} = 0, \quad \frac{d}{dt}\left(\frac{\partial L}{\partial \dot{y}}\right) - \frac{\partial L}{\partial y} = 0,$$

$$\left. \begin{array}{l} \ddot{x} - 2n\dot{y} + (2E - n^2)x + Gy = \frac{\partial L_3}{\partial x} + \frac{\partial L_4}{\partial x} \\ \ddot{x} + 2n\dot{x} + (2F - n^2)y + Gx = \frac{\partial L_3}{\partial y} + \frac{\partial L_4}{\partial y} \end{array} \right\}, \tag{3.2}$$

Since x and y are double d'Alembert series, $x^j x^k (j \geq 0, k \geq 0, j + k \geq 0)$ is also a double d'Alembert series, the time derivatives $\dot{x}, \dot{y}, \ddot{x}, \ddot{y}$ are also double d'Alembert series. We can write

$$\dot{x} = \sum_{n\geq 1} \dot{x}_n, \quad \dot{y} = \sum_{n\geq 1} \dot{y}_n, \quad \ddot{x} = \sum_{n\geq 1} \ddot{x}_n, \quad \ddot{y} = \sum_{n\geq 1} \ddot{y}_n,$$

$$r_5 = \frac{1}{\omega_1\omega_2(2\omega_1+\omega_2)(4\omega_1+2\omega_2)}\Bigg\{(\omega_1+\omega_2)^3 \times$$

$$\left[\Big\{J_{13}J_{22}(\frac{\omega_1}{\omega_2})^{1/2}-J_{14}J_{21}(\frac{\omega_2}{\omega_1})^{1/2}\Big\}F_1'\right.$$

$$\left.-2\Big\{J_{21}J_{24}(\frac{\omega_2}{\omega_1})^{1/2}-J_{22}J_{23}(\frac{\omega_1}{\omega_2})^{1/2}\Big\}F_1''\right]$$

$$-(\omega_1+\omega_2)^2\left[\Big\{2\{J_{13}J_{14}F_2\right.$$

$$+(J_{13}J_{24}+J_{14}J_{23})F_2'\}(\omega_1\omega_2)^{1/2}+\Big\{\frac{J_{21}J_{22}}{(\omega_1\omega_2)^{1/2}}$$

$$\left.+J_{23}J_{24}(\omega_1\omega_2)^{1/2}\Big\}F_2''\right]$$

$$-(\omega_1+\omega_2)\left[\Big\{J_{13}J_{22}(\frac{\omega_1}{\omega_2})^{1/2}-J_{14}J_{21}(\frac{\omega_2}{\omega_1})^{1/2}\Big\}F_3'\right.$$

$$\left.-2\Big\{J_{21}J_{24}(\frac{\omega_2}{\omega_1})^{1/2}-J_{22}J_{23}(\frac{\omega_1}{\omega_2})^{1/2}\Big\}F_3''\right]$$

$$+\left[\Big\{2\{J_{13}J_{14}F_4+(J_{13}J_{24}+J_{14}J_{23})F_4'\}(\omega_1\omega_2)^{1/2}\right.$$

$$\left.+2\Big\{\frac{J_{21}J_{22}}{(\omega_1\omega_2)^{1/2}}+J_{23}J_{24}(\omega_1\omega_2)^{1/2}\Big\}F_4''\right]\Bigg\}$$

$$r_6 = \frac{-1}{\omega_1\omega_2(2\omega_1-\omega_2)(4\omega_1-2\omega_2)}\Bigg\{(\omega_1-\omega_2)^3 \times$$

$$\left[\Big\{J_{13}J_{22}(\frac{\omega_1}{\omega_2})^{1/2}-J_{14}J_{21}(\frac{\omega_2}{\omega_1})^{1/2}\Big\}F_1'\right.$$

$$\left.+2\Big\{J_{21}J_{24}(\frac{\omega_2}{\omega_1})^{1/2}+J_{22}J_{23}(\frac{\omega_1}{\omega_2})^{1/2}\Big\}F_1''\right]$$

$$+(\omega_1-\omega_2)^2\left[\Big\{2\{J_{13}J_{14}F_2+(J_{13}J_{24}+J_{14}J_{23})F_2'\}(\omega_1\omega_2)^{1/2}\right.$$

$$\left.-2\Big\{\frac{J_{21}J_{22}}{(\omega_1\omega_2)^{1/2}}-J_{23}J_{24}(\omega_1\omega_2)^{1/2}\Big\}F_2''\right]-(\omega_1-\omega_2)\left[\Big\{J_{13}J_{22}(\frac{\omega_1}{\omega_2})^{1/2}\right.$$

$$\left.-J_{14}J_{21}(\frac{\omega_2}{\omega_1})^{1/2}\Big\}F_3'+2\Big\{J_{21}J_{22}(\frac{\omega_2}{\omega_1})^{1/2}+J_{22}J_{23}(\frac{\omega_1}{\omega_2})^{1/2}\Big\}F_3''\right]$$

$$-\left[\Big\{2\{J_{13}J_{14}F_4+(J_{13}J_{24}+J_{14}J_{23})F_4'\}(\omega_1\omega_2)^{1/2}\right.$$

$$\left.-2\Big\{\frac{J_{21}J_{22}}{(\omega_1\omega_2)^{1/2}}-J_{23}J_{24}(\omega_1\omega_2)^{1/2}\Big\}F_4''\right]\Bigg\}$$

$$r_7 = \frac{1}{3\omega_1^2(4\omega_1^2 - \omega_2^2)}\left\{8\omega_1^3\left[J_{13}(J_{13}F_1 + J_{23}F_1')\omega_1 - \left(\frac{J_{21}^2}{\omega_1} - J_{23}^2\omega_1\right)F_1''\right]\right.$$

$$-2\omega_1\left[\omega_1 J_{13}(J_{13}F_3 + J_{23}F_3') - \left(\frac{J_{21}^2}{\omega_1} - J_{23}^2\omega_1\right)F_3''\right]$$

$$\left.-4\omega_1^2 J_{21}(J_{13}F_2 + J_{23}F_2'')\omega_1 + J_{21}(J_{13}F_4' + 2J_{23}F_4'')\right\}$$

$$r_8 = \frac{-1}{3\omega_2^2(4\omega_2^2 - \omega_1^2)}\left\{8\omega_2^3\left[J_{14}(J_{14}F_1 + J_{24}F_1')\omega_2 - \left(\frac{J_{22}^2}{\omega_2} - J_{24}^2\omega_2\right)F_1''\right]\right.$$

$$+ 4\omega_2^2 J_{22}(J_{14}F_2 + 2J_{24}F_2'')\omega_2 - 2\omega_2\left[\omega_2 J_{14}(J_{14}F_3 + J_{24}F_3')\right.$$

$$\left.\left.-\left(\frac{J_{22}^2}{\omega_2} - J_{24}^2\omega_2\right)F_3''\right] - J_{22}(J_{14}F_4' + 2J_{24}F_4'')\right\}$$

$$r_9 = \frac{1}{\omega_1\omega_2(2\omega_1 + \omega_2)(\omega_1 + 2\omega_2)}\left\{(\omega_1 + \omega_2)^3\left[\left\{2J_{13}J_{14}F_1\right.\right.\right.$$

$$+ (J_{13}J_{24} + J_{14}J_{23})F_1'\}(\omega_1\omega_2)^{1/2} + 2\left\{\frac{J_{21}J_{22}}{(\omega_1\omega_2)^{1/2}} + J_{23}J_{24}(\omega_1\omega_2)^{1/2}\right\}F_1''\right]$$

$$-(\omega_1 + \omega_2)^2\left[\left\{J_{13}J_{22}(\frac{\omega_1}{\omega_2})^{1/2} - J_{14}J_{21}(\frac{\omega_2}{\omega_1})^{1/2}\right\}F_2'\right.$$

$$\left.-2\left\{J_{21}J_{24}(\frac{\omega_2}{\omega_1})^{1/2} - J_{22}J_{23}(\frac{\omega_1}{\omega_2})^{1/2}\right\}F_2''\right]$$

$$-(\omega_1 + \omega_2)\left[\left\{2\left\{J_{13}J_{14}F_3 + (J_{13}J_{24} + J_{14}J_{23})F_3'\right\}(\omega_1\omega_2)^{1/2}\right.\right.$$

$$+ 2\left\{\frac{J_{21}J_{22}}{(\omega_1\omega_2)^{1/2}} + J_{23}J_{24}(\omega_1\omega_2)^{1/2}\right\}F_3''\right]$$

$$-\left[\left\{J_{13}J_{22}(\frac{\omega_1}{\omega_2})^{1/2}\right.\right.$$

$$-J_{14}J_{21}(\frac{\omega_2}{\omega_1})^{1/2}\}F_4' - 2\left\{J_{21}J_{24}(\frac{\omega_2}{\omega_1})^{1/2}\right.$$

$$\left.\left.\left.-J_{22}J_{23}(\frac{\omega_1}{\omega_2})^{1/2}\right\}F_4''\right]\right]\right\}$$

$$r_{10} = \frac{1}{\omega_1\omega_2(2\omega_1 - \omega_2)(2\omega_2 - \omega_1)}\left\{(\omega_1 - \omega_2)^3\left[\left\{2J_{13}J_{14}F_1\right.\right.\right.$$

$$+ (J_{13}J_{24} + J_{14}J_{23})F_1'\right\}(\omega_1\omega_2)^{1/2} - 2\left\{\frac{J_{21}J_{22}}{(\omega_1\omega_2)^{1/2}}\right.$$

$$\left.\left.-J_{23}J_{24}(\omega_1\omega_2)^{1/2}\right\}F_1''\right]$$

$$-(\omega_1 - \omega_2)^2\left[\left\{J_{13}J_{22}(\frac{\omega_1}{\omega_2})^{1/2} - J_{14}J_{21}(\frac{\omega_2}{\omega_1})^{1/2}\right\}F_2'\right.$$

$$\left.+ 2\left\{J_{21}J_{24}(\frac{\omega_2}{\omega_1})^{1/2} + J_{22}J_{23}(\frac{\omega_1}{\omega_2})^{1/2}\right\}F_2''\right]$$

$$-(\omega_1 - \omega_2)\left[\left\{2\left\{J_{13}J_{14}F_3 + (J_{13}J_{24} + J_{14}J_{23})F_3'\right\}(\omega_1\omega_2)^{1/2}\right.\right.$$

$$\left.-2\left\{\frac{J_{21}J_{22}}{(\omega_1\omega_2)^{1/2}} - J_{23}J_{24}(\omega_1\omega_2)^{1/2}\right\}F_3''\right]$$

$$+ \left[\left\{J_{13}J_{22}(\frac{\omega_1}{\omega_2})^{1/2} - J_{14}J_{21}(\frac{\omega_2}{\omega_1})^{1/2}\right\}F_4'\right.$$

$$\left.\left.+ 2\left\{J_{21}J_{24}(\frac{\omega_2}{\omega_1})^{1/2} - J_{22}J_{23}(\frac{\omega_1}{\omega_2})^{1/2}\right\}F_4''\right]\right\}.$$

For more details on the restricted 3–body problem, see [KI04, KSI06].

3.1.2 Scaling and Reduction in the 3–Body Problem

Now, following [MS00], we make a series of symplectic changes of variables in the 3–body problem which show how to look at the restricted problem as the limit of the reduced problem with one small mass. To the first approximation the reduced problem with one small mass is separable, i.e., the Hamiltonian of the reduced problem to the first approximation is the sum of the Hamiltonian of the restricted problem and the Hamiltonian of a harmonic oscillator. This result was established in [Mey81a, MH92] for the planar problem, so we shall consider only the spatial case here.

The 3–body problem in 3D space has 9DOF. By placing the center of mass at the origin and setting linear momentum equal to zero the problem reduces one with six degrees of freedom. This is easily done using Jacobi coordinates. The Hamiltonian of the 3–body problem in rotating (about the $z-$axis) Jacobi coordinates $(u_0, u_1, u_2, v_0, v_1, v_2)$ is

$$H = \frac{\|v_0\|^2}{2M_0} - u_0^T J v_0 + \frac{\|v_1\|^2}{2M_1} - u_1^T J v_1 - \frac{m_0 m_1}{\|u_1\|}$$
$$+ \frac{\|v_2\|^2}{2M_2} - u_2^T J v_2 - \frac{m_1 m_2}{\|u_2 - \alpha_0 u_1\|} - \frac{m_2 m_0}{\|u_2 + \alpha_1 u_1\|}, \qquad \text{where} \quad u_i, v_i \in \mathbb{R}^3, \tag{3.8}$$

$$M_0 = m_0 + m_1 + m_2, \qquad M_1 = m_0 m_1 / (m_0 + m_1),$$
$$M_2 = m_2(m_0 + m_1)/(m_0 + m_1 + m_2),$$
$$\alpha_0 = m_0/(m_0 + m_1), \qquad \alpha_1 = m_1/(m_0 + m_1), \qquad \text{and}$$

$$J = \begin{pmatrix} 0 & 1 & 0 \\ -1 & 0 & 0 \\ 0 & 0 & 0 \end{pmatrix}.$$

In these coordinates u_0 is the center of mass, v_0 is total linear momentum, and total angular momentum is [MH92]

$$A = u_0 \times v_0 + u_1 \times v_1 + u_2 \times v_2.$$

The set where $u_0 = v_0 = 0$ is invariant, and setting these two coordinates to zero effects the first reduction. Setting $u_0 = v_0 = 0$ reduces the problem by three degrees of freedom.

We consider angular momentum to be nonzero. One way to reduce the problem by two more degrees is to hold the vector A fixed and eliminate the rotational symmetry about the A axis. Another way to reduce the problem is to note that A_z, the z--component of angular momentum, and $\mathbf{A} = \| A \|$, the magnitude of angular momentum, are integrals in involution. Two independent integrals in involution can be used to reduce a system by two degrees of freedom, see [Whi27]. In either case we pass to the reduced space as defined in [Mey73] or [MW74].

Assume that one of the particles has small mass by setting $m_2 = \varepsilon^2$, where ε is to be considered as a small parameter. Also set $m_0 = \mu, m_1 = 1 - \mu$ and $\nu = \mu(1 - \mu)$, so that

$$M_1 = \nu = \mu(1 - \mu), \qquad M_2 = \varepsilon^2/(1 + \varepsilon^2) = \varepsilon^2 - \varepsilon^4 + \cdots .$$

$$\alpha_0 = \mu, \qquad \alpha_1 = 1 - \mu.$$

The Hamiltonian becomes

$$H = K + \tilde{H}, \qquad \text{where}$$
$$K = \frac{1}{2\nu} \| v_1 \|^2 - u_1^T J v_1 - \frac{\nu}{\| u_1 \|} \qquad \text{and}$$
$$\tilde{H} = \frac{(1 + \varepsilon^2)}{2\varepsilon^2} \| v_2 \|^2 - u_2^T J v_2 - \frac{\varepsilon^2(1 - \mu)}{\| u_2 - \mu u_1 \|}$$
$$- \frac{\varepsilon^2 \mu}{\| u_2 + (1 - \mu)u_1 \|}.$$

K is the Hamiltonian of the Kepler problem in rotating coordinates. We can simplify K by making the scaling $u_i \to u_i, v_i \to \nu v_i, K \to \nu^{-1}K, \tilde{H} \to \nu^{-1}\tilde{H}, \varepsilon^2 \nu^{-1} \to \varepsilon^2$, so that

$$K = \frac{1}{2} \| v_1 \|^2 - u_1^T J v_1 - \frac{1}{\| u_1 \|}, \qquad \text{and} \qquad (3.9)$$

$$\tilde{H} = \frac{(1 + \nu \varepsilon^2)}{2\varepsilon^2} \parallel v_2 \parallel^2 - u_2^T J v_2 - \frac{\varepsilon^2 (1 - \mu)}{\parallel u_2 - \mu u_1 \parallel} - \frac{\varepsilon^2 \mu}{\parallel u_2 + (1 - \mu) u_1 \parallel}. \quad (3.10)$$

K has a critical point at $u_1 = a = (1,0,0)^T$, $v_1 = b = (0,1,0)^T$; it corresponds to a circular orbit of the Kepler problem. Expand K in a Taylor series about this point, ignore the constant term, and make the scaling

$$u_1 \to a + \varepsilon q, \qquad v_1 \to b + \varepsilon p, \qquad K \to \varepsilon^{-2} K$$

to get $K = K_0 + O(\varepsilon)$, where

$$K_0 = \frac{1}{2} \left(p_1^2 + p_2^2 + p_3^2 \right) + q_2 p_1 - q_1 p_2 + \frac{1}{2} \left(-2 q_1^2 + q_2^2 + q_3^2 \right). \quad (3.11)$$

Now scale $\tilde{H}$ by the above and

$$u_2 = \xi, \quad v_2 = \varepsilon^2 \eta, \quad \tilde{H} \longrightarrow \varepsilon^{-2} \tilde{H}.$$

The totality is a symplectic scaling with multiplier ε^{-2}, and so the Hamiltonian of the 3–body problem becomes $H_R + K_0 + O(\varepsilon)$, where K_0 is given in (3.11) and H_R is the Hamiltonian of the restricted problem, i.e.,

$$H_R = \frac{1}{2} \parallel \eta \parallel^2 - \xi^T J \eta - \frac{(1 - \mu)}{\parallel \xi - (\mu, 0, 0) \parallel} - \frac{\mu}{\parallel \xi + (1 - \mu, 0, 0) \parallel}. \quad (3.12)$$

To get the expansions above, recall that $u_1 = (1,0,0) + O(\varepsilon)$.

We have already reduced the problem by using the transitional invariance and the conservation of linear momentum, so now we will complete the reduction by using the rotational invariance and the conservation of angular momentum.

Recall that angular momentum in the original coordinates is $A = u_1 \times v_1 + u_2 \times v_2$ and in the scaled coordinates it becomes

$$A = (a + \varepsilon q) \times (b + \varepsilon p) + \varepsilon^2 \xi \times \eta, \quad (3.13)$$

and so holding angular momentum fixed by setting $A = a \times b$ imposes the constraint

$$0 = a \times p + q \times b + O(\varepsilon) = (-q_3, -p_3, p_2 + q_1) + O(\varepsilon). \quad (3.14)$$

Now let us do the reduction when $\varepsilon = 0$, so that the Hamiltonian is $H = H_R + K_0$ and holding angular momentum fixed is equivalent to $q_3 = p_3 = p_2 + q_1 = 0$. Notice that the angular momentum constraint is only on the q, p variables. Make the symplectic change of variables

$$\begin{array}{lll} r_1 = q_1 + p_2, & R_1 = p_1 & \\ r_2 = q_2 + p_1, & R_2 = p_2, & \text{so that} \qquad (3.15) \\ r_3 = q_3, & R_3 = p_3 & \end{array}$$

$$K_0 = \frac{1}{2}(r_2^2 + R_2^2) + \frac{1}{2}(r_3^2 + R_3^2) + r_1 R_2 - r_1^2. \tag{3.16}$$

Notice that holding angular momentum fixed in these coordinates is equivalent to $r_1 = r_3 = R_3 = 0$, that R_1 is an ignorable coordinate, and that r_1 is an integral. Thus passing to the reduced space reduces K_0 to

$$K_0 = \frac{1}{2}(r_2^2 + R_2^2). \tag{3.17}$$

Thus when $\varepsilon = 0$ the Hamiltonian of the reduced 3–body problem becomes

$$H = H_R + \frac{1}{2}(r^2 + R^2), \tag{3.18}$$

which is the sum of the Hamiltonian of the restricted 3–body problem and a harmonic oscillator. Here in (3.18) and henceforth we have dropped the subscript 2. The equations and integrals all depend smoothly on ε, and so for small ε the Hamiltonian becomes

$$H = H_R + \frac{1}{2}(r^2 + R^2) + O(\varepsilon). \tag{3.19}$$

We can also introduce action–angle variables (I, ι) by

$$r = \sqrt{2I} \cos \iota, \qquad R = \sqrt{2I} \sin \iota, \qquad \text{to get}$$

$$H = H_R + I + O(\varepsilon). \tag{3.20}$$

The reduced 3–body problem in two or three dimensions with one small mass is approximately the product of the restricted problem and a harmonic oscillator.

3.1.3 Periodic Solutions of the 3–Body Problem

A periodic solution of a conservative Hamiltonian system always has the characteristic multiplier $+1$ with algebraic multiplicity at least 2. If the periodic solution has the characteristic multiplier $+1$ with algebraic multiplicity exactly equal to 2, then the periodic solution is called *non–degenerate* or sometimes *elementary*. A non–degenerate periodic solution lies in a smooth cylinder of periodic solutions which are parameterized by the Hamiltonian. Moreover, if the Hamiltonian depends smoothly on parameters, then the periodic solution persists for small variations of the parameters (see [MH92] for details).

A nondegenerate periodic solution of the planar or spatial restricted 3–body problem whose period is not a multiple of 2π can be continued into the reduced 3–body problem. More precisely [MS00]: Let $\eta = \phi(t), \xi = \psi(t)$ be a periodic solution with period T of the restricted problem whose Hamiltonian is (3.12). Let its multipliers be $+1, +1, \beta, \beta^{-1}$ in the planar case or $+1, +1, \beta_1, \beta_1^{-1}, \beta_2, \beta_2^{-1}$ in the spatial case. Assume that $T \neq n2\pi$ for all

$n \in \mathbb{Z}$, and $\beta \neq +1$ in the planar case or $\beta_1 \neq +1$ and $\beta_2 \neq +1$ in the spatial case. Then the reduced 3–body problem, the system with Hamiltonian (3.19), has a periodic solution of the form

$$\eta = \phi(t) + O(\varepsilon), \qquad \xi = \psi(t) + O(\varepsilon), \qquad r = O(\varepsilon), \qquad R = O(\varepsilon),$$

whose period is $T + O(\varepsilon)$. Moreover, its multipliers are $+1, +1, \beta + O(\varepsilon), \beta^{-1} + O(\varepsilon), e^{iT} + O(\varepsilon), e^{-iT} + O(\varepsilon)$ in the planar case, or $+1, +1, \beta_1 + O(\varepsilon), \beta_1^{-1} + O(\varepsilon), \beta_2 + O(\varepsilon), \beta_2^{-1} + O(\varepsilon), e^{iT} + O(\varepsilon), e^{-iT} + O(\varepsilon)$ in the spatial case.

The planar version of this theorem is due to [Had75]. There are similar theorems about non–degenerate symmetric periodic solutions (see [Mey81a]).

There are three classes of non–degenerate periodic solutions of the planar restricted problem that are obtained by continuation of the circular orbits of the Kepler problem using a small parameter. The small parameter might be μ, the mass ratio parameter, giving the periodic solutions of the first kind of Poincaré [SM71, Poi899], a small distance giving Hill's lunar orbits [Bir15, Bir27, Con63, SM71], or a large distance giving the comet orbits [Mey81b, Mou12]. All these papers cited except [Mey81b] use a symmetry argument, and so do not calculate the multipliers. However, in Meyer and Hall [MH92] a unified treatment of all three cases is given, and the multipliers are computed and found to be nondegenerate. Thus, there are three corresponding families of periodic solutions of the reduced problem. The corresponding results with independent proofs for the reduced problem are found in [Mey94, Mey81a, Mou06, Mou12, Per37, Sie50].

One of the most interesting families of nondegenerate periodic solution of the spatial restricted problem can be found in [Bel81]. He regularized double collisions when $\mu = 0$, and showed that some spatial collision orbits are non–degenerate periodic solutions in the regularized coordinates. Thus, they can be continued into the spatial restricted problem as nondegenerate periodic solutions for $\mu \neq 0$. Now these same orbits can be continued into the reduced 3–body problem.

3.1.4 Bifurcating Periodic Solutions of the 3–Body Problem

Many families of periodic solutions of the restricted problem have been studied, and numerous bifurcations have been observed. Most of these bifurcations are 'generic one parameter bifurcations' as defined in [Mey70, MH92]. Other bifurcations seem to be generic in either the class of symmetric solutions or generic two–parameter bifurcations. We claim that these bifurcations can be carried over to the reduced 3–body problem, *mutatis mutandis*. Since there are a multitude of different bifurcations and they are all generalize in a similar manner, we shall illustrate only one simple case: the 3–bifurcation of [Mey70], called the phantom kiss in [AM78].

Let $p(t, h)$ be a smooth family of non–degenerate periodic solutions of the restricted problem parameterized by H_R , i.e., $H_R(p(t, h)) = h$, with period

$\tau(h)$. When $h = h_0$ let the periodic solution be $p_0(t)$ with period τ_0, so $p_0(t) = p(t, h_0)$ and $\tau_0 = \tau(h_0)$. We will say that the τ_0–periodic solution $p_0(t)$ of the restricted problem is a *3–bifurcation orbit* if the cross section map $(\psi, \Psi) \longrightarrow (\psi', \Psi')$ in the surface $H_R = h$ for this periodic orbit can be put into the normal form

$$\psi' = \psi + (2\pi k/3) + \alpha(h - h_0) + \beta\Psi^{1/2}\cos(3\psi) + \cdots ,$$
$$\Psi' = \Psi - 2\beta\Psi^{3/2}\sin(3\psi) + \cdots , \qquad T = \tau_0 + \cdots ,$$

and $k = 1, 2$, and α and β are non–zero constants. In the above ψ, Ψ are normalized action–angle coordinates in the cross section intersect $H_R = h$, and T is the first return time for the cross section. The periodic solution, $p(t, h)$, corresponds to the point $\Psi = 0$. The multipliers of the periodic solution $p_0(t)$ are $+1, +1, \mathrm{e}^{+2k\pi i/3}, \mathrm{e}^{-2k\pi i/3}$ (cube roots of unity) so the periodic solution is a nondegenerate elliptic periodic solution. Thus, this family of periodic solutions can be continued into the reduced problem, provided τ_0 is not a multiple of 2π, by the result of the last subsection.

The above assumptions imply that the periodic solution $p(t, h)$ of the restricted problem undergoes a bifurcation. In particular, there is a one parameter family $p_3(t, h)$ of hyperbolic periodic solution of period $3\tau_0 + \cdots$ whose limit is $p_0(t)$ as $h \longrightarrow h_0$ (see [Mey71, MH92] for details).

Let $p_0(t)$ be a 3–bifurcation orbit of the restricted problem that is not in resonance with the harmonic oscillator, i.e., assume that $3\tau_0 \neq 2n\pi$, for $n \in \mathbb{Z}$. Let $\tilde{p}(t, h, \varepsilon)$ be the $\tilde{\tau}(h, \varepsilon)$–periodic solution which is the continuation into the reduced problem of the periodic solution $p(t, h)$ for small ε. Thus $\tilde{p}(t, h, \varepsilon) \longrightarrow (p(t, h), 0, 0)$ and $\tilde{\tau}(h, \varepsilon) \longrightarrow \tau(h)$ as $\varepsilon \longrightarrow 0$. Then there is a smooth function $\tilde{h}_0(\varepsilon)$ with $\tilde{h}_0(0) = h_0$ such that $\tilde{p}(t, \tilde{h}_0(\varepsilon), \varepsilon)$ has multipliers $+1, +1, \mathrm{e}^{+2k\pi i/3}, \mathrm{e}^{-2k\pi i/3}, \mathrm{e}^{+\tau i} + O(\varepsilon), \mathrm{e}^{-\tau i} + O(\varepsilon)$, i.e., exactly one pair of multipliers are cube roots of unity. Moreover, there is a family of periodic solutions of the reduced problem, $\tilde{p}_3(t, h, \varepsilon)$ with period $3\tilde{\tau}(h, \varepsilon) + \cdots$ such that $\tilde{p}_3(t, h, \varepsilon) \longrightarrow (p_3(t, h), 0, 0)$ as $\varepsilon \longrightarrow 0$ and $\tilde{p}_3(t, h, \varepsilon) \longrightarrow \tilde{p}(t, \tilde{h}_0(\varepsilon), \varepsilon)$ as $h \longrightarrow \tilde{h}_0(\varepsilon)$. The periodic solutions of the family $\tilde{p}_3(t, h, \varepsilon)$ are hyperbolic–elliptic, i.e., they have two multipliers equal to $+1$, two multipliers which are of unit modulus, and two multipliers which are real and not equal to ± 1 [MS00].

There are many other types of generic bifurcations, e.g., Hamiltonian saddle–node bifurcation, period doubling, k–bifurcations with $k > 3$, etc., as listed in [Mey70, MH92]. If such a bifurcation occurs in the restricted problem and the period of the basic periodic orbit is not a multiple of 2π, then a similar bifurcation takes place in the reduced problem also. The proofs will be essentially the same as the proof given above.

3.1.5 Bifurcations in Lagrangian Equilibria

The restricted 3–body problem has five equilibrium points, of which three lie on the line joining the masses (the *Eulerian equilibria*) and two at the vertices of the two equilateral triangles whose base is the segment joining the primaries (the *Lagrangian equilibria*) [MH92]. These later two equilibria are usually denoted by $\mathcal{L}_4$, $\mathcal{L}_5$. By the symmetry of the problem it is enough to consider just $\mathcal{L}_4$. For $0 < \mu \leq \mu_1 = 27\mu(1-\mu)/4$ the linearized equations at $\mathcal{L}_4$ have pure imaginary eigenvalues $\pm i\omega_1$, $\pm i\omega_2$. These pure imaginary eigenvalues give rise to two families of periodic solutions emanating from the equilibria for most values of μ, $0 < \mu < \mu_1$. These families are known as the Lyapunov families.

As an application of the previous result, consider the Lyapunov families of periodic solutions emanating from $\mathcal{L}_4$ in the restricted problem. Except as noted below, these two families have continuations into the reduced problem as Lyapunov families emanating from $\mathcal{L}_4$.

There are a multitude of interesting bifurcations that occur at the Lagrange equilateral triangular equilibria, $\mathcal{L}_4$, in the restricted problem as the mass ratio parameter μ is varied. We shall look at just a few of these interesting bifurcations to illustrate how many of the results that have been established in the restricted 3–body problem can be carried over to the reduced 3–body problem. Here we shall consider only the planar problem, since there is a wealth of literature on the planar restricted problem. Again we hope that the reader will realize that similar results too numerous to expound will follow in a similar manner for both the planar and spatial problems.

Now, recall [MH92] that the characteristic polynomial of the linearized equations at the equilibrium point $\mathcal{L}_4$ in the restricted problem is

$$p_R(\lambda, \mu) = \lambda^4 + \lambda^2 + \frac{27}{4}\mu(1-\mu).$$

For $\mu > \mu_1$ the eigenvalues are complex, two with positive real part and two with negative real part; for $\mu < \mu_1$ the eigenvalues are $\pm i\omega_1 = \pm i\omega_1(\mu)$, $\pm i\omega_2 = \pm i\omega_2(\mu)$, where $\omega_1^2 + \omega_2^2 = 1$, $\omega_1^2 \omega_2^2 = 27\mu(1-\mu)/4$ and $\omega_1 < \omega_2$; for $\mu = \mu_1$ the eigenvalues are $\pm\sqrt{2}/2i$, $\pm\sqrt{2}/2i$, where $\mu_1 = \frac{1}{2}(1 - \sqrt{69}/9) = 0.0385\ldots$ is the critical mass ratio of Routh.

Let μ_r denote the mass ratio at which $\omega_1/\omega_2 = r$. These mass ratios are given as a function of r by

$$\mu_r = \frac{1}{2} \pm \frac{1}{2}\left\{1 - \frac{16r^2}{27(r^2+1)^2}\right\}^{1/2}, \tag{3.21}$$

where both signs are admissible since $0 < \mu < 1$. μ_r is called a *critical mass ratio of order* k if $r = k$ is a positive integer and a *resonance mass ratio of order* (p, q) if $r = p/q$, $p > q > 1$. The frequencies are $\omega_1 = k/(1+k^2)^{1/2}$, $\omega_2 = 1/(1+k^2)^{1/2}$ at μ_k and $\omega_1 = p/(p^2+q^2)^{1/2}$, $\omega_2 = q/(p^2+q^2)^{1/2}$ at $\mu_{p/q}$.

The reduced problem has a relative equilibrium at the Lagrange triangular configuration – recall the Hamiltonian of the reduced problem is (3.19) in rotating coordinates. From (3.19) the characteristic polynomial of the linearized equations at the Lagrange relative equilibrium is

$$p_F(\lambda, \mu, \varepsilon) = p_R(\lambda, \mu)(\lambda^2 + 1) + O(\varepsilon).$$

But $(\lambda^2 + 1)$ is always a factor of the characteristic polynomial of a relative equilibrium on the reduced space, and so $\pm i$ are always characteristic exponents of the relative equilibrium, see [Mey94]. Thus we can write

$$p_F(\lambda, \mu, \varepsilon) = c(\lambda, \mu, \varepsilon)(\lambda^2 + 1), \qquad \text{where} \qquad c(\lambda, \mu, \varepsilon) = p_R(\lambda, \mu) + O(\varepsilon).$$

These polynomials are polynomials in λ^2, and so c and p_R are quadratics in λ^2. The discriminant is

$$\Delta(\mu, \varepsilon) = 1 - 27\mu(1 - \mu) + O(\varepsilon).$$

By definition $\Delta(\mu_1, 0) = 0$, and also $\partial \Delta(\mu_1, 0)/\partial \mu = 54\mu_1 - 27 \neq 0$. Thus by the implicit function theorem there is a smooth function $\tilde{\mu}_1(\varepsilon) = \mu_1 + O(\varepsilon)$ such that $\Delta(\tilde{\mu}_1(\varepsilon), \varepsilon) = 0$ for small ε. Along the curve

$$\mathcal{C}_1 = \{(\tilde{\mu}_1(\varepsilon), \varepsilon) : \varepsilon \text{ small}\}$$

in the μ, ε--parameter plane the linearized equations of the reduced problem at the Lagrange relative equilibrium has one pair of pure imaginary eigenvalues with multiplicity two. That is, along this curve the eigenvalues are $\pm i, \pm \tilde{\omega}(\varepsilon)i, \pm \tilde{\omega}(\varepsilon)i$, where $\tilde{\omega}(\varepsilon) = \sqrt{2}i/2 + O(\varepsilon)$. For $\mu < \tilde{\mu}_1(\varepsilon)$ and small ε the relative equilibrium has eigenvalues of the form $\pm i, \pm \tilde{\omega}_1 = \pm \tilde{\omega}_1(\mu, \varepsilon), \pm \tilde{\omega}_2 = \pm \tilde{\omega}_2(\mu, \varepsilon)$.

The frequencies have ratio r, $\tilde{\omega}_1/\tilde{\omega}_2 = r$, iff $p_F(\lambda, \mu, \varepsilon)$ and $p_F(r\lambda, \mu, \varepsilon)$ have a common root, that is, iff

$$S_r(\mu, \varepsilon) = \text{resultant}(z^2 + z + \frac{27}{4}\mu(1 - \mu) + O(\varepsilon), r^4 z^2 + r^2 z$$

$$+ \frac{27}{4}\mu(1 - \mu) + O(\varepsilon)) = \frac{27}{16}\mu(\mu - 1)(r^2 - 1)^2(27\mu^2 r^4 - 27\mu r^4$$

$$+ 54\mu^2 r^2 - 54\mu r^2 + 4r^2 + 27\mu^2 - 27\mu) + O(\varepsilon)$$

By definition $S_r(\mu_r, 0) = 0$, and Mathematica computes

$$\frac{\partial S(\mu_r, 0)}{\partial \mu} = \frac{3\sqrt{3}r^2(r^2 - 1)\sqrt{27r^4 + 38r^2 + 27}}{4(r^2 + 1)} \neq 0,$$

for $r \neq 1$. Again the implicit function theorem gives smooth functions $\tilde{\mu}_r(\varepsilon) = \mu_r + O(\varepsilon)$ such that $S_r(\tilde{\mu}_r(\varepsilon), \varepsilon) = 0$. On the curve $\mathcal{C}_r = \{(\tilde{\mu}_r(\varepsilon), \varepsilon) : \varepsilon \text{ small}\}$ the linearized equations of the reduced problem at the Lagrange relative equilibrium have eigenvalues $\pm i, \pm \tilde{\omega}_1 i, \pm \tilde{\omega}_2 i$ with $\pm \tilde{\omega}_1 i = \tilde{\omega}_1(\tilde{\mu}_r(\varepsilon), \varepsilon)$, $\pm \tilde{\omega}_2 i = \tilde{\omega}_2(\tilde{\mu}_r(\varepsilon), \varepsilon)$ and $\tilde{\omega}_1/\tilde{\omega}_2 = r$.

For small ε the linearized equations of the reduced problem have three frequencies at the Lagrangian relative equilibrium for $0 < \mu < \tilde{\mu}_1$; namely,

$$1 > \tilde{\omega}_1 > \tilde{\omega}_2, \qquad \text{with periods}$$

$$\tilde{T}_0 = 2\pi < \tilde{T}_1 = \frac{2\pi}{\tilde{\omega}_1} < \tilde{T}_2 = \frac{2\pi}{\tilde{\omega}_2}.$$

Since $\frac{1}{2} < \tilde{\omega}_1 < 1$ for small ε, the period $\tilde{T}_1$ is never an integral multiple of $\tilde{T}_0 = 2\pi$. But $T_2 = kT_0$ when $\varepsilon = 0$, and

$$\mu_k^* = \frac{1}{2} - \frac{\sqrt{3(16 - 16k^2 + 27k^4)}}{18k^2}.$$

Again, we apply the implicit function theorem to find functions $\tilde{\mu}_k^*(\varepsilon) = \mu_k^* + O(\varepsilon)$ such that $\tilde{T}_2(\varepsilon) = k2\pi$ when $\mu = \tilde{\mu}_k^*$.

Lyapunov's Center Theorem

For fixed small ε and $\mu < \mu_1(\varepsilon)$ the linearized equations of the reduced problem at the Lagrange relative equilibrium have three families of periodic solutions with periods $\tilde{T}_0 = 2\pi$, $\tilde{T}_1(\mu, \varepsilon) = 2\pi/\tilde{\omega}_1(\mu, \varepsilon)$, $\tilde{T}_2(\mu, \varepsilon) = 2\pi/\tilde{\omega}_2(\mu, \varepsilon)$ with $\tilde{T}_2 > \tilde{T}_1 > \tilde{T}_0$. Lyapunov's center theorem [Lya47, MH92] can be applied to yield three families of periodic solutions emanating from the Lagrange relative equilibrium in the reduced problem for $\mu < \tilde{\mu}_1(\varepsilon)$ for ε small, except in some special cases discussed here.

First of all there is always the family with the shortest period, $\tilde{T}_0 = 2\pi$, emanating from the relative equilibrium for $\mu < \tilde{\mu}_1(\varepsilon)$, ε small. But this family is always at a relative equilibrium on the reduced space [Mey94, SM71]. Next, since $\tilde{\omega}_1(\mu, \varepsilon)$ is not a multiple of 2π for $\mu < \tilde{\mu}_1(\varepsilon)$ for ε small, there is always a family emanating from the relative equilibrium whose limit period at the relative equilibrium is $\tilde{T}_1(\mu, \varepsilon)$. This family is the continuation of the *short period family* in the restricted problem.

Finally, there is a family whose limit period at the relative equilibrium is $\tilde{T}_2(\mu, \varepsilon)$ provided $0 < \mu < \tilde{\mu}_1(\varepsilon)$, $\mu \neq \tilde{\mu}_k(\varepsilon)$, $\mu \neq \tilde{\mu}_k^*(\varepsilon)$, $k \notin \mathbb{Z}$, for ε small. This is the continuation (where it exists) of the *long period family* in the restricted problem. It would be interesting to see a numerical investigation of the evolution of this last family as the parameters are varied–provided the investigation is carried out with the same care as found in the classic studies [DH68, Pal67].

Hamiltonian Hopf Bifurcation

One of the most interesting bifurcations occurs in the restricted problem at the equilibrium point $\mathcal{L}_4$ as the mass ratio parameter passes through the Routh critical mass ratio μ_1. The linearized equations at $\mathcal{L}_4$ have two pairs of pure imaginary eigenvalues, $\pm\omega_1 i$, $\pm\omega_2 i$ for $0 < \mu < \mu_1$, eigenvalues $\pm i\sqrt{2}/2$ of multiplicity two for $\mu = \mu_1$, and eigenvalues $\pm\alpha \pm \beta i$, $\alpha \neq 0$, $\beta \neq 0$, for $\mu_1 < \mu \leq 1/2$. For $\mu < \mu_1$ and μ near μ_1, Lyapunov's Center Theorem establishes the existence of two families of periodic solutions emanating from the equilibrium point $\mathcal{L}_4$, and for $\mu_1 < \mu \leq 1/2$, the stable manifold theorem asserts that there are no periodic solutions near $\mathcal{L}_4$. What happens to these periodic solutions as μ passes through μ_1?

Buchanan [Buc41] proved, up to a small computation, that there are still two families of periodic solutions emanating from the libration point $\mathcal{L}_4$ even when $\mu = \mu_1$. This is particularly interesting, because the linearized equations have only one family. The small computation of a coefficient of a higher order term was completed by Deprit and Henrard [DH68], thus showing that Buchanan's theorem did indeed apply to the restricted problem. Palmore [Pal67] investigated the question numerically and was led to the conjecture that the two families detach as a unit from the libration point and recede as μ increases from μ_1. Finally, Meyer and Schmidt [MS71] established the general theorem which has become known as the Hamiltonian Hopf Bifurcation Theorem, and then established Palmore's conjecture using the calculation of Deprit and Henrard [DH68]. Unfortunately, a spurious factor of $\sqrt{2}$ occurred in the application of Deprit's calculation, but the result holds. Also see [MH92], pages 218–24. Subsequently, this theorem has been re–proved by several authors by essentially the same method – see for example [GMS95] for a complete contemporary treatment of this and related problems.

First we must set up the general theorem along the lines found in [MH92, MS71]. We will drop the explicit ε dependence until it is necessary to display it, but we will indicate those variables which will depend on ε with a tilde. The normal form for a quadratic Hamiltonian (linear Hamiltonian system) with eigenvalues $\pm \tilde{\omega} i$ with multiplicity two, which is non–simple, and $\pm i$ with multiplicity one, is

$$Q_0 = \tilde{\omega}(\xi_2 \eta_1 - \xi_1 \eta_2) + (\delta/2)(\xi_1^2 + \xi_2^2) + \tfrac{1}{2}(r^2 + R^2),$$

where $\delta = \pm 1$, which gives rise to the linear system of equations $\dot{z} = A_0 z$, where

$$z = \begin{pmatrix} \xi_1 \\ \xi_2 \\ r \\ \eta_1 \\ \eta_2 \\ R \end{pmatrix} \quad \text{and} \quad A_0 = \begin{pmatrix} 0 & \tilde{\omega} & 0 & 0 & 0 & 0 \\ -\tilde{\omega} & 0 & 0 & 0 & 0 & 0 \\ 0 & 0 & 0 & 0 & 0 & 1 \\ -\delta & 0 & 0 & 0 & \tilde{\omega} & 0 \\ 0 & -\delta & 0 & -\tilde{\omega} & 0 & 0 \\ 0 & 0 & -1 & 0 & 0 & 0 \end{pmatrix}.$$

Consider a smooth quadratic perturbation of Q_0, i.e., a quadratic Hamiltonian of the form $Q(\nu) = Q_0 + \nu Q_1 + \cdots$, where ν is the perturbation parameter. By Sokol'skii [Sok78] (also see the discussion in [MH92]) there are four quantities that are important in the theory of normal forms for this problem, namely

$$\Gamma_1 = \xi_2 \eta_1 - \xi_1 \eta_2, \quad \Gamma_2 = (\xi_1^2 + \xi_2^2)/2, \quad \Gamma_3 = (\eta_1^2 + \eta_2^2)/2, \quad I = \frac{1}{2}(r^2 + R^2).$$

The higher order terms in $Q(\nu)$ are in normal form if they are functions of Γ_1, Γ_3 and I only. Assume that $Q(\nu)$ is normalized through terms in ν, so that

$Q_1 = a\Gamma_1 + b\Gamma_3 + cI$ or

$$Q(\nu) = \tilde{\omega}\Gamma_1 + \delta\Gamma_2 + I + \nu(\tilde{a}\Gamma_1 + \tilde{b}\Gamma_3 + \tilde{c}I) + \cdots .$$

In the case under consideration here $\tilde{c} = 0$, since $\pm i$ is always an exponent of a relative equilibrium. The characteristic polynomial of the linear system defined by $Q(\nu)$ is

$$(\lambda^2 + 1)(\{\lambda^2 + (\omega + \nu\tilde{a})^2\}^2 + 2\nu\tilde{b}\delta\{\lambda^2 - (\omega + \nu\tilde{a})^2\} + \nu^2\tilde{b}^2\delta^2 + \cdots),$$

which has roots

$$\lambda = \pm i, \qquad \lambda = \pm(\tilde{\omega} + \nu\tilde{a})i \pm \sqrt{-\tilde{b}\delta\nu} + \cdots .$$

So the coefficient '$\tilde{a}$" controls the way the eigenvalues move in the imaginary direction, and the coefficient "$\tilde{b}$" controls the way the eigenvalues split off the imaginary axis. The assumption that $\tilde{b} \neq 0$ means that the eigenvalues move off the imaginary axis when $\tilde{b}\delta\nu < 0$.

Now consider a nonlinear Hamiltonian system depending on the parameter ν which has $Q(\nu)$ as its quadratic part and when $\nu = 0$ has been put in Sokol'skii's normal form through the fourth order terms (see [MH92]), i.e., consider

$$\begin{aligned}
\tilde{H}(\nu) = {}& \tilde{\omega}\Gamma_1 + \delta\Gamma_2 + I + \nu(\tilde{a}\Gamma_1 + \tilde{b}\Gamma_3) \\
& + \frac{1}{2}(\tilde{d}_1\Gamma_1^2 + \tilde{d}_2\Gamma_3^2 + \tilde{d}_3 I^2 + \tilde{d}_4\Gamma_1\Gamma_3 + \tilde{d}_5\Gamma_2 I + \tilde{d}_6\Gamma_3 I) + \cdots ,
\end{aligned}$$
$$(3.22)$$

where here the ellipsis stands for terms which are at least second order in ν or fifth order in the rectangular variables.

The slight generalization of the main bifurcation in [MS71] is as follows (see [MH92, GMS95]).

Consider a Hamiltonian of the form (3.22) with $\tilde{\omega}$ not equal to a small rational number, $\delta = \pm 1, \tilde{b} \neq 0, \tilde{d}_2 \neq 0$. Then there is always a Lyapunov family of periodic solutions emanating from the origin with period close to 2π. Case A: $\delta\tilde{d}_2 > 0$. The two additional Lyapunov families emanate from the origin when $\delta\tilde{b}\nu$ is small and positive. These families persist when $\nu = 0$ as two distinct families of periodic orbits emanating from the origin. As $\delta\tilde{b}\nu$ becomes negative, the two families detach from the origin as a single family and recede from the origin. Case B: $\delta\tilde{d}_2 < 0$. The two additional Lyapunov families emanate from the origin when $\delta\tilde{b}\nu$ is small and positive, and the families are globally connected. This global family shrinks to the origin as $\delta\tilde{b}\nu$ tends to zero through positive values. When $\delta\tilde{b}\nu$ is small and negative, there are no periodic solutions close to the origin.

One can compute the multipliers approximately to show that in Case A the periodic solutions are elliptic. In Case B, the periodic solutions are initially elliptic as they emanate from the origin, but go through extremal bifurcations to become hyperbolic–elliptic.

To apply this theorem to the reduced problem let $\nu = \mu - \tilde{\mu}_1(\varepsilon)$, so for fixed small ε the Lagrange relative equilibrium of the reduced problem has exponents $\pm i$ with multiplicity one and $\pm i\tilde{\omega}(\varepsilon)$ with multiplicity two, where $\tilde{\omega} = \sqrt{2}/2 + O(\varepsilon)$. So for fixed small ε the Hamiltonian of the reduced problem at the Lagrange relative equilibrium can be put into Sokol'skii's normal form (3.22). When $\varepsilon = 0$ the important coefficients have been calculated in the restricted problem, namely $\tilde{a}(0) = -3\sqrt{2}\sqrt{69}/16$, $\tilde{b}(0) = -3\sqrt{69}/8$ (see [Sch90]), and $\tilde{d}_2(0) = 59/108$ (see [Sch94]). So for small ε, $\tilde{b}(\varepsilon)$ and $\tilde{d}_2(\varepsilon)$ are non-zero and have the same sign. Thus: For small ε the reduced problem undergoes the bifurcation given in Case A of the above theorem at the Lagrange relative equilibrium as μ passes through $\tilde{\mu}_1(\varepsilon)$ [MS00].

Bridges and Natural Centers

There are many different bifurcations from the two Lyapunov families of periodic solutions which emanate from $\mathcal{L}_4$ as the parameter μ is varied. In particular, very interesting bifurcations occur as the parameter μ passes through the values $\mu_{p/q}$, where p/q is a rational number.

The careful numerical work of Deprit, Henrard and Palmore [DH68, Pal67] found families of periodic solutions, called *bridges*, connecting the two Lyapunov families at $\mathcal{L}_4$. In general, a bridge consisted of two families of periodic solutions, of which one is elliptic and the other hyperbolic. As the parameter μ is varied, a particular bridge would collapse into the equilibrium $\mathcal{L}_4$ as $\mu \longrightarrow \mu_{p/q}$.

Using the computations of the normal form at $\mathcal{L}_4$ by Deprit and Deprit–Bartholomé [DD67], Meyer and Palmore [MP70] were able to establish the existence of these bridges using a variation of Birkhoff's fixed–point theorem. As with most fixed–point arguments, there were no uniqueness or continuity conclusions. Finally, Schmidt [Sch74] proved a series of theorems for general Hamiltonian systems which explain almost all the local bifurcations observed in the restricted problem in [DH68, Pal67]. These theorems have been re-proved many times; see [GMS95] for additional references.

Using the same line of argument, one can show that for small ε the reduced problem has bridges and natural centers which are continuations of the bridges and natural centers found in the restricted problem. There are surely many more bifurcations in the spatial restricted problem, due to the third Lyapunov family in the third direction, but that analysis remains to be done.

3.1.6 Continuation of KAM–Tori

There are several examples where the classical KAM theorems have been applied to the restricted problem, but we know of only one case where our method applies at present. That one case is in the planar problem, so we will only consider the planar problem now. Because the planar restricted problem has two degrees of freedom, one gets not only the existence of invariant tori, but some stability information also. For the three degree of freedom reduced problem the existence of the invariant tori will not imply any stability information.

We will say that a nondegenerate $\tau-$-periodic solution $p(t)$ of the restricted problem with $H_R(p(t)) = h_0$ is of *general elliptic type* if the cross section map for this periodic orbit is of the form

$$\Psi' = \Psi + \cdots , \qquad \psi' = \psi + \omega + \alpha h + \beta \Psi + \cdots , \qquad T = \tau + \gamma h + \delta \Psi + \cdots ,$$

where the constants $\alpha, \beta, \gamma, \delta$ satisfy

$$\alpha\delta - \beta\gamma \neq 0,$$

τ is not a multiple of 2π, and ω is irrational. In the above $h = H_R - h_0$, Ψ, ψ are action–angle coordinates in the cross section in the level set $H_R = h_0 + h$, and T is the first return time for the cross section. The periodic solution, p, corresponds to the point $\Psi = 0$ when $h = 0$.

Since the periodic solution is nondegenerate ($\tau\omega \neq 2\pi k$, $k \in \mathbb{Z}$) and τ is not a multiple of 2π, the periodic solution can be continued into the reduced problem for small ε by above results. As we shall see, additional assumptions are needed to apply one of the standard KAM theorems to the continuation of this periodic solution.

Let $\tilde{p}(t, h, \varepsilon)$ be the continuation into the reduced problem of the general elliptic periodic solution $p(t)$ of the restricted problem. For ε small and h near h_0 the periodic solution $\tilde{p}(t, h, \varepsilon)$ is the limit of invariant three–dimensional tori (KAM tori) [MS00].

Using the normal form calculation of Deprit and Deprit–Bartholomé [DD67], Meyer and Palmore [MP70] computed the cross section map to the Lyapunov periodic solutions and showed that there was a 'twist'. Let μ_n denote the mass ratio at which $\omega_1/\omega_2 = n$ and $\mu_n \leq 1/2$, where n is a positive integer. The explicit formula for μ_n is given in (3.21), but for now note that $\mu_{n+1} < \mu_n$. Provided $0 < \mu < \mu_1$, $\mu \neq \mu_2, \mu_3, \mu_4$, the Hamiltonian of the restricted problem at $\mathcal{L}_4$ can be brought into the normal form

$$H(J_1, J_2, \psi_1, \psi_2) = \omega_1 J_1 - \omega_2 J_2 + \frac{1}{2}\{AJ_1^2 + 2BJ_1 J_2 + CJ_2^2\} + \cdots ,$$

where J_1, J_2, ψ_1, ψ_2 are action–angle variables and

$$0 < \omega_2 < \frac{\sqrt{2}}{2} < \omega_1, \qquad \omega_1^2 + \omega_2^2 = 1, \qquad \omega_1^2\omega_2^2 = 27\mu(1 - \mu)/4.$$

When $\mu = \mu_2, \mu_3, \mu_4$ there are other terms in the normal form that appear at the same level of truncation (see [MH92]).

In the classic paper by Deprit and Deprit–Bartholomé [DD67] the constants of the normal form were found to be

$$A = \frac{\omega_2^2(81 - 696\omega_1^2 + 124\omega_1^4)}{72(1 - 2\omega_1^2)^2(1 - 5\omega_1^2)}, \qquad B = -\frac{\omega_1\omega_2(43 + 64\omega_1^2\omega_2^2)}{6(4\omega_1^2\omega_2^2 - 1)(25\omega_1^2\omega_2^2 - 4)},$$

$C(\omega_1, \omega_2) = A(\omega_2, \omega_1).$

3.1.7 Parametric Resonance and Chaos in Cosmology

Recall that *parametric resonance* is a phenomenon that occurs in various cosmological and high energy physics models. It manifests itself as a *rapid growth of a physical field at an exponential rate*. Recently this phenomenon has been used to explain some physical processes such as reheating in the early universe [KLS94, Kai97] and in phase transitions in disordered chiral condensates [Kai99]. At the same time a lot of attention has been given to the study of chaotic systems, i.e., systems whose trajectories in phase–space diverge exponentially, but at the same time remain within a bounded region.

Here, following [KP05], we explore the relationship between parametric resonance and chaos in low–dimensional dynamical systems.

Floquet Index

From the general theory of differential equations we know that any second–order linear ODE

$$\ddot{y} + f(t)\dot{y} + g(t)y = 0, \tag{3.23}$$

will have two linearly independent solutions. According to the *Floquet's theorem* [MF53], if $f(t)$ and $g(t)$ are functions periodic in t with a period T, then those solutions will have form:

$$y(t) = e^{\mu t}P(t), \tag{3.24}$$

where $P(t)$ is periodic function with period T, as well. Therefore, stability of the solution (3.24) is entirely determined by the exponent μ, which is also called *Floquet exponent*, or *Floquet index*. There is no general procedure for estimating Floquet exponent, however there are a lot of particular cases such as the Mathieu equation where an extensive analysis of the Floquet indices has been done.

Lyapunov Exponents

Recall that Lyapunov exponents are a quantitative indication of chaos, and are used to measure the average rate at which initially close trajectories in phase–space diverge from each other. The Lyapunov exponent is usually defined as:

$$\lambda_i = \lim_{t \to \infty} \frac{1}{t} \ln \frac{\epsilon_i(t)}{\epsilon_i(0)}, \tag{3.25}$$

where $\epsilon_i(t)$ denotes the separation of two trajectories, and the index i denotes the direction of growth in phase–space or Lyapunov directions [WSS85]. If at least one Lyapunov exponent is positive, then trajectories of the system will separate at an exponential rate and we say that *system is chaotic*. This will manifest itself in a high sensitivity to a change in the initial conditions. Since the motion of chaotic systems is performed within a bounded region in phase–space, besides an exponential stretch some kind of fold of the trajectory must occur. Because of that, the Lyapunov directions i in phase–space are not constant in time. They tend to rotate, and at the same time to collapse towards the direction of the most rapid growth. A powerful and very general algorithm for determining Lyapunov exponents is presented in [BGG80, SN79], and its implementation, which we use for our calculations, in [WSS85].

The algorithm is based on an analysis of the evolution of an orthonormal basis in phase–space along a trajectory. In the linear approximation directions of growth in phase–space will remain orthogonal, therefore allowing one to measure the growth directly. In [WSS85] it is suggested to describe evolution of the orthonormal basis by linearized equations and use a small enough time step so this approximation remains valid, and so we can neglect the effect of the Lyapunov directions collapsing. After the growth was measured the basis is re-orthonormalized, and the procedure is repeated for the next small enough time interval. If the limit of (3.25) exists, then the average of the measured growths should converge to its asymptotic value. This is a robust numerical algorithm and it works for almost every problem whose governing equation is known. It is not suitable though for unbounded systems that diverge exponentially, since the limits of numerical accuracy may be exceeded before the average growth starts converging.

Mathieu Equation

Perhaps the simplest model that exhibits parametric resonance is the *Mathieu equation* (which was originally used to describe small oscillations of a pendulum with vertically driven base)

$$\ddot{y} + (A - 2q\cos 2t)y = 0 \qquad (3.26)$$

This is a *Floquet–type equation* with two parameters A and q, and it has a solution of the form of (3.24). The value of the Floquet index μ depends on the equation's parameters. For certain values of A and q (e.g., $A = 2.5$ and $q = 1$) Floquet exponents would be purely imaginary, meaning that the solutions of (3.24) will both be periodic. Therefore, the solution will be stable, and its trajectory in phase–space will remain within a bounded region. Otherwise, both Floquet exponents will be purely real, and one of them hence positive. A solution with a positive Floquet exponent is unstable and grows exponentially. In some physical models such growth of the field y can be interpreted as a massive production of certain particles. This is also referred to as parametric resonance.

The rate of exponential growth of the solution (i.e., a positive Floquet exponent) can be determined from the graph for $\log |y|^2$ plotted against time t. Since the term in the solution containing a positive exponent is dominant, the slope of envelope of the graph will yield numerical value of 2μ. Regions of stability in parameter space of the Mathieu equation have been very well studied (see, for example, [AS72]). There are bands of stability and instability in the parameter space, and their boundaries are continuous curves.

In analogy with interpreting y as an angle, we impose suitable 'winding' conditions on the solution of (3.26) so that it always stays within segment $[-1, 1]$. There is no physical motivation to interpret y as an angle though, unless it stays within the limits of a small angle approximation. With this additional restriction imposed, both stable and unstable solutions are bounded, so parametric resonance does not occur. The stable solution remains periodic and exhibits the same behavior as before. The unstable solution, on the other hand, instead of parametric resonance, exhibits chaotic behavior, which manifests in high sensitivity in change of initial conditions. For this solution we estimated the Lyapunov spectrum, and found the positive Lyapunov exponent to be $\lambda_1 = 0.453 \pm 0.001$, which is the same as the Floquet exponent. This result could be anticipated because the first Lyapunov direction always point to the direction of the fastest growth in phase–space. For the Mathieu equation this growth is entirely described by the solution with a positive Floquet exponent. The linearization procedure in the algorithm for the Lyapunov exponents calculation will in a sense 'unwind' the trajectory, so that the exponential divergence measured by the Lyapunov exponent has to be the same as that described by the Floquet index.

Parametric Resonance Model

If the parameters of the Mathieu equation (3.26) are not constant, but rather some functions of time, then the solution will eventually switch between regions of stability and instability in parameter space, and therefore phases of quasiperiodicity and exponential growth will interchange during that time. Here is a somewhat simplified system of equations which illustrates such behavior. (3.27, 3.28) may be used to describe the decay of ϕ−particles into χ−particles [KP05]

$$\ddot{\chi} + H\dot{\chi} + (m^2 + g\phi^2)\chi = 0 \qquad (3.27)$$

$$\ddot{\phi} + (A - 2q\cos 2t)\phi = 0 \qquad (3.28)$$

(3.27) is a Floquet–type of equation, with parameters m and g set near the boundary between the stability and instability regions. (3.28) is a Mathieu equation which is coupled to (3.27). If we set the parameters A and q so that (3.28) has a periodic solution, then the term $g\phi^2$ will periodically drive (3.27) between its stability and instability regions, and therefore its Floquet exponent will change periodically in time. Although the Floquet exponent changes periodically in time, the system spends more time in a region of instability, and parametric resonance occurs. The average value of the Floquet exponent has a positive real part.

Now, let us impose the same 'winding' conditions like that for the Mathieu equation, so that $\chi \in [-1, 1]$. As before, parametric resonance will not occur, but the field χ will exhibit chaotic–like behavior instead. In order to find the Lyapunov exponent spectrum for this system we need to perform our calculation in 5D phase–space $(\dot{\chi}, \chi, \dot{\phi}, \phi, t)$. We found two exponents in the spectrum to be positive, and their sum to be $\lambda_1 + \lambda_2 = 0.0973 \pm 0.0006$, which agrees with the value for average Floquet exponent.

Again, this is an expected result, considering the algorithm for estimation of the Lyapunov exponents. The sum of all positive Lyapunov exponents is an average rate of exponential divergence of the solution of (3.27).

If we, however, substitute a chaotic solution of the Mathieu equation into (3.27), the system will exhibit a very complex behavior. The system will chaotically switch between stability and instability regions so it will be impossible to predict any kind of resonant behavior due to a high sensitivity to change in initial conditions. Furthermore, (3.27) will not be a Floquet equation any more, and its solution will not have the simple form of (3.24).

3.2 Elements of Chaos Control

3.2.1 Feedback and Non–Feedback Algorithms for Chaos Control

Although the presence of chaotic behavior is generic and robust for suitable nonlinearities, ranges of parameters and external forces, there are practical situations where one wishes to avoid or control chaos so as to improve the performance of the dynamical system. Also, although chaos is sometimes useful as in a mixing process or in heat transfer, it is often unwanted or undesirable. For example, increased drag in flow systems, erratic fibrillations of heart beats, extreme weather patterns and complicated circuit oscillations are situations where chaos is harmful. Clearly, the ability to control chaos, that is to convert chaotic oscillations into desired regular ones with a periodic time dependence would be beneficial in working with a particular system. The possibility of purposeful selection and stabilization of particular orbits in a normally chaotic system, using minimal, predetermined efforts, provides a unique opportunity to maximize the output of a dynamical system. It is thus of great practical importance to develop suitable control methods and to analyze their efficacy.

Let us consider a general nD nonlinear dynamical system,

$$\dot{x} = F(x, p, t), \tag{3.29}$$

where $x = (x_1, x_2, x_3, ..., x_n)$ represents the n state variables and p is a control or external parameter. Let $x(t)$ be a chaotic solution of (3.29). Different control algorithms are essentially based on the fact that one would like to effect the most minimal changes to the original system so that it will not be grossly deformed. From this point of view, controlling methods or algorithms can be broadly classified into two categories:

(i) feedback methods, and

(ii) non–feedback algorithms.

Feedback methods essentially make use of the intrinsic properties of chaotic systems, including their sensitivity to initial conditions, to stabilize orbits already existing in the systems. Some of the prominent methods are the following (see, [Lak97, Lak03, Sch88, II06b]):

1. Adaptive control algorithm;
2. Nonlinear control algorithm;
3. Ott–Grebogi–Yorke (OGY) method of stabilizing unstable periodic orbits;
4. Singer's method of stabilizing unstable periodic orbits; and
5. Various control engineering approaches.

In contrast to feedback control techniques, non–feedback methods make use of a small perturbing external force such as a small driving force, a small noise term, a small constant bias or a weak modulation to some system parameter. These methods modify the underlying chaotic dynamical system weakly so that stable solutions appear. Some of the important controlling methods of this type are the following.

1. Parametric perturbation method
2. Addition of a weak periodic signal, constant bias or noise
3. Entrainment–open loop control
4. Oscillator absorber method.

Here is a typical example of adaptive control algorithm. We can control the chaotic orbit $X_s = (x_s, y_s)$ of the *Van der Pol oscillator* (1.35) by introducing the following dynamics on the parameter A_1:

$$\dot{x} = x - \frac{x^3}{3} - y + A_0 + A_1 \cos \omega t, \qquad \dot{y} = c(x + a - by),$$

$$\dot{A}_1 = -\epsilon[(x - x_s) - (y - y_s)], \qquad \epsilon << 1.$$

On the other hand, recall from [II06b] that a generic SISO nonlinear system

$$\dot{x} = f(x) + g(x)\, u \qquad y = h(x) \tag{3.30}$$

is said to have *relative degree* r at a point x^o if

(i) $L_g L_f^k h(x) = 0$ for all x in a neighborhood of x^o and all $k < r - 1$

(ii) $L_g L_f^{r-1} h(x^o) \neq 0$, where L_g denotes the *Lie derivative* in the direction of the vector–field g.

Now, the Van der Pol oscillator (1.30) has the state space form

$$\dot{x} = f(x) + g(x)\, u = \begin{bmatrix} x_2 \\ 2\omega\zeta\,(1 - \mu x_1^2)\, x_2 - \omega^2 x_1 \end{bmatrix} + \begin{bmatrix} 0 \\ 1 \end{bmatrix} u. \tag{3.31}$$

Suppose the output function is chosen as

$$y = h(x) = x_1. \tag{3.32}$$

In this case we have

$$L_g h(x) = \frac{\partial h}{\partial x} g(x) = \begin{bmatrix} 1 & 0 \end{bmatrix} \begin{bmatrix} 0 \\ 1 \end{bmatrix} = 0, \qquad \text{and} \tag{3.33}$$

$$L_f h(x) = \frac{\partial h}{\partial x} f(x) = \begin{bmatrix} 1 & 0 \end{bmatrix} \begin{bmatrix} x_2 \\ 2\omega\zeta \left(1 - \mu x_1^2\right) x_2 - \omega^2 x_1 \end{bmatrix} = x_2. \tag{3.34}$$

Moreover

$$L_g L_f h(x) = \frac{\partial (L_f h)}{\partial x} g(x) = \begin{bmatrix} 0 & 1 \end{bmatrix} \begin{bmatrix} 0 \\ 1 \end{bmatrix} = 1 \tag{3.35}$$

and thus we see that the Van der Pol oscillator system has relative degree 2 at any point x^o.

However, if the output function is, for instance

$$y = h(x) = \sin x_2 \tag{3.36}$$

then $L_g h(x) = \cos x_2$. The system has relative degree 1 at any point x^o, provided that $(x^o)_2 \neq (2k+1)\pi/2$. If the point x^o is such that this condition is violated, no relative degree can be defined.

Both adaptive and nonlinear control methods can be naturally extended to other chaotic systems, e.g., *Lorenz attractor* (see Figure 3.2).

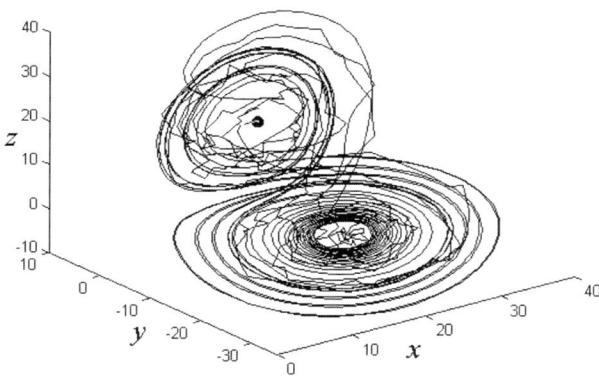

Fig. 3.2. Nonlinear control of the Lorenz system: targeting of unstable upper and lower states in the Lorenz attractor (after applying random perturbations, see [Pet96]), using a MIMO nonlinear controller (see [II06b]); simulated using $Matlab^{TM}$.

Hybrid Systems and Homotopy ODEs

Consider a *hybrid dynamical system of variable structure*, given by an nD ODE–system (see [MWH01])

$$\dot{x} = f(t, x), \tag{3.37}$$

where $x = x(t) \in \mathbb{R}^n$ and $f = f(t, x) : \mathbb{R}^+ \times \mathbb{R}^n \rightarrow \mathbb{R}^n$. Let the domain $G \subset \mathbb{R}^+ \times \mathbb{R}^n$, on which the vector–field $f(t, x)$ is defined, be divided into two subdomains, G^+ and G^-, by means of a smooth $(n-1)$–manifold M. In $G^+ \cup M$, let there be given a vector–field $f^+(t, x)$, and in $G^- \cup M$, let there be given a vector–field $f^-(t, x)$. Assume that both $f^+ = f^+(t, x)$ and $f^- = f^-(t, x)$ are continuous in t and smooth in x. For the system (3.37), let

$$f = \begin{cases} f^+ & \text{when} \ \ x \in G^+ \\ f^- & \text{when} \ \ x \in G^- \end{cases}.$$

Under these conditions, a solution $x(t)$ of ODE (3.37) is well–defined while passing through G until the manifold M is reached.

Upon reaching the manifold M, in physical systems with inertia, the transition

$$\text{from} \ \ \dot{x} = f^-(t, x) \ \ \ \text{to} \ \ \dot{x} = f^+(t, x)$$

does not take place instantly on reaching M, but after some delay. Due to this delay, the solution $x(t)$ oscillates about M, $x(t)$ being displaced along M with some mean velocity.

As the delay tends to zero, the limiting motion and velocity along M are determined by the *linear homotopy ODE*

$$\dot{x} = f^0(t, x) \equiv (1 - \alpha) f^-(t, x) + \alpha f^+(t, x), \tag{3.38}$$

where $x \in M$ and $\alpha \in [0, 1]$ is such that the *linear homotopy segment* $f^0(t, x)$ is tangential to M at the point x, i.e., $f^0(t, x) \in T_x M$, where $T_x M$ is the tangent space to the manifold M at the point x.

The vector–field $f^0(t, x)$ of the system (3.38) can be constructed as follows: at the point $x \in M$, $f^-(t, x)$ and $f^+(t, x)$ are given and their ends are joined by the linear homotopy segment. The point of intersection between this segment and $T_x M$ is the end of the required vector–field $f^0(t, x)$. The vector function $x(t)$ which satisfies (3.37) in G^- and G^+, and (3.38) when $x \in M$, can be considered as a *solution* of (3.37) in a *general sense*.

However, there are cases in which the solution $x(t)$ cannot consist of a finite or even countable number of arcs, each of which passes through G^- or G^+ satisfying (3.37), or moves along the manifold M and satisfies the homotopic ODE (3.38). To cover such cases, assume that the vector–field $f = f(t, x)$ in ODE (3.37) is a Lebesgue–measurable function in a domain $G \subset \mathbb{R}^+ \times \mathbb{R}^n$, and that for any closed bounded domain $D \subset G$ there exists a summable function $K(t)$ such that almost everywhere in D we have $|f(t, x)| \leq K(t)$. Then the absolutely continuous vector function $x(t)$ is called the *generalized solution* of the ODE (3.37) *in the sense of Filippov* (see [MWH01]) if for almost all t, the vector $\dot{x} = \dot{x}(t)$ belongs to the least convex closed set containing all the limiting values of the vector–field $f(t, x^*)$, where x^* tends towards x in an arbitrary manner, and the values of the function $f(t, x^*)$ on a set of measure zero in $\mathbb{R}^n$ are ignored.

Such *hybrid systems* of variable structure occur in the study of nonlinear electric networks (endowed with electronic switches, relays, diodes, rectifiers, etc.), in models of both natural and artificial neural networks, as well as in feedback control systems (usually with continuous–time plants and digital controllers/filters).

3.2.2 Exploiting Critical Sensitivity

The fact that some dynamical systems showing the necessary conditions for chaotic behavior possess such a critical dependence on the initial conditions was known since the end of the last century. However, only in the last thirty years, experimental observations have pointed out that, in fact, chaotic systems are common in nature. They can be found, e.g., in chemistry (*Belouzov–Zhabotinski reaction*), in nonlinear optics (lasers), in electronics (*Chua–Matsumoto circuit*), in fluid dynamics (*Rayleigh–Bénard convection*), etc. Many natural phenomena can also be characterized as being chaotic. They can be found in meteorology, solar system, heart and brain of living organisms and so on.

Due to their critical dependence on the initial conditions, and due to the fact that, in general, experimental initial conditions are never known perfectly, these systems are intrinsically unpredictable. Indeed, the prediction trajectory emerging from an initial condition and the real trajectory emerging from the real initial condition diverge exponentially in course of time, so that the error in the prediction (the distance between prediction and real trajectories) grows exponentially in time, until making the system's real trajectory completely different from the predicted one at long times.

For many years, this feature made chaos undesirable, and most experimentalists considered such characteristic as something to be strongly avoided. Besides their critical sensitivity to initial conditions, chaotic systems exhibit two other important properties. Firstly, there is an infinite number of unstable periodic orbits embedded in the underlying chaotic set. In other words, the skeleton of a chaotic attractor is a collection of an infinite number of periodic orbits, each one being unstable. Secondly, the dynamics in the chaotic attractor is *ergodic*, which implies that during its temporal evolution the system ergodically visits small neighborhood of every point in each one of the unstable periodic orbits embedded within the chaotic attractor.

A relevant consequence of these properties is that a chaotic dynamics can be seen as shadowing some periodic behavior at a given time, and erratically jumping from one to another periodic orbit. The idea of controlling chaos is then when a trajectory approaches ergodically a desired periodic orbit embedded in the attractor, one applies small perturbations to stabilize such an orbit. If one switches on the stabilizing perturbations, the trajectory moves to the neighborhood of the desired periodic orbit that can now be stabilized. This fact has suggested the idea that the critical sensitivity of a chaotic system to changes (perturbations) in its initial conditions may be, in fact, very desirable in practical experimental situations. Indeed, if it is true that a small perturbation can give rise to a very large response in the course of time, it is also true that a judicious choice of such a perturbation can direct the trajectory to wherever one wants in the attractor, and to produce a series of desired dynamical states. This is exactly the idea of targeting [BGL00].

The important point here is that, because of chaos, one is able to produce an infinite number of desired dynamical behaviors (either periodic and not periodic) using the same chaotic system, with the only help of tiny perturbations chosen properly. We stress that this is not the case for a non–chaotic dynamics, wherein the perturbations to be done for producing a desired behavior must, in general, be of the same order of magnitude as the un–perturbed evolution of the dynamical variables.

The *idea* of *chaos control* was enunciated in 1990 at the University of Maryland, by E. Ott, C. Grebogi and J.A. Yorke [OGY90], widely referred to as Ott–Grebogi–Yorke (OGY, for short). In OGY–paper [OGY90], the ideas for controlling chaos were outlined and a method for stabilizing an unstable periodic orbit was suggested, as a proof of principle. The main idea consisted in waiting for a natural passage of the chaotic orbit close to the desired periodic behavior, and then applying a small judiciously chosen perturbation, in order to stabilize such periodic dynamics (which would be, in fact, unstable for the un–perturbed system). Through this mechanism, one can use a given laboratory system for producing an infinite number of different periodic behavior (the infinite number of its unstable periodic orbits), with a great flexibility in switching from one to another behavior. Much more, by constructing appropriate goal dynamics, compatible with the chaotic attractor, an operator may apply small perturbations to produce any kind of desired dynamics, even not periodic, with practical application in the coding process of signals.

A branch of the theory of dynamical systems has been developed with the aim of formalizing and quantitatively characterizing the sensitivity to initial conditions. The *largest Lyapunov exponent* λ (together with the related *Kaplan–Yorke dimension* d_{KY}) and the *Kolmogorov–Sinai entropy* h_{KS} are the two indicators for measuring the *rate of error growth* and *information* produced by the dynamical system [ER85].

3.2.3 Lyapunov Exponents and Kaplan–Yorke Dimension

The characteristic Lyapunov exponents are somehow an extension of the linear stability analysis to the case of aperiodic motions. Roughly speaking, they measure the typical rate of exponential divergence of nearby trajectories. In this sense they give information on the rate of growth of a very small error on the initial state of a system [BCF02].

Consider an nD dynamical system given by the set of ODEs of the form

$$\dot{x} = f(x), \qquad (3.39)$$

where $x = (x_1, \dots, x_n) \in \mathbb{R}^n$ and $f : \mathbb{R}^n \to \mathbb{R}^n$. Recall that since the r.h.s of equation (3.39) does not depend on t explicitly, the system is called *autonomous*. We assume that f is smooth enough that the evolution is well–defined for time intervals of arbitrary extension, and that the motion occurs in a bounded region R of the system phase–space M. We intend to study the separation between two trajectories in M, $x(t)$ and $x'(t)$, starting from two close initial conditions, $x(0)$ and $x'(0) = x(0) + \delta x(0)$ in $R_0 \subset M$, respectively.

As long as the difference between the trajectories, $\delta x(t) = x'(t) - x(t)$, remains infinitesimal, it can be regarded as a vector, $z(t)$, in the tangent space $T_x M$ of M. The time evolution of $z(t)$ is given by the linearized differential equations:

$$\dot{z}_i(t) = \left. \frac{\partial f_i}{\partial x_j} \right|_{x(t)} z_j(t).$$

Under rather general hypothesis, Oseledets [Ose68] proved that for almost all initial conditions $x(0) \in R$, there exists an orthonormal basis $\{e_i\}$ in the tangent space $T_x M$ such that, for large times,

$$z(t) = c_i e_i \exp(\lambda_i t), \qquad (3.40)$$

where the coefficients $\{c_i\}$ depend on $z(0)$. The exponents $\lambda_1 \geq \lambda_2 \geq \cdots \geq \lambda_d$ are called *characteristic Lyapunov exponents*. If the dynamical system has an ergodic invariant measure on M, the spectrum of LEs $\{\lambda_i\}$ does not depend on the initial conditions, except for a set of measure zero with respect to the natural invariant measure.

Equation (3.40) describes how a dD spherical region $R = S^n \subset M$, with radius ϵ centered in $x(0)$, deforms, with time, into an ellipsoid of semi–axes $\epsilon_i(t) = \epsilon \exp(\lambda_i t)$, directed along the e_i vectors. Furthermore, for a generic small perturbation $\delta x(0)$, the distance between the reference and the perturbed trajectory behaves as

$$|\delta x(t)| \sim |\delta x(0)| \exp(\lambda_1 t) \left[1 + O\left(\exp -(\lambda_1 - \lambda_2)t \right) \right].$$

If $\lambda_1 > 0$ we have a rapid (exponential) amplification of an error on the initial condition. In such a case, the system is chaotic and, unpredictable on the long times. Indeed, if the initial error amounts to $\delta_0 = |\delta x(0)|$, and we purpose to predict the states of the system with a certain tolerance Δ, then the prediction is reliable just up to a *predictability time* given by

$$T_p \sim \frac{1}{\lambda_1} \ln \left(\frac{\Delta}{\delta_0} \right).$$

This equation shows that T_p is basically determined by the *positive leading Lyapunov exponent*, since its dependence on δ_0 and Δ is logarithmically weak. Because of its preeminent role, λ_1 is often referred as 'the leading positive Lyapunov exponent', and denoted by λ.

Therefore, Lyapunov exponents are average rates of expansion or contraction along the principal axes. For the ith principal axis, the corresponding Lyapunov exponent is defined as (compare with (4.226) above)

$$\lambda_i = \lim_{t \to \infty} \{(1/t) \ln[L_i(t)/L_i(0)]\}, \tag{3.41}$$

where $L_i(t)$ is the radius of the ellipsoid along the ith principal axis at time t. For technical details on calculating Lyapunov exponents from any time series data, see [WSS85].

An initial volume V_0 of the phase–space region R_0 evolves on average as

$$V(t) = V_0 e^{(\lambda_1 + \lambda_2 + \cdots + \lambda_{2n})t}, \tag{3.42}$$

and therefore the rate of change of $V(t)$ is simply

$$\dot{V}(t) = \sum_{i=1}^{2n} \lambda_i V(t).$$

In the case of a 2D phase area A, evolving as $A(t) = A_0 e^{(\lambda_1 + \lambda_2)t}$, a *Lyapunov dimension* d_L is defined as

$$d_L = \lim_{\epsilon \to 0} \left[\frac{d(\ln(N(\epsilon)))}{d(\ln(1/\epsilon))} \right],$$

where $N(\epsilon)$ is the number of squares with sides of length ϵ required to cover $A(t)$, and d represents an ordinary *capacity dimension*,

$$d_c = \lim_{\epsilon \to 0} \left(\frac{\ln N}{\ln(1/\epsilon)} \right).$$

Lyapunov dimension can be extended to the case of nD phase–space by means of the *Kaplan–Yorke dimension* [Kap00, YAS96, OGY90]) as

$$d_{KY} = j + \frac{\lambda_1 + \lambda_2 + \cdots + \lambda_j}{|\lambda_{j+1}|},$$

where the λ_i are ordered (λ_1 being the largest) and j is the index of the smallest nonnegative Lyapunov exponent.

3.2.4 Kolmogorov–Sinai Entropy

The LE, λ, gives a first quantitative information on how rapidly we loose the ability of predicting the evolution of a system [BCF02]. A state, initially determined with an error $\delta x(0)$, after a time enough larger than $1/\lambda$, may be found almost everywhere in the region of motion $R \in M$. In this respect, the *Kolmogorov–Sinai* (KS) *entropy*, h_{KS}, supplies a more refined information. The error on the initial state is due to the maximal resolution we use for observing the system. For simplicity, let us assume the same resolution ϵ for each degree of freedom. We build a partition of the phase–space M with cells of volume ϵ^d, so that the state of the system at $t = t_0$ is found in a region R_0 of volume $V_0 = \epsilon^d$ around $x(t_0)$. Now we consider the trajectories starting from V_0 at t_0 and sampled at discrete times $t_j = j\,\tau$ $(j = 1, 2, 3, \ldots, t)$. Since we are considering motions that evolve in a bounded region $R \subset M$, all the trajectories visit a finite number of different cells, each one identified by a symbol. In this way a unique sequence of symbols $\{s(0), s(1), s(2), \ldots\}$ is associated with a given trajectory $x(t)$. In a chaotic system, although each evolution $x(t)$ is univocally determined by $x(t_0)$, a great number of different symbolic sequences originates by the same initial cell, because of the divergence of nearby trajectories. The total number of the admissible symbolic sequences, $\widetilde{N}(\epsilon, t)$, increases exponentially with a rate given by the topological entropy

$$h_T = \lim_{\epsilon \to 0} \lim_{t \to \infty} \frac{1}{t} \ln \widetilde{N}(\epsilon, t)\,.$$

However, if we consider only the number of sequences $N_{eff}(\epsilon, t) \leq \widetilde{N}(\epsilon, t)$ which appear with very high probability in the long time limit – those that can be numerically or experimentally detected and that are associated with the natural measure – we arrive at a more physical quantity, namely the *Kolmogorov–Sinai entropy* [ER85]:

$$h_{KS} = \lim_{\epsilon \to 0} \lim_{t \to \infty} \frac{1}{t} \ln N_{eff}(\epsilon, t) \leq h_T. \tag{3.43}$$

h_{KS} quantifies the long time exponential rate of growth of the number of the effective coarse-grained trajectories of a system. This suggests a link with information theory where the Shannon entropy measures the mean asymptotic growth of the number of the typical sequences – the ensemble of which has probability almost one – emitted by a source.

We may wonder what is the number of cells where, at a time $t > t_0$, the points that evolved from R_0 can be found, i.e., we wish to know how big is the coarse–grained volume $V(\epsilon, t)$, occupied by the states evolved from the volume V_0 of the region R_0, if the minimum volume we can observe is $V_{min} = \epsilon^d$. As stated above (3.42), we have

$$V(t) \sim V_0 \exp(t \sum_{i=1}^{d} \lambda_i).$$

However, this is true only in the limit $\epsilon \to 0$. In this (unrealistic) limit, $V(t) = V_0$ for a conservative system (where $\sum_{i=1}^{d} \lambda_i = 0$) and $V(t) < V_0$ for a dissipative system (where $\sum_{i=1}^{d} \lambda_i < 0$). As a consequence of limited resolution power, in the evolution of the volume $V_0 = \epsilon^d$ the effect of the contracting directions (associated with the negative Lyapunov exponents) is completely lost. We can experience only the effect of the expanding directions, associated with the positive Lyapunov exponents. As a consequence, in the typical case, the coarse grained volume behaves as

$$V(\epsilon, t) \sim V_0 \, e^{(\sum_{\lambda_i > 0} \lambda_i) \, t},$$

when V_0 is small enough. Since $N_{eff}(\epsilon, t) \propto V(\epsilon, t)/V_0$, one has

$$h_{KS} = \sum_{\lambda_i > 0} \lambda_i.$$

This argument can be made more rigorous with a proper mathematical definition of the metric entropy. In this case one derives the Pesin relation [Pes77, ER85]

$$h_{KS} \leq \sum_{\lambda_i > 0} \lambda_i. \tag{3.44}$$

Because of its relation with the Lyapunov exponents – or by the definition (3.43) – it is clear that also h_{KS} is a fine-grained and global characterization of a dynamical system.

The metric entropy is an invariant characteristic quantity of a dynamical system, i.e., given two systems with invariant measures, their KS–entropies exist and they are equal iff the systems are isomorphic [Bil65].

3.2.5 Chaos Control by Ott, Grebogi and Yorke (OGY)

Besides the occurrence of chaos in a large variety of natural processes, chaos may also occur because one may wish to design a physical, biological or chemical experiment, or to project an industrial plant to behave in a chaotic manner. The OGY–idea is that chaos may indeed be desirable since it can be controlled by using small perturbation to some accessible parameter.

The major key ingredient for the OGY–control of chaos is the observation that a chaotic set, on which the trajectory of the chaotic process lives, has embedded within it a large number of unstable low–period periodic orbits. In addition, because of ergodicity, the trajectory visits or accesses the neighborhood of each one of these periodic orbits. Some of these periodic orbits may correspond to a desired system's performance according to some criterion. The second ingredient is the realization that chaos, while signifying sensitive dependence on small changes to the current state and henceforth rendering unpredictable the system state in the long time, also implies that the system's behavior can be altered by using small perturbations. Then, the accessibility of the chaotic systems to many different periodic orbits combined with its sensitivity to small perturbations allows for the control and the manipulation of the chaotic process. Specifically, the OGY approach is then as follows. One first determines some of the unstable low–period periodic orbits that are embedded in the chaotic set. One then examines the location and the stability of these orbits and chooses one which yields the desired system performance. Finally, one applies small control to stabilize this desired periodic orbit. However, all this can be done from data by using nonlinear time series analysis for the observation, understanding and control of the system. This is particularly important since chaotic systems are rather complicated and the detailed knowledge of the equations of the process is often unknown [BGL00].

Simple Example of Chaos Control: a 1D Map. The basic idea of controlling chaos can be understood [Lai94] by considering May's classical *logistic map* [May76] (1.27)

$$x_{n+1} = f(x_n, r) = rx_n(1 - x_n),$$

where x is restricted to the unit interval $[0, 1]$, and r is a control parameter. It is known that this map develops chaos via the *period–doubling bifurcation* route. For $0 < r < 1$, the asymptotic state of the map (or the attractor of the map) is $x = 0$; for $1 < r < 3$, the attractor is a nonzero fixed–point $x_F = 1 - 1/r$; for $3 < r < 1 + \sqrt{6}$, this fixed–point is unstable and the attractor is a stable period-2 orbit. As r is increased further, a sequence of period–doubling bifurcations occurs in which successive period–doubled orbits become stable. The period–doubling cascade accumulates at $r = r_\infty \approx 3.57$, after which chaos can arise.

Consider the case $r = 3.8$ for which the system is apparently chaotic. An important characteristic of a chaotic attractor is that there exists an infinite number of unstable periodic orbits embedded within it. For example, there are a fixed–point $x_F \approx 0.7368$ and a period-2 orbit with components $x(1) \approx 0.3737$ and $x(2) = 0.8894$, where $x(1) = f(x(2))$ and $x(2) = f(x(1))$.

Now suppose we want to avoid chaos at $r = 3.8$. In particular, we want trajectories resulting from a randomly chosen initial condition x_0 to be as close as possible to the period-2 orbit, assuming that this period-2 orbit gives the best system performance. Of course, we can choose the desired asymptotic state of the map to be any of the infinite number of unstable periodic orbits. Suppose that the parameter r can be finely tuned in a small range around the value $r_0 = 3.8$, i.e., r is allowed to vary in the range $[r_0 - \delta, r_0 + \delta]$, where $\delta \ll 1$. Due to the nature of the chaotic attractor, a trajectory that begins from an arbitrary value of x_0 will fall, with probability one, into the neighborhood of the desired period-2 orbit at some later time. The trajectory would diverge quickly from the period-2 orbit if we do not intervene. Our task is to program the variation of the control parameter so that the trajectory stays in the neighborhood of the period-2 orbit as long as the control is present. In general, the small parameter perturbations will be time dependent [BGL00].

The logistic map in the neighborhood of a periodic orbit can be approximated by a linear equation expanded around the periodic orbit. Denote the target period-m orbit to be controlled as $x(i)$, $i = 1, ..., m$, where

$x(i + 1) = f(x(i))$ and $x(m + 1) = x(1)$. Assume that at time n, the trajectory falls into the neighborhood of component i of the period-m orbit. The linearized dynamics in the neighborhood of component $i + 1$ is then

$$x_{n+1} - x(i + 1) = \frac{\partial f}{\partial x}[x_n - x(i)] + \frac{\partial f}{\partial r}\Delta r_n$$
$$= r_0[1 - 2x(i)][x_n - x(i)] + x(i)[1 - x(i)]\Delta r_n,$$

where the partial derivatives are evaluated at $x = x(i)$ and $r = r_0$. We require x_{n+1} to stay in the neighborhood of m. Hence, we set $x_{n+1} - x(i + 1) = 0$, which gives

$$\Delta r_n = r_0\frac{[2x(i) - 1][x_n - x(i)]}{x(i)[1 - x(i)]}. \tag{3.45}$$

Equation (3.45) holds only when the trajectory x_n enters a small neighborhood of the period-m orbit, i.e., when $|x_n - x(i)| << 1$, and hence the required parameter perturbation Δr_n is small. Let the length of a small interval defining the neighborhood around each component of the period-m orbit be 2ε. In general, the required maximum parameter perturbation δ is proportional to ε. Since ε can be chosen to be arbitrarily small, δ also can be made arbitrarily small. The average transient time before a trajectory enters the neighborhood of the target periodic orbit depends on ε (or δ). When the trajectory is outside the neighborhood of the target periodic orbit, we do not apply any parameter perturbation, so the system evolves at its nominal parameter value r_0. Hence we set $\Delta r_n = 0$ when $\Delta r_n > \delta$. The parameter perturbation Δr_n depends on x_n and is time–dependent.

The above strategy for controlling the orbit is very flexible for stabilizing different periodic orbits at different times. Suppose we first stabilize a chaotic trajectory around a period-2 orbit. Then we might wish to stabilize the fixed–point of the logistic map, assuming that the fixed–point would correspond to a better system performance at a later time. To achieve this change of control, we simply turn off the parameter control with respect to the period-2 orbit. Without control, the trajectory will diverge from the period-2 orbit exponentially. We let the system evolve at the parameter value r_0. Due to the nature of chaos, there comes a time when the chaotic trajectory enters a small neighborhood of the fixed–point. At this time we turn on a new set of parameter perturbations calculated with respect to the fixed–point. The trajectory can then be stabilized around the fixed–point [Lai94].

In the presence of external noise, a controlled trajectory will occasionally be 'kicked' out of the neighborhood of the periodic orbit. If this behavior occurs, we turn off the parameter perturbation and let the system evolve by itself. With probability one the chaotic trajectory will enter the neighborhood of the target periodic orbit and be controlled again. The effect of the noise is to turn a controlled periodic trajectory into an intermittent one in which chaotic phases (uncontrolled trajectories) are interspersed with laminar phases (controlled periodic trajectories). It is easy to verify that the averaged length of the laminar phase increases as the noise amplitude decreases [Lai94].

3.2.6 Floquet Stability Analysis and OGY Control

Controlling chaos, or stabilization of unstable periodic orbits of chaotic systems, has established to a field of large interest since the seed paper of Ott, Grebogi, Yorke [OGY90]. The idea is to stabilize by a feedback calculated at each *Poincaré section*, which reduces the control problem to stabilization of an unstable fixed–point of an iterated map. The feedback can, as in OGY scheme, be chosen proportional to the distance to the desired fixed–point, or proportional to the difference in phase–space position between actual and last but one Poincaré section. This difference control scheme [BDG93], being a time–discrete counterpart of the Pyragas approach [Pyr92, Pyr95], allows for stabilization of inaccurately known fixed–points, and can be extended by a memory term to overcome stability restrictions and to allow for for tracking of drifting fixed–points [CMP98a].

In this section the stability of perturbations $x(t)$ around an unstable periodic orbit being subject to a Poincaré–based control scheme is analyzed by means of *Floquet theory* [HL93]. This approach allows to investigate viewpoints that have not been accessible by considering only the iteration dynamics between the Poincaré sections. Among these are primary the discussion of small measurement delays and variable impulse lengths. The impulse length is for both OGY and difference control usually a fixed parameter; and the iterated dynamics is uniquely defined only as long as this impulse length is not varied. The influence of the impulse length has not been point of consideration before; if reported at all, usually for both OGY and difference control a relative length of approximately $1/3$ is chosen without any reported sensitivity [Cla02b].

The linearized ODEs of both schemes are invariant under translation in time, $t \to t + T$. Therefore we can expand the solutions after periodic solutions $u(t + T) = u(t)$ according to

$$x(t) = e^{\gamma T} u_\gamma(t).$$

The necessary condition for stability of the solution is $\mathrm{Re}(\gamma) < 0$; and $x(t) \equiv 0$ refers to motion along the orbit.

Whereas for the *Pyragas control* method (in which the delayed state feedback enforces a time–continuous description) a *Floquet stability analysis* is known [JBO97], here the focus is on the time–discrete control schemes.

Time–Continuous Stability Analysis of OGY Control

Due to the mathematically elegant and practical convenient description and application of OGY control in the Poincaré section up to now there seems to have been no need to calculate explicitly the *Floquet multiplicator* for a stability analysis. However, this allows a novel viewpoint on the differences between the local dynamics around an instable periodic orbit of a dynamical system being subject to Pyragas and OGY control.

For the 1D case, one has the dynamical system [Cla02b]

$$\dot{x}(t) = \lambda x(t) + \mu\varepsilon x(t - (t \bmod T)).$$

In the first time interval between $t = 0$ and $t = T$ the differential equation reads

$$\dot{x}(t) = \lambda x(t) - \mu\varepsilon x(0), \qquad \text{for} \qquad 0 < t < T.$$

Integration of this differential equation yields

$$x(t) = \left((1 - \frac{\mu\varepsilon}{\lambda})e^{\lambda t} + \frac{\mu\varepsilon}{\lambda} \right) x(0).$$

This gives us an iterated dynamics (here we label the beginning of the time period again with t)

$$x(t + T) = \left((1 - \frac{\mu\varepsilon}{\lambda})e^{\lambda T} + \frac{\mu\varepsilon}{\lambda} \right) x(t).$$

The Floquet multiplier of an orbit therefore is

$$e^{\gamma T} = (1 - \frac{\mu\varepsilon}{\lambda})e^{\lambda T} + \frac{\mu\varepsilon}{\lambda}.$$

Influence of the Duration of the Control Impulse on OGY Control

The time–discrete viewpoint now allows to investigate the influence of timing questions on control. First we consider the case that the control impulse is applied timely in the Poincaré section, but only for a finite period $T \cdot p$ within the orbit period $(0 < p < 1)$.

This situation is described by the differential equation [Cla02b]

$$\dot{x}(t) = \lambda x(t)\mu\varepsilon x(t - (t \bmod T)) \cdot \Theta((t \bmod T) - p).$$

Here Θ is a step function ($\Theta(x) = 1$ for $x > 0$ and $\Theta(x) = 0$ elsewhere). In the first time interval between $t = 0$ and $t = T \cdot p$ the differential equation reads

$$\dot{x}(t) = \lambda x(t) + \mu\varepsilon x(0), \qquad \text{for} \qquad 0 < t < T \cdot p.$$

Integration of this differential equation yields

$$x(t) = \left((1 + \frac{\mu\varepsilon}{\lambda})e^{\lambda t} - \frac{\mu\varepsilon}{\lambda} \right) x(0), \qquad x(T \cdot p) = \left((1 + \frac{\mu\varepsilon}{\lambda})\epsilon^{\lambda T \cdot p} - \frac{\mu\varepsilon}{\lambda} \right) x(0).$$

In the second interval between $t = T \cdot p$ and $t = T$ the differential equation is the same as without control,

$$\dot{x}(t) = \lambda x(t), \qquad \text{for} \qquad T \cdot p < t < T.$$

From this one has immediately

$$x(t) = e^{\lambda(t - T \cdot p)}x(T \cdot p).$$

If the beginning of the integration period again is denoted by t, this defines an iteration dynamics,

$$x(t+T) = e^{\lambda(1-p)T}\left((1+\frac{\mu\varepsilon}{\lambda})e^{\lambda T \cdot p} - \frac{\mu\varepsilon}{\lambda}\right)x(t) = \left(\left(1+\frac{\mu\varepsilon}{\lambda}\right)e^{\lambda T} - \frac{\mu\varepsilon}{\lambda}e^{\lambda(1-p)T}\right),$$

and the Floquet multiplier of an orbit is given by

$$e^{\gamma T} = (1-\frac{\mu\varepsilon}{\lambda})e^{\lambda T} + \frac{\mu\varepsilon}{\lambda}e^{\lambda(1-p)T} = e^{\lambda T}\left(1 - \frac{\mu\varepsilon}{\lambda}(1-e^{-\lambda pT})\right). \qquad (3.46)$$

One finds that in zero order the 'strength' of control is given by the product $p \cdot \mu\varepsilon$; in fact there is a weak linear correction in p. For $\lambda pT \leq 1$ one has

$$e^{\gamma T} = e^{\lambda T}(1 + \mu\varepsilon pT + \frac{1}{2}\mu\varepsilon\lambda p^2 T^2 + o(p^3)) = e^{\lambda T}(1 + \mu\varepsilon pT(1 - \frac{1}{2}\lambda pT + o(p^2))),$$

i.e., to get a constant strength of control, one has to fulfill the condition

$$\mu\varepsilon pT = \frac{1}{1 - \frac{\lambda T}{2}p} = 1 + \frac{\lambda T}{2}p + o(p^2).$$

The result is, apart from a weak linear correction for OGY control the length of the impulse can be chosen arbitrarily, and the 'strength' of control in zero order is given by the time integral over the control impulse.

Floquet Stability Analysis of Difference Control

Again the starting point is the linearized equation of motion around the periodic orbit when control is applied. For difference control now there is a dependency on two past time steps,

$$\dot{x}(t) = \lambda x(t) + \mu\varepsilon x(t - (t \bmod T)) - \mu\varepsilon x(t - T - (t \bmod T)). \qquad (3.47)$$

Although the r.h.s of (3.47) depends on x at three different times, it can be nevertheless integrated exactly, which is mainly due to the fact that the two past times (of the two last Poincaré crossings) have a fixed time difference being equal to the orbit length. This allows not only for an exact solution, but also offers a correspondence to the time–discrete dynamics and the matrix picture used in time–delayed coordinates [CMP98a, CS98, CMP98b].

Stability Analysis of Difference Control

Now also for difference control the experimentally more common situation of a finite but small measurement delay $T \cdot s$ is considered, together with a finite impulse length $T \cdot p$ (here $0 < p < 1$ and $0 < (s+p) < 1$) [Cla02b].

In the first time interval between $t = 0$ and $t = T \cdot s$ the ODE reads

$$\dot{x}(t) = \lambda x(t), \qquad \text{for} \qquad 0 < t < T \cdot s.$$

The integration gives $x(T \cdot s) = e^{\lambda T \cdot s} x(0)$.

For the second interval between $t = T \cdot s$ and $t = T \cdot (s + p)$ we have

$$\dot{x}(t) = \lambda x(t) - \mu \varepsilon x(0) = \lambda x(t) + \mu \varepsilon (x(0) - x(-T)), \quad \text{for} \quad T \cdot s < t < T \cdot (s+p).$$

Integration of this ODE yields

$$x(t) = -\frac{\mu \varepsilon}{\lambda}(x(0) - x(-T)) + \frac{\mu \varepsilon}{\lambda}(x(0) - x(-T)) + e^{\lambda s T} x(0) e^{\lambda(t - sT)}$$

$$x(T(s+p)) = -\frac{\mu \varepsilon}{\lambda}(x(0) - x(-T))\frac{\mu \varepsilon}{\lambda}(x(0) - x(-T)) + e^{\lambda p T} + e^{\lambda(s+p)T}x(0).$$

For the third interval, the ODE is homogeneous again and one has

$$x(t) = e^{\lambda(t - (s+p)T)} x(T \cdot (s+p)), \quad \text{for} \quad T \cdot (s+p) < t < T.$$

Insertion gives

$$x(T) = x(0) e^{\lambda T} \left(1 + \frac{\mu \varepsilon}{\lambda} e^{-\lambda s T}(1 - e^{-\lambda p T}) \right) - x(-T) e^{\lambda T} \frac{\mu \varepsilon}{\lambda} e^{-\lambda s T}(1 - e^{-\lambda p T})$$

or, in time–delayed coordinates of the last and last but one Poincaré crossing [Cla02b]

$$\begin{pmatrix} x_{n+1} \\ x_n \end{pmatrix} = \begin{pmatrix} e^{\lambda T}\left(1 + \frac{\mu \varepsilon (1 - e^{-\lambda p T})}{\lambda e^{\lambda s T}}\right) & -e^{\lambda T}\frac{\mu \varepsilon (1 - e^{-\lambda p T})}{\lambda e^{\lambda s T}} \\ 1 & 0 \end{pmatrix} \begin{pmatrix} x_n \\ x_{n-1} \end{pmatrix}$$

If we identify with the coefficients of the time–discrete case, $\lambda_{\mathrm{d}} = e^{\lambda T}$ and $\mu_{\mathrm{d}}\varepsilon_{\mathrm{d}} = e^{-\lambda s T}(1 - e^{\lambda p T})\frac{\mu \varepsilon}{\lambda}$, the dynamics in the Poincaré iteration $t = nT$ becomes identical with the pure discrete description; this again illustrates the power of the concept of the Poincaré map. Due to the low degree of the characteristic polynomial, one in principle can explicitly diagonalize the iteration matrix, allowing for a closed expression for the nth power of the iteration matrix. As for the stability analysis only the eigenvalues are needed, this straightforward calculation is excluded here.

For the Floquet multiplier one has [Cla02b]

$$e^{2\gamma T} = e^{\gamma T} e^{\lambda T} \left(1 + \frac{\mu \varepsilon}{\lambda} e^{-\lambda s T}(1 - e^{-\lambda p T}) \right) - e^{\lambda T}\frac{\mu \varepsilon}{\lambda} e^{-\lambda s T}(1 - e^{-\lambda p T}).$$

This quadratic equation yields two Floquet multipliers,

$$e^{\gamma T} = \frac{1}{2} e^{\lambda T}\left(1 + \frac{\mu \varepsilon}{\lambda} e^{-\lambda s T}(1 - e^{-\lambda p T})\right)$$

$$\pm \frac{1}{2}\sqrt{\left(e^{\lambda T}\left(1 + \frac{\mu \varepsilon}{\lambda} e^{-\lambda s T}(1 - e^{-\lambda p T})\right)\right)^2 + 4e^{\lambda T}\frac{\mu \varepsilon}{\lambda} e^{-\lambda s T}(1 - e^{-\lambda p T})}.$$

For $s = 0$ one gets the special cases discussed above.

3.2.7 Blind Chaos Control

One of the most surprising successes of chaos theory has been in biology: the experimentally demonstrated ability to control the timing of spikes of electrical activity in complex and apparently chaotic systems such as heart tissue [GSD92] and brain tissue [SJD94]. In these experiments, PPF control, a modified formulation of OGY control [OGY90], was applied to set the timing of external stimuli; the controlled system showed stable periodic trajectories instead of the irregular inter–spike intervals seen in the uncontrolled system. The mechanism of control in these experiments was interpreted originally as analogous to that of OGY control: unstable periodic orbits riddle the chaotic attractor and the electrical stimuli place the system's state on the stable manifold of one of these periodic orbits [Kap00].

Alternative possible mechanisms for the experimental observations have been described by Zeng and Glass [GZ94] and Christini and Collins [CC95]. These authors point out that the controlling external stimuli serve to truncate the inter–spike interval to a maximum value. When applied, the control stimulus sets the next interval s_{n+1} to be on the line

$$s_{n+1} = As_n + C. \qquad (3.48)$$

We will call this relationship the 'control line.' Zeng and Glass showed that if the uncontrolled relationship between inter–spike intervals is a chaotic 1D function, $s_{n+1} = f(s_n)$, then the control system effectively flattens the top of this map and the controlled dynamics may have fixed points or other periodic orbits [GZ94]. Christini and Collins showed that behavior analogous to the fixed–point control seen in the biological experiments can be accomplished even in completely random systems [CC95]. Since neither chaotic 1D systems nor random systems have a stable manifold, the interval–truncation interpretation of the biological experiments is different than the OGY interpretation. The interval–truncation method differs also from OGY and related control methods in that the perturbing control input is a fixed–size stimulus whose timing can be treated as a continuous parameter. This type of input is conventional in cardiology (e.g., [HCT97]).

Kaplan demonstrated in [Kap00] that the state–truncation interpretation was applicable in cases where there was a stable manifold of a periodic orbit as well as in cases where there were only unstable manifolds. He found that superior control could be achieved by intentionally placing the system's state off of any stable manifold. That suggested a powerful scheme for the rapid experimental identification of fixed points and other periodic orbits in systems where inter–spike intervals were of interest.

The chaos control in [GSD92] and [SJD94] was implemented in two stages. First, inter–spike intervals s_n from the uncontrolled, 'natural' system were observed. Modelling the system as a function of two variables

$$s_{n+1} = f(s_n, s_{n-1}),$$

the location $s^\star$ of a putative unstable flip–saddle type fixed–point and the corresponding stable eigenvalue λ_s were estimated from the data[3] [CK97]. The linear approximation to the stable manifold lies on a line given by (3.48) with

$$A = \lambda_s \qquad \text{and} \qquad C = (1 - \lambda_s)s^\star.$$

Second, using estimated values of A and C, the control system was turned on. Following each observed interval s_n, the maximum allowed value of the next inter–spike interval was computed as

$$S_{n+1} = As_n + C.$$

If the next interval naturally was shorter than S_{n+1} no control stimulus was applied to the system. Otherwise, an external stimulus was provided to truncate the inter–spike interval at $s_{n+1} = S_{n+1}$.

In practice, the values of $s^\star$ and λ_s for a real fixed–point of the natural system are known only imperfectly from the data. Insofar as the estimates are inaccurate, the control system does not place the state on the true stable manifold. Therefore, we will analyze the controlled system without presuming that A and C in (3.48) correspond to the stable manifold.

If the natural dynamics of the system is modelled by

$$s_{n+1} = f(s_n, s_{n-1}),$$

then the dynamics of the controlled system is given by [Kap0C]

$$s_{n+1} = \min \begin{cases} f(s_n, s_{n-1}) : & \text{Natural Dynamics}, \\ As_n + C & : \text{Control Line}. \end{cases} \qquad (3.49)$$

We can study the dynamics of the controlled system close to a natural fixed–point, $s^\star$, by approximating the natural dynamics linearly as[4]

$$s_{n+1} = f(s_n, s_{n-1}) = (\lambda_s + \lambda_u)s_n - \lambda_s\lambda_u s_{n-1} + s^\star(1 + \lambda_s\lambda_u - \lambda_s - \lambda_u).$$

Since the controlled system (3.49) is nonlinear even when $f()$ is linear, it is difficult to analyze its behavior by algebraic iteration. Nonetheless, the controlled system can be studied in terms of 1D maps.

Following any inter–spike interval when the controlling stimulus has been applied, the system's state (s_n, s_{n-1}) will lie somewhere on the control line. From this time onward the state will lie on an image of the control line even if additional stimuli are applied during future inter–spike intervals.

The stability of the controlled dynamics fixed–point and the size of its basin of attraction can be analyzed in terms of the control line and its image.

---

[3] Since the fixed–point is unstable, there is also an unstable eigenvalue λ_u.

[4] Equation (3.50) is simply the linear equation $s_{n+1} = as_n + bs_{n-1} + c$ with a, b, and c set to give eigenvalues λ_s and λ_u and fixed–point $s^\star$.

the flip–saddle value of $s^\star = 0.884$. A different type bifurcation occurs at the non–flip saddle fixed–point at $s^\star = -1.584$. To the left of the bifurcation point, the iterates are diverging to $-\infty$ and are not plotted.

Adding gaussian dynamical noise (of standard deviation 0.05) does not substantially alter the bifurcation diagram, suggesting that examination of the truncation control bifurcation diagram may be a practical way to read off the location of the unstable fixed points in an experimental preparation.

Unstable periodic orbits can be difficult to find in uncontrolled dynamics because there is typically little data near such orbits. Application of PPF control, even blindly, can stabilize such orbits and dramatically improve the ability to locate them. This, and the robustness of the control, may prove particularly useful in biological experiments where orbits may drift in time as the properties of the system change [Kap00].

3.2.8 Jerk Functions of Simple Chaotic Flows

Recall that the celebrated *Lorenz equations* (1.21) can be rewritten as

$$\dot{x} = -ax + ay, \qquad \dot{y} = -xz + bx - y, \qquad \dot{z} = xy - cz. \qquad (3.53)$$

Note that there are seven terms in the phase–flow of these equations, two of which are nonlinear (xz and xy); also, there are three parameters, for which Lorenz found chaos with $a = 10$, $b = 28$, and $c = 8/3$. The number of independent parameters is generally $d + 1$ less than the number of terms for a $d-$dimensional system, since each of the variables (x, y, and z in this case) and time (t) can be arbitrarily rescaled [SL00]. The Lorenz system has been extensively studied, and there is an entire book [Spa82] devoted to it.

Although the Lorenz system is often taken as the prototypical chaotic flow, it is not the algebraically simplest such system [SL00]. Recall that in 1976, *Rössler* [Ros76] proposed his equations (1.38), rewritten here as

$$\dot{x} = -y - z, \qquad \dot{y} = x + ay, \qquad \dot{z} = b + xz - cz. \qquad (3.54)$$

Rössler phase–flow also has seven terms and two parameters, which Rössler took as $a = b = 0.2$ and $b = 5.7$, but only a single quadratic nonlinearity (xz).

In 1994, Sprott [Spr94] embarked on an extensive search for autonomous three–dimensional chaotic systems with fewer than seven terms and a single quadratic nonlinearity and systems with fewer than six terms and two quadratic nonlinearities. The *brute–force* method [Spr93a, Spr93b] involved the numerical solution of a huge number (about 10^8) systems of autonomous ODEs with randomly chosen real coefficients and initial conditions. The criterion for chaos was the existence of a leading Lyapunov exponent. He found fourteen algebraically distinct cases with six terms and one nonlinearity, and five cases with five terms and two nonlinearities. One case was volume conserving (conservative), and all the others were volume–contracting (dissipative), implying the existence of a strange attractor. Sprott provided a table of the spectrum of Lyapunov exponents, the related Kaplan–Yorke dimension, and the types and eigenvalues of the unstable fixed–points for each of the nineteen cases [SL00].

Subsequently, Hoover [Hoo95] pointed out that the conservative case found by Sprott in [Spr94]

$$\dot{x} = y, \qquad \dot{y} = -x + yz, \qquad \dot{z} = 1 - y^2, \tag{3.55}$$

is a special case of the Nosé–Hoover thermostated dynamic system that had earlier been shown [PHV86] to exhibit time–reversible *Hamiltonian chaos*.

In response to Sprott's work, Gottlieb [Got96] pointed out that the conservative system (3.55) could be recast in the *explicit third–order form*

$$\dddot{x} = -\dot{x} + \ddot{x}(x + \ddot{x})/\dot{x},$$

which he called a '*jerk function*' since it involves a third derivative of $\ddot{x}$, which in a mechanical system is the time rate of change of the acceleration, also called the 'jerk' [Sch78]. It is known that any explicit ODE can be cast in the form of a system of coupled first–order ODEs, but the converse does not hold in general. Even if one can reduce the dynamical system to a jerk form for each of the phase–space variables, the resulting differential equations may look quite different. Gottlieb asked the provocative question 'What is the simplest jerk function that gives chaos?'

One response was provided by Linz [Lin97] who showed that both the original Rössler model and the Lorenz model can be reduced to jerk forms. The Rössler model (3.54) can be rewritten (in a slightly modified form) as

$$\dddot{x} + (c - \varepsilon + \varepsilon x - \dot{x})\ddot{x} + [1 - \varepsilon c - (1 + \varepsilon^2)x + \varepsilon\dot{x}]\dot{x} + (\varepsilon x + c)x + \varepsilon = 0,$$

where $\varepsilon = 0.2$ and $c = 5.7$ gives chaos. Note that the jerk form of the Rössler equation is a rather complicated quadratic polynomial with 10 terms.

The Lorenz model in (3.53) can be written as

$$\dddot{x} + (1 + \sigma + b - \dot{x}/x)\ddot{x} + [b(1 + \sigma + x^2) - (1 + \sigma)\dot{x}/x]\dot{x} - b\sigma(r - 1 - x^2)x = 0.$$

The jerk form of the Lorenz equation is not a polynomial since it contains terms proportional to $\dot{x}/x$ as is typical of dynamical systems with multiple nonlinearities. Its jerk form contains eight terms.

Linz [Lin97] showed that Sprott's case R model (see [Spr94]) can be written as a polynomial with only five terms and a single quadratic nonlinearity

$$\dddot{x} + \ddot{x} - x\dot{x} + ax + b = 0,$$

with chaos for $a = 0.9$ and $b = 0.4$.

Sprott [Spr97] also took up Gottlieb's challenge and embarked on an extensive numerical search for chaos in systems of the explicit form

$$\dddot{x} = J(\ddot{x}, \dot{x}, x),$$

where the jerk function J is a simple quadratic or cubic polynomial. He found a variety of cases, including two with three terms and two quadratic nonlinearities in their jerk function,

$$\dddot{x} + ax\ddot{x} - \dot{x}^2 + x = 0,$$

with $a = 0.645$ and

$$\dddot{x} + ax\ddot{x} - x\dot{x} + x = 0,$$

with $a = -0.113$, and a particularly simple case with three terms and a single quadratic nonlinearity

$$\dddot{x} + a\ddot{x} \pm \dot{x}^2 + x = 0, \qquad (3.56)$$

with $a = 2.017$. For this value of a, the Lyapunov exponents are (0.0550, 0, -2.0720) and the Kaplan–Yorke dimension is $d_{KY} = 2.0265$.

Equation (3.56) is simpler than any previously discovered case. The range of the parameter a over which chaos occurs is quite narrow ($2.0168\ldots < a < 2.0577\ldots$). It also has a relatively small basin of attraction, so that initial conditions must be chosen carefully. One choice of initial conditions that lies in the basin of attraction is $(x, y, z) = (0, 0, \pm 1)$, where the sign is chosen according to the sign of the $\pm\dot{x}^2$ term.

All above systems share a common *route to chaos*. The control parameter a can be considered a *damping rate* for the nonlinear oscillator. For large values of a, there are one or more stable equilibrium points. As a decreases, a *Hopf bifurcation* (see [CD98]) occurs in which the equilibrium becomes unstable, and a stable limit cycle is born. The limit cycle grows in size until it bifurcates into a more complicated limit cycle with two loops, which then bifurcates into four loops, and so forth, in a *sequence of period doublings* until chaos finally onsets. A further decrease in a causes the *chaotic attractor* to grow in size, passing through infinitely many periodic windows, and finally becoming unbounded when the attractor grows to touch the boundary of its *basin of attraction* (a 'crisis').

Recently, Malasoma [Mal00] joined the search for simple chaotic jerk functions and found a cubic case as simple as (3.56) but of a different form

$$\dddot{x} + a\ddot{x} - x\dot{x}^2 + x = 0,$$

which exhibits chaos for $a = 2.05$. For this value of a, the Lyapunov exponents are $(0.0541, 0, -2.1041)$, and the Kaplan–Yorke dimension is $d_{KY} = 2.0257$. This case follows the usual period–doubling route to chaos, culminating in a boundary crisis and unbounded solutions as a is lowered. The range of a over which chaos occurs is very narrow, ($2.0278\ldots < a < 2.0840\ldots$). There is also a second extraordinarily small window of chaos for ($0.0753514\ldots < a < 0.0753624\ldots$), which is five thousand times smaller than the previous case. Malasoma points out that this system is invariant under the parity transformation $x \to -x$ and speculates that this system is the simplest such example.

Both Linz and Sprott pointed out that if the jerk function is considered the time derivative of an acceleration of a particle of mass m, then *Newton's second law* implies a *force* F whose time derivative is $\dot{F} = mJ$. If the force has an explicit dependence on only $\dot{x}$, x, and time, it is considered to be 'Newtonian jerky'. The condition for $F = F(\dot{x}, x, t)$ is that J depends only linearly on $\ddot{x}$. In such a case the force in general includes a *memory term* of the form

$$M = \int_0^t G(x(\tau))\, d\tau,$$

which depends on the dynamical history of the motion.

The jerk papers by Linz [Lin97] and Sprott [Spr97] appeared in the same issue of the American Journal of Physics and prompted von Baeyer [Bae98] to comment: "The articles with those funny titles are not only perfectly serious, but they also illustrate in a particularly vivid way the revolution that is transforming the ancient study of mechanics into a new science – one that is not just narrowly concerned with the motion of physical bodies, but that deals with changes of all kinds." He goes on to say that the method of searching for chaos in a large class of systems "is not just empty mathematical formalism. Rather it illustrates the arrival of a new level of abstraction in physical science... At that higher level of abstraction, dynamics has returned to the classical Aristotelian goal of trying to understand all change.'

3.2.9 Example: Chaos Control in Molecular Dynamics

Recall that classically modelled molecular dynamics are often characterized by the presence of chaotic behavior. A central issue in these studies is the role of a control variable. A small variation of this variable can cause a transition from the periodic or quasi–periodic regime to the chaotic regime. For example, the molecular energy can serve as a control variable. It was shown in [BRR95] that the isotopic mass could be viewed as a discrete control variable and related effects were also evident in quantum calculations.

In this approach the variation of the control variable or parameters changes the route taken by the dynamics of the system. It was shown that a small time–dependent perturbation of the control parameters could convert a chaotic attractor to any of a large number of possible attracting periodic orbits [SOG90, SDG92, RGO92]. The control parameter could stabilize the chaotic dynamics about some periodic orbit.

In [BRR95], a general method to control molecular systems has been proposed by employing optimally designed laser pulses. This approach is capable of designing the time–dependent laser pulse shapes to permit the steering of intramolecular dynamics to desired physical objectives. Such objectives include selective bond breaking through infrared excitation [SR91, SR90, SR91] and through electronic excitation between multiple surfaces [TR85, KRG89], control of curve-crossing reactions [GNR92, CBS91], selective electronic excitation in condensed media [GNR91, GNR93, SB89], control of the electric susceptibility of a molecular gas [SSR93] and selective rotational excitation [JLR90]. The pulse shape design is obtained by minimization of a design cost functional with respect to the control field pulse which corresponds to the optimization of the particular objective. Also, multiple solutions of the optimal control equations exist, which gives flexibility for adding further terms in the cost functional or changing certain design parameters so that the particular objective is better reached.

In this section, following [BRR95, BRR95], we will explore optimal control properties of nonlinear classical Hamiltonians under certain constraints in order to suppress their chaotic behavior. This approach to molecular dynamics simulations reduces the computational cost and instabilities especially associated with the treatment of the Lagrange multipliers.

Conventional optimal control with an ensemble of N trajectories would call for the introduction of rN Lagrange constraint functions where r is the dimension of the phase–space. In contrast the new method only requires the same number r of Lagrange multipliers as the dimension of the phase–space of a single molecule. The reduction of Lagrange multipliers is achieved by only following the control of the average trajectory. Here we present the application of these ideas to the classical control of chaotic Hamiltonian dynamics, where control is realized by a small external interaction. An important application of this approach is to finding regular orbits amongst a dense set of irregular trajectories.

Hamiltonian Chaotic Dynamics

The methodology introduced below is quit general, but to aid its presentation we develop it in the context of controlling the motion of a particular Hamiltonian [BRR95]

$$H = \frac{1}{2}(P_1^2 + P_2^2 + R_1^4 + R_2^4) - KR_1^2R_2^2, \qquad (3.57)$$

where K is a coupling constant between R_1 and R_2. The Hamiltonian in (4.33) presents certain characteristics such as scaling, where the motion of the system at any energy can be determined from $E = 1$ by simple scaling relations.

This latter scaling property can be shown through Poincaré sections and from this approach it is also possible to see the transition to chaos. The *Poincaré section* is constructed as follows. The Hamiltonian is associated with a 4D phase–space (R_1, R_2, P_1, P_2). We can reduce this to a two dimensional phase–space by fixing the energy and one variable, for instance $R_1 = 0$. With this method, the Poincaré sections permit us to establish the character of the system motion at each desired energy and value of the parameter K. When we increase K, the cross term in this Hamiltonian increases its impact over the un–coupled quartic terms. Thus, for a fixed energy, K can be used as a perturbation control parameter, showing the scaling properties of the energy at different values of K.

Optimal Control Algorithm

The dynamics equation for the Hamiltonian at very strong coupling constant, in the chaotic regime, can be modified by the a small perturbation from an external interaction field. This external perturbation can guide the chaotic motion towards the average trajectory of an ensemble or draw the system towards a periodic or quasi–periodic orbit(if it exists). Based on this control idea, we introduce an interaction term for controlling the chaotic motion as [BRR95]

$$H_{int}(R(t), \epsilon(t)) = (R_1(t) - \eta R_2(t))\epsilon(t) \tag{3.58}$$

where η is a parameter. This term is similar to previous work in the control of chaos, where the parameter η is zero [SR91]. It is physically desirable to keep this external interaction small; an intense external interaction could itself lead to chaotic motion.

Let us define a set of N trajectories corresponding to the initial conditions as a vector $w(t) = [w_1(t), w_2(t)...w_{4N}(t)]$ where the first N values are the positions for $R_1(t)$, $N + 1$ to $2N$ the positions of $R_2(t)$, $2N + 1$ to $3N$ the values for $P_1(t)$ and the last values $3N + 1$ to $4N$ for $P_2(t)$. With this definition the classical dynamics is governed by Hamiltonian's equations,

$$\dot{w}_1(t) = g_1(w(t), \epsilon(t)),$$
$$\dot{w}_2(t) = g_2(w(t), \epsilon(t)),$$
$$... \qquad ... \tag{3.59}$$
$$\dot{w}_{4N}(t) = g_{4N}(\mathbf{w(t)}, \epsilon(t)),$$

which depend on the dynamical observables and the time dependent field $\epsilon(t)$. The functions $g_i(w(t), \epsilon(t))$ may be readily identified as momentum or coordinate derivatives of the Hamiltonian. Alteration of the external control interaction $\epsilon(t)$ can steer about the classical dynamics and change its characteristic behavior.

The classical average of the coordinates and momenta are [BRR95]

$$\langle z_i(t) \rangle = \frac{1}{N} \sum_{j=(i-1)N+1}^{iN} w_j(t), \qquad (i = 1..4), \tag{3.60}$$

and the average approximate energy as a function of time is

$$E_l(t) = \frac{1}{N} \sum_{j=(l+1)N+1}^{(l+2)N} \frac{w_j^2(t)}{2} + \frac{1}{N} \sum_{j=(l-1)N+1}^{lN} V(w_j(t)), \tag{3.61}$$

where $l = 1$ and 2 labels the two degrees of freedom and $V(w_j(t))$ represents the un–coupled quartic term in the (4.33). This energy is approximate because it does not account for the nonlinear coupling term. With these definitions, we can proceed to define the control cost functional.

In the formulation of the classical optimal control problem, we first identify the physical objective and constraints. The objective at time T is to control the ensemble of trajectories by requiring that they are confined to a small region in the phase–space. In addition to the final time goal, another physical objective is to require that the chaotic trajectories evolve to either be close to the average trajectory $Z_i(t) = \langle z_i(t) \rangle$ or be confined around an imposed fiducial trajectory $Z_i(t)$. This fiducial trajectory may be a periodic or quasi–periodic trajectory. The cost functional, in quadratic form, that represents the objective and cost is [BRR95]

$$J[w, \epsilon] = \sum_{i=1}^{4} \sigma_i(\langle z_i(T) \rangle - \gamma_i)^2 + \frac{1}{N} \sum_{i=1}^{4} \sum_{j=(i-1)N+1}^{iN} \varrho_i(w_j(T) - Z_i(T))^2$$

$$+ \frac{1}{N} \int_0^T dt \sum_{i=1}^{4} \sum_{j=(i-1)N+1}^{iN} W_i(w_j(t) - Z_i(t))^2 + w_e \int_0^T dt \epsilon(t)^2, \qquad (3.62)$$

where W_i, ω_e, σ_i and ϱ_i are positive weights, which balance the importance of each term in the cost functional. The notation $J[w, \epsilon]$ indicated the functional dependence on $w(t), \epsilon(t)$ for $0 \le t \le T$. γ_i is the a specified constant target for each average variable in the phase–space. This cost functional can be decomposed into a sum of four terms where each of them represents the component average in the phase–space. $J[w, \epsilon]$ is to be minimized with respect to the external interaction subject to satisfying the equations of motion. This latter constraint can be satisfied by introducing Lagrange multipliers, $\lambda_i(t)$, to get an unconstrained functional

$$\bar{J}[w, \epsilon] = J[w, \epsilon] - \frac{1}{N} \int_0^T dt \sum_{i=1}^{4} \lambda_i(t)[\sum_{j=(i-1)N+1}^{iN} \dot{w}_j(t) - g_j(w(t), \epsilon(t))]. \quad (3.63)$$

Note that there are four Lagrange multipliers corresponding to constraints on the average equation of motion. These four Lagrange multipliers are introduced to be consistent with the cost functional (3.62) that represent the sum of four terms over each average component on the phase–space. The classical variational problem for the N initial conditions is given by

$$\epsilon(t) \to \epsilon(t) + \delta\epsilon(t), \qquad w_j(t) \to w_j(t) + \delta w_j(t), \qquad \lambda_i(t) \to \lambda_i(t) + \delta\lambda_i(t).$$

The variation of $\lambda_i(t)$ yields the average Hamiltonian equations of motion. The variation of $w_j(t)$ and $\epsilon(t)$, gives the following equations

$$\delta \bar{J}[w, \epsilon] = \sum_{i=1}^{4} \lambda_i(T)\delta\langle z_i(T)\rangle - 2\sum_{i=1}^{4} \sigma_i[\langle z_i(T)\rangle - \gamma_i]\delta\langle z_i(T)\rangle \qquad (3.64)$$

$$-\frac{2}{N}\sum_{i=1}^{4}\sum_{j=(i-1)N+1}^{iN} \varrho_i[w_j(T) - Z_i(T)][\delta w_j(T) - \delta Z_i(T)]$$

$$+\int_0^T dt \sum_{i=1}^{4}\sum_{j=(i-1)N+1}^{iN} \{\dot{\lambda}_i(t)\delta\langle z_i(t)\rangle + \frac{2W_i}{N}[w_j(t) - Z_i(t)][\delta w_j(t) - \delta Z_i(t)]$$

$$+\frac{\lambda_i(t)}{N}\sum_{k=1}^{4N} \frac{\partial g_j(w(t), \epsilon(t))}{\partial w_k(t)}\delta w_k(t)\}$$

$$+\int_0^T dt[2\omega_e\epsilon(t) + \frac{1}{N}\sum_{i=1}^{4}\sum_{j=(i-1)N+1}^{iN} \lambda_i(t)\frac{\partial g_j(w(t), \epsilon(t))}{\partial \epsilon(t)}]\delta\epsilon(t),$$

where $\delta\langle z_i(t)\rangle$ is given by

$$\delta\langle z_i(t)\rangle = \frac{1}{N}\sum_{j=(i-1)N+1}^{iN} \delta w_j(t),$$

and take account all the variations for each initial condition at each instant of time t. The variation of $Z_i(t)$ is equal to zero if the target trajectory is a fiducial trajectory (constant trajectory in the control process) and $\delta Z_i(t) = \delta\langle z_i(t)\rangle$ if we consider the control evolution over its average trajectory. These two methods to control the chaotic behavior will be explained below.

The final conditions at time T,

$$\lambda_i(T)\delta\langle z_i(T)\rangle = 2\sigma_i[\langle z_i(T)\rangle - \gamma_i]\delta\langle z_i(T)\rangle \qquad (3.65)$$

$$+\frac{2}{N}\sum_{j=(i-1)N+1}^{iN} \varrho_i[w_j(T) - Z_i(T)][\delta w_j(T) - \delta Z_i(T)]$$

are obtained through (3.64). The second variational equation for the Lagrange multipliers is derived from the second part of (3.64),

$$\int_0^T dt \sum_{i=1}^{4}\sum_{j=(i-1)N+1}^{iN} \{\dot{\lambda}_i(t)\delta\langle z_i(t)\rangle + \frac{2W_i}{N}[w_j(t) - Z_i(t)][\delta w_j(t) - \delta Z_i(t)]$$

$$+\frac{\lambda_i(t)}{N}\sum_{k=1}^{4N} \frac{\partial g_j(w(t), \epsilon(t))}{\partial w_k(t)}\delta w_k(t)\} = 0. \qquad (3.66)$$

The gradient with respect to the field is given by

$$\frac{\delta \bar{J}[w, \epsilon]}{\delta\epsilon(t)} = 2\omega_e\epsilon(t) + \frac{1}{N}\sum_{i=1}^{4}\sum_{j=(i-1)N+1}^{iN} \lambda_i(t)\frac{\partial g_j(w(t), \epsilon(t))}{\partial \epsilon(t)}, \qquad (3.67)$$

and the minimum field solution is

$$\epsilon(t) = -\frac{1}{2N\omega_e} \sum_{i=1}^{4} \sum_{j=(i-1)N+1}^{iN} \lambda_i(t) \frac{\partial g_j(w(t), \epsilon(t))}{\partial \epsilon(t)}. \qquad (3.68)$$

This minimum solution links Hamilton's equations with (3.66) for the Lagrange multipliers. In order to integrate the equations for the Lagrange multipliers we need to know the classical evolution for each component $w_j(t)$ and the final condition of the Lagrange multipliers at time T. These final conditions for the Lagrange multipliers in (3.65) are given by the objective in the control dynamics. The solution (if it exists) of these coupled equations prescribes the external control field interaction $\epsilon(t)$.

In order to find a solution of the (3.66), the Lagrange multipliers were chosen to satisfy the following equation

$$\sum_{i=1}^{4} \dot{\lambda}_i(t)\delta\langle z_i(t)\rangle = -\frac{1}{N} \sum_{i=1}^{4} \sum_{j=(i-1)N+1}^{iN} \{2W_i[w_j(t) - Z_i(t)][\delta w_j(t) - \delta Z_i(t)]$$

$$+ \lambda_i(t) \sum_{k=1}^{4N} \frac{\partial g_j(w(t), \epsilon(t))}{\partial w_k(t)} \delta w_k(t)\}, \qquad (3.69)$$

where each Lagrange multiplier gives information about the status of the average or fiducial trajectory $Z_i(t)$. The values of the Lagrange multipliers at each instant of time directs how the external field must evolve. The generalization of these equations is obvious when we increase the dimensionality of the system. Equations (3.69) are over–specified, and a physically motivated assumption on the nature of the solutions must be introduced to close the equations.

There is considerable flexibility in the choice of Lagrange multipliers and within this control formalism we will address two different ways to control the chaotic trajectories. The basic role of the Lagrange multipliers is to guide the physical system towards a desirable solution. We shall take advantage of these observations to develop the two approaches towards control of chaotic trajectories. The first method represents the desire to achieve control around the mean trajectory. In this case the control dynamics searches the phase–space, using the variational principle, finding the mean trajectory and the control field works to keep the ensemble around this trajectory. The second method directs the formerly chaotic trajectories towards a periodic or quasi–periodic orbit or other specified fiducial trajectory.

Following the Mean Trajectory

In this case the fiducial trajectory $Z_i(t)$ is not fixed, and can change at each cycle of cost functional minimization. Here the choice is the average trajectory,

$Z_i(t) = \langle z_i(t)\rangle$. Substituting the equations of motion into (3.69) and (3.67), we have [BRR95]

$$\dot{\lambda}_1(t)\delta\langle z_1(t)\rangle = \frac{1}{N}\sum_{j=1}^{N}\delta w_j(t)\{2\lambda_3(t)[3w_j(t)^2 - Kw_{j+N}^2(t)]$$

$$-4K\lambda_4(t)w_j(t)w_{j+N}(t)\}$$

$$-\frac{2W_1}{N}\sum_{j=1}^{N}[w_j(t) - \langle z_1(t)\rangle][\delta w_j(t) - \delta\langle z_1(t)\rangle], \qquad (3.70)$$

$$\dot{\lambda}_2(t)\delta\langle z_2(t)\rangle = \frac{1}{N}\sum_{j=1}^{N}\delta w_{j+N}(t)\{2\lambda_4(t)[3w_{j+N}^2(t) - Ku_j^2(t)]$$

$$-4K\lambda_3(t)w_j(t)w_{j+N}(t)\}$$

$$-\frac{2W_2}{N}\sum_{j=1}^{N}[w_{j+N}(t) - \langle z_2(t)\rangle][\delta w_{j+N}(t) - \delta\langle z_2(t)\rangle],$$

$$\dot{\lambda}_3(t)\delta\langle z_3(t)\rangle = -\lambda_1(t)\delta\langle z_3(t)\rangle - \frac{2W_3}{N}\sum_{j=1}^{N}[w_{j+2N}(t)$$

$$-\langle z_3(t)\rangle][\delta w_{j+2N}(t) - \delta\langle z_3(t)\rangle],$$

$$\dot{\lambda}_4(t)\delta\langle z_4(t)\rangle = -\lambda_2(t)\delta\langle z_4(t)\rangle - \frac{2W_4}{N}\sum_{j=1}^{N}[w_{j+3N}(t)$$

$$-\langle z_4(t)\rangle][\delta w_{j+3N}(t) - \delta\langle z_4(t)\rangle],$$

and the gradient of the cost functional with respect of the external field is

$$\frac{\delta\bar{J}[w(t),\epsilon(t)]}{\delta\epsilon(t)} = 2\omega_e\epsilon(t) - \lambda_3(t) + \eta\lambda_4(t). \qquad (3.71)$$

This equation only depends on the values of the Lagrange multipliers, $\lambda_3(t)$, $\lambda_4(t)$, that represent the driving force. In this approach the control dynamics equations to be solved result from (3.71), (3.70) and the final conditions for the Lagrange multipliers are given by (3.65).

When the classical trajectories are very close or are attracted to a regular orbit, then these control equations for the Lagrange multipliers can be closed and written as

$$\dot{\lambda}_1(t) = 2\lambda_3(t)[3\langle z_1(t)\rangle^2 - K\langle z_2(t)\rangle^2] - 4K\lambda_4(t)\langle z_1(t)\rangle\langle z_2(t)\rangle,$$
$$\dot{\lambda}_2(t) = 2\lambda_4(t)[3\langle z_2(t)\rangle^2 - K\langle z_1(t)\rangle^2] - 4K\lambda_3(t)\langle z_1(t)\rangle\langle z_2(t)\rangle,$$
$$\dot{\lambda}_3(t) = -\lambda_1(t), \qquad \dot{\lambda}_4(t) = -\lambda_2(t). \qquad (3.72)$$

This gives the control dynamics equations for one effective classical trajectory, in this case the average trajectory. Thus, the system to solve is (3.71) and (3.72) with boundary conditions (3.65).

Following a Fixed Trajectory

This case treats the desire to draw the chaotic trajectories around a specified fiducial trajectory. This trajectory is the target trajectory, $Z_i(t)$. In the present case the choice of this trajectory is the average trajectory produced from the original ensemble evolving from the initial conditions. The control dynamics equations are [BRR95]

$$\dot{\lambda}_1(t)\delta\langle z_1(t)\rangle = \frac{1}{N}\sum_{j=1}^{N}\delta w_j(t)\{2\lambda_3(t)[3w_j(t)^2 - Kw_{j+N}^2(t)]$$

$$-4K\lambda_4(t)w_j(t)w_{j+N}(t)\}$$

$$-\frac{2W_1}{N}\sum_{j=1}^{N}[w_j(t) - Z_1(t)]\delta w_j(t),$$

$$\dot{\lambda}_2(t)\delta\langle z_2(t)\rangle = \frac{1}{N}\sum_{j=1}^{N}\delta w_{j+N}(t)\{2\lambda_4(t)[3w_{j+N}^2(t) - Kw_j^2(t)]$$

$$-4K\lambda_3(t)w_j(t)w_{j+N}(t)\}$$

$$-\frac{2W_2}{N}\sum_{j=1}^{N}[w_{j+N}(t) - Z_2(t)]\delta w_{j+N}(t),$$

$$\dot{\lambda}_3(t)\delta\langle z_3(t)\rangle = -\lambda_1(t)\delta\langle z_3(t)\rangle - \frac{2W_3}{N}\sum_{j=1}^{N}[w_{j+2N}(t) - Z_3(t)]\delta w_{j+2N} \quad (3.73)$$

$$\dot{\lambda}_4(t)\delta\langle z_4(t)\rangle = -\lambda_2(t)\delta\langle z_4(t)\rangle - \frac{2W_4}{N}\sum_{j=1}^{N}[w_{j+3N}(t) - Z_4(t)]\delta w_{j+3N},$$

with the final conditions for the Lagrange multipliers as

$$\lambda_i(T)\delta\langle z_i(T)\rangle = 2\sigma_i[\langle z_i(T)\rangle - \gamma_i]\delta\langle z_i(T)\rangle + \frac{2}{N}\sum_{j=(i-1)N+1}^{iN}\varrho_i[w_j(T) - Z_i(T)]\delta w_j(T).$$

$$(3.74)$$

In this case the system of equations to solve are (3.71), (3.73) and the boundary conditions (3.74).

Once again, the self consistency of (3.73) needs to be addressed and here we seek the non-chaotic result of each trajectory closely following the corresponding initially specified mean $Z_i(t)$. Then for this one classical trajectory, we get the following equations

$$\dot{\lambda}_1(t) = 2\lambda_3(t)[3\langle z_1(t)\rangle^2 - K\langle z_2(t)\rangle^2]$$
$$-4K\lambda_4(t)\langle z_1(t)\rangle\langle z_2(t)\rangle - 2W_1[\langle z_1(t)\rangle - Z_1(t)],$$
$$\dot{\lambda}_2(t) = 2\lambda_4(t)[3\langle z_2(t)\rangle^2 - K\langle z_1(t)\rangle^2]$$
$$-4K\lambda_3(t)\langle z_1(t)\rangle\langle z_2(t)\rangle - 2W_2[\langle z_2(t)\rangle - Z_2(t)],$$
$$\dot{\lambda}_3(t) = -\lambda_1(t) - 2W_3[\langle z_3(t)\rangle - Z_3(t)], \tag{3.75}$$
$$\dot{\lambda}_4(t) = -\lambda_2(t) - 2W_4[\langle z_4(t)\rangle - Z_4(t)].$$

This case expresses the desire and assumption that the formally chaotic trajectories are tightly drawn around the fiducial trajectory. The system to solve is (3.75), (3.71) and (3.74).

Utilizing these two methods to control the irregular motion using optimal control theory, we present two different algorithms in order to find the optimal solution $\epsilon(t)$. The first algorithm returns to the variational principle over all classical trajectories as in (3.69). This evaluation of the Lagrange multipliers takes account the single variation of each component in the phase–space. The procedure is implemented by evaluating the trajectory variations as

$$\delta w_j(t) \simeq w_j(t) - w_j^{old}(t) \tag{3.76}$$

at each instant of time from two successive steps of the minimization process ($w_j^{old}(t)$ is the value in the previous minimization step). This first algorithm is a rigorous test (accepting the finite difference nature of (3.76)) of the arguments leading to the approximate Lagrange equations (3.72) and (3.75).

The second algorithm accepts the approximate form of (3.72) and (3.75) to ultimately yield the field $\epsilon(t)$. Although the average trajectory is used in this approach, the full dynamical equations for $w_j(t)$ are followed and used to evaluate the cost functional $J[\mathbf{w}, \epsilon]$ attesting to the quality of the results.

Computational Method

The optimal control of N classical trajectories is given by the solution of the (3.70), (3.65) and (3.71) for following the average trajectory, and (3.73), (3.74) and (3.71) for a fixed fiducial trajectory.

The following iterative scheme is adopted to find the external control field $\epsilon(t)$ that meets the physical objectives, for N classical trajectories:

a) Make an initial guess for the external interaction $\epsilon(t)$.

b) Integrate the equation of motion (3.59) forward in time for all initial conditions.

c) Calculate the cost functional and boundary conditions for the Lagrange multipliers $\lambda(T)$.

d) Integrate the Lagrange multiplier $\lambda(t)$ equation backwards in time.

f) Calculate the gradient and save all the classical trajectories as the 'old' configuration.

g) Upgrade the new external interaction as [BRR95]

$$\epsilon^{new}(t) = \epsilon(t) - \alpha \frac{\delta \bar{J}}{\delta \epsilon(t)},$$

where α is a suitable small positive constant. This process is repeated until the cost functional and field converge. In one of the algorithmic approaches we need to calculate the variation of $w_j(t)$ as

$$\delta w_j(t) \simeq w_j(t) - w_j^{old}(t),$$

where $w_j^{old}(t)$ is the previous value. In this case at the beginning of the minimization procedure, we choose one classical trajectory at random from the set of initial conditions as the 'old' configuration.

The procedure starts by examining a Poincaré section and choosing one point and giving a random distribution around this point (e.g., $R_2 \pm |\delta|$, $P_2 \pm |\delta|$), where δ is a random number between 0 and 1×10^{-3} (or another small number). After two cycles (one cycle is defined as the apparent quasi–period of the $R_1(t)$ motion at the beginning of the simulation), the system starts its chaotic behavior and the motion is bounded. The approximate energy for each degree of freedom reflects the irregular behavior with strong coupling between them. In five cycles the system is completely irregular, and the objective is to control this irregular motion by keeping the classical trajectories close during a time interval of 5 cycles; no demand is made on the behavior greater than 5 cycles, although, it is also possible to apply this method for more that 5 cycles [BRR95].

4

Phase Transitions and Synergetics

Like the previous Chapter, this one also addresses dynamical *sources of chaos* and the means of *chaos control*. However, here we are explicitly dealing with a high–dimensional chaos (in terms of phase transitions, both classical–Landau's and modern–topological), and a high–dimensional control (in terms of Haken's synergetics and related techniques).

4.1 Phase Transitions, Partition Function and Noise

In this section we present the basic principles of equilibrium phase transitions, which are used in *synergetics*, *quantum biodynamics* and *topological biodynamics*. We also give a brief on *partition function*, the most important quantity in statistical mechanics.

4.1.1 Equilibrium Phase Transitions

In *thermodynamics*, a *phase transition* represents the transformation of a system from one phase to another. Here the therm *phase* denotes a set of states of a macroscopic physical system that have relatively uniform chemical composition and physical properties (i.e., density, crystal structure, index of refraction, and so forth.) The most familiar examples of phases are solids, liquids, and gases. Less familiar phases include plasmas, Bose–Einstein condensates and fermionic condensates and the paramagnetic and ferromagnetic phases of magnetic materials.

In essence, all thermodynamic properties of a system (entropy, heat capacity, magnetization, compressibility, and so forth) may be expressed in terms of the *free energy potential* $\mathcal{F}$ and its partial derivatives. For example, the *entropy* S is the first derivative of the free energy $\mathcal{F}$ with respect to the temperature T, i.e., $S = -\partial \mathcal{F}/\partial T$, while the *specific heat capacity* C is the second derivative, $C = T\,\partial S/\partial T$. As long as the free energy $\mathcal{F}$ remains analytic, all the thermodynamic properties will be well–behaved.

Now, the distinguishing characteristic of a phase transition is an *abrupt sudden change in one or more physical properties*, in particular the specific heat c, with a small change in a thermodynamic variable such as the temperature T. Standard examples of phase transitions are (see e.g., [LL78, Wik05]):

1. The transitions between the solid, liquid, and gaseous phases (boiling, melting, sublimation, etc.)
2. The transition between the ferromagnetic and paramagnetic phases of magnetic materials at the Curie point.
3. The emergence of superconductivity in certain metals when cooled below a critical temperature.
4. Quantum condensation of bosonic fluids, such as Bose–Einstein condensation and the superfluid transition in liquid helium.
5. The breaking of symmetries in the laws of physics during the early history of the universe as its temperature cooled.

When a system goes from one phase to another, there will generally be a stage where the free energy is non–analytic. This is known as a phase transition. Familiar examples of phase transitions are melting (solid to liquid), freezing (liquid to solid), boiling (liquid to gas), and condensation (gas to liquid). Due to this *non–analyticity*, the free energies on either side of the transition are two different functions, so one or more thermodynamic properties will behave very differently after the transition. The property most commonly examined in this context is the heat capacity. During a transition, the heat capacity may become infinite, jump abruptly to a different value, or exhibit a 'kink' or *discontinuity* in its derivative[1].

Therefore, phase transitions come about when the free energy of a system is non–analytic for some choice of thermodynamic variables. This non–analyticity generally stems from the interactions of an extremely large number of particles in a system, and does not appear in systems that are too small.

4.1.2 Classification of Phase Transitions

Ehrenfest Classification

The first attempt at classifying phase transitions was the *Ehrenfest classification scheme*, which grouped phase transitions based on the degree of non-

---

[1] In practice, each type of phase is distinguished by a handful of relevant thermodynamic properties. For example, the distinguishing feature of a solid is its rigidity; unlike a liquid or a gas, a solid does not easily change its shape. Liquids are distinct from gases because they have much lower compressibility: a gas in a large container fills the container, whereas a liquid forms a puddle in the bottom. Not all the properties of solids, liquids, and gases are distinct; for example, it is not useful to compare their magnetic properties. On the other hand, the ferromagnetic phase of a magnetic material is distinguished from the paramagnetic phase by the presence of bulk magnetization without an applied magnetic field.

analyticity involved. Though useful, Ehrenfest's classification is flawed, as we will discuss in the next section.

Under this scheme, phase transitions were labelled by the lowest partial derivative of the free energy that is discontinuous at the transition. *First-order phase transitions* exhibit a discontinuity in the first derivative of the free energy with respect to a thermodynamic variable. The various solid⇒liquid⇒gas transitions are classified as first–order transitions, as the density, which is the first partial derivative of the free energy with respect to chemical potential, changes discontinuously across the transitions[2]. *Second-order phase transitions* have a discontinuity in a second derivative of the free energy. These include the ferromagnetic phase transition in materials such as iron, where the magnetization, which is the first derivative of the free energy with the applied magnetic field strength, increases continuously from zero as the temperature is lowered below the Curie temperature. The magnetic susceptibility, the second derivative of the free energy with the field, changes discontinuously. Under the *Ehrenfest classification scheme*, there could in principle be third, fourth, and higher–order phase transitions.

Modern Classification

The Ehrenfest scheme is an inaccurate method of classifying phase transitions, for it is based on the *mean–field theory* of phases, which is inaccurate in the vicinity of phase transitions, as it neglects the role of thermodynamic fluctuations. For instance, it predicts a finite discontinuity in the heat capacity at the ferromagnetic transition, which is implied by Ehrenfest's definition of second–order transitions. In real ferromagnets, the heat capacity diverges to infinity at the transition.

In the modern classification scheme, phase transitions are divided into two broad categories, named similarly to the Ehrenfest classes:

- The *first–order phase transitions*, or, *discontinuous phase transitions*, are those that involve a latent heat. During such a transition, a system either absorbs or releases a fixed (and typically large) amount of energy. Because energy cannot be instantaneously transferred between the system and its environment, first–order transitions are associated with *mixed–phase regimes* in which some parts of the system have completed the transition and others have not. This phenomenon is familiar to anyone who has boiled a pot of water: the water does not instantly turn into gas, but forms a turbulent mixture of water and water vapor bubbles. Mixed–phase systems are difficult to study, because their dynamics are violent and hard to control. However, many important phase transitions fall in this category, including the solid⇒liquid⇒gas transitions.
- The *second–order phase transitions* are the *continuous phase transitions*. These have no associated latent heat. Examples of second–order phase

---

[2] The pressure must be continuous across the phase boundary in equilibrium.

transitions are the ferromagnetic transition, the superfluid transition, and Bose–Einstein condensation.

4.1.3 Basic Properties of Phase Transitions

Critical Points

In systems containing liquid and gaseous phases, there exist a special combination of pressure and temperature, known as the *critical point*, at which the transition between liquid and gas becomes a second–order transition. Near the critical point, the fluid is sufficiently hot and compressed that the distinction between the liquid and gaseous phases is almost non–existent.

This is associated with the phenomenon of critical opalescence, a milky appearance of the liquid, due to density fluctuations at all possible wavelengths (including those of visible light).

Symmetry

Phase transitions often (but not always) take place between phases with different symmetry. Consider, for example, the transition between a fluid (i.e., liquid or gas) and a crystalline solid. A fluid, which is composed of atoms arranged in a disordered but homogenous manner, possesses continuous translational symmetry: each point inside the fluid has the same properties as any other point. A crystalline solid, on the other hand, is made up of atoms arranged in a regular lattice. Each point in the solid is *not* similar to other points, unless those points are displaced by an amount equal to some lattice spacing.

Generally, we may speak of one phase in a phase transition as being more symmetrical than the other. The transition from the more symmetrical phase to the less symmetrical one is a *symmetry–breaking* process. In the fluid–solid transition, for example, we say that continuous translation symmetry is broken.

The ferromagnetic transition is another example of a symmetry–breaking transition, in this case the symmetry under reversal of the direction of electric currents and magnetic field lines. This symmetry is referred to as 'up–down symmetry' or 'time–reversal symmetry'. It is broken in the ferromagnetic phase due to the formation of magnetic domains containing aligned magnetic moments. Inside each domain, there is a magnetic field pointing in a fixed direction chosen spontaneously during the phase transition. The name *time–reversal symmetry* comes from the fact that electric currents reverse direction when the time coordinate is reversed.

The presence of symmetry–breaking (or nonbreaking) is important to the behavior of phase transitions. It was pointed out by Landau that, given any state of a system, one may unequivocally say whether or not it possesses a given symmetry [LL78]. Therefore, it cannot be possible to analytically deform a state in one phase into a phase possessing a different symmetry. This means, for example, that it is impossible for the solid–liquid phase boundary to end in a critical point like the liquid–gas boundary. However, symmetry–breaking transitions can still be either first or second order.

Typically, the more symmetrical phase is on the high–temperature side of a phase transition, and the less symmetrical phase on the low–temperature side. This is certainly the case for the solid–fluid and ferromagnetic transitions. This happens because the Hamiltonian of a system usually exhibits all the possible symmetries of the system, whereas the low–energy states lack some of these symmetries (this phenomenon is known as spontaneous symmetry breaking.) At low temperatures, the system tends to be confined to the low–energy states. At higher temperatures, thermal fluctuations allow the system to access states in a broader range of energy, and thus more of the symmetries of the Hamiltonian.

When symmetry is broken, one needs to introduce one or more extra variables to describe the state of the system. For example, in the ferromagnetic phase one must provide the net magnetization, whose direction was spontaneously chosen when the system cooled below the Curie point. Such variables are instances of *order parameters*. However, note that order parameters can also be defined for symmetry–nonbreaking transitions.

Symmetry–breaking phase transitions play an important role in cosmology. It has been speculated that, in the hot early universe, the vacuum (i.e., the various quantum fields that fill space) possessed a large number of symmetries. As the universe expanded and cooled, the vacuum underwent a series of symmetry–breaking phase transitions. For example, the electroweak transition broke the $SU(2) \times U(1)$ symmetry of the electroweak field into the $U(1)$ symmetry of the present–day electromagnetic field. This transition is important to understanding the asymmetry between the amount of matter and antimatter in the present–day universe.

Critical Exponents and Universality Classes

Continuous phase transitions are easier to study than first–order transitions due to the absence of latent heat, and they have been discovered to have many interesting properties. The phenomena associated with continuous phase transitions are called *critical phenomena*, due to their association with critical points.

It turns out that continuous phase transitions can be characterized by parameters known as critical exponents. For instance, let us examine the behavior of the heat capacity near such a transition. We vary the temperature T of the system while keeping all the other thermodynamic variables fixed, and find that the transition occurs at some critical temperature T_c. When T is near T_c, the heat capacity C typically has a *power law* behavior: $C \sim |T_c - T|^{-\alpha}$. Here, the constant α is the critical exponent associated with the heat capacity. It is not difficult to see that it must be less than 1 in order for the transition to have no latent heat. Its actual value depends on the type of phase transition we are considering. For $-1 < \alpha < 0$, the heat capacity has a 'kink' at the transition temperature. This is the behavior of liquid helium at the 'lambda transition' from a normal state to the superfluid state, for which experiments have found $\alpha = -0.013 \pm 0.003$. For $0 < \alpha < 1$, the heat capacity diverges at the transition temperature (though, since $\alpha < 1$, the divergence is not strong enough to produce a latent heat.) An example of such behavior is the 3D ferromagnetic phase transition. In the $3D$ Ising model for uniaxial magnets, detailed theoretical studies have yielded the exponent $\alpha \sim 0.110$.

Some model systems do not obey this power law behavior. For example, mean–field theory predicts a finite discontinuity of the heat capacity at the transition temperature, and the 2D Ising model has a logarithmic divergence. However, these systems are an exception to the rule. Real phase transitions exhibit power law behavior.

Several other critical exponents – β, γ, δ, ν, and η – are defined, examining the power law behavior of a measurable physical quantity near the phase transition.

It is a remarkable fact that phase transitions arising in different systems often possess the same set of critical exponents. This phenomenon is known as *universality*. For example, the critical exponents at the liquid–gas critical point have been found to be independent of the chemical composition of the fluid. More amazingly, they are an exact match for the critical exponents of the ferromagnetic phase transition in uniaxial magnets. Such systems are said to be in the same *universality class*. Universality is a prediction of the renormalization group theory of phase transitions, which states that the thermodynamic properties of a system near a phase transition depend only on a small number of features, such as dimensionality and symmetry, and is insensitive to the underlying microscopic properties of the system.

4.1.4 Landau's Theory of Phase Transitions

Landau's theory of phase transitions is a simple but powerful empirical thermodynamic theory by which the behavior of crystals at phase transitions can be described. It is based simply on a power series expansion of the free energy of the crystal with respect to one or a few prominent parameters distorting the symmetry of the crystal. The symmetry of the distortion decides which terms may be present, and which not. For example, odd terms on the power series expansion often are not allowed because the energy of the system is symmetric with respect to positive or negative distortion. With Landau's theory, the thermodynamics of the crystal (free energy, entropy, heat capacity) can be directly linked to it's structural state (volume, deviation from high symmetry, etc.), and both can be described as they change as a function of temperature or pressure.

More precisely, in Landau's theory, the probability (density) distribution function f is exponentially related to the potential $\mathcal{F}$,

$$f \approx e^{-\mathcal{F}(T)}, \tag{4.1}$$

if $\mathcal{F}$ is considered as a function of the order parameter o. Therefore, the most probable order parameter is determined by the requirement $\mathcal{F} = \min$.

When $M_\uparrow$ elementary magnets point upwards and $M_\downarrow$ elementary magnets point downwards, the magnetization order parameter o is given by

$$o = (M_\uparrow - M_\downarrow)\, m, \tag{4.2}$$

where m is the magnetic moment of a single elementary magnet.

We expand the potential $\mathcal{F} = \mathcal{F}(o, T)$ into a power series of o,

$$\mathcal{F}(o, T) = \mathcal{F}(0, T) + \mathcal{F}'(0, T)o + \cdots + \frac{1}{4!}\mathcal{F}''''(0, T)o^4 + \cdots, \tag{4.3}$$

and discus $\mathcal{F}$ as a function of o. In a number of cases $\mathcal{F}' = \mathcal{F}''' = 0$, due to inversion symmetry. In this case, $\mathcal{F}$ has the form

$$\mathcal{F}(o, T) = \mathcal{F}(0, T) + \frac{\sigma}{2}o^2 + \frac{\beta}{4}o^4, \tag{4.4}$$

where $\beta > 0$, and $\sigma = a(T - T_c)$, $(a > 0)$, i.e., it changes its sign at the critical temperature $T = T_c$.

Recall that the (negative) first partial derivative of the free energy potential $\mathcal{F}$ with respect to the *control parameter* – temperature T is the entropy

$$S = -\frac{\partial \mathcal{F}(q, T)}{\partial T}. \tag{4.5}$$

For $T > T_c$, $\sigma > 0$, and the minimum of $\mathcal{F}$ lies at $o = o_0 = 0$, and

$$S = S_0 = -\frac{\partial \mathcal{F}(0, T)}{\partial T}.$$

Also recall that the second partial derivative of $\mathcal{F}$ with respect to T is the specific heat capacity (besides the factor T)

$$C = T\frac{\partial S}{\partial T}. \tag{4.6}$$

One may readily check that S is continuous at $T = T_c$ for $\sigma = 0$. However, when we calculate the specific heat we get two different expressions above and below the critical temperature and thus a discontinuity at $T = T_c$.

Closely related to the Landau's theory of phase transitions is *Ginzburg–Landau model of superconductivity* (named after *Nobel Laureates Vitaly L. Ginzburg and Lev D. Landau*). It does not purport to explain the microscopic mechanisms giving rise to superconductivity. Instead, it examines the macroscopic properties of a superconductor with the aid of general thermodynamic arguments. Based on Landau's previously–established theory of second–order phase transitions, Landau and Ginzburg argued that the free energy F of a superconductor near the superconducting transition can be expressed in terms of a complex *order parameter* ψ, which describes how deep into the superconducting phase the system is. By minimizing the free energy with respect to fluctuations in the order parameter and the vector potential, one arrives at the *Ginzburg–Landau equation*, which is a generalization of the *nonlinear Schrödinger equation*.

4.1.5 Partition Function

Recall that in *statistical mechanics*, the *partition function Z* is used for statistical description of a system in thermodynamic equilibrium. Z depends on the physical system under consideration and is a function of temperature T as well as other parameters (such as volume V enclosing a gas etc.). The partition function forms the basis for most calculations in statistical mechanics. It is most easily formulated in *quantum statistical mechanics* (see e.g., [Fey72]).

Classical Partition Function

A system subdivided into N subsystems, where each subsystem (e.g., a particle) can attain any of the energies ϵ_j $(j = 1, ..., N)$, has the partition function given by the sum of its Boltzmann factors,

$$\zeta = \sum_{j=0}^{\infty} e^{-\beta \epsilon_j},$$

where $\beta = \frac{1}{k_B T}$ and k_B is *Boltzmann constant*. The interpretation of ζ is that the probability that the subsystem will have energy ϵ_j is $e^{-\beta \epsilon_j}/\zeta$. When the number of energies ϵ_j is definite (e.g., particles with spin in a crystal lattice under an external magnetic field), then the indefinite sum is replaced with a definite sum. However, the total partition function for the system containing N subsystems is of the form

$$Z = \prod_{j=1}^{N} \zeta_j = \zeta_1 \zeta_2 \zeta_3 \cdot ...,$$

where ζ_j is the partition function for the jth subsystem. Another approach is to sum over all system's total energy states,

$$Z = \sum_{r=1}^{N} e^{-\beta E_r}, \qquad \text{where} \qquad E_j = n_1^{(j)} \epsilon_1 + n_2^{(j)} \epsilon_2 + ...$$

In case of a system containing N non–interacting subsystems (e.g., a real gas), the system's partition function is given by

$$Z = \frac{1}{N!}\zeta^N.$$

This equation also has the more general form

$$Z = \frac{1}{N!h^{3N}} \int \prod_{i=1}^{N} d^3q^i d^3p_i \sum_{i=1}^{N} e^{-\beta H_i},$$

where $H_i = H_i(q^i, p_i)$ is the ith subsystem's Hamiltonian, while h^{3N} is a normalization factor.

Given the partition function Z, the system's *free energy* F is defined as

$$F = -k_B T \ln Z,$$

while the *average energy* U is given by

$$U = \frac{1}{Z} E_i e^{-\frac{E_i}{k_B T}} = -\frac{d}{d\beta}(\ln Z).$$

Liner Harmonic Oscillators in Thermal Equilibrium. The partition function Z, free energy F, and average energy U of the system of M oscillators can be found as follows: The oscillators do not interact with each other, but only with the *heat bath*. Since each oscillator is independent, one can find F_i of the ith oscillator and then $F = \sum_{i=1}^{M} F_i$. For each ith oscillator (that can be in one of N states) we have [Fey72]

$$Z_i = \sum_{n=1}^{N} e^{-\frac{E_n^i}{k_B T}}, \qquad F_i = -k_B T \ln Z_i, \qquad U_i = \frac{1}{Z} \sum_{n=1}^{N} E_n^i e^{-\frac{E_n^i}{k_B T}}.$$

Quantum Partition Function

Partition function Z of a quantum–mechanical system may be written as a trace over all states (which may be carried out in any basis, as the trace is basis–independent),

$$Z = \mathrm{Tr}(e^{-\beta \hat{H}}),$$

where $\hat{H}$ is the system's Hamiltonian operator. If $\hat{H}$ contains a dependence on a parameter λ, as in $\hat{H} = \hat{H}_0 + \lambda \hat{A}$, then the statistical average over $\hat{A}$ may be found from the dependence of the partition function on the parameter, by differentiation,

$$< \hat{A} > = -\beta^{-1} \frac{d}{d\lambda} \ln Z(\beta, \lambda).$$

However, if one is interested in the average of an operator that does not appear in the Hamiltonian, one often adds it artificially to the Hamiltonian, calculates Z as a function of the extra new parameter and sets the parameter equal to zero after differentiation.

More general, in *quantum field theory*, we have a *generating functional J* of the *field $\phi(q)$* and the partition function is usually expressed by the *Feynman path integral* [Fey72]

$$Z[J] = \int \mathcal{D}\phi e^{i(S[\phi] + \int d^N q \, J(q)\phi(q))},$$

where $S = S[\phi]$ is the *field action functional*.

Vibrations of Coupled Oscillators

In this subsection, following [Fey72], we give both classical and quantum analysis of vibrations of coupled oscillators. R. Feynman used this method as a generic model for the crystal lattice.

Consider a *crystal lattice* with A atoms per unit cell, such that $3A$ coordinates α must be given to locate each atom. Also let $Q_{\alpha,N}$ denote the *displacement* from equilibrium of the coordinate α in the Nth cell. $Q_{\alpha,N+M}$ is the displacement of an atom in a cell close to N.

The *kinetic energy* of the lattice is given by

$$T = \frac{1}{2} Q_{\alpha,N} Q_{\alpha,N},$$

while its *potential energy* (in linear approximation) is given by

$$V = \frac{1}{2} C^M_{\alpha\beta} Q_{\alpha,N} Q_{\beta,N+M}.$$

Classical Problem

The so–called *original Hamiltonian* is given by

$$\bar{H} = \sum_i \frac{p_i\prime^2}{2m_i} + \frac{1}{2} C_{ij}\prime q^i\prime q^j\prime,$$

where the $q^i\prime$ are the coordinates of the amount of the lattice displacement from its equilibrium, $p_i\prime = m_i \dot{q}^i\prime$ are the canonical momenta, and $C_{ij}\prime = C_{ji}\prime$ are constants. To eliminate the mass constants m_i, let

$$q^i = q^i\prime \sqrt{m_i} \qquad \text{and} \qquad C_{ij} = \frac{C_{ij}\prime}{\sqrt{m_i m_j}}.$$

Then

$$p_i = \frac{\partial L}{\partial \dot{q}^i} = \frac{p_i\prime}{\sqrt{m_i}}, \qquad (L \text{ is the Lagrangian of the system})$$

and we get the *simplified Hamiltonian*

$$H = \frac{1}{2}\sum_i p_i^2 + \frac{1}{2}C_{ij}q^i q^j.$$

The Hamilton's equations of motion now read

$$\dot{q}^i = \partial_{p_i} H = p_i, \qquad \dot{p}_i = -\partial_{q^i} H = -C_{ij}q^j = \ddot{q}^i.$$

We now break the motion of the system into *modes*, each of which has its *own frequency* ω. *The total motion of the system is a sum of the motions of the modes.* Let the αth mode have frequency ω_α so that

$$q^i_{(\alpha)} = e^{-i\omega_\alpha t}a_i^{(\alpha)}$$

for the motion of the αth mode, with $a_i^{(\alpha)}$ independent of time. Then

$$\omega_\alpha^2 a_i^{(\alpha)} = C_{ij}a_j^{(\alpha)}.$$

In this way, the classical *problem of vibrations of coupled oscillators* has been reduced to the *problem of finding eigenvalues* and *eigenvectors* of the real, symmetric matrix $\|C_{ij}\|$. In order to get the ω_α we must solve the *characteristic equation*

$$\det \|C_{ij} - \omega^2 \delta_{ij}\| = 0.$$

Then the eigenvectors $a_i^{(\alpha)}$ can be found. It is possible to choose the $a_i^{(\alpha)}$ so that

$$a_i^{(\alpha)}a_i^{(\beta)} = \delta_{\alpha\beta}.$$

The general solution for q^i is

$$q^i = C_\alpha q^i_{(\alpha)},$$

where the C_α are arbitrary constants. If we take

$$Q_\alpha = C_\alpha e^{-i\omega_\alpha t},$$

we get

$$q^i = a_i^{(\alpha)}Q_\alpha.$$

From this it follows that

$$a_i^{(j)}q^i = a_i^{(j)}a_i^{(\alpha)}Q_\alpha = \delta_{\alpha j}Q_\alpha = Q_j.$$

Making the change of variables, $Q_j = a_i^{(j)}q^i$, we get $H = \sum_\alpha H_\alpha$, where

$$H_\alpha = \frac{1}{2}p_\alpha^2 + \frac{1}{2}\omega_\alpha^2 Q_\alpha.$$

This has the expected solutions: $Q_\alpha = C_\alpha e^{-i\omega_\alpha t}.$

Quantum–Mechanical Problem

Again we have the original Hamiltonian

$$H = \sum_i \frac{p_i'^2}{2m_i} + \frac{1}{2}C_{ij}'q^i{}'q^j{}',$$

where this time

$$p_i' = \frac{1}{\mathrm{i}}\frac{\partial}{\partial q^i{}'} \qquad \text{(in normal units } \hbar = 1\text{).}$$

Making the same change of variables as before, we get

$$Q_\alpha = a_i^{(\alpha)}q^i = a_i^{(\alpha)}\sqrt{m_i}q^i{}',$$

$$H = \sum_\alpha H_\alpha, \qquad \text{where} \qquad H_\alpha = -\frac{1}{2}\frac{\partial^2}{\partial Q_\alpha^2} + \frac{1}{2}\omega_\alpha^2 Q_\alpha.$$

It follows immediately that the eigenvalues of our original Hamiltonian are

$$E = \sum_\alpha (N_\alpha + \frac{1}{2})\omega_\alpha.$$

The solution of a quantum–mechanical system of coupled oscillators is trivial once we have solved the characteristic equation

$$0 = \det \left\| C_{ij} - \omega^2\delta_{ij} \right\| = \det \left\| \frac{C_{ij}'}{\sqrt{m_i m_j}} - \omega^2\delta_{ij} \right\|.$$

If we have a solid with $\frac{1}{3}(10^{23})$ atoms we must apparently find the eigenvalues of a 10^{23} by 10^{23} matrix. But if the solid is *crystal*, the problem is enormously simplified. The *classical Hamiltonian for a crystal* is

$$H = \frac{1}{2}\sum_{\alpha,N}\dot{Q}_{\alpha,N}^2 + \frac{1}{2}\sum_{\alpha,\beta,N,M}C_{\alpha\beta}^M Q_{\alpha,N}Q_{\beta,N+M},$$

and the *classical equation of motion for a crystal lattice* is (using $C_{\alpha\beta}^M = C_{\beta\alpha}^{-M}$)

$$\ddot{Q}_{\alpha,N} = -\sum_{M,\beta}C_{\alpha\beta}^M Q_{\beta,N+M}.$$

In a given mode, if one cell of the crystal is vibrating in a certain manner, it is reasonable to expect all cells to vibrate the same way, but with different phases. So we try

$$Q_{\alpha,N} = a_\alpha(K)\mathrm{e}^{-\mathrm{i}\omega t}\mathrm{e}^{\mathrm{i}K\cdot N},$$

where K expresses the relative phase between cells. The $\mathrm{e}^{\mathrm{i}K\cdot N}$ factor allows for wave motion. We now want to find the dispersion relations, or $\omega = \omega(K)$.

$$\omega^2 a_\alpha e^{iK\cdot N} = \sum_{M,\beta} (C^M_{\alpha\beta} a_\beta e^{iK\cdot M}) e^{iK\cdot N}.$$

Let

$$\gamma_{\alpha\beta}(K) = \sum_M C^M_{\alpha\beta} e^{iK\cdot M}$$

(note that $\gamma_{\alpha\beta}(K)$ is Hermitian).

Then $\omega^2 a_\alpha = \sum_\beta \gamma_{\alpha\beta} a_\beta$, and we must solve the characteristic equation of a $3A-$by$-3A$ matrix:

$$\det |\gamma_{\alpha\beta} - \omega^2 \delta_{\alpha\beta}| = 0.$$

The solutions of the characteristic equation are

$$\omega^{(r)}(K) = \omega(^r_K),$$

where r runs from 1 to $3A$. The motion of a particular mode can be written

$$Q^{(r)}_{\alpha,N}(K) = a^r_\alpha(K) e^{-i\omega^{(r)}(K)^r} e^{iK\cdot N},$$

where

$$a^r_\alpha a^{*r'}_\alpha = \delta_{rr'}.$$

Then the general motion can be described by

$$Q_{\alpha,N} = \sum_{K,r} C_r(K) a^r_\alpha(K) e^{-i\omega^{(r)}(K)^r} e^{iK\cdot N},$$

where $C_r(K)$ are arbitrary constants.

Let $Q_r(K) = C_r(K) e^{-i\omega^{(r)}(K)^r}$. $Q_r(K)$ describe the motion of a particular mode. Then we have

$$Q_{\alpha,N} = \sum_{K,r} Q_r(K) a^r_\alpha(K) e^{iK\cdot N}.$$

It follows that

$$Q_r(K) \propto \sum_{\alpha,N} Q_{\alpha,N} a^{*r}_\alpha(K) e^{-iK\cdot N}, \qquad (\propto \text{ means 'proportional to'}),$$

and the Hamiltonian for the system is

$$H = \frac{1}{2} \sum_{\alpha,N} \left[\dot{Q}^2_{\alpha,N} + \sum_{\beta,M} C^M_{\alpha\beta} Q_{\alpha,N} Q_{\beta,N+M} \right]$$

$$= \frac{1}{2} \sum_{K,r} \left[|\dot{Q}_r(K)|^2 + \omega^{2(r)}(K) |Q_r(K)|^2 \right].$$

A Cubic Lattice of Harmonic Oscillators

Assume the unit cell to be a cubic lattice with one atom per cell. Each atom behaves as an harmonic oscillator, with spring constants k_A (nearest neighbors), and k_B (diagonal–, or next–nearest neighbors). This case is fairly simple, and we can simplify the notation: $\alpha = 1, 2, 3$.

$$Q_{1,N} = X_N, \qquad Q_{2,N} = Y_N, \qquad Q_{3,N} = Q_N.$$

We wish to find the 3 natural frequencies associated with each k of the crystal. To do this, we must find $C_{\alpha\beta}^M$ and then $\gamma_{\alpha\beta}$. In complex coordinates,

$$V = \sum_{\alpha,\beta} V_{\alpha\beta}, \qquad \text{where} \qquad V_{\alpha\beta} = \sum_{N,M} C_{\alpha\beta}^M Q_{\alpha,N}^* Q_{\beta,N+M},$$

where * denotes complex conjugation. For example,

$$V_{11} = \sum_{N,M} C_{11}^M X_N^* X_{N+M}.$$

If we express the displacement of atom N from its normal position as X_N, then the *potential energy* from the distortion of the spring between atoms N and M is

$$\frac{1}{2} k_M \left[(X_N - X_{N+M}) \cdot \frac{M}{|M|} \right]^2,$$

where $k_M = k_A$ for $N + M$ a nearest neighbor to N, $k_M = k_B$ for $N + M$ a next–nearest neighbor.

In summing over N and M to get the total potential energy we must *divide V by two*, for we count each spring twice. If we use complex coordinates, however, we *multiply V by two* to get the correct equations of motion:

$$V = \frac{1}{2} \sum_{N,M} k_M \left[(X_N - X_{N+M}) \cdot \frac{M}{|M|} \right]^2,$$

$$V_{11} = \frac{1}{2} \sum_{N,M} k_M \left(\frac{M_X}{|M|} \right)^2 (X_N^* - X_{N+M}^*)(X_N - X_{N+M})$$

$$= \frac{1}{2} \sum_{N,M} k_M \left(\frac{M_X}{|M|} \right)^2 \left[(X_N^* X_N + X_{N+M}^* X_{N+M}) - (X_N^* X_{N+M} + X_N X_{N+M}^*) \right]$$

$$= \sum_{N,M} k_M \left(\frac{M_X}{|M|} \right)^2 \left[X_N^* X_N - X_N X_{N+M}^* \right].$$

Comparing the above expressions we see that

$$C_{11}^0 = 2k_A + 4k_B, \qquad C_{11}^{\pm(1,0,0)} = -k_A, \qquad \text{and so on.}$$

In this way, all the $C_{\alpha\beta}^{M}$ can be found. We can then calculate

$$\gamma_{\alpha\beta}(K) = \sum_{M} C_{\alpha\beta}^{M} e^{iK \cdot M}.$$

We wish to solve

$$\det |\gamma_{\alpha\beta} - \omega^2 \delta_{\alpha\beta}| = 0.$$

For each relative phase K, there are 3 solutions for ω. Thus we get $3N$ values of ω and $\omega^{(r)}(K)$.

4.1.6 Noise–Induced Non–equilibrium Phase Transitions

Noise is usually thought of as a phenomenon which perturbs the observation and creates disorder (see section 4.6.1). This idea is based mainly on our day to day experience and, in the context of physical theories, on the study of equilibrium systems. The effect of noise can, however, be quite different in *nonlinear non–equilibrium systems*. Several situations have been documented in the literature, in which the noise actually participates in the creation of ordered states or is responsible for surprising phenomena through its interaction with the nonlinearities of the system [HL84]. Recently, a quite spectacular phenomenon was discovered in a specific model of a spatially distributed system with multiplicative noise, white in space and time. It was found that the noise generates an *ordered symmetry–breaking state* through a genuine *second–order phase transition*, whereas no such transition is observed in the absence of noise [BPT94, BPT97].

Recently it has been shown that a white and *Gaussian multiplicative noise* can lead an *extended* dynamical system (fulfilling appropriate conditions) to undergo a *phase* transition towards an *ordered* state, characterized by a nonzero order parameter and by the breakdown of ergodicity [BPT94]. This result–first got within a Curie–Weiss–like *mean–field approximation*, and further extended to consider the simplest correlation function approach–has been confirmed through extensive numerical simulations [BPT97]. In addition to its *critical* nature as a function of the noise intensity σ, the newly found noise–induced phase transition has the noteworthy feature of being *reentrant*: for each value of D above a threshold one, the ordered state exists only inside a window $[\sigma_1, \sigma_2]$. At variance with the known case of *equilibrium* order⇒disorder transitions that are induced (in the simplest lattice models) by the nearest–neighbor coupling constant D and rely on the bi–stability of the local potential, the transition in the case at hand is led by the *combined effects* of D and σ through the nonlinearities of the system. Neither the zero–dimensional system (corresponding to the $D = 0$ limit) nor the deterministic one ($\sigma = 0$) show any transition.

General Zero–Dimensional System

To smoothly introduce the subject, we will start from the well–known *logistic equation*, and add to it a multiplicative white noise.

Noisy Logistic Equation

Recall that the logistic equation (also called the *Verhulst model* or *logistic growth curve*) is a model of population growth first published by P. Verhulst in 1845 (see [Wei05]). The model is continuous in time, but a modification of the continuous equation to a discrete quadratic recurrence equation known as the *logistic map* is widely used in *chaos theory*. The standard logistic equation

$$\dot{x} = \lambda x - x^2, \tag{4.7}$$

where the parameter λ is usually constrained to be positive, has a solution

$$x(t) = \frac{1}{1 + \left(\frac{1}{x_0} - 1\right) e^{-\lambda t}}.$$

Now, if we add a *multiplicative zero–mean Gaussian white noise* $\xi = \xi(t)$ with *noise intensity* σ to (4.7), we get the *Langevin SDE* (stochastic differential equation)

$$\dot{x} = \lambda x - x^2 + x\,\xi. \tag{4.8}$$

If we apply the *Stratonovitch interpretation* to the Langevin equation (4.8), we get the corresponding *Fokker–Planck equation*

$$\partial_t P(x,t) = -\partial_t \left(\lambda x - x^2\right) P(x,t) + \frac{\sigma^2}{2} \partial_x x \, \partial_x P(x,t) \tag{4.9}$$

derermining the *probability density* $P(x,t)$ for the variable $x(t)$. The equation (4.9) has the *stationary probability density*

$$P_{st}(x) = \frac{1}{Z} x^{\frac{2\lambda}{\sigma^2} - 1} \exp\left(-\frac{2x}{\sigma^2}\right)$$

(where Z is a normalization constant), with *two extrema*:

$$x_1 = 0, \qquad x_2 = \lambda - \frac{\sigma^2}{2}.$$

General Zero–Dimensional Model

Now, following [BPT94, BPT97], we consider the following SDE that generalizes noisy logistic equation (4.8),

$$\dot{x} = f(x) + g(x)\,\xi, \tag{4.10}$$

where, as above, $\xi = \xi(t)$ denotes the Gaussian white noise with first two moments

$$\langle \xi(t) \rangle = 0, \qquad \langle \xi(t)\,\xi(t') \rangle = \sigma^2 \delta(t - t').$$

If we interpret equation (4.10) according to the Stratonovitch interpretation, we get the corresponding Fokker–Planck equation

$$\partial_t P(x,t) = -\partial_x \left[f(x) + P(x,t) \right] + \frac{\sigma^2}{2} \partial_x \left(g(x)\,\partial_x \left[g(x)\,P(x,t) \right] \right),$$

with the *steady–state solution*

$$P_{st}(x) = \frac{1}{Z} \exp \left(\int_0^x \frac{f(y) - \frac{\sigma^2}{2} g(y) g'(y)}{\frac{\sigma^2}{2} g^2(y)} dy \right), \qquad (4.11)$$

where $g'(x)$ stands for the derivative of $g(x)$ with respect to its argument. The extrema of the steady–state probability density obey the following equation

$$f(x) - \frac{\sigma^2}{2} g(x) g'(x) = 0. \qquad (4.12)$$

Note that this equation is not identical to the equation $f(x) = 0$ for the steady states in the absence of multiplicative noise. As a result, the most probable states need not coincide with the deterministic stationary states. More importantly, solutions can appear or existing solutions can be destabilized by the noise. These changes in the asymptotic behavior of the system have been generally named noise–induced phase transitions [HL84].

To illustrate this phenomenon, consider the case of a deterministically stable steady state at $x = 0$, e.g.,

$$f(x) = -x + o(x),$$

perturbed by a multiplicative noise. As is clear from equations (4.11–4.12), a noise term of the form

$$g(x) = 1 + x^2 + o(x^2)$$

will have a stabilizing effect, since

$$-(\sigma^2/2)g(x)g'(x) = -\sigma^2 x + o(x^2),$$

and it makes the coefficient of x more negative. On the other hand, noise of the form

$$g(x) = 1 - x^2 + o(x^2)$$

i.e., with maximal amplitude at the reference state $x = 0$, has the tendency to 'destabilize' the reference state. In fact, above a critical intensity $\sigma^2 > \sigma_c^2 = 1$, the stationary probability density will no longer have a maximum at $x = 0$, and 'noise–induced' maxima can appear. This phenomenon remains possible even if the deterministic steady–state equation, got by fixing the random value

by a Fokker–Planck equation. The resulting equation, being *linear* in $\dot{x}$ (but not in x), can be immediately solved for $\dot{x}$, giving

$$\dot{x} = Q(x;\bar{x}) + S(x;\bar{x})\xi, \tag{4.20}$$

with

$$Q(x;\bar{x}) \equiv (f + \bar{\Delta})\theta, \tag{4.21}$$

$$S(x;\bar{x}) \equiv \sigma g \theta, \tag{4.22}$$

$$\theta(x;\bar{x}) \equiv \{1 - \tau g[(f + \bar{\Delta})/g]'\}^{-1}. \tag{4.23}$$

The Fokker–Planck equation associated to the SDE (8.45) is

$$\partial_t P(x,t;\bar{x}) = -\partial_x \left[R_1(x;\bar{x})P(x,t;\bar{x})\right] + \frac{1}{2}\partial_x^2 \left[R_2(x;\bar{x})P(x,t;\bar{x})\right], \tag{4.24}$$

with *drift* and *diffusion* coefficients given by

$$R_1(x;\bar{x}) = Q + \frac{1}{4}(S^2)' \tag{4.25}$$

$$R_2(x;\bar{x}) = S^2. \tag{4.26}$$

The solution of the time–independent Fokker–Planck equation leads to the stationary probability density

$$P_{st}(x;\bar{x}) = \frac{1}{Z}\exp\left[\int_0^x dx' \frac{2R_1(x';\bar{x}) - \partial_{x'}R_2(x';\bar{x})}{R_2(x';\bar{x})}\right]. \tag{4.27}$$

The value of $\bar{x}$ arises from a *self–consistency relation*, once we equate it to the average value of the random variable x_i in the stationary state

$$\bar{x} = \langle x \rangle \equiv \int_{-\infty}^{\infty} dx\, x\, P_{st}(x;\bar{x}) \equiv F(\bar{x}). \tag{4.28}$$

Now, the condition

$$\left.\frac{dF}{d\bar{x}}\right|_{\bar{x}=0} = 1 \tag{4.29}$$

allows us to find the *transition line* between the *ordered* and the *disordered* phases.

Results

The mean–field approximation of the general dD extended system are the following (see [MDT00]):

A. As in the white–noise case $\tau = 0$, the ordering phase transition is *reentrant with respect to* σ: for a range of values of D that depends on τ, ordered states can only exist within a window $[\sigma_1, \sigma_2]$. The fact that this window shifts to the right *for small* τ means that, for fixed D, color *destroys* order just above σ_1 but *creates* it just above σ_2.

B. For fixed $\sigma > 1$ and $\tau \neq 0$, ordered states exist *only within a window* of values for D. Thus the ordering phase transition is *also reentrant with respect to D*. For τ small enough the maximum value of D compatible with the ordered phase increases rather steeply with σ, reaching a maximum around $\sigma \sim 5$ and then decreases gently. For $\tau \geq 0.1$ it becomes evident (in the ranges of D and σ analyzed) that the region sustaining the ordered phase is *closed*, and shrinks to a point for a value slightly larger than $\tau = 0.123$.

C. For fixed values of $\sigma > 1$ and D larger than its minimum for $\tau = 0$, the system *always* becomes disordered for τ large enough. The maximum value of τ consistent with order altogether corresponds to $\sigma \sim 5$ and $D \sim 32$. In other words, ordering is possible *only* if the multiplicative noise inducing it has short memory.

D. The fact that the region sustaining the ordered phase finally shrinks to a point means that even for that small region in the $\sigma-D$ plane for which order is induced by color, a further increase in τ destroys it. In other words, the phase transition is *also reentrant with respect to τ*. For D large enough there may exist even *two* such windows.

Order Parameter

As already mentioned above, the *order parameter* in this system is $m \equiv |\bar{x}|$, namely, the positive solution of the consistency equation (4.28). Consistently with what has been discussed in (A) and (C), we see that as τ increases the window of σ values where ordering occurs shrinks until it disappears. One also notices that at least for this D, the value of σ corresponding to the maximum order parameter varies very little with τ.

The *short–time evolution* of $\langle x \rangle$ can be obtained multiplying (8.49) by x and integrating:

$$\frac{d\langle x \rangle}{dt} = \int_{-\infty}^{\infty} dx\, R_1(x; \bar{x}) P(x, t; \bar{x}). \tag{4.30}$$

Let us assume an initial condition such that at early times $P(x, t \sim 0; \bar{x}) = \delta(x - \bar{x})$. Equating $\bar{x} = \langle x \rangle$ as before, we get the *order parameter equation*

$$\frac{d\langle x \rangle}{dt} = R_1(\bar{x}, \bar{x}). \tag{4.31}$$

The solution of (4.31) has an initial *rising* period (it is initially *unstable*) reaching very soon a maximum and tending to zero afterwards.

For $D/\sigma^2 \rightarrow \infty$, equation (4.31) is valid also in the *asymptotic regime* since $P_{st}(x) = \delta(x - \bar{x})$ [BPT97]. According to this criterion, in the $D/\sigma^2 \rightarrow \infty$ limit the system undergoes a second–order phase transition *if* the corresponding zero-dimensional model presents *a linear instability in its short–time dynamics*, i.e., if after linearizing (4.31):

$$\langle \dot{x} \rangle = -\alpha \langle x \rangle, \tag{4.32}$$

one finds that $\alpha < 0$. We then see that the trivial (disordered) solution $\langle x \rangle = 0$ is stable only for $\alpha > 0$. For $\alpha < 0$ other stable solutions with $\langle x \rangle \neq 0$ appear, and the system develops order through a genuine *phase* transition. In this case, $\langle x \rangle$ can be regarded as the *order parameter*. In the white noise limit $\tau = 0$ this is known to be the case for sufficiently large values of the coupling D and for a window of values for the noise amplitude $\sigma \in [\sigma_1, \sigma_2]$.

In summary, we have:

A. Multiplicative noise can shift or induce phase transitions in 0D systems.
B. Multiplicative noise can induce phase transitions in spatially extended systems.
C. Mean–field approximation predicts a minimal coupling strength for the appearance of noise induced phase transitions.
D. Mean–field approximation predicts, that the phase transition is reentrant, i.e., the ordered phase is destroyed by even higher noise intensity.
E. Appearance of an ordered phase results from a nontrivial cooperative effect between multiplicative noise, nonlinearity and diffusion.

4.2 Elements of Haken's Synergetics

In this section we present the basics of the most powerful tool for high–dimensional chaos control, which is the *synergetics*. This powerful scientific tool to *extract order from chaos* has been developed outside of chaos theory, with intention to deal with much more complex, high–dimensional, hierarchical systems, in the realm of *synergetics*. Synergetics is an interdisciplinary field of research that was founded by H. Haken in 1969 (see [Hak83, Hak93, Hak96, Hak00]). Synergetics deals with complex systems that are composed of many individual parts (components, elements) that interact with each other and are able to produce spatial, temporal or functional structures by self–organization. In particular, synergetics searches for general principles governing self–organization irrespective of the nature of the individual parts of the systems that may belong to a variety of disciplines such as physics (lasers, fluids, plasmas), meteorology, chemistry (pattern formation by chemical reactions, including flames), biology (morphogenesis, evolution theory) movement science, brain activities, computer sciences (synergetic computer), sociology (e.g., city growth) psychology and psychiatry (including Gestalt psychology).

The aim of synergetics has been to describe processes of *spontaneous self–organization and cooperation* in complex systems built from many subsystems which themselves can be complicated nonlinear objects (like many individual neuro–muscular components of the human motion system, having their own excitation and contraction dynamics, embedded in a synergistic way to produce coordinated human movement). General properties of the subsystems are their own nonlinear/chaotic dynamics as well as mutual nonlinear/chaotic interactions. Furthermore, the systems of synergetics are *open*. The influence from outside is measured by a certain set of *control parameters* $\{\sigma\}$ (like amplitudes, frequencies and time characteristics of neuro–muscular driving forces). Processes of self-organization in synergetics, (like musculo–skeletal coordination in human motion dynamics) are observed as temporal macroscopic patterns. They are described by a small set of *order parameters* $\{o\}$, similar to those in Landau's *phase–transition theory* (named after *Nobel Laureate Lev D. Landau*) of physical systems in *thermal equilibrium* [Hak83].

Now, recall that the *measure for the degree of disorder* in any isolated, or conservative, system (such a system that does not interact with its surrounding, i.e., does neither dissipate nor gain energy) is *entropy*. The *second law of thermodynamics*[3] states that in every conservative irreversible system the entropy ever increases to its maximal value, i.e., to the total disorder of the system (or remains constant for a reversible system).

Example of such a system is conservative Hamiltonian dynamics of human skeleton in the phase–space Γ defined by all joint angles q^i and momenta p_i[4], defined by ordinary (conservative) Hamilton's equations

$$\dot{q}^i = \partial_{p_i} H, \qquad \dot{p}_i = -\partial_{q^i} H. \tag{4.33}$$

The basic fact of the conservative Hamiltonian system is that its phase–flow, the time evolution of equations (4.33), preserves the phase–space volume (the so–called *Liouville measure*), as proposed by the *Liouville theorem*. This might look fine at first sight, however, the preservation of phase–space volume causes *structural instability* of the conservative Hamiltonian system, i.e., the phase–space spreading effect, by which small phase regions R_t will tend to get distorted from the initial one R_0 during the system evolution. The problem is much more serious in higher dimensions than in lower dimensions, since there are so many 'directions' in which the region can locally spread. Here we see the work of the second law of thermodynamics on an irreversible process: the increase of entropy towards the total disorder/chaos [Pen89]. In this way, the conservative Hamiltonian systems of the form (4.33) cover the wide range of dynamics, from completely integrable, to completely ergodic. Biodynamics of human–like movement is probably somewhere in the middle of this range, the more DOF included in the model, the closer to the ergodic case. One can easily imagine that the conservative skeleton–like system with 300 DOF, which means 600–D system of the form (4.33), which is full of trigonometry (coming from its noncommutative rotational matrices), is probably closer to the ergodic than to the completely integrable case.

On the other hand, when we manipulate a system from the outside, by the use of certain *control parameters* $\{\sigma\}$, we can change its *degree of order* (see [Hak83, Hak93]). Consider for example *water vapor*. At elevated temperature

---

[3] This is the only physical law that implies the arrow of time.

[4] If we neglect joints dissipation and muscular driving forces, we are dealing with pure skeleton conservative dynamics.

its molecules move freely without mutual correlation. When temperature is lowered, a liquid drop is formed, the molecules now keep a mean distance between each other. Their motion is thus highly correlated. Finally, at still lower temperature, at the freezing point, water is transformed into ice crystals. The transitions between the different aggregate states, also called phases, are quite abrupt. Though the same kind of molecules are involved all the time, the macroscopic features of the three phases differ drastically.

Similar type of ordering, but not related to the thermal equilibrium conditions, occurs in *lasers*, mathematically given by Lorenz–like attractor equations. Lasers are certain types of lamps which are capable of emitting coherent light. A typical laser consists of a crystal rod filled with gas, with the following features important from the synergetics point of view: when the atoms the laser material consists of are excited or 'pumped' from the outside, they emit light waves. So, the pump power, or pump rate represents the control parameter σ. At low pump power, the waves are entirely uncorrelated as in a usual lamp. Could we hear light, it would sound like noise to us [Hak83].

When we increase the pump rate to a critical value σ_c, the noise disappears and is replaced by a pure tone. This means that the atoms emit a pure sinusoidal light wave which in turn means that the individual atoms act in a perfectly correlated way – they become self–organized. When the pump rate is increased beyond a second critical value, the laser may periodically emit very intense and short pulses. In this way the following *instability sequence* occurs [Hak83]:

$$\text{noise} \mapsto \{\text{coherent oscillation at frequency} \, \omega_1\} \mapsto$$

periodic pulses at frequency ω_2 which modulate oscillation at frequency ω_1

i.e., no oscillation $\mapsto$ first frequency $\mapsto$ second frequency.

Under different conditions the light emission may become *chaotic* or even *turbulent*. The frequency spectrum becomes broadened.

The laser played a crucial role in the development of synergetics for various reasons [Hak83]. In particular, it allowed detailed theoretical and experimental study of the phenomena occurring within the transition region: *lamp* $\leftrightarrow$ *laser*, where a surprising and far–reaching analogy with phase transitions of systems in thermal equilibrium was discovered. This analogy includes all basic *phase–transition effects*: a *symmetry breaking instability*, *critical slowing down* and *hysteresis effect*.

4.2.1 Phase Transitions

Besides water vapor, a typical example is a *ferromagnet* [Hak83]. When a ferromagnet is heated, it suddenly loses its magnetization. When temperature is lowered, the magnet suddenly regains its magnetization. What happens on a microscopic, atomic level, is this: We may visualize the magnet as being composed of many, elementary (atomic) magnets (called spins). At elevated temperature, the elementary magnets point in random directions. Their magnetic moments, when added up, cancel each other and no macroscopic magnetization results. Below a critical value of temperature T_c, the elementary magnets are lined up, giving rise to a macroscopic magnetization. Thus the *order on the microscopic level* is a cause of a *new feature* of the material *on the macroscopic level*. The change of one phase to the other one is called *phase transition*.

A thermodynamical description of ferromagnet is based on analysis of its *free energy potential* (in thermal equilibrium conditions). The free energy $\mathcal{F}$, depends on the *control parameter* $\sigma = T$, the temperature. We seek the minimum of the potential $\mathcal{F}$ for a fixed value of magnetization o, which is called *order parameter* in Landau's theory of phase transitions (see Appendix).

This phenomenon is called a *phase transition of second order* because the second derivative (specific heat) of the free energy potential $\mathcal{F}$ is discontinuous. On the other hand, the entropy S (the first derivative of $\mathcal{F}$) itself is continuous so that this transition is also referred to as a *continuous phase transition*.

In statistical physics one also investigates the temporal change of the order parameter – magnetization o. Usually, in a more or less phenomenological manner, one assumes that o obeys an equation of the form

$$\dot{o} = -\frac{\partial \mathcal{F}}{\partial o} = -\sigma o - \beta o^3. \tag{4.34}$$

For $\sigma \to 0$ we observe a phenomenon called *critical slowing down*, because the 'particle' with coordinate o falls down the slope of the 'potential well' more and more slowly. Simple relation (4.34) is called *order parameter equation*.

We now turn to the case where the free energy potential has the form

$$\mathcal{F}(o, T) = \frac{\sigma}{2} o^2 + \frac{\gamma}{3} o^3 + \frac{\beta}{4} o^4, \tag{4.35}$$

(β and γ – positive but σ may change its sign according to $\sigma = a(T - T_c)$, ($a > 0$)). When we change the control parameter – temperature T, i.e., the parameter σ, we pass through a sequence of deformations of the potential curve.

When lowering temperature, the local minimum first remains at $o_0 = 0$. When lowering temperature, the 'particle' may fall down from o_0 to the new (global) minimum of $\mathcal{F}$ at o_1. The entropies of the two states, o_0 and o_1, differ. This phenomenon is called a *phase transition of first order* because the first derivative of the potential $\mathcal{F}$ with respect to the control parameter T is discontinuous. Since the entropy S is discontinuous this transition is also referred to as a *discontinuous phase transition*. When we now increase the temperature, is apparent that the system stays at o_1 longer than it had been before when lowering the control parameter. This represents *hysteresis effect*.

In the case of the potential (4.35) the order parameter equation gets the form

$$\dot{o} = -\sigma o - \gamma o^2 - \beta o^3.$$

Similar *disorder* $\Rightarrow$ *order* transitions occur also in various non–equilibrium systems of physics, chemistry, biology, psychology, sociology, as well as in human motion dynamics. The analogy is subsumed in Table 1.

Table 1. Phase transition analogy

System in thermal equilibrium	Non–equilibrium system
Free energy potential $\mathcal{F}$	Generalized potential V
Order parameters o_i	Order parameters o_i
$\dot{o}_i = -\frac{\partial \mathcal{F}}{\partial o_i}$	$\dot{o}_i = -\frac{\partial V}{\partial o_i}$
Temperature T	Control input u
Entropy S	System output y
Specific Heat c	System efficiency e

In the case of human motion dynamics, natural control inputs u_i are muscular torques F_i, natural system outputs y_i are joint coordinates q^i and momenta p_i, while the system efficiencies e_i represent the changes of coordinates and momenta with changes of corresponding muscular torques for the ith joint,

$$
e_i^q = \frac{\partial q^i}{\partial F_i}, \qquad e_i^p = \frac{\partial p_i}{\partial F_i}.
$$

Order parameters o_i represent certain important qualities of the human motion system, depending on muscular torques as control inputs, similar to *magnetization*, and usually defined by equations similar to (4.34) or

$$
\dot{o}_i = -\sigma o - \gamma o^2 - \beta o^3,
$$

with nonnegative parameters σ, β, γ, and corresponding to the second and first order phase transitions, respectively. The choice of actual order parameters is a matter of *expert knowledge* and *purpose of macroscopic system modelling* [Hak83].

4.2.2 Mezoscopic Derivation of Order Parameters

Basic Hamiltonian equations (1.156–1.157) are in general quite complicated and can hardly be solved completely in the whole locomotor phase–space Γ, spanned by the set of possible joint vectors $\{q^i(t), p_i(t)\}$. We therefore have to restrict ourselves to local concepts for analyzing the behavior of our locomotor system. To this end we shall consider a reference musculo–skeletal state $\{q_0, p_0\}$ and its neighborhood. Following the procedures of the mezoscopic synergetics (see [Hak83, Hak93]), we assume that the reference state has the properties of an attractor and is a comparably low–dimensional object in Γ. In order to explore the behavior of our locomotor system (dependent on the set of control parameters σ) in the neighborhood of $\{q_0, p_0\}$ we look for the time development of small deviations from the reference state (to make the formalism as simple as possible, we drop the joint index in this section)

$$q(t) = q_0 + \delta q(t), \qquad p(t) = p_0 + \delta p(t),$$

and consider $\delta q(t)$ and $\delta p(t)$ as small entities. As a result we may linearize the equations of δq and δp in the vicinity of the reference state $\{q_0, p_0\}$ We get

$$\partial_t \delta q(t) = L[q_0, p_0, \sigma]\,\delta q(t), \qquad \partial_t \delta p(t) = K[q_0, p_0, \sigma]\,\delta p(t),$$

where $L[.]$ and $K[.]$ are linear matrices independent of $\delta q(t)$ and $\delta p(t)$, which can be derived from the basic Hamiltonian vector–field (1.156–1.157) by standard synergetics methods [Hak83, Hak93, Hak96, Hak00]. We now assume that we can construct a complete set of eigenvectors $\{l^{(j)}(t), k^{(j)}(t)\}$ corresponding to (4.34). These eigenvectors allow us to decompose arbitrary deviations $\delta q(t)$ and $\delta p(t)$ into elementary collective deviations along the directions of the eigenvectors

$$\delta q(t) = \xi_j(t)\, l^j(t), \qquad \delta p(t) = \zeta_j(t)\, k^j(t), \tag{4.36}$$

where $\xi_j(t)$ and $\zeta_j(t)$ represent the excitations of the system along the directions in the phase–space Γ prescribed by the eigenvectors $l^j(t)$ and $k^j(t)$, respectively. These amplitudes are still dependent on the set of control parameters $\{\sigma\}$. We note that the introduction of the eigenvectors $\{l^j(t), k^j(t)\}$ is of crucial importance. In the realm of synergetics they are considered as the *collective modes or patterns* of the system. Whereas the basic Hamiltonian equation (1.156–1.157) is formulated on the basis of the human locomotor–system variables (coordinates and momenta) of the single subsystems (joints), we can now give a new formulation which is based on these collective patterns and describes the dynamical behavior of the locomotor system in terms of these different collective patterns. Inserting relations (4.36) into the basic system (1.156–1.157) we get equations for the amplitudes $\xi_j(t)$ and $\zeta_j(t)$,

$$\dot{\xi}_i(t) = A_{ij} \cdot \xi_j(t) + \text{nonlinear terms}, \qquad \dot{\zeta}_j(t) = B_{ij} \cdot \zeta_j(t) + \text{nonlinear terms},$$

where $\cdot$ denotes the scalar product, and it is assumed that the time dependence of the linear matrices L and K is carried out by the eigenvectors leaving us with constant matrices A and B.

We now summarize the results by discussing the following time–evolution formulas for joint coordinates $q(t)$ and momenta $p(t)$,

$$q(t) = q_0 + \xi_j(t)\, l^j(t), \qquad p(t) = p_0 + \zeta_j(t)\, k^j(t), \tag{4.37}$$

which describes the time dependence of the phase vectors $q(t)$ and $p(t)$ through the evolution of the collective patterns. Obviously, the reference musculo–skeletal state $\{q_0(t), p_0(t)\}$ can be called stable when all the possible excitations $\{\xi_j(t), \zeta_j(t)\}$ decay during the curse of time. When we now change the control parameters $\{\sigma\}$ some of the $\{\xi_j(t), \zeta_j(t)\}$ can become unstable and start to grow in time. The border between decay and growth in parameter space is called a Tablecritical region. Haken has shown that the few unstable amplitudes, denoted by u_q and u_p, change very slowly in the vicinity of a critical region, whereas the damped amplitudes, denoted by s_q and s_p, quickly decay to values which are completely prescribed by the unstable modes. This fact is expressed as the Tableslaving principle of synergetics [Hak83, Hak93, Hak96, Hak00], in our case reading as

$$s_q = s_q(u_q), \qquad s_p = s_p(u_p).$$

These relations allow us to eliminate the stable modes in (4.37), and leave us with a low–dimensional set of equations for the unstable modes which play the role of the order parameters. These Tableorder parameter equations then completely rule the behavior of our microscopic nD musculo–skeletal system on macroscopic scales near an instability.

The fundamental result of synergetics consists in the observation that on macroscopic scales new laws can be discovered which exist in their own right [Hak83]. These laws which are expressed by the order parameter equations turn out to be independent of the detailed nature of the subsystems and their interactions. As a consequence this allows us to introduce the concept of Tablenormal forms [Arn88] as a method to discus instabilities and qualitative dynamics in the neighborhood of the critical regions. This method of phenomenological synergetics allows us to start qualitative analysis from purely macroscopic considerations.

Using the so–called Tableadiabatic elimination of fast variables [Hak83], one tries to identify macroscopic quantities related to global musculo–skeletal dynamics (similar but different from the *mean–field* center–of–mass dynamics) – from experience and classifies them according to time–scale arguments. The slowest variables are usually identified with the control parameters which are assumed to be quasi static quantities. The slow macroscopic dynamics of the system has to be attributed to the order parameters. Very quickly relaxing variables have to be considered as enslaved modes.

4.2.3 Example: Synergetic Control of Biodynamics

Recall from [II05, II06a, II06b] that the basic microscopic synergetic level of human musculo–skeletal dynamics (1.156–1.157), can be viewed on the highest, *macroscopic synergetic center–of–mass organization level* of human motion dynamics as a simple *Hamilton oscillator*, physically representing the damped, sinusoidally driven pendulum (1.29) of the unit mass and length l

$$l^2 \ddot{q} + \gamma \dot{q} + lg \sin q = A \cos(p_D t).$$

This equation expresses Newtonian second law of motion with the various terms on the left representing acceleration, damping, and gravitation. The angular momentum of the forcing p_D, may be different from the natural frequency of the pendulum. In order to minimize the number of adjustable parameters the equation may be rewritten in dimensionless form as

$$\ddot{q} + (1/\nu)\dot{q} + \sin q = \epsilon \cos(p_D t),$$

where ν is the damping or quality parameter, ϵ is the forcing amplitude, and p_D is the drive frequency. The low–amplitude natural angular frequency of the pendulum is unity, and time is regarded as dimensionless. This equation satisfies the necessary conditions for chaos when it is written as an extended Hamiltonian system

$$\dot{q} = p, \qquad \dot{p} = -(1/\nu)p - \sin q + \epsilon \cos \phi, \qquad \dot{\phi} = p_D. \qquad (4.38)$$

The variable ϕ is introduced as the phase of the drive term. Three variables are evident and also two nonlinear coupling terms. Whether the motion is chaotic depends upon the values of the three parameters: damping, forcing amplitude and drive frequency. For some values the pendulum locks onto the driving force, oscillating in a periodic motion whose frequency is the driving frequency, possibly with some harmonics or subharmonics. But for other choices of the parameters the pendulum motion is chaotic. One may view the chaos as resulting from a subtle interplay between the tendency of the pendulum to oscillate at its 'natural' frequency and the action of the forcing term. The transitions between non–chaotic and chaotic states, due to changes in the parameters, occur in several ways and depend delicately upon the values of the parameters.

To include (in the simplest possible way) the muscle excitation–contraction dynamics, and thus make the damped, driven Hamilton oscillator (4.38) a more realistic macroscopic model for human motion dynamics, we assume that the time–dependent forcing amplitude $\epsilon = \epsilon(t)$ has the form of a low pass filter, a characteristic feature of biological systems, given by first–order transfer function $\frac{K}{Ts+1}$. Here K denotes gain of the filter and T its time constant.

Therefore, macroscopic mechanical model of human motion dynamics gets the fully–functional form

$$\ddot{q} + (1/\nu)\dot{q} + \sin q = K(1 - e^{-t/T}) \cos(p_D t).$$

which can be rewritten in the form of *extended Hamilton oscillator*

$$\dot{q} = p, \qquad \dot{p} = -(1/\nu)p - \sin q + K(1 - e^{-t/T}) \cos \phi, \qquad \dot{\phi} = p_D. \qquad (4.39)$$

Now, to effectively control the macroscopic HML model (4.39), we can use two standard nonlinear–control techniques:

A. Adaptive Lie–derivative based geometric control; and
B. Adaptive fuzzy–logic based AI control.

4.2.4 Example: Chaotic Psychodynamics of Perception

Perceptual alternation phenomena of ambiguous figures have been studied for a long time. Figure–ground, perspective (depth) and semantic ambiguities are

The two competitive interpretations are embedded in the network as minima of the energy map (see Figure 4.2):

$$E = -\frac{1}{2} w_{ij} X_i X_j,$$

at HNP. This is done by using a iterative perception learning rule for $p(< N)$ patterns $\{\xi_i^\mu\} \equiv (\xi_1^\mu, \cdots, \xi_N^\mu), (\mu = 1, \cdots, p; \xi_i^\mu = +1 \, or -1)$ in the form :

$$w_{ij}^{new} = w_{ij}^{old} + \sum_\mu \delta w_{ij}^\mu, \qquad \text{with} \qquad \delta w_{ij}^\mu = \frac{1}{N} \theta(1 - \gamma_i^\mu) \xi_i^\mu \xi_j^\mu,$$

$$\text{where} \qquad \gamma_i^\mu \equiv \xi_i^\mu w_{ij} \xi_j^\mu,$$

and $\theta(h)$ is the unit step function. The learning mode is separated from the performance mode by (4.46).

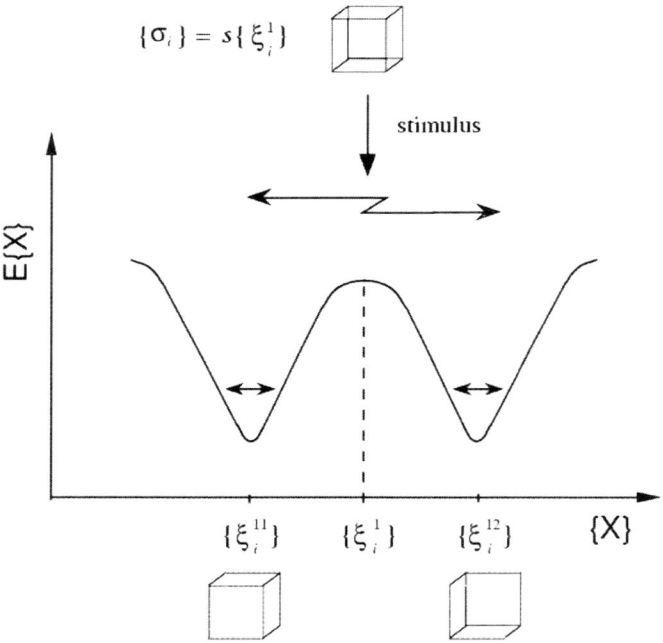

Fig. 4.2. Conceptual psychodynamic model of [NNM00], illustrating state transitions induced by chaotic activity.

Simulations of the CNN have shown that the neural chaos leads to perceptual alternations as responses to ambiguous stimuli in the chaotic neural network. Its emergence is based on the simple process in a realistic bottom–up framework. In the same stage, similar results can not be obtained by the stochastic activity. This simulation suggests functional usefulness of the chaotic activity in perceptual systems even at higher cognitive levels. The perceptual alternation appears to be an inherent feature built in the chaotic neuron assembly. It may be interesting to study the brain with the experimental technique (e.g., functional MRI) under the circumstance where the perceptual alternation is running [NNM00].

4.2.5 Kick Dynamics and Dissipation–Fluctuation Theorem

Deterministic Delayed Kicks

Following [Hak02], we consider the mechanical example of a soccer ball that is kicked by a soccer player and rolls over grass, whereby its motion will be slowed down. We start with the Newton's (second) law of motion, $m\dot{v} = force$, and in order to get rid of superfluous constants, we put temporarily $m = 1$. The $force$ on the r.h.s. consists of the damping force $-\gamma v(t)$ of the grass (where γ is the damping constant) and the sharp force $F(t) = s\delta(t - \sigma)$ of the individual kick occurring at time $t = \sigma$ (where s is the strength of the kick, and δ is the Dirac's 'delta' function). In this way, the (single) *kick equation* of the ball motion becomes

$$\dot{v} = -\gamma v(t) + s\delta(t - \sigma), \qquad (4.47)$$

with the general solution

$$v(t) = sG(t - \sigma),$$

where $G(t - \sigma)$ is the *Green's function*[5]

$$G(t - \sigma) = \begin{cases} 0 & \text{for } t < \sigma \\ e^{-\gamma(t-\sigma)} & \text{for } t \geq \sigma \end{cases}.$$

Now, we can generalize the above to N kicks with individual strengths s_j, occurring at a sequence of times $\{\sigma_j\}$, so that the total kicking force becomes

$$F(t) = \sum_{j=1}^{N} s_j \delta(t - \sigma_j).$$

In this way, we get the *multi–kick equation* of the ball motion

---

[5] This is the Green's function of the first order system (4.47). Similarly, the Green's function

$$G(t - \sigma) = \begin{cases} 0 & \text{for } t < \sigma \\ (t - \sigma)e^{-\gamma(t-\sigma)} & \text{for } t \geq \sigma \end{cases}$$

corresponds to the second order system

$$\left(\frac{d}{dt} + \gamma\right)^2 G(t - \sigma) = \delta(t - \sigma).$$

$$\dot{v} = -\gamma v(t) + \sum_{j=1}^{N} s_j \delta(t - \sigma_j),$$

with the general solution

$$v(t) = \sum_{j=1}^{N} s_j G(t - \sigma_j). \tag{4.48}$$

As a final generalization, we would imagine that the kicks are continuously exerted on the ball, so that kicking force becomes

$$F(t) = \int_{t_0}^{T} s(\sigma)\delta(t - \sigma)d\sigma \equiv \int_{t_0}^{T} d\sigma F(\sigma)\delta(t - \sigma),$$

so that the continuous multi–kick equation of the ball motion becomes

$$\dot{v} = -\gamma v(t) + \int_{t_0}^{T} s(\sigma)\delta(t - \sigma)d\sigma \equiv -\gamma v(t) + \int_{t_0}^{T} d\sigma F(\sigma)\delta(t - \sigma),$$

with the general solution

$$v(t) = \int_{t_0}^{T} d\sigma F(\sigma)G(t - \sigma) = \int_{t_0}^{T} d\sigma F(\sigma)e^{-\gamma(t-\sigma)}. \tag{4.49}$$

Random Kicks and Langevin Equation

We now denote the times at which kicks occur by t_j and indicate their direction in a one–dimensional game by $(\pm 1)_j$, where the choice of the plus or minus sign is random (e.g., throwing a coin). Thus the kicking force can be written in the form [Hak02]

$$F(t) = s \sum_{j=1}^{N} \delta(t - t_j)(\pm 1)_j, \tag{4.50}$$

where for simplicity we assume that all kicks have the same strength s. When we observe many games, then we may perform an average $< ... >$ over all these different *performances*,

$$< F(t) >= s < \sum_{j=1}^{N} \delta(t - t_j)(\pm 1)_j > . \tag{4.51}$$

Since the direction of the kicks is assumed to be independent of the time at which the kicks happen, we may split (4.51) into the product

$$< F(t) >= s < \sum_{j=1}^{N} \delta(t - t_j) >< (\pm 1)_j > .$$

As the kicks are assumed to happen with equal frequency in both directions, we get the cancellation

$$< (\pm 1)_j >= 0,$$

which implies that the average kicking force also vanishes,

$$< F(t) >= 0.$$

In order to characterize the strength of the force (4.50), we consider a quadratic expression in F, e.g., by calculating the *correlation function* for two times t, t',

$$< F(t)F(t') >= s^2 < \sum_j \delta(t - t_j)(\pm 1)_j \sum_k \delta(t' - t_k)(=1)_k > .$$

As the ones for $j \neq k$ will cancel each other and for $j = k$ will become 1, the correlation function becomes a single sum

$$< F(t)F(t') >= s^2 < \sum_j \delta(t - t_j)\delta(t' - t_k) >, \tag{4.52}$$

which is usually evaluated by assuming the *Poisson process* for the times of the kicks.

Now, proper description of random motion is given by *Langevin rate equation*, which describes the *Brownian motion*: when a particle is immersed in a fluid, the velocity of this particle is slowed down by a force proportional to its velocity and the particle undergoes a zig–zag motion (the particle is steadily pushed by much smaller particles of the liquid in a random way). In physical terminology, we deal with the behavior of a system (particle) which is coupled to a *heat bath* or reservoir (namely the liquid). The heat bath has two effects [Hak02]:

A. It decelerates the mean motion of the particle; and
B. It causes statistical fluctuation.

The standard Langevin equation has the form

$$\dot{v} = -\gamma v(t) + F(t), \tag{4.53}$$

where $F(t)$ is a *fluctuating force* with the following properties:

A. Its statistical average (4.51) vanishes; and
B. Its correlation function (6.20) is given by

$$< F(t)F(t') >= Q\delta(t - t_0), \tag{4.54}$$

where $t_0 = T/N$ denotes the mean free time between kicks, and $Q = s^2/t_0$ is the *random fluctuation*.

The general solution of the Langevin equation (4.53) is given by (4.49).

The average velocity vanishes, $< v(t) >= 0$, as both directions are possible and cancel each other. Using the integral solution (4.49) we get

$$< v(t)v(t') >=< \int_{t_0}^{t} d\sigma \int_{t_0}^{t'} d\sigma' F(\sigma)F(\sigma')e^{-\gamma(t-\sigma)}e^{-\gamma(t'-\sigma')} >,$$

which, in the steady–state, reduces to

$$< v(t)v(t') >= \frac{Q}{2\gamma}e^{-\gamma(t-\sigma)},$$

and for equal times

$$< v(t)^2 >= \frac{Q}{2\gamma}.$$

If we now repeat all the steps performed so far with $m \neq 1$, the final result reads

$$< v(t)^2 >= \frac{Q}{2\gamma m}. \tag{4.55}$$

Now, according to thermodynamics, the *mean kinetic energy* of a particle is given by

$$\frac{m}{2} < v(t)^2 >= \frac{1}{2}k_B T, \tag{4.56}$$

where T is the (absolute) temperature, and k_B is the Boltzman's constant. Comparing (4.55) and (4.56), we get the important Einstein's result

$$Q = 2\gamma k_B T,$$

which says that whenever there is damping, i.e., $\gamma \neq 0$, then there are random fluctuations (or noise) Q. In other words, fluctuations or noise are inevitable in any physical system. For example, in a resistor (with the resistance R) the electric field E fluctuates with a correlation function (similar to (4.54))

$$< \bar{E}(t)E(t') >= 2Rk_B T\delta(t - t_0).$$

This is the simplest example of the *fluctuation–dissipation theorem*.

4.3 Synergetics of Recurrent and Attractor Neural Networks

Recall that *recurrent neural networks* are neural networks with synaptic feedback loops. Provided that we restrict ourselves to large neural systems, we can apply to their analysis tools from statistical mechanics. Here, we have two possibilities. Under the common conditions of synaptic symmetry, the stochastic process of evolving neuron states leads towards an equilibrium situation where the microscopic state probabilities are known, where the classical techniques of *equilibrium statistical mechanics* can be applied. On the other hand, for non–symmetric networks, where the asymptotic (stationary) statistics are not known, synergetic techniques from *non–equilibrium statistical mechanics* are the only tools available for analysis. Here, the 'natural' set of macroscopic *order parameters* to be calculated can be defined in practice as the smallest set which will obey closed deterministic equations in the limit of an infinitely large network.

Being high–dimensional nonlinear systems with extensive feedback, the dynamics of recurrent neural networks are generally dominated by a wealth of different attractors, and the practical use of recurrent neural networks (in both biology and engineering) lies in the potential for creation and manipulation of these attractors through adaptation of the network parameters (synapses and thresholds). Input fed into a recurrent neural network usually serves to induce a specific initial configuration (or firing pattern) of the neurons, which serves as a cue, and the 'output' is given by the (static or dynamic) attractor which has been triggered by this cue. The most familiar types of recurrent neural network models, where the idea of creating and manipulating attractors has been worked out and applied explicitly, are the so–called *attractor neural networks* for associative memory, designed to store and retrieve information in the form of neuronal firing patterns and/or sequences of neuronal firing patterns. Each pattern to be stored is represented as a microscopic state vector. One then constructs synapses and thresholds such that the dominant attractors of the network are precisely the pattern vectors (in the case of static recall), or where, alternatively, they are trajectories in which the patterns are successively generated microscopic system states. From an initial configuration (the 'cue', or input pattern to be recognized) the system is allowed to evolve in time autonomously, and the final state (or trajectory) reached can be interpreted as the pattern (or pattern sequence) recognized by network from the input. For such programmes to work one clearly needs recurrent neural networks with extensive *ergodicity breaking*: the state vector will during the course of the dynamics (at least on finite time–scales) have to be confined to a restricted region of state space (an 'ergodic component'), the location of which is to depend strongly on the initial conditions. Hence our interest will mainly be in systems with many attractors. This, in turn, has implications at a theoretical/mathematical level: solving models of recurrent neural networks with extensively many attractors requires advanced tools from disordered systems theory, such as statical replica theory and dynamical partition function analysis.

The *equilibrium* statistical mechanical techniques can provide much detailed quantitative information on the behavior of recurrent neural networks, but they obviously have serious restrictions. The first one is that, by definition, they will only provide information on network properties in the stationary state. For associative memories, for instance, it is not clear how one can calculate quantities like sizes of domains of attraction without solving the dynamics. The second, and more serious, restriction is that for equilibrium statistical mechanics to apply the dynamics of the network under study must obey detailed balance, i.e., absence of microscopic probability currents in the stationary state. For recurrent networks in which the dynamics take the form of a stochastic alignment of neuronal firing rates to post–synaptic potentials which, in turn, depend linearly on the firing rates, this requirement of detailed balance usually implies symmetry of the synaptic matrix. From a physiological point of view this requirement is clearly unacceptable, since it is violated in any network that obeys Dale's law as soon as an excitatory neuron is connected to an inhibitory one. Worse still, in any network of graded–response neurons detailed balance will always be violated, even when the synapses are symmetric. The situation will become even worse when we turn to networks of yet more realistic (spike–based) neurons, such as integrate-and-fire ones. In contrast to this, *non–equilibrium* statistical mechanical techniques, it will turn out, do not impose such biologically non–realistic restrictions on neuron types and synaptic symmetry, and they are consequently the more appropriate avenue for future theoretical research aimed at solving biologically more realistic models (for details, see [Coo01, SC00, SC01, CKS05]).

4.3.1 Stochastic Dynamics of Neuronal Firing States

Recall that the simplest non–trivial definition of a recurrent neural network is that where N binary neurons $\sigma_i \in \{-1, 1\}$ (in which the states '1' and '-1' represent firing and rest, respectively) respond iteratively and synchronously to post–synaptic potentials (or local fields) $h_i(\boldsymbol{\sigma})$, with $\boldsymbol{\sigma} = (\sigma_1, \ldots, \sigma_N)$. The fields are assumed to depend linearly on the instantaneous neuron states (summation convention upon repeated indices is always used):

Parallel Dynamics: $\sigma_i(\ell+1) = \mathrm{sgn}\left[h_i(\boldsymbol{\sigma}(\ell)) + T\eta_i(\ell)\right], \quad h_i(\sigma) = J_{ij}\sigma_j + \theta_i.$
$$(4.57)$$

The stochasticity is in the independent random numbers $\eta_i(\ell) \in \mathbb{R}$ (representing threshold noise), which are all drawn according to some distribution $w(\eta)$. The parameter T is introduced to control the amount of noise. For $T = 0$ the process (4.57) is deterministic: $\sigma_i(\ell+1) = \mathrm{sgn}[h_i(\sigma(\ell))]$. The opposite extreme is choosing $T = \infty$, here the system evolution is fully random. The external fields θ_i represent neural thresholds and/or external stimuli, J_{ij} represents the synaptic efficacy at the junction $j \to i$ ($J_{ij} > 0$ implies excitation, $J_{ij} < 0$ inhibition). Alternatively we could decide that at each iteration step ℓ only a single randomly drawn neuron σ_{i_ℓ} is to undergo an update of the type (4.57):

Sequential Dynamics: $\begin{array}{l} i \neq i_\ell : \sigma_i(\ell + 1) = \sigma_i(\ell), \\ i = i_\ell : \sigma_i(\ell + 1) = \mathrm{sgn}\left[h_i(\boldsymbol{\sigma}(\ell)) + T\eta_i(\ell)\right], \end{array}$ (4.58)

with the local fields as in (4.57). The stochasticity is now both in the independent random numbers $\eta_i(\ell)$ (the threshold noise) and in the site i_ℓ to be updated, drawn randomly from the set $\{1, \ldots, N\}$. For simplicity we assume $w(-\eta) = w(\eta)$, and define

$$g(z) = 2 \int_0^z d\eta \, w(\eta) : \qquad g(-z) = -g(z), \qquad \lim_{z \to \pm\infty} g(z) = \pm 1, \qquad \partial_z g(z) \geq 0.$$

Popular choices for the threshold noise distributions are

$$w(\eta) = (2\pi)^{-\frac{1}{2}} e^{-\frac{1}{2}\eta^2} : \qquad g(z) = \mathrm{Erf}[z/\sqrt{2}],$$

$$w(\eta) = \frac{1}{2}[1 - \tanh^2(\eta)] : \qquad g(z) = \tanh(z).$$

Now, from the microscopic equations (4.57,4.58), which are suitable for numerical simulations, we can derive an equivalent but mathematically more convenient description in terms of microscopic state probabilities $p_\ell(\boldsymbol{\sigma})$. Equations (4.57,4.58) state that, if the system state $\boldsymbol{\sigma}(\ell)$ is given, a neuron i to be updated will obey

$$\mathrm{Prob}\left[\sigma_i(\ell+1)\right] = \frac{1}{2}\left[1 + \sigma_i(\ell+1) \, g[\beta h_i(\boldsymbol{\sigma}(\ell))]\right], \qquad (4.59)$$

with $\beta = T^{-1}$. In the case (4.57) this rule applies to all neurons, and thus we simply get $p_{\ell+1}(\boldsymbol{\sigma}) = \prod_{i=1}^N \frac{1}{2}[1 + \sigma_i \, g[\beta h_i(\boldsymbol{\sigma}(\ell))]]$. If, on the other hand, instead of $\boldsymbol{\sigma}(\ell)$ only the probability distribution $p_\ell(\boldsymbol{\sigma})$ is given, this expression for $p_{\ell+1}(\boldsymbol{\sigma})$ is to be averaged over the possible states at time ℓ:

$$\text{Parallel Dynamics} : \qquad p_{\ell+1}(\boldsymbol{\sigma}) = \sum_{\boldsymbol{\sigma}'} W\left[\boldsymbol{\sigma}; \boldsymbol{\sigma}'\right] p_\ell(\boldsymbol{\sigma}'), \qquad (4.60)$$

$$W\left[\boldsymbol{\sigma}; \boldsymbol{\sigma}'\right] = \prod_{i=1}^N \frac{1}{2}\left[1 + \sigma_i \, g[\beta h_i(\boldsymbol{\sigma}')]\right].$$

This is the standard representation of a *Markov chain*. Also the sequential process (4.58) can be formulated in terms of probabilities, but here expression (4.59) applies only to the randomly drawn candidate i_ℓ. After averaging over all possible realisations of the sites i_ℓ we get

$$p_{\ell+1}(\boldsymbol{\sigma}) = \frac{1}{N}\left\{ [\prod_{j \neq i} \delta_{\sigma_j, \sigma_j(\ell)}] \frac{1}{2}[1 + \sigma_i \, g[\beta h_i(\boldsymbol{\sigma}(\ell))]] \right\},$$

with the Kronecker symbol: $\delta_{ij} = 1$, if $i = j$ and $\delta_{ij} = 0$, otherwise. If, instead of $\boldsymbol{\sigma}(\ell)$, the probabilities $p_\ell(\boldsymbol{\sigma})$ are given, this expression is to be averaged over the possible states at time ℓ, with the result:

$$p_{\ell+1}(\boldsymbol{\sigma}) = \frac{1}{2N} [1 + \sigma_i \, g[\beta h_i(\boldsymbol{\sigma})]] \, p_\ell(\boldsymbol{\sigma})$$

$$+ \frac{1}{2N} [1 + \sigma_i \, g[\beta h_i(F_i \boldsymbol{\sigma})]] \, p_\ell(F_i \boldsymbol{\sigma}).$$

with the state-flip operators $F_i \Phi(\boldsymbol{\sigma}) = \Phi(\sigma_1, \ldots, \sigma_{i-1}, -\sigma_i, \sigma_{i+1}, \ldots, \sigma_N)$. This equation can again be written in the standard form

$$p_{\ell+1}(\boldsymbol{\sigma}) = \sum_{\boldsymbol{\sigma}'} W[\boldsymbol{\sigma}; \boldsymbol{\sigma}'] p_\ell(\boldsymbol{\sigma}'),$$

but now with the transition matrix

Sequential Dynamics: $W[\boldsymbol{\sigma}; \boldsymbol{\sigma}'] = \delta_{\boldsymbol{\sigma}, \boldsymbol{\sigma}'} + \dfrac{1}{N} \{ w_i(F_i \boldsymbol{\sigma}) \delta_{\boldsymbol{\sigma}, F_i \boldsymbol{\sigma}'} - w_i(\boldsymbol{\sigma}) \delta_{\boldsymbol{\sigma}, \boldsymbol{\sigma}'} \},$

$$\text{(4.61)}$$

where $\delta_{\boldsymbol{\sigma}, \boldsymbol{\sigma}'} = \prod_i \delta_{\sigma_i, \sigma_i'}$ and $w_i(\boldsymbol{\sigma}) = \dfrac{1}{2} [1 - \sigma_i \tanh [\beta h_i(\boldsymbol{\sigma})]].$

$$\text{(4.62)}$$

Note that, as soon as $T > 0$, the two transition matrices $W[\boldsymbol{\sigma}; \boldsymbol{\sigma}']$ in (4.60,4.61) both describe *ergodic systems*: from any initial state $\boldsymbol{\sigma}'$ one can reach any final state $\boldsymbol{\sigma}$ with nonzero probability in a finite number of steps (being one in the parallel case, and N in the sequential case). It now follows from the standard theory of stochastic processes (see e.g., [Kam92, Gar94]) that in both cases the system evolves towards a unique stationary distribution $p_\infty(\boldsymbol{\sigma})$, where all probabilities $p_\infty(\boldsymbol{\sigma})$ are non–zero [Coo01, SC00, SC01, CKS05].

The above processes have the (mathematically and biologically) less appealing property that time is measured in discrete units. For the sequential case we will now assume that the *duration* of each of the iteration steps is a continuous random number (for parallel dynamics this would make little sense, since all updates would still be made in full synchrony). The statistics of the durations are described by a function $\pi_\ell(t)$, defined as the probability that at time t precisely ℓ updates have been made. Upon denoting the previous discrete-time probabilities as $\hat{p}_\ell(\boldsymbol{\sigma})$, our new process (which now includes the randomness in step duration) will be described by

$$p_t(\boldsymbol{\sigma}) = \sum_{\ell \geq 0} \pi_\ell(t) \hat{p}_\ell(\boldsymbol{\sigma}) = \sum_{\ell \geq 0} \pi_\ell(t) \sum_{\boldsymbol{\sigma}'} W^\ell[\boldsymbol{\sigma}; \boldsymbol{\sigma}'] p_0(\boldsymbol{\sigma}'),$$

and time has become a continuous variable. For $\pi_\ell(t)$ we make the Poisson choice,

$$\pi_\ell(t) = \frac{1}{\ell+} (\frac{t}{\Delta})^\ell e^{-t/\Delta}.$$

From $\langle \ell \rangle_\pi = t/\Delta$ and $\langle \ell^2 \rangle_\pi = t/\Delta + t^2/\Delta^2$ it follows that Δ is the average duration of an iteration step, and that the relative deviation in ℓ at a given t vanishes for $\Delta \to 0$ as $\sqrt{\langle \ell^2 \rangle_\pi - \langle \ell \rangle_\pi^2}/\langle \ell \rangle_\pi = \sqrt{\Delta/t}$. The nice properties of the Poisson distribution under temporal derivation allow us to derive:

$$\Delta \dot{p}_t(\boldsymbol{\sigma}) = \sum_{\boldsymbol{\sigma}'} W[\boldsymbol{\sigma}; \boldsymbol{\sigma}'] p_t(\boldsymbol{\sigma}') - p_t(\boldsymbol{\sigma}).$$

For sequential dynamics, we choose $\Delta = \frac{1}{N}$ so that, as in the parallel case, in one time unit each neuron will on average be updated once. The master equation corresponding to (4.61) acquires the form $w_i(\boldsymbol{\sigma})$; (4.62) now play the role of *transition rates*. The choice $\Delta = \frac{1}{N}$ implies $\sqrt{\langle \ell^2 \rangle_\pi - \langle \ell \rangle_\pi^2} / \langle \ell \rangle_\pi = \sqrt{1/Nt}$, so we will still for $N \to \infty$ no longer have uncertainty in where we are on the t axis.

Alternatively, we could start with continuous neuronal variables σ_i (representing e.g., firing frequencies or oscillator phases), where $i = 1, \ldots, N$, and with stochastic equations of the form

$$\sigma_i(t + \Delta) = \sigma_i(t) + \Delta f_i(\boldsymbol{\sigma}(t)) + \sqrt{2T\Delta}\xi_i(t). \tag{4.63}$$

Here, we have introduced (as yet unspecified) deterministic state–dependent forces $f_i(\boldsymbol{\sigma})$, and uncorrelated Gaussian distributed random forces $\xi_i(t)$ (the noise), with

$$\langle \xi_i(t) \rangle = 0 \qquad \text{and} \qquad \langle \xi_i(t)\xi_j(t') \rangle = \delta_{ij}\delta_{t,t'}.$$

As before, the parameter T controls the amount of noise in the system, ranging from $T = 0$ (deterministic dynamics) to $T = \infty$ (completely random dynamics). If we take the limit $\Delta \to 0$ in (4.63) we find a Langevin equation (with a continuous time variable) [Coo01, SC00, SC01, CKS05]:

$$\dot{\sigma}_i(t) = f_i(\boldsymbol{\sigma}(t)) + \eta_i(t). \tag{4.64}$$

This equation acquires its meaning only as the limit $\Delta \to 0$ of (4.63). The moments of the new noise variables $\eta_i(t) = \xi_i(t)\sqrt{2T/\Delta}$ in (4.64) are given by

$$\langle \eta_i(t) \rangle = 0 \qquad \text{and} \qquad \langle \eta_i(t)\eta_j(t') \rangle = 2T\delta_{ij}\delta(t - t').$$

This can be derived from the moments of the $\xi_i(t)$. For instance:

$$\langle \eta_i(t)\eta_j(t') \rangle = \lim_{\Delta \to 0} \frac{2T}{\Delta} \langle \xi_i(t)\xi_j(t') \rangle$$

$$= 2T\delta_{ij} \lim_{\Delta \to 0} \frac{1}{\Delta}\delta_{t,t'} = 2TC\delta_{ij}\delta(t - t').$$

The constant C is found by summing over t', before taking the limit $\Delta \to 0$, in the above equation:

$$\int dt' \, \langle \eta_i(t)\eta_j(t') \rangle = \lim_{\Delta \to 0} 2T \sum_{t'=-\infty}^{\infty} \langle \xi_i(t)\xi_j(t') \rangle$$

$$= 2T\delta_{ij} \lim_{\Delta \to 0} \sum_{t'=-\infty}^{\infty} \delta_{t,t'} = 2T\delta_{ij}.$$

Thus $C = 1$, which indeed implies $\langle \eta_i(t)\eta_j(t') \rangle = 2T\delta_{ij}\delta(t - t')$. More directly, one can also calculate the moment partition function:

$$\langle e^{i \int dt \psi_i(t) \eta_i(t)} \rangle = \lim_{\Delta \to 0} \prod_{i,t} \int \frac{dz}{\sqrt{2\pi}} \, e^{-\frac{1}{2}z^2 + iz\psi_i(t)\sqrt{2T\Delta}} \qquad (4.65)$$

$$= \lim_{\Delta \to 0} \prod_{i,t} e^{-T\Delta\psi_i^2(t)} = e^{-T \int dt \sum_i \psi_i^2(t)}. \qquad (4.66)$$

On the other hand, a mathematically more convenient description of the process (4.64) is provided by the Fokker–Planck equation for the microscopic state probability density $p_t(\boldsymbol{\sigma}) = \langle \delta[\boldsymbol{\sigma} - \boldsymbol{\sigma}(t)] \rangle$, which we will now derive. For the discrete–time process (4.63) we expand the δ–distribution in the definition of $p_{t+\Delta}(\boldsymbol{\sigma})$ (in a distributional sense) [Coo01, SC00, SC01, CKS05]:

$$p_{t+\Delta}(\boldsymbol{\sigma}) - p_t(\boldsymbol{\sigma}) = \langle \delta \left[\boldsymbol{\sigma} - \boldsymbol{\sigma}(t) - \Delta \mathbf{f}(\boldsymbol{\sigma}(t)) - \sqrt{2T\Delta} \boldsymbol{\xi}(t) \right] \rangle - \langle \delta[\boldsymbol{\sigma} - \boldsymbol{\sigma}(t)] \rangle$$

$$= -\frac{\partial}{\partial \sigma_i} \langle \delta[\boldsymbol{\sigma} - \boldsymbol{\sigma}(t)] \left[\Delta f_i(\boldsymbol{\sigma}(t)) + \sqrt{2T\Delta} \xi_i(t) \right] \rangle$$

$$+ T\Delta \frac{\partial^2}{\partial \sigma_i \partial \sigma_j} \langle \delta[\boldsymbol{\sigma} - \boldsymbol{\sigma}(t)] \xi_i(t) \xi_j(t) \rangle + \mathcal{O}(\Delta^{\frac{3}{2}}).$$

The variables $\boldsymbol{\sigma}(t)$ depend only on noise variables $\xi_j(t')$ with $t' < t$, so that for any function A,

$$\langle A[\boldsymbol{\sigma}(t)]\xi_i(t) \rangle = \langle A[\boldsymbol{\sigma}(t)] \rangle \langle \xi_i(t) \rangle = 0, \qquad \text{and}$$
$$\langle A[\boldsymbol{\sigma}(t)]\xi_i(t)\xi_j(t) \rangle = \delta_{ij} \langle A[\boldsymbol{\sigma}(t)] \rangle.$$

As a consequence, we have:

$$\frac{1}{\Delta} [p_{t+\Delta}(\boldsymbol{\sigma}) - p_t(\boldsymbol{\sigma})] = -\frac{\partial}{\partial \sigma_i} \langle \delta[\boldsymbol{\sigma} - \boldsymbol{\sigma}(t)] f_i(\boldsymbol{\sigma}(t)) \rangle$$

$$+ T\frac{\partial^2}{\partial \sigma_i^2} \langle \delta[\boldsymbol{\sigma} - \boldsymbol{\sigma}(t)] \rangle + \mathcal{O}(\Delta^{\frac{1}{2}})$$

$$= -\frac{\partial}{\partial \sigma_i} [p_t(\boldsymbol{\sigma}) f_i(\boldsymbol{\sigma})] + T\frac{\partial^2}{\partial \sigma_i^2} p_t(\boldsymbol{\sigma}) + \mathcal{O}(\Delta^{\frac{1}{2}}).$$

By taking the limit $\Delta \to 0$ we then arrive at the Fokker–Planck equation:

$$\dot{p}_t(\boldsymbol{\sigma}) = -\frac{\partial}{\partial \sigma_i} [p_t(\boldsymbol{\sigma}) f_i(\boldsymbol{\sigma})] + T\frac{\partial^2}{\partial \sigma_i^2} p_t(\boldsymbol{\sigma}). \qquad (4.67)$$

In the case of graded–response neurons, the continuous variable σ_i represents the membrane potential of neuron i, and (in their simplest form) the deterministic forces are given by

$$f_i(\boldsymbol{\sigma}) = J_{ij} \tanh[\gamma \sigma_j] - \sigma_i + \theta_i, \qquad \text{with} \qquad \gamma > 0,$$

and with the θ_i representing injected currents. Conventional notation is restored by putting $\sigma_i \to u_i$. Thus equation (4.64) specializes to

$$\dot{u}_i(t) = J_{ij} \tanh[\gamma u_j(t)] - u_i(t) + \theta_i + \eta_i(t). \tag{4.68}$$

One often chooses $T = 0$ (i.e., $\eta_i(t) = 0$), the rationale being that threshold noise is already assumed to have been incorporated via the nonlinearity in (4.68).

In our second example the variables σ_i represent the phases of coupled neural oscillators, with forces of the form

$$f_i(\boldsymbol{\sigma}) = J_{ij} \sin(\sigma_j - \sigma_i) + \omega_i.$$

Individual synapses J_{ij} now try to enforce either pair–wise synchronization ($J_{ij} > 0$), or pair–wise anti–synchronization ($J_{ij} < 0$), and the ω_i represent the natural frequencies of the individual oscillators. Conventional notation dictates $\sigma_i \to \xi_i$, giving [Coo01, SC00, SC01, CKS05]

$$\dot{\xi}_i(t) = \omega_i + J_{ij} \sin[\xi_j(t) - \xi_i(t)] + \eta_i(t). \tag{4.69}$$

4.3.2 Synaptic Symmetry and Lyapunov Functions

In the deterministic limit $T \to 0$ the rules (4.57) for networks of synchronously evolving binary neurons reduce to the deterministic map

$$\sigma_i(\ell + 1) = \mathrm{sgn}\left[h_i(\boldsymbol{\sigma}(\ell))\right]. \tag{4.70}$$

It turns out that for systems with symmetric interactions, $J_{ij} = J_{ji}$ for all (ij), one can construct a Lyapunov function, i.e., a function of $\boldsymbol{\sigma}$ which during the dynamics decreases monotonically and is bounded from below (see e.g., [Kha92]):

$$\text{Binary \& Parallel Dynamics:} \quad L[\boldsymbol{\sigma}] = -\sum_i |h_i(\boldsymbol{\sigma})| - \sigma_i \theta_i. \tag{4.71}$$

Clearly, $L \geq -\sum_i [\sum_j |J_{ij}| + |\theta_i|] - \sum_i |\theta_i|$. During iteration of (4.70) we find:

$$
\begin{aligned}
L[\boldsymbol{\sigma}(\ell+1)] - L[\boldsymbol{\sigma}(\ell)] &= -\sum_i |h_i(\boldsymbol{\sigma}(\ell+1))| \\
&\quad + \sigma_i(\ell+1)[J_{ij}\sigma_j(\ell) + \theta_i] - \theta_i \left[\sigma_i(\ell+1) - \sigma_i(\ell)\right] \\
&= -\sum_i |h_i(\boldsymbol{\sigma}(\ell+1))| + \sigma_i(\ell)h_i(\boldsymbol{\sigma}(\ell+1)) \\
&= -\sum_i |h_i(\boldsymbol{\sigma}(\ell+1))| \left[1 - \sigma_i(\ell+2)\sigma_i(\ell)\right] \ \leq \ 0,
\end{aligned}
$$

where we have used (4.70) and $J_{ij} = J_{ji}$. So L decreases monotonically until a stage is reached where $\sigma_i(\ell + 2) = \sigma_i(\ell)$ for all i. Thus, with symmetric interactions this system will in the deterministic limit always end up in a limit cycle with period ≤ 2. A similar result is found for networks with binary neurons and sequential dynamics. In the limit $T \to 0$ the rules (4.58) reduce to the map

$$\sigma_i(\ell + 1) = \delta_{i,i_\ell} \operatorname{sgn}[h_i(\boldsymbol{\sigma}(\ell))] + [1 - \delta_{i,i_\ell}]\sigma_i(\ell). \qquad (4.72)$$

(in which we still have randomness in the choice of site to be updated). For systems with symmetric interactions and without self–interactions, i.e., $J_{ii} = 0$ for all i, we again find a Lyapunov function:

Binary & Sequential Dynamics: $L[\boldsymbol{\sigma}] = -\dfrac{1}{2}\sigma_i J_{ij}\sigma_j - \sigma_i\theta_i.$ (4.73)

This quantity is bounded from below, $L \geq -\frac{1}{2}\sum_{ij}|J_{ij}| - \sum_i|\theta_i|$. Upon calling the site i_ℓ selected for update at step ℓ simply i, the change in L during iteration of (4.72) can be written as [Coo01, SC00, SC01, CKS05]:

$$
\begin{aligned}
L[\boldsymbol{\sigma}(\ell + 1)] - L[\boldsymbol{\sigma}(\ell)] &= -\theta_i[\sigma_i(\ell + 1) - \sigma_i(\ell)] \\
&\quad - \frac{1}{2}J_{ik}[\sigma_i(\ell + 1)\sigma_k(\ell + 1) - \sigma_i(\ell)\sigma_k(\ell)] \\
&\quad - \frac{1}{2}J_{ji}[\sigma_j(\ell + 1)\sigma_i(\ell + 1) - \sigma_j(\ell)\sigma_i(\ell)] \\
&= [\sigma_i(\ell) - \sigma_i(\ell + 1)][J_{ij}\sigma_j(\ell) + \theta_i] \\
&= -|h_i(\boldsymbol{\sigma}(\ell))|\,[1 - \sigma_i(\ell)\sigma_i(\ell + 1)] \;\leq\; 0.
\end{aligned}
$$

Here we used (4.72), $J_{ij} = J_{ji}$, and absence of self–interactions. Thus L decreases monotonically until $\sigma_i(t + 1) = \sigma_i(t)$ for all i. With symmetric synapses, but without diagonal terms, the sequentially evolving binary neurons system will in the deterministic limit always end up in a stationary state.

Now, one can derive similar results for models with continuous variables. Firstly, in the deterministic limit the graded–response equations (4.68) simplify to

$$\dot{u}_i(t) = J_{ij}\tanh[\gamma u_j(t)] - u_i(t) + \theta_i. \qquad (4.74)$$

Symmetric networks again admit a Lyapunov function (without a need to eliminate self-interactions):

Graded–Response Dynamics : $L[\mathbf{u}] = -\dfrac{1}{2}J_{ij}\tanh(\gamma u_i)\tanh(\gamma u_j) +$

$$\sum_i \left[\gamma \int_0^{u_i} dv\; v[1 - \tanh^2(\gamma v)] - \theta_i \tanh(\gamma u_i)\right].$$

Clearly, $L \geq -\frac{1}{2}\sum_{ij}|J_{ij}| - \sum_i|\theta_i|$; the term in $L[\mathbf{u}]$ with the integral is non–negative. During the noise–free dynamics (4.74) one can use the identity

$$\frac{\partial L}{\partial u_i} = -\gamma\dot{u}_i[1 - \tanh^2(\gamma u_i)],$$

valid only when $J_{ij} = J_{ji}$, to derive [Coo01, SC00, SC01, CKS05]

$$\dot{L} = \frac{\partial L}{\partial u_i} \dot{u}_i = -\gamma \sum_i [1 - \tanh^2(\gamma u_i)] \, \dot{u}_i^2 \leq 0.$$

Again L is found to decrease monotonically, until $\dot{u}_i = 0$ for all i, i.e., until we are at a fixed–point.

The coupled oscillator equations (4.69) reduce in the *noise–free limit* to

$$\dot{\xi}_i(t) = \omega_i + J_{ij} \sin[\xi_j(t) - \xi_i(t)]. \tag{4.75}$$

Note that self-interactions J_{ii} always drop out automatically. For symmetric oscillator networks, a construction of the type followed for the graded–response equations would lead us to propose

Coupled Oscillators Dynamics: $L[\boldsymbol{\xi}] = -\frac{1}{2} J_{ij} \cos[\xi_i - \xi_j] - \omega_i \xi_i.$ (4.76)

This function decreases monotonically, due to $\partial L / \partial \xi_i = -\dot{\xi}_i$:

$$\dot{L} = \frac{\partial L}{\partial \xi_i} \dot{\xi}_i = -\sum_i \dot{\xi}_i^2 \leq 0.$$

Actually, (4.75) describes gradient descent on the surface $L[\boldsymbol{\xi}]$. However, due to the term with the natural frequencies ω_i the function $L[\boldsymbol{\xi}]$ is not bounded, so it cannot be a Lyapunov function. This could have been expected; when $J_{ij} = 0$ for all (i, j), for instance, one finds continually increasing phases, $\xi_i(t) = \xi_i(0) + \omega_i t$. Removing the ω_i, in contrast, gives the bound $L \geq -\sum_j |J_{ij}|$. Now the system must go to a fixed-point. In the special case $\omega_i = \omega$ (N identical natural frequencies) we can transform away the ω_i by putting $\xi(t) = \tilde{\xi}_i(t) + \omega t$, and find the relative phases $\tilde{\xi}_i$ to go to a fixed-point.

4.3.3 Detailed Balance and Equilibrium Statistical Mechanics

The results got above indicate that networks with symmetric synapses are a special class [Coo01, SC00, SC01, CKS05]. We now show how synaptic symmetry is closely related to the detailed balance property, and derive a number of consequences. An ergodic Markov chain of the form (4.60,4.61), i.e.,

$$p_{\ell+1}(\boldsymbol{\sigma}) = \sum_{\boldsymbol{\sigma}'} W[\boldsymbol{\sigma}; \boldsymbol{\sigma}'] p_{\ell}(\boldsymbol{\sigma}'), \tag{4.77}$$

is said to obey detailed balance if its (unique) stationary solution $p_\infty(\boldsymbol{\sigma})$ has the property

$$W[\boldsymbol{\sigma}; \boldsymbol{\sigma}'] p_\infty(\boldsymbol{\sigma}') = W[\boldsymbol{\sigma}'; \boldsymbol{\sigma}] p_\infty(\boldsymbol{\sigma}), \qquad \text{(for all } \boldsymbol{\sigma}, \boldsymbol{\sigma}'). \tag{4.78}$$

All $p_\infty(\boldsymbol{\sigma})$ which satisfy (4.78) are stationary solutions of (4.77), this is easily verified by substitution. The converse is not true. Detailed balance states that, in addition to $p_\infty(\boldsymbol{\sigma})$ being stationary, one has *equilibrium*: there is no net probability current between any two microscopic system states.

It is not a trivial matter to investigate systematically for which choices of the threshold noise distribution $w(\eta)$ and the synaptic matrix $\{J_{ij}\}$ detailed balance holds. It can be shown that, apart from trivial cases (e.g., systems with self–interactions only) a Gaussian distribution $w(\eta)$ will not support detailed balance. Here we will work out details only for the choice $w(\eta) = \frac{1}{2}[1 - \tanh^2(\eta)]$, and for $T > 0$ (where both discrete systems are ergodic). For parallel dynamics the transition matrix is given in (4.60), now with $g[z] = \tanh[z]$, and the detailed balance condition (4.78) becomes

$$\frac{e^{\beta\sigma_i h_i(\boldsymbol{\sigma}')}p_\infty(\boldsymbol{\sigma}')}{\prod_i \cosh[\beta h_i(\boldsymbol{\sigma}')]} = \frac{e^{\beta\sigma_i' h_i(\boldsymbol{\sigma})}p_\infty(\boldsymbol{\sigma})}{\prod_i \cosh[\beta h_i(\boldsymbol{\sigma})]}, \qquad \text{(for all } \boldsymbol{\sigma}, \boldsymbol{\sigma}'\text{).} \qquad (4.79)$$

All $p_\infty(\boldsymbol{\sigma})$ are non-zero (ergodicity), so we may safely put

$$p_\infty(\boldsymbol{\sigma}) = e^{\beta[\theta_i\sigma_i + K(\boldsymbol{\sigma})]} \prod_i \cosh[\beta h_i(\boldsymbol{\sigma})],$$

which, in combination with definition (4.57), simplifies the detailed balance condition to:

$$K(\boldsymbol{\sigma}) - K(\boldsymbol{\sigma}') = \sigma_i\,[J_{ij} - J_{ji}]\,\sigma_j', \qquad \text{(for all } \boldsymbol{\sigma}, \boldsymbol{\sigma}'\text{).} \qquad (4.80)$$

Averaging (4.80) over all possible $\boldsymbol{\sigma}'$ gives $K(\boldsymbol{\sigma}) = \langle K(\boldsymbol{\sigma}')\rangle_{\boldsymbol{\sigma}'}$ for all $\boldsymbol{\sigma}$, i.e., K is a constant, whose value follows from normalizing $p_\infty(\boldsymbol{\sigma})$. So, if detailed balance holds the equilibrium distribution must be [Coo01, SC00, SC01, CKS05]:

$$p_{eq}(\boldsymbol{\sigma}) \sim e^{\beta\theta_i\sigma_i} \prod_i \cosh[\beta h_i(\boldsymbol{\sigma})]. \qquad (4.81)$$

For symmetric systems detailed balance indeed holds: (4.81) solves (4.79), since $K(\boldsymbol{\sigma}) = K$ solves the reduced problem (4.80). For non-symmetric systems, however, there can be no equilibrium. For $K(\boldsymbol{\sigma}) = K$ the condition (4.80) becomes $\sum_{ij}\sigma_i\,[J_{ij} - J_{ji}]\,\sigma_j' = 0$ for all $\boldsymbol{\sigma}, \boldsymbol{\sigma}' \in \{-1, 1\}^N$. For $N \geq 2$ the vector pairs $(\boldsymbol{\sigma}, \boldsymbol{\sigma}')$ span the space of all $N \times N$ matrices, so $J_{ij} - J_{ji}$ must be zero. For $N = 1$ there simply exists no non-symmetric synaptic matrix. In conclusion: for binary networks with parallel dynamics, interaction symmetry implies detailed balance, and vice versa.

For sequential dynamics, with $w(\eta) = \frac{1}{2}[1 - \tanh^2(\eta)]$, the transition matrix is given by (4.61) and the detailed balance condition (4.78) simplifies to

$$\frac{e^{\beta\sigma_i h_i(F_i\boldsymbol{\sigma})}p_\infty(F_i\boldsymbol{\sigma})}{\cosh[\beta h_i(F_i\boldsymbol{\sigma})]} = \frac{e^{-\beta\sigma_i h_i(\boldsymbol{\sigma})}p_\infty(\boldsymbol{\sigma})}{\cosh[\beta h_i(\boldsymbol{\sigma})]}, \qquad \text{(for all } \boldsymbol{\sigma}, i\text{).}$$

Self–interactions J_{ii}, inducing $h_i(F_i\boldsymbol{\sigma}) \neq h_i(\boldsymbol{\sigma})$, complicate matters. Therefore we first consider systems where all $J_{ii} = 0$. All stationary probabilities $p_\infty(\boldsymbol{\sigma})$ being non–zero (ergodicity), we may write:

$$p_\infty(\boldsymbol{\sigma}) = e^{\beta[\theta_i \sigma_i + \frac{1}{2}\sigma_i J_{ij}\sigma_j + K(\boldsymbol{\sigma})]}. \tag{4.82}$$

Using relations like

$$J_{kl}F_i(\sigma_k \sigma_l) = J_{kl}\sigma_k\sigma_l - 2\sigma_i \left[J_{ik} + J_{ki}\right]\sigma_k,$$

we can simplify the detailed balance condition to

$$K(F_i\boldsymbol{\sigma}) - K(\boldsymbol{\sigma}) = \sigma_i \left[J_{ik} - J_{ki}\right]\sigma_k, \qquad \text{(for all } \boldsymbol{\sigma}, i\text{)}.$$

If to this expression we apply the general identity

$$[1 - F_i]\, f(\boldsymbol{\sigma}) = 2\sigma_i \langle \sigma_i f(\boldsymbol{\sigma}) \rangle_{\sigma_i},$$

we find for $i \neq j$ [Coo01, SC00, SC01, CKS05]:

$$K(\boldsymbol{\sigma}) = -2\sigma_i \sigma_j \left[J_{ij} - J_{ji}\right], \qquad \text{(for all } \boldsymbol{\sigma} \text{ and all } i \neq j\text{)}.$$

The left–hand side is symmetric under permutation of the pair (i, j), which implies that the interaction matrix must also be symmetric: $J_{ij} = J_{ji}$ for all (i, j). We now find the trivial solution $K(\boldsymbol{\sigma}) = K$ (constant), detailed balance holds and the corresponding equilibrium distribution is

$$p_{eq}(\boldsymbol{\sigma}) \sim e^{-\beta H(\boldsymbol{\sigma})}, \qquad H(\boldsymbol{\sigma}) = -\frac{1}{2}\sigma_i J_{ij}\sigma_j - \theta_i \sigma_i.$$

In conclusion: for binary networks with sequential dynamics, but without self–interactions, interaction symmetry implies detailed balance, and vice versa. In the case of self–interactions the situation is more complicated. However, here one can still show that non-symmetric models with detailed balance must be pathological, since the requirements can be met only for very specific choices for the $\{J_{ij}\}$.

Now, let us turn to the question of when we find microscopic equilibrium (stationarity without probability currents) in continuous models described by a Fokker–Planck equation (4.67). Note that (4.67) can be seen as a continuity equation for the density of a conserved quantity:

$$\dot{p}_t(\boldsymbol{\sigma}) + \frac{\partial}{\partial \sigma_i} J_i(\boldsymbol{\sigma}, t) = 0.$$

The components $J_i(\boldsymbol{\sigma}, t)$ of the current density are given by

$$J_i(\boldsymbol{\sigma}, t) = [f_i(\boldsymbol{\sigma}) - T\frac{\partial}{\partial \sigma_i}]p_t(\boldsymbol{\sigma}).$$

Stationary distributions $p_\infty(\boldsymbol{\sigma})$ are those which give $\sum_i \frac{\partial}{\partial \sigma_i} J_i(\boldsymbol{\sigma}, \infty) = 0$ (divergence-free currents). Detailed balance implies the stronger statement $J_i(\boldsymbol{\sigma}, \infty) = 0$ for all i (zero currents), so

$$f_i(\boldsymbol{\sigma}) = T \frac{\partial \log p_\infty(\boldsymbol{\sigma})}{\partial \sigma_i}, \qquad \text{or}$$

$$f_i(\boldsymbol{\sigma}) = -\frac{\partial H(\boldsymbol{\sigma})}{\partial \sigma_i}, \qquad p_\infty(\boldsymbol{\sigma}) \sim e^{-\beta H(\boldsymbol{\sigma})}, \qquad (4.83)$$

for some $H(\boldsymbol{\sigma})$, i.e., the forces $f_i(\boldsymbol{\sigma})$ must be conservative. However, one can have conservative forces without a normalizable equilibrium distribution. Just take $H(\boldsymbol{\sigma}) = 0$, i.e., $f_i(\boldsymbol{\sigma}, t) = 0$: here we have $p_{\mathrm{eq}}(\boldsymbol{\sigma}) = C$, which is not normalizable for $\boldsymbol{\sigma} \in \mathbb{R}^N$. For this particular case equation (4.67) is solved easily:

$$p_t(\boldsymbol{\sigma}) = [4\pi T t]^{-N/2} \int d\boldsymbol{\sigma}' \, p_0(\boldsymbol{\sigma}') e^{-[\boldsymbol{\sigma}-\boldsymbol{\sigma}']^2/4Tt},$$

so the limit $\lim_{t\to\infty} p_t(\boldsymbol{\sigma})$ does not exist. One can prove the following (see e.g., [Zin93]). If the forces are conservative and if $p_\infty(\boldsymbol{\sigma}) \sim e^{-\beta H(\boldsymbol{\sigma})}$ is normalizable, then it is the unique stationary solution of the Fokker–Planck equation, to which the system converges for all initial distributions $p_0 \in L^1[\mathbb{R}^N]$ which obey $\int_{\mathbb{R}^N} d\boldsymbol{\sigma} \, e^{\beta H(\boldsymbol{\sigma})} p_0^2(\boldsymbol{\sigma}) < \infty$.

Note that conservative forces must obey [Coo01, SC00, SC01, CKS05]

$$\frac{\partial f_i(\boldsymbol{\sigma})}{\partial \sigma_j} - \frac{\partial f_j(\boldsymbol{\sigma})}{\partial \sigma_i} = 0, \qquad \text{(for all } \boldsymbol{\sigma} \text{ and all } i \neq j\text{).} \qquad (4.84)$$

In the graded–response equations (4.74) the deterministic forces are

$$f_i(\mathbf{u}) = J_{ij} \tanh[\gamma u_j] - u_i + \theta_i, \qquad \text{where}$$

$$\frac{\partial f_i(\mathbf{u})}{\partial u_j} - \frac{\partial f_j(\mathbf{u})}{\partial u_i} = \gamma\{J_{ij}[1 - \tanh^2[\gamma u_j]] - J_{ji}[1 - \tanh^2[\gamma u_i]]\}.$$

At $\mathbf{u} = \mathbf{0}$ this reduces to $J_{ij} - J_{ji}$, i.e., the interaction matrix must be symmetric. For symmetric matrices we find away from $\mathbf{u} = \mathbf{0}$:

$$\frac{\partial f_i(\mathbf{u})}{\partial u_j} - \frac{\partial f_j(\mathbf{u})}{\partial u_i} = \gamma J_{ij}\{\tanh^2[\gamma u_i] - \tanh^2[\gamma u_j]\}.$$

The only way for this to be zero for any $\mathbf{u}$ is by having $J_{ij} = 0$ for all $i \neq j$, i.e., all neurons are disconnected (in this trivial case the system (4.74) does indeed obey detailed balance). Network models of interacting graded–response neurons of the type (4.74) apparently never reach equilibrium, they will always violate detailed balance and exhibit microscopic probability currents. In the case of coupled oscillators (4.69), where the deterministic forces are

$$f_i(\boldsymbol{\xi}) = J_{ij} \sin[\xi_j - \xi_i] + \omega_i,$$

one finds the left-hand side of condition (4.84) to give

$$\frac{\partial f_i(\boldsymbol{\xi})}{\partial \xi_j} - \frac{\partial f_j(\boldsymbol{\xi})}{\partial \xi_i} = [J_{ij} - J_{ji}] \cos[\xi_j - \xi_i].$$

Requiring this to be zero for any $\boldsymbol{\xi}$ gives the condition $J_{ij} = J_{ji}$ for any $i \neq j$. We have already seen that symmetric oscillator networks indeed have conservative forces:

$$f_i(\boldsymbol{\xi}) = -\partial H(\boldsymbol{\xi})/\partial \xi_i, \qquad \text{with} \qquad H(\boldsymbol{\xi}) = -\frac{1}{2} J_{ij} \cos[\xi_i - \xi_j] - \omega_i \xi_i.$$

If in addition we choose all $\omega_i = 0$ the function $H(\boldsymbol{\sigma})$ will also be bounded from below, and, although $p_\infty(\boldsymbol{\xi}) \sim \mathrm{e}^{-\beta H(\boldsymbol{\xi})}$ is still not normalizable on $\boldsymbol{\xi} \in \mathbb{R}^N$, the full 2π−periodicity of the function $H(\boldsymbol{\sigma})$ now allows us to identify $\xi_i - 2\pi \equiv \xi_i$ for all i, so that now $\boldsymbol{\xi} \in [-\pi, \pi]^N$ and $\int d\boldsymbol{\xi}\ \mathrm{e}^{-\beta H(\boldsymbol{\xi})}$ does exist. Thus symmetric coupled oscillator networks with zero natural frequencies obey detailed balance. In the case of non-zero natural frequencies, in contrast, detailed balance does not hold.

The above results establish the link with *equilibrium statistical mechanics* (see e.g., [Yeo92, PB94]). For binary systems with symmetric synapses (in the sequential case: without self-interactions) and with threshold noise distributions of the form

$$w(\eta) = \frac{1}{2}[1 - \tanh^2(\eta)],$$

detailed balance holds and we know the equilibrium distributions. For sequential dynamics it has the Boltzmann form (4.83) and we can apply standard equilibrium statistical mechanics. The parameter β can formally be identified with the inverse 'temperature' in equilibrium, $\beta = T^{-1}$, and the function $H(\boldsymbol{\sigma})$ is the usual *Ising–spin Hamiltonian*. In particular we can define the *partition function Z* and the *free energy F* [Coo01, SC00, SC01, CKS05]:

$$p_{\mathrm{eq}}(\boldsymbol{\sigma}) = \frac{1}{Z} \mathrm{e}^{-\beta H(\boldsymbol{\sigma})}, \qquad H(\boldsymbol{\sigma}) = -\frac{1}{2}\sigma_i J_{ij} \sigma_j - \theta_i \sigma_i, \qquad (4.85)$$

$$Z = \sum_{\boldsymbol{\sigma}} \mathrm{e}^{-\beta H(\boldsymbol{\sigma})}, \qquad F = -\beta^{-1} \log Z. \qquad (4.86)$$

The free energy can be used as the partition function for equilibrium averages. Taking derivatives with respect to external fields θ_i and interactions J_{ij}, for instance, produces $\langle \sigma_i \rangle = -\partial F/\partial \theta_i$ and $\langle \sigma_i \sigma_j \rangle = -\partial F/\partial J_{ij}$, whereas equilibrium averages of arbitrary state variable $f(\boldsymbol{\sigma})$ can be obtained by adding suitable partition terms to the Hamiltonian:

$$H(\boldsymbol{\sigma}) \to H(\boldsymbol{\sigma}) + \lambda f(\boldsymbol{\sigma}), \qquad \langle f \rangle = \lim_{\lambda \to 0} \frac{\partial F}{\partial \lambda}.$$

In the parallel case (4.81) we can again formally write the equilibrium probability distribution in the Boltzmann form [Per84] and define a corresponding partition function $\tilde{Z}$ and a free energy $\tilde{F}$:

$$p_{\mathrm{eq}}(\boldsymbol{\sigma}) = \frac{1}{Z} \mathrm{e}^{-\beta \tilde{H}(\boldsymbol{\sigma})}, \qquad \tilde{H}(\boldsymbol{\sigma}) = -\theta_i \sigma_i - \frac{1}{\beta} \sum_i \log 2 \cosh[\beta h_i(\boldsymbol{\sigma})], \quad (4.87)$$

$$\tilde{Z} = \sum_{\sigma} e^{-\beta \tilde{H}(\sigma)}, \qquad \tilde{F} = -\beta^{-1} \log \tilde{Z}, \qquad (4.88)$$

which again serve to generate averages: $\tilde{H}(\sigma) \to \tilde{H}(\sigma) + \lambda f(\sigma)$, $\langle f \rangle = \lim_{\lambda \to 0} \partial \tilde{F} / \partial \lambda$. However, standard thermodynamic relations involving derivation with respect to β need no longer be valid, and derivation with respect to fields or interactions generates different types of averages, such as [Coo01, SC00, SC01, CKS05]

$$-\frac{\partial \tilde{F}}{\partial \theta_i} = \langle \sigma_i \rangle + \langle \tanh[\beta h_i(\sigma)] \rangle, \qquad -\frac{\partial \tilde{F}}{\partial J_{ii}} = \langle \sigma_i \tanh[\beta h_i(\sigma)] \rangle,$$

$$i \neq j: \qquad \frac{\partial \tilde{F}}{\partial J_{ii}} = \langle \sigma_i \tanh[\beta h_j(\sigma)] \rangle + \langle \sigma_j \tanh[\beta h_i(\sigma)] \rangle.$$

One can use $\langle \sigma_i \rangle = \langle \tanh[\beta h_i(\sigma)] \rangle$, which can be derived directly from the equilibrium equation $p_{eq}(\sigma) = \sum_{\sigma'} W[\sigma; \sigma'] p_{eq}(\sigma)$, to simplify the first of these identities.

A connected network of graded–response neurons can never be in an equilibrium state, so our only model example with continuous neuronal variables for which we can set up the equilibrium statistical mechanics formalism is the system of coupled oscillators (4.69) with symmetric synapses and absent (or uniform) natural frequencies ω_i. If we define the phases as $\xi_i \in [-\pi, \pi]$ we have again an equilibrium distribution of the Boltzmann form, and we can define the standard thermodynamic quantities:

$$p_{eq}(\boldsymbol{\xi}) = \frac{1}{Z} e^{-\beta H(\boldsymbol{\xi})}, \qquad H(\boldsymbol{\xi}) = -\frac{1}{2} J_{ij} \cos[\xi_i - \xi_j], \qquad (4.89)$$

$$Z = \int_{-\pi}^{\pi} \cdots \int_{-\pi}^{\pi} d\boldsymbol{\xi} \, e^{-\beta H(\boldsymbol{\xi})}, \qquad F = -\beta^{-1} \log Z. \qquad (4.90)$$

These generate equilibrium averages in the usual manner. For instance

$$\langle \cos[\xi_i - \xi_j] \rangle = -\frac{\partial F}{\partial J_{ij}},$$

whereas averages of arbitrary state variables $f(\boldsymbol{\xi})$ follow, as before, upon introducing suitable partition terms:

$$H(\boldsymbol{\xi}) \to H(\boldsymbol{\xi}) + \lambda f(\boldsymbol{\xi}), \qquad \langle f \rangle = \lim_{\lambda \to 0} \frac{\partial F}{\partial \lambda}.$$

4.3.4 Simple Recurrent Networks with Binary Neurons

Networks with Uniform Synapses

We now turn to a simple toy model to show how equilibrium statistical mechanics is used for solving neural network models, and to illustrate similarities and differences between the different dynamics types [Coo01, SC00, SC01, CKS05]. We choose uniform infinite-range synapses and zero external fields, and calculate the free energy for the binary systems (4.57,4.58), parallel and sequential, and with threshold–noise distribution $w(\eta) = \frac{1}{2}[1 - \tanh^2(\eta)]$:

$$J_{ij} = J_{ji} = J/N, \qquad (i \neq j), \qquad J_{ii} = \theta_i = 0, \qquad \text{(for all } i\text{)}.$$

The free energy is an extensive object, $\lim_{N \to \infty} F/N$ is finite. For the models (4.57,4.58) we now get:

Binary & Sequential Dynamics:

$$\lim_{N \to \infty} F/N = - \lim_{N \to \infty} (\beta N)^{-1} \log \sum_{\sigma} e^{\beta N\left[\frac{1}{2} Jm^2(\sigma)\right]},$$

Binary & Parallel Dynamics:

$$\lim_{N \to \infty} \tilde{F}/N = - \lim_{N \to \infty} (\beta N)^{-1} \log \sum_{\sigma} e^{N[\log 2 \cosh[\beta Jm(\sigma)]]},$$

with the average activity $m(\sigma) = \frac{1}{N} \sum_k \sigma_k$. We have to count the number of states σ with a prescribed average activity $m = 2n/N - 1$ (n is the number of neurons i with $\sigma_i = 1$), in expressions of the form

$$\frac{1}{N} \log \sum_{\sigma} e^{NU[m(\sigma)]} = \frac{1}{N} \log \sum_{n=0}^{N} \binom{N}{n} e^{NU[2n/N-1]}$$

$$= \frac{1}{N} \log \int_{-1}^{1} dm \; e^{N[\log 2 - c^*(m) + U[m]]},$$

$$\lim_{N \to \infty} \frac{1}{N} \log \sum_{\sigma} e^{NU[m(\sigma)]} = \log 2 + \max_{m \in [-1,1]} \{U[m] - c^*(m)\},$$

with the *entropic function*

$$c^*(m) = \frac{1}{2}(1+m)\log(1+m) + \frac{1}{2}(1-m)\log(1-m).$$

In order to get there we used Stirling's formula to get the leading term of the factorials (only terms which are exponential in N survive the limit $N \to \infty$), we converted (for $N \to \infty$) the summation over n into an integration over $m = 2n/N - 1 \in [-1,1]$, and we carried out the integral over m via *saddle–point integration* (see e.g., [Per92]). This leads to a saddle–point problem whose solution gives the free energies [Coo01, SC00, SC01, CKS05]:

$$\lim_{N \to \infty} F/N = \min_{m \in [-1,1]} f_{\text{seq}}(m), \qquad \beta f_{\text{seq}}(m) = c^*(m) - \log 2 - \frac{1}{2}\beta Jm^2.$$
(4.91)

$$\lim_{N \to \infty} \tilde{F}/N = \min_{m \in [-1,1]} f_{\text{par}}(m), \qquad \beta f_{\text{par}}(m) = c^*(m) - 2\log 2 - \log \cosh[\beta Jm].$$
(4.92)

The equations from which to solve the minima are easily got by differentiation, using $\frac{d}{dm}c^*(m) = \tanh^{-1}(m)$. For sequential dynamics we find

$$\text{Binary \& Sequential Dynamics:} \qquad m = \tanh[\beta Jm], \qquad (4.93)$$

which is the so–called *Curie–Weiss law*. For parallel dynamics we find

$$m = \tanh\left[\beta J \tanh[\beta Jm]\right].$$

One finds that the solutions of the latter equation again obey a Curie–Weiss law. The definition $\hat{m} = \tanh[\beta|J|m]$ transforms it into the coupled equations $m = \tanh[\beta|J|\hat{m}]$ and $\hat{m} = \tanh[\beta|J|m]$, from which we derive

$$0 \le [m - \hat{m}]^2 = [m - \hat{m}]\left[\tanh[\beta|J|\hat{m}] - \tanh[\beta|J|m]\right] \le 0.$$

Since $\tanh[\beta|J|m]$ is a monotonically increasing function of m, this implies $\hat{m} = m$, so

$$\text{Binary \& Parallel Dynamics:} \qquad m = \tanh[\beta|J|m]. \qquad (4.94)$$

Our study of the toy models has thus been reduced to analyzing the nonlinear equations (4.93) and (4.94). If $J \ge 0$ (excitation) the two types of dynamics lead to the same behavior. At high noise levels, $T > J$, both minimisation problems are solved by $m = 0$, describing a disorganized (paramagnetic) state. This can be seen upon writing the right–hand side of (4.93) in integral form [Coo01, SC00, SC01, CKS05]:

$$m^2 = m\tanh[\beta Jm] = \beta Jm^2 \int_0^1 dz\,[1 - \tanh^2[\beta Jmz]] \le \beta Jm^2.$$

So $m^2[1 - \beta J] \le 0$, which gives $m = 0$ as soon as $\beta J < 1$. A phase transition occurs at $T = J$ (a bifurcation of non–trivial solutions of (4.93)), and for $T < J$ the equations for m are solved by the two non-zero solutions of (4.93), describing a state where either all neurons tend to be firing ($m > 0$) or where they tend to be quiet ($m < 0$). This becomes clear when we expand (4.93) for small m: $m = \beta Jm + \mathcal{O}(m^3)$, so precisely at $\beta J = 1$ one finds a de–stabilization of the trivial solution $m = 0$, together with the creation of (two) stable non–trivial ones. Furthermore, using the identity $c^*(\tanh x) = x\tanh x - \log\cosh x$, we get from (4.91,4.92) the relation $\lim_{N\to\infty} \tilde{F}/N = 2\lim_{N\to\infty} F/N$. For $J < 0$ (inhibition), however, the two types of dynamics give quite different results. For sequential dynamics the relevant minimum is located at $m = 0$ (the paramagnetic state). For parallel dynamics, the minimization problem is invariant under $J \to -J$, so the behavior is again of the Curie-Weiss type, with a paramagnetic state for $T > |J|$, a phase transition at $T = |J|$, and order for $T < |J|$. This difference between the two types of dynamics for $J < 0$ is explained by studying dynamics. For the present (toy) model in the limit $N \to \infty$ the average activity evolves in time according to the deterministic laws [Coo01, SC00, SC01, CKS05]

$$\dot{m} = \tanh[\beta J m] - m, \qquad m(t+1) = \tanh[\beta J m(t)],$$

for sequential and parallel dynamics, respectively. For $J < 0$ the sequential system always decays towards the trivial state $m = 0$, whereas for sufficiently large β the parallel system enters the stable limit–cycle $m(t) = M_\beta(-1)^t$, where M_β is the non-zero solution of (4.94). The concepts of 'distance' and 'local minima' are quite different for the two dynamics types; in contrast to the sequential case, parallel dynamics allows the system to make the transition $m \to -m$ in equilibrium.

Phenomenology of Hopfield Models

Recall that the Hopfield model [Hop82] represents a network of binary neurons of the type (4.57,4.58), with threshold noise $w(\eta) = \frac{1}{2}[1 - \tanh^2(\eta)]$, and with a specific recipe for the synapses J_{ij} aimed at storing patterns, motivated by suggestions made in the late nineteen-forties [Heb49]. The original model was in fact defined more narrowly, as the zero noise limit of the system (4.58), but the term has since then been accepted to cover a larger network class. Let us first consider the simplest case and try to store a single pattern $\boldsymbol{\xi} \in \{-1,1\}^N$ in noise–less infinite–range binary networks. Appealing candidates for interactions and thresholds would be $J_{ij} = \xi_i \xi_j$ and $\theta_i = 0$ (for sequential dynamics we put $J_{ii} = 0$ for all i). With this choice the Lyapunov function (4.73) becomes:

$$L_{\text{seq}}[\boldsymbol{\sigma}] = \frac{1}{2}N - \frac{1}{2}[\xi_i \sigma_i]^2.$$

This system indeed reconstructs dynamically the original pattern $\boldsymbol{\xi}$ from an input vector $\boldsymbol{\sigma}(0)$, at least for sequential dynamics. However, *en passant* we have created an additional attractor: the state -$\boldsymbol{\xi}$. This property is shared by all binary models in which the external fields are zero, where the Hamiltonians $H(\boldsymbol{\sigma})$ (4.85) and $\tilde{H}(\boldsymbol{\sigma})$ (4.87) are invariant under an overall sign change $\boldsymbol{\sigma} \to -\boldsymbol{\sigma}$. A second feature common to several (but not all) attractor neural networks is that *each* initial state will lead to pattern reconstruction, even nonsensical (random) ones.

The Hopfield model is got by generalizing the previous simple one-pattern recipe to the case of an arbitrary number p of binary patterns $\boldsymbol{\xi}^\mu = (\xi_1^\mu, \ldots, \xi_N^\mu) \in \{-1,1\}^N$ [Coo01, SC00, SC01, CKS05]:

$$J_{ij} = \frac{1}{N}\xi_i^\mu \xi_j^\mu, \qquad \theta_i = 0 \qquad (\text{for all } i; \quad \mu = 1, ..., p), \qquad (4.95)$$

$$(\text{sequential dynamics}: \ J_{ii} \to 0, \quad \text{for all } i).$$

The prefactor N^{-1} has been inserted to ensure that the limit $N \to \infty$ will exist in future expressions. The process of interest is that where, triggered by correlation between the initial state and a stored pattern $\boldsymbol{\xi}^\lambda$, the state vector $\boldsymbol{\sigma}$ evolves towards $\boldsymbol{\xi}^\lambda$. If this happens, pattern $\boldsymbol{\xi}^\lambda$ is said to be recalled.

The similarity between a state vector and the stored patterns is measured by so–called *Hopfield overlaps*

$$m_\mu(\boldsymbol{\sigma}) = \frac{1}{N} \xi_i^\mu \sigma_i. \tag{4.96}$$

The Hopfield model represents as an associative memory, in which the recall process is described in terms of overlaps.

Analysis of Hopfield Models Away From Saturation

A binary Hopfield network with parameters given by (4.95) obeys detailed balance, and the Hamiltonian $H(\boldsymbol{\sigma})$ (4.85) (corresponding to sequential dynamics) and the pseudo-Hamiltonian $\tilde{H}(\boldsymbol{\sigma})$ (4.87) (corresponding to parallel dynamics) become [Coo01, SC00, SC01, CKS05]

$$H(\boldsymbol{\sigma}) = -\frac{1}{2} N \sum_{\mu=1}^{p} m_\mu^2(\boldsymbol{\sigma}) + \frac{1}{2} p, \tag{4.97}$$

$$\tilde{H}(\boldsymbol{\sigma}) = -\frac{1}{\beta} \sum_i \log 2 \cosh[\beta \xi_i^\mu m_\mu(\boldsymbol{\sigma})],$$

with the overlaps (4.96). Solving the statics implies calculating the free energies F and $\tilde{F}$:

$$F = -\frac{1}{\beta} \log \sum_{\boldsymbol{\sigma}} e^{-\beta H(\boldsymbol{\sigma})}, \qquad \tilde{F} = -\frac{1}{\beta} \log \sum_{\boldsymbol{\sigma}} e^{-\beta \tilde{H}(\boldsymbol{\sigma})}.$$

Upon introducing the short-hand notation $\mathbf{m} = (m_1, \ldots, m_p)$ and $\boldsymbol{\xi}_i = (\xi_i^1, \ldots, \xi_i^p)$, both free energies can be expressed in terms of the density of states $\mathcal{D}(\mathbf{m}) = 2^{-N} \sum_{\boldsymbol{\sigma}} \delta[\mathbf{m} - \mathbf{m}(\boldsymbol{\sigma})]$:

$$F/N = -\frac{1}{\beta} \log 2 - \frac{1}{\beta N} \log \int d\mathbf{m} \, \mathcal{D}(\mathbf{m}) \, e^{\frac{1}{2} \beta N \mathbf{m}^2} + \frac{p}{2N}, \tag{4.98}$$

$$\tilde{F}/N = -\frac{1}{\beta} \log 2 - \frac{1}{\beta N} \log \int d\mathbf{m} \, \mathcal{D}(\mathbf{m}) \, e^{\sum_{i=1}^{N} \log 2 \cosh[\beta \boldsymbol{\xi}_i \cdot \mathbf{m}]}, \tag{4.99}$$

using $\int d\mathbf{m} \, \delta[\mathbf{m} - \mathbf{m}(\boldsymbol{\sigma})] = 1$. In order to proceed, we need to specify how the number of patterns p scales with the system size N. In this section we will follow [AGS85] (equilibrium analysis following sequential dynamics) and [FK88] (equilibrium analysis following parallel dynamics), and assume p to be finite. One can now easily calculate the leading contribution to the density of states, using the integral representation of the δ−function and keeping in mind that according to (4.98,4.99) only terms exponential in N will retain statistical relevance for $N \to \infty$:

$$\lim_{N\to\infty} \frac{1}{N} \log \mathcal{D}(\mathbf{m}) = \lim_{N\to\infty} \frac{1}{N} \log \int d\mathbf{x}\; \mathrm{e}^{iN\mathbf{x}\cdot\mathbf{m}} \langle \mathrm{e}^{-i\sigma_i \boldsymbol{\xi}_i \cdot \mathbf{x}} \rangle_{\boldsymbol{\sigma}}$$

$$= \lim_{N\to\infty} \frac{1}{N} \log \int d\mathbf{x}\; \mathrm{e}^{N[i\mathbf{x}\cdot\mathbf{m} + \langle \log \cos[\boldsymbol{\xi}\cdot\mathbf{x}] \rangle_{\boldsymbol{\xi}}]},$$

with the abbreviation $\langle \Phi(\boldsymbol{\xi}) \rangle_{\boldsymbol{\xi}} = \lim_{N\to\infty} \frac{1}{N} \sum_{i=1}^{N} \Phi(\boldsymbol{\xi}_i)$. The leading contribution to both free energies can be expressed as a finite-dimensional integral, for large N dominated by that saddle–point (extremum) for which the extensive exponent is real and maximal [Coo01, SC00, SC01, CKS05]:

$$\lim_{N\to\infty} F/N = -\frac{1}{\beta N} \log \int d\mathbf{m}d\mathbf{x}\; \mathrm{e}^{-N\beta f(\mathbf{m},\mathbf{x})} = \mathrm{extr}_{\mathbf{x},\mathbf{m}}\; f(\mathbf{m},\mathbf{x}),$$

$$\lim_{N\to\infty} \tilde{F}/N = -\frac{1}{\beta N} \log \int d\mathbf{m}d\mathbf{x}\; \mathrm{e}^{-N\beta \tilde{f}(\mathbf{m},\mathbf{x})} = \mathrm{extr}_{\mathbf{x},\mathbf{m}}\; \tilde{f}(\mathbf{m},\mathbf{x}), \qquad \text{with}$$

$$f(\mathbf{m},\mathbf{x}) = -\tfrac{1}{2}\mathbf{m}^2 - i\mathbf{x}\cdot\mathbf{m} - \beta^{-1}\langle \log 2\cos\left[\beta\boldsymbol{\xi}\cdot\mathbf{x}\right]\rangle_{\boldsymbol{\xi}},$$

$$\tilde{f}(\mathbf{m},\mathbf{x}) = -\beta^{-1}\langle \log 2\cosh\left[\beta\boldsymbol{\xi}\cdot\mathbf{m}\right]\rangle_{\boldsymbol{\xi}} - i\mathbf{x}\cdot\mathbf{m} - \beta^{-1}\langle \log 2\cos\left[\beta\boldsymbol{\xi}\cdot\mathbf{x}\right]\rangle_{\boldsymbol{\xi}}.$$

The saddle–point equations for f and $\tilde{f}$ are given by:

$$f : \mathbf{x} = i\mathbf{m}, \qquad\qquad i\mathbf{m} = \langle \boldsymbol{\xi}\tan\left[\beta\boldsymbol{\xi}\cdot\mathbf{x}\right]\rangle_{\boldsymbol{\xi}},$$

$$\tilde{f} : \mathbf{x} = i\langle \boldsymbol{\xi}\tanh\left[\beta\boldsymbol{\xi}\cdot\mathbf{m}\right]\rangle_{\boldsymbol{\xi}},\; i\mathbf{m} = \langle \boldsymbol{\xi}\tan\left[\beta\boldsymbol{\xi}\cdot\mathbf{x}\right]\rangle_{\boldsymbol{\xi}}.$$

In saddle-points $\mathbf{x}$ turns out to be purely imaginary. However, after a shift of the integration contours, putting $\mathbf{x} = i\mathbf{x}^\star(\mathbf{m}) + \mathbf{y}$ (where $i\mathbf{x}^\star(\mathbf{m})$ is the imaginary saddle–point, and where $\mathbf{y} \in \mathbb{R}^p$) we can eliminate $\mathbf{x}$ in favor of $\mathbf{y} \in \mathbb{R}^p$ which does have a real saddle–point, by construction.(Our functions to be integrated have no poles, but strictly speaking we still have to verify that the integration segments linking the original integration regime to the shifted one will not contribute to the integrals. This is generally a tedious and distracting task, which is often skipped. For simple models, however (e.g., networks with uniform synapses), the verification can be carried out properly, and all is found to be safe.) We then get

Sequential Dynamics: $\mathbf{m} = \langle \boldsymbol{\xi}\tanh[\beta\boldsymbol{\xi}\cdot\mathbf{m}]\rangle_{\boldsymbol{\xi}}$,

Parallel Dynamics: $\mathbf{m} = \langle \boldsymbol{\xi}\tanh[\beta\boldsymbol{\xi}\cdot[\langle \boldsymbol{\xi}'\tanh[\beta\boldsymbol{\xi}'\cdot\mathbf{m}]\rangle_{\boldsymbol{\xi}'}]]\rangle_{\boldsymbol{\xi}}$,

(compare to e.g., (4.93,4.94)). The solutions of the above two equations will in general be identical. To see this, let us denote $\hat{\mathbf{m}} = \langle \boldsymbol{\xi}\tanh\left[\beta\boldsymbol{\xi}\cdot\mathbf{m}\right]\rangle_{\boldsymbol{\xi}}$, with which the saddle point equation for $\tilde{f}$ decouples into:

$$\mathbf{m} = \langle \boldsymbol{\xi}\tanh\left[\beta\boldsymbol{\xi}\cdot\hat{\mathbf{m}}\right]\rangle_{\boldsymbol{\xi}}, \qquad \hat{\mathbf{m}} = \langle \boldsymbol{\xi}\tanh\left[\beta\boldsymbol{\xi}\cdot\mathbf{m}\right]\rangle_{\boldsymbol{\xi}}, \qquad \text{so}$$

$$[\mathbf{m} - \hat{\mathbf{m}}]^2 = \langle [(\boldsymbol{\xi}\cdot\mathbf{m}) - (\boldsymbol{\xi}\cdot\hat{\mathbf{m}})]\left[\tanh(\beta\boldsymbol{\xi}\cdot\hat{\mathbf{m}}) - \tanh(\beta\boldsymbol{\xi}\cdot\mathbf{m})\right]\rangle_{\boldsymbol{\xi}}.$$

Since tanh is a monotonicaly–increasing function, we must have $[\mathbf{m} - \hat{\mathbf{m}}]\cdot\boldsymbol{\xi} = 0$ for each $\boldsymbol{\xi}$ that contributes to the averages $\langle\ldots\rangle_{\boldsymbol{\xi}}$. For all choices of patterns where the covariance matrix $C_{\mu\nu} = \langle \xi_\mu \xi_\nu \rangle_{\boldsymbol{\xi}}$ is positive definite, we thus get $\mathbf{m} = \hat{\mathbf{m}}$. The final result is: for both types of dynamics (sequential and parallel) the overlap order parameters in equilibrium are given by the solution $\mathbf{m}^*$ of

$$\mathbf{m} = \langle \boldsymbol{\xi} \tanh [\beta \boldsymbol{\xi} \cdot \mathbf{m}] \rangle_{\boldsymbol{\xi}}, \qquad \text{which minimises} \qquad (4.100)$$

$$f(\mathbf{m}) = \frac{1}{2}\mathbf{m}^2 - \frac{1}{\beta}\langle \log 2 \cosh [\beta \boldsymbol{\xi} \cdot \mathbf{m}] \rangle_{\boldsymbol{\xi}}. \qquad (4.101)$$

The free energies of the ergodic components are $\lim_{N\to\infty} F/N = f(\mathbf{m}^*)$ and $\lim_{N\to\infty} \tilde{F}/N = 2f(\mathbf{m}^*)$. Adding partition terms of the form $H \to H + \lambda g[\mathbf{m}(\boldsymbol{\sigma})]$ to the Hamiltonians allows us identify $\langle g[\mathbf{m}(\boldsymbol{\sigma})]\rangle_{\text{eq}} = \lim_{\lambda\to 0} \partial F/\partial\lambda = g[\mathbf{m}^*]$. Thus, in equilibrium the fluctuations in the overlap order parameters $\mathbf{m}(\boldsymbol{\sigma})$ (4.96) vanish for $N \to \infty$. Their deterministic values are simply given by $\mathbf{m}^*$. Note that in the case of sequential dynamics we could also have used linearization with Gaussian integrals (as used previously for coupled oscillators with uniform synapses) to arrive at this solution, with p auxiliary integrations, but that for parallel dynamics this would not have been possible.

Now, in analysis of *order parameter equations*, we will restrict our further discussion to the case of randomly drawn patterns, so [Coo01, SC00, SC01, CKS05]

$$\langle \Phi(\boldsymbol{\xi}) \rangle_{\boldsymbol{\xi}} = 2^{-p} \sum_{\boldsymbol{\xi}\in\{-1,1\}^p} \Phi(\boldsymbol{\xi}), \qquad \langle \xi_\mu \rangle_{\boldsymbol{\xi}} = 0, \qquad \langle \xi_\mu \xi_\nu \rangle_{\boldsymbol{\xi}} = \delta_{\mu\nu}.$$

We first establish an upper bound for the temperature for where non–trivial solutions $\mathbf{m}^*$ could exist, by writing (4.100) in integral form:

$$m_\mu = \beta \langle \xi_\mu (\boldsymbol{\xi} \cdot \mathbf{m}) \int_0^1 d\lambda [1 - \tanh^2[\beta\lambda\boldsymbol{\xi} \cdot \mathbf{m}]] \rangle_{\boldsymbol{\xi}},$$

from which we deduce

$$0 = \mathbf{m}^2 - \beta\langle (\boldsymbol{\xi} \cdot \mathbf{m})^2 \int_0^1 d\lambda [1 - \tanh^2 [\beta\lambda\boldsymbol{\xi} \cdot \mathbf{m}]] \rangle_{\boldsymbol{\xi}}$$

$$\geq \mathbf{m}^2 - \beta\langle (\boldsymbol{\xi} \cdot \mathbf{m})^2 \rangle_{\boldsymbol{\xi}} = \mathbf{m}^2(1 - \beta),$$

For $T > 1$ the only solution of (4.100) is the paramagnetic state $\mathbf{m} = 0$, which gives for the free energy per neuron $-T\log 2$ and $-2T\log 2$ (for sequential and parallel dynamics, respectively). At $T = 1$ a phase transition occurs, which follows from expanding (4.100) for small $|\mathbf{m}|$ in powers of $\tau = \beta - 1$:

$$m_\mu = (1+\tau)m_\mu - \frac{1}{3}m_\nu m_\rho m_\lambda \langle \xi_\mu \xi_\nu \xi_\rho \xi_\lambda \rangle_{\boldsymbol{\xi}}$$

$$+ \mathcal{O}(\mathbf{m}^5, \tau\mathbf{m}^3) = m_\mu[1 + \tau - \mathbf{m}^2 + \frac{2}{3}m_\mu^2] + \mathcal{O}(\mathbf{m}^5, \tau\mathbf{m}^3).$$

The new saddle–point scales as $m_\mu = \tilde{m}_\mu \tau^{1/2} + \mathcal{O}(\tau^{3/2})$, with for each μ: $\tilde{m}_\mu = 0$ or $0 = 1 - \tilde{\mathbf{m}}^2 + \frac{2}{3}\tilde{m}_\mu^2$.

The solutions are of the form $\tilde{m}_\mu \in \{-\tilde{m}, 0, \tilde{m}\}$. If we denote with n the number of non-zero components in the vector $\hat{\mathbf{m}}$, we derive from the above identities: $\tilde{m}_\mu = 0$ or $\tilde{m}_\mu = \pm\sqrt{3}/\sqrt{3n-2}$. These saddle-points are called *mixture states*, since they correspond to microscopic configurations correlated equally with a finite number n of the stored patterns (or their negatives). Without loss of generality we can always perform gauge transformations on the set of stored patterns (permutations and reflections), such that the mixture states acquire the form [Coo01, SC00, SC01, CKS05]

$$\mathbf{m} = m_n(\overbrace{1,\ldots,1}^{n \text{ times}}, \overbrace{0,\ldots,0}^{p-n \text{ times}}), \qquad (4.102)$$
$$m_n = [\frac{3}{3n-2}]^{\frac{1}{2}}(\beta - 1)^{1/2} + \ldots$$

These states are in fact saddle–points of the surface $f(\mathbf{m})$ (4.101) for any finite temperature, as can be verified by substituting (4.102) as an *ansatz* into (4.100):

$$\mu \leq n: \qquad m_n = \langle \xi_\mu \tanh[\beta m_n \sum_{\nu \leq n} \xi_\nu]\rangle_{\boldsymbol{\xi}},$$

$$\mu > n: \qquad 0 = \langle \xi_\mu \tanh[\beta m_n \sum_{\nu \leq n} \xi_\nu]\rangle_{\boldsymbol{\xi}}.$$

The second equation is automatically satisfied since the average factorizes. The first equation leads to a condition determining the amplitude m_n of the mixture states:

$$m_n = \langle [\frac{1}{n} \sum_{\mu \leq n} \xi_\mu] \tanh[\beta m_n \sum_{\nu \leq n} \xi_\nu]\rangle_{\boldsymbol{\xi}}. \qquad (4.103)$$

The corresponding values of $f(\mathbf{m})$, to be denoted by f_n, are

$$f_n = \frac{1}{2}nm_n^2 - \frac{1}{\beta}\langle \log 2 \cosh[\beta m_n \sum_{\nu \leq n} \xi_\nu]\rangle_{\boldsymbol{\xi}}. \qquad (4.104)$$

The relevant question at this stage is whether or not these saddle-points correspond to local minima of the surface $f(\mathbf{m})$ (4.101). The second derivative of $f(\mathbf{m})$ is given by

$$\frac{\partial^2 f(\mathbf{m})}{\partial m_\mu \partial m_\nu} = \delta_{\mu\nu} - \beta\langle \xi_\mu \xi_\nu [1 - \tanh^2[\beta\boldsymbol{\xi}\cdot\mathbf{m}]]\rangle_{\boldsymbol{\xi}}, \qquad (4.105)$$

where a local minimum corresponds to a positive definite second derivative. In the trivial saddle–point $\mathbf{m} = 0$ this gives simply $\delta_{\mu\nu}(1-\beta)$, so at $T = 1$ this state destabilizes. In a mixture state of the type (4.102) the second derivative becomes:

$$D_{\mu\nu}^{(n)} = \delta_{\mu\nu} - \beta\langle \xi_\mu \xi_\nu [1 - \tanh^2[\beta m_n \sum_{\rho \leq n} \xi_\rho]]\rangle_{\boldsymbol{\xi}}.$$

Due to the symmetries in the problem the spectrum of the matrix $D^{(n)}$ can be calculated. One finds the following eigen–spaces, with

$$Q = \langle \tanh^2[\beta m_n \sum_{\rho \leq n} \xi_\rho] \rangle_{\boldsymbol{\xi}} \qquad \text{and} \qquad R = \langle \xi_1 \xi_2 \tanh^2[\beta m_n \sum_{\rho \leq n} \xi_\rho] \rangle_{\boldsymbol{\xi}},$$

Eigenspace :	Eigenvalue :
$I:$ $\mathbf{x} = (0, \ldots, 0, x_{n+1}, \ldots, x_p)$,	$1 - \beta[1 - Q]$,
$II:$ $\mathbf{x} = (1, \ldots, 1, 0, \ldots, 0)$,	$1 - \beta[1 - Q + (1-n)R]$,
$III: \mathbf{x} = (x_1, \ldots, x_n, 0, \ldots, 0)$, $\sum_\mu x_\mu = 0$,	$1 - \beta[1 - Q + R]$.

The eigen–space III and the quantity R only come into play for $n > 1$. To find the smallest eigenvalue we need to know the sign of R. With the abbreviation $M_{\boldsymbol{\xi}} = \sum_{\rho \leq n} \xi_\rho$ we find [Coo01, SC00, SC01, CKS05]:

$$n(n-1)R = \langle M_{\boldsymbol{\xi}}^2 \tanh^2[\beta m_n M_{\boldsymbol{\xi}}] \rangle_{\boldsymbol{\xi}} - n\langle \tanh^2[\beta m_n M_{\boldsymbol{\xi}}] \rangle_{\boldsymbol{\xi}}$$

$$= \langle [M_{\boldsymbol{\xi}}^2 - \langle M_{\boldsymbol{\xi}'}^2 \rangle_{\boldsymbol{\xi}'}] \tanh^2[\beta m_n |M_{\boldsymbol{\xi}}|] \rangle_{\boldsymbol{\xi}}$$

$$= \langle [M_{\boldsymbol{\xi}}^2 - \langle M_{\boldsymbol{\xi}'}^2 \rangle_{\boldsymbol{\xi}'}] \left\{ \tanh^2[\beta m_n \sqrt{M_{\boldsymbol{\xi}}^2}] - \tanh^2[\beta m_n \sqrt{\langle M_{\boldsymbol{\xi}'}^2 \rangle_{\boldsymbol{\xi}'}}] \right\} \rangle_{\boldsymbol{\xi}} \geq 0.$$

We may now identify the conditions for an $n-$mixture state to be a local minimum of $f(\mathbf{m})$. For $n = 1$ the relevant eigenvalue is I, now the quantity Q simplifies considerably. For $n > 1$ the relevant eigenvalue is III, here we can combine Q and R into one single average:

$$n = 1 : 1 - \beta[1 - \tanh^2[\beta m_1]] > 0$$
$$n = 2 : 1 - \beta > 0$$
$$n \geq 3 : 1 - \beta[1 - \langle \tanh^2[\beta m_n \sum_{\rho=3}^n \xi_\rho] \rangle_{\boldsymbol{\xi}}] > 0$$

The $n = 1$ states, correlated with one pattern only, are the desired solutions. They are stable for all $T < 1$, since partial differentiation with respect to β of the $n = 1$ amplitude equation (4.103) gives

$$m_1 = \tanh[\beta m_1] \quad \rightarrow \quad 1 - \beta[1 - \tanh^2[\beta m_1]]$$
$$= m_1[1 - \tanh^2[\beta m_1]](\partial m_1 / \partial \beta)^{-1},$$

so that clearly $\text{sgn}[m_1] = \text{sgn}[\partial m_1 / \partial \beta]$. The $n = 2$ mixtures are always unstable. For $n \geq 3$ we have to solve the amplitude equations (4.103) numerically to evaluate their stability. It turns out that only for odd n will there be a critical temperature below which the $n-$mixture states are local minima of $f(\mathbf{m})$.

We have now solved the model in equilibrium for finite p and $N \rightarrow \infty$. For non–random patterns one simply has to study the bifurcation properties of equation (4.100) for the new pattern statistics at hand; this is only qualitatively different from the random pattern analysis explained above. The occurrence of multiple saddle–points corresponding to local minima of the free energy signals ergodicity breaking. Although among these only the *global minimum* will correspond to the thermodynamic equilibrium state, the non–global minima correspond to true ergodic components, i.e., on finite time–scales they will be just as relevant as the global minimum.

4.3.5 Simple Recurrent Networks of Coupled Oscillators

Coupled Oscillators with Uniform Synapses

Models with continuous variables involve integration over states, rather than summation. For a coupled oscillator network (4.69) with uniform synapses $J_{ij} = J/N$ and zero frequencies $\omega_i = 0$ (which is a simple version of the model in [Kur84]) we get for the free energy per oscillator [Coo01, SC00, SC01, CKS05]:

$$\lim_{N \to \infty} F/N = -\lim_{N \to \infty} \frac{1}{\beta N} \log \int_{-\pi}^{\pi} \cdots \int_{-\pi}^{\pi} d\boldsymbol{\xi} \times$$
$$\times e^{(\beta J/2N)\left[[\sum_i \cos(\xi_i)]^2 + [\sum_i \sin(\xi_i)]^2 \right]}.$$

We would now have to 'count' microscopic states with prescribed average cosines and sines. A faster route exploits auxiliary Gaussian integrals, via the identity

$$e^{\frac{1}{2} y^2} = \int Dz \, e^{yz}, \tag{4.106}$$

with the short–hand $Dx = (2\pi)^{-\frac{1}{2}} e^{-\frac{1}{2} x^2} dx$ (this alternative would also have been open to us in the binary case; my aim in this section is to explain both methods):

$$\lim_{N \to \infty} F/N = -\lim_{N \to \infty} \frac{1}{\beta N} \log \int_{-\pi}^{\pi} \cdots \int_{-\pi}^{\pi} d\boldsymbol{\xi} \int DxDy \times$$
$$\times e^{\sqrt{\beta J/N} \left[x \sum_i \cos(\xi_i) + y \sum_i \sin(\xi_i) \right]}$$
$$= -\lim_{N \to \infty} \frac{1}{\beta N} \log \int DxDy \left[\int_{-\pi}^{\pi} d\xi \, e^{\cos(\xi)\sqrt{\beta J(x^2+y^2)/N}} \right]^N$$
$$= -\lim_{N \to \infty} \frac{1}{\beta N} \log \int_0^{\infty} dq \, q e^{-\frac{1}{2} N \beta |J| q^2} \times$$
$$\times \left[\int_{-\pi}^{\pi} d\xi \, e^{\beta |J| q \cos(\xi) \sqrt{\mathrm{rmsgn}(J)}} \right]^N,$$

where we have transformed to polar coordinates, $(x, y) = q\sqrt{\beta |J| N}(\cos\theta, \sin\theta)$, and where we have already eliminated (constant) terms which will not survive the limit $N \to \infty$. Thus, saddle–point integration gives us, quite similar to the previous cases (4.91,4.92):

$$\lim_{N\to\infty} F/N = \min_{q\geq 0} f(q), \qquad \begin{array}{l} J > 0 : \beta f(q) = \frac{1}{2}\beta|J|q^2 - \log[2\pi I_0(\beta|J|q)] \\ J < 0 : \beta f(q) = \frac{1}{2}\beta|J|q^2 - \log[2\pi I_0(i\beta|J|q)] \end{array},$$

$$(4.107)$$

in which the $I_n(z)$ are the Bessel functions (see e.g., [AS72]). The equations from which to solve the minima are got by differentiation, using $\frac{d}{dz}I_0(z) = I_1(z)$:in which the $I_n(z)$ are the *Bessel functions* (see e.g., [AS72]). The equations from which to solve the minima are got by differentiation, using $\frac{d}{dz}I_0(z) = I_1(z)$:

$$J > 0 : \qquad q = \frac{I_1(\beta|J|q)}{I_0(\beta|J|q)}, \qquad J < 0 : \qquad q = i\,\frac{I_1(i\beta|J|q)}{I_0(i\beta|J|q)}. \qquad (4.108)$$

Again, in both cases the problem has been reduced to studying a single non-linear equation. The physical meaning of the solution follows from the identity $-2\partial F/\partial J = \langle N^{-1}\sum_{i\neq j}\cos(\xi_i - \xi_j)\rangle$:

$$\lim_{N\to\infty}\langle[\frac{1}{N}\sum_i \cos(\xi_i)]^2\rangle + \lim_{N\to\infty}\langle[\frac{1}{N}\sum_i \sin(\xi_i)]^2\rangle = \operatorname{sgn}(J)\,q^2.$$

From this equation it also follows that $q \leq 1$. Note: since $\partial f(q)/\partial q = 0$ at the minimum, one only needs to consider the explicit derivative of $f(q)$ with respect to J. If the synapses induce anti-synchronization, $J < 0$, the only solution of (4.108) (and the minimum in (4.107)) is the trivial state $q = 0$. This also follows immediately from the equation which gave the physical meaning of q. For synchronizing forces, $J > 0$, on the other hand, we again find the trivial solution at high noise levels, but a globally synchronized state with $q > 0$ at low noise levels. Here a phase transition occurs at $T = \frac{1}{2}J$ (a bifurcation of non–trivial solutions of (4.108)), and for $T < \frac{1}{2}J$ the minimum of (4.107) is found at two non-zero values for q. The critical noise level is again found upon expanding the saddle–point equation, using

$$I_0(z) = 1 + \mathcal{O}(z^2) \qquad \text{and} \qquad I_1(z) = \frac{1}{2}z + \mathcal{O}(z^3) : q = \frac{1}{2}\beta Jq + \mathcal{O}(q^3).$$

Precisely at $\beta J = 2$ one finds a de-stabilization of the trivial solution $q = 0$, together with the creation of (two) stable non–trivial ones. Note that, in view of (4.107), we are only interested in non–negative values of q. One can prove, using the properties of the Bessel functions, that there are no other (discontinuous) bifurcations of non–trivial solutions of the saddle–point equation. Note, finally, that the absence of a state with global anti-synchronization for $J < 0$ has the same origin as the absence of an anti-ferromagnetic state for $J < 0$ in the previous models with binary neurons. Due to the long-range nature of the synapses $J_{ij} = J/N$ such states simply cannot exist: whereas any set of oscillators can be in a fully synchronized state, if two oscillators are in anti-synchrony it is already impossible for a third to be simultaneously in anti-synchrony with the first two (since anti-synchrony with one implies synchrony with the other) [Coo01, SC00, SC01, CKS05].

Coupled Oscillator Attractor Networks

Let us now turn to an alternative realisation of information storage in a recurrent network based upon the creation of attractors [Coo01, SC00, SC01, CKS05]. We will solve models of coupled neural oscillators of the type (4.69), with zero natural frequencies (since we wish to use equilibrium techniques), in which real-valued patterns are stored as stable configurations of oscillator phases, following [Coo89]. Let us, however, first find out how to store a single pattern $\boldsymbol{\xi} \in [-\pi, \pi]^N$ in a noise-less infinite-range oscillator network. For simplicity we will draw each component ξ_i independently at random from $[-\pi, \pi]$, with uniform probability density. This allows us to use asymptotic properties such as $|N^{-1} \sum_j e^{i\ell\xi_j}| = \mathcal{O}(N^{-\frac{1}{2}})$ for any integer ℓ. A sensible choice for the synapses would be $J_{ij} = \cos[\xi_i - \xi_j]$. To see this we work out the corresponding Lyapunov function (4.76):

$$L[\boldsymbol{\xi}] = -\frac{1}{2N^2} \cos[\xi_i - \xi_j] \cos[\xi_i - \xi_j],$$

$$L[\boldsymbol{\xi}] = -\frac{1}{2N^2} \cos^2[\xi_i - \xi_j] = -\frac{1}{4} + \mathcal{O}(N^{-\frac{1}{2}}),$$

where the factors of N have been inserted to achieve appropriate scaling in the $N \to \infty$ limit. The function $L[\boldsymbol{\xi}]$, which is obviously bounded from below, must decrease monotonically during the dynamics. To find out whether the state $\boldsymbol{\xi}$ is a stable fixed-point of the dynamics we have to calculate L and derivatives of L at $\boldsymbol{\xi} = \boldsymbol{\xi}$:

$$\left.\frac{\partial L}{\partial \xi_i}\right|_{\boldsymbol{\xi}} = \frac{1}{2N^2} \sum_j \sin[2(\xi_i - \xi_j)],$$

$$\left.\frac{\partial^2 L}{\partial \xi_i^2}\right|_{\boldsymbol{\xi}} = \frac{1}{N^2} \sum_j \cos^2[\xi_i - \xi_j],$$

$$i \neq j : \quad \left.\frac{\partial^2 L}{\partial \xi_i \partial \xi_j}\right|_{\boldsymbol{\xi}} = -\frac{1}{N^2} \cos^2[\xi_i - \xi_j].$$

Clearly $\lim_{N \to \infty} L[\boldsymbol{\xi}] = -\frac{1}{4}$. Putting $\boldsymbol{\xi} = \boldsymbol{\xi} + \Delta\boldsymbol{\xi}$, with $\Delta\xi_i = \mathcal{O}(N^0)$, we find

$$L[\boldsymbol{\xi} + \Delta\boldsymbol{\xi}] - L[\boldsymbol{\xi}] = \Delta\xi_i \frac{\partial L}{\partial \xi_i}|_{\boldsymbol{\xi}} \qquad (4.109)$$

$$+ \frac{1}{2} \Delta\xi_i \Delta\xi_j \frac{\partial^2 L}{\partial \xi_i \partial \xi_j}|_{\boldsymbol{\xi}} + \mathcal{O}(\Delta\boldsymbol{\xi}^3)$$

$$= \frac{1}{4N} \sum_i \Delta\xi_i^2 - \frac{1}{2N^2} \Delta\xi_i \Delta\xi_j \cos^2[\xi_i - \xi_j] + \mathcal{O}(N^{-\frac{1}{2}}, \Delta\boldsymbol{\xi}^3)$$

$$= \frac{1}{4}\{\frac{1}{N}\sum_i \Delta\xi_i^2 - [\frac{1}{N}\sum_i \Delta\xi_i]^2 - [\frac{1}{N}\Delta\xi_i \cos(2\xi_i)]^2$$

$$-[\frac{1}{N}\sum_i \Delta\xi_i \sin(2\xi_i)]^2\} + \mathcal{O}(N^{-\frac{1}{2}}, \Delta\boldsymbol{\xi}^3).$$

In leading order in N the following three vectors in $\mathbb{R}^N$ are normalized and orthogonal:

$$\mathbf{e}_1 = \frac{1}{\sqrt{N}}(1, 1, \ldots, 1), \qquad \mathbf{e}_2 = \frac{\sqrt{2}}{\sqrt{N}}(\cos(2\xi_1), \ldots, \cos(2\xi_N)),$$

$$\mathbf{e}_2 = \frac{\sqrt{2}}{\sqrt{N}}(\sin(2\xi_1), \ldots, \sin(2\xi_N)).$$

We may therefore use

$$\Delta\boldsymbol{\xi}^2 \geq (\Delta\boldsymbol{\xi}\cdot_1)^2 + (\Delta\boldsymbol{\xi}\cdot_2)^2 + (\Delta\boldsymbol{\xi}\cdot_3)^2,$$

insertion of which into (4.109) leads to

$$L[\boldsymbol{\xi} + \Delta\boldsymbol{\xi}] - L[\boldsymbol{\xi}] \geq [\frac{1}{2N}\sum_i \Delta\xi_i \cos(2\xi_i)]^2$$

$$+ [\frac{1}{2N}\sum_i \Delta\xi_i \sin(2\xi_i)]^2 + \mathcal{O}(N^{-\frac{1}{2}}, \Delta\boldsymbol{\xi}^3).$$

Thus for large N the second derivative of L is non-negative at $\boldsymbol{\xi} = \boldsymbol{\xi}$, and the phase pattern $\boldsymbol{\xi}$ has indeed become a fixed–point attractor of the dynamics of the noise-free coupled oscillator network. The same is found to be true for the states $\boldsymbol{\xi} = \pm\boldsymbol{\xi} + \alpha(1, \ldots, 1)$ (for any α).

We next follow the strategy of the Hopfield model and attempt to simply extend the above recipe for the synapses to the case of having a finite number p of phase patterns $\boldsymbol{\xi}^\mu = (\xi_1^\mu, \ldots, \xi_N^\mu) \in [-\pi, \pi]^N$, giving

$$J_{ij} = \frac{1}{N}\sum_{\mu=1}^p \cos[\xi_i^\mu - \xi_j^\mu], \tag{4.110}$$

where the factor N, as before, ensures a proper limit $N \to \infty$ later. In analogy with our solution of the Hopfield model we define the following averages over pattern variables:

$$\langle g[\boldsymbol{\xi}]\rangle_\xi = \lim_{N\to\infty}\sum_i g[\boldsymbol{\xi}_i], \qquad \boldsymbol{\xi}_i = (\xi_i^1, \ldots, \xi_i^p) \in [-\pi, \pi]^p.$$

We can write the Hamiltonian $H(\boldsymbol{\xi})$ of (4.89) in the form [Coo01, SC00, SC01, CKS05]

$$H(\boldsymbol{\xi}) = -\frac{1}{2N} \sum_{\mu=1}^{p} \cos[\xi_i^\mu - \xi_j^\mu] \cos[\xi_i - \xi_j]$$

$$= -\frac{N}{2} \sum_{\mu=1}^{p} \left\{ m_{cc}^\mu(\boldsymbol{\xi})^2 + m_{cs}^\mu(\boldsymbol{\xi})^2 + m_{sc}^\mu(\boldsymbol{\xi})^2 + m_{ss}^\mu(\boldsymbol{\xi})^2 \right\},$$

in which

$$m_{cc}^\mu(\boldsymbol{\xi}) = \frac{1}{N} \cos(\xi_i^\mu) \cos(\xi_i), \qquad\qquad (4.111)$$

$$m_{cs}^\mu(\boldsymbol{\xi}) = \frac{1}{N} \cos(\xi_i^\mu) \sin(\xi_i),$$

$$m_{sc}^\mu(\boldsymbol{\xi}) = \frac{1}{N} \sin(\xi_i^\mu) \cos(\xi_i), \qquad\qquad (4.112)$$

$$m_{ss}^\mu(\boldsymbol{\xi}) = \frac{1}{N} \sin(\xi_i^\mu) \sin(\xi_i).$$

The free energy per oscillator can now be written as

$$F/N = -\frac{1}{\beta N} \log \int \cdots \int d\boldsymbol{\xi} \; \mathrm{e}^{-\beta H(\boldsymbol{\xi})} =$$

$$-\frac{1}{\beta N} \log \int \cdots \int d\boldsymbol{\xi} \; \mathrm{e}^{\frac{1}{2}\beta N \sum_\mu \sum_{\star\star} m_{\star\star}^\mu(\boldsymbol{\xi})^2},$$

with $\star\star \in \{cc, ss, cs, sc\}$. Upon introducing the notation $\mathbf{m}_{\star\star} = (m_{\star\star}^1, \ldots, m_{\star\star}^p)$ we can again express the free energy in terms of the density of states $\mathcal{D}(\{\mathbf{m}_{\star\star}\}) = (2\pi)^{-N} \int \cdots \int d\boldsymbol{\xi} \prod_{\star\star} \delta[\mathbf{m}_{\star\star} - \mathbf{m}_{\star\star}(\boldsymbol{\sigma})]$:

$$F/N = -\frac{1}{\beta} \log(2\pi) - \frac{1}{\beta N} \log \int \prod_{\star\star} d\mathbf{m}_{\star\star} \; \mathcal{D}(\{\mathbf{m}_{\star\star}\}) \mathrm{e}^{\frac{1}{2}\beta N \sum_{\star\star} \mathbf{m}_{\star\star}^2}. \quad (4.113)$$

Since p is finite, the leading contribution to the density of states (as $N \to \infty$), which will give us the entropy, can be calculated by writing the δ−functions in integral representation:

$$\lim_{N \to \infty} \frac{1}{N} \log \mathcal{D}(\{\mathbf{m}_{\star\star}\}) = \lim_{N \to \infty} \frac{1}{N} \log \int \prod_{\star\star} \left[d\mathbf{x}_{\star\star} \; \mathrm{e}^{iN\mathbf{x}_{\star\star} \cdot \mathbf{m}_{\star\star}} \right] \; \times$$

$$\int \cdots \int \frac{d\boldsymbol{\xi}}{(2\pi)^N} \; \times$$

$$\mathrm{e}^{-i[x_{cc}^\mu \cos(\xi_i^\mu) \cos(\xi_i) + x_{cs}^\mu \cos(\xi_i^\mu) \sin(\xi_i) + x_{sc}^\mu \sin(\xi_i^\mu) \cos(\xi_i) + x_{ss}^\mu \sin(\xi_i^\mu) \sin(\xi_i)]}$$

$$= \mathrm{extr}_{\{\mathbf{x}_{\star\star}\}} \{ i \sum_{\star\star} \mathbf{x}_{\star\star} \cdot \mathbf{m}_{\star\star} + \langle \log \int \frac{d\xi}{2\pi} \; \times$$

$$\mathrm{e}^{-i[x_{cc}^\mu \cos(\xi_\mu) \cos(\xi) + x_{cs}^\mu \cos(\xi_\mu) \sin(\xi) + x_{sc}^\mu \sin(\xi_\mu) \cos(\xi) + x_{ss}^\mu \sin(\xi_\mu) \sin(\xi)]} \rangle_{\boldsymbol{\xi}} \}.$$

The relevant extremum is purely imaginary so we put $\mathbf{x}_{\star\star} = i\beta\mathbf{y}_{\star\star}$ (see our previous discussion for the Hopfield model) and, upon inserting the density of states into our original expression for the free energy per oscillator, arrive at

$$\lim_{N\to\infty} F/N = \mathrm{extr}_{\{\mathbf{m}_{\star\star},\mathbf{y}_{\star\star}\}}\, f(\{\mathbf{m}_{\star\star},\mathbf{y}_{\star\star}\}),$$

$$f(\{\mathbf{m}_{\star\star},\mathbf{y}_{\star\star}\}) = -\frac{1}{\beta}\log(2\pi) - \frac{1}{2}\sum_{\star\star}\mathbf{m}_{\star\star}^2 + \sum_{\star\star}\mathbf{y}_{\star\star}\cdot\mathbf{m}_{\star\star}$$

$$-\frac{1}{\beta}\left\langle\log\int\frac{d\xi}{2\pi}e^{\beta[y^\mu_{cc}\cos(\xi_\mu)\cos(\xi)+y^\mu_{cs}\cos(\xi_\mu)\sin(\xi)+y^\mu_{sc}\sin(\xi_\mu)\cos(\xi)+y^\mu_{ss}\sin(\xi_\mu)\sin(\xi)]}\right\rangle_{\boldsymbol{\xi}}$$

Taking derivatives with respect to the order parameters $\mathbf{m}_{\star\star}$ gives us $\mathbf{y}_{\star\star} = \mathbf{m}_{\star\star}$, with which we can eliminate the $\mathbf{y}_{\star\star}$. Derivation with respect to the $\mathbf{m}_{\star\star}$ subsequently gives the saddle–point equations

$$m^\mu_{cc} = \tag{4.114}$$
$$\left\langle\cos[\xi_\mu]\frac{\int d\xi\,\cos[\xi]e^{\beta\cos[\xi][m^\nu_{cc}\cos[\xi_\nu]+m^\nu_{sc}\sin[\xi_\nu]]+\beta\sin[\xi][m^\nu_{cs}\cos[\xi_\nu]+m^\nu_{ss}\sin[\xi_\nu]]}}{\int d\xi\,e^{\beta\cos[\xi][m^\nu_{cc}\cos[\xi_\nu]+m^\nu_{sc}\sin[\xi_\nu]]+\beta\sin[\xi][m^\nu_{cs}\cos[\xi_\nu]+m^\nu_{ss}\sin[\xi_\nu]]}}\right\rangle_{\boldsymbol{\xi}},$$

$$m^\mu_{cs} = \tag{4.115}$$
$$\left\langle\cos[\xi_\mu]\frac{\int d\xi\,\sin[\xi]e^{\beta\cos[\xi][m^\nu_{cc}\cos[\xi_\nu]+m^\nu_{sc}\sin[\xi_\nu]]+\beta\sin[\xi][m^\nu_{cs}\cos[\xi_\nu]+m^\nu_{ss}\sin[\xi_\nu]]}}{\int d\xi\,e^{\beta\cos[\xi][m^\nu_{cc}\cos[\xi_\nu]+m^\nu_{sc}\sin[\xi_\nu]]+\beta\sin[\xi][m^\nu_{cs}\cos[\xi_\nu]+m^\nu_{ss}\sin[\xi_\nu]]}}\right\rangle_{\boldsymbol{\xi}},$$

$$m^\mu_{sc} = \tag{4.116}$$
$$\left\langle\sin[\xi_\mu]\frac{\int d\xi\,\cos[\xi]e^{\beta\cos[\xi][m^\nu_{cc}\cos[\xi_\nu]+m^\nu_{sc}\sin[\xi_\nu]]+\beta\sin[\xi][m^\nu_{cs}\cos[\xi_\nu]+m^\nu_{ss}\sin[\xi_\nu]]}}{\int d\xi\,e^{\beta\cos[\xi][m^\nu_{cc}\cos[\xi_\nu]+m^\nu_{sc}\sin[\xi_\nu]]+\beta\sin[\xi][m^\nu_{cs}\cos[\xi_\nu]+m^\nu_{ss}\sin[\xi_\nu]]}}\right\rangle_{\boldsymbol{\xi}},$$

$$m^\mu_{ss} = \tag{4.117}$$
$$\left\langle\sin[\xi_\mu]\frac{\int d\xi\,\sin[\xi]e^{\beta\cos[\xi][m^\nu_{cc}\cos[\xi_\nu]+m^\nu_{sc}\sin[\xi_\nu]]+\beta\sin[\xi][m^\nu_{cs}\cos[\xi_\nu]+m^\nu_{ss}\sin[\xi_\nu]]}}{\int d\xi\,e^{\beta\cos[\xi][m^\nu_{cc}\cos[\xi_\nu]+m^\nu_{sc}\sin[\xi_\nu]]+\beta\sin[\xi][m^\nu_{cs}\cos[\xi_\nu]+m^\nu_{ss}\sin[\xi_\nu]]}}\right\rangle_{\boldsymbol{\xi}},$$

The equilibrium values of the observables $\mathbf{m}_{\star\star}$, as defined in (4.111,4.112), are now given by the solution of the coupled equations (4.114-4.117) which minimizes

$$f(\{\mathbf{m}_{\star\star}\}) = \frac{1}{2}\sum_{\star\star}\mathbf{m}_{\star\star}^2 - \frac{1}{\beta}\langle\log\int d\xi\;\times \tag{4.118}$$
$$e^{\beta\cos[\xi][m^\nu_{cc}\cos[\xi_\nu]+m^\nu_{sc}\sin[\xi_\nu]]+\beta\sin[\xi][m^\nu_{cs}\cos[\xi_\nu]+m^\nu_{ss}\sin[\xi_\nu]]}\rangle_{\boldsymbol{\xi}}.$$

We can confirm that the relevant saddle–point must be a minimum by inspecting the $\beta = 0$ limit (infinite noise levels):

$$\lim_{\beta\to 0} f(\{\mathbf{m}_{\star\star}\}) = \frac{1}{2}\sum_{\star\star}\mathbf{m}_{\star\star}^2 - \frac{1}{\beta}\log(2\pi).$$

From now on we will restrict our analysis to phase pattern components ξ_i^μ which have all been drawn independently at random from $[-\pi, \pi]$, with uniform probability density, so that $\langle g[\boldsymbol{\xi}] \rangle_{\boldsymbol{\xi}} = (2\pi)^{-p} \int_{-\pi}^{\pi} \cdots \int_{-\pi}^{\pi} d\boldsymbol{\xi} \; g[\boldsymbol{\xi}]$. At $\beta = 0$ $(T = \infty)$ one finds only the trivial state $m_{\star\star}^\mu = 0$. It can be shown that there will be no discontinuous transitions to a non–trivial state as the noise level (temperature) is reduced. The continuous ones follow upon expansion of the equations (4.114-4.117) for small $\{\mathbf{m}_{\star\star}\}$, which is found to give (for each μ and each combination $\star\star$):

$$m_{\star\star}^\mu = \frac{1}{4}\beta m_{\star\star}^\mu + \mathcal{O}(\{\mathbf{m}_{\star\star}^2\}).$$

Thus a continuous transition to recall states occurs at $T = \frac{1}{4}$. Full classification of all solutions of (4.114-4.117) is ruled out. Here we will restrict ourselves to the most relevant ones, such as the pure states, where $m_{\star\star}^\mu = m_{\star\star}\delta_{\mu\lambda}$ (for some pattern label λ). Here the oscillator phases are correlated with only one of the stored phase patterns (if at all). Insertion into the above expression for $f(\{\mathbf{m}_{\star\star}\})$ shows that for such solutions we have to minimize [Coo01, SC00, SC01, CKS05]

$$f(\{m_{\star\star}\}) = \frac{1}{2}\sum_{\star\star} m_{\star\star}^2 - \frac{1}{\beta}\int \frac{d\xi}{2\pi} \; \log \int d\xi \; \times \qquad (4.119)$$

$$\mathrm{e}^{\beta \cos[\xi][m_{cc}\cos[\xi]+m_{sc}\sin[\xi]]+\beta \sin[\xi][m_{cs}\cos[\xi]+m_{ss}\sin[\xi]]}.$$

We anticipate solutions corresponding to the (partial) recall of the stored phase pattern $\boldsymbol{\xi}^\lambda$ or its mirror image (modulo overall phase shifts $\xi_i \to \xi_i + \delta$, under which the synapses are obviously invariant). Insertion into (4.114-4.117) of the state

$$\xi_i = \xi_i^\lambda + \delta \quad \text{gives} \quad (m_{cc}, m_{sc}, m_{cs}, m_{ss}) = \frac{1}{2}(\cos\delta, -\sin\delta, \sin\delta, \cos\delta).$$

Similarly, insertion into (4.114-4.117) of

$$\xi_i = -\xi_i^\lambda + \delta \quad \text{gives} \quad (m_{cc}, m_{sc}, m_{cs}, m_{ss}) = \frac{1}{2}(\cos\delta, \sin\delta, \sin\delta, -\cos\delta).$$

Thus we can identify retrieval states as those solutions which are of the form

(i) retrieval of $\boldsymbol{\xi}^\lambda$: $(m_{cc}, m_{sc}, m_{cs}, m_{ss}) = m(\cos\delta, -\sin\delta, \sin\delta, \cos\delta)$,

(ii) retrieval of $-\boldsymbol{\xi}^\lambda$: $(m_{cc}, m_{sc}, m_{cs}, m_{ss}) = m(\cos\delta, \sin\delta, \sin\delta, -\cos\delta)$,

with full recall corresponding to $m = \frac{1}{2}$. Insertion into the saddle–point equations and into (4.119), followed by an appropriate shift of the integration variable ξ, shows that the free energy is independent of δ (so the above two ansätzes solve the saddle–point equations for any δ) and that

$$m = \frac{1}{2}\frac{\int d\xi \; \cos[\xi]\mathrm{e}^{\beta m \cos[\xi]}}{\int d\xi \; \mathrm{e}^{\beta m \cos[\xi]}}, \qquad f(m) = m^2 - \frac{1}{\beta}\log \int d\xi \; \mathrm{e}^{\beta m \cos[\xi]}.$$

Expansion in powers of m, using $\log(1 + z) = z - \frac{1}{2}z^2 + \mathcal{O}(z^3)$, reveals that non-zero minima m indeed bifurcate continuously at $T = \beta^{-1} = \frac{1}{4}$:

$$f(m) + \frac{1}{\beta} \log[2\pi] = (1 - \frac{1}{4}\beta)m^2 + \frac{1}{64}\beta^3 m^4 + \mathcal{O}(m^6). \qquad (4.120)$$

Retrieval states are obviously not the only pure states that solve the saddle–point equations. The function (4.119) is invariant under the following discrete (non-commuting) transformations:

$$\text{I}: \ (m_{cc}, m_{sc}, m_{cs}, m_{ss}) \rightarrow (m_{cc}, m_{sc}, -m_{cs}, -m_{ss}),$$
$$\text{II}: \ (m_{cc}, m_{sc}, m_{cs}, m_{ss}) \rightarrow (m_{cs}, m_{ss}, m_{cc}, m_{sc}).$$

We expect these to induce solutions with specific symmetries. In particular we anticipate the following symmetric and antisymmetric states:

symmetric under I : $(m_{cc}, m_{sc}, m_{cs}, m_{ss}) = \sqrt{2}m(\cos\delta, \sin\delta, 0, 0)$,
antisymmetric under I : $(m_{cc}, m_{sc}, m_{cs}, m_{ss}) = \sqrt{2}m(0, 0, \cos\delta, \sin\delta)$,
symmetric under II : $(m_{cc}, m_{sc}, m_{cs}, m_{ss}) = m(\cos\delta, \sin\delta, \cos\delta, \sin\delta)$,
antisymmetric under II : $(m_{cc}, m_{sc}, m_{cs}, m_{ss}) = m(\cos\delta, \sin\delta, -\cos\delta, -\sin\delta)$.

Insertion into the saddle–point equations and into (4.119) shows in all four cases the parameter δ is arbitrary and that always

$$m = \frac{1}{\sqrt{2}} \int \frac{d\xi}{2\pi} \ \cos[\xi] \ \frac{\int d\xi \ \cos[\xi] e^{\beta m \sqrt{2} \cos[\xi] \cos[\xi]}}{\int d\xi \ e^{\beta m \sqrt{2} \cos[\xi] \cos[\xi]}},$$
$$f(m) = m^2 - \frac{1}{\beta} \int \frac{d\xi}{2\pi} \ \log \int d\xi \ e^{\beta m \sqrt{2} \cos[\xi] \cos[\xi]}.$$

Expansion in powers of m reveals that non-zero solutions m here again bifurcate continuously at $T = \frac{1}{4}$:

$$f(m) + \frac{1}{\beta} \log[2\pi] = (1 - \frac{1}{4}\beta)m^2 + \frac{3}{2} \cdot \frac{1}{64}\beta^3 m^4 + \mathcal{O}(m^6). \qquad (4.121)$$

However, comparison with (4.120) shows that the free energy of the pure recall states is lower. Thus the system will prefer the recall states over the above solutions with specific symmetries.

Note, that the free energy and the order parameter equation for the pure recall states can be written in terms of *Bessel functions* as follows:

$$m = \frac{1}{2} \frac{I_1(\beta m)}{I_0(\beta m)}, \qquad f(m) = m^2 - \frac{1}{\beta} \log[2\pi I_0(\beta m)]$$

4.3.6 Attractor Neural Networks with Binary Neurons

The simplest non–trivial recurrent neural networks consist of N binary neurons $\sigma_i \in \{-1, 1\}$, which respond stochastically to post-synaptic potentials (or local fields) $h_i(\boldsymbol{\sigma})$, with $\boldsymbol{\sigma} = (\sigma_1, \ldots, \sigma_N)$. The fields depend linearly on the instantaneous neuron states,

$$h_i(\boldsymbol{\sigma}) = J_{ij}\sigma_j + \theta_i,$$

with the J_{ij} representing synaptic efficacies, and the θ_i representing external stimuli and/or neural thresholds [Coo01, SC00, SC01, CKS05].

Closed Macroscopic Laws for Sequential Dynamics

Here, we will first show how for sequential dynamics (where neurons are updated one after the other) one can calculate, from the microscopic stochastic laws, differential equations for the probability distribution of suitably defined macroscopic observables. For mathematical convenience our starting point will be the continuous-time master equation for the microscopic probability distribution $p_t(\boldsymbol{\sigma})$

$$\dot{p}_t(\boldsymbol{\sigma}) = \{w_i(F_i\boldsymbol{\sigma})p_t(F_i\boldsymbol{\sigma}) - w_i(\boldsymbol{\sigma})p_t(\boldsymbol{\sigma})\}, \tag{4.122}$$

$$w_i(\boldsymbol{\sigma}) = \frac{1}{2}[1 - \sigma_i \tanh[\beta h_i(\boldsymbol{\sigma})]],$$

with $F_i\Phi(\boldsymbol{\sigma}) = \Phi(\sigma_1, \ldots, \sigma_{i-1}, -\sigma_i, \sigma_{i+1}, \ldots, \sigma_N)$.

Let us illustrate the basic ideas with the help of a simple (infinite range) toy model: $J_{ij} = (J/N)\eta_i\xi_j$ and $\theta_i = 0$ (the variables η_i and ξ_i are arbitrary, but may not depend on N). For $\eta_i = \xi_i = 1$ we get a network with uniform synapses. For $\eta_i = \xi_i \in \{-1, 1\}$ and $J > 0$ we recover the Hopfield [Hop82] model with one stored pattern. Note: the synaptic matrix is non-symmetric as soon as a pair (ij) exists such that $\eta_i\xi_j \neq \eta_j\xi_i$, so in general equilibrium statistical mechanics will not apply. The local fields become $h_i(\boldsymbol{\sigma}) = J\eta_i m(\boldsymbol{\sigma})$ with $m(\boldsymbol{\sigma}) = \frac{1}{N}\sum_k \xi_k\sigma_k$. Since they depend on the microscopic state $\boldsymbol{\sigma}$ only through the value of m, the latter quantity appears to constitute a natural macroscopic level of description. The probability density of finding the macroscopic state $m(\boldsymbol{\sigma}) = m$ is given by $\mathcal{P}_t[m] = \sum_{\boldsymbol{\sigma}} p_t(\boldsymbol{\sigma})\delta[m - m(\boldsymbol{\sigma})]$. Its time derivative follows upon inserting (4.122):

$$\dot{\mathcal{P}}_t[m] = \sum_{\boldsymbol{\sigma}} p_t(\boldsymbol{\sigma})w_k(\boldsymbol{\sigma}) \left\{ \delta[m - m(\boldsymbol{\sigma}) + \frac{2}{N}\xi_k\sigma_k] - \delta[m - m(\boldsymbol{\sigma})] \right\}$$

$$= \frac{\partial}{\partial m} \left\{ \sum_{\boldsymbol{\sigma}} p_t(\boldsymbol{\sigma})\delta[m - m(\boldsymbol{\sigma})] \frac{2}{N}\xi_k\sigma_k w_k(\boldsymbol{\sigma}) \right\} + \mathcal{O}(\frac{1}{N}).$$

Inserting our expressions for the transition rates $w_i(\boldsymbol{\sigma})$ and the local fields $h_i(\boldsymbol{\sigma})$ gives:

$$\dot{\mathcal{P}}_t[m] = \frac{\partial}{\partial m} \left\{ \mathcal{P}_t[m] \left[m - \frac{1}{N}\xi_k \tanh[\eta_k\beta Jm] \right] \right\} + \mathcal{C}(N^{-1}).$$

In the limit $N \to \infty$ only the first term survives. The general solution of the resulting Liouville equation is

$$\mathcal{P}_t[m] = \int dm_0 \, \mathcal{P}_0[m_0]\delta\left[m - m(t|m_0)\right],$$

where $m(t|m_0)$ is the solution of

$$\dot{m} = \lim_{N \to \infty} \frac{1}{N}\xi_k \tanh[\eta_k \beta J m] - m, \qquad m(0) = m_0. \qquad (4.123)$$

This describes deterministic evolution; the only uncertainty in the value of m is due to uncertainty in initial conditions. If at $t = 0$ the quantity m is known exactly, this will remain the case for finite time–scales; m turns out to evolve in time according to (4.123).

Let us now allow for less trivial choices of the synaptic matrix $\{J_{ij}\}$ and try to calculate the evolution in time of a given set of macroscopic observables $\mathbf{\Omega}(\boldsymbol{\sigma}) = (\Omega_1(\boldsymbol{\sigma}), \ldots, \Omega_n(\boldsymbol{\sigma}))$ in the limit $N \to \infty$. There are no restrictions yet on the form or the number n of these state variables; these will, however, arise naturally if we require the observables $\mathbf{\Omega}$ to obey a closed set of deterministic laws, as we will see. The probability density of finding the system in macroscopic state $\mathbf{\Omega}$ is given by [Coo01, SC00, SC01, CKS05]:

$$\mathcal{P}_t[\mathbf{\Omega}] = \sum_{\boldsymbol{\sigma}} p_t(\boldsymbol{\sigma})\delta\left[\mathbf{\Omega} - \mathbf{\Omega}(\boldsymbol{\sigma})\right]. \qquad (4.124)$$

Its time derivative is got by inserting (4.122). If in those parts of the resulting expression which contain the operators F_i we perform the transformations $\boldsymbol{\sigma} \to F_i\boldsymbol{\sigma}$, we arrive at

$$\dot{\mathcal{P}}_t[\mathbf{\Omega}] = \sum_{\boldsymbol{\sigma}} p_t(\boldsymbol{\sigma})w_i(\boldsymbol{\sigma})\left\{\delta\left[\mathbf{\Omega} - \mathbf{\Omega}(F_i\boldsymbol{\sigma})\right] - \delta\left[\mathbf{\Omega} - \mathbf{\Omega}(\boldsymbol{\sigma})\right]\right\}.$$

If we define

$$\Omega_\mu(F_i\boldsymbol{\sigma}) = \Omega_\mu(\boldsymbol{\sigma}) + \Delta_{i\mu}(\boldsymbol{\sigma})$$

and make a Taylor expansion in powers of $\{\Delta_{i\mu}(\boldsymbol{\sigma})\}$, we finally get the so–called *Kramers–Moyal expansion*:[6]

$$\dot{\mathcal{P}}_t[\mathbf{\Omega}] = \sum_{\ell \geq 1} \frac{(-1)^\ell}{\ell+} \sum_{\mu_1=1}^{n} \cdots \sum_{\mu_\ell=1}^{n} \frac{\partial^\ell}{\partial\Omega_{\mu_1} \cdots \partial\Omega_{\mu_\ell}}\left\{\mathcal{P}_t[\mathbf{\Omega}] \, F^{(\ell)}_{\mu_1 \cdots \mu_\ell}[\mathbf{\Omega}; t]\right\}. \qquad (4.125)$$

----

[6] The Kramers–Moyal expansion (4.125) is to be interpreted in a distributional sense, i.e., only to be used in expressions of the form $\int d\mathbf{\Omega}_t(\mathbf{\Omega})G(\mathbf{\Omega})$ with smooth functions $G(\mathbf{\Omega})$, so that all derivatives are well–defined and finite. Furthermore, (4.125) will only be useful if the $\Delta_{j\mu}$, which measure the sensitivity of the macroscopic quantities to single neuron state changes, are sufficiently small. This is to be expected: for finite N any observable can only assume a finite number of possible values; only for $N \to \infty$ may we expect smooth probability distributions for our macroscopic quantities.

It involves conditional averages $\langle f(\boldsymbol{\sigma}) \rangle_{\boldsymbol{\Omega};t}$ and the 'discrete derivatives'

$$\Delta_{j\mu}(\boldsymbol{\sigma}) = \Omega_\mu(F_j\boldsymbol{\sigma}) - \Omega_\mu(\boldsymbol{\sigma}),$$

$$F^{(l)}_{\mu_1\cdots\mu_l}[\boldsymbol{\Omega};t] = \langle w_j(\boldsymbol{\sigma})\Delta_{j\mu_1}(\boldsymbol{\sigma})\cdots\Delta_{j\mu_l}(\boldsymbol{\sigma})\rangle_{\boldsymbol{\Omega};t},$$

$$\langle f(\boldsymbol{\sigma}) \rangle_{\boldsymbol{\Omega};t} = \frac{\sum_{\boldsymbol{\sigma}} p_t(\boldsymbol{\sigma})\delta\left[\boldsymbol{\Omega} - \boldsymbol{\Omega}(\boldsymbol{\sigma})\right] f(\boldsymbol{\sigma})}{\sum_{\boldsymbol{\sigma}} p_t(\boldsymbol{\sigma})\delta\left[\boldsymbol{\Omega} - \boldsymbol{\Omega}(\boldsymbol{\sigma})\right]}. \tag{4.126}$$

Retaining only the $\ell = 1$ term in (4.125) would lead us to a Liouville equation, which describes deterministic flow in $\boldsymbol{\Omega}$ space. Including also the $\ell = 2$ term leads us to a Fokker–Planck equation which, in addition to flow, describes diffusion of the macroscopic probability density. Thus a sufficient condition for the observables $\boldsymbol{\Omega}(\boldsymbol{\sigma})$ to evolve in time deterministically in the limit $N \to \infty$ is:

$$\lim_{N\to\infty} \sum_{\ell\geq 2} \frac{1}{\ell+} \sum_{\mu_1=1}^{n} \cdots \sum_{\mu_\ell=1}^{n} \langle |\Delta_{j\mu_1}(\boldsymbol{\sigma})\cdots\Delta_{j\mu_\ell}(\boldsymbol{\sigma})|\rangle_{\boldsymbol{\Omega};t} = 0. \tag{4.127}$$

In the simple case where all observables Ω_μ scale similarly in the sense that all 'derivatives' $\Delta_{j\mu} = \Omega_\mu(F_i\boldsymbol{\sigma}) - \Omega_\mu(\boldsymbol{\sigma})$ are of the same order in N (i.e., there is a monotonic function $\tilde{\Delta}_N$ such that $\Delta_{j\mu} = \mathcal{O}(\tilde{\Delta}_N)$ for all j,μ), for instance, criterion (4.127) becomes:

$$\lim_{N\to\infty} n\tilde{\Delta}_N\sqrt{N} = 0. \tag{4.128}$$

If for a given set of observables condition (4.127) is satisfied, we can for large N describe the evolution of the macroscopic probability density by a Liouville equation:

$$\dot{\mathcal{P}}_t[\boldsymbol{\Omega}] = -\frac{\partial}{\partial \Omega_\mu}\left\{\mathcal{P}_t[\boldsymbol{\Omega}]\, F^{(1)}_\mu[\boldsymbol{\Omega};t]\right\},$$

whose solution describes deterministic flow,

$$\mathcal{P}_t[\boldsymbol{\Omega}] = \int d\boldsymbol{\Omega}_0 \mathcal{P}_0[\boldsymbol{\Omega}_0]\delta[\boldsymbol{\Omega} - \boldsymbol{\Omega}(t|\boldsymbol{\Omega}_0)],$$

with $\boldsymbol{\Omega}(t|\boldsymbol{\Omega}_0)$ given, in turn, as the solution of

$$\dot{\boldsymbol{\Omega}}(t) = \mathbf{F}^{(1)}[\boldsymbol{\Omega}(t);t], \qquad \boldsymbol{\Omega}(0) = \boldsymbol{\Omega}_0. \tag{4.129}$$

In taking the limit $N \to \infty$, however, we have to keep in mind that the resulting deterministic theory is got by taking this limit for *finite* t. According to (4.125) the $\ell > 1$ terms do come into play for sufficiently large times t; for $N \to \infty$, however, these times diverge by virtue of (4.127).

Now, equation (4.129) will in general not be autonomous; tracing back the origin of the explicit time dependence in the right–hand side of (4.129) one finds that to calculate $\mathbf{F}^{(1)}$ one needs to know the microscopic probability density $p_t(\boldsymbol{\sigma})$. This, in turn, requires solving equation (4.122) (which is exactly what one tries to avoid). We will now discuss a mechanism via which to eliminate the offending explicit time dependence, and to turn the observables $\boldsymbol{\Omega}(\boldsymbol{\sigma})$ into an autonomous level of description, governed by *closed* dynamic laws. The idea is to choose the observables $\boldsymbol{\Omega}(\boldsymbol{\sigma})$ in such a way that there is no explicit time dependence in the flow field $\mathbf{F}^{(1)}[\boldsymbol{\Omega};t]$ (if possible). According to (4.126) this implies making sure that there exist functions $\varPhi_\mu[\boldsymbol{\Omega}]$ such that

$$\lim_{N\to\infty} w_j(\boldsymbol{\sigma})\Delta_{j\mu}(\boldsymbol{\sigma}) = \varPhi_\mu[\boldsymbol{\Omega}(\boldsymbol{\sigma})], \tag{4.130}$$

in which case the time dependence of $\mathbf{F}^{(1)}$ indeed drops out and the macroscopic state vector simply evolves in time according to [Coo01, SC00, SC01, CKS05]:

$$\dot{\boldsymbol{\Omega}} = \boldsymbol{\Phi}[\boldsymbol{\Omega}], \qquad \boldsymbol{\Phi} = (\varPhi_1[\boldsymbol{\Omega}], \dots, \varPhi_n[\boldsymbol{\Omega}]).$$

Clearly, for this closure method to apply, a suitable separable structure of the synaptic matrix is required. If, for instance, the macroscopic observables Ω_μ depend linearly on the microscopic state variables $\boldsymbol{\sigma}$ (i.e., $\Omega_\mu(\boldsymbol{\sigma}) = \frac{1}{N}\sum_{j=1}^N \omega_{\mu j}\sigma_j$), we get with the transition rates defined in (4.122):

$$\dot{\Omega}_\mu = \lim_{N\to\infty} \frac{1}{N}\omega_{\mu j}\tanh(\beta h_j(\boldsymbol{\sigma})) - \Omega_\mu, \tag{4.131}$$

in which case the only further condition for (4.130) to hold is that all local fields $h_k(\boldsymbol{\sigma})$ must (in leading order in N) depend on the microscopic state $\boldsymbol{\sigma}$ only through the values of the observables $\boldsymbol{\Omega}$; since the local fields depend linearly on $\boldsymbol{\sigma}$ this, in turn, implies that the synaptic matrix must be separable: if $J_{ij} = \sum_\mu K_{i\mu}\omega_{\mu j}$ then indeed $h_i(\boldsymbol{\sigma}) = \sum_\mu K_{i\mu}\Omega_\mu(\boldsymbol{\sigma}) + \theta_i$. Next we will show how this approach can be applied to networks for which the matrix of synapses has a separable form (which includes most symmetric and non–symmetric Hebbian type attractor models). We will restrict ourselves to models with $\theta_i = 0$; introducing non–zero thresholds is straightforward and does not pose new problems.

Application to Separable Attractor Networks

We consider the following class of models, in which the interaction matrices have the form

$$J_{ij} = \frac{1}{N}Q(\boldsymbol{\xi}_i; \boldsymbol{\xi}_j), \qquad \boldsymbol{\xi}_i = (\xi_i^1, \dots, \xi_i^p). \tag{4.132}$$

The components ξ_i^μ, representing the information ('patterns') to be stored or processed, are assumed to be drawn from a finite discrete set Λ, containing n_Λ elements (they are not allowed to depend on N). The Hopfield model [Hop82] corresponds to choosing $Q(\mathbf{x}; \mathbf{y}) = \mathbf{x}\cdot\mathbf{y}$ and $\Lambda \equiv \{-1, 1\}$. One now introduces a partition of the system $\{1, \dots, N\}$ into n_Λ^p so–called sublattices $I_{\boldsymbol{\eta}}$:

$$I_{\eta} = \{i \mid \boldsymbol{\xi}_i = \boldsymbol{\eta}\}, \qquad \{1, \ldots, N\} = \bigcup_{\eta} I_{\eta}, \qquad (\boldsymbol{\eta} \in \Lambda^p).$$

The number of neurons in sublattice I_{η} is denoted by $|I_{\eta}|$ (this number will have to be large). If we choose as our macroscopic observables the average activities ('magnetizations') within these sublattices, we are able to express the local fields h_k solely in terms of macroscopic quantities:

$$m_{\boldsymbol{\eta}}(\boldsymbol{\sigma}) = \frac{1}{|I_{\eta}|} \sum_{i \in I_{\eta}} \sigma_i, \qquad h_k(\boldsymbol{\sigma}) = p_{\eta} Q\left(\boldsymbol{\xi}_k; \boldsymbol{\eta}\right) m_{\boldsymbol{\eta}}, \qquad (4.133)$$

with the relative sublattice sizes $p_{\eta} = |I_{\eta}|/N$. If all p_{η} are of the same order in N (which, for example, is the case if the vectors $\boldsymbol{\xi}_i$ have been drawn at random from the set Λ^p) we may write $\Delta_{j\eta} = \mathcal{O}(n_{\Lambda}^p N^{-1})$ and use (4.128). The evolution in time of the sublattice activities is then found to be deterministic in the $N \to \infty$ limit if $\lim_{N \to \infty} p/\log N = 0$. Furthermore, condition (4.130) holds, since

$$w_j(\boldsymbol{\sigma})\Delta_{j\eta}(\boldsymbol{\sigma}) = \tanh[\beta p_{\eta'} Q\left(\boldsymbol{\eta}; \boldsymbol{\eta}'\right) m_{\boldsymbol{\eta}'}] - m_{\boldsymbol{\eta}}.$$

This situation is described by (4.131), and that the evolution in time of the sublattice activities is governed by the following autonomous set of differential equations [RKH88]:

$$\dot{m}_{\boldsymbol{\eta}} = \tanh[\beta p_{\eta'} Q\left(\boldsymbol{\eta}; \boldsymbol{\eta}'\right) m_{\boldsymbol{\eta}'}] - m_{\boldsymbol{\eta}}. \qquad (4.134)$$

We see that, in contrast to the equilibrium techniques as described above, here there is no need at all to require symmetry of the interaction matrix or absence of self-interactions. In the symmetric case $Q(\mathbf{x}; \mathbf{y}) = Q(\mathbf{y}; \mathbf{x})$ the system will approach equilibrium; if the kernel Q is positive definite this can be shown, for instance, by inspection of the *Lyapunov function* $\mathcal{L}\{n_{\eta}\}$:[7]

$$\mathcal{L}\{m_{\boldsymbol{\eta}}\} = \frac{1}{2} p_{\eta} m_{\boldsymbol{\eta}} Q(\boldsymbol{\eta}; \boldsymbol{\eta}') m_{\boldsymbol{\eta}'} p_{\eta'}$$
$$- \frac{1}{\beta} \sum_{\eta} p_{\eta} \log \cosh[\beta Q(\boldsymbol{\eta}; \boldsymbol{\eta}') m_{\boldsymbol{\eta}'} p_{\eta'}],$$

which is bounded from below and obeys:

$$\dot{\mathcal{L}} = -\left[p_{\eta}\dot{m}_{\boldsymbol{\eta}}\right] Q(\boldsymbol{\eta}; \boldsymbol{\eta}') \left[p_{\eta'}\dot{m}_{\boldsymbol{\eta}'}\right] \leq 0. \qquad (4.135)$$

Note that from the sublattice activities, in turn, follow the *Hopfield overlaps* $m_{\mu}(\boldsymbol{\sigma})$,

---

[7] Recall that *Lyapunov function* is a function of the state variables which is bounded from below and whose value decreases monotonically during the dynamics, see e.g., [Kha92]. Its existence guarantees evolution towards a stationary state (under some weak conditions).

$$m_\mu(\boldsymbol{\sigma}) = \frac{1}{N}\xi_i^\mu \sigma_i = p_{\boldsymbol{\eta}}\eta_\mu m_{\boldsymbol{\eta}}. \tag{4.136}$$

Simple examples of relevant models of the type (4.132), the dynamics of which are for large N described by equation (4.134), are for instance the ones where one applies a nonlinear operation Φ to the standard Hopfield–type [Hop82] (or Hebbian) interactions . This nonlinearity could result from e.g., a clipping procedure or from retaining only the *sign* of the Hebbian values:

$$J_{ij} = \frac{1}{N}\Phi(\xi_i^\mu \xi_j^\mu) : \qquad \text{e.g.,}$$

$$\Phi(x) = \begin{cases} -K & \text{for} \quad x \le K \\ x & \text{for} \quad -K < x < K \\ K & \text{for} \quad x \ge K \end{cases}, \qquad \text{or} \qquad \Phi(x) = \mathrm{sgn}(x).$$

The effect of introducing such nonlinearities is found to be of a quantitative nature, giving rise to little more than a re-scaling of critical noise levels and storage capacities. We will illustrate this statement by working out the $p = 2$ equations for randomly drawn pattern bits $\xi_i^\mu \in \{-1, 1\}$, where there are only four sub-lattices, and where $p_{\boldsymbol{\eta}} = \frac{1}{4}$ for all $\boldsymbol{\eta}$ (details can be found in e.g., [DHS91]). Using $\Phi(0) = 0$ and $\Phi(-x) = -\Phi(x)$ (as with the above examples) we get from (4.134):

$$\dot{m}_{\boldsymbol{\eta}} = \tanh[\frac{1}{4}\beta\Phi(2)(m_{\boldsymbol{\eta}} - m_{-\boldsymbol{\eta}})] - m_{\boldsymbol{\eta}}. \tag{4.137}$$

Here the choice made for $\Phi(x)$ shows up only as a re-scaling of the temperature. From (4.137) we further get $\frac{d}{dt}(m_{\boldsymbol{\eta}} + m_{-\boldsymbol{\eta}}) = -(m_{\boldsymbol{\eta}} + m_{-\boldsymbol{\eta}})$. The system decays exponentially towards a state where, according to (4.136), $m_{\boldsymbol{\eta}} = -m_{-\boldsymbol{\eta}}$ for all $\boldsymbol{\eta}$. If at $t = 0$ this is already the case, we find (at least for $p = 2$) decoupled equations for the sub-lattice activities.

Now, equations (4.134.4.136) suggest that at the level of overlaps there will be, in turn, closed laws if the kernel Q is bilinear, $Q(\mathbf{x}; \mathbf{y}) = \sum_{\mu\nu} x_\mu A_{\mu\nu} y_\nu$, or [Coo01, SC00, SC01, CKS05]:

$$J_{ij} = \frac{1}{N}\xi_i^\mu A_{\mu\nu}\xi_i^\nu, \qquad \boldsymbol{\xi}_i = (\xi_i^1, \dots, \xi_i^p). \tag{4.138}$$

We will see that now the ξ_i^μ need not be drawn from a finite discrete set (as long as they do not depend on N). The Hopfield model corresponds to $A_{\mu\nu} = \delta_{\mu\nu}$ and $\xi_i^\mu \in \{-1, 1\}$. The fields h_k can now be written in terms of the overlaps m_μ:

$$h_k(\boldsymbol{\sigma}) = \boldsymbol{\xi}_k \cdot A\mathbf{m}(\boldsymbol{\sigma}), \qquad \mathbf{m} = (m_1, \dots, m_p), \qquad m_\mu(\boldsymbol{\sigma}) = \frac{1}{N}\xi_i^\mu \sigma_i.$$

For this choice of macroscopic variables we find $\Delta_{j\mu} = \mathcal{O}(N^{-1})$, so the evolution of the vector $\mathbf{m}$ becomes deterministic for $N \to \infty$ if, according to (4.128), $\lim_{N\to\infty} p/\sqrt{N} = 0$. Again (4.130) holds, since

$$w_j(\boldsymbol{\sigma})\Delta_{j\mu}(\boldsymbol{\sigma}) = \frac{1}{N}\boldsymbol{\xi}_k \tanh\left[\beta\boldsymbol{\xi}_k \cdot A\mathbf{m}\right] - \mathbf{m}.$$

Thus the evolution in time of the overlap vector $\mathbf{m}$ is governed by a closed set of differential equations:

$$\dot{\mathbf{m}} = \langle\boldsymbol{\xi}\tanh\left[\beta\boldsymbol{\xi}\cdot A\mathbf{m}\right]\rangle_{\boldsymbol{\xi}} - \mathbf{m}, \qquad \langle\Phi(\boldsymbol{\xi})\rangle_{\boldsymbol{\xi}} = \int d\boldsymbol{\xi}\;\rho(\boldsymbol{\xi})\Phi(\boldsymbol{\xi}), \qquad (4.139)$$

with $\rho(\boldsymbol{\xi}) = \lim_{N\to\infty} N^{-1}\sum_i \delta[\boldsymbol{\xi} - \boldsymbol{\xi}_i]$. Symmetry of the synapses is not required. For certain non–symmetric matrices A one finds stable limit–cycle solutions of (4.139). In the symmetric case $A_{\mu\nu} = A_{\nu\mu}$ the system will approach equilibrium; the Lyapunov function (4.135) for positive definite matrices A now becomes:

$$\mathcal{L}\{\mathbf{m}\} = \frac{1}{2}\mathbf{m}\cdot A\mathbf{m} - \frac{1}{\beta}\langle\log\cosh\left[\beta\boldsymbol{\xi}\cdot A\mathbf{m}\right]\rangle_{\boldsymbol{\xi}}.$$

As a second simple application of the flow equations (4.139) we turn to the relaxation times corresponding to the attractors of the Hopfield model (where $A_{\mu\nu} = \delta_{\mu\nu}$). Expanding (4.139) near a stable fixed–point $\mathbf{m}^*$, i.e., $\mathbf{m}(t) = \mathbf{m}^* + \mathbf{x}(t)$ with $|\mathbf{x}(t)| \ll 1$, gives the linearized equation

$$\dot{x}_\mu = [\beta\langle\xi_\mu\xi_\nu\tanh[\beta\boldsymbol{\xi}\cdot\mathbf{m}^*]\rangle_{\boldsymbol{\xi}} - \delta_{\mu\nu}]x_\nu + \mathcal{O}(\mathbf{x}^2). \qquad (4.140)$$

The Jacobian of (4.139), which determines the linearized equation (4.140), turns out to be *minus* the curvature matrix of the free energy surface at the fixed-point. The asymptotic relaxation towards any stable attractor is generally exponential, with a characteristic time τ given by the inverse of the smallest eigenvalue of the curvature matrix. If, in particular, for the fixed–point $\mathbf{m}^*$ we substitute an n–mixture state, i.e., $m_\mu = r_n$ ($\mu \leq n$) and $m_\mu = 0$ ($\mu > n$), and transform (4.140) to the basis where the corresponding curvature matrix $\mathbf{D}^{(n)}$ (with eigenvalues D_λ^n) is diagonal, $\mathbf{x} \to \tilde{\mathbf{x}}$, we get $\tilde{x}_\lambda(t) = \tilde{x}_\lambda(0)\mathrm{e}^{-tD_\lambda^n} + \ldots$ so $\tau^{-1} = \min_\lambda D_\lambda^n$, which we have already calculated in determining the character of the saddle–points of the free-energy surface. The relaxation time for the n–mixture attractors decreases monotonically with the degree of mixing n, for any noise level. At the transition where a macroscopic state $\mathbf{m}^*$ ceases to correspond to a local minimum of the free energy surface, it also de–stabilizes in terms of the linearized dynamic equation (4.140) (as it should). The Jacobian develops a zero eigenvalue, the relaxation time diverges, and the long–time behavior is no longer got from the linearized equation. This gives rise to critical slowing down (i.e., power law relaxation as opposed to exponential relaxation). For instance, at the transition temperature $T_c = 1$ for the $n = 1$ (pure) state, we find by expanding (4.139):

$$\dot{m}_\mu = m_\mu[\frac{2}{3}m_\mu^2 - \mathbf{m}^2] + \mathcal{O}(\mathbf{m}^5),$$

which gives rise to a relaxation towards the trivial fixed–point of the form $\mathbf{m} \sim t^{-\frac{1}{2}}$.

If one is willing to restrict oneself to the limited class of models (4.138) (as opposed to the more general class (4.132)) and to the more global level of description in terms of p overlap parameters m_μ instead of n_A^p sublattice activities $m_{\boldsymbol{\eta}}$, then there are two rewards. Firstly there will be no restrictions on the stored pattern components ξ_i^μ (for instance, they are allowed to be real-valued); secondly the number p of patterns stored can be much larger for the deterministic autonomous dynamical laws to hold ($p \ll \sqrt{N}$ instead of $p \ll \log N$, which from a biological point of view is not impressive [Coo01, SC00, SC01, CKS05].

Closed Macroscopic Laws for Parallel Dynamics

We now turn to the parallel dynamics counterpart of (4.122), i.e., the Markov chain

$$p_{\ell+1}(\boldsymbol{\sigma}) = \sum_{\boldsymbol{\sigma}'} W\left[\boldsymbol{\sigma}; \boldsymbol{\sigma}'\right] p_\ell(\boldsymbol{\sigma}'), \qquad (4.141)$$

$$W\left[\boldsymbol{\sigma}; \boldsymbol{\sigma}'\right] = \prod_{i=1}^N \frac{1}{2}\left[1 + \sigma_i \tanh[\beta h_i(\boldsymbol{\sigma}')]\right],$$

with $\sigma_i \in \{-1, 1\}$, and with local fields $h_i(\boldsymbol{\sigma})$ defined in the usual way. The evolution of macroscopic probability densities will here be described by discrete maps, in stead of differential equations.

Let us first see what happens to our previous toy model: $J_{ij} = (J/N)\eta_i \xi_j$ and $\theta_i = 0$. As before we try to describe the dynamics at the (macroscopic) level of the quantity $m(\boldsymbol{\sigma}) = \frac{1}{N}\sum_k \xi_k \sigma_k$. The evolution of the macroscopic probability density $\mathcal{P}_t[m]$ is got by inserting (4.141):

$$\mathcal{P}_{t+1}[m] = \sum_{\boldsymbol{\sigma}\boldsymbol{\sigma}'} \delta\left[m - m(\boldsymbol{\sigma})\right] W\left[\boldsymbol{\sigma}; \boldsymbol{\sigma}'\right] p_t(\boldsymbol{\sigma}')$$

$$= \int dm'\, \tilde{W}_t\left[m, m'\right] \mathcal{P}_t[m'], \qquad \text{with} \qquad (4.142)$$

$$\tilde{W}_t\left[m, m'\right] = \frac{\sum_{\boldsymbol{\sigma}\boldsymbol{\sigma}'} \delta\left[m - m(\boldsymbol{\sigma})\right] \delta\left[m' - m(\boldsymbol{\sigma}')\right] W\left[\boldsymbol{\sigma}; \boldsymbol{\sigma}'\right] p_t(\boldsymbol{\sigma}')}{\sum_{\boldsymbol{\sigma}'} \delta\left[m' - m(\boldsymbol{\sigma}')\right] p_t(\boldsymbol{\sigma}')}.$$

We now insert our expression for the transition probabilities $W[\boldsymbol{\sigma}; \boldsymbol{\sigma}']$ and for the local fields. Since the fields depend on the microscopic state $\boldsymbol{\sigma}$ only through $m(\boldsymbol{\sigma})$, the distribution $p_t(\boldsymbol{\sigma})$ drops out of the above expression for $\tilde{W}_t$ which thereby loses its explicit time–dependence, $\tilde{W}_t\left[m, m'\right] \to \tilde{W}\left[m, m'\right]$:

$$\tilde{W}\left[m, m'\right] = \mathrm{e}^{-\sum_i \log\cosh(\beta J m' \eta_i)} \langle \delta\left[m - m(\boldsymbol{\sigma})\right] \mathrm{e}^{\beta J m' \eta_i \sigma_i}\rangle_{\boldsymbol{\sigma}}$$

with $\langle \dots \rangle_{\boldsymbol{\sigma}} = 2^{-N} \sum_{\boldsymbol{\sigma}} \dots$

Inserting the integral representation for the δ−function allows us to perform the average:

$$\tilde{W}[m,m'] = \left[\frac{\beta N}{2\pi}\right]\int dk\; e^{N\Psi(m,m',k)},$$

$$\Psi = i\beta km + \langle \log\cosh\beta[J\eta m' - ik\xi]\rangle_{\eta,\xi} - \langle \log\cosh\beta[J\eta m']\rangle_{\eta}.$$

Since $\tilde{W}[m,m']$ is (by construction) normalized, $\int dm\;\tilde{W}[m,m'] = 1$, we find that for $N\to\infty$ the expectation value with respect to $\tilde{W}[m,m']$ of any sufficiently smooth function $f(m)$ will be determined only by the value $m^*(m')$ of m in the relevant saddle–point of Ψ:

$$\int dm\; f(m)\tilde{W}[m,m'] = \frac{\int dmdk\; f(m)e^{N\Psi(m,m',k)}}{\int dmdk\; e^{N\Psi(m,m',k)}} \to f(m^*(m')),\quad (N\to\infty).$$

Variation of Ψ with respect to k and m gives the two saddle–point equations:

$$m = \langle \xi\tanh\beta[J\eta m' - \xi k]\rangle_{\eta,\xi},\qquad k = 0.$$

We may now conclude that $\lim_{N\to\infty}\tilde{W}[m,m'] = \delta[m - m^*(m')]$ with $m^*(m') = \langle \xi\tanh(\beta J\eta m')\rangle_{\eta,\xi}$, and that the macroscopic equation (4.142) becomes [Coo01, SC00, SC01, CKS05]:

$$\mathcal{P}_{t+1}[m] = \int dm'\;\delta\left[m - \langle\xi\tanh(\beta J\eta m')\rangle_{\eta\xi}\right]\mathcal{P}_t[m'],\qquad (N\to\infty).$$

This describes deterministic evolution. If at $t=0$ we know m exactly, this will remain the case for finite time-scales, and m will evolve according to a discrete version of the sequential dynamics law (4.123):

$$m_{t+1} = \langle\xi\tanh[\beta J\eta m_t]\rangle_{\eta,\xi}.$$

We now try to generalize the above approach to less trivial classes of models. As for the sequential case we will find in the limit $N\to\infty$ closed deterministic evolution equations for a more general set of intensive macroscopic state variables $\boldsymbol{\Omega}(\boldsymbol{\sigma}) = (\Omega_1(\boldsymbol{\sigma}),\dots,\Omega_n(\boldsymbol{\sigma}))$ if the local fields $h_i(\boldsymbol{\sigma})$ depend on the microscopic state $\boldsymbol{\sigma}$ only through the values of $\boldsymbol{\Omega}(\boldsymbol{\sigma})$, and if the number n of these state variables necessary to do so is not too large. The evolution of the ensemble probability density (4.124) is now got by inserting the Markov equation (4.141):

$$\mathcal{P}_{t+1}[\boldsymbol{\Omega}] = \int d\boldsymbol{\Omega}'\;\tilde{W}_t[\boldsymbol{\Omega},\boldsymbol{\Omega}']\,\mathcal{P}_t[\boldsymbol{\Omega}'],\tag{4.143}$$

$$\tilde{W}_t[\boldsymbol{\Omega},\boldsymbol{\Omega}'] = \frac{\sum_{\boldsymbol{\sigma}\boldsymbol{\sigma}'}\delta[\boldsymbol{\Omega}-\boldsymbol{\Omega}(\boldsymbol{\sigma})]\,\delta[\boldsymbol{\Omega}'-\boldsymbol{\Omega}(\boldsymbol{\sigma}')]\,W[\boldsymbol{\sigma};\boldsymbol{\sigma}']\,p_t(\boldsymbol{\sigma}')}{\sum_{\boldsymbol{\sigma}'}\delta[\boldsymbol{\Omega}'-\boldsymbol{\Omega}(\boldsymbol{\sigma}')]\,p_t(\boldsymbol{\sigma}')}\tag{4.144}$$

$$= \langle\delta[\boldsymbol{\Omega}-\boldsymbol{\Omega}(\boldsymbol{\sigma})]\,\langle e^{[\beta\sigma_i h_i(\boldsymbol{\sigma}') - \log\cosh(\beta h_i(\boldsymbol{\sigma}'))]}\rangle_{\boldsymbol{\Omega}';\bar{z}}\rangle_{\boldsymbol{\sigma}}.$$

with $\langle\dots\rangle_{\boldsymbol{\sigma}} = 2^{-N}\sum_{\boldsymbol{\sigma}}\dots$, and with the conditional (or sub-shell) average defined as in (4.126). It is clear from (4.144) that in order to find autonomous macroscopic laws, i.e., for the distribution $p_t(\boldsymbol{\sigma})$ to drop out, the

terms if $\mathbf{m}(\boldsymbol{\sigma})$ only, as for the models (4.138) (or, more generally, for all models in which the interactions are of the form $J_{ij} = \sum_{\mu \leq p} f_{i\mu} \xi_j^{\mu}$), and with the following restriction on the number p of embedded patterns: $\lim_{N \to \infty} p/\sqrt{N} = 0$ (as with sequential dynamics). For the bilinear models (4.138), the evolution in time of the overlap vector $\mathbf{m}$ (which depends linearly on the σ_i) is governed by (4.146), which now translates into the iterative map:

$$\mathbf{m}(t+1) = \langle \boldsymbol{\xi} \tanh[\beta \boldsymbol{\xi} \cdot A\mathbf{m}(t)] \rangle_{\boldsymbol{\xi}}, \tag{4.148}$$

with $\rho(\boldsymbol{\xi})$ as defined in (4.139). Again symmetry of the synapses is not required. For parallel dynamics it is far more difficult than for sequential dynamics to construct Lyapunov functions, and prove that the macroscopic laws (4.148) for symmetric systems evolve towards a stable fixed–point (as one would expect), but it can still be done. For non–symmetric systems the macroscopic laws (4.148) can in principle display all the interesting, but complicated, phenomena of non–conservative nonlinear systems. Nevertheless, it is also not uncommon that the equations (4.148) for non–symmetric systems can be mapped by a time–dependent transformation onto the equations for related symmetric systems (mostly variants of the original Hopfield model).

Note that the fixed–points of the macroscopic equations (4.134) and (4.139) (derived for sequential dynamics) are identical to those of (4.147) and (4.148) (derived for parallel dynamics). The stability properties of these fixed–points, however, need not be the same, and have to be assessed on a case–by–case basis. For the Hopfield model, i.e., equations (4.139,4.148) with $A_{\mu\nu} = \delta_{\mu\nu}$, they are found to be the same, but already for $A_{\mu\nu} = -\delta_{\mu\nu}$ the two types of dynamics would behave differently [Coo01, SC00, SC01, CKS05].

4.3.7 Attractor Neural Networks with Continuous Neurons

Closed Macroscopic Laws

We have seen above that models of recurrent neural networks with continuous neural variables (e.g., graded–response neurons or coupled oscillators) can often be described by a Fokker–Planck equation for the microscopic state probability density $p_t(\boldsymbol{\sigma})$:

$$\dot{p}_t(\boldsymbol{\sigma}) = -\frac{\partial}{\partial \sigma_i}[p_t(\boldsymbol{\sigma})f_i(\boldsymbol{\sigma})] + T\sum_i \frac{\partial^2}{\partial \sigma_i^2}p_t(\boldsymbol{\sigma}).$$

Averages over $p_t(\boldsymbol{\sigma})$ are denoted by $\langle G \rangle = \int d\boldsymbol{\sigma}\, p_t(\boldsymbol{\sigma})G(\boldsymbol{\sigma}, t)$. From (4.67) one gets directly (through integration by parts) an equation for the time derivative of averages [Coo01, SC00, SC01, CKS05]:

$$\frac{d}{dt}\langle G \rangle = \langle \frac{\partial G}{\partial t} \rangle + \langle \left[f_i(\boldsymbol{\sigma}) + T\frac{\partial}{\partial \sigma_i} \right] \frac{\partial G}{\partial \sigma_i} \rangle. \tag{4.149}$$

In particular, if we apply (4.149) to $G(\boldsymbol{\sigma}, t) = \delta[\boldsymbol{\Omega} - \boldsymbol{\Omega}(\boldsymbol{\sigma})]$ for any set of macroscopic observables $\boldsymbol{\Omega}(\boldsymbol{\sigma}) = (\Omega_1(\boldsymbol{\sigma}), \ldots, \Omega_n(\boldsymbol{\sigma}))$ (in the spirit of the previous section), we get a dynamic equation for the macroscopic probability density $P_t(\boldsymbol{\Omega}) = \langle \delta[\boldsymbol{\Omega} - \boldsymbol{\Omega}(\boldsymbol{\sigma})] \rangle$, which is again of the Fokker–Planck form:

$$
\dot{P}_t(\boldsymbol{\Omega}) = -\frac{\partial}{\partial \Omega_\mu} \left\{ P_t(\boldsymbol{\Omega}) \, \langle \left[f_i(\boldsymbol{\sigma}) + T \frac{\partial}{\partial \sigma_i} \right] \frac{\partial}{\partial \sigma_i} \Omega_\mu(\boldsymbol{\sigma}) \rangle_{\boldsymbol{\Omega};t} \right\} \tag{4.150}
$$
$$
+ T \frac{\partial^2}{\partial \Omega_\mu \partial \Omega_\nu} \left\{ P_t(\boldsymbol{\Omega}) \, \langle \left[\frac{\partial}{\partial \sigma_i} \Omega_\mu(\boldsymbol{\sigma}) \right] \left[\frac{\partial}{\partial \sigma_i} \Omega_\nu(\boldsymbol{\sigma}) \right] \rangle_{\boldsymbol{\Omega};t} \right\},
$$

with the conditional (or sub–shell) averages:

$$
\langle G(\boldsymbol{\sigma}) \rangle_{\boldsymbol{\Omega},t} = \frac{\int d\boldsymbol{\sigma} \, p_t(\boldsymbol{\sigma}) \delta[\boldsymbol{\Omega} - \boldsymbol{\Omega}(\boldsymbol{\sigma})] G(\boldsymbol{\sigma})}{\int d\boldsymbol{\sigma} \, p_t(\boldsymbol{\sigma}) \delta[\boldsymbol{\Omega} - \boldsymbol{\Omega}(\boldsymbol{\sigma})]}.
$$

From (4.150) we infer that a sufficient condition for the observables $\boldsymbol{\Omega}(\boldsymbol{\sigma})$ to evolve in time deterministically (i.e., for having vanishing diffusion matrix elements in (4.150)) in the limit $N \to \infty$ is

$$
\lim_{N \to \infty} \langle \sum_i \left[\sum_\mu \left| \frac{\partial}{\partial \sigma_i} \Omega_\mu(\boldsymbol{\sigma}) \right| \right]^2 \rangle_{\boldsymbol{\Omega};t} = 0. \tag{4.151}
$$

If (4.151) holds, the macroscopic Fokker–Planck equation (4.150) reduces for $N \to \infty$ to a Liouville equation, and the observables $\boldsymbol{\Omega}(\boldsymbol{\sigma})$ will evolve in time according to the coupled deterministic equations:

$$
\dot{\Omega}_\mu = \lim_{N \to \infty} \langle \left[f_i(\boldsymbol{\sigma}) + T \frac{\partial}{\partial \sigma_i} \right] \frac{\partial}{\partial \sigma_i} \Omega_\mu(\boldsymbol{\sigma}) \rangle_{\boldsymbol{\Omega};t}. \tag{4.152}
$$

The deterministic macroscopic equation (4.152), together with its associated condition for validity (4.151) will form the basis for the subsequent analysis.

The general derivation given above went smoothly. However, the equations (4.152) are not yet closed. It turns out that to achieve closure even for simple continuous networks we can no longer get away with just a finite (small) number of macroscopic observables (as with binary neurons). This we will now illustrate with a simple toy network of graded–response neurons:

$$
\dot{u}_i(t) = J_{ij} \, g[u_j(t)] - u_i(t) + \eta_i(t),
$$

with $g[z] = \frac{1}{2}[\tanh(\gamma z) + 1]$ and with the standard Gaussian white noise $\eta_i(t)$. In the language of (4.67) this means

$$
f_i(\mathbf{u}) = J_{ij} g[u_j] - u_i.
$$

We choose uniform synapses $J_{ij} = J/N$, so $f_i(\mathbf{u}) \to (J/N) \sum_j g[u_j] - u_i$. If (4.151) were to hold, we would find the deterministic macroscopic laws

$$\dot{\Omega}_\mu = \lim_{N \to \infty} \langle \sum_i [\frac{J}{N} \sum_j g[u_j] - u_i + T\frac{\partial}{\partial u_i}]\frac{\partial}{\partial u_i} \Omega_\mu(\mathbf{u}) \rangle_{\mathbf{\Omega};t}. \qquad (4.153)$$

In contrast to similar models with binary neurons, choosing as our macroscopic level of description $\mathbf{\Omega}(\mathbf{u})$ again simply the average $m(\mathbf{u}) = N^{-1}\sum_i u_i$ now leads to an equation which fails to close [Coo01, SC00, SC01, CKS05]:

$$\dot{m} = \lim_{N \to \infty} J\langle \frac{1}{N}\sum_j g[u_j]\rangle_{m;t} - m.$$

The term $N^{-1}\sum_j g[u_j]$ cannot be written as a function of $N^{-1}\sum_i u_i$. We might be tempted to try dealing with this problem by just including the offending term in our macroscopic set, and choose $\mathbf{\Omega}(\mathbf{u}) = (N^{-1}\sum_i u_i, N^{-1}\sum_i g[u_i])$. This would indeed solve our closure problem for the $m-$equation, but we would now find a new closure problem in the equation for the newly introduced observable. The only way out is to choose an observable *function*, namely the distribution of potentials

$$\rho(u; \mathbf{u}) = \frac{1}{N}\sum_i \delta[u - u_i], \qquad (4.154)$$

$$\rho(u) = \langle \rho(u; \mathbf{u})\rangle = \langle \frac{1}{N}\sum_i \delta[u - u_i]\rangle.$$

This is to be done with care, in view of our restriction on the number of observables: we evaluate (4.154) at first only for n specific values u_μ and take the limit $n \to \infty$ only after the limit $N \to \infty$. Thus we define $\Omega_\mu(\mathbf{u}) = \frac{1}{N}\sum_i \delta[u_\mu - u_i]$, condition (4.151) reduces to the familiar expression $\lim_{N \to \infty} n/\sqrt{N} = 0$, and we get for $N \to \infty$ and $n \to \infty$ (taken in that order) from (4.153) a diffusion equation for the distribution of membrane potentials (describing a so–called '*time–dependent Ornstein–Uhlenbeck process*' [Kam92, Gar94]):

$$\dot{\rho}(u) = -\frac{\partial}{\partial u}\left\{\rho(u)\left[J\int du'\,\rho(u')g[u'] - u\right]\right\} + T\frac{\partial^2}{\partial u^2}\rho(u). \qquad (4.155)$$

The natural solution of (4.155) [8] is the Gaussian distribution

$$\rho_t(u) = [2\pi\Sigma^2(t)]^{-\frac{1}{2}}e^{-\frac{1}{2}[u-\overline{u}(t)]^2/\Sigma^2(t)}, \qquad (4.156)$$

in which $\Sigma = [T + (\Sigma_0^2 - T)e^{-2t}]^{\frac{1}{2}}$, and $\overline{u}$ evolves in time according to

$$\frac{d}{dt}\overline{u} = J\int Dz\, g[\overline{u} + \Sigma z] - \overline{u},$$

---

[8] For non-Gaussian initial conditions $\rho_0(u)$ the solution of (4.155) would in time converge towards the Gaussian solution.

with $Dz = (2\pi)^{-\frac{1}{2}} e^{-\frac{1}{2}z^2} dz$. We can now also calculate the distribution $p(s)$ of neuronal firing activities $s_i = g[u_i]$ at any time,

$$p(s) = \int du \, \rho(u) \, \delta[s - g[u]] = \frac{\rho(g^{\mathrm{inv}}[s])}{\int_0^1 ds' \, \rho(g^{\mathrm{inv}}[s'])}.$$

For our choice $g[z] = \frac{1}{2} + \frac{1}{2}\tanh[\gamma z]$ we have $g^{\mathrm{inv}}[s] = \frac{1}{2\gamma}\log[s/(1-s)]$, so in combination with (4.156)[9]

$$0 < s < 1: \qquad p(s) = \frac{e^{-\frac{1}{2}[(2\gamma)^{-1}\log[s/(1-s)]-\bar{u}]^2/\Sigma^2}}{\int_0^1 ds' \, e^{-\frac{1}{2}[(2\gamma)^{-1}\log[s'/(1-s')]-\bar{u}]^2/\Sigma^2}}.$$

Application to Graded–Response Attractor Networks

We will now turn to attractor networks with graded–response neurons of the type (4.68), in which p binary patterns $\boldsymbol{\xi}^\mu = (\xi_1^\mu, \ldots, \xi_N^\mu) \in \{-1, 1\}^N$ have been stored via separable Hebbian synapses (4.138): $J_{ij} = (2/N)\xi_i^\mu A_{\mu\nu}\xi_j^\nu$ (the extra factor 2 is inserted for future convenience). Adding suitable thresholds $\theta_i = -\frac{1}{2}\sum_j J_{ij}$ to the right–hand sides of (4.68), and choosing the nonlinearity $g(z) = \frac{1}{2}(1 + \tanh[\gamma z])$ would then give us [Coo01, SC00, SC01, CKS05]

$$\dot{u}_i(t) = \sum_{\mu\nu} \xi_i^\mu A_{\mu\nu} \frac{1}{N} \sum_j \xi_j^\nu \tanh[\gamma u_j(t)] - u_i(t) + \eta_i(t),$$

so the deterministic forces are $f_i(\mathbf{u}) = N^{-1}\sum_{\mu\nu}\xi_i^\mu A_{\mu\nu}\sum_j \xi_j^\nu \tanh[\gamma u_j] - u_i$. Choosing our macroscopic observables $\boldsymbol{\Omega}(\mathbf{u})$ such that (4.151) holds, would lead to the deterministic macroscopic laws

$$\dot{\Omega}_\mu = \lim_{N\to\infty}\sum_{\mu\nu} A_{\mu\nu}\left\langle \left[\frac{1}{N}\xi_j^\nu \tanh[\gamma u_j]\right]\left[\xi_i^\mu \frac{\partial}{\partial u_i}\Omega_\mu(\mathbf{u})\right]\right\rangle_{\boldsymbol{\Omega};t} \quad (4.157)$$

$$+ \lim_{N\to\infty}\left\langle \left[T\frac{\partial}{\partial u_i} - u_i\right]\frac{\partial}{\partial u_i}\Omega_\mu(\mathbf{u})\right\rangle_{\boldsymbol{\Omega};t}.$$

As with the uniform synapses case, the main problem to be dealt with is how to choose the $\Omega_\mu(\mathbf{u})$ such that (4.157) closes. It turns out that the canonical choice is to turn to the distributions of membrane potentials within each of the 2^p sub–lattices, as introduced above:

---

[9] None of the above results (not even those on the stationary state) could have been got within equilibrium statistical mechanics, since any network of connected graded–response neurons will violate detailed balance. Secondly, there appears to be a qualitative difference between simple networks (e.g., $J_{ij} = J/N$) of binary neurons versus those of continuous neurons, in terms of the types of macroscopic observables needed for deriving closed deterministic laws: a single number $m = N^{-1}\sum_i \sigma_i$ versus a distribution $\rho(\sigma) = N^{-1}\sum_i \delta[\sigma - \sigma_i]$.

$$I_{\boldsymbol{\eta}} = \{i| \; \boldsymbol{\xi}_i = \boldsymbol{\eta}\} : \qquad \rho_{\boldsymbol{\eta}}(u; \mathbf{u}) = \frac{1}{|I_{\boldsymbol{\eta}}|} \sum_{i \in I_{\boldsymbol{\eta}}} \delta[u - u_i], \qquad (4.158)$$

$$\rho_{\boldsymbol{\eta}}(u) = \langle \rho_{\boldsymbol{\eta}}(u; \mathbf{u}) \rangle,$$

with $\boldsymbol{\eta} \in \{-1, 1\}^p$ and $\lim_{N \to \infty} |I_{\boldsymbol{\eta}}|/N = p_{\boldsymbol{\eta}}$. Again we evaluate the distributions in (4.158) at first only for n specific values u_{μ} and send $n \to \infty$ after $N \to \infty$. Now condition (4.151) reduces to $\lim_{N \to \infty} 2^p/\sqrt{N} = 0$. We will keep p finite, for simplicity. Using identities such as $\sum_i \cdots = \sum_{\boldsymbol{\eta}} \sum_{i \in I_{\boldsymbol{\eta}}} \cdots$ and $i \in I_{\boldsymbol{\eta}}$:

$$\frac{\partial}{\partial u_i} \rho_{\boldsymbol{\eta}}(u; \mathbf{u}) = -|I_{\boldsymbol{\eta}}|^{-1} \frac{\partial}{\partial u} \delta[u - u_i], \qquad \frac{\partial^2}{\partial u_i^2} \rho_{\boldsymbol{\eta}}(u; \mathbf{u}) = |I_{\boldsymbol{\eta}}|^{-1} \frac{\partial^2}{\partial u^2} \delta[u - u_i],$$

we then get for $N \to \infty$ and $n \to \infty$ (taken in that order) from equation (4.157) 2^p coupled diffusion equations for the distributions $\rho_{\boldsymbol{\eta}}(u)$ of membrane potentials in each of the 2^p sub-lattices $I_{\boldsymbol{\eta}}$:

$$\dot{\rho}_{\boldsymbol{\eta}}(u) = -\frac{\partial}{\partial u} \{ \rho_{\boldsymbol{\eta}}(u) [\eta_{\mu} A_{\mu\nu} p_{\boldsymbol{\eta}'} \eta_{\nu}' \int du' \; \rho_{\boldsymbol{\eta}'}(u') \tanh[\gamma u'] - u] \} + T \frac{\partial^2}{\partial u^2} \rho_{\boldsymbol{\eta}}(u). \qquad (4.159)$$

Equation (4.159) is the basis for our further analysis. It can be simplified only if we make additional assumptions on the system's initial conditions, such as δ−distributed or Gaussian distributed $\rho_{\boldsymbol{\eta}}(u)$ at $t = 0$ (see below); otherwise it will have to be solved numerically.

It is clear that (4.159) is again of the time–dependent Ornstein–Uhlenbeck form, and will thus again have Gaussian solutions as the natural ones [Coo01, SC00, SC01, CKS05]:

$$\rho_{t,\boldsymbol{\eta}}(u) = [2\pi \Sigma_{\boldsymbol{\eta}}^2(t)]^{-\frac{1}{2}} e^{-\frac{1}{2} [u - \bar{u}_{\boldsymbol{\eta}}(t)]^2 / \Sigma_{\boldsymbol{\eta}}^2(t)},$$

in which $\Sigma_{\boldsymbol{\eta}}(t) = [T + (\Sigma_{\boldsymbol{\eta}}^2(0) - T) e^{-2t}]^{\frac{1}{2}}$, and with the $\bar{u}_{\boldsymbol{\eta}}(t)$ evolving in time according to

$$\frac{d}{dt} \bar{u}_{\boldsymbol{\eta}} = p_{\boldsymbol{\eta}'}(\boldsymbol{\eta} \cdot \mathbf{A} \boldsymbol{\eta}') \int Dz \; \tanh[\gamma(\bar{u}_{\boldsymbol{\eta}'} + \Sigma_{\boldsymbol{\eta}'} z)] - \bar{u}_{\boldsymbol{\eta}}. \qquad (4.160)$$

Our problem has thus been reduced successfully to the study of the 2^p coupled scalar equations (4.160). We can also measure the correlation between the firing activities $s_i(u_i) = \frac{1}{2}[1 + \tanh(\gamma u_i)]$ and the pattern components (similar to the overlaps in the case of binary neurons). If the pattern bits are drawn at random, i.e., $\lim_{N \to \infty} |I_{\boldsymbol{\eta}}|/N = p_{\boldsymbol{\eta}} = 2^{-p}$ for all $\boldsymbol{\eta}$, we can define a 'graded–response' equivalent $m_{\mu}(\mathbf{u}) = 2N^{-1} \xi_i^{\mu} s_i(u_i) \in [-1, 1]$ of the Hopfield pattern overlaps:

$$m_{\mu}(\mathbf{u}) = \frac{2}{N} \xi_i^{\mu} s_i(\mathbf{u}) = \frac{1}{N} \xi_i^{\mu} \tanh(\gamma u_i) + \mathcal{O}(N^{-\frac{1}{2}})$$

$$= p_{\boldsymbol{\eta}} \; \eta_{\mu} \int du \; \rho_{\boldsymbol{\eta}}(u; \mathbf{u}) \tanh(\gamma u) + \mathcal{O}(N^{-\frac{1}{2}}).$$

Full recall of pattern μ implies $s_i(u_i) = \frac{1}{2}[\xi_i^\mu + 1]$, giving $m_\mu(\mathbf{u}) = 1$. Since the distributions $\rho_{\boldsymbol\eta}(u)$ obey deterministic laws for $N \to \infty$, the same will be true for the overlaps $\mathbf{m} = (m_1, \ldots, m_p)$. For the Gaussian solutions (4.160) of (4.159) we can now proceed to replace the 2^p macroscopic laws (4.160), which reduce to $\frac{d}{dt}\bar{u}_{\boldsymbol\eta} = \boldsymbol\eta \cdot \mathbf{Am} - \bar{u}_{\boldsymbol\eta}$ and give $\bar{u}_{\boldsymbol\eta} = \bar{u}_{\boldsymbol\eta}(0)e^{-t} + \boldsymbol\eta \cdot \mathbf{A}\int_0^t ds\, \epsilon^{s-t}\mathbf{m}(s)$, by p integral equations in terms of overlaps only:

$$m_\mu(t) = p_{\boldsymbol\eta}\, \eta_\mu \int Dz\ \tanh[\gamma(\bar{u}_{\boldsymbol\eta}(0)e^{-t} \tag{4.161}$$

$$+ \boldsymbol\eta \cdot \mathbf{A}\int_0^t ds\ e^{s-t}\mathbf{m}(s) + z\sqrt{T + (\Sigma_{\boldsymbol\eta}^2(0) - T)e^{-2t}}],$$

with $Dz = (2\pi)^{-\frac{1}{2}}e^{-\frac{1}{2}z^2}dz$. Here the sub–lattices only come in via the initial conditions.

The equations describing the asymptotic (stationary) state can be written entirely without sub-lattices,[10]

$$m_\mu = \langle \xi_\mu \int Dz\ \tanh[\gamma(\boldsymbol\xi \cdot \mathbf{Am} + z\sqrt{T})]\rangle_{\boldsymbol\xi}, \tag{4.162}$$

$$\rho_{\boldsymbol\eta}(u) = [2\pi T]^{-\frac{1}{2}}e^{-\frac{1}{2}[u - \boldsymbol\eta\cdot\mathbf{Am}]^2/T},$$

by taking the $t \to \infty$ limit in (4.161), using $\bar{u}_{\boldsymbol\eta} \to \boldsymbol\eta \cdot \mathbf{Am}$, $\Sigma_{\boldsymbol\eta} \to \sqrt{T}$, and the familiar notation

$$\langle g(\boldsymbol\xi)\rangle_{\boldsymbol\xi} = \lim_{N\to\infty}\frac{1}{N}\sum_i g(\boldsymbol\xi_i) = 2^{-p}\sum_{\boldsymbol\xi\in\{-1,1\}^p} g(\boldsymbol\xi).$$

For the simplest non–trivial choice, $A_{\mu\nu} = \delta_{\mu\nu}$ (i.e., $J_{ij} = (2/N)\sum_\mu \xi_i^\mu \xi_j^\mu$, as in the Hopfield [Hop82] model) equation (4.162) yields the familiar pure and mixture state solutions. For $T = 0$ we find a continuous phase transition from non–recall to pure states of the form $m_\mu = m\delta_{\mu\nu}$ (for some ν) at $\gamma_c = 1$. For $T > 0$ we have in (4.162) an additional Gaussian noise, absent in the models with binary neurons. Again the pure states are the first non–trivial solutions to enter the stage. Substituting $m_\mu = m\delta_{\mu\nu}$ into (4.162) gives

$$m = \int Dz\ \tanh[\gamma(m + z\sqrt{T})]. \tag{4.163}$$

Writing (4.163) as $m^2 = \gamma m\int_0^m dk[1 - \int Dz\ \tanh^2[\gamma(k + z\sqrt{T})]] \leq \gamma m^2$, reveals that $m = 0$ as soon as $\gamma < 1$. A continuous transition to an $m > 0$ state occurs when

---

[10] Note the appealing similarity with previous results on networks with binary neurons in equilibrium. For $T = 0$ the overlap equations (4.162) become identical to those found for attractor networks with binary neurons and finite p (hence our choice to insert an extra factor 2 in defining the synapses), with γ replacing the inverse noise level β in the former.

$$\gamma^{-1} = 1 - \int Dz \ \tanh^2[\gamma z \sqrt{T}].$$

A parametrization of this transition line in the (γ, T)−plane is given by

$$\gamma^{-1}(x) = 1 - \int Dz \ \tanh^2(zx),$$

$$T(x) = x^2/\gamma^2(x), \qquad x \geq 0.$$

Discontinuous transitions away from $m = 0$ (for which there is no evidence) would have to be calculated numerically. For $\gamma = \infty$ we get the equation $m = \mathrm{erf}[m/\sqrt{2T}]$, giving a continuous transition to $m > 0$ at $T_c = 2/\pi \approx 0.637$. Alternatively the latter number can also be found by taking $\lim_{x \to \infty} T(x)$ in the above parametrization:

$$T_c(\gamma = \infty) = \lim_{x \to \infty} x^2 [1 - \int Dz \ \tanh^2(zx)]^2$$

$$= \lim_{x \to \infty} [\int Dz \ \frac{\partial}{\partial z} \tanh(zx)]^2 = [2 \int Dz \ \delta(z)]^2 = 2/\pi.$$

Let us now turn to dynamics. It follows from (4.162) that the 'natural' initial conditions for $\bar{u}_\eta$ and Σ_η are of the form: $\bar{u}_\eta(0) = \boldsymbol{\eta} \cdot \mathbf{k}_0$ and $\Sigma_\eta(0) = \Sigma_0$ for all $\boldsymbol{\eta}$. Equivalently:

$$t = 0: \qquad \rho_\eta(u) = [2\pi\Sigma_0^2]^{-\frac{1}{2}} e^{-\frac{1}{2}[u - \boldsymbol{\eta} \cdot \mathbf{k}_0]^2/\Sigma_0^2},$$

$$\mathbf{k}_0 \in \mathbb{R}^p, \ \Sigma_0 \in \mathbb{R}.$$

These would also be the typical and natural statistics if we were to prepare an initial firing state $\{s_i\}$ by hand, via manipulation of the potentials $\{u_i\}$. For such initial conditions we can simplify the dynamical equation (4.161) to [Coo01, SC00, SC01, CKS05]

$$m_\mu(t) = \langle \ \xi_\mu \int Dz \ \tanh[\gamma(\boldsymbol{\xi} \cdot [\mathbf{k}_0 e^{-t} \qquad (4.164)$$

$$+ \mathbf{A} \int_0^t ds \ e^{s-t} \mathbf{m}(s)] + z\sqrt{T + (\Sigma_0^2 - T)e^{-2t}})]\rangle_{\boldsymbol{\xi}}. \qquad (4.165)$$

For the special case of the Hopfield synapses, i.e., $A_{\mu\nu} = \delta_{\mu\nu}$, it follows from (4.164) that recall of a given pattern ν is triggered upon choosing $k_{0,\mu} = k_0 \delta_{\mu\nu}$ (with $k_0 > 0$), since then equation (4.164) generates $m_\mu(t) = m(t)\delta_{\mu\nu}$ at any time, with the amplitude $m(t)$ following from

$$m(t) = \int Dz \ \tanh[\gamma[k_0 e^{-t} + \int_0^t ds \ e^{s-t} m(s) + z\sqrt{T + (\Sigma_0^2 - T)e^{-2t}}]],$$

$$(4.166)$$

which is the dynamical counterpart of equation (4.163) (to which indeed it reduces for $t \to \infty$).

We finally specialize further to the case where our Gaussian initial conditions are not only chosen to trigger recall of a single pattern $\boldsymbol{\xi}^\nu$, but in addition describe uniform membrane potentials within the sub-lattices, i.e., $k_{0,\mu} = k_0 \delta_{\mu\nu}$ and $\Sigma_0 = 0$, so $\rho_{\boldsymbol{\eta}}(u) = \delta[u - k_0 \eta_\nu]$. Here we can derive from (4.166) at $t = 0$ the identity $m_0 = \tanh[\gamma k_0]$, which enables us to express k_0 as $k_0 = (2\gamma)^{-1} \log[(1 + m_0)/(1 - m_0)]$, and find (4.166) reducing to

$$m(t) = \int Dz \ \tanh[e^{-t} \log[\frac{1 + m_0}{1 - m_0}]^{\frac{1}{2}} + \gamma[\int_0^t ds \ e^{s-t} m(s) + z \sqrt{T(1 - e^{-2t})}]].$$

Compared to the overlap evolution in large networks of binary networks (away from saturation) one can see richer behavior, e.g., non–monotonicity [Coo01, SC00, SC01, CKS05].

4.3.8 Correlation– and Response–Functions

We now turn to correlation functions $C_{ij}(t, t')$ and response functions $G_{ij}(t, t')$. These will become the language in which the partition function methods are formulated, which will enable us to solve the dynamics of recurrent networks in the (complex) regime near saturation (we take $t > t'$):

$$C_{ij}(t, t') = \langle \sigma_i(t) \sigma_j(t') \rangle, \qquad G_{ij}(t, t') = \partial \langle \sigma_i(t) \rangle / \partial \theta_j(t') \qquad (4.167)$$

The $\{\sigma_i\}$ evolve in time according to equations of the form (4.122) (binary neurons, sequential updates), (4.141) (binary neurons, parallel updates) or (4.67) (continuous neurons). The θ_i represent thresholds and/or external stimuli, which are added to the local fields in the cases (4.122,4.141), or added to the deterministic forces in the case of a Fokker–Planck equation (4.67). We retain $\theta_i(t) = \theta_i$, except for a perturbation $\delta\theta_j(t')$ applied at time t' in defining the response function. Calculating averages such as (4.167) requires determining joint probability distributions involving neuron states at different times.

Fluctuation–Dissipation Theorems

For networks of binary neurons with discrete time dynamics of the form $p_{\ell+1}(\boldsymbol{\sigma}) = \sum_{\boldsymbol{\sigma}'} W[\boldsymbol{\sigma}; \boldsymbol{\sigma}'] p_\ell(\boldsymbol{\sigma}')$, the probability of observing a given 'path' $\boldsymbol{\sigma}(\ell') \to \boldsymbol{\sigma}(\ell' + 1) \to \ldots \to \boldsymbol{\sigma}(\ell - 1) \to \boldsymbol{\sigma}(\ell)$ of successive configurations between step ℓ' and step ℓ is given by the product of the corresponding transition matrix elements (without summation):

$$\text{Prob}[\boldsymbol{\sigma}(\ell'), \ldots, \boldsymbol{\sigma}(\ell)] = W[\boldsymbol{\sigma}(\ell); \boldsymbol{\sigma}(\ell - 1)] W[\boldsymbol{\sigma}(\ell - 1);$$
$$\boldsymbol{\sigma}(\ell - 2)] \ldots W[\boldsymbol{\sigma}(\ell' + 1); \boldsymbol{\sigma}(\ell')] p_{\ell'}(\boldsymbol{\sigma}(\ell')).$$

This allows us to write [Coo01, SC00, SC01, CKS05]

$$C_{ij}(\ell, \ell') = \sum_{\boldsymbol{\sigma}(\ell')} \cdots \sum_{\boldsymbol{\sigma}(\ell)} \text{Prob}[\boldsymbol{\sigma}(\ell'), \ldots, \boldsymbol{\sigma}(\ell)]\sigma_i(\ell)\sigma_j(\ell') \qquad (4.168)$$

$$= \sum_{\boldsymbol{\sigma}\boldsymbol{\sigma}'} \sigma_i \sigma'_j W^{\ell-\ell'}[\boldsymbol{\sigma}; \boldsymbol{\sigma}']p_{\ell'}(\boldsymbol{\sigma}'),$$

$$G_{ij}(\ell, \ell') = \sum_{\boldsymbol{\sigma}\boldsymbol{\sigma}'\boldsymbol{\sigma}''} \sigma_i W^{\ell-\ell'-1}[\boldsymbol{\sigma}; \boldsymbol{\sigma}''] \left[\frac{\partial}{\partial \theta_j} W[\boldsymbol{\sigma}''; \boldsymbol{\sigma}'] \right] p_{\ell'}(\boldsymbol{\sigma}'). \qquad (4.169)$$

From (4.168) and (4.169) it follows that both $C_{ij}(\ell, \ell')$ and $G_{ij}(\ell, \ell')$ will in the stationary state, i.e., upon substituting $p_{\ell'}(\boldsymbol{\sigma}') = p_\infty(\boldsymbol{\sigma}')$, only depend on $\ell - \ell'$: $C_{ij}(\ell, \ell') \to C_{ij}(\ell - \ell')$ and $G_{ij}(\ell, \ell') \to G_{ij}(\ell - \ell')$. For this we do not require detailed balance. Detailed balance, however, leads to a simple relation between the response function $G_{ij}(\tau)$ and the temporal derivative of the correlation function $C_{ij}(\tau)$.

We now turn to equilibrium systems, i.e., networks with symmetric synapses (and with all $J_{ii} = 0$ in the case of sequential dynamics). We calculate the derivative of the transition matrix that occurs in (4.169) by differentiating the equilibrium condition $p_{\text{eq}}(\boldsymbol{\sigma}) = \sum_{\boldsymbol{\sigma}'} W[\boldsymbol{\sigma}; \boldsymbol{\sigma}']p_{\text{eq}}(\boldsymbol{\sigma}')$ with respect to external fields:

$$\frac{\partial}{\partial \theta_j} p_{\text{eq}}(\boldsymbol{\sigma}) = \sum_{\boldsymbol{\sigma}'} \{ \frac{\partial W[\boldsymbol{\sigma}; \boldsymbol{\sigma}']}{\partial \theta_j} p_{\text{eq}}(\boldsymbol{\sigma}') + W[\boldsymbol{\sigma}; \boldsymbol{\sigma}'] \frac{\partial}{\partial \theta_j} p_{\text{eq}}(\boldsymbol{\sigma}') \}.$$

Detailed balance implies $p_{\text{eq}}(\boldsymbol{\sigma}) = Z^{-1} e^{-\beta H(\boldsymbol{\sigma})}$ (in the parallel case we simply substitute the appropriate Hamiltonian $H \to \tilde{H}$), giving $\partial p_{\text{eq}}(\boldsymbol{\sigma})/\partial \theta_j = -[Z^{-1}\partial Z/\partial \theta_j + \beta \partial H(\boldsymbol{\sigma})/\partial \theta_j]p_{\text{eq}}(\boldsymbol{\sigma})$, so that

$$\sum_{\boldsymbol{\sigma}'} \frac{\partial W[\boldsymbol{\sigma}; \boldsymbol{\sigma}']}{\partial \theta_j} p_{\text{eq}}(\boldsymbol{\sigma}') = \beta \{ \sum_{\boldsymbol{\sigma}'} W[\boldsymbol{\sigma}; \boldsymbol{\sigma}'] \frac{\partial H(\boldsymbol{\sigma}')}{\partial \theta_j} p_{\text{eq}}(\boldsymbol{\sigma}') - \frac{\partial H(\boldsymbol{\sigma})}{\partial \theta_j} p_{\text{eq}}(\boldsymbol{\sigma}) \},$$

in which the term containing Z drops out. We now get for the response function (4.169) in equilibrium the following result:

$$G_{ij}(\ell) = \qquad\qquad (4.170)$$

$$\beta \sum_{\boldsymbol{\sigma}\boldsymbol{\sigma}'} \sigma_i W^{\ell-1}[\boldsymbol{\sigma}; \boldsymbol{\sigma}'] \left\{ \sum_{\boldsymbol{\sigma}''} W[\boldsymbol{\sigma}'; \boldsymbol{\sigma}''] \frac{\partial H(\boldsymbol{\sigma}'')}{\partial \theta_j} p_{\text{eq}}(\boldsymbol{\sigma}'') - \frac{\partial H(\boldsymbol{\sigma}')}{\partial \theta_j} p_{\text{eq}}(\boldsymbol{\sigma}') \right\}.$$

The structure of (4.170) is similar to what follows upon calculating the evolution of the equilibrium correlation function (4.168) in a single iteration step:

$$C_{ij}(\ell) - C_{ij}(\ell - 1) = \sum_{\boldsymbol{\sigma}\boldsymbol{\sigma}'} \sigma_i W^{\ell-1}[\boldsymbol{\sigma}; \boldsymbol{\sigma}'] \times \qquad (4.171)$$

$$\times \{ \sum_{\boldsymbol{\sigma}''} W[\boldsymbol{\sigma}'; \boldsymbol{\sigma}''] \sigma''_j p_{\text{eq}}(\boldsymbol{\sigma}'') - \sigma'_j p_{\text{eq}}(\boldsymbol{\sigma}') \}.$$

Finally we calculate the relevant derivatives of the two Hamiltonians

$$H(\boldsymbol{\sigma}) = -J_{ij}\sigma_i\sigma_j + \theta_i\sigma_i, \qquad \text{and}$$
$$\tilde{H}(\boldsymbol{\sigma}) = -\theta_i\sigma_i - \beta^{-1}\sum_i \log 2\cosh[\beta h_i(\boldsymbol{\sigma})]$$

(with $h_i(\boldsymbol{\sigma}) = J_{ij}\sigma_j + \theta_i$),

$$\frac{\partial H(\boldsymbol{\sigma})}{\partial\theta_j} = -\sigma_j, \qquad \frac{\partial\tilde{H}(\boldsymbol{\sigma})}{\partial\theta_j} = -\sigma_j - \tanh[\beta h_j(\boldsymbol{\sigma})].$$

For sequential dynamics we hereby arrive directly at a *fluctuation–dissipation theorem*. For parallel dynamics we need one more identity (which follows from the definition of the transition matrix in (4.141) and the detailed balance property) to transform the *tanh* occurring in the derivative of $\tilde{H}$:

$$\tanh[\beta h_j(\boldsymbol{\sigma}')]p_{\text{eq}}(\boldsymbol{\sigma}') = \sum_{\boldsymbol{\sigma}''}\sigma_j''W\left[\boldsymbol{\sigma}'';\boldsymbol{\sigma}'\right]p_{\text{eq}}(\boldsymbol{\sigma}')$$
$$= \sum_{\boldsymbol{\sigma}''}W\left[\boldsymbol{\sigma}';\boldsymbol{\sigma}''\right]\sigma_j''p_{\text{eq}}(\boldsymbol{\sigma}'').$$

For parallel dynamics ℓ and ℓ' are the real time labels t and t', and we get, with $\tau = t - t'$:

$$G_{ij}(\tau > 0) = -\beta[C_{ij}(\tau+1) - C_{ij}(\tau-1)], \qquad G_{ij}(\tau \leq 0) = 0. \qquad (4.172)$$

For the continuous-time version (4.122) of sequential dynamics the time t is defined as $t = \ell/N$, and the difference equation (4.171) becomes a differential equation. For perturbations at time t' in the definition of the response function (4.169) to retain a non-vanishing effect at (re-scaled) time t in the limit $N \to \infty$, they will have to be re-scaled as well: $\delta\theta_j(t') \to N\delta\theta_j(t')$. As a result:

$$G_{ij}(\tau) = -\beta\theta(\tau)\frac{d}{d\tau}C_{ij}(\tau). \qquad (4.173)$$

The need to re–scale perturbations in making the transition from discrete to continuous times has the same origin as the need to re-scale the random forces in the derivation of the continuous-time Langevin equation from a discrete-time process. Going from ordinary derivatives to functional derivatives (which is what happens in the continuous–time limit), implies replacing Kronecker delta's $\delta_{t,t'}$ by Dirac delta-functions according to $\delta_{t,t'} \to \Delta\delta(t - t')$, where Δ is the average duration of an iteration step. Equations (4.172) and (4.173) are examples of so–called *fluctuation–dissipation theorems* (FDT).

For systems described by a Fokker–Planck equation (4.67) the simplest way to calculate correlation- and response-functions is by first returning to the underlying discrete-time system and leaving the continuous time limit $\Delta \to 0$ until the end. We saw above that for small but finite time-steps Δ the underlying discrete-time process is described by [Coo01, SC00, SC01, CKS05]

$$t = \ell\Delta, \qquad p_{\ell\Delta+\Delta}(\boldsymbol{\sigma}) = [1 + \Delta\mathcal{L}_{\boldsymbol{\sigma}} + \mathcal{O}(\Delta^{\frac{3}{2}})]p_{\ell\Delta}(\boldsymbol{\sigma}),$$

with $\ell = 0, 1, 2, \ldots$ and with the differential operator

$$\mathcal{L}_{\boldsymbol{\sigma}} = -\frac{\partial}{\partial\sigma_i}[f_i(\boldsymbol{\sigma}) - T\frac{\partial}{\partial\sigma_i}].$$

From this it follows that the conditional probability density $p_{\ell\Delta}(\boldsymbol{\sigma}|\boldsymbol{\sigma}', \ell'\Delta)$ for finding state $\boldsymbol{\sigma}$ at time $\ell\Delta$, given the system was in state $\boldsymbol{\sigma}'$ at time $\ell'\Delta$, must be

$$p_{\ell\Delta}(\boldsymbol{\sigma}|\boldsymbol{\sigma}', \ell'\Delta) = [1 + \Delta\mathcal{L}_{\boldsymbol{\sigma}} + \mathcal{O}(\Delta^{\frac{3}{2}})]^{\ell-\ell'}\delta[\boldsymbol{\sigma} - \boldsymbol{\sigma}']. \tag{4.174}$$

Equation (4.174) will be our main building block. Firstly, we will calculate the correlations:

$$C_{ij}(\ell\Delta, \ell'\Delta) = \langle\sigma_i(\ell\Delta)\sigma_j(\ell'\Delta)\rangle$$

$$= \int d\boldsymbol{\sigma}d\boldsymbol{\sigma}' \; \sigma_i\sigma_j' \; p_{\ell\Delta}(\boldsymbol{\sigma}|\boldsymbol{\sigma}', \ell'\Delta)p_{\ell'\Delta}(\boldsymbol{\sigma}')$$

$$= \int d\boldsymbol{\sigma} \; \sigma_i[1 + \Delta\mathcal{L}_{\boldsymbol{\sigma}} + \mathcal{O}(\Delta^{\frac{3}{2}})]^{\ell-\ell'} \int d\boldsymbol{\sigma}' \; \sigma_j'\delta[\boldsymbol{\sigma} - \boldsymbol{\sigma}']p_{\ell'\Delta}(\boldsymbol{\sigma}')$$

$$= \int d\boldsymbol{\sigma} \; \sigma_i[1 + \Delta\mathcal{L}_{\boldsymbol{\sigma}} + \mathcal{O}(\Delta^{\frac{3}{2}})]^{\ell-\ell'} \left[\sigma_j \; p_{\ell'\Delta}(\boldsymbol{\sigma})\right].$$

At this stage, we can take the limits $\Delta \to 0$ and $\ell, \ell' \to \infty$, with $t = \ell\Delta$ and $t' = \ell'\Delta$ finite, using $\lim_{\Delta\to 0}[1 + \Delta A]^{k/\Delta} = e^{kA}$:

$$C_{ij}(t, t') = \int d\boldsymbol{\sigma} \; \sigma_i \; e^{(t-t')\mathcal{L}_{\boldsymbol{\sigma}}} \left[\sigma_j \; p_{t'}(\boldsymbol{\sigma})\right]. \tag{4.175}$$

Next we turn to the response function. A perturbation applied at time $t' = \ell'\Delta$ to the Langevin forces $f_i(\boldsymbol{\sigma})$ comes in at the transition $\boldsymbol{\sigma}(\ell'\Delta) \to \boldsymbol{\sigma}(\ell'\Delta + \Delta)$. As with sequential dynamics binary networks, the perturbation is re-scaled with the step size Δ to retain significance as $\Delta \to 0$:

$$G_{ij}(\ell\Delta, \ell'\Delta) = \frac{\partial\langle\sigma_i(\ell\Delta)\rangle}{\Delta\partial\theta_j(\ell'\Delta)} = \frac{\partial}{\Delta\partial\theta_j(\ell'\Delta)} \int d\boldsymbol{\sigma}d\boldsymbol{\sigma}' \; \sigma_i \; p_{\ell\Delta}(\boldsymbol{\sigma}|\boldsymbol{\sigma}', \ell'\Delta)p_{\ell'\Delta}(\boldsymbol{\sigma}')$$

$$= \int d\boldsymbol{\sigma}d\boldsymbol{\sigma}'d\boldsymbol{\sigma}'' \; \sigma_i \; p_{\ell\Delta}(\boldsymbol{\sigma}|\boldsymbol{\sigma}'', \ell'\Delta + \Delta) \left[\frac{\partial p_{\ell''\Delta+\Delta}(\boldsymbol{\sigma}|\boldsymbol{\sigma}', \ell'\Delta)}{\Delta\partial\theta_j}\right] p_{\ell'\Delta}(\boldsymbol{\sigma}')$$

$$= \int d\boldsymbol{\sigma}d\boldsymbol{\sigma}'d\boldsymbol{\sigma}'' \; \sigma_i[1 + \Delta\mathcal{L}_{\boldsymbol{\sigma}} + \mathcal{O}(\Delta^{\frac{3}{2}})]^{\ell-\ell'-1}\delta[\boldsymbol{\sigma} - \boldsymbol{\sigma}''] \times$$

$$\left[\frac{1}{\Delta}\frac{\partial}{\partial\theta_j}[1 + \Delta\mathcal{L}_{\boldsymbol{\sigma}''} + \mathcal{O}(\Delta^{\frac{3}{2}})]\delta[\boldsymbol{\sigma}'' - \boldsymbol{\sigma}']\right] p_{\ell'\Delta}(\boldsymbol{\sigma}')$$

$$= -\int d\boldsymbol{\sigma}d\boldsymbol{\sigma}'d\boldsymbol{\sigma}'' \; \sigma_i[1 + \Delta\mathcal{L}_{\boldsymbol{\sigma}} + \mathcal{O}(\Delta^{\frac{3}{2}})]^{\ell-\ell'-1} \times$$

$$\delta[\boldsymbol{\sigma} - \boldsymbol{\sigma}'']\delta[\boldsymbol{\sigma}'' - \boldsymbol{\sigma}'][\frac{\partial}{\partial\sigma_j'} + \mathcal{O}(\Delta^{\frac{1}{2}})] \; p_{\ell'\Delta}(\boldsymbol{\sigma}')$$

$$= -\int d\boldsymbol{\sigma} \; \sigma_i[1 + \Delta\mathcal{L}_{\boldsymbol{\sigma}} + \mathcal{O}(\Delta^{\frac{3}{2}})]^{\ell-\ell'-1}[\frac{\partial}{\partial\sigma_j} + \mathcal{O}(\Delta^{\frac{1}{2}})] \; p_{\ell'\Delta}(\boldsymbol{\sigma}).$$

We take the limits $\Delta \to 0$ and $\ell, \ell' \to \infty$, with $t = \ell\Delta$ and $t' = \ell'\Delta$ finite:

$$G_{ij}(t, t') = -\int d\boldsymbol{\sigma} \; \sigma_i \; \mathrm{e}^{(t-t')\mathcal{L}_\sigma} \frac{\partial}{\partial \sigma_j} p_{t'}(\boldsymbol{\sigma}). \qquad (4.176)$$

Equations (4.175) and (4.176) apply to arbitrary systems described by Fokker–Planck equations. In the case of conservative forces, i.e., $f_i(\boldsymbol{\sigma}) = -\partial H(\boldsymbol{\sigma})/\partial \sigma_i$, and when the system is in an equilibrium state at time t' so that $C_{ij}(t, t') = C_{ij}(t - t')$ and $G_{ij}(t, t') = G_{ij}(t - t')$, we can take a further step using $p_{t'}(\boldsymbol{\sigma}) = p_{\mathrm{eq}}(\boldsymbol{\sigma}) = Z^{-1}\mathrm{e}^{-\beta H(\boldsymbol{\sigma})}$. In that case, taking the time derivative of expression (4.175) gives

$$\frac{\partial}{\partial \tau} C_{ij}(\tau) = \int d\boldsymbol{\sigma} \; \sigma_i \; \mathrm{e}^{\tau \mathcal{L}_\sigma} \mathcal{L}_\sigma \left[\sigma_j \; p_{\mathrm{eq}}(\boldsymbol{\sigma}) \right].$$

Working out the key term in this expression gives

$$\mathcal{L}_\sigma[\sigma_j \; p_{\mathrm{eq}}(\boldsymbol{\sigma})] = -\sum_i \frac{\partial}{\partial \sigma_i} [f_i(\boldsymbol{\sigma}) - T \frac{\partial}{\partial \sigma_i}][\sigma_j \; p_{\mathrm{eq}}(\boldsymbol{\sigma})]$$

$$= T \frac{\partial}{\partial \sigma_j} p_{\mathrm{eq}}(\boldsymbol{\sigma}) - \sum_i \frac{\partial}{\partial \sigma_i} [\sigma_j J_i(\boldsymbol{\sigma})],$$

with the components of the probability current density $J_i(\boldsymbol{\sigma}) = [f_i(\boldsymbol{\sigma}) - T\frac{\partial}{\partial \sigma_i}]p_{\mathrm{eq}}(\boldsymbol{\sigma})$. In equilibrium, however, the current is zero by definition, so only the first term in the above expression survives. Insertion into our previous equation for $\partial C_{ij}(\tau)/\partial \tau$, and comparison with (4.176) leads to the FDT for continuous systems [Coo01, SC00, SC01, CKS05]:

$$\text{Continuous Dynamics:} \qquad G_{ij}(\tau) = -\beta \theta(\tau) \frac{d}{d\tau} C_{ij}(\tau).$$

We will now calculate the correlation and response functions explicitly, and verify the validity or otherwise of the FDT relations, for attractor networks away from saturation.

Simple Attractor Networks with Binary Neurons

We will consider the continuous time version (4.122) of the sequential dynamics, with the local fields $h_i(\boldsymbol{\sigma}) = J_{ij}\sigma_j + \theta_i$, and the separable interaction matrix (4.138). We already solved the dynamics of this model for the case with zero external fields and away from saturation (i.e., $p \ll \sqrt{N}$). Having non-zero, or even time–dependent, external fields does not affect the calculation much; one adds the external fields to the internal ones and finds the macroscopic laws (4.139) for the overlaps with the stored patterns being replaced by [Coo01, SC00, SC01, CKS05]

$$\dot{\mathbf{m}}(t) = \lim_{N \to \infty} \frac{1}{N} \boldsymbol{\xi}_i \tanh \left[\beta \boldsymbol{\xi}_i \cdot \mathbf{A}\mathbf{m}(t) + \theta_i(t) \right] - \mathbf{m}(t), \qquad (4.177)$$

Fluctuations in the local fields are of vanishing order in N (since the fluctuations in $\mathbf{m}$ are), so that one can easily derive from the master equation (4.122) the following expressions for spin averages:

$$\frac{d}{dt}\langle \sigma_i(t) \rangle = \tanh \beta[\boldsymbol{\xi}_i \cdot \mathbf{Am}(t) + \theta_i(t)] - \langle \sigma_i(t) \rangle, \tag{4.178}$$

$$i \neq j: \quad \frac{d}{dt}\langle \sigma_i(t)\sigma_j(t) \rangle = \tanh \beta[\boldsymbol{\xi}_i \cdot \mathbf{Am}(t) + \theta_i(t)]\langle \sigma_j(t) \rangle$$
$$+ \tanh \beta[\boldsymbol{\xi}_j \cdot \mathbf{Am}(t) + \theta_j(t)]\langle \sigma_i(t) \rangle - 2\langle \sigma_i(t)\sigma_j(t) \rangle.$$

Correlations at different times are calculated by applying (4.178) to situations where the microscopic state at time t' is known exactly, i.e., where $p_{t'}(\boldsymbol{\sigma}) = \delta_{\boldsymbol{\sigma},\boldsymbol{\sigma}'}$ for some $\boldsymbol{\sigma}'$:

$$\langle \sigma_i(t) \rangle|_{\boldsymbol{\sigma}(t')=\boldsymbol{\sigma}'} = \sigma'_i e^{-(t-t')} + \int_{t'}^t ds\ e^{s-t} \tanh \beta \times \tag{4.179}$$
$$\times [\boldsymbol{\xi}_i \cdot \mathbf{Am}(s; \boldsymbol{\sigma}', t') + \theta_i(s)],$$

with $\mathbf{m}(s; \boldsymbol{\sigma}', t')$ denoting the solution of (4.177) following initial condition $\mathbf{m}(t') = \frac{1}{N}\sigma'_i \boldsymbol{\xi}_i$. If we multiply both sides of (4.179) by σ'_j and average over all possible states $\boldsymbol{\sigma}'$ at time t' we get in leading order in N:

$$\langle \sigma_i(t)\sigma_j(t') \rangle = \langle \sigma_i(t')\sigma_j(t') \rangle e^{-(t-t')} +$$
$$\int_{t'}^t ds\ e^{s-t} \langle \tanh \beta[\boldsymbol{\xi}_i \cdot \mathbf{Am}(s; \boldsymbol{\sigma}(t'), t') + \theta_i(s)]\sigma_j(t') \rangle.$$

Because of the existence of deterministic laws for the overlaps $\mathbf{m}$ in the $N \to \infty$ limit, we know with probability one that during the stochastic process the actual value $\mathbf{m}(\boldsymbol{\sigma}(t'))$ must be given by the solution of (4.177), evaluated at time t'. As a result we get, with $C_{ij}(t, t') = \langle \sigma_i(t)\sigma_j(t') \rangle$:

$$C_{ij}(t, t') = C_{ij}(t', t')e^{-(t-t')} + \tag{4.180}$$
$$\int_{t'}^t ds\ e^{s-t} \tanh \beta[\boldsymbol{\xi}_i \cdot \mathbf{Am}(s) + \theta_i(s)]\langle \sigma_j(t') \rangle.$$

Similarly we get from the solution of (4.178) an equation for the leading order in N of the response functions, by derivation with respect to external fields:

$$\frac{\partial \langle \sigma_i(t) \rangle}{\partial \theta_j(t')} = \beta \theta(t - t') \int_{-\infty}^t ds\ e^{s-t} \left[1 - \tanh^2 \beta[\boldsymbol{\xi}_i \cdot \mathbf{Am}(s) + \theta_i(s)]\right] \times$$
$$\times \left[\frac{1}{N}(\boldsymbol{\xi}_i \cdot \mathbf{A}\boldsymbol{\xi}_k)\frac{\partial \langle \sigma_k(s) \rangle}{\partial \theta_j(t')} + \delta_{ij}\delta(s - t')\right], \qquad \text{or}$$

$$G_{ij}(t,t') = \beta\delta_{ij}\theta(t-t')\mathrm{e}^{-(t-t')}\left[1 - \tanh^2\beta[\boldsymbol{\xi}_i \cdot \mathbf{Am}(t') + \theta_i(t')]\right] \quad (4.181)$$

$$+ \beta\theta(t-t')\int_{t'}^{t} ds\ \mathrm{e}^{s-t} \times$$

$$\times\left[1 - \tanh^2\beta[\boldsymbol{\xi}_i \cdot \mathbf{Am}(s) + \theta_i(s)]\right]\frac{1}{N}(\boldsymbol{\xi}_i \cdot \mathbf{A}\boldsymbol{\xi}_k)G_{kj}(s,t').$$

For $t = t'$ we retain in leading order in N only the instantaneous single-site contribution

$$\lim_{t'\uparrow t} G_{ij}(t,t') = \beta\delta_{ij}\left[1 - \tanh^2\beta[\boldsymbol{\xi}_i \cdot \mathbf{Am}(t) + \theta_i(t)]\right]. \qquad (4.182)$$

This leads to the following ansatz for the scaling with N of the $G_{ij}(t,t')$, which can be shown to be correct by insertion into (4.181), in combination with the correctness at $t = t'$ following from (4.182):

$$i = j: \quad G_{ii}(t,t') = \mathcal{O}(1), \qquad i \neq j: \quad G_{ij}(t,t') = \mathcal{O}(N^{-1})$$

Note that this implies $\frac{1}{N}(\boldsymbol{\xi}_i \cdot \mathbf{A}\boldsymbol{\xi}_k)G_{kj}(s,t') = \mathcal{O}(\frac{1}{N})$. In leading order in N we now find

$$G_{ij}(t,t') = \beta\delta_{ij}\theta(t-t')\mathrm{e}^{-(t-t')}\left[1 - \tanh^2\beta[\boldsymbol{\xi}_i \cdot \mathbf{Am}(t') + \theta_i(t')]\right]. \quad (4.183)$$

For those cases where the macroscopic laws (4.177) describe evolution to a stationary state $\mathbf{m}$, obviously requiring stationary external fields $\theta_i(t) = \theta_i$, we can take the limit $t \to \infty$, with $t-t' = \tau$ fixed, in (4.180) and (4.183). Using the $t \to \infty$ limits of (4.178) we subsequently find time translation invariant expressions: $\lim_{t\to\infty} C_{ij}(t,t-\tau) = C_{ij}(\tau)$ and $\lim_{t\to\infty} G_{ij}(t,t-\tau) = G_{ij}(\tau)$, with in leading order in N

$$C_{ij}(\tau) = \tanh\beta[\boldsymbol{\xi}_i \cdot \mathbf{Am} + \theta_i]\tanh\beta[\boldsymbol{\xi}_j \cdot \mathbf{Am} + \theta_j]$$

$$+ \delta_{ij}\mathrm{e}^{-\tau}\left[1 - \tanh^2\beta[\boldsymbol{\xi}_i \cdot \mathbf{Am} + \theta_i]\right],$$

$$G_{ij}(\tau) = \beta\delta_{ij}\theta(\tau)\mathrm{e}^{-\tau}\left[1 - \tanh^2\beta[\boldsymbol{\xi}_i \cdot \mathbf{Am} + \theta_i]\right].$$

for which the fluctuation–dissipation theorem (4.173) holds [Coo01, SC00, SC01, CKS05]:

$$G_{ij}(\tau) = -\beta\theta(\tau)\frac{d}{d\tau}C_{ij}(\tau).$$

We now turn to the parallel dynamical rules (4.141), with the local fields $h_i(\boldsymbol{\sigma}) = J_{ij}\sigma_j + \theta_i$, and the interaction matrix (4.138). As before, having time–dependent external fields amounts simply to adding these fields to the internal ones, and the dynamic laws (4.148) are found to be replaced by

$$\mathbf{m}(t+1) = \lim_{N\to\infty}\frac{1}{N}\boldsymbol{\xi}_i\tanh\left[\beta\boldsymbol{\xi}_i \cdot \mathbf{Am}(t) + \theta_i(t)\right]. \qquad (4.184)$$

Fluctuations in the local fields are again of vanishing order in N, and the parallel dynamics versions of equations (4.178), to be derived from (4.141), are found to be

$$\langle \sigma_i(t+1) \rangle = \tanh \beta [\boldsymbol{\xi}_i \cdot A\mathbf{m}(t) + \theta_i(t)], \tag{4.185}$$

$$i \neq j: \qquad \langle \sigma_i(t+1) \sigma_j(t+1) \rangle = \tag{4.186}$$
$$\tanh \beta [\boldsymbol{\xi}_i \cdot A\mathbf{m}(t) + \theta_i(t)] \tanh \beta [\boldsymbol{\xi}_j \cdot A\mathbf{m}(t) + \theta_j(t)].$$

With $\mathbf{m}(t; \boldsymbol{\sigma}', t')$ denoting the solution of the map (4.184) following initial condition $\mathbf{m}(t') = \frac{1}{N} \sigma_i' \boldsymbol{\xi}_i$, we immediately get from equations (4.185,4.186) the correlation functions:

$$C_{ij}(t,t) = \delta_{ij} + [1 - \delta_{ij}] \tanh \beta [\boldsymbol{\xi}_i \cdot A\mathbf{m}(t-1) + \theta_i(t-1)] \times$$
$$\tanh \beta [\boldsymbol{\xi}_j \cdot A\mathbf{m}(t-1) + \theta_j(t-1)],$$

$$t > t': \qquad C_{ij}(t,t') = \langle \tanh \beta [\boldsymbol{\xi}_i \cdot A\mathbf{m}(t-1; \boldsymbol{\sigma}(t'), t') + \theta_i(t-1)] \sigma_j(t') \rangle$$
$$= \tanh \beta [\boldsymbol{\xi}_i \cdot A\mathbf{m}(t-1) + \theta_i(t-1)] \tanh \beta [\boldsymbol{\xi}_j \cdot A\mathbf{m}(t'-1) + \theta_j(t'-1)].$$

From (4.185) also follow equations determining the leading order in N of the response functions $G_{ij}(t,t')$, by derivation with respect to the external fields $\theta_j(t')$:

$$t' > t - 1 : G_{ij}(t,t') = 0,$$
$$t' = t - 1 : G_{ij}(t,t') = \beta \delta_{ij} \left[1 - \tanh^2 \beta [\boldsymbol{\xi}_i \cdot A\mathbf{m}(t-1) + \theta_i(t-1)]\right],$$
$$t' < t - 1 : G_{ij}(t,t') = \beta \left[1 - \tanh^2 \beta [\boldsymbol{\xi}_i \cdot A\mathbf{m}(t-1) + \theta_i(t-1)]\right] \times$$
$$\times \frac{1}{N} (\boldsymbol{\xi}_i \cdot \mathbf{A}\boldsymbol{\xi}_k) G_{kj}(t-1,t').$$

It now follows iteratively that all off–diagonal elements must be of vanishing order in N: $G_{ij}(t,t-1) = \delta_{ij} G_{ii}(t,t-1) \quad \rightarrow \quad G_{ij}(t,t-2) = \delta_{ij} G_{ii}(t,t-2) \quad \rightarrow \quad \ldots$, so that in leading order

$$G_{ij}(t,t') = \beta \delta_{ij} \delta_{t,t'+1} \left[1 - \tanh^2 \beta [\boldsymbol{\xi}_i \cdot A\mathbf{m}(t') + \theta_i(t')]\right].$$

For those cases where the macroscopic laws (4.184) describe evolution to a stationary state $\mathbf{m}$, with stationary external fields, we can take the limit $t \rightarrow \infty$, with $t - t' = \tau$ fixed above. We find time translation invariant expressions: $\lim_{t \rightarrow \infty} C_{ij}(t, t - \tau) = C_{ij}(\tau)$ and $\lim_{t \rightarrow \infty} G_{ij}(t, t - \tau) = G_{ij}(\tau)$, with in leading order in N:

$$C_{ij}(\tau) = \tanh \beta [\boldsymbol{\xi}_i \cdot A\mathbf{m} + \theta_i] \tanh \beta [\boldsymbol{\xi}_j \cdot A\mathbf{m} + \theta_j]$$
$$+ \delta_{ij} \delta_{\tau,0} \left[1 - \tanh^2 \beta [\boldsymbol{\xi}_i \cdot A\mathbf{m} + \theta_i]\right],$$
$$G_{ij}(\tau) = \beta \delta_{ij} \delta_{\tau,1} \left[1 - \tanh^2 \beta [\boldsymbol{\xi}_i \cdot A\mathbf{m} + \theta_i]\right],$$

obeying the Fluctuation-Dissipation Theorem (4.172):

$$G_{ij}(\tau > 0) = -\beta [C_{ij}(\tau + 1) - C_{ij}(\tau - 1)].$$

Graded–Response Neurons with Uniform Synapses

Let us finally find out how to calculate correlation and response function for the simple network (4.68) of graded–response neurons, with (possibly time–dependent) external forces $\theta_i(t)$, and with uniform synapses $J_{ij} = J/N$ [Coo01, SC00, SC01, CKS05]:

$$\dot{u}_i(t) = \frac{J}{N} \sum_j g[\gamma u_j(t)] - u_i(t) + \theta_i(t) + \eta_i(t). \qquad (4.187)$$

For a given realisation of the external forces and the Gaussian noise variables $\{\eta_i(t)\}$ we can formally integrate (4.187) and find

$$u_i(t) = u_i(0)e^{-t} + \int_0^t ds \; e^{s-t} \left[J \int du \, \rho(u; \mathbf{u}(s)) \, g[\gamma u] + \theta_i(s) + \eta_i(s) \right], \qquad (4.188)$$

with the distribution of membrane potentials $\rho(u; \mathbf{u}) = N^{-1} \sum_i \delta[u - u_i]$. The correlation function $C_{ij}(t, t') = \langle u_i(t) u_j(t') \rangle$ immediately follows from (4.188). Without loss of generality we can define $t \geq t'$. For absent external forces (which were only needed in order to define the response function), and upon using $\langle \eta_i(s) \rangle = 0$ and $\langle \eta_i(s)\eta_j(s') \rangle = 2T\delta_{ij}\delta(s - s')$, we arrive at

$$C_{ij}(t, t') = T\delta_{ij}(e^{t'-t} - e^{-t'-t}) +$$

$$\left\langle \left[u_i(0)e^{-t} + J \int du \, g[\gamma u] \int_0^t ds \; e^{s-t}\rho(u; \mathbf{u}(s)) \right] \times \right.$$

$$\left. \times \left[u_j(0)e^{-t'} + J \int du \, g[\gamma u] \int_0^{t'} ds' \; e^{s'-t'}\rho(u; \mathbf{u}(s')) \right] \right\rangle.$$

For $N \to \infty$, however, we know the distribution of potentials to evolve deterministically: $\rho(u; \mathbf{u}(s)) \to \rho_s(u)$ where $\rho_s(u)$ is the solution of (4.155). This allows us to simplify the above expression to

$$N \to \infty: \qquad C_{ij}(t, t') = T\delta_{ij}(e^{t'-t} - e^{-t'-t}) \qquad (4.189)$$

$$+ \; \left\langle \left[u_i(0)e^{-t} + J \int du \, g[\gamma u] \int_0^t ds \; e^{s-t}\rho_s(u) \right] \times \right.$$

$$\left. \times \left[u_j(0)e^{-t'} + J \int du \, g[\gamma u] \int_0^{t'} ds' \; e^{s'-t'}\rho_{s'}(u) \right] \right\rangle.$$

Next we turn to the response function $G_{ij}(t, t') = \delta \langle u_i(t) \rangle / \delta \xi_j(t')$ (its definition involves functional rather than scalar differentiation, since time is continuous). After this differentiation the forces $\{\theta_i(s)\}$ can be put to zero. Functional differentiation of (4.188), followed by averaging, then leads us to

$$G_{ij}(t,t') = \theta(t-t')\,\delta_{ij}\,\mathrm{e}^{t'-t} - J \int du\ g[\gamma u]\frac{\partial}{\partial u} \times$$

$$\int_0^t ds\ \mathrm{e}^{s-t}\,\frac{1}{N}\sum_k \lim_{\boldsymbol{\theta}\to\mathbf{0}}\langle\ \delta[u-u_k(s)]\frac{\delta u_k(s)}{\delta\theta_j(t')}\ \rangle.$$

In view of (4.188) we make the self-consistent ansatz $\delta u_k(s)/\delta\xi_j(s') = \mathcal{O}(N^{-1})$ for $k\ne j$. This produces

$$N\to\infty:\qquad G_{ij}(t,t') = \theta(t-t')\,\delta_{ij}\,\mathrm{e}^{t'-t}.$$

Since equation (4.155) evolves towards a stationary state, we can also take the limit $t\to\infty$, with $t-t' = \tau$ fixed, in (4.189). Assuming non–pathological decay of the distribution of potentials allows us to put $\lim_{t\to\infty}\int_0^t ds\ \mathrm{e}^{s-t}\rho_s(u) = \rho(u)$ (the stationary solution of (4.155)), with which we find also (4.189) reducing to time translation invariant expressions for $N\to\infty$, $\lim_{t\to\infty}C_{ij}(t,t-\tau) = C_{ij}(\tau)$ and $\lim_{t\to\infty}G_{ij}(t,t-\tau) = G_{ij}(\tau)$, in which

$$C_{ij}(\tau) = T\delta_{ij}\mathrm{e}^{-\tau} + J^2\left\{\int du\ \rho(u)g[\gamma u]\right\}^2,$$

$$G_{ij}(\tau) = \theta(\tau)\delta_{ij}\mathrm{e}^{-\tau}.$$

Clearly the leading orders in N of these two functions obey the fluctuation-dissipation theorem:

$$G_{ij}(\tau) = -\beta\theta(\tau)\frac{d}{d\tau}C_{ij}(\tau).$$

As with the binary neuron attractor networks for which we calculated the correlation and response functions earlier, the impact of detailed balance violation (occurring when $A_{\mu\nu}\ne A_{\nu\mu}$ in networks with binary neurons and synapses (4.138), and in all networks with graded–response neurons on the validity of the fluctuation-dissipation theorems, vanishes for $N\to\infty$, provided our networks are relatively simple and evolve to a stationary state in terms of the macroscopic observables (the latter need not necessarily happen. Detailed balance violation, however, would be noticed in the finite size effects [CCV98].

4.3.9 Path–Integral Approach for Complex Dynamics

Let us return to the simplest setting in which to study the problem: single pattern recall in an attractor neural network with N binary neurons and $p = \alpha N$ stored patterns in the non–trivial regime, where $\alpha > 0$. We choose parallel dynamics, i.e., (4.141), with Hebbian synapses of the form (4.138) with $A_{\mu\nu} = \delta_{\mu\nu}$, i.e., $J_{ij} = N^{-1}\sum_\mu^p \xi_i^\mu\xi_j^\mu$, giving us the parallel dynamics version of the Hopfield model [Hop82]. Our interest is in the recall overlap $m(\boldsymbol{\sigma}) = N^{-1}\sum_i \sigma_i\xi_i^1$ between system state and pattern one. We saw above that for $N\to\infty$ the fluctuations in the values of the recall overlap m will vanish, and that for initial states where all $\sigma_i(0)$ are drawn independently the overlap m will obey:

$$m(t+1) = \int dz \; P_t(z) \tanh[\beta(m(t)+z)] : \qquad (4.190)$$

$$P_t(z) = \lim_{N\to\infty} \frac{1}{N} \sum_i \langle \delta[z - \frac{1}{N}\xi_i^1 \xi_i^\mu \xi_j^\mu \sigma_j(t)] \rangle,$$

and that all complications in a dynamical analysis of the $\alpha > 0$ regime are concentrated in the calculation of the distribution $P_t(z)$ of the (generally non–trivial) interference noise.

As a simple *Gaussian approximation* one could just assume [Ama77, Ama78] that the σ_i remain uncorrelated at all times, i.e., $\mathrm{Prob}[\sigma_i(t) = \pm\xi_i^1] = \frac{1}{2}[1 \pm m(t)]$ for all $t \geq 0$, such that the argument given above for $t = 0$ (leading to a Gaussian $P(z)$) would hold generally, and where the map (4 190) would describe the overlap evolution at all times:

$$P_t(z) = [2\pi\alpha]^{-\frac{1}{2}} e^{-\frac{1}{2}z^2/\alpha} :$$

$$m(t+1) = \int Dz \; \tanh[\beta(m(t)+z\sqrt{\alpha})],$$

with the *Gaussian measure* $Dz = (2\pi)^{-\frac{1}{2}} e^{-\frac{1}{2}z^2} dz$. This equation, however, must be generally incorrect. Rather than taking all σ_i to be independent, a weaker assumption would be to just assume the interference noise distribution $P_t(z)$ to be a zero-average Gaussian one, at any time, with statistically independent noise variables z at different times. One can then derive (for $N \to \infty$ and fully connected networks) an evolution equation for the width $\Sigma(t)$, giving [AM88]:

$$m(t+1) = \int Dz \; \tanh[\beta(m(t)+z\Sigma(t))] :$$

$$P_t(z) = [2\pi\Sigma^2(t)]^{-\frac{1}{2}} e^{-\frac{1}{2}z^2/\Sigma^2(t)} :$$

$$\Sigma^2(t+1) = \alpha + 2\alpha m(t+1)m(t)h[m(t), \Sigma(t)]$$
$$+ \Sigma^2(t)h^2[m(t), \Sigma(t)], \qquad \text{with}$$

$$h[m, \Sigma] = \beta\left[1 - \int Dz \; \tanh^2[\beta(m+z\Sigma)]\right].$$

These equations describe correctly the qualitative features of recall dynamics, and are found to work well when retrieval actually occurs. A final refinement of the Gaussian approach [Oka95] consisted in allowing for correlations between the noise variables z at different times (while still describing them by Gaussian distributions). This results in a hierarchy of macroscopic equations, which improve upon the previous Gaussian theories and even predict the correct stationary state and phase diagrams, but still fail to be correct at intermediate times.

In view of the non–Gaussian shape of the interference noise distribution, several attempts have been made at constructing non–Gaussian approximations. In all cases the aim is to arrive at a theory involving only macroscopic observables with a *single* time-argument. For a fully connected network with binary neurons and parallel dynamics a more accurate ansatz for $P_t(z)$ would be the sum of two Gaussian functions. In [HO90] the following choice was proposed:

$$P_t(z) = P_t^+(z) + P_t^-(z),$$

$$P_t^\pm(z) = \lim_{N \to \infty} \frac{1}{N} \sum_i \delta_{\sigma_i(t), \pm \xi_i^1} \langle \delta[z - \frac{1}{N} \xi_i^1 \xi_i^\mu \xi_j^\mu \sigma_j(t)] \rangle$$

$$P_t^\pm(z) = \frac{1 \pm m(t)}{2\Sigma(t)\sqrt{2\pi}} e^{-\frac{1}{2}[z \mp d(t)]^2 / \Sigma^2(t)},$$

followed by a self–consistent calculation of $d(t)$ (representing an effective 'retarded self–interaction', since it has an effect equivalent to adding $h_i(\boldsymbol{\sigma}(t)) \to h_i(\boldsymbol{\sigma}(t)) + d(t)\sigma_i(t)$), and of the width $\Sigma(t)$ of the two distributions $P_t^\pm(z)$, together with

$$m(t+1) = \frac{1}{2}[1 + m(t)] \int Dz \ \tanh[\beta(m(t) + d(t) + z\Sigma(t))]$$

$$+ \frac{1}{2}[1 - m(t)] \int Dz \ \tanh[\beta(m(t) - d(t) + z\Sigma(t))].$$

The resulting three–parameter theory, in the form of closed dynamic equations for $\{m, d, \Sigma\}$, is found to give a nice (but not perfect) agreement with numerical simulations.

A different philosophy was followed in [CS94] (for sequential dynamics). First (as yet exact) equations are derived for the evolution of the two macroscopic observables $m(\boldsymbol{\sigma}) = m_1(\boldsymbol{\sigma})$ and $r(\boldsymbol{\sigma}) = \alpha^{-1} \sum_{\mu>1} m_\mu^2(\boldsymbol{\sigma})$, with $m_\mu(\boldsymbol{\sigma}) = N^{-1} \sum_i \xi_i^1 \sigma_i$, which are both found to involve $P_t(z)$:

$$\dot{m} = \int dz \ P_t(z) \tanh[\beta(m+z)],$$

$$\dot{r} = \frac{1}{\alpha} \int dz \ P_t(z) z \tanh[\beta(m+z)] + 1 - r.$$

Next one closes these equations *by hand*, using a maximum–entropy (or, 'Occam's Razor') argument: instead of calculating $P_t(z)$ from (4.190) with the real (unknown) microscopic distribution $p_t(\boldsymbol{\sigma})$, it is calculated upon assigning equal probabilities to all states $\boldsymbol{\sigma}$ with $m(\boldsymbol{\sigma}) = m$ and $r(\boldsymbol{\sigma}) = r$, followed by averaging over all realisations of the stored patterns with $\mu > 1$. In order words: one assumes (i) that the microscopic states visited by the system are 'typical' within the appropriate (m, r) sub-shells of state space, and (ii) that one can average over the disorder. Assumption (ii) is harmless, the most important step is (i). This procedure results in an explicit (non-Gaussian) expression for the noise distribution in terms of (m, r) only, a closed 2–parameter theory which is exact for short times and in equilibrium, accurate predictions of the macroscopic flow in the (m, r)–plane, but (again) deviations in predicted time-dependencies at intermediate times. This theory, and its performance, was later improved by applying the same ideas to a derivation of a dynamic equation for the function $P_t(z)$ itself (rather than for m and r only).

If we now use the powerful *path–integral formalism* (see [II06b]), instead of working with the probability $p_t(\boldsymbol{\sigma})$ of finding a microscopic state $\boldsymbol{\sigma}$ at time t in order to calculate the statistics of a set of macroscopic observables $\boldsymbol{\Omega}(\boldsymbol{\sigma})$ at time t, we turn to the probability $\mathrm{Prob}[\boldsymbol{\sigma}(0), \dots, \boldsymbol{\sigma}(t_m)]$ of finding a microscopic *path* $\boldsymbol{\sigma}(0) \to \boldsymbol{\sigma}(1) \to \dots \to \boldsymbol{\sigma}(t_m)$. W also add time–dependent external sources to the local fields, $h_i(\boldsymbol{\sigma}) \to h_i(\boldsymbol{\sigma}) + \theta_i(t)$, in order to probe the networks via perturbations and define a response function. The idea is to concentrate on the moment partition function $Z[\boldsymbol{\psi}]$, which, like $\mathrm{Prob}[\boldsymbol{\sigma}(0), \dots, \boldsymbol{\sigma}(t_m)]$, fully captures the statistics of paths:

$$Z[\boldsymbol{\psi}] = \left\langle e^{-i \sum_{t=0}^{t_m} \psi_i(t)\sigma_i(t)} \right\rangle.$$

It generates averages of the relevant observables, including those involving neuron states at different times, such as correlation functions $C_{ij}(t, t') = \langle \sigma_i(t)\sigma_j(t') \rangle$ and response functions $G_{ij}(t, t') = \partial \langle \sigma_i(t) \rangle / \partial \theta_j(t')$, upon differentiation with respect to the dummy variables $\{\psi_i(t)\}$:

$$\langle \sigma_i(t) \rangle = i \lim_{\boldsymbol{\psi} \to \mathbf{0}} \frac{\partial Z[\boldsymbol{\psi}]}{\partial \psi_i(t)}, \qquad C_{ij}(t, t') = - \lim_{\boldsymbol{\psi} \to \mathbf{0}} \frac{\partial^2 Z[\boldsymbol{\psi}]}{\partial \psi_i(t) \partial \psi_j(t')},$$

$$G_{ij}(t, t') = i \lim_{\boldsymbol{\psi} \to \mathbf{0}} \frac{\partial^2 Z[\boldsymbol{\psi}]}{\partial \psi_i(t) \partial \theta_j(t')}.$$

Next one assumes (correctly) that for $N \to \infty$ only the statistical properties of the stored patterns will influence the macroscopic quantities, so that the partition function $Z[\boldsymbol{\psi}]$ can be averaged over all pattern realisations, i.e., $Z[\boldsymbol{\psi}] \to \overline{Z[\boldsymbol{\psi}]}$. As in replica theories (the canonical tool to deal with complexity in equilibrium) one carries out the disorder average *before* the average over the statistics of the neuron states, resulting for $N \to \infty$ in what can be interpreted as a theory describing a single 'effective' binary neuron $\sigma(t)$, with an effective local field $h(t)$ and the dynamics $\mathrm{Prob}[\sigma(t + 1) = \pm 1] = \frac{1}{2}[1 \pm \tanh[\beta h(t)]]$. However, this effective local field is found to generally depend on past states of the neuron, and on zero-average but temporally correlated Gaussian noise contributions $\xi(t)$:

$$h(t | \{\sigma\}, \{\xi\}) = m(t) + \theta(t) + \alpha \sum_{t' < t} R(t, t')\sigma(t') + \sqrt{\alpha}\xi(t). \qquad (4.191)$$

The first comprehensive neural network studies along these lines, dealing with fully connected networks, were applied to asymmetrically and symmetrically extremely diluted networks [KZ91, WS91]. More recent applications include sequence processing networks [DCS98].[11] For $N \to \infty$ the differences between different models are found to show up only in the actual form taken by the effective local field (4.191), i.e., in the dependence of the 'retarded self-interaction' kernel $R(t, t')$ and the covariance matrix $\langle \xi(t)\xi(t') \rangle$ of the interference–induced Gaussian noise on the macroscopic objects $\mathbf{C} = \{C(s, s') = \lim_{N \to \infty} \frac{1}{N} C_{ii}(s, s')\}$ and $\mathbf{G} = \{G(s, s') = \lim_{N \to \infty} \frac{1}{N} G_{ii}(s, s')\}$. For instance [Coo01, SC00, SC01, CKS05]:

model	synapses J_{ij}	$R(t, t')$	$\langle \xi(t)\xi(t') \rangle$
fully connected, static patterns	$\frac{1}{N}\xi_i^\mu \xi_j^\mu$	$[(\mathbf{1} - \mathbf{G})^{-1}\mathbf{G}](t, t')$	$[(\mathbf{1} - \mathbf{G})^{-1}\mathbf{C}(\mathbf{1} - \mathbf{G}^\dagger)^{-1}](t, t')$
fully connected, pattern sequence	$\frac{1}{N}\xi_i^{\mu+1} \xi_j^\mu$	0	$\sum_{n \geq 0}[(\mathbf{G}^\dagger)^n \mathbf{C}\mathbf{G}^n](t, t')$
symm extr diluted, static patterns	$\frac{c_{ij}}{c}\xi_i^\mu \xi_j^\mu$	$G(t, t')$	$C(t, t')$
asymm extr diluted, static patterns	$\frac{c_{ij}}{c}\xi_i^\mu \xi_j^\mu$	0	$C(t, t')$

with the c_{ij} drawn at random according to $P(c_{ij}) = \frac{c}{N}\delta_{c_{ij},1} + (1 - \frac{c}{N})\delta_{c_{ij},0}$ (either symmetrically, i.e., $c_{ij} = c_{ji}$, or independently) and where $c_{ii} = 0$, $\lim_{N \to \infty} c/N = 0$, and $c \to \infty$. In all cases the observables (overlaps and correlation– and response–functions) are to be solved from the following closed equations, involving the statistics of the single effective neuron experiencing the field (4.191):

$$m(t) = \langle \sigma(t) \rangle, \qquad C(t, t') = \langle \sigma(t)\sigma(t') \rangle, \qquad G(t, t') = \partial \langle \sigma(t) \rangle / \partial \theta(t').$$

It is now clear that Gaussian theories can at most produce exact results for asymmetric networks. Any degree of symmetry in the synapses is found to induce a non–zero retarded self–interaction, via the kernel $K(t, t')$, which constitutes a non–Gaussian contribution to the local fields. Exact closed macroscopic theories apparently require a number of macroscopic observables which grows as $\mathcal{O}(t^2)$ in order to predict the dynamics up to time t. In the case of sequential dynamics the picture is found to be very similar to the one above; instead of discrete time labels $t \in \{0, 1, \ldots, t_m\}$, path summations and matrices, there one has a real time variable $t \in [0, t_m]$, path–integrals, integral operators, and partition–functions.

Partition–Function Analysis for Binary Neurons

First we will define parallel dynamics, i.e., (4.141), driven as usual by local fields of the form $h_i(\boldsymbol{\sigma}; t) = J_{ij}\sigma_j + \theta_i(t)$, but with a more general choice of *Hebbian synapses*, in which we allow for a possible random dilution (to reduce repetition in our subsequent derivations):

---

[11] In the case of sequence recall the overlap m is defined with respect to the 'moving' target, i.e., $m(t) = \frac{1}{N}\sigma_i(t)\xi_i^t$

$$J_{ij} = \frac{c_{ij}}{c}\xi_i^\mu \xi_j^\mu, \qquad \text{with} \qquad p = \alpha c. \qquad (4.192)$$

Architectural properties are reflected in the variables $c_{ij} \in \{0,1\}$, whereas information storage is to be effected by the remainder in (4.192), involving p randomly and independently drawn patterns $\boldsymbol{\xi}^\mu = (\xi_1^\mu, \ldots, \xi_N^\mu) \in \{-1,1\}^N$. I will deal both with symmetric and with asymmetric architectures (always putting $c_{ii} = 0$), in which the variables c_{ij} are drawn randomly according to

$$c_{ij} = c_{ji}, \qquad \text{(for all } i < j), \qquad P(c_{ij}) = \frac{c}{N}\delta_{c_{ij},1} + (1 - \frac{c}{N})\delta_{c_{ij},0},$$

$$\text{(for all } i = j), \qquad P(c_{ij}) = \frac{c}{N}\delta_{c_{ij},1} + (1 - \frac{c}{N})\delta_{c_{ij},0}.$$

Thus c_{kl} is statistically independent of c_{ij} as soon as $(k,l) \notin \{(i,j),(j,i)\}$. In leading order in N one has $\langle \sum_j c_{ij} \rangle = c$ for all i, so c gives the average number of neurons contributing to the field of any given neuron. In view of this, the number p of patterns to be stored can be expected to scale as $p = \alpha c$. The connectivity parameter c is chosen to diverge with N, i.e., $\lim_{N\to\infty} c^{-1} = 0$. If $c = N$ we get the fully connected (parallel dynamics) Hopfield model. Extremely diluted networks are got when $\lim_{N\to\infty} c/N = 0$.

For simplicity, we make the so–called 'condensed ansatz': we assume that the system state has an $\mathcal{O}(N^0)$ overlap only with a single pattern, say $\mu = 1$. This situation is induced by initial conditions: we take a randomly drawn $\boldsymbol{\sigma}(0)$, generated by

$$p(\boldsymbol{\sigma}(0)) = \prod_i \left\{ \frac{1}{2}[1 + m_0]\delta_{\sigma_i(0),\xi_i^1} + \frac{1}{2}[1 - m_0]\delta_{\sigma_i(0),-\xi_i^1} \right\},$$

so $\qquad \frac{1}{N}\xi_i^1 \langle \sigma_i(0) \rangle = m_0.$

The patterns $\mu > 1$, as well as the architecture variables c_{ij}, are viewed as disorder. One assumes that for $N \to \infty$ the macroscopic behavior of the system is 'self–averaging', i.e., only dependent on the statistical properties of the disorder (rather than on its microscopic realisation). Averages over the disorder are written as $\overline{\cdots}$. We next define the *disorder–averaged partition function*:

$$\overline{Z[\boldsymbol{\psi}]} = \overline{\langle e^{-i\sum_t \psi_i(t)\sigma_i(t)} \rangle}, \qquad (4.193)$$

in which the time t runs from $t = 0$ to some (finite) upper limit t_m. Note that $\overline{Z[\mathbf{0}]} = 1$. With a modest amount of foresight we define the macroscopic site-averaged and disorder–averaged objects $m(t) = N^{-1}\xi_i^1 \overline{\langle \sigma_i(t) \rangle}$, $C(t,t') = N^{-1}\overline{\langle \sigma_i(t)\sigma_i(t') \rangle}$ and $G(t,t') = N^{-1}\partial \overline{\langle \sigma_i(t) \rangle}/\partial \theta_i(t')$. They can be obtained from (4.193) as follows:

$$m(t) = \lim_{\boldsymbol{\psi}\to\mathbf{0}} \frac{i}{N}\xi_j^1 \frac{\partial \overline{Z[\boldsymbol{\psi}]}}{\partial \psi_j(t)}, \qquad C(t,t') = -\lim_{\boldsymbol{\psi}\to\mathbf{0}} \frac{1}{N}\frac{\partial^2 \overline{Z[\boldsymbol{\psi}]}}{\partial \psi_j(t)\partial \psi_j(t')},$$

$$G(t,t') = \lim_{\boldsymbol{\psi}\to\mathbf{0}} \frac{i}{N}\frac{\partial^2 \overline{Z[\boldsymbol{\psi}]}}{\partial \psi_j(t)\partial \theta_j(t')}.$$

Now, as in equilibrium replica calculations, the hope here is that progress can be made by carrying out the disorder averages first. In equilibrium calculations we use the replica trick to convert our disorder averages into feasible ones; here the idea is to isolate the local fields at different times and different sites by inserting appropriate $\delta-$distributions:

$$1 = \prod_{it} \int dh_i(t)\delta[h_i(t) - J_{ij}\sigma_j(t) - \theta_i(t)]$$

$$= \int \{d\mathbf{h}d\hat{\mathbf{h}}\}e^{i\sum_{it}\hat{h}_i(t)[h_i(t) - J_{ij}\sigma_j(t) - \theta_i(t)]},$$

with $\{d\mathbf{h}d\hat{\mathbf{h}}\} = \prod_{it}[d\hat{h}_i(t)dh_i(t)/2\pi]$, giving

$$\overline{Z[\boldsymbol{\psi}]} = \int \{d\mathbf{h}d\hat{\mathbf{h}}\}e^{i\sum_{it}\hat{h}_i(t)[h_i(t) - \theta_i(t)]} \times$$

$$\times \left\langle e^{-i\sum_{it}\psi_i(t)\sigma_i(t)} \overline{\left[e^{-i\sum_{it}\hat{h}_i(t)J_{ij}\sigma_j(t)}\right]}\right\rangle_{\mathrm{pf}},$$

in which $\langle \ldots \rangle_{\mathrm{pf}}$ refers to averages over a constrained stochastic process of the type (4.141), but with prescribed fields $\{h_i(t)\}$ at all sites and at all times. Note that with such prescribed fields the probability of partition a path $\{\boldsymbol{\sigma}(0), \ldots, \boldsymbol{\sigma}(t_m)\}$ is given by

$$\mathrm{Prob}[\boldsymbol{\sigma}(0), \ldots, \boldsymbol{\sigma}(t_m)|\{h_i(t)\}] =$$
$$p(\boldsymbol{\sigma}(0))e^{\sum_{it}[\beta\sigma_i(t+1)h_i(t) - \log 2\cosh[\beta h_i(t)]]}, \qquad \text{so}$$

$$\overline{Z[\boldsymbol{\psi}]} = \int \{d\mathbf{h}d\hat{\mathbf{h}}\} \sum_{\boldsymbol{\sigma}(0)} \cdots \sum_{\boldsymbol{\sigma}(t_m)} p(\boldsymbol{\sigma}(0))e^{N\mathcal{F}[\{\boldsymbol{\sigma}\},\{\hat{\mathbf{h}}\}]} \times$$

$$\times \prod_{it} e^{i\hat{h}_i(t)[h_i(t) - \theta_i(t)] - i\psi_i(t)\sigma_i(t) + \beta\sigma_i(t+1)h_i(t) - \log 2\cosh[\beta h_i(t)]},$$

$$\text{with} \qquad \mathcal{F}[\{\boldsymbol{\sigma}\}, \{\hat{\mathbf{h}}\}] = \frac{1}{N}\log \overline{\left[e^{-i\sum_{it}\hat{h}_i(t)J_{ij}\sigma_j(t)}\right]}. \qquad (4.194)$$

We concentrate on the term $\mathcal{F}[\ldots]$ (with the disorder), of which we need only know the limit $N \to \infty$, since only terms inside $\overline{Z[\boldsymbol{\psi}]}$ which are exponential in N will retain statistical relevance. In the disorder–average of (4.194) every site i plays an equivalent role, so the leading order in N of (4.194) should depend only on site–averaged functions of the $\{\sigma_i(t), \hat{h}_i(t)\}$, with no reference to any special direction except the one defined by pattern $\boldsymbol{\xi}^1$. The simplest such functions with a single time variable are

$$a(t; \{\boldsymbol{\sigma}\}) = \frac{1}{N}\xi_i^1\sigma_i(t), \qquad k(t; \{\hat{\mathbf{h}}\}) = \frac{1}{N}\xi_i^1\hat{h}_i(t),$$

whereas the simplest ones with two time variables would appear to be

$$q(t, t'; \{\boldsymbol{\sigma}\}) = \frac{1}{N} \sigma_i(t) \sigma_i(t'), \qquad Q(t, t'; \{\hat{\mathbf{h}}\}) = \frac{1}{N} \hat{h}_i(t) \hat{h}_\cdot(t'),$$

$$K(t, t'; \{\boldsymbol{\sigma}, \hat{\mathbf{h}}\}) = \frac{1}{N} \hat{h}_i(t) \sigma_i(t').$$

It will turn out that all models of the type (4.192), have the crucial property that above are in fact the *only* functions to appear in the leading order of (4.194):

$$\mathcal{F}[\ldots] = \Phi[\{a(t; \ldots), k(t; \ldots), q(t, t'; \ldots), Q(t, t'; \ldots), K(t, t'; \ldots)\}] + \ldots,$$

for $N \to \infty$ and some as yet unknown function $\Phi[\ldots]$. This allows us to proceed with the evaluation of $\overline{Z[\boldsymbol{\psi}]}$. We can introduce suitable δ-distributions (taking care that all exponents scale linearly with N, to secure statistical relevance). Thus we insert

$$1 = \prod_{t=0}^{t_m} \int da(t) \ \delta[a(t) - a(t; \{\boldsymbol{\sigma}\})]$$

$$= \left[\frac{N}{2\pi}\right]^{t_m+1} \int d\mathbf{a} d\hat{\mathbf{a}} \ e^{iN \sum_t \hat{a}(t)[a(t) - \frac{1}{N} \xi_j^1 \sigma_j(\cdot)]},$$

$$1 = \prod_{t=0}^{t_m} \int dk(t) \ \delta[k(t) - k(t; \{\hat{\mathbf{h}}\})]$$

$$= \left[\frac{N}{2\pi}\right]^{t_m+1} \int d\mathbf{k} d\hat{\mathbf{k}} \ e^{iN \sum_t \hat{k}(t)[k(t) - \frac{1}{N} \xi_j^1 \hat{h}_j(t)]},$$

$$1 = \prod_{t,t'=0}^{t_m} \int dq(t, t') \ \delta[q(t, t') - q(t, t'; \{\boldsymbol{\sigma}\})]$$

$$= \left[\frac{N}{2\pi}\right]^{(t_m+1)^2} \int d\mathbf{q} d\hat{\mathbf{q}} \ e^{iN \sum_{t,t'} \hat{q}(t,t')[q(t,t') - \frac{1}{N} \sigma_j(t) \sigma_j(t')]},$$

$$1 = \prod_{t,t'=0}^{t_m} \int dQ(t, t') \ \delta[Q(t, t') - Q(t, t'; \{\hat{\mathbf{h}}\})]$$

$$= \left[\frac{N}{2\pi}\right]^{(t_m+1)^2} \int d\mathbf{Q} d\hat{\mathbf{Q}} \ e^{iN \sum_{t,t'} \hat{Q}(t,t')[Q(t,t') - \frac{1}{N} \hat{h}_j(t) \hat{h}_j(t')]},$$

$$1 = \prod_{t,t'=0}^{t_m} \int dK(t, t') \ \delta[K(t, t') - K(t, t'; \{\boldsymbol{\sigma}, \hat{\mathbf{h}}\})]$$

$$= \left[\frac{N}{2\pi}\right]^{(t_m+1)^2} \int d\mathbf{K} d\hat{\mathbf{K}} \ e^{iN \sum_{t,t'} \hat{K}(t,t')[K(t,t') - \frac{1}{N} \hat{h}_j(t) \sigma_\cdot(t')]}.$$

Using the short–hand

$$\Psi[\mathbf{a}, \hat{\mathbf{a}}, \mathbf{k}, \hat{\mathbf{k}}, \mathbf{q}, \hat{\mathbf{q}}, \mathbf{Q}, \hat{\mathbf{Q}}, \mathbf{K}, \hat{\mathbf{K}}] = i \sum_t [\hat{a}(t)a(t) + \hat{k}(t)k(t)]$$

$$+ \ i \sum_{t,t'} [\hat{q}(t,t')q(t,t') + \hat{Q}(t,t')Q(t,t') + \hat{K}(t,t')K(t,t')]$$

then leads us to $\overline{Z[\psi]} = \int d\mathbf{a} d\hat{\mathbf{a}} d\mathbf{k} d\hat{\mathbf{k}} d\mathbf{q} d\hat{\mathbf{q}} d\mathbf{Q} d\hat{\mathbf{Q}} d\mathbf{K} d\hat{\mathbf{K}} \ \times$

$$e^{N\Psi[\mathbf{a}, \hat{\mathbf{a}}, \mathbf{k}, \hat{\mathbf{k}}, \mathbf{q}, \hat{\mathbf{q}}, \mathbf{Q}, \hat{\mathbf{Q}}, \mathbf{K}, \hat{\mathbf{K}}] + N\Phi[\mathbf{a}, \mathbf{k}, \mathbf{q}, \mathbf{Q}, \mathbf{K}] + \mathcal{O}(...)}$$

$$\times \int \{d\mathbf{h} d\hat{\mathbf{h}}\} \sum_{\boldsymbol{\sigma}(0)} \cdots \sum_{\boldsymbol{\sigma}(t_m)} p(\boldsymbol{\sigma}(0)) \ \times$$

$$\prod_{it} e^{i\hat{h}_i(t)[h_i(t) - \theta_i(t)] - i\psi_i(t)\sigma_i(t) + \beta\sigma_i(t+1)h_i(t) - \log 2\cosh[\beta h_i(t)]} \ \times$$

$$\prod_i e^{-i\xi_i^1 \sum_t [\hat{a}(t)\sigma_i(t) + \hat{k}(t)\hat{h}_i(t)] - i\sum_{t,t'} [\hat{q}(t,t')\sigma_i(t)\sigma_i(t') + \hat{Q}(t,t')\hat{h}_i(t)\hat{h}_i(t') + \hat{K}(t,t')\hat{h}_i(t)\sigma_i(t')]},$$

in which the term denoted as $\mathcal{O}(...)$ covers both the non–dominant orders in (4.194) and the $\mathcal{O}(\log N)$ relics of the various pre–factors $[N/2\pi]$ in the above integral representations of the δ–distributions (note: t_m was assumed fixed). We now see explicitly that the summations and integrations over neuron states and local fields fully factorize over the N sites. A simple transformation

$$\{\sigma_i(t), h_i(t), \hat{h}_i(t)\} \longrightarrow \{\xi_i^1 \sigma_i(t), \xi_i^1 h_i(t), \xi_i^1 \hat{h}_i(t)\}$$

brings the result into the form [Coo01, SC00, SC01, CKS05]

$$e^{N \ \Xi[\hat{\mathbf{a}}, \hat{\mathbf{k}}, \hat{\mathbf{q}}, \hat{\mathbf{Q}}, \hat{\mathbf{K}}]} \ = \ \int \{d\mathbf{h} d\hat{\mathbf{h}}\} \sum_{\boldsymbol{\sigma}(0)} \cdots \sum_{\boldsymbol{\sigma}(t_m)} p(\boldsymbol{\sigma}(0)) \ \times$$

$$\times \prod_{it} e^{i\hat{h}_i(t)[h_i(t) - \xi_i^1 \theta_i(t)] - i\xi_i^1 \psi_i(t)\sigma_i(t) + \beta\sigma_i(t+1)h_i(t) - \log 2\cosh[\beta h_i(t)]} \ \times$$

$$\prod_i e^{-i\xi_i^1 \sum_t [\hat{a}(t)\sigma_i(t) + \hat{k}(t)\hat{h}_i(t)] - i\sum_{t,t'} [\hat{q}(t,t')\sigma_i(t)\sigma_i(t') + \hat{Q}(t,t')\hat{h}_i(t)\hat{h}_i(t') + \hat{K}(t,t')\hat{h}_i(t)\sigma_i(t')]},$$

in which

$$\{d\mathbf{h} d\hat{\mathbf{h}}\} = \prod_t [dh(t)d\hat{h}(t)/2\pi], \qquad \pi_0(\sigma) = \frac{1}{2}[1 + m_0]\delta_{\sigma,1} + \frac{1}{2}[1 - m_0]\delta_{\sigma,-1}.$$

At this stage $\overline{Z[\psi]}$ acquires the form of an integral to be evaluated via the saddle–point (or 'steepest descent') method,

$$\overline{Z[\{\psi(t)\}]} = \int d\mathbf{a} d\hat{\mathbf{a}} d\mathbf{k} d\hat{\mathbf{k}} d\mathbf{q} d\hat{\mathbf{q}} d\mathbf{Q} d\hat{\mathbf{Q}} d\mathbf{K} d\hat{\mathbf{K}} \ e^{N\{\Psi[...] + \Phi[...] + \Xi[...]\} + \mathcal{O}(...)}.$$

$$(4.195)$$

The disorder–averaged partition function (4.195) is for $N \to \infty$ dominated by the physical saddle–point of the macroscopic surface

$$\Psi[\mathbf{a}, \hat{\mathbf{a}}, \mathbf{k}, \hat{\mathbf{k}}, \mathbf{q}, \hat{\mathbf{q}}, \mathbf{Q}, \hat{\mathbf{Q}}, \mathbf{K}, \hat{\mathbf{K}}] + \Phi[\mathbf{a}, \mathbf{k}, \mathbf{q}, \mathbf{Q}, \mathbf{K}] + \Xi[\hat{\mathbf{a}}, \hat{\mathbf{k}}, \hat{\mathbf{q}}\ \hat{\mathbf{Q}}, \hat{\mathbf{K}}]. \quad (4.196)$$

It will be advantageous at this stage to define the following effective measure:

$$\langle f[\{\sigma\}, \{h\}, \{\hat{h}\}] \rangle_{\star} = (4.197)$$

$$\frac{1}{N} \left\{ \frac{\int \{dh d\hat{h}\} \sum_{\sigma(0)\cdots\sigma(t_m)} M_i[\{\sigma\}, \{h\}, \{\hat{h}\}] \, f[\{\sigma\}, \{h\}, \{\hat{h}\}]}{\int \{dh d\hat{h}\} \sum_{\sigma(0)\cdots\sigma(t_m)} M_i[\{\sigma\}, \{h\}, \{\hat{h}\}]} \right\},$$

with

$$M_i[\{\sigma\}, \{h\}, \{\hat{h}\}] =$$

$$\pi_0(\sigma(0)) \, e^{\sum_t \{i\hat{h}(t)[h(t) - \xi_i^1 \theta_i(t)] - i\xi_i^1 \psi_i(t)\sigma(t) + \beta\sigma(t+1)h(t) - \log 2\cosh[\beta h(t)]\}}$$

$$\times e^{-i\sum_t [\hat{a}(t)\sigma(t) + \hat{k}(t)\hat{h}(t)] - i\sum_{t,t'} [\hat{q}(t,t')\sigma(t)\sigma(t') + \hat{Q}(t,t')\hat{h}(t)\hat{h}(t') + \hat{K}(t,t')\hat{h}(t)\sigma(t')]},$$

in which the values to be inserted for $\{\hat{m}(t), \hat{k}(t), \hat{q}(t,t'), \hat{Q}(t,t'), \hat{K}(t,t')\}$ are given by the saddle–point of (4.196). Variation of (4.196) with respect to all the original macroscopic objects occurring as arguments (those without the 'hats') gives the following set of saddle–point equations:

$$\hat{a}(t) = i\frac{\partial\Phi}{\partial a(t)}, \qquad \hat{k}(t) = i\frac{\partial\Phi}{\partial k(t)}, \qquad \hat{q}(t,t') = i\frac{\partial\Phi}{\partial q(t,t')},$$

$$\hat{Q}(t,t') = i\frac{\partial\Phi}{\partial Q(t,t')}, \qquad \hat{K}(t,t') = i\frac{\partial\Phi}{\partial K(t,t')}.$$

Variation of (4.196) with respect to the conjugate macroscopic objects (those with the 'hats'), in turn, and usage of our newly introduced short-hand notation $\langle \ldots \rangle_{\star}$, gives:

$$a(t) = \langle \sigma(t) \rangle_{\star}, \qquad k(t) = \langle \hat{h}(t) \rangle_{\star}, \qquad q(t,t') = \langle \sigma(t)\sigma(t') \rangle_{\star},$$

$$Q(t,t') = \langle \hat{h}(t)\hat{h}(t') \rangle_{\star}, \qquad K(t,t') = \langle \hat{h}(t)\sigma(t') \rangle_{\star}$$

The above coupled equations have to be solved simultaneously, once we have calculated the term $\Phi[\ldots]$ that depends on the synapses. This appears to be a formidable task; it can, however, be simplified considerably upon first deriving the physical meaning of the above macroscopic quantities. We use identities such as

$$\frac{\partial\Xi[\ldots]}{\partial\psi_j(t)} = -\frac{i}{N}\xi_j^1 \left[\frac{\int \{dh d\hat{h}\} \sum_{\sigma(0)\cdots\sigma(t_m)} M_j[\{\sigma\}, \{h\}, \{\hat{h}\}]\sigma(t)}{\int \{dh d\hat{h}\} \sum_{\sigma(0)\cdots\sigma(t_m)} M_j[\{\sigma\}, \{h\}, \{\hat{n}\}]} \right],$$

$$\frac{\partial\Xi[\ldots]}{\partial\theta_j(t)} = -\frac{i}{N}\xi_j^1 \left[\frac{\int \{dh d\hat{h}\} \sum_{\sigma(0)\cdots\sigma(t_m)} M_j[\{\sigma\}, \{h\}, \{\hat{h}\}]\hat{h}(t)}{\int \{dh d\hat{h}\} \sum_{\sigma(0)\cdots\sigma(t_m)} M_j[\{\sigma\}, \{h\}, \{\hat{h}\}]} \right],$$

$$\frac{\partial^2 \Xi[\ldots]}{\partial \psi_j(t) \partial \psi_j(t')} = -\frac{1}{N} \left[\frac{\int \{dh d\hat{h}\} \sum_{\sigma(0) \cdots \sigma(t_m)} M_j[\{\sigma\}, \{h\}, \{\hat{h}\}] \sigma(t) \sigma(t')}{\int \{dh d\hat{h}\} \sum_{\sigma(0) \cdots \sigma(t_m)} M_j[\{\sigma\}, \{h\}, \{\hat{h}\}]} \right]$$
$$- N \left[\frac{\partial \Xi[\ldots]}{\partial \psi_j(t)} \right] \left[\frac{\partial \Xi[\ldots]}{\partial \psi_j(t')} \right],$$

$$\frac{\partial^2 \Xi[\ldots]}{\partial \theta_j(t) \partial \theta_j(t')} = -\frac{1}{N} \left[\frac{\int \{dh d\hat{h}\} \sum_{\sigma(0) \cdots \sigma(t_m)} M_j[\{\sigma\}, \{h\}, \{\hat{h}\}] \hat{h}(t) \hat{h}(t')}{\int \{dh d\hat{h}\} \sum_{\sigma(0) \cdots \sigma(t_m)} M_j[\{\sigma\}, \{h\}, \{\hat{h}\}]} \right]$$
$$- N \left[\frac{\partial \Xi[\ldots]}{\partial \theta_j(t)} \right] \left[\frac{\partial \Xi[\ldots]}{\partial \theta_j(t')} \right],$$

$$\frac{\partial^2 \Xi[\ldots]}{\partial \psi_j(t) \partial \theta_j(t')} = -\frac{i}{N} \left[\frac{\int \{dh d\hat{h}\} \sum_{\sigma(0) \cdots \sigma(t_m)} M_j[\{\sigma\}, \{h\}, \{\hat{h}\}] \sigma(t) \hat{h}(t')}{\int \{dh d\hat{h}\} \sum_{\sigma(0) \cdots \sigma(t_m)} M_j[\{\sigma\}, \{h\}, \{\hat{h}\}]} \right]$$
$$- N \left[\frac{\partial \Xi[\ldots]}{\partial \psi_j(t)} \right] \left[\frac{\partial \Xi[\ldots]}{\partial \theta_j(t')} \right],$$

and using the short–hand notation (4.197) wherever possible. Note that the external fields $\{\psi_i(t), \theta_i(t)\}$ occur only in the function $\Xi[\ldots]$, not in $\Psi[\ldots]$ or $\Phi[\ldots]$, and that overall constants in $\overline{Z[\psi]}$ can always be recovered *a posteriori*, using $\overline{Z[\mathbf{0}]} = 1$:

$$m(t) = \lim_{\psi \to 0} \frac{i}{N} \sum_i \xi_i^1 \frac{\int d\mathbf{a} \ldots d\hat{\mathbf{K}} \left[\frac{N \partial \Xi}{\partial \psi_i(t)} \right] e^{N[\Psi + \Phi + \Xi] + \mathcal{O}(\ldots)}}{\int d\mathbf{a} \ldots d\hat{\mathbf{K}} \; e^{N[\Psi + \Phi + \Xi] + \mathcal{O}(\ldots)}}$$
$$= \lim_{\psi \to 0} \langle \sigma(t) \rangle_\star,$$

$$C(t, t') =$$
$$- \lim_{\psi \to 0} \frac{1}{N} \sum_i \frac{\int d\mathbf{a} \ldots d\hat{\mathbf{K}} \left[\frac{N \partial^2 \Xi}{\partial \psi_i(t) \partial \psi_i(t')} + \frac{N \partial \Xi}{\partial \psi_i(t)} \frac{N \partial \Xi}{\partial \psi_i(t')} \right] e^{N[\Psi + \Phi + \Xi] + \mathcal{O}(\ldots)}}{\int d\mathbf{a} \ldots d\hat{\mathbf{K}} \; e^{N[\Psi + \Phi + \Xi] + \mathcal{O}(\ldots)}}$$
$$= \lim_{\psi \to 0} \langle \sigma(t) \sigma(t') \rangle_\star,$$

$$iG(t, t') =$$
$$- \lim_{\psi \to 0} \frac{1}{N} \sum_i \frac{\int d\mathbf{a} \ldots d\hat{\mathbf{K}} \left[\frac{N \partial^2 \Xi}{\partial \psi_i(t) \partial \theta_i(t')} + \frac{N \partial \Xi}{\partial \psi_i(t)} \frac{N \partial \Xi}{\partial \theta_i(t')} \right] e^{N[\Psi + \Phi + \Xi] + \mathcal{O}(\ldots)}}{\int d\mathbf{a} \ldots d\hat{\mathbf{K}} \; e^{N[\Psi + \Phi + \Xi] + \mathcal{O}(\ldots)}}$$
$$= \lim_{\psi \to 0} \langle \sigma(t) \hat{h}(t') \rangle_\star.$$

Finally we get useful identities from the seemingly trivial statements $N^{-1} \sum_i \xi_i^1 \partial \overline{Z[\mathbf{0}]} / \partial \theta_i(t) = 0$ and $N^{-1} \sum_i \partial^2 \overline{Z[\mathbf{0}]} / \partial \theta_i(t) \partial \theta_i(t') = 0$,

$$0 = \lim_{\psi \to 0} \frac{i}{N} \sum_i \xi_i^1 \frac{\int d\mathbf{a} \dots d\hat{\mathbf{K}} \left[\frac{N \partial \Xi}{\partial \theta_i(t)} \right] e^{N[\Psi + \Phi + \Xi] + \mathcal{O}(\dots)}}{\int d\mathbf{a} \dots d\hat{\mathbf{K}} \, e^{N[\Psi + \Phi + \Xi] + \mathcal{O}(\dots)}} = \lim_{\psi \to 0} \langle \hat{h}(t) \rangle_\star,$$

$$0 = -\lim_{\psi \to 0} \frac{1}{N} \sum_i \frac{\int d\mathbf{a} \dots d\hat{\mathbf{K}} \left[\frac{N \partial^2 \Xi}{\partial \theta_i(t) \partial \theta_i(t')} + \frac{N \partial \Xi}{\partial \theta_i(t)} \frac{N \partial \Xi}{\partial \theta_i(t')} \right] e^{N[\Psi + \Phi + \Xi] + \mathcal{O}(\dots)}}{\int d\mathbf{a} \dots d\hat{\mathbf{K}} \, e^{N[\Psi + \Phi + \Xi] + \mathcal{O}(\cdot\cdot)}}$$

$$= \lim_{\psi \to 0} \langle \hat{h}(t) \hat{h}(t') \rangle_\star.$$

The above identities simplify our problem considerably. The dummy fields $\psi_i(t)$ have served their purpose and will now be put to zero, as a result we can now identify our macroscopic observables *at the relevant saddle–point* as:

$$a(t) = m(t), \qquad k(t) = 0, \qquad q(t,t') = C(t,t'),$$
$$Q(t,t') = 0, \qquad K(t,t') = iG(t',t).$$

Finally we make a convenient choice for the external fields, $\theta_i(t) = \xi_i^1 \theta(t)$, with which the effective measure $\langle \dots \rangle_\star$ simplifies to

$$\langle f[\{\sigma\}, \{h\}, \{\hat{h}\}] \rangle_\star = \frac{\int \{dh d\hat{h}\} \sum_{\sigma(0) \dots \sigma(t_m)} M[\{\sigma\}, \{h\}, \{\hat{h}\}] \, f[\{\sigma\}, \{h\}, \{\hat{h}\}]}{\int \{dh d\hat{h}\} \sum_{\sigma(0) \dots \sigma(t_m)} M[\{\sigma\}, \{h\}, \{\hat{h}\}]},$$

(4.198)

with $\quad M[\{\sigma\}, \{h\}, \{\hat{h}\}] =$

$$\pi_0(\sigma(0)) \, e^{\sum_t \{i\hat{h}(t)[h(t) - \theta(t)] + \beta \sigma(t+1)h(t) - \log 2\cosh[\beta h(t)]\} - i \sum_t [\hat{a}(t)\sigma(t) + \hat{k}(t)\hat{h}(t)]}$$

$$\times e^{-i \sum_{t,t'} [\hat{q}(t,t')\sigma(t)\sigma(t') + \hat{Q}(t,t')\hat{h}(t)\hat{h}(t') - \hat{K}(t,t')\hat{h}(t)\sigma(t')]}.$$

Our final task is calculating the leading order of

$$\mathcal{F}[\{\boldsymbol{\sigma}\}, \{\hat{\mathbf{h}}\}] == \frac{1}{N} \log \overline{\left[e^{-i \sum_{it} \hat{h}_i(t) J_{ij} \sigma_j(t)} \right]}.$$

(4.199)

Parallel Hopfield Model Near Saturation

The fully connected Hopfield [Hop82] network (here with parallel dynamics) is got upon choosing $c = N$ in the recipe (4.192), i.e., $c_{ij} = 1 - \delta_{ij}$ and $p = \alpha N$. The disorder average thus involves only the patterns with $\mu > 1$. Now (4.199) gives

$$\mathcal{F}[\dots] = \frac{1}{N} \log \overline{\left[e^{-iN^{-1} \sum_t \xi_i^\mu \xi_j^\mu \hat{h}_i(t) \sigma_j(t)} \right]}$$

(4.200)

$$= i\alpha \sum_t K(t,t; \{\boldsymbol{\sigma}, \hat{\mathbf{h}}\}) - i \sum_t a(t) k(t) +$$

$$\alpha \log \overline{\left[e^{-i \sum_t [\xi_i \hat{h}_i(t)/\sqrt{N}][\xi_i \sigma_i(t)/\sqrt{N}]} \right]} + \mathcal{O}(N^{-1}).$$

We concentrate on the last term:

$$\overline{\left[e^{-i\sum_t[\xi_i\hat{h}_i(t)/\sqrt{N}][\xi_i\sigma_i(t)/\sqrt{N}]}\right]} = \int d\mathbf{x}d\mathbf{y} \; e^{-i\mathbf{x}\cdot\mathbf{y}} \; \times$$

$$\overline{\times\prod_t\left\{\delta[x(t) - \frac{\xi_i\sigma_i(t)}{\sqrt{N}}\cdot\delta[y(t) - \frac{\xi_i\hat{h}_i(t)}{\sqrt{N}}]\right\}}$$

$$= \int \frac{d\mathbf{x}d\mathbf{y}\,d\hat{\mathbf{x}}d\hat{\mathbf{y}}}{(2\pi)^{2(t_m+1)}} \; e^{i[\hat{\mathbf{x}}\cdot\mathbf{x}+\hat{\mathbf{y}}\cdot\mathbf{y}-\mathbf{x}\cdot\mathbf{y}]} \; \overline{\left[e^{-\frac{i}{\sqrt{N}}\xi_i\sum_t[\hat{x}(t)\sigma_i(t)+\hat{y}(t)\hat{h}_i(t)]}\right]}$$

$$= \int \frac{d\mathbf{x}d\mathbf{y}\,d\hat{\mathbf{x}}d\hat{\mathbf{y}}}{(2\pi)^{2(t_m+1)}} \; e^{i[\hat{\mathbf{x}}\cdot\mathbf{x}+\hat{\mathbf{y}}\cdot\mathbf{y}-\mathbf{x}\cdot\mathbf{y}]+\sum_i\log\cos\left[\frac{1}{\sqrt{N}}\sum_t[\hat{x}(t)\sigma_i(t)+\hat{y}(t)\hat{h}_i(t)]\right]}$$

$$= \int \frac{d\mathbf{x}d\mathbf{y}\,d\hat{\mathbf{x}}d\hat{\mathbf{y}}}{(2\pi)^{2(t_m+1)}} \; e^{i[\hat{\mathbf{x}}\cdot\mathbf{x}+\hat{\mathbf{y}}\cdot\mathbf{y}-\mathbf{x}\cdot\mathbf{y}]-\frac{1}{2N}\sum_i\{\sum_t[\hat{x}(t)\sigma_i(t)+\hat{y}(t)\hat{h}_i(t)]\}^2+\mathcal{O}(N^{-1})} =$$

$$\int \frac{d\mathbf{x}d\mathbf{y}\,d\hat{\mathbf{x}}d\hat{\mathbf{y}}}{(2\pi)^{2(t_m+1)}} \; e^{i[\hat{\mathbf{x}}\cdot\mathbf{x}+\hat{\mathbf{y}}\cdot\mathbf{y}-\mathbf{x}\cdot\mathbf{y}]-\frac{1}{2}\sum_{t,t'}[\hat{x}(t)\hat{x}(t')q(t,t')+2\hat{x}(t)\hat{y}(t')K(t',t)+\hat{y}(t)\hat{y}(t')Q(t,t')]}.$$

Together with (4.200) we have now shown that the disorder average (4.199) is indeed, in leading order in N, with

$$\Phi[\mathbf{a},\mathbf{k},\mathbf{q},\mathbf{Q},\mathbf{K}] =$$

$$i\alpha\sum_t K(t,t) - i\mathbf{a}\cdot\mathbf{k} + \alpha\log\int \frac{d\mathbf{x}d\mathbf{y}\,d\hat{\mathbf{x}}d\hat{\mathbf{y}}}{(2\pi)^{2(t_m+1)}} \; e^{i[\hat{\mathbf{x}}\cdot\mathbf{x}+\hat{\mathbf{y}}\cdot\mathbf{y}-\mathbf{x}\cdot\mathbf{y}]-\frac{1}{2}[\hat{\mathbf{x}}\cdot\mathbf{q}\hat{\mathbf{x}}+2\hat{\mathbf{y}}\cdot\mathbf{K}\hat{\mathbf{x}}+\hat{\mathbf{y}}\cdot\mathbf{Q}\hat{\mathbf{y}}]}$$

$$= i\alpha\sum_t K(t,t) - i\mathbf{a}\cdot\mathbf{k} + \alpha\log\int \frac{d\mathbf{u}d\mathbf{v}}{(2\pi)^{t_m+1}} \; e^{-\frac{1}{2}[\mathbf{u}\cdot\mathbf{q}\mathbf{u}+2\mathbf{v}\cdot\mathbf{K}\mathbf{u}-2i\mathbf{u}\cdot\mathbf{v}+\mathbf{v}\cdot\mathbf{Q}\mathbf{v}]}.$$

Now, for the single–time observables, this gives $\hat{a}(t) = k(t)$ and $\hat{k}(t) = a(t)$, and for the two–time ones:

$$\hat{q}(t,t') = -\frac{1}{2}\alpha i \; \frac{\int d\mathbf{u}d\mathbf{v} \; u(t)u(t')e^{-\frac{1}{2}[\mathbf{u}\cdot\mathbf{q}\mathbf{u}+2\mathbf{v}\cdot\mathbf{K}\mathbf{u}-2i\mathbf{u}\cdot\mathbf{v}+\mathbf{v}\cdot\mathbf{Q}\mathbf{v}]}}{\int d\mathbf{u}d\mathbf{v} \; e^{-\frac{1}{2}[\mathbf{u}\cdot\mathbf{q}\mathbf{u}+2\mathbf{v}\cdot\mathbf{K}\mathbf{u}-2i\mathbf{u}\cdot\mathbf{v}+\mathbf{v}\cdot\mathbf{Q}\mathbf{v}]}},$$

$$\hat{Q}(t,t') = -\frac{1}{2}\alpha i \; \frac{\int d\mathbf{u}d\mathbf{v} \; v(t)v(t')e^{-\frac{1}{2}[\mathbf{u}\cdot\mathbf{q}\mathbf{u}+2\mathbf{v}\cdot\mathbf{K}\mathbf{u}-2i\mathbf{u}\cdot\mathbf{v}+\mathbf{v}\cdot\mathbf{Q}\mathbf{v}]}}{\int d\mathbf{u}d\mathbf{v} \; e^{-\frac{1}{2}[\mathbf{u}\cdot\mathbf{q}\mathbf{u}+2\mathbf{v}\cdot\mathbf{K}\mathbf{u}-2i\mathbf{u}\cdot\mathbf{v}+\mathbf{v}\cdot\mathbf{Q}\mathbf{v}]}},$$

$$\hat{K}(t,t') = -\alpha i \; \frac{\int d\mathbf{u}d\mathbf{v} \; v(t)u(t')e^{-\frac{1}{2}[\mathbf{u}\cdot\mathbf{q}\mathbf{u}+2\mathbf{v}\cdot\mathbf{K}\mathbf{u}-2i\mathbf{u}\cdot\mathbf{v}+\mathbf{v}\cdot\mathbf{Q}\mathbf{v}]}}{\int d\mathbf{u}d\mathbf{v} \; e^{-\frac{1}{2}[\mathbf{u}\cdot\mathbf{q}\mathbf{u}+2\mathbf{v}\cdot\mathbf{K}\mathbf{u}-2i\mathbf{u}\cdot\mathbf{v}+\mathbf{v}\cdot\mathbf{Q}\mathbf{v}]}} - \alpha\delta_{t,t'}.$$

At the physical saddle–point we can now express all non–zero objects in terms of the observables $m(t)$, $C(t,t')$ and $G(t,t')$, with a clear physical meaning. Thus we find $\hat{a}(t) = 0$, $\hat{k}(t) = m(t)$, and

$$\hat{q}(t,t') = -\frac{1}{2}\alpha i\, \frac{\int d\mathbf{u}d\mathbf{v}\ u(t)u(t')e^{-\frac{1}{2}[\mathbf{u}\cdot\mathbf{C}\mathbf{u}-2i\mathbf{u}\cdot[\mathbf{1}-\mathbf{G}]\mathbf{v}]}}{\int d\mathbf{u}d\mathbf{v}\ e^{-\frac{1}{2}[\mathbf{u}\cdot\mathbf{C}\mathbf{u}-2i\mathbf{u}\cdot[\mathbf{1}-\mathbf{G}]\mathbf{v}]}} = 0,$$

$$\hat{Q}(t,t') = -\frac{1}{2}\alpha i\, \frac{\int d\mathbf{u}d\mathbf{v}\ v(t)v(t')e^{-\frac{1}{2}[\mathbf{u}\cdot\mathbf{C}\mathbf{u}-2i\mathbf{u}\cdot[\mathbf{1}-\mathbf{G}]\mathbf{v}]}}{\int d\mathbf{u}d\mathbf{v}\ e^{-\frac{1}{2}[\mathbf{u}\cdot\mathbf{C}\mathbf{u}-2i\mathbf{u}\cdot[\mathbf{1}-\mathbf{G}\ \mathbf{v}]}}$$

$$= -\frac{1}{2}\alpha i\left[(\mathbf{1}-\mathbf{G})^{-1}\mathbf{C}(\mathbf{1}-\mathbf{G}^{\dagger})^{-1}\right](t,t'),$$

$$\hat{K}(t,t') + \alpha\delta_{t,t'} = -\alpha i\, \frac{\int d\mathbf{u}d\mathbf{v}\ v(t)u(t')e^{-\frac{1}{2}[\mathbf{u}\cdot\mathbf{C}\mathbf{u}-2i\mathbf{u}\cdot[\mathbf{1}-\mathbf{G}]\mathbf{v}]}}{\int d\mathbf{u}d\mathbf{v}\ e^{-\frac{1}{2}[\mathbf{u}\cdot\mathbf{C}\mathbf{u}-2i\mathbf{u}\cdot[\mathbf{1}-\mathbf{G}]\mathbf{v}]}}$$

$$= \alpha(\mathbf{1}-\mathbf{G})^{-1}(t,t'),$$

with $G^{\dagger}(t,t') = G(t',t)$, and using standard manipulations of Gaussian integrals. Note that we can use the identity $(\mathbf{1}-\mathbf{G})^{-1} - \mathbf{1} = \sum_{\ell\geq 0}\mathbf{G}^{\ell} - \mathbf{1} = \sum_{\ell>0}\mathbf{G}^{\ell} = \mathbf{G}(\mathbf{1}-\mathbf{G})^{-1}$ to compactify the last equation to

$$\hat{K}(t,t') = \alpha[\mathbf{G}(\mathbf{1}-\mathbf{G})^{-1}](t,t'). \tag{4.201}$$

We have now expressed all our objects in terms of the disorder–averaged recall Hopfield overlap $\mathbf{m} = \{m(t)\}$ and the disorder–averaged single–site correlation– and response–functions $\mathbf{C} = \{C(t,t')\}$ and $\mathbf{G} = \{G(t,t')\}$. We can next simplify the effective measure (4.198), which plays a crucial role in the remaining saddle–point equations. Inserting $\hat{a}(t) = \hat{q}(t,t') = 0$ and $\hat{k}(t) = m(t)$ into (4.198), first of all, gives us

$$M[\{\sigma\},\{h\},\{\hat{h}\}] = \pi_0(\sigma(0))\ \times \tag{4.202}$$

$$e^{\sum_t\{i\hat{h}(t)[h(t)-m(t)-\theta(t)-\sum_{t'}\hat{K}(t,t')\sigma(t')]+\beta\sigma(t+1)h(t)-\log 2\cosh[\beta h(t)]\} -i\sum_{t,t'}\hat{Q}(t,t')\hat{h}(t)\hat{h}(t')}.$$

Secondly, causality ensures that $G(t,t') = 0$ for $t \leq t'$, from which, in combination with (4.201), it follows that the same must be true for the kernel $\hat{K}(t,t')$, since

$$\hat{K}(t,t') = \alpha[\mathbf{G}(\mathbf{1}-\mathbf{G})^{-1}](t,t') = \alpha\left\{\mathbf{G}+\mathbf{G}^2+\mathbf{G}^3+\ldots\right\}(t,t').$$

This, in turn, guarantees that the function $M[\ldots]$ in (4.202) is already normalised:

$$\int\{dhd\hat{h}\}\sum_{\sigma(0)\cdots\sigma(t_m)}M[\{\sigma\},\{h\},\{\hat{h}\}] = 1.$$

One can prove this iteratively. After summation over $\sigma(t_m)$ (which due to causality cannot occur in the term with the kernel $\hat{K}(t,t')$) one is left with just a single occurrence of the field $h(t_m)$ in the exponent, integration over which reduces to $\delta[\hat{h}(t_m)]$, which then eliminates the conjugate field $\hat{h}(t_m)$. This cycle of operations is next applied to the variables at time $t_m - 1$, etc. The effective measure (4.198) can now be written simply as

$$\langle f[\{\sigma\}, \{h\}, \{\hat{h}\}]\rangle_{\star} = \sum_{\sigma(0)\cdots\sigma(t_m)} \int \{dh d\hat{h}\} \, M[\{\sigma\}, \{h\}, \{\hat{h}\}] \, f[\{\sigma\}, \{h\}, \{\hat{h}\}],$$

with $M[\dots]$ as given in (4.202). The remaining saddle–point equations to be solved, which can be slightly simplified by using the identity

$$\langle \sigma(t)\hat{h}(t')\rangle_{\star} = i\partial\langle\sigma(t)\rangle_{\star}/\partial\theta(t'), \qquad \text{are}$$
$$m(t) = \langle\sigma(t)\rangle_{\star}, \qquad C(t,t') = \langle\sigma(t)\sigma(t')\rangle_{\star}, \qquad G(t,t') = \partial\langle\sigma(t)\rangle_{\star}/\partial\theta(t').$$

Here we observe that we only need to insert functions of spin states into the effective measure $\langle\dots\rangle_{\star}$ (rather than fields or conjugate fields), so the effective measure can again be simplified. We get

$$\langle f[\{\sigma\}]\rangle_{\star} = \sum_{\sigma(0)\cdots\sigma(t_m)} \text{Prob}[\{\sigma\}] \, f[\{\sigma\}], \qquad \text{with}$$

$$\text{Prob}[\{\sigma\}] = \pi_0(\sigma(0)) \int \{d\phi\} \, P[\{\phi\}] \times \qquad (4.203)$$
$$\times \prod_t \left[\frac{1}{2}[1 + \sigma(t+1)\tanh[\beta h(t|\{\sigma\}, \{\phi\})]]\right],$$

in which $\pi_0(\sigma(0)) = \frac{1}{2}[1 + \sigma(0)m_0]$, and

$$h(t|\{\sigma\}, \{\phi\}) = m(t) + \theta(t) + \alpha \sum_{t'<t}[\mathbf{G}(\mathbf{1} - \mathbf{G})^{-1}](t,t')\sigma(t') + \alpha^{\frac{1}{2}}\phi(t), (4.204)$$

$$P[\{\phi\}] = \frac{e^{-\frac{1}{2}\sum_{t,t'}\phi(t)[(\mathbf{1}-\mathbf{G}^{\dagger})\mathbf{C}^{-1}(\mathbf{1}-\mathbf{G})](t,t')\phi(t')}}{(2\pi)^{(t_m+1)/2}\det^{-\frac{1}{2}}[(\mathbf{1} - \mathbf{G}^{\dagger})\mathbf{C}^{-1}(\mathbf{1} - \mathbf{G})]}.$$

We recognize (4.203) as describing an effective single neuron, with the usual dynamics $\text{Prob}[\sigma(t+1) = \pm 1] = \frac{1}{2}[1 \pm \tanh[\beta h(t)]]$, but with the fields (4.204). This result is indeed of the form (4.191), with a retarded self–interaction kernel $R(t,t')$ and covariance matrix $\langle\phi(t)\phi(t')\rangle$ of the Gaussian $\phi(t)$ given by

$$R(t,t') = [\mathbf{G}(\mathbf{1} - \mathbf{G})^{-1}](t,t'),$$
$$\langle\phi(t)\phi(t')\rangle = [(\mathbf{1} - \mathbf{G})^{-1}\mathbf{C}(\mathbf{1} - \mathbf{G}^{\dagger})^{-1}](t,t').$$

For $\alpha \to 0$ we loose all the complicated terms in the local fields, and recover the type of simple expression we found earlier for finite p: $m(t+1) = \tanh[\beta(m(t) + \theta(t))]$.

Note that always $C(t,t) = \langle\sigma^2(t)\rangle_{\star} = 1$ and $G(t,t') = R(t,t') = 0$ for $t \leq t'$. As a result the covariance matrix of the Gaussian fields can be written as

$$\langle \phi(t)\phi(t')\rangle = [(\mathbf{1}-\mathbf{G})^{-1}\mathbf{C}(\mathbf{1}-\mathbf{G}^\dagger)^{-1}](t,t')$$

$$= \sum_{s,s'\geq 0}[\delta_{t,s}+R(t,s)]C(s,s')[\delta_{s',t'}+R(t',s')]$$

$$= \sum_{s=0}^{t}\sum_{s'=0}^{t'}[\delta_{t,s}+R(t,s)]C(s,s')[\delta_{s',t'}+R(t',s')].$$

Considering arbitrary positive integer powers of the response function immediately shows that

$$(\mathbf{G}^\ell)(t,t') = 0, \qquad \text{if} \qquad t' > t - \ell, \qquad \text{which gives}$$

$$R(t,t') = \sum_{\ell > 0}(\mathbf{G}^\ell)(t,t') = \sum_{\ell=1}^{t-t'}(\mathbf{G}^\ell)(t,t').$$

Similarly we get from $(\mathbf{1}-\mathbf{G})^{-1} = \mathbf{1} + \mathbf{R}$ that for $t' \geq t$: $(\mathbf{1}-\mathbf{G})^{-1}(t,t') = \delta_{t,t'}$. To suppress notation we will simply put $h(t|..)$ instead of $h(t|\{\sigma\},\{\phi\})$; this need not cause any ambiguity. We notice that summation over neuron variables $\sigma(s)$ and integration over Gaussian variables $\phi(s)$ with time arguments s higher than than those occurring in the function to be averaged can always be carried out immediately, giving (for $t > 0$ and $t' < t$):

$$m(t) = \sum_{\sigma(0)...\sigma(t-1)} \pi_0(\sigma(0)) \int \{d\phi\} P[\{\phi\}] \, \tanh[\beta h(t-1|.\,)] \times$$

$$\times \prod_{s=0}^{t-2} \frac{1}{2}[1+\sigma(s+1)\tanh[\beta h(s|..)]],$$

$$G(t,t') = \beta\{C(t,t'+1) - \sum_{\sigma(0)...\sigma(t-1)} \pi_0(\sigma(0)) \times$$

$$\times \int \{d\phi\} P[\{\phi\}] \, \tanh[\beta h(t-1|..)]\tanh[\beta h(t'|..)]$$

$$\times \prod_{s=0}^{t-2} \frac{1}{2}[1+\sigma(s+1)\tanh[\beta h(s|..)]]\},$$

which we get directly for $t' = t - 1$, and which follows for times $t' < t - 1$ upon using the identity

$$\sigma[1-\tanh^2(x)] = [1+\sigma\tanh(x)][\sigma-\tanh(x)].$$

For the correlations we distinguish between $t' = t - 1$ and $t' < t - 1$,

$$C(t,t-1) = \sum_{\sigma(0)...\sigma(t-2)} \pi_0(\sigma(0)) \int \{d\phi\} P[\{\phi\}] \, \tanh[\beta h(t-1|..)] \times$$

$$\times \tanh[\beta h(t-2|..)] \prod_{s=0}^{t-3} \frac{1}{2}[1+\sigma(s+1)\tanh[\beta h(s|..)]],$$

whereas for $t' < t - 1$ we have

$$C(t, t') = \sum_{\sigma(0)...\sigma(t-1)} \pi_0(\sigma(0)) \int \{d\phi\} P[\{\phi\}] \ \tanh[\beta h(t-1|..)] \times$$

$$\times \ \sigma(t') \prod_{s=0}^{t-2} \frac{1}{2} [1 + \sigma(s+1) \tanh[\beta h(s|..)]] \,.$$

Now, the field at $t = 0$ is $h(0|..) = m_0 + \theta(0) + \alpha^{\frac{1}{2}}\phi(0)$, since the retarded self–interaction does not yet come into play. The distribution of $\phi(0)$ is fully characterized by its variance $\langle \phi^2(0) \rangle = C(0,0) = 1$. Therefore, with $Dz = (2\pi)^{-\frac{1}{2}}e^{-\frac{1}{2}z^2}dz$, we immediately find

$$m(1) = \int Dz \ \tanh[\beta(m_0 + \theta(0) + z\sqrt{\alpha})], \qquad C(1,0) = m_0 m(1),$$

$$G(1,0) = \beta \left\{ 1 - \int Dz \ \tanh^2[\beta(m_0 + \theta(0) + z\sqrt{\alpha})] \right\}.$$

For the self–interaction kernel this implies that $R(1,0) = G(1,0)$. We now move on to $t = 2$,

$$m(2) = \frac{1}{2} \sum_{\sigma(0)} \int d\phi(0)d\phi(1) P[\phi(0), \phi(1)] \ \tanh[\beta h(1|..)][1 + \sigma(0)m_0],$$

$$C(2,1) = \frac{1}{2} \sum_{\sigma(0)} \int d\phi(1)d\phi(0) P[\phi(0), \phi(1)] \ \times$$

$$\times \tanh[\beta h(1|..)] \tanh[\beta h(0|..)][1 + \sigma(0)m_0]$$

$$C(2,0) = \frac{1}{2} \sum_{\sigma(0)\sigma(1)} \int \{d\phi\} P[\{\phi\}] \ \tanh[\beta h(1|..)] \times$$

$$\sigma(0)\frac{1}{2} [1 + \sigma(1) \tanh[\beta h(0|..)]] [1 + \sigma(0)m_0],$$

$$G(2,1) = \beta\{1 - \frac{1}{2} \sum_{\sigma(0)} \int d\phi(0)d\phi(1) P[\phi(0), \phi(1)] \ \times$$

$$\times \tanh^2[\beta h(1|..)][1 + \sigma(0)m_0]\},$$

$$G(2,0) = \beta\{C(2,1) - \frac{1}{2} \sum_{\sigma(0)} \int d\phi(0)d\phi(1) P[\phi(0), \phi(1)] \ \times$$

$$\times \tanh[\beta h(1|..)] \tanh[\beta h(0|..)][1 + \sigma(0)m_0]\} = 0.$$

We already know that $\langle\phi^2(0)\rangle = 1$; the remaining two moments we need in order to determine $P[\phi(0), \phi(1)]$ read [Coo01, SC00, SC01, CKS05]

$$\langle\phi(1)\phi(0)\rangle = \sum_{s=0}^{1}[\delta_{1,s} + \delta_{0,s}R(1,0)]C(s,0) = C(1,0) + G(1,0),$$

$$\langle\phi^2(1)\rangle = \sum_{s=0}^{1}\sum_{s'=1}^{1}[\delta_{1,s} + \delta_{0,s}R(1,0)]C(s,s')[\delta_{s',1} + \delta_{s' 0}R(1,0)]$$
$$= G^2(1,0) + 2C(0,1)G(1,0) + 1.$$

We now know $P[\phi(0), \phi(1)]$ and can work out all macroscopic objects with $t = 2$ explicitly, if we wish. I will not do this here in full, but only point at the emerging pattern of all calculations at a given time t depending only on macroscopic quantities that have been calculated at times $t' < t$, which allows for iterative solution. Let us just work out $m(2)$ explicitly, in order to compare the first two recall overlaps $m(1)$ and $m(2)$ with the values found in simulations and in approximate theories. We note that calculating $m(2)$ only requires the field $\phi(1)$, for which we found $\langle\phi^2(1)\rangle = G^2(1,0) + 2C(0,1)G(1,0) + 1$:

$$m(2) = \frac{1}{2}\sum_{\sigma(0)}\int d\phi(1)P[\phi(1)] \; \tanh[\beta(m(1) + \theta(1)$$
$$+ \alpha G(1,0)\sigma(0) + \alpha^{\frac{1}{2}}\phi(1))][1 + \sigma(0)m_0]$$
$$= \frac{1}{2}[1 + m_0]\int Dz \; \tanh[\beta(m(1) + \theta(1) + \alpha G(1,0)$$
$$+ z\sqrt{\alpha[G^2(1,0) + 2m_0 \; m(1) \; G(1,0) + 1]})]$$
$$+ \frac{1}{2}[1 - m_0]\int Dz \; \tanh[\beta(m(1) + \theta(1) - \alpha G(1,0)$$
$$+ z\sqrt{\alpha[G^2(1,0) + 2m_0 \; m(1) \; G(1,0) + 1]})]$$

Here we give a comparison of some of the approximate theories, the (exact) partition function (i.e., path–integral) formalism, and numerical simulations, for the case $\theta(t) = 0$ on the fully connected networks. The evolution of the recall overlap in the first two time–steps has been described as follows:

Naive Gaussian Approx: $m(1) = \int Dz \; \tanh[\beta(m(0) + z\sqrt{\alpha})]$

$$m(2) = \int Dz \; \tanh[\beta(m(1) + z\sqrt{\alpha})]$$

Amari–Maginu Theory: $\quad m(1) = \int Dz \; \tanh[\beta(m(0) + z\sqrt{\alpha})]$

$$m(2) = \int Dz \; \tanh[\beta(m(1) + z\Sigma\sqrt{\alpha})]$$

$$\Sigma^2 = \quad 1 + 2m(0)m(1)G + G^2$$
$$G = \quad \beta\left[1 - \int Dz \; \tanh^2[\beta(m(0) + z\sqrt{\alpha})]\right]$$

Exact Solution: $\quad m(1) = \int Dz \; \tanh[\beta(m(0) + z\sqrt{\alpha})]$

$$m(2) = \tfrac{1}{2}[1 + m_0] \int Dz \; \tanh[\beta(m(1) + \alpha G + z\Sigma\sqrt{\alpha})]$$
$$+ \quad \tfrac{1}{2}[1 - m_0] \int Dz \; \tanh[\beta(m(1) - \alpha G + z\Sigma\sqrt{\alpha})]$$

$$\Sigma^2 = \quad 1 + 2m(0)m(1)G + G^2$$
$$G = \quad \beta\left[1 - \int Dz \; \tanh^2[\beta(m(0) + z\sqrt{\alpha})]\right]$$

We can now appreciate why the more advanced Gaussian approximation (Amari–Maginu theory, [AM88]) works well when the system state is close to the target attractor. This theory gets the moments of the Gaussian part of the interference noise distribution at $t = 1$ exactly right, but not the discrete part, whereas close to the attractor both the response function $G(1,0)$ and one of the two pre–factors $\tfrac{1}{2}[1 \pm m_0]$ in the exact expression for $m(2)$ will be very small, and the latter will therefore indeed approach a Gaussian shape. One can also see why the non–Gaussian approximation of [HO90] made sense: in the calculation of $m(2)$ the interference noise distribution can indeed be written as the sum of two Gaussian ones (although for $t > 2$ this will cease to be true).

Extremely Diluted Attractor Networks Near Saturation

The extremely diluted attractor networks were first studied in [DGZ87] (asymmetric dilution) and [WS91] (symmetric dilution). These models are got upon choosing $\lim_{N\to\infty} c/N = 0$ (while still $c \to \infty$) in definition (4.192) of the Hebbian synapses. The disorder average now involves both the patterns with $\mu > 1$ and the realisation of the 'wiring' variables $c_{ij} \in \{0, 1\}$. Again, in working out the key function (4.199) we will show that for $N \to \infty$ the outcome can be written in terms of the above macroscopic quantities. We carry out the average over the spatial structure variables $\{c_{ij}\}$ first:

$$\mathcal{F}[\ldots] = \frac{1}{N} \log \overline{\left[e^{-\frac{i}{c}c_{ij}\xi_i^\mu \xi_j^\mu \sum_t \hat{h}_i(t)\sigma_j(t)}\right]}$$

$$= \frac{1}{N} \log \overline{\prod_{i<j} e^{-\frac{i}{c}\xi_i^\mu \xi_j^\mu [c_{ij}\sum_t \hat{h}_i(t)\sigma_j(t) + c_{ji}\sum_t \hat{h}_j(t)\sigma_i(t)]}}.$$

Now we have to distinguish between *symmetric* and *asymmetric dilution*. First we deal with the case of *symmetric dilution*: $c_{ij} = c_{ji}$ for all $i \neq j$. The average over the c_{ij} is trivial:

$$\overline{\prod_{i<j} e^{-\frac{i}{c}c_{ij}\sum_{\mu}\xi_i^{\mu}\xi_j^{\mu}\sum_t[\hat{h}_i(t)\sigma_j(t)+\hat{h}_j(t)\sigma_i(t)]}}$$

$$=\overline{\prod_{i<j}\left\{1+\frac{c}{N}[e^{-\frac{i}{c}\xi_i^{\mu}\xi_j^{\mu}\sum_t[\hat{h}_i(t)\sigma_j(t)+\hat{h}_j(t)\sigma_i(t)]}-1]\right\}}$$

$$=\prod_{i<j}e^{-\frac{i}{N}\xi_i^{\mu}\xi_j^{\mu}\sum_t[\hat{h}_i(t)\sigma_j(t)+\hat{h}_j(t)\sigma_i(t)]-\frac{1}{2cN}[\xi_i^{\mu}\xi_j^{\mu}\sum_t[\hat{h}_i(t)\sigma_j(t)+\hat{h}_j(t)\sigma_i(t)]]^2+\mathcal{O}(\frac{1}{N\sqrt{c}})+\mathcal{O}(\frac{c}{N^2})}.$$

We separate in the exponent the terms where $\mu = \nu$ in the quadratic term (being of the form $\sum_{\mu\nu}\ldots$), and the terms with $\mu = 1$. We get (note: $p = \alpha c$):

$$\mathcal{F}[\ldots]=-i\sum_t a(t)k(t)-\frac{1}{2}\alpha\sum_{st}[q(s,t)Q(s,t)+K(s,t)K(t,s)]+\mathcal{O}(c^{-\frac{1}{2}})+\mathcal{O}(c/N)+$$

$$\overline{\frac{1}{N}\log\{e^{-\frac{i}{N}\sum_t[\xi_i^{\mu}\hat{h}_i(t)][\xi_j^{\mu}\sigma_j(t)]-\frac{1}{4cN}\sum_{st}\xi_i^{\mu}\xi_j^{\mu}\xi_i^{\nu}\xi_j^{\nu}[\hat{h}_i(s)\sigma_j(s)+\hat{h}_j(s)\sigma_i(s)][\hat{h}_i(t)\sigma_j(t)+\hat{h}_j(t)\sigma_i(t)]}\}}.$$

Our 'condensed ansatz' implies that for $\mu > 1$: $N^{-\frac{1}{2}}\xi_i^{\mu}\sigma_i(t) = \mathcal{O}(1)$ and $N^{-\frac{1}{2}}\xi_i^{\mu}\hat{h}_i(t) = \mathcal{O}(1)$. Thus the first term in the exponent containing the disorder is $\mathcal{O}(c)$, contributing $\mathcal{O}(c/N)$ to $\mathcal{F}[\ldots]$. We therefore retain only the second term in the exponent. However, the same argument applies to the second term. There all contributions can be seen as uncorrelated in leading order, so that $\sum_{i\neq j}\sum_{\mu\neq\nu}\ldots = \mathcal{O}(Np)$, giving a non-leading $\mathcal{O}(N^{-1})$ cumulative contribution to $\mathcal{F}[\ldots]$. Thus, provided $\lim_{N\to\infty}c^{-1} = \lim_{N\to\infty}c/N = 0$ (which we assumed), we have shown that the disorder average (4.199) is again, in leading order in N, with

Symmetric Case: $\Phi[\mathbf{a},\mathbf{k},\mathbf{q},\mathbf{Q},\mathbf{K}]=-i\mathbf{a}\cdot\mathbf{k}-\frac{1}{2}\alpha\sum_{st}[q(s,t)Q(s,t)+K(s,t)K(t,s)]$.

Next, we deal with the asymmetric case, where c_{ij} and c_{ji} are independent. Again, the average over the c_{ij} is trivial; here it gives

$$\overline{\prod_{i<j}\left\{e^{-\frac{i}{c}c_{ij}\xi_i^{\mu}\xi_j^{\mu}\sum_t\hat{h}_i(t)\sigma_j(t)}e^{-\frac{i}{c}c_{ji}\xi_i^{\mu}\xi_j^{\mu}\sum_t\hat{h}_j(t)\sigma_i(t)}\right\}}$$

$$=\prod_{i<j}\left\{1+\frac{c}{N}[e^{-\frac{i}{c}\xi_i^{\mu}\xi_j^{\mu}\sum_t\hat{h}_i(t)\sigma_j(t)}-1]\right\}\left\{1+\frac{c}{N}[e^{-\frac{i}{c}\xi_i^{\mu}\xi_j^{\mu}\sum_t\hat{h}_j(t)\sigma_i(t)}-1]\right\}$$

$$=\prod_{i<j}\left\{1-\frac{c}{N}[\frac{i}{c}\xi_i^{\mu}\xi_j^{\mu}\sum_t\hat{h}_i(t)\sigma_j(t)+\frac{1}{2c^2}[\xi_i^{\mu}\xi_j^{\mu}\sum_t\hat{h}_i(t)\sigma_j(t)]^2+\mathcal{O}(c^{-\frac{3}{2}})]\right\}$$

$$\times\left\{1-\frac{c}{N}[\frac{i}{c}\xi_i^{\mu}\xi_j^{\mu}\sum_t\hat{h}_j(t)\sigma_i(t)+\frac{1}{2c^2}[\xi_i^{\mu}\xi_j^{\mu}\sum_t\hat{h}_j(t)\sigma_i(t)]^2+\mathcal{O}(c^{-\frac{3}{2}})]\right\}$$

$$=\prod_{i<j}e^{-\frac{i}{N}\xi_i^{\mu}\xi_j^{\mu}\sum_t[\hat{h}_i(t)\sigma_j(t)+\hat{h}_j(t)\sigma_i(t)]-\frac{1}{2cN}[\xi_i^{\mu}\xi_j^{\mu}\sum_t\hat{h}_i(t)\sigma_j(t)]^2-\frac{1}{2cN}[\xi_i^{\mu}\xi_j^{\mu}\sum_t\hat{h}_j(t)\sigma_i(t)]^2+\mathcal{O}(\frac{1}{N\sqrt{c}})+\mathcal{O}(\frac{c}{N^2})}$$

Again, we separate in the exponent the terms where $\mu = \nu$ in the quadratic term and the terms with $\mu = 1$ and get

$$
\mathcal{F}[\ldots] = -i \sum_t a(t)k(t) - \frac{1}{2}\alpha \sum_{st} q(s,t)Q(s,t) + \mathcal{O}(c^{-\frac{1}{2}}) + \mathcal{O}(c/n)
$$
$$
+ \frac{1}{N} \log \left\{ e^{-\frac{i}{N}\sum_t [\xi_i^\mu \hat{h}_i(t)][\xi_j^\mu \sigma_j(t)] - \frac{1}{2cN}\xi_i^\mu \xi_j^\mu \xi_i^\nu \xi_j^\nu \sum_{st} \hat{h}_i(s)\sigma_j(s)\hat{h}_i(t)\sigma_j(t)} \right\}.
$$

The scaling arguments given in the symmetric case, based on our 'condensed ansatz', apply again, and tell us that the remaining terms with the disorder are of vanishing order in N. We have again shown that the disorder average (4.199) is, in leading order in N with

Asymmetric Case: $\qquad \Phi[\mathbf{a}, \mathbf{k}, \mathbf{q}, \mathbf{Q}, \mathbf{K}] = -i\mathbf{a} \cdot \mathbf{k} - \frac{1}{2}\alpha \sum_{st} q(s,t)Q(s,t).$

Now, asymmetric dilution corresponds to $\Delta = 0$, i.e., there is no retarded self–interaction, and the response function no longer plays a role. We now only retain $h(t|\ldots) = m(t) + \theta(t) + \alpha^{\frac{1}{2}}\phi(t)$, with $\langle \phi^2(t)\rangle = C(1,1) = 1$. We now get

$$
m(t+1) = \sum_{\sigma(0)\ldots\sigma(t)} \pi_0(\sigma(0)) \int \{d\phi\}P[\{\phi\}] \ \tanh[\beta h(t|\ldots)] \times
$$
$$
\times \prod_{s=0}^{t-1} \frac{1}{2}[1 + \sigma(s+1)\tanh[\beta h(s|\ldots)]] = \int Dz \ \tanh[\beta(m(t) + \theta(t) + z\sqrt{\alpha})].
$$

Similarly, for $t > t'$ equations for correlation and response functions reduce to

$$
C(t,t') = \int \frac{d\phi_a d\phi_b}{2\pi\sqrt{1 - C^2(t-1,t'-1)}} e^{-\frac{1}{2}\frac{\phi_a^2 + \phi_b^2 - 2C(t-1,t'-1)\phi_a\phi_b}{1 - C^2(t-1,t'-1)}} \times
$$
$$
\times \tanh[\beta(m(t-1) + \theta(t-1) + \phi_a\sqrt{\alpha})] \times
$$
$$
\times \tanh[\beta(m(t'-1) + \theta(t'-1) + \phi_b\sqrt{\alpha})],
$$
$$
G(t,t') = \beta\delta_{t,t'+1}\left\{1 - \int Dz \ \tanh^2[\beta(m(t-1) + \theta(t-1) + z\sqrt{\alpha})]\right\}.
$$

Let us also inspect the stationary state $m(t) = m$, for $\theta(t) = 0$. One easily proves that $m = 0$ as soon as $T > 1$, using

$$
m^2 = \beta m \int_0^m dk[1 - \int Dz \tanh^2[\beta(k + z\sqrt{\alpha})]] \le \beta m^2.
$$

A continuous bifurcation occurs from the $m = 0$ state to an $m > 0$ state when $T = 1 - \int Dz \ \tanh^2[\beta z\sqrt{\alpha}]$. A parametrization of this transition line in the (α, T)−plane is given by

$$T(x) = 1 - \int Dz \, \tanh^2(zx), \qquad \alpha(x) = x^2 T^2(x), \qquad \text{for } x \geq 0.$$

For $\alpha = 0$ we just jet $m = \tanh(\beta m)$ so $T_c = 1$. For $T = 0$ we get the equation $m = \text{erf}[m/\sqrt{2\alpha}]$, giving a continuous transition to $m > 0$ solutions at $\alpha_c = 2/\pi \approx 0.637$. The remaining question concerns the nature of the $m = 0$ state. Inserting $m(t) = \theta(t) = 0$ (for all t) into $C(t, t')$ tells us that $C(t, t') = f[C(t-1, t'-1)]$ for $t > t' > 0$, with 'initial conditions' $C(t, 0) = m(t)m_0$, where

$$f[C] = \int \frac{d\phi_a d\phi_b}{2\pi\sqrt{1 - C^2}} \, e^{-\frac{1}{2}\frac{\phi_a^2 + \phi_b^2 - 2C\phi_a\phi_b}{1 - C^2}} \tanh[\beta\sqrt{\alpha}\phi_a] \tanh[\beta\sqrt{\alpha}\phi_b].$$

In the $m = 0$ regime we have $C(t, 0) = 0$ for any $t > 0$, inducing $C(t, t') = 0$ for any $t > t'$, due to $f[0] = 0$. Thus we conclude that $C(t, t') = \delta_{t,t'}$ in the $m = 0$ phase, i.e., this phase is para–magnetic rather than of a spin–glass type.

On the other hand, physics of networks with symmetric dilution is more complicated situation. In spite of the extreme dilution, the interaction symmetry makes sure that the spins still have a sufficient number of common ancestors for complicated correlations to build up in finite time. We have

$$h(t|\{\sigma\}, \{\phi\}) = m(t) + \theta(t) + \alpha \sum_{t' < t} G(t, t')\sigma(t') + \alpha^{\frac{1}{2}}\phi(t),$$

$$P[\{\phi\}] = \frac{e^{-\frac{1}{2}\sum_{t,t'} \phi(t)\mathbf{C}^{-1}(t,t')\phi(t')}}{(2\pi)^{(t_m+1)/2}\det^{\frac{1}{2}}\mathbf{C}}.$$

Thus the effective single neuron problem is found to be exactly of the form found also for the Gaussian model defined above (which, in turn, maps onto the parallel dynamics *Sherrington–Kirkpatrick model* (SK model) [SK75]) with the synapses $J_{ij} = J_0\xi_i\xi_j/N + Jz_{ij}/\sqrt{N}$ (in which the z_{ij} are symmetric zero–average and unit–variance Gaussian variables, and $J_{ii} = 0$ for all i), with the identification:

$$J \to \sqrt{\alpha}, \qquad \text{with} \qquad J_0 \to 1.$$

Since one can show that for $J_0 > 0$ the parallel dynamics SK model gives the same equilibrium state as the sequential one, we can now immediately write down the stationary solution of our dynamic equations which corresponds to the FDT regime, with $q = \lim_{\tau \to \infty}\lim_{t \to \infty} C(t, t + \tau)$:

$$q = \int Dz \tanh^2[\beta(m + z\sqrt{\alpha q})], \qquad m = \int Dz \tanh[\beta(m + z\sqrt{\alpha q})].$$

4.3.10 Hierarchical Self–Programming in Neural Networks

The *hierarchical self–programming* in recurrent neural networks was studied in [UC02]. The authors studied recurrent networks of binary (Ising–spin) neuronal state variables with symmetric interactions J_{ij}, taken to be of infinite range. In contrast to most standard neural network models, not only the neuron states but also the interactions were allowed to evolve in time (simultaneously), driven by correlations in the states of the neurons (albeit slowly compared to the dynamics of the latter), reflecting the effect of 'learning' or 'long–term potentiation' in real nervous tissue. Since the interactions represent the 'programme' of the system, and since the slow interaction dynamics are driven by the states of the neurons (the 'processors'), such models can be regarded as describing self–programming information–processing systems, which can be expected to exhibit highly complex dynamical behavior.

The first papers in which self–programming recurrent neural networks were studied appear to be [Shi87, DH85]. In the language of self-programming systems one could say that these authors were mostly concerned with the stability properties of embedded 'programmes' (usually taken to be those implementing content-addressable or associative memories). In both [Shi87, DH85] the programme dynamics, i.e., that of the $\{J_{ij}\}$, was defined to be adiabatically slow compared to the neuronal dynamics, and fully deterministic. However, the authors already made the important observation that the natural type of (deterministic) programme dynamics (from a biological point of view), so–called Hebbian learning, could be written as a gradient descent of the interactions $\{J_{ij}\}$ on the free energy surface of a symmetric recurrent neural network equipped with these interactions.

In order to study more generally the potential of such self–programming systems, several authors (simultaneously and independently) took the natural next step [CPS93, PCS93, DFM94, PS94, FD94, Cat94]: they generalized the interaction dynamics by adding Gaussian white noise to the deterministic laws, converting the process into one described by conservative Langevin equations, and were thus able to set up an equilibrium statistical mechanics of the self–programming process. This was (surprisingly) found to take the form of a replica theory with finite replica dimension, whose value was given by the ratio of the noise levels in the neuronal dynamics and the interaction dynamics, respectively. Furthermore, adding explicit quenched disorder to the problem in the form of additional random (but frozen) forces in the interaction dynamics, led to theories with two nested levels of replicas, one representing the disorder (with zero replica dimension) and one representing the adiabatically slow dynamics of the interactions [PS94, JBC98, JAB00] (with nonzero replica dimension). The structure of these latter theories was found to be more or less identical to those of ordinary disordered spin systems such as the SK model [SK75], with fixed interactions but quenched disorder, when described by replica theories with one step replica symmetry breaking (RSB) [MPV87]. The only (yet crucial) difference was that in ordinary disordered spin systems the size m of the level-1 block in the Parisi solution is determined by extermination of the free energy, which forces m to lie in the interval $[0, 1]$, whereas in the self-programming neural networks of [PS94, JBC98, JAB00] m was an independent control parameters, given by the ratio of two temperatures, which can take any non-zero value. As a consequence one can observe in the latter systems, for sufficiently large values of such dimensions, much more complicated scenarios of (generally discontinuous) phase transitions.

In contrast to the previous models involving coupled dynamics of fast neurons and slow interactions, [UC02] studied systems in which the interactions did not evolve on a single time-scale, but where they were divided into a hierarchy of L different groups, each with their own characteristic time-scale τ_ℓ and noise level T_ℓ ($\ell = 1, \ldots, L$), describing a hierarchy of increasingly non-volatile programming levels. This appeared to be a much more realistic representation of self-programming systems; conventional programmes generally take the form of hierarchies of routines, sub-routines and so on, and it would appear appropriate to allow low-level sub-routines to be more easily modifiable than high-level ones. In order to retain analytical solvability we choose the different groups of interactions randomly (prescribing only their sizes).

The authors of [UC02] solved the model in equilibrium, and found, upon making the replica–symmetric (i.e., ergodic) ansatz within each level of our hierarchy, a theory which resembled, but was not identical to Parisi's $L-$level replica symmetry–breaking solution for spin systems with frozen disorder. They referred to the binary neurons as spins and to the synaptic interactions as couplings. They wrote the $N-$spin state vector as $\boldsymbol{\sigma} = (\sigma_1, \ldots, \sigma_N) \in \{-1, 1\}^N$, and the matrix of interactions as $\mathbf{J} = \{J_{ij}\}$. The spins were taken to have a stochastic Glauber-type dynamics such that for *stationary* choices of the couplings the microscopic spin probability density would evolve towards a Boltzmann distribution

$$p_\infty(\boldsymbol{\sigma}) = \frac{\mathrm{e}^{-\beta H(\boldsymbol{\sigma})}}{Z}, \qquad Z = \sum_{\boldsymbol{\sigma}} \mathrm{e}^{-\beta H(\boldsymbol{\sigma})}, \tag{4.205}$$

with the conventional Hamiltonian

$$H(\boldsymbol{\sigma}) = -J_{ij}\sigma_i\sigma_j, \tag{4.206}$$

where $i, j \in \{1, \ldots, N\}$, and with the inverse temperature $\beta = T^{-1}$.

The couplings J_{ij} also evolve in a stochastic manner, in response to the states of the spins, but adiabatically slowly compared to the spins, such that on the time-scales of the couplings the spins are always in an equilibrium state described by (4.205). For the coupling dynamics the following Langevin equations are proposed:

$$\tau_{ij}\dot{J}_{ij} = \frac{1}{N}\langle\sigma_i\sigma_j\rangle_{\mathrm{sp}} - \mu_{ij}J_{ij} + \eta_{ij}(t), \qquad (i < j = 1 \ldots N), \tag{4.207}$$

with $\tau_{ij} \gg 1$. In the adiabatic limit $\tau_{ij} \to \infty$ the term $\langle\sigma_i\sigma_j\rangle_{\mathrm{sp}}$, representing spin correlations associated with the coupling J_{ij}, becomes an average over the Boltzmann distribution (4.205) of the spins, given the instantaneous couplings $\mathbf{J}$. The $\eta_{ij}(t)$ represent Gaussian white noise contributions, of zero mean and covariance

$$\langle \eta_{ij}(t)\eta_{kl}(t') \rangle = 2T_{ij}\delta_{ik}\delta_{jl}\delta(t-t'),$$

with associated temperature $T_{ij} = \beta_{ij}^{-1}$. Appropriate factors of N have been introduced in order to ensure non–trivial behavior in the limit $N \to \infty$. We classify the spin pairs (i,j) according to the characteristic time-scale τ_{ij} and the control parameters (T_{ij}, μ_{ij}) associated with their interactions J_{ij}. In contrast to papers such as [CPS93, PCS93, PS94, JBC98, JAB00], where $\tau_{ij} = \tau$, $T_{ij} = \tilde{T}$ and $\mu_{ij} = \mu$ for all (i,j), here the various time-scales, temperatures and decay rates are no longer assumed to be identical, but to come in L distinct adiabatically separated groups I_ℓ (always with $i < j$):

$$I_\ell = \{(i,j)|\ \tau_{ij} = \tau_\ell, \qquad T_{ij} = T_\ell,\ \mu_{ij} = \mu_\ell\}, \qquad (l = 1, \cdots, L),$$

with $1 \ll \tau_1 \ll \tau_2 \ll \ldots \ll \tau_L$. Thus $\{(i,j)\} = \bigcup_{\ell \le L} I_\ell$. We will write the set of spin-interactions with time–scale τ_ℓ as $\mathbf{J}^\ell = \{J_{ij}|\ (i,j) \in I_\ell\}$. The interactions in group I_2 are adiabatically slow compared to those in group I_1, and so on. The rationale of this set–up is that, in information processing terms, this would represent a stochastic self–programming neural information processing system equipped with a program which consists of a hierarchy of increasingly less volatile and less easily modifiable sub-routines.

Finally we have to define the detailed partitioning of the $\frac{1}{2}N(N-1)$ interactions into the L volatility groups. We introduce $\epsilon_{ij}(\ell) \in \{0,1\}$ such that $\epsilon_{ij}(\ell) = 1$ iff $(i,j) \in I_\ell$, so $\sum_{\ell=1}^{L} \epsilon_{ij}(\ell) = 1$ for all (i,j). In order to arrive at a solvable mean–field problem, with full equivalence of the N sites, we will choose the $\epsilon_{ij}(\ell)$ independently at random for each pair (i,j) with $i < j$, with probabilities [UC02]

$$\mathrm{Prob}[\epsilon_{ij}(\ell) = 1] = \epsilon_\ell, \qquad \sum_{\ell=1}^{L} \epsilon_\ell = 1. \qquad (4.208)$$

These time-scale and temperature allocation variables $\{\epsilon_{ij}(\ell)\}$ thus introduce quenched disorder into the problem in hand.

Let us denote averages over the probability distribution of the couplings at level ℓ in the hierarchy (for which $\tau_{ij} = \tau_\ell$, $T_{ij} = T_\ell$ and $\mu_{ij} = \mu_\ell$) as $\langle \ldots \rangle_\ell$. At every level ℓ, the stochastic equation (4.207) for those couplings which evolve on that particular time-scale τ_ℓ has now become conservative:

$$\tau_\ell \dot{J}_{ij} = \frac{1}{N} \langle \ldots \langle\ \langle \sigma_i \sigma_j \rangle_{\mathrm{sp}} \rangle_1 \ldots \rangle_{\ell-1} - \mu_\ell J_{ij} + \eta_{ij}(t)\sqrt{\frac{\tau_\ell}{N}}$$

$$= -\frac{1}{N}\frac{\partial}{\partial J_{ij}}H_\ell(\mathbf{J}^\ell, \ldots, \mathbf{J}^L) + \eta_{ij}(t)\sqrt{\frac{\tau_\ell}{N}}, \qquad (4.209)$$

with the following effective Hamiltonian for the couplings at level ℓ:

$$H_1(\mathbf{J}^1, \ldots, \mathbf{J}^L) = -\beta^{-1}\log Z[\mathbf{J}^1, \ldots, \mathbf{J}^L], \qquad (4.210)$$

$$H_{\ell+1}(\mathbf{J}^{\ell+1},\ldots,\mathbf{J}^L) = -\beta_\ell^{-1} \log Z_\ell[\mathbf{J}^{\ell+1},\ldots,\mathbf{J}^L] \qquad (1 \le \ell < L), \quad (4.211)$$

and with the partition functions

$$Z[\mathbf{J}^1,\ldots,\mathbf{J}^L] = \sum_{\boldsymbol{\sigma}} e^{-\beta H(\boldsymbol{\sigma},\mathbf{J})}, \qquad (4.212)$$

$$Z_\ell[\mathbf{J}^{\ell+1},\ldots,\mathbf{J}^L] = \int d\mathbf{J}^\ell \, e^{-\beta_\ell H_\ell(\mathbf{J}^l,\ldots,\mathbf{J}^L)}, \qquad (1 \le \ell < L), \qquad (4.213)$$

$$Z_L = \int d\mathbf{J}^L \, e^{-\beta_L H_L(\mathbf{J}^L)}, \qquad \text{in which} \qquad (4.214)$$

$$H(\boldsymbol{\sigma},\mathbf{J}) = -J_{ij}\sigma_i\sigma_j + \frac{1}{2}N\mu_{ij}J_{ij}^2 \qquad (4.215)$$

This describes a hierarchy of nested equilibrations. At each time-scale τ_ℓ the interactions $\mathbf{J}^\ell$ equilibrate to a Boltzmann distribution, with an effective Hamiltonian H_ℓ which is the free energy of the previous (faster) level $\ell+1$ in the hierarchy, starting from the overall Hamiltonian (4.215) (for spins and couplings) at the fastest (spin) level. As a result of having different effective temperatures T_ℓ associated with each level, the partition functions of subsequent levels are found to generate replica theories with replica dimensions $m_\ell \ge 0$ which represent the ratios of the effective temperatures of the two levels involved. This follows from substitution of (4.211) into (4.213):

$$Z_\ell[\mathbf{J}^{l+1},\ldots,\mathbf{J}^L] = \int d\mathbf{J}^\ell \, \{Z_{\ell-1}[\mathbf{J}^l,\ldots,\mathbf{J}^L]\}^{m_\ell}, \qquad (4.216)$$

$$m_\ell = \beta_\ell/\beta_{\ell-1} \qquad m_1 = \beta_1/\beta \qquad (4.217)$$

The statics of the system, including the effect of the quenched disorder, are governed by the disorder-averaged free energy $\mathcal{F}$ associated with the partition function Z_L in (4.214), where the slowest variables have finally been integrated out [UC02]:

$$\mathcal{F} = -\frac{1}{\beta_L}\overline{\log Z_L} = -\lim_{m_{L+1}\to 0}\frac{1}{m_{L+1}\beta_L}\log\overline{Z_L^{m_{L+1}}}. \qquad (4.218)$$

This function is found to act as the general generator of equilibrium values for observables at any level in the hierarchy, since upon adding suitable partition terms to the Hamiltonian (4.215), i.e., $H(\boldsymbol{\sigma},\mathbf{J}) \to H(\boldsymbol{\sigma},\mathbf{J}) + \lambda\Phi(\boldsymbol{\sigma},\mathbf{J})$, one finds:

$$\overline{\langle\ldots\langle\,\langle\Phi(\boldsymbol{\sigma},\mathbf{J})\rangle_{\mathrm{sp}}\rangle_1\ldots\rangle_L} = \lim_{\lambda\to 0}\frac{\partial}{\partial\lambda}\mathcal{F}.$$

We can now combine our previous results and write down an explicit expression for $\mathcal{F}$, involving multiple replications due to (4.216,4.218). We find $\mathcal{F} = \lim_{m_{L+1}\to 0} F[m_1,\ldots,m_{L+1}]$, with $F[m_1,\ldots,m_{L+1}] = -(m_{L+1}\beta_L)^{-1}\log\overline{Z_1^{m_{L+1}}}$, and where $Z_L^{m_{L+1}}$ is written as [UC02]

$$Z_L^{m_{L+1}} = \int \left[\prod_{\alpha_{L+1}} d\mathbf{J}^{L,\alpha_{L+1}} \right] \prod_{\alpha_{L+1}} \left\{ Z_{L-1}[\mathbf{J}^{L,\alpha_{L+1}}] \right\}^{m_L}$$

$$= \int \left[\prod_{\alpha_{L+1}} d\mathbf{J}^{L,\alpha_{L+1}} \dots \prod_{\alpha_2,\dots,\alpha_{L+1}} d\mathbf{J}^{1,\alpha_2,\dots,\alpha_{L+1}} \right] \times$$

$$\times \prod_{\alpha_2,\dots,\alpha_{L+1}} \left\{ Z[\mathbf{J}^{1,\alpha_2,\dots,\alpha_{L+1}}, \dots, \mathbf{J}^{L,\alpha_{L+1}}] \right\}^{m_1}$$

$$= \sum_{\{\sigma^{\alpha_1,\dots,\alpha_{L+1}}\}} \int \left[\prod_{\alpha_{L+1}} d\mathbf{J}^{L,\alpha_{L+1}} \dots \prod_{\alpha_2,\dots,\alpha_{L+1}} d\mathbf{J}^{1,\alpha_2,\dots,\alpha_{L+1}} \right] \times$$

$$\times \prod_{\alpha_1,\dots,\alpha_{L+1}} e^{-\beta H(\sigma^{\alpha_1,\dots,\alpha_{L+1}}, \sigma^{1,\alpha_2,\dots,\alpha_{L+1}}, \dots, \sigma^{L,\alpha_{L+1}})}$$

$$= \sum_{\{\sigma^{\alpha_1,\dots,\alpha_{L+1}}\}} \prod_{\ell=1}^{L} \prod_{(i<j)\in I_\ell} \prod_{\alpha_{\ell+1},\dots,\alpha_{L+1}} \times$$

$$\times \int dz \; e^{-\frac{1}{2}\beta\mu_\ell N[m_1\dots m_\ell z^2 - 2(\mu_\ell N)^{-1} z \sum_{\alpha_1\dots\alpha_\ell} \sigma_i^{\alpha_1,\dots,\alpha_{L+1}} \sigma_j^{\alpha_1,\dots,\alpha_{L+1}}]}$$

$$= \sum_{\{\sigma^{\alpha_1,\dots,\alpha_{L+1}}\}} e^{\frac{\beta}{2N} \sum_{\ell \le L} \frac{1}{m_1\dots m_\ell \mu_\ell} \sum_{(i<j)\in I_\ell} \sum_{\alpha_{\ell+1},\dots,\alpha_{L+1}} [\sum_{\alpha_1\dots\alpha_\ell} \sigma_i^{\alpha_1,\dots,\alpha_{L+1}} \sigma_j^{\alpha_1,\dots,\alpha_{L+1}}]}.$$

(modulo irrelevant constants), in which always $\alpha_\ell = 1, \dots, m_\ell$.

Now, in order to average the last expression for $Z_L^{m_{L+1}}$ over the disorder, we note that the spin summation in the exponent can also be written in terms of the allocation variables $\epsilon_{ij}(\ell)$:

$$\sum_{\ell \le L} \frac{1}{m_1 \cdots m_\ell \mu_\ell} \sum_{(i<j)\in I_\ell} \sum_{\alpha_{l+1},\dots,\alpha_{L+1}} \dots =$$

$$\sum_{i<j} \sum_{\ell \le L} \frac{\beta}{\beta_\ell} \frac{\epsilon_{ij}(\ell)}{\mu_\ell} \sum_{\alpha_{\ell+1},\dots,\alpha_{L+1}} [\sum_{\alpha_1\dots\alpha_\ell} \sigma_i^{\alpha_1,\dots,\alpha_{L+1}} \sigma_j^{\alpha_1,\dots,\alpha_{L+1}}]^2,$$

where we used $m_1 \dots m_\ell = \beta_\ell/\beta$. This allows us to carry out the average over the (independent) $\epsilon_{ij}(\ell)$ in:

$$F[\ldots] = -\frac{1}{m_{L+1}\beta_L}\log\sum_{\{\sigma^{\alpha_1,\ldots,\alpha_{L+1}}\}}$$

$$\times\prod_{i<j}\left[e^{\frac{\beta^2}{2N}\sum_{\ell\le L}\frac{\epsilon_{ij}(\ell)}{\beta_\ell\mu_\ell}\sum_{\alpha_{\ell+1},\ldots,\alpha_{L+1}}[\sum_{\alpha_1\ldots\alpha_\ell}\sigma_i^{\alpha_1,\ldots,\alpha_{L-1}}\sigma_j^{\alpha_1,\ldots,\alpha_{L+1}}]^2}\right]$$

$$= -\frac{1}{m_{L+1}\beta_L}\log\sum_{\{\sigma^{\alpha_1,\ldots,\alpha_{L+1}}\}}e^{\mathcal{O}(N^0)}$$

$$\times\prod_{i<j}\left[1+\frac{\beta^2}{2N}\sum_{\ell\le L}\frac{\epsilon_{ij}(\ell)}{\beta_\ell\mu_\ell}\sum_{\alpha_{\ell+1},\ldots,\alpha_{L+1}}[\sum_{\alpha_1\ldots\alpha_\ell}\sigma_i^{\alpha_1,\ldots,\alpha_{L+1}}\sigma_j^{\alpha_1,\ldots,\alpha_{L+1}}]^2\right]$$

$$= -\frac{1}{m_{L+1}\beta_L}\log\sum_{\{\sigma^{\alpha_1,\ldots,\alpha_{L+1}}\}}e^{\mathcal{O}(N^0)}$$

$$\times\exp\left\{\frac{N\beta^2}{4}\sum_{\alpha_1,\ldots,\alpha_{L+1}}\sum_{\beta_1,\ldots,\beta_{L+1}}\sum_{\ell\le L}\frac{\epsilon_\ell}{\beta_\ell\mu_\ell}\delta_{(\alpha_{\ell+1},\ldots,\alpha_{L+1}),(\beta_{\ell+1},\ldots,\beta_{L+1})}\right.$$

$$\left.\times[\frac{1}{N}\sum_i\sigma_i^{\alpha_1,\ldots,\alpha_{L+1}}\sigma_i^{\beta_1,\ldots,\beta_{L+1}}]^2\right\},$$

(again modulo irrelevant constants). We abbreviate $a = (\alpha_1,\ldots,\alpha_{L+1})$, and introduce the spin-glass replica order parameters $q_{ab} = \frac{1}{N}\sigma_i^{\alpha_1,\ldots,\alpha_{L+1}}\sigma_i^{\beta_1,\ldots,\beta_{L+1}}$ by inserting appropriate integrals over $\delta-$distributions, and arrive at

$$F[\ldots] = -\frac{1}{m_{L+1}\beta_L}\log\int\{dq_{ab}d\hat{q}_{ab}\}e^{NG[\{q_{ab},\hat{q}_{ab}\}]+\mathcal{O}(N^0)}$$

$$G[\{q_{ab},\hat{q}_{ab}\}] = i\sum_{ab}\hat{q}_{ab}q_{ab} + \frac{\beta^2}{4}\sum_{\ell\le L}\frac{\epsilon_\ell}{\beta_\ell\mu_\ell}\sum_{ab}\delta_{(\alpha_{\ell+1},\ldots,\alpha_{L+1}),(\beta_{\ell+1},\ldots,\beta_{L+1})}q_{ab}^2$$

$$+\log\sum_{\{\sigma_{\alpha_1,\ldots,\alpha_{L+1}}\}}e^{-i\sum_{ab}\hat{q}_{ab}\sigma_{\alpha_1,\ldots,\alpha_{L+1}}\sigma_{\beta_1,\ldots,\beta_{L+1}}}.$$

For $N\to\infty$ the above integral can be evaluated by steepest descent, and upon elimination of the conjugate order parameters $\{\hat{q}_{ab}\}$ by variation of $\{q_{ab}\}$, the disorder–averaged free energy per spin $f = \lim_{N\to\infty}\mathcal{F}/N$ is found to be

$$f = \lim_{m_{L+1}\to 0}\frac{1}{m_{L+1}\beta_L}\left\{\frac{\beta^2}{4}\sum_{\ell\le L}\frac{\epsilon_\ell}{\beta_\ell\mu_\ell}\sum_{ab}\delta_{(\alpha_{\ell+1},\ldots,\alpha_{L+1}),(\beta_{\ell+1},\ldots,\beta_{L+1})}q_{ab}^2\right.$$

$$\left.-\log\sum_{\{\sigma_a\}}e^{\frac{\beta^2}{2}\sum_{\ell\le L}\frac{\epsilon_\ell}{\beta_\ell\mu_\ell}\sum_{ab}\delta_{(\alpha_{\ell+1},\ldots,\alpha_{L+1}),(\beta_{\ell+1},\ldots,\beta_{L-1})}q_{ab}\sigma_a\sigma_b}\right\}.(4.219)$$

The saddle–point equations from which to solve the $\{q_{ab}\}$ are given by [UC02]

$$q_{cd} = \frac{\sum_{\{\sigma_a\}} \sigma_c \sigma_d \, e^{\frac{\beta^2}{2} \sum_{\ell \leq L} \epsilon_\ell / (\beta_\ell \mu_\ell) \sum_{ab} \delta_{(\alpha_{\ell+1},\ldots,\alpha_{L+1}),(\beta_{\ell+1},\ldots,\beta_{L+1})} q_{ab} \sigma_a \sigma_b}}{\sum_{\{\sigma_a\}} e^{\frac{\beta^2}{2} \sum_{\ell \leq L} \epsilon_\ell / (\beta_\ell \mu_\ell) \sum_{ab} \delta_{(\alpha_{\ell+1},\ldots,\alpha_{L+1}),(\beta_{\ell+1},\ldots,\beta_{L+1})} q_{ab} \sigma_a \sigma_b}}.$$

with our short-hand $a = (\alpha_1, \ldots, \alpha_{L+1})$, and with $\alpha_\ell \in \{1, \ldots, m_\ell\}$ for all ℓ, where the dimensions m_ℓ are given by the ratios of the temperatures of subsequent programming levels in the hierarchy according to (4.217).

Our full order parameters are $q_{ab} = \frac{1}{N} \sigma_i^{\alpha_1, \ldots, \alpha_{L+1}} \sigma_i^{\beta_1, \ldots, \beta_{L+1}}$. Note that $\alpha_{L+1} = \beta_{L+1}$. Spin variables with $\alpha_L = \beta_L$ have identical level–L bonds. Those with $\alpha_L = \beta_L$ and $\alpha_{L-1} = \beta_{L-1}$ have identical level L bonds *and* identical level $L-1$ bonds, and so on. Hence, for the present model the replica-symmetric (RS) ansatz (describing ergodicity at each level of the hierarchy of time-scales) takes the following form [UC02]:

$$
\begin{aligned}
& & \alpha_L \neq \beta_L : & \quad q_{ab} = q_L, \\
(\alpha_{\ell+1}, \ldots, \alpha_{L+1}) = (\beta_{\ell+1}, \ldots, \beta_{L+1}), & \quad \alpha_\ell \neq \beta_\ell : & \quad q_{ab} = q_\ell, \\
& & a = b : & \quad q_{ab} = 1,
\end{aligned}
$$

$$\text{or} \quad q_{ab} = \sum_{\ell=1}^{L} q_\ell \, \delta_{(\alpha_{\ell+1},\ldots,\alpha_{L+1})(\beta_{\ell+1},\ldots,\beta_{L+1})} \overline{\delta}_{\alpha_\ell \beta_\ell} + \delta_{ab}, \qquad (4.220)$$

where $\overline{\delta}_{ij} = 1 - \delta_{ij}$, and where $0 \leq q_L \leq \ldots \leq q_1 \leq 1$.

4.4 Topological Phase Transitions and Hamiltonian Chaos

4.4.1 Phase Transitions in Hamiltonian Systems

Recall that *phase transitions* (PTs) are phenomena which bring about *qualitative* physical changes at the macroscopic level in presence of the same microscopic forces acting among the constituents of a system. Their mathematical description requires to translate into *quantitative* terms the mentioned qualitative changes. The standard way of doing this is to consider how the values of thermodynamic observables, get in laboratory experiments, vary with temperature, or volume, or an external field, and then to associate the experimentally observed discontinuities at a PT to the appearance of some kind of singularity entailing a loss of analyticity. Despite the smoothness of the statistical measures, after the *Yang–Lee theorem* [YL52] we know that in the $N \to \infty$ limit non–analytic behaviors of thermodynamic functions are possible whenever the analyticity radius in the complex fugacity plane shrinks to zero, because this entails the loss of *uniform convergence* in N (number of degrees of freedom) of any sequence of real–valued thermodynamic functions, and all this depends on the distribution of the zeros of the grand canonical partition function. Also the other developments of the rigorous theory of PTs [Geo88, Rue78], identify PTs with the loss of analyticity.

In this subsection we will address a recently proposed geometric approach to thermodynamic phase transitions (see [CCC97, FCS99 FPS00, FP04]). Given any Hamiltonian system, the configuration space can be equipped with a metric, in order to get a Riemannian geometrization of the dynamics. At the beginning, several numerical and analytical studies of a variety of models showed that the fluctuation of the curvature becomes singular at the transition point. Then the following conjecture was proposed in [CCC97]: The phase transition is determined by a change in the topology of the configuration space, and the loss of analyticity in the thermodynamic observables is nothing but a consequence of such topological change. The latter conjecture is also known as the *topological hypothesis*.

The topological hypothesis states that suitable topology changes of equipotential submanifolds of the Hamiltonian system's configuration manifold can entail thermodynamic phase transitions [FPS00]. The authors of the topological hypothesis gave both a theoretical argument and numerical demonstration in case of $2d$ lattice φ^4 model. They considered classical many–particle (or many–subsystem) systems described by standard *mechanical Hamiltonians*

$$H(p,q) = \sum_{i=1}^{N} \frac{p_i^2}{2m} + V(q), \qquad (4.221)$$

where the coordinates $q^i = q^i(t)$ and momenta $p_i = p_i(t)$, $(i = 1, ..., N)$, have continuous values and the system's potential energy $V(q)$ is bounded below.

Now, assuming a large number of subsystems N, the statistical behavior of physical systems described by Hamiltonians of the type (4.221) is usually encompassed, in the system's canonical ensemble, by the *partition function* in the system's phase–space

$$Z_N(\beta) = \int \prod_{i=1}^{N} dp_i dq^i e^{-\beta H(p,q)} = \left(\frac{\pi}{\beta}\right)^{\frac{N}{2}} \int \prod_{i=1}^{N} dq^i e^{-\beta V(q)}$$

$$= \left(\frac{\pi}{\beta}\right)^{\frac{N}{2}} \int_0^{\infty} dv\, e^{-\beta v} \int_{M_v} \frac{d\sigma}{\|\nabla V\|}, \qquad (4.222)$$

where the last term is written using a *co–area formula* [Fed69], and v labels the *equipotential hypersurfaces* M_v of the system's configuration manifold M,

$$M_v = \{(q^1, \ldots, q^N) \in \mathbb{R}^N | V(q^1, \ldots, q^N) = v\}. \qquad (4.223)$$

Equation (4.222) shows that for Hamiltonians (4.221) the relevant statistical information is contained in the *canonical configurational partition function*

$$Z_N^C = \int \prod_{i=1}^{N} dq^i \exp[-\beta V(q)].$$

Therefore, partition function Z_N^C is decomposed – in the last term of equation (4.222) – into an infinite summation of geometric integrals, $\int_{M_v} d\sigma / \|\nabla V\|$, defined on the $\{M_v\}_{v \in \mathbb{R}}$. Once the microscopic interaction potential $V(q)$ is given, the configuration space of the system is automatically foliated into the family $\{M_v\}_{v \in \mathbb{R}}$ of these equipotential hypersurfaces. Now, from standard statistical mechanical arguments we know that, at any given value of the inverse temperature β, the larger the number N of particles the closer to $M_v \equiv M_{u_\beta}$ are the microstates that significantly contribute to the averages – computed through $Z_N(\beta)$ – of thermodynamic observables. The hypersurface M_{u_β} is the one associated with the average potential energy computed at a given β,

$$
u_\beta = (Z_N^C)^{-1} \int \prod_{i=1}^{N} dq^i \, V(q) \exp[-\beta V(q)].
$$

Thus, at any β, if N is very large the effective support of the canonical measure shrinks very close to a single $M_v = M_{u_\beta}$.

Explicitly, the topological hypothesis reads: the basic origin of a phase transition lies in a suitable topology change of the $\{M_v\}$, occurring at some v_c. This topology change induces the singular behavior of the thermodynamic observables at a phase transition. By change of topology we mean that $\{M_v\}_{v<v_c}$ are not diffeomorphic to the $\{M_v\}_{v>v_c}$. In other words, canonical measure should 'feel' a big and sudden change of the topology of the equipotential hypersurfaces of its underlying support, the consequence being the appearance of the typical signals of a phase transition.

This point of view has the interesting consequence that – also at finite N – in principle *different* mathematical objects, i.e., manifolds of different cohomology type, could be associated to *different* thermodynamical phases, whereas from the point of view of measure theory [YL52] the only mathematical property available to signal the appearance of a phase transition is the loss of analyticity of the grand–canonical and canonical averages, a fact which is compatible with analytic statistical measures only in the mathematical $N \to \infty$ limit.

As it is conjectured that the counterpart of a phase transition is a breaking of diffeomorphicity among the surfaces M_v, it is appropriate to choose a *diffeomorphism invariant* to probe if and how the topology of the M_v changes as a function of v. This is a very challenging task because we have to deal with high dimensional manifolds. Fortunately a topological invariant exists whose computation is feasible, yet demands a big effort. Recall that this is the *Euler characteristic*, a diffeomorphism invariant of the system's configuration manifold, expressing its fundamental topological information.

4.4.2 Geometry of the Largest Lyapunov Exponent

Now, the topological hypothesis has recently been promoted into a *topological theorem* [FP04]. The new theorem says that non–analyticity is the 'shadow' of a more fundamental phenomenon occurring in the system's configuration manifold: a *topology change* within the family of equipotential hypersurfaces (4.223). This topological approach to PTs stems from the numerical study of the Hamiltonian dynamical counterpart of phase transitions, and precisely from the observation of *discontinuous* or *cuspy patterns*, displayed by the *largest Lyapunov exponent* at the transition energy (or temperature).

Recall that the *Lyapunov exponents* measure the strength of *dynamical chaos* and cannot be measured in laboratory experiments, at variance with thermodynamic observables, thus, being genuine dynamical observables they are only measurable in numerical simulations of the microscopic dynamics. To get a hold of the reason why the largest Lyapunov exponent λ_1 should probe configuration space topology, let us first remember that for standard Hamiltonian systems, λ_1 is computed by solving the *tangent dynamics equation* (or, Jacobi equation of geodesic deviation), for Hamiltonian systems,

$$\ddot{\xi}_i + \left(\frac{\partial^2 V}{\partial q^i \partial q^j}\right)_{q(t)} \xi^j = 0, \tag{4.224}$$

which, for the nonlinear Hamiltonian system

$$\dot{q}^1 = p_1, \qquad\qquad \dot{p}_1 = -\partial_{q^1} V,$$
$$\dots \qquad\qquad\qquad \dots$$
$$\dot{q}^N = p_N, \qquad\qquad \dot{p}_N = -\partial_{q^N} V,$$

expands into *linearized Hamiltonian dynamics*

$$\dot{\xi}_1 = \xi_{N+1}, \qquad\qquad \dot{\xi}_{N+1} = -\sum_{j=1}^{N}\left(\frac{\partial^2 V}{\partial q_1 \partial q_j}\right)_{q(t)}\xi_j,$$
$$\dots \qquad\qquad\qquad \dots \tag{4.225}$$
$$\dot{\xi}_n = \xi_{2N}, \qquad\qquad \dot{\xi}_{2N} = -\sum_{j=1}^{N}\left(\frac{\partial^2 V}{\partial q_N \partial q_j}\right)_{q(t)}\xi_j.$$

Using (4.224) we can get the analytical expression for the largest Lyapunov exponent

$$\lambda_1 = \lim_{t\to\infty}\frac{1}{t}\log\frac{\left[\xi_1^2(t)+\cdots+\xi_N^2(t)+\dot{\xi}_1^2(t)+\cdots+\dot{\xi}_N^2(t)\right]^{1/2}}{\left[\xi_1^2(0)+\cdots+\xi_N^2(0)+\dot{\xi}_1^2(0)+\cdots+\dot{\xi}_N^2(0)\right]^{1/2}}. \tag{4.226}$$

If there are critical points of V in configuration space, that is points $q_c = [\bar{q}^1, \dots, \bar{q}^N]$ such that $\nabla V(q)|_{q=q_c} = 0$, according to the *Morse lemma* (see e.g., [Hir76]), in the neighborhood of any critical point q_c there always exists a coordinate system $\tilde{q}(t) = [\bar{q}^1(t), \dots, \bar{q}^N(t)]$ for which

$$V(\tilde{q}) = V(q_c) - \left(\overline{q}^1\right)^2 - \cdots - \left(\overline{q}^k\right)^2 + \left(\overline{q}^{k+1}\right)^2 + \cdots + \left(\overline{q}^N\right)^2 , \qquad (4.227)$$

where k is the index of the critical point, i.e., the number of negative eigenvalues of the Hessian of V. In the neighborhood of a critical point, equation (4.227) yields

$$\partial^2 V/\partial q^i \partial q^j = \pm \delta_{ij},$$

which, substituted into equation (4.224), gives k *unstable directions* which contribute to the exponential growth of the norm of the tangent vector $\xi = \xi(t)$. This means that the strength of dynamical chaos, measured by the largest Lyapunov exponent λ_1, is affected by the existence of critical points of V. In particular, let us consider the possibility of a sudden variation, with the potential energy v, of the number of critical points (or of their indexes) in configuration space at some value v_c, it is then reasonable to expect that the pattern of $\lambda_1(v)$ – as well as that of $\lambda_1(E)$ since $v = v(E)$ – will be consequently affected, thus displaying *jumps* or *cusps* or other *singular patterns* at v_c.

On the other hand, recall that *Morse theory* teaches us that the existence of critical points of V is associated with topology changes of the hypersurfaces $\{M_v\}_{v\in\mathbb{R}}$, provided that V is a good *Morse function* (that is: bounded below, with no vanishing eigenvalues of its Hessian matrix). Thus the existence of critical points of the potential V makes possible a conceptual link between dynamics and configuration space topology, which, on the basis of both direct and indirect evidence for a few particular models, has been formulated as a topological hypothesis about the relevance of topology for PTs phenomena (see [FPS00, FP04, GM04]).

Here we give two simple examples of standard Hamiltonian systems of the form (4.221), namely Peyrard–Bishop system and mean–field XY model.

Peyrard–Bishop Hamiltonian System

The *Peyrard–Bishop system* [PB89][12] exhibits a *second–order phase transition*. It is defined by the following potential energy

$$V(q) = \sum_{i=1}^{N} \left[\frac{K}{2}(q^{i+1} - q^i)^2 + D(e^{-aq^i} - 1)^2 + Dhaq^i \right], \qquad (4.228)$$

which represents the energy of a string of N base pairs of reduced mass m. Each hydrogen bond is characterized by the stretching q^i and its conjugate momentum $p_i = m\dot{q}^i$. The elastic transverse force between neighboring pairs is tuned by the constant K, while the energy D and the inverse length a determine, respectively, the plateau and the narrowness of the on–site potential well that mimics the interaction between bases in each pair. It is understood

---

[12] The Peyrard–Bishop system has been proposed as a simple model for describing the DNA thermally induced denaturation [GM04].

that K, D, and a are all positive parameters. The transverse, external stress $h \geq 0$ is a computational tool useful in the evaluation of the susceptibility. Our interest in it lies in the fact that a phase transition can occur only when $h = 0$. We assume periodic boundary conditions.

The transfer operator technique [DTP02] maps the problem of computing the classical partition function into the easier task of evaluating the lowest energy eigenvalues of a 'quantum' mechanical Morse oscillator (no real quantum mechanics is involved, since the temperature plays the role of $\hbar$). One can then observe that, as the temperature increases, the number of levels belonging to the discrete spectrum decreases, until for some critical temperature $T_c = 2\sqrt{2KD}/(ak_B)$ only the continuous spectrum survives. This passage from a localized ground state to an unnormalizable one corresponds to the second–order phase transition of the statistical model. Various critical exponents can be analytically computed and all applicable scaling laws can be checked. The simplicity of this model permits an analytical computation of the largest Lyapunov exponent by exploiting the geometric method proposed in [CCC97].

Mean–Field XY Hamiltonian System

The mean–field XY model describes a system of N equally coupled planar classical rotators (see [AR95, CCP99]). It is defined by a Hamiltonian of the class (4.221) where the potential energy is

$$V(\varphi) = \frac{J}{2N} \sum_{i,j=1}^{N} \left[1 - \cos(\varphi_i - \varphi_j) \right] - h \sum_{i=1}^{N} \cos\varphi_i. \qquad (4.229)$$

Here $\varphi_i \in [0, 2\pi]$ is the rotation angle of the i−th rotator and h is an external field. Defining at each site i a classical spin vector $\mathbf{s}_i = (\cos\varphi_i, \sin\varphi_i)$ the model describes a planar (XY) Heisenberg system with interactions of equal strength among all the spins. We consider only the ferromagnetic case $J > 0$; for the sake of simplicity, we set $J = 1$. The equilibrium statistical mechanics of this system is exactly described, in the thermodynamic limit, by the *mean–field theory* [AR95]. In the limit $h \to 0$, the system has a continuous phase transition, with classical critical exponents, at $T_c = 1/2$, or $\varepsilon_c = 3/4$, where $\varepsilon = E/N$ is the energy per particle.

The Lyapunov exponent λ_1 of this system is extremely sensitive to the phase transition. According to reported numerical simulations (see [CCP99]), $\lambda_1(\varepsilon)$ is positive for $0 < \varepsilon < \varepsilon_c$, shows a sharp maximum immediately below the critical energy, and drops to zero at ε_c in the thermodynamic limit, where it remains zero in the whole region $\varepsilon > \varepsilon_c$, which corresponds to the thermodynamic disordered phase. In fact in this phase the system is integrable, reducing to an assembly of un–coupled rotators.

4.4.3 Euler Characteristics of Hamiltonian Systems

Recall that *Euler characteristic* χ is a number that is a characterization of the various classes of geometric figures based only on the topological relationship between the numbers of vertices V, edges E, and faces F, of a geometric Figure. This number, $\chi = F - E + V$, is the same for all figures the boundaries of which are composed of the same number of connected pieces. Therefore, the Euler characteristic is a *topological invariant*, i.e., any two geometric figures that are homeomorphic to each other have the same Euler characteristic.

More specifically, a standard way to analyze a geometric Figure is to fragment it into other more familiar objects and then to examine how these pieces fit together. Take for example a surface M in the Euclidean 3D space. Slice M into pieces that are curved triangles (this is called a triangulation of the surface). Then count the number F of faces of the triangles, the number E of edges, and the number V of vertices on the tesselated surface. Now, no matter how we triangulate a compact surface Σ, its Euler characteristic, $\chi(\Sigma) = F - E + V$, will always equal a constant which is characteristic of the surface and which is invariant under diffeomorphisms $\phi : \Sigma \rightarrow \Sigma'$.

At higher dimensions this can be again defined by using higher dimensional generalizations of triangles (simplexes) and by defining the Euler characteristic $\chi(M)$ of the nD manifold M to be the alternating sum:

{number of points} − {number of 2-simplices}   +

{number of 3-simplices} − {number of 4-simplices}   +   ...

i.e.,

$$\chi(M) = \sum_{k=0}^{n} (-1)^k (\text{number of faces of dimension } k). \qquad (4.230)$$

and then define the Euler characteristic of a manifold as the Euler characteristic of any simplicial complex homeomorphic to it. With this definition, circles and squares have Euler characteristic 0 and solid balls have Euler characteristic 1.

The Euler characteristic χ of a manifold is closely related to its *genus g* as $\chi = 2 - 2g$.[13]

Recall that in differential topology a more standard definition of $\chi(M)$ is

---

[13] Recall that the *genus* of a topological space such as a surface is a topologically invariant property defined as the largest number of nonintersecting simple closed curves that can be drawn on the surface without separating it, i.e., an integer representing the maximum number of cuts that can be made through it without rendering it disconnected. This is roughly equivalent to the number of holes in it, or handles on it. For instance: a point, line, and a sphere all have genus 0; a torus has genus 1, as does a coffee cup as a solid object (solid torus), a Möbius strip, and the symbol 0; the symbols 8 and B have genus 2; etc.

$$\chi(M) = \sum_{k=0}^{n} (-1)^k b_k(M), \qquad (4.231)$$

where b_k are the kth *Betti numbers* of M.

In general, it would be hopeless to try to practically calculate $\chi(M)$ from (4.231) in the case of non–trivial physical models at large dimension. Fortunately, there is a possibility given by the *Gauss–Bonnet formula*, that relates $\chi(M)$ with the total *Gauss–Kronecker curvature* of the manifold,

$$\chi(M) = \gamma \int_M K_G \, d\sigma, \qquad (4.232)$$

which is valid for even dimensional hypersurfaces of Euclidean spaces $\mathbb{R}^N$ [here $\dim(M) = n \equiv N - 1$], and where:

$$\gamma = 2/\operatorname{vol}(S_1^n)$$

is twice the inverse of the volume of an $n-$dimensional sphere of unit radius S_1^n; K_G is the Gauss–Kronecker curvature of the manifold;

$$d\sigma = \sqrt{\det(g)} \, dx^1 dx^2 \cdots dx^n$$

is the invariant volume measure of M and g is its Riemannian metric (induced from $\mathbb{R}^N$). Let us briefly sketch the meaning and definition of the Gauss–Kronecker curvature. The study of the way in which an $n-$surface M curves around in $\mathbb{R}^N$ is measured by the way the normal direction changes as we move from point to point on the surface. The rate of change of the normal direction ξ at a point $x \in M$ in direction v is described by the *shape operator*

$$L_x(v) = -\mathcal{L}_v \xi = [v, \xi],$$

where v is a tangent vector at x and $\mathcal{L}_v$ is the Lie derivative, hence

$$L_x(v) = -(\nabla \xi_1 \cdot v, \dots, \nabla \xi_{n+1} \cdot v);$$

gradients and vectors are represented in $\mathbb{R}^N$. As L_x is an operator of the tangent space at x into itself, there are n independent eigenvalues $\kappa_1(x), \dots, \kappa_n(x)$ which are called the principal curvatures of M at x [Tho79]. Their product is the Gauss–Kronecker curvature:

$$K_G(x) = \prod_{i=1}^{n} \kappa_i(x) = \det(L_x).$$

Alternatively, recall that according to the *Morse theory*, it is possible to understand the topology of a given manifold by studying the regular critical points of a smooth *Morse function* defined on it. In our case, the manifold M is the configuration space $\mathbb{R}^N$ and the natural choice for the Morse function is the potential $V(q)$. Hence, one is lead to define the family M_v (4.223) of submanifolds of M.

A full characterization of the topological properties of M_v generally requires the critical points of $V(q)$, which means solving the equations

$$\partial_{q^i} V = 0, \qquad (i = 1, \dots, N). \tag{4.233}$$

Moreover, one has to calculate the indexes of all the critical points, that is the number of negative eigenvalues of the Hessian $\partial^2 V/(\partial q^i \partial q_j)$. Then the Euler characteristic $\chi(M_v)$ can be computed by means of the formula

$$\chi(M_v) = \sum_{k=0}^{N} (-1)^k \mu_k(M_v), \tag{4.234}$$

where $\mu_k(M_v)$ is the total number of critical points of $V(q)$ on M_v which have index k, i.e., the so–called *Morse numbers* of a manifold M, which happen to be upper bounds of the Betti numbers,

$$b_k(M) \leq \mu_k(M) \qquad (k = 0, \dots, n). \tag{4.235}$$

Among all the Morse functions on a manifold M, there is a special class, called *perfect Morse functions*, for which the Morse inequalities (4.235) hold as equalities. Perfect Morse functions characterize completely the topology of a manifold.

Now, we continue with our two examples started before.

Peyrard–Bishop System

If applied to any generic model, calculation of (4.234) turns out to be quite formidable, but the exceptional simplicity of the Peyrard–Bishop model (4.228) makes it possible to carry on completely the topological analysis without invoking equation (4.234).

For the potential in exam, equation (4.233) results in the nonlinear system

$$\frac{a}{R}(q^{i+1} - 2q^i + q^{i-1}) = h - 2(e^{-2aq^i} - e^{-aq^i}),$$

where $R = Da^2/K$ is a dimensionless ratio. It is easy to verify that a particular solution is given by

$$q^i = -\frac{1}{a} \ln \frac{1 + \sqrt{1 + 2h}}{2}, \qquad (i = 1, \dots, N).$$

The corresponding minimum of potential energy is

$$V_{\min} = ND \left(\frac{1 + h - \sqrt{1 + 2h}}{2} - h \ln \frac{1 + \sqrt{1 + 2h}}{2} \right).$$

Mean–Field XY Model

In the case of the mean–field XY model (4.229) it is possible to show analytically that a topological change in the configuration space exists and that it can be related to the thermodynamic phase transition. Consider again the family M_v of submanifolds of the configuration space defined in (4.223); now the potential energy per degree of freedom is that of the mean–field XY model, i.e.,

$$\mathcal{V}(\varphi) = \frac{V(\varphi)}{N} = \frac{J}{2N^2} \sum_{i,j=1}^{N} \left[1 - \cos(\varphi_i - \varphi_j) \right] - h \sum_{i=1}^{N} \cos \varphi_i,$$

where $\varphi_i \in [0, 2\pi]$. Such a function can be considered a Morse function on M, so that, according to Morse theory, all these manifolds have the same topology until a critical level $\mathcal{V}^{-1}(v_c)$ is crossed, where the topology of M_v changes.

A change in the topology of M_v can only occur when v passes through a critical value of $\mathcal{V}$. Thus in order to detect topological changes in M_v we have to find the critical values of $\mathcal{V}$, which means solving the equations

$$\partial_{\varphi_i} \mathcal{V}(\varphi) = 0, \qquad (i = 1, \dots, N). \tag{4.236}$$

For a general potential energy function $\mathcal{V}$, the solution of (4.236) would be a formidable task, but in the case of the mean–field XY model, the mean–field character of the interaction greatly simplifies the analysis, allowing an analytical treatment of (4.236); moreover, a projection of the configuration space onto a 2D plane is possible [CCP99, CPC03].

4.4.4 Pathways to Self–Organization in Human Biodynamics

Here we explore the pathways to self–organization in human biodynamics[14] (see Figure 4.4.4), following the general principles of synergetics [Hak96, Hak00].

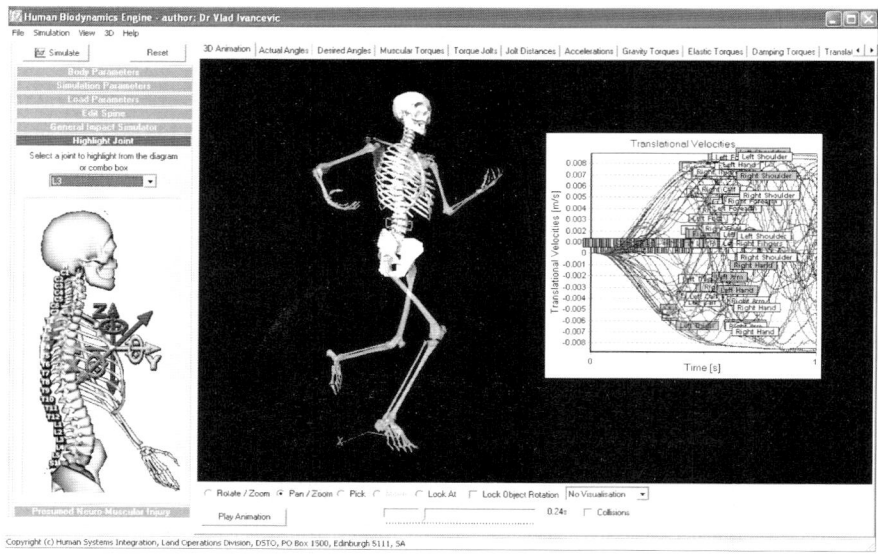

Fig. 4.3. Sample output from the Human Biodynamics Engine: running with the speed of 5 m/s.

---

[14] *Human Biodynamics Engine* (HBE) is a generalized Hamiltonian system with 264 DOF, including 132 rotational DOF (considered active) and 132 translational DOF (considered passive). Passive joint dynamics models visco–elastic properties of intervertebral discs, joint tendons and muscle ligaments as a nonlinear spring–damper system. Active joint dynamics is driven by 264 *nonlinear muscular actuators*, each with its own excitation–contraction dynamics (following traditional biomechanical models) and two–level neural–like control. The lower control level resembles spinal–reflex positive and negative force feedbacks (stretch and Golgi reflexes). The higher control level resembles cerebellum's postural stabilization and velocity control (modelled as a high–dimensional Lie–derivative controller). The HBE includes over 2000 body parameters, all derived from individual user data, using standard biomechanical tables. It models stabilizing and reaching body movements at a spot, walking and running with any speed and a generic crash simulator. The HBE incorporates a new theory of soft neuro–musculo–skeletal injuries, much more sensitive than the traditional Principal Loading Hypothesis (of tension, compression, bending and shear) for spinal and other neuro–musculo–skeletal injuries. It is based on the concept of the local Jolts and Torque–Jolts, which are the time derivatives of the total forces and torques localized in each joint at a particular time instant.

Self–Organization Through Change of Control Parameters

We may change the global impact of the surroundings on a certain biody-namic system, as expressed by its control parameters $\{\sigma\}$ (e.g., changing a surrounding temperature $\sigma = T$ causes faster muscular contractions [Hil38]). When we slowly change the impact of the surroundings on the system, at certain points the system may acquire new states of higher order or structure (e.g., a 'warm up' of the human muscular system [II06a]).

Self–Organization Through Change of Components

Self–organization can be caused by a mere increase of number of components of a certain biodynamic system (e.g., activating more muscular–like actua-tors enables more sophisticated motion patterns [II06a]). Even if we put the same microscopic components together, entirely new behavior can arise on the macroscopic level (e.g., reducing the basic biodynamics model (1.156–1.157) to the macro–model (4.39)).

Self–Organization Through Transients

Self–organization can be also caused by a sudden change of control parameters when a certain biodynamic system (e.g., muscular–like actuator) tries to relax to a new state under the new conditions (constraints). Patterns in this system may be formed in a self–organized fashion when it passes from an initial disordered (or homogenous) state to another final state which we need not specify or which even does not exist. For example, if we have a musculo–skeletal–like state vector of the form

$$q(x, t) = u(t)v(x),$$

where $v(x)$ describes some muscular–like spatial order, and the order param-eter equation

$$\dot{o} = \lambda o.$$

When we change a control parameter σ quickly, so that $\lambda < 0$ is quickly replaced by $\lambda > 0$, we get a transient state vector of the form

$$q(x, t) = \exp(\lambda t)\, v(x). \tag{4.237}$$

It clearly describes some structure, but it does not tend to a new stable state.

Now, to get the solution (4.237) started, some *fluctuations* must be present. Otherwise, the solution $o \equiv 0$, and thus $q \equiv 0$ will persist forever.

Self–Organization and Neural Networks

The basic synergetic idea of self–organization by reduction of the system's dimension has been fully exploited in T. Kohonen's neural network model, the so–called *self–organizing map* (SOM). The SOM transforms the network's input of arbitrary dimension into a one or 2D discrete map subject to a topo-logical (neighborhood preserving) constraint. The SOM is computed using Kohonen's special form of *unsupervised learning*. The output of the SOM can be used as input to a supervised classification feedforward network. This net-work's key advantage is the clustering produced by the SOM which reduces the input space into representative features using a self–organizing process. Hence the underlying structure of the input space is kept, while the dimensionality of the space is reduced.

5

Phase Synchronization in Chaotic Systems

This Chapter deals with phase *synchronization* in high–dimensional chaotic systems.

Recall that synchronization phenomena occur abundantly in nature and in day to day life. A few well known examples are the observations in coupled systems such as pendulum clocks, radio circuits, swarms of light-emitting fireflies, groups of neurons and neuronal ensembles in sensory systems, chemical systems, Josephson junctions, cardiorespiratory interactions, etc. Starting from the observation of pendulum clocks by Huygens, a vast literature already exists which studies synchronization in coupled nonlinear systems – in systems of coupled maps as well as in oscillators and networks (see [PRK01] and references therein). In recent times, different kinds of synchronization have been classified – mutual synchronization, lag synchronization, phase synchronization and complete synchronization (see [RPK96, Bal06]).

5.1 Lyapunov Vectors and Lyapunov Exponents

In this section, following [PGY06], we discuss a method to determine *Lyapunov exponents* (LEs) from suitable ensemble averages. It is easy to write down a formal meaningful definition, but the problem lies in translating it into a workable procedure. With reference to an ND discrete–time system, given by the *map*

$$\mathbf{x}_{t+1} = \mathbf{f}_d(\mathbf{x}_t), \qquad (\mathbf{x} \in \mathbb{R}^N), \tag{5.1}$$

one can express the ith LE (as usual, LE are supposed to be ordered from the largest to the smallest one) as

$$\lambda^{(i)} = \frac{1}{2} \int d\mathbf{x} P(\mathbf{x}) \ln \left[\frac{||\partial_x \mathbf{f}_d \mathbf{V}^{(i)}(\mathbf{x})||^2}{||\mathbf{V}^{(i)}(\mathbf{x})||^2} \right] \tag{5.2}$$

where $P(\mathbf{x})$ is the corresponding invariant measure, $\partial_x \mathbf{f}_d$ is the Jacobian of the transformation, and the Lyapunov vector $\mathbf{V}^{(i)}(\mathbf{x})$ identifies the ith most expanding direction in $\mathbf{x}$.

With reference to a continuous–time system, ruled by the ODE

$$\dot{\mathbf{x}} = \mathbf{f}_c(\mathbf{x}), \qquad (\mathbf{x} \in \mathbb{R}^N). \tag{5.3}$$

the ith LE is defined by

$$\lambda^{(i)} = \int d\mathbf{x} P(\mathbf{x}) \frac{[\partial_x \mathbf{f}_c \mathbf{V}^{(i)}(\mathbf{x})] \cdot \mathbf{V}^{(i)}(\mathbf{x})}{||\mathbf{V}^{(i)}(\mathbf{x})||^2}, \tag{5.4}$$

where $\cdot$ denotes the scalar product.

Unless a clear procedure to determine the LV is given, (5.2,5.4) are nothing but formal statements. As anticipated in the introduction, $\mathbf{V}^{(i)}(\mathbf{x})$ can be obtained by following a two–step procedure. We start with a generic set of i linearly independent vectors lying in the tangent space and let them evolve in time. This is the standard procedure to determine LEs, and it is well known that the hyper–volume $\mathbf{Y}^{(i)}$ identified by such vectors contains for, large enough times, the i most expanding directions. Furthermore, with reference to the set of orthogonal cordinates got by implementing the *Gram–Schmidt procedure*, the component v_k of a generic vector $\mathbf{v}$ evolves according to the following ODE [EP98]

$$\dot{v}_k = \sum_{j=k}^{\imath} \sigma_{k,j}(\mathbf{x}) v_j, \qquad (1 \le k \le i), \tag{5.5}$$

where $\sigma_{k,j}$ does not explicitly depend on time, but only through the position $\mathbf{x}$ in the phase–space. As a result, the ith Lyapunov exponent can be formally expressed as the ensemble average of the local expansion rate $\sigma_{i,i}$, i.e.,

$$\lambda^{(i)} = \int d\mathbf{x} P(\mathbf{x}) \sigma_{i,i}(\mathbf{x}). \tag{5.6}$$

By comparing with (5.4), one finds the obvious equality

$$\sigma_{i,i} = \frac{[\partial_x \mathbf{f}_c \mathbf{V}^{(i)}(\mathbf{x})] \cdot \mathbf{V}^{(i)}(\mathbf{x})}{||\mathbf{V}^{(i)}(\mathbf{x})||^2}. \tag{5.7}$$

In subsection 5.1.2 below, we will apply this formalism to a phase–synchronization problem, and we will find that the only workable way to get an analytic expression for $\sigma_{i,i}$ passes through the determination of the direction of the corresponding LV vector $\mathbf{V}^{(i)}(\mathbf{x})$.

Let us now consider the *backward evolution* of a generic vector $\mathbf{V}^{(i)} \in \mathbf{Y}^{(i)}$. Its direction is identified by the $(i-1)$D vector

$$\mathbf{u} \equiv (u_1, u_2, \ldots, u_{i-1}), \tag{5.8}$$

where $u_k = v_k/v_i$. From (5.5) and the definition of $\mathbf{u}$, it follows that the backward evolution follows the equation

$$\dot{u}_k = (\sigma_{i,i} - \sigma_{k,k})u_k - \sum_{j=k+1}^{i-1} \sigma_{k,j}(t)u_j - \sigma_{k,i}, \qquad (1 \leq k < i). \qquad (5.9)$$

This is a cascade of skew–product linear stable equations (they are stable because the Lyapunov exponents are organized in descending order). The overall stability is basically determined by the smallest $(\sigma_{k,k} - \sigma_{i,i})$ that is got for $k = i - 1$. It is, therefore, sufficient to turn our attention to the last $(i - 1)$ component of the vector $\mathbf{V}$. Its equation has the following structure

$$\dot{u}(t) = \gamma u + \sigma(t), \qquad (5.10)$$

where $\gamma = \lambda_i - \lambda_{i-1} < 0$ and we have dropped the subscript i for simplicity. The value of the direction u is got by integrating this equation. By neglecting the temporal fluctuations of γ (it is not difficult to include them, but this is not important for our final goal), the formal solution of (5.10) reads

$$u(\mathbf{x}(t)) = \int_{-\infty}^{t} e^{\gamma(t-\tau)}\sigma(\mathbf{x})\, d\tau. \qquad (5.11)$$

This equation does not simply tell us the value of u at time t, but the value of u when the trajectory sits in $\mathbf{x}(\mathbf{t})$. It is in fact important to investigate the dependence of u on $\mathbf{x}$. We proceed by determining the deviation $\delta_j u$ induced by a perturbation δx_j of $\mathbf{x}$ along the jth direction,

$$\delta_j u = \int_{-\infty}^{t} e^{\gamma(t-\tau)}\delta_j\sigma(\tau)\, d\tau, \qquad (5.12)$$

where, assuming a smooth dependence of σ on $\mathbf{x}$,

$$\delta_j\sigma(\tau) \approx \sigma_x(\tau)\delta x_j(\tau) = \sigma_x(\tau)\delta x_j(t)e^{\lambda_j(t-\tau)} \qquad (5.13)$$

(notice that the dynamics is flowing backward). If the Lyapunov exponent λ_j is negative, $\delta_j\sigma(\tau)$ decreases for $\tau \to -\infty$ and the integral over τ in (5.12) converges. As a result, $\delta_j u$ is proportional to δx_j, indicating that the direction of the LV is smooth along the jth direction. If λ_j is positive, $\delta_j\sigma(\tau)$ diverges, and below time t_0, where

$$\delta x_j(t)e^{\lambda_j(t-t_0)} = 1, \qquad (5.14)$$

linearization breaks down. In this case, $\delta\sigma(\tau)$ for $\tau < t_0$ is basically uncorrelated with its 'initial value' $\delta_j\sigma(t)$ and one can estimate $\delta_j u$, by limiting the integral to the range $[t_0, t]$

$$\delta_j u(t) = \delta x_j(t) \int_{t_0}^{t} d\tau e^{(\lambda_j+\gamma)(t-\tau)}\sigma_x(\tau). \qquad (5.15)$$

where t_0 is given by (5.14). By bounding σ_x with constant functions and thereby performing the integral in (5.15), we finally get

$$\delta_j u(t) \approx \delta x_j(t) + \delta x_j(t)^{-\gamma/\lambda_j}. \qquad (5.16)$$

The scaling behavior is finally got as the smallest number between 1 and -γ/λ_j. If we now introduce the exponent η_j to identify the scaling behavior of the deviation of the LV direction when the point of reference is moved along the jth direction in phase–space, the results are summarized in the following way

$$\eta_j = \begin{cases} 1, & \text{for } \lambda_j \le -\gamma, \\ -\gamma/\lambda_j, & \text{for } \lambda_j > -\gamma. \end{cases} \qquad (5.17)$$

The former case corresponds to a smooth behavior (the derivative is finite), while the latter one reveals a singular behavior that is the signature of a generalized synchronization.

5.1.1 Forced Rössler Oscillator

The first model where phase synchronization has been explored is the periodically *forced Rössler oscillator* [RPK96]. In this section we derive a discrete–time map describing a forced Rössler system in the limit of weak coupling [PZR97]. We start with the ODE,

$$\begin{aligned} \dot{x} &= -y - z + \varepsilon y \cos(\Omega t + \psi_0), \\ \dot{y} &= x + a_0 y - \varepsilon x \sin(\Omega t + \psi_0), \\ \dot{z} &= a_1 + z(x - a_2), \end{aligned} \qquad (5.18)$$

where ψ_0 fixes the phase of the forcing term at time 0. It is convenient to introduce cylindrical coordinates, namely $\mathbf{u} = (\varphi, r, z)$, ($x = r\cos\phi$, $y = r\sin\phi$). For the future sake of clarity, let us denote with $\mathbf{S_c}$ the 3D space parametrized by such coordinates, so that (5.18) reads

$$\dot{\mathbf{u}} = \mathbf{F}(\mathbf{u}) + \varepsilon \mathbf{G}(\mathbf{u}, \Omega t + \psi_0), \qquad \text{where} \qquad (5.19)$$

$$\begin{aligned} \mathbf{F} &= [1 + \frac{z}{r}\sin\phi + \frac{a_0}{2}\sin 2\phi, \, a_0 r \sin^2\phi - z\cos\phi, \, a_1 + z(r\cos\phi - a_2),] \\ \mathbf{G} &= [-\sin^2\phi \cos(\Omega t + \psi_0) - \cos^2\phi \sin(\Omega t + \psi_0), \\ &\quad \frac{r}{\sqrt{2}}\sin 2\phi \cos(\Omega t + \psi_0 + \pi/4), \, 0]. \end{aligned}$$

Note that system (5.19) can be written in the equivalent autonomous form

$$\dot{\mathbf{u}} = \mathbf{F}(\mathbf{u}) + \varepsilon \mathbf{G}(\mathbf{u}, \psi), \qquad \dot{\psi} = \Omega,$$

where ψ denotes the phase of the forcing term.

We pass to a discrete–time description, by monitoring the system each time the phase ϕ is a multiple of 2π. In the new framework, the relevant variables are r, z, and ψ, all measured when the Poincaré section is crossed. The task is to determine the transformation map the state (r, z, ψ) onto (r', z', ψ').

In order to get the expression of the map, it is necessary to formally integrate the equations of motion from one to the next section. This can be done, by expanding around the un–perturbed solution for $\varepsilon = 0$ (which must nevertheless be obtained numerically). The task is anyhow worth, because it allows determining the structure of the resulting map, which turns out to be [PGY06]

$$\psi' = \psi + \langle T^{(0)} \rangle \, \Omega + A_1 + \varepsilon \left(B_1^c \cos \psi + B_1^s \sin \psi \right),$$
$$r' = A_2 + \varepsilon \left(B_2^c \cos \psi + B_2^s \sin \psi \right), \tag{5.20}$$
$$z' = A_3 + \varepsilon \left(B_3^c \cos \psi + B_3^s \sin \psi \right),$$

where $\langle T^{(0)} \rangle$ is the average period of the un–perturbed Rössler oscillator and A_m's and B_m's are functions of z and r. They can be numerically determined by integrating the appropriate set of equations. Up to first order in ε, the structure of the model is fairly general as it is got for a generic periodically forced oscillator represented in cylindrical coordinates (as long the phase of the attractor can be unambiguously identified).

For the usual parameter values, the Rössler attractor is characterized by a strong contraction along one direction [YML00]. As a result, one can neglect the z dependence since this variable is basically a function of r, and thus write

$$\psi' = \psi + \langle T^{(0)} \rangle \, \Omega + A_1(r) + \varepsilon \left(B_1^c(r) \cos \psi + B_1^s(r) \sin \psi \right),$$
$$r' = A_2(r) + \varepsilon \left(B_2^c(r) \cos \psi + B_2^s(r) \sin \psi \right), \tag{5.21}$$

where all the functions can be obtained by integrating numerically the equations of motion of the single Rössler oscillator.

To simplify further manipulations, we finally recast equation (5.21) in the form

$$\psi' = \psi + K + A_1(r) + \varepsilon g_1(r) \cos \left(\psi + \beta_1(r) \right), \tag{5.22}$$
$$r' = A_2(r) + \varepsilon \, g_2(r) \cos \left(\psi + \beta_2(r) \right), \qquad \text{where}$$
$$B_i^c(r) = g_i(r) \cos \beta_i(r), \qquad B_i^s(r) = -g_i(r) \sin \beta_i(r)$$

for $i = 1, 2$. The parameter $K = \langle T^{(0)} \rangle \, \Omega - 2\pi$ represents the detuning between the original Rössler–system average frequency and the forcing frequency Ω.

The GSF (5.22) generalizes the model introduced in [PZR97], where the effect of the phase on the r dynamics was not included. This implies that the GSF looses the skew–product structure. This has important consequences on the orientation of the second Lyapunov vector that we determine in the next sections. Notice also that the GSF (5.22) generalizes and justifies the model invoked in [POR97].

For the sake of simplicity, we have analyzes the following model,

$$r' = f(r) + 2\varepsilon c g(r)\cos(\psi + \alpha)$$
$$\psi' = \psi + K + \Delta r + \varepsilon b \cos\psi, \qquad \text{where} \qquad (5.23)$$
$$f(r) = 1 - 2|r|, \quad g(r) = r^2 - |r|,$$

with $r \in [-1,1]$. The tent–map choice for r ensures that $[-1,0]$ and $[0,1]$ are the two atoms of a Markov partition. Moreover, since $g(r)$ is equal to 0 for $r = 0$ and $r = \pm 1$, this remains true also when the perturbation is switched on.

In this 2D setup, the formal expression of the ith LE (5.2) reads

$$\lambda^{(i)} = \frac{1}{2}\int_{-1}^{1} dr \int_{0}^{2\pi} d\psi P(r,\psi)\ln\left[\frac{||\mathbf{J}(r,\psi)\mathbf{V}^{(i)}(r,\psi)||^2}{||\mathbf{V}^{(i)}(r,\psi)||^2}\right], \qquad (5.24)$$

and the Jacobian is given by

$$\mathbf{J}(r,\psi) = \begin{pmatrix} f_r(r) + 2\varepsilon c g_r(r)\cos(\psi + \alpha) & -2\varepsilon c g(r)\sin(\psi + \alpha) \\ \Delta & 1 - \varepsilon b \sin\psi \end{pmatrix},$$

where the subscript r denotes the derivative with respect to r. The computation of the Lyapunov exponent therefore, requires determining both the invariant measure $P(r,\psi)$ and the local direction of the Lyapunov vector $\mathbf{V}^{(i)}$.

5.1.2 Second Lyapunov Exponent: Perturbative Calculation

Here we derive a perturbative expression for the second LE of the GSF (5.23), by expanding (5.24). One of the key ingredients is the second LV, whose direction can be identified by writing $\mathbf{V} = (V, 1)$ (for the sake of clarity, from now on, we omit the superscript $i = 2$ in $\mathbf{V}$ and λ, as we shall refer only to the second direction). Due to the skew-product structure of the un–perturbed map (5.23), the second LV is, for $\varepsilon = 0$, aligned along the ψ direction (i.e., $V = 0$). It is therefore natural to expand V in powers of ε

$$V \approx \varepsilon v_1(r,\psi) + \varepsilon^2 v_2(r,\psi). \qquad (5.25)$$

Accordingly, the logarithm of the norm of $\mathbf{V}$ is

$$\ln ||\mathbf{V}||^2 = \ln(1 + \varepsilon^2 v_1^2) = \varepsilon^2 v_1^2,$$

while its forward iterate writes as (including only those terms that contribute up to second order in the norm),

$$\mathbf{JV} = \begin{pmatrix} \varepsilon f_r(r)v_1 - 2c\varepsilon g(r)\sin(\psi + \alpha) \\ 1 - \varepsilon(\Delta v_1 - b\sin\psi) + \varepsilon^2 \Delta v_2) \end{pmatrix}, \qquad (5.26)$$

Notice that we have omitted the (r,ψ) dependence of v_1 and v_2 to keep the notation compact [PGY06].

The Euclidean norm of the forward iterate is

$$\|\mathbf{JV}\|^2 = 1 + 2\varepsilon(\Delta v_1 - b\sin\psi) + \varepsilon^2\{(\Delta v_1 - b\sin\psi)^2$$
$$+ 2\Delta v_2 + [f_r(r)v_1 - 2cg(r)\sin(\psi + \alpha)]^2\},$$

and its logarithm is

$$\ln\|\mathbf{JV}\|^2 = 2\varepsilon(\Delta v_1 - b\sin\psi) - \varepsilon^2\{(\Delta v_1 - b\sin\psi)^2$$
$$- 2\Delta v_2 - [f_r(r)v_1 - 2cg(r)\sin(\psi + \alpha)]^2\}.$$

We now proceed by formally expanding the invariant measure in powers of ε

$$P(r, \psi) \approx p_0(\psi) + \varepsilon p_1(r, \psi) + \varepsilon^2 p_2(r, \psi). \tag{5.27}$$

The determination of the p_i coefficients is presented in the next section, but here we anticipate that, as a consequence of the skew-product structure for $\varepsilon = 0$, the zeroth-order component of the invariant measure does not depend on the phase ψ. Moreover, because of the structure of the tent-map, p_0 is also independent of r, i.e., $p_0 = 1/4\pi$. The second Lyapunov exponent can thus be written as

$$\lambda = \int_{-1}^{1} dr \int_{0}^{2\pi} d\psi (\frac{1}{4\pi} + \varepsilon p_1(r, \psi))\{2\varepsilon(\Delta v_1(r, \psi) \tag{5.28}$$
$$- b\sin\psi) - \varepsilon^2[(\Delta v_1(r, \psi) - b\sin\psi)^2 - 2\Delta v_2(r, \psi)$$
$$+ [f_r(r)v_1(r, \psi) + 2cg(r)\sin(\psi + \alpha)]^2 + v_1^2(r, \psi)]\} + o(\varepsilon^2).$$

As the variable ψ is a phase, it is not a surprise that some simplifications can be found by expanding the relevant functions into Fourier components. We start writing the first component of the invariant measure as

$$p_1(r, \psi) = \frac{1}{2\pi} \sum_n q_i(r) e^{in\psi}. \tag{5.29}$$

We then consider the first order component $v_1(r, \psi)$ of the second LV (5.25). Due to the sinusoidal character of the forcing term in the GSF (5.23), it is easy to verify (see the next section) that $v_1(r, \psi)$ contains just the first Fourier component,

$$v_1(r, \psi) = c[L(r)\sin(\psi + \alpha) + R(r)\cos(\psi + \alpha)]. \tag{5.30}$$

By now, inserting (5.29–5.30) into (5.28) and performing the integration over ψ, we get

$$\lambda = \varepsilon^2 \int_{-1}^{1} dr \left\{ \Delta c[q_1^r[L(r)\sin\alpha + R(r)\cos\alpha] - q_1^i[L(r)\cos\alpha - R(r)\sin\alpha]] \right.$$
$$+ bq_1^i - \frac{b^2}{8} + \Delta\frac{bc}{4}[L(r)\cos\alpha - R(r)\sin\alpha] + \frac{c^2}{8}(3 - \Delta^2)[L^2(r) + R^2(r)]$$
$$+ \frac{c^2}{2}g^2(r) + c^2\frac{|r|}{r}g(r)L(r)\right\} + \frac{\Delta I_2}{4\pi}, \tag{5.31}$$

where we have further decomposed $q_1(r)$ in its real and imaginary parts

$$q_1(r) = q_1^r(r) + iq_1^i(r),$$

and we have defined

$$I_2 = \int_{-1}^{1} dr \int_0^{2\pi} d\psi\, v_2(r, \psi), \qquad (5.32)$$

which accounts for the contribution arising from the second order correction to the LV. This expansion shows that the highest–order contribution to the second Lyapunov exponent of the GSF scales quadratically with the perturbation amplitude. This is indeed a general result that does not depend on the particular choice of the functions used to define the GSF, but only on the skew–product structure of the un–perturbed time evolution and on the validity of the expansion assumed in (5.27).

According to [PGY06], we finally get the perturbative expression for the second LE,

$$\lambda = \varepsilon^2 \{ \frac{c^2}{30} - \frac{b^2}{4} + \int_{-1}^{1} dr[bq_1^i(r) + \frac{c^2}{16}(6 - \Delta^2)[L^2(r) + R^2(r)] \qquad (5.33)$$

$$+ \Delta c q_1^r(r)[L(r) \sin \alpha + R(r) \cos \alpha] + \Delta c \left(\frac{b}{4} - q_1^i(r) \right) [L(r) \cos \alpha - R(r) \sin \alpha]$$

$$+ c^2 \frac{|r|}{r} g(r)L(r) + \frac{\Delta c^2}{4} r \sin \left(\frac{\Delta(1 - r)}{2} \right) [L(r) \cos K - R(r) \sin K]]\}.$$

Accordingly, the numerical value of the second LE can be obtained by performing integrals which involve the four functions $q_1^r(r)$, $q_1^i(r)$, $L(r)$, and $R(r)$.

5.2 Phase Synchronization in Coupled Chaotic Oscillators

Over the past decade or so, *synchronization in chaotic oscillators* [FY83, PC90] has received much attention because of its fundamental importance in nonlinear dynamics and potential applications to laser dynamics [DBO01], electronic circuits [KYR98], chemical and biological systems [ESH98], and secure communications [KP95]. Synchronization in chaotic oscillators is characterized by the loss of exponential instability in the transverse direction through interaction. In coupled chaotic oscillators, it is known, various types of synchronization are possible to observe, among which are *complete synchronization* (CS) [FY83, PC90], *phase synchronization* (PS) [RPK96, ROH98], *lag synchronization* (LS) [RPK97] and *generalized synchronization* (GS) [KP96].

One of the noteworthy synchronization phenomena in this regard is PS which is defined by the phase locking between nonidentical chaotic oscillators whose amplitudes remain chaotic and uncorrelated with each other:

$|\theta_1 - \theta_2| \leq$ const. Since the first observation of PS in mutually coupled chaotic oscillators [RPK96], there have been extensive studies in theory [ROH98] and experiments [DBO01]. The most interesting recent development in this regard is the report that the interdependence between physiological systems is represented by PS and *temporary phase–locking* (TPL) states, e.g., (a) *human heart beat and respiration* [SRK98], (b) a certain brain area and the tremor activity [TRW98, RGL99]. Application of the concept of PS in these areas sheds light on the analysis of non–stationary bivariate data coming from biological systems which was thought to be impossible in the conventional statistical approach. And this calls new attention to the PS phenomenon [KK00, KLR03].

Accordingly, it is quite important to elucidate a detailed transition route to PS in consideration of the recent observation of a TPL state in biological systems. What is known at present is that TPL[ROH98] transits to PS and then transits to LS as the coupling strength increases. On the other hand, it is noticeable that the phenomenon from non–synchronization to PS have hardly been studied, in contrast to the wide observations of the TPL states in the biological systems.

In this section, following [KK00, KLR03], we study the characteristics of TPL states observed in the regime from non–synchronization to PS in coupled chaotic oscillators. We report that there exists a special locking regime in which a TPL state shows maximal periodicity, which phenomenon we would call *periodic phase synchronization* (PPS). We show this PPS state leads to local negativeness in one of the vanishing Lyapunov exponents, taking the measure by which we can identify the maximal periodicity in a TPL state. We present a qualitative explanation of the phenomenon with a nonuniform oscillator model in the presence of noise.

We consider here the unidirectionally coupled non–identical Rössler oscillators for first example:

$$\dot{x}_1 = -\omega_1 y_1 - z_1, \quad \dot{y}_1 = \omega_1 x_1 + 0.15 y_1, \quad \dot{z}_1 = 0.2 + z_1(x_1 - 10.0),$$
$$\dot{x}_2 = -\omega_2 y_2 - z_2, \quad \dot{y}_2 = \omega_2 x_2 + 0.165 y_2 + \epsilon(y_1 - y_2), \tag{5.34}$$
$$\dot{z}_2 = 0.2 + z_2(x_2 - 10.0),$$

where the subscripts imply the oscillators 1 and 2, respectively, $\omega_{1,2}$ ($= 1.0 \pm 0.015$) is the overall frequency of each oscillator, and ϵ is the coupling strength. It is known that PS appears in the regime $\epsilon \geq \epsilon_c$ and that 2π phase jumps arise when $\epsilon < \epsilon_c$. Lyapunov exponents play an essential role in the investigation of the transition phenomenon with coupled chaotic oscillators and as generally understood that PS transition is closely related to the transition to the negative value in one of the vanishing Lyapunov exponents [PC90].

A vanishing Lyapunov exponent corresponds to a phase variable of an oscillator and it exhibits the neutrality of an oscillator in the phase direction. Accordingly, the local negativeness of an exponent indicates this neutrality is locally broken [RPK96]. It is important to define an appropriate phase variable in order to study the TPL state more thoroughly. In this regard, several methods have been proposed methods of using linear interpolation at a Poincaré section [RPK96], phase–space projection [RPK96, ROH98], tracing of the center of rotation in phase–space [YL97], Hilbert transformation [RPK96], or wavelet transformation [KK00, KLR03]. Among these we take the method of phase–space projection onto the $x_1 - y_1$ and $x_2 - y_2$ planes with the geometrical relation

$$\theta_{1,2} = \arctan(y_{1,2}/x_{1,2}),$$

and get *phase difference* $\varphi = \theta_1 - \theta_2$.

The system of coupled oscillators is said to be in a TPL state (or laminar state) when $\langle \varphi \rangle < \Lambda_c$ where $\langle ... \rangle$ is the running average over appropriate short time scale and Λ_c is the cutoff value to define a TPL state. The locking length of the TPL state, τ, is defined by time interval between two adjacent peaks of $\langle \varphi \rangle$.

In order to study the characteristics of the locking length τ, we introduce a measure [KK00, KLR03]: $P(\epsilon) = \sqrt{\mathrm{var}(\tau)}/\langle \tau \rangle$, which is the ratio between the average value of time lengths of TPL states and their standard deviation. In terminology of stochastic resonance, it can be interpreted as noise–to–signal ratio [PK97, Jun93]. The measure would be minimized where the periodicity is maximized in TPL states.

To validate the argument, we explain the phenomenon in simplified dynamics. From (5.34), we get the equation of motion in terms of phase difference:

$$\dot{\varphi} = \Delta\omega + A(\theta_1, \theta_2, \epsilon) \sin\varphi + \xi(\theta_1, \theta_2, \epsilon), \qquad \text{where} \qquad (5.35)$$

$$A(\theta_1, \theta_2, \epsilon) = (\epsilon + 0.15)\cos(\theta_1 + \theta_2) - \frac{\epsilon}{2}(\frac{R_1}{R_2}),$$

$$\xi(\theta_1, \theta_2, \epsilon) = \frac{\epsilon}{2}\frac{R_1}{R_2}\sin(\theta_1 + \theta_2) + \frac{z_1}{R_1}\sin(\theta_1) - \frac{z_2}{R_2}\sin(\theta_2)$$
$$+ (\epsilon + 0.015)\cos(\theta_2)\sin(\theta_2).$$

$$\text{Here,} \qquad \Delta\omega = \omega_1 - \omega_2, \qquad R_{1,2} = \sqrt{x_{1,2}^2 + y_{1,2}^2}.$$

And from (5.35) we get the simplified equation to describe the phase dynamics:

$$\dot{\varphi} = \Delta\omega + \langle A \rangle \sin(\varphi) + \xi,$$

where $\langle A \rangle$ is the time average of $A(\theta_1, \theta_2, \epsilon)$. This is a nonuniform oscillator in the presence of noise where ξ plays a role of effective noise [Str94] and the value of $\langle A \rangle$ controls the width of bottleneck (i.e, non–uniformity of the flow). If the bottleneck is wide enough, (i.e., faraway from the saddle–node bifurcation point: $\Delta\omega \gg -\langle A \rangle$), the effective noise hardly contributes to the phase dynamics of the system. So the passage time is wholly governed by the width of the bottleneck as follows:

$$\langle \tau \rangle \sim 1/\sqrt{\Delta\omega^2 - \langle A \rangle^2} \sim 1/\sqrt{\Delta\omega^2 - \epsilon^2/4},$$

which is a slowly increasing function of ϵ. In this region while the standard deviation of TPL states is nearly constant (because the widely opened bottlenecks periodically appears and those lead to small standard deviation), the average value of locking length of TPL states is relatively short and the ratio between them is still large.

On the contrary as the bottleneck becomes narrower (i.e., near the saddle–node bifurcation point: $\Delta\omega \geq -\langle A \rangle$) the effective noise begins to perturb the process of bottleneck passage and regular TPL states develop into intermittent ones [ROH98, KK00]. It makes the standard deviation increase very rapidly and this trend overpowers that of the average value of locking lengths of the TPL states. Thus we understand that the competition between width of bottleneck and amplitude of effective noise produces the crossover at the minimum point of $P(\epsilon)$ which shows the maximal periodicity of TPL states.

Rosenblum *et al.* firstly observed the dip in mutually coupled chaotic oscillators [RPK96]. However the origin and the dynamical characteristics of the dip have been left unclarified. We argue that the dip observed in mutually coupled chaotic oscillators has the same origin as observed above in unidirectionally coupled systems.

Common apprehension is that near the border of synchronization the phase difference in coupled regular oscillators is periodic [RPK96] whereas in coupled chaotic oscillators it is irregular [ROH98]. On the contrary, we report that the special locking regime exhibiting the maximal periodicity of a TPL state also exists in the case of coupled chaotic oscillators. In general, the phase difference of coupled chaotic oscillators is described by the 1D Langevin equation,

$$\dot{\varphi} = F(\varphi) + \xi,$$

where ξ is the effective noise with finite amplitude. The investigation with regard to PS transition is the study of scaling of the laminar length around the virtual fixed–point φ^* where $F(\varphi^*) = 0$ [KK00, KT01] and PS transition is established when

$$|\int_{\varphi}^{\varphi^*} F(\varphi)d\phi| > \max |\xi|.$$

Consequently, the crossover region, from which the value of P grows exponentially, exists because intermittent series of TPL states with longer locking length τ appears as PS transition is nearer. Eventually it leads to an exponential growth of the standard deviation of the locking length. Thus we argue that PPS is the generic phenomenon mostly observed in coupled chaotic oscillators prior to PS transition.

In conclusion, analyzing the dynamic behaviors in coupled chaotic oscillators with slight parameter mismatch we have completed the whole transition route to PS. We find that there exists a special locking regime called PPS in which a TPL state shows maximal periodicity and that the periodicity leads to local negativeness in one of the vanishing Lyapunov exponents. We have also made a qualitative description of this phenomenon with the nonuniform oscillator model in the presence of noise. Investigating the characteristics of TPL states between non–synchronization and PS, we have clarified the transition route before PS. Since PPS appears in the intermediate regime between non–synchronization and PS, we expect that the concept of PPS can be used as a tool for analyzing weak interdependences, i.e., those not strong enough to develop to PS, between non–stationary bivariate data coming from biological systems, for instance [KK00, KLR03]. Moreover PPS could be a possible mechanism of the chaos regularization phenomenon [Har92, Rul01] observed in neurobiological experiments.

5.3 Oscillatory Phase Neurodynamics

In coupled oscillatory neuronal systems, under suitable conditions, the original dynamics can be reduced theoretically to a simpler phase dynamics. The state of the ith neuronal oscillatory system can be then characterized by a single phase variable φ_i representing the timing of the neuronal firings. The typical dynamics of *oscillator neural networks* are described by the *Kuramoto model* [Kur84, HI97, Str00], consisting of N equally weighted, all–to–all, phase–coupled limit–cycle oscillators, where each oscillator has its own natural frequency ω_i drawn from a prescribed distribution function:

$$\dot{\varphi}_i = \omega_i + \frac{K}{N} \sum_{i=1}^{N} J_{ij} \sin(\varphi_j - \varphi_i + \beta_{ij}). \qquad (5.36)$$

Here, J_{ij} and β_{ij} are parameters representing the effect of the interaction, while $K \geq 0$ is the coupling strength. For simplicity, we assume that all natural frequencies ω_i are equal to some fixed value ω_0. We can then eliminate ω_0 by applying the transformation

$$\varphi_i \rightarrow \varphi_i + \omega_0 t.$$

Using the complex representation

$$W_i = \exp(i\varphi_i) \qquad \text{and} \qquad C_{ij} = J_{ij} \exp(i\beta_{ij})$$

in (5.36), it is easily found that all neurons relax toward their stable equilibrium states, in which the relation $W_i = h_i/|h_i|$ ($h_i = \sum_{j=1}^{N} C_{ij} W_j$) is satisfied. Following this line of reasoning, as a synchronous update version of the oscillator neural network we can consider the alternative discrete form [AN99],

$$W_i(t+1) = \frac{h_i(t)}{|h_i(t)|}, \qquad h_i(t) = C_{ij} W_j(t). \qquad (5.37)$$

Now we will attempt to construct an extended model of the oscillator neural networks to retrieve sparsely coded phase patterns. In equation (5.37), the complex quantity h_i can be regarded as the local field produced by all other neurons. We should remark that the phase of this field, h_i, determines the timing of the ith neuron at the next time step, while the amplitude $|h_i|$ has no effect on the retrieval dynamics (5.37). It seems that the amplitude can be thought of as the strength of the local field with regard to emitting spikes. Pursuing this idea, as a natural extension of the original model we stipulate that the system does not fire and stays in the resting state if the amplitude is smaller than a certain value. Therefore, we consider a network of N oscillators whose dynamics are governed by

$$W_i(t+1) = f(|h_i(t)|)\frac{h_i(t)}{|h_i(t)|}, \qquad h_i(t) = C_{ij}W_j(t). \qquad (5.38)$$

We assume that $f(x) = \Theta(x - H)$, where the real variable H is a threshold parameter and $\Theta(x)$ is the unit step function; $\Theta(x) = 1$ for $x \geq 0$ and 0 otherwise. Therefore, the amplitude $|W_i^t|$ assumes a value of either 1 or 0, representing the state of the ith neuron as firing or non–firing. Consequently, the neuron can emit spikes when the amplitude of the local field $h_i(t)$ is greater than the threshold parameter H.

Now, let us define a set of P patterns to be memorized as $\xi_i^\mu = A_i^\mu \exp(i\theta_i^\mu)$ $(\mu = 1, 2, \ldots, P)$, where θ_i^μ and A_i^μ represent the phase and the amplitude of the ith neuron in the μth pattern, respectively. For simplicity, we assume that the θ_i^μ are chosen at random from a uniform distribution between 0 and 2π. The amplitudes A_i^μ are chosen independently with the probability distribution

$$P(A_i^\mu) = a\delta(A_i^\mu - 1) + (1-a)\delta(A_i^\mu),$$

where a is the mean activity level in the patterns. Note that, if $H = 0$ and $a = 1$, this model reduces to (5.37).

For the synaptic efficacies, to realize the function of the associative memory, we adopt the *generalized Hebbian rule* in the form

$$C_{ij} = \frac{1}{aN}\xi_i^\mu \tilde{\xi}_j^\mu, \qquad (5.39)$$

where $\tilde{\xi}_j^\mu$ denotes the complex conjugate of ξ_j^μ. The overlap $M_\mu(t)$ between the state of the system and the pattern μ at time t is given by

$$M_\mu(t) = m_\mu(t)\, e^{i\varphi_\mu(t)} = \frac{1}{aN}\tilde{\xi}_j^\mu W_j(t), \qquad (5.40)$$

In practice, the rotational symmetry forces us to measure the correlation of the system with the pattern μ in terms of the amplitude component $m_\mu(t) = |M_\mu(t)|$.

Let us consider the situation in which the network is recalling the pattern ξ_i^1; that is, $m_1(t) = m(t) \sim O(1)$ and $m_\mu(t) \sim O(1/\sqrt{N})(\mu \neq 1)$. The local field $h_i(t)$ in (5.38) can then be separated as

$$h_i(t) = C_{ij}W_j(t) = m_t e^{i\varphi_1(t)}\xi_i^1 + z_i(t), \tag{5.41}$$

where $z_i(t)$ is defined by

$$z_i(t) = \frac{1}{aN}\xi_i^\mu \tilde{\xi}_j^\mu W_j(t). \tag{5.42}$$

The first term in (5.41) acts to recall the pattern, while the second term can be regarded as the noise arising from the other learned patterns. The essential point in this analysis is the treatment of the second term as *complex Gaussian noise* characterized by

$$< z_i(t) > = 0, \qquad < |z_i(t)|^2 > = 2\sigma(t)^2. \tag{5.43}$$

We also assume that $\varphi_1(t)$ remains a constant, that is, $\varphi_1(t) = \varphi_0$. By applying the method of statistical neurodynamics to this model under the above assumptions [AN99], we can study the retrieval properties analytically. As a result of such analysis we have found that the retrieval process can be characterized by some macroscopic order parameters, such as $m(t)$ and $\sigma(t)$.

From (5.40), we find that the overlap at time $t + 1$ is given by

$$m(t+1) = \left\langle\!\!\left\langle f(|m(t) + z(t)|)\frac{m(t) + z(t)}{|m(t) + z(t)|}\right\rangle\!\!\right\rangle, \tag{5.44}$$

where $\langle\!\langle\cdots\rangle\!\rangle$ represents an average over the complex Gaussian $z(t)$ with mean 0 and variance $2\sigma(t)^2$. For the noise $z(t+1)$, in the limit $N \to \infty$ we get [AN99]

$$
\begin{aligned}
z_i(t+1) \sim &\frac{1}{aN}\sum_{j=1}^{N}\sum_{\mu=2}^{P}\xi_i^\mu \tilde{\xi}_j^\mu f(|h_{j,\mu}(t)|)\frac{h_{j,\mu}(t)}{|h_{j,\mu}(t)|} \\
&+ z_i(t)\left(\frac{f'(|h_{j,\mu}(t)|)}{2} + \frac{f(|h_{j,\mu}(t)|)}{2|h_{j,\mu}(t)|}\right),
\end{aligned} \tag{5.45}
$$

where $h_{j,\mu}(t) = 1/aN\sum_{k=1}^{N}\sum_{\nu\neq mu,1}^{P}\xi_j^\nu \tilde{\xi}_k^\nu W_k(t)$.

5.3.1 Kuramoto Synchronization Model

The microscopic individual level dynamics of the Kuramoto model (5.36) is easily visualized by imagining oscillators as points running around on the unit circle. Due to rotational symmetry, the average frequency $\Omega = \sum_{i=1}^{N}\omega_i/N$ can be set to 0 without loss of generality; this corresponds to observing dynamics in the co–rotating frame at frequency Ω.

The governing equation (5.36) for the ith oscillator phase angle φ_i can be simplified to

$$\dot{\varphi}_i = \omega_i + \frac{K}{N}\sum_{i=1}^{N}\sin(\varphi_j - \varphi_i), \quad 1 \leq i \leq N. \tag{5.46}$$

It is known that as K is increased from 0 above some critical value K_c, more and more oscillators start to get synchronized (or phase–locked) until all the oscillators get fully synchronized at another critical value of K_{tp}. In the choice of $\Omega = 0$, the fully synchronized state corresponds to an exact steady state of the 'detailed', fine–scale problem in the co–rotating frame.

Such synchronization dynamics can be conveniently summarized by considering the fraction of the synchronized (phase–locked) oscillators, and conventionally described by a *complex–valued order parameter* [Kur84, Str00], $re^{i\psi} = \frac{1}{N}e^{i\varphi_J}$, where the radius r measures the phase coherence, and ψ is the average phase angle.

Transition from Full to Partial Synchronization

Following [MK05], here we restate certain facts about the nature of the second transition mentioned above, a transition between the full and the partial synchronization regime at $K = K_{tp}$, in the direction of decreasing K.

A fully synchronized state in the continuum limit corresponds to the solution to the mean–field type alternate form of equation (5.46),

$$\dot{\varphi}_i = \Omega = \omega_i + rK \sum_{i=1}^{N} \sin(\psi - \varphi_i), \qquad (5.47)$$

where Ω is the common angular velocity of the fully synchronized oscillators (which is set to 0 in our case). Equation (5.47) can be further rewritten as

$$\frac{\Omega - \omega_i}{rK} = \sum_{i=1}^{N} \sin(\psi - \varphi_i), \qquad (5.48)$$

where the absolute value of the r.h.s is bounded by unity.

As K approaches K_{tp} from above, the l.h.s for the 'extreme' oscillator (the oscillator in a particular family that has the maximum value of $|\Omega - \omega_i|$) first exceeds unity, and a real–valued solution to (5.48) ceases to exist. Different random draws of ω_i's from $g(\omega)$ for a finite number of oscillators result in slightly different values of K_{tp}. K_{tp} appears to follow the Gumbel type extreme distribution function [KN00], just as the maximum values of $|\Omega - \omega_i|$ do:

$$p(K_{tp}) = \sigma^{-1}e^{-(K_{tp}-\mu)/\sigma} \exp[-e^{-(K_{tp}-\mu)/\sigma}],$$

where σ and μ are parameters.

5.3.2 Lyapunov Chaotic Synchronization

The notion of *conditional Lyapunov exponents* was introduced by Pecora and Carroll in their study of synchronization of chaotic systems. First, in [PC91], they generalized the idea of driving a stable system to the situation when the drive signal is chaotic. This leaded to the concept of conditional Lyapunov exponents and also generalized the usual criteria of the linear stability theorem. They showed that driving with chaotic signals can be done in a robust fashion, rather insensitive to changes in system parameters. The calculation of the stability criteria leaded naturally to an estimate for the convergence of the driven system to its stable state. The authors focussed on a homogeneous driving situation that leaded to the construction of synchronized chaotic subsystems. They applied these ideas to the Lorenz and Rössler systems, as well as to an electronic circuit and its numerical model. Later, in [PC98], they showed that many coupled oscillator array configurations considered in the literature could be put into a simple form so that determining the stability of the synchronous state could be done by a master stability function, which could be tailored to one's choice of stability requirement. This solved, once and for all, the problem of synchronous stability for any linear coupling of that oscillator.

It turns out, that, like the full Lyapunov exponent, the conditional exponents are well defined ergodic invariants, which are reliable quantities to quantify the relation of a global dynamical system to its constituent parts and to characterize dynamical self–organization [Men98].

Given a dynamical system defined by a map $f : M \rightarrow M$, with $M \subset R^m$ the conditional exponents associated to the splitting $R^k \times R^{m-k}$ are the eigenvalues of the limit

$$\lim_{n \to \infty} (D_k f^{n*}(x) D_k f^n(x))^{\frac{1}{2n}},$$

where $D_k f^n$ is the $k \times k$ diagonal block of the full Jacobian.

Mendes [Men98] proved that existence of the conditional Lyapunov exponents as well–defined ergodic invariants was guaranteed under the same conditions that established the existence of the Lyapunov exponents.

Recall that for measures μ that are absolutely continuous with respect to the Lebesgue measure of M or, more generally, for measures that are smooth along unstable directions (SBR measures) Pesin's [Pes77] identity holds

$$h(\mu) = \sum_{\lambda_i > 0} \lambda_i,$$

relating *Kolmogorov–Sinai entropy* $h(\mu)$ to the sum of the Lyapunov exponents. By analogy we may define the *conditional exponent entropies* [Men98] associated to the splitting $R^k \times R^{m-k}$ as the sum of the positive conditional exponents counted with their multiplicity

$$h_k(\mu) = \sum_{\xi_i^{(k)} > 0} \xi_i^{(k)}, \qquad h_{m-k}(\mu) = \sum_{\xi_i^{(m-k)} > 0} \xi_i^{(m-k)}.$$

The Kolmogorov–Sinai entropy of a dynamical system measures the rate of information production per unit time. That is, it gives the amount of randomness in the system that is not explained by the defining equations (or the minimal model [CY89]). Hence, the conditional exponent entropies may be interpreted as a measure of the randomness that would be present if the two parts $S^{(k)}$ and $S^{(m-k)}$ were un–coupled. The difference $h_k(\mu)+h_{m-k}(\mu)-h(\mu)$ represents the effect of the coupling.

Given a dynamical system S composed of N parts $\{S_k\}$ with a total of m degrees of freedom and invariant measure μ, one defines a *measure of dynamical self–organization* $I(S, \Sigma, \mu)$ as

$$I(S, \Sigma, \mu) = \sum_{k=1}^{N} \{h_k(\mu) + h_{m-k}(\mu) - h(\mu)\} .$$

For each system S, this quantity will depend on the partition Σ into N parts that one considers. $h_{m-k}(\mu)$ always denotes the conditional exponent entropy of the complement of the subsystem S_k. Being constructed out of ergodic invariants, $I(S, \Sigma, \mu)$ is also a well–defined ergodic invariant for the measure μ. $I(S, \Sigma, \mu)$ is formally similar to a mutual information. However, not being strictly a mutual information, in the information theory sense, $I(S, \Sigma, \mu)$ may take negative values.

5.4 Synchronization Geometry

In this section, mainly following [Bal06], we present a general geometry of noisy synchronization.

5.4.1 Geometry of Coupled Nonlinear Oscillators

In a series of papers [SW87, SW89a, SW89b], Shapere and Wilczek established a geometric framework to discuss the motion of deformable objects in the absence of applied external force. Following [Bal06], we adopt these methods to understand a fundamental phenomenon in nonlinear dynamics, namely phase synchronization in coupled dynamical systems. Elaborating along these lines, we study a general system of coupled nonlinear oscillators under the influence of additive Gaussian white noise and we find the conditions under which the coupled units within the full system can exhibit phase-locked behavior and phase synchronization.

We consider a system of n coupled generalized oscillators $q(x, t)$ where the state variables q could in general be functions not only of time, but could also depend on a set of additional variables x, say spatial variables when there is a metric structure associated with the variables:

$$\dot{q}^i = f_i(q^1, q^2, \ldots, q^n, \mu_j) \ , \qquad (i = 1, \ldots, n \ , \ j = 1, \ldots, p). \qquad (5.49)$$

Thus q^i include also extended systems where the individual elements could mutually influence each other through a distance–dependent interaction between the elements. For instance in the case of coupled chemical oscillators

[FHI00], q^i would denote concentrations which have spatial dependence. Another example occurs in biological information processing where neurons interact in ensembles.

We study the simplest case in which, in the absence of couplings, each of the n subsystems admits oscillatory solutions for some parameter values μ_j. Switching on the mutual coupling between these oscillators results in the emergence of a collective behavior. It is then appealing to view the collective behavior as having arisen as a result of some sort of communication between different points in the configuration space. Thus one is led to a geometrical description of the system's dynamical evolution. For simplicity, we restrict ourselves to the case in which the collective dynamics also exhibits limit cycle behavior emerging via one or more *Hopf bifurcations*. We define the configuration space of the full system as the space of all possible flow lines and closed paths. We consider the situation when there is no external driving force, so that the space of all possible contours is the space of oriented contours centered at the origin.

In the absence of mutual couplings, the space of contours consists of oriented closed orbits, each orbit inclined at an angle with respect to the other. If we now turn on the mutual couplings between these n subsystems gradually, each of the orbits would gradually get deformed, going through a sequence of shape changes, and resulting subsequently in a net rotation for it. The problem of interest is to link the dynamical variables of the system with the net rotation induced by a change of shape of the orbits for each of the n oscillators in phase–space.

The relative orientations of any two contour shapes can be compared by fixing coordinate axes for each. Since there exists a degeneracy in the possible choice of axes one can make at each point in the space of contour shapes, each set of reference frame we can choose from being isomorphic to E_n, a *gauge structure* is therefore induced in this space which facilitates going from one particular choice of axes to another.

In [SW87, SW89a, SW89b], the problem of self–propulsion at low *Reynold's number* made possible solely through shape deformations was discussed. Each choice of reference frame fixed to a shape, which assigned a 'standard location' in space for each shape, was associated with the motion and location of any arbitrary shape in relation to its standard location. We follow their methods closely to discuss deformations of the oriented contours in the space of contour shapes.

Following [SW87, SW89a, SW89b], the sequence of oriented contours $S(t)$ can be similarly related to the sequence of the corresponding chosen reference standard contour shapes $S_0(t)$ by a rigid displacement $\mathcal{R}(t)$:

$$S(t) = \mathcal{R}(t)S_0(t), \tag{5.50}$$

where in general, an $n-$dimensional motion $\mathcal{R}$ includes both rotations R and a translation l:

$$[R, l] = \begin{pmatrix} R & l \\ 0 & 1 \end{pmatrix}, \tag{5.51}$$

where $R(t)$ is an $n \times n$ rotation matrix and stands for a sequence of time–dependent rigid motions. The contour boundaries are parametrized by the control parameters μ_i, for each of which the rigid motion $\mathcal{R}$ acts on the vector $[S_0(\mu), 1]^T$. The physical contours $S(t)$ are invariant under a local change

$$\tilde{S}_0 = \Omega[S_0]S_0, \tag{5.52}$$

made in the choice of standard contours S_0. Then the contour shape evolution can be written by combining (5.52) with (5.50) as :

$$\tilde{S}(t) = \mathcal{R}(t)\Omega^{-1}(S_0(t))\tilde{S}_0 = \tilde{\mathcal{R}}(t)\tilde{S}_0(t), \qquad \text{or} \tag{5.53}$$
$$\tilde{\mathcal{R}}(t) = \mathcal{R}(t)\Omega^{-1}(S_0(t)). \tag{5.54}$$

The temporal change in the sequence of rigid motions can be written as:

$$\frac{d\mathcal{R}}{dt} = \mathcal{R}(\mathcal{R}^{-1}\frac{d\mathcal{R}}{dt}) \equiv \mathcal{R}A, \tag{5.55}$$

where A can be identified with the infinitesimal rotation arising from an infinitesimal deformation of $S_0(t)$. Equation (5.55) can be integrated to get the full motion for finite t:

$$\mathcal{R}(t_2) = \mathcal{R}(t_1)\mathcal{P}e^{\int_{t_1}^{t_2} A(t)dt}, \tag{5.56}$$

where $\mathcal{P}$ stands for the *Wilson line integral* W:

$$W_{21} = \mathcal{P}e^{\int_{t_1}^{t_2} A(t)dt} = 1 + \int_{t_1 < t < t_2} A(t)dt + \int_{t_1 < t < t' < t_2} \int A(t)A(t')dtdt' + \dots \tag{5.57}$$

in which the matrices are ordered such that the ones occurring at earlier times are on the left. It can be seen from (5.53–5.55), that A transforms like a *gauge potential*,

$$\tilde{A} = \Omega A \Omega^{-1} + \Omega \frac{d\Omega^{-1}}{dt}, \tag{5.58}$$

and the Wilson integral transforms as

$$\tilde{W}_{21} = \Omega_1 W_{21} \Omega_2^{-1}. \tag{5.59}$$

Shapere and Wilczek exploited the invariance of (5.57) under rescaling of time, $t \rightarrow \tau(t)$, the measure scaling as $dt \rightarrow \dot{\tau}dt$, $A \rightarrow A/\dot{\tau}$, to rewrite it in a time–independent geometric form. This was done in [SW87, SW89a, SW89b] by defining an abstract vector–field A on the tangent space to S_0. The projection $A(t)$ of A at the contour shape $S_0(t)$ is evaluated in the direction $\frac{\delta S_0}{\delta t}$ in which the shape is changing:

$$A(t) \equiv A_{\dot{S}_0}[S_0(t)]. \tag{5.60}$$

In terms of these projected vector–fields, (5.56) was rewritten in a time–independent form for a given path and independent of the manner in which the path is parametrized in the contour shape space as:

$$R(t_2) = R(t_1)\mathcal{P}e^{\int_{S_0(t_1)}^{S_0(t_2)} A(S_0)dS_0}. \tag{5.61}$$

Each of the components $A_i[S_0]$ of A coming from each direction in the contour space generates a rigid motion and can be defined in terms of a fixed basis of tangent vectors $\{w_i\}$ at S_0:

$$A_i[S_0] \equiv A_{w_i}[S_0]. \tag{5.62}$$

An infinitesimal deformation $s(t)$ of a contour $S_0(t)$ can be represented as:

$$S_0(t) = S_0 + s(t), \tag{5.63}$$

where an expansion of $s(t)$ can be made:

$$s(t) = \sum_i \alpha_i(t)w_i. \tag{5.64}$$

It was shown in [SW87, SW89a, SW89b] that for the particular case $S_0(t_1) = S_0(t_2)$, i,e., for a closed cycle in which the sequence of deformations returns the system to the original contour shape in its configuration space, the line integral in (5.57) becomes the closed *Wilson loop* which can be simplified to

$$W = \mathcal{P}e^{\oint A(t)dt} = 1 + \frac{1}{2} \oint \sum_{i,j} F_{ij}\alpha_i\dot{\alpha}_j dt, \tag{5.65}$$

$$\text{where} \quad F_{ij} = \frac{\partial A_{w_i}}{\partial w_j} - \frac{\partial A_{w_j}}{\partial w_i} + [A_{w_i}, A_{w_j}]. \tag{5.66}$$

The field strength tensor F_{ij} gives the resultant net displacement when a sequence of successive deformations is made of S_0 around a closed path and is thus the curvature associated with the gauge potential.

In the configuration space of contour shapes, the orbit of each of the n subsystems of the full coupled systems of oscillators undergoes the shape deformations described above. Because of the mutual couplings, the motion in phase–space of any one oscillator coordinate is inseparably linked with that of any other phase–space point which may be the coordinate of another oscillator. The deformation and motion in the configuration space of the various flow lines and closed paths of the entire coupled system can thus be viewed as those on the surface of a solid deformable body which is undergoing motion solely due to these deformations.

The full system of n oscillators can be represented by an $n-$component vector ψ^i, $(i = 1, \ldots n)$ in an abstract complex vector space:

$$\psi = \begin{pmatrix} q^1 \\ q^2 \\ \cdot \\ \cdot \\ \cdot \\ q^n \end{pmatrix}. \tag{5.67}$$

A rotation through an angle Λ with respect to a chosen axis in this internal vector space does not change the state of the full system, but just takes one oscillator state q^i to another:

$$q^i \to \tilde{q}_i = U(\Lambda)q^i = \mathrm{e}^{it^\alpha \Lambda^\alpha} q^i, \tag{5.68}$$

where t^k are k number of $n \times n$ matrices and are representations of the generators of the transformation group. Each of the q^is represents the state of the ith oscillator at time t.

There are n independent gauge potentials A_{w_i} corresponding to the n independent internal rotations. Any two rotation matrices $U(\Lambda^a)$ and $U(\Lambda^b)$ do not commute unless Λ^a and Λ^b point in the same direction. On application of a common input to the full system, all the different n oscillators respond to it. In this case emergence of a collective behavior is determined by the same gauge potential, although perhaps by different amounts or strengths.

5.4.2 Noisy Coupled Nonlinear Oscillators

Now, following [Bal06], we consider the system of n coupled nonlinear oscillators q^i subject to additive Gaussian white noise ξ^i in the limit of weak noise:

$$\dot{q}^i = f_i(q^1, q^2, \dots, q^n, \mu) + \xi^i, \qquad (i = 1, \dots, n, \quad \mu \in \mathbb{R}^v), \tag{5.69}$$

where the noise correlations are defined as:

$$\langle \xi^i(t)\xi_j(t') \rangle = Q\delta_{ij}\delta(t - t'). \tag{5.70}$$

The eigenvalues of the linear stability matrix of the coupled deterministic system determine the route through which the full system moves towards a collective behavior. A pure imaginary complex conjugate pair of eigenvalues at the bifurcation point with the remaining $(n-2)$ eigenvalues having nonzero real parts signals a *Hopf bifurcation*. The orbit structure near the non–hyperbolic fixed–points (q^0, μ_0) of (5.49) is determined by the *center manifold theorem*. When the system described by (5.70) undergoes a change in stability through a Hopf bifurcation, one gets a $p-$parameter family of vector–fields on a 2D center manifold.

In this case, one observes an emergent common frequency of oscillation for the coupled system. Such a situation automatically realises frequency synchronization also since the Hopf oscillator rotates with a characteristic frequency.

If there are more than one Hopf bifurcations, clearly it indicates more than one common frequency of oscillation and one expects to observe a clustering of the various n coupled oscillators around these common characteristic frequencies.

The full system of n oscillators changes stability as the parameter under consideration takes on different values, and at some parameter values undergoes bifurcations. We wil. study the system in the close neighborhood of the bifurcation points where the system exhibits critical behavior. It is in these regimes that the behaviors of the individual oscillators gives way to the collective behavior of the entire coupled system of the n oscillators. We employ center manifold reduction techniques for the system in the presence of fluctuations and perform a separation of variables in terms of fast and slow variables as in [KW83, BMB82a, BMB82b], exploiting their dynamical evolution on different time scales. A drastic simplification can then be made of the system's dynamics and one can write the probability $P(q^i, t)$ for the system to be in a certain configuration at time t in the weak noise limit as the product:

$$P(q^i, t) = p(q^f | q^s) P(q^s, t) \tag{5.71}$$

where q^s and q^f are the slow and the fast variables respectively of the system. The probability $P(q^s, t)$ for the critical variables occurs on a slow time scale and is non–Gaussian in nature.

The properties of the fast variables depend upon the nonlinearities in the system. For instance in the case when the coupled system exhibits a cusp bifurcation the fast variable could exhibit non–Gaussian fluctuations as it is coupled to the critical variable. It can be shown [KW83] that the joint probability density $p(q^f | q^s)$ is confined to a narrow strip peaked about the center manifold. We are interested in the case when the coupled system also exhibits self–sustained oscillatory behavior and makes a transition to limit cycle behavior in the presence of fluctuations. It was shown in [KW83, BMB82a, BMB82b], that for a transition via a Hopf bifurcation, $p(q^f | q^s)$ has the time–independent Gaussian form in the q^f variables with width which depends upon the slow variables q^s:

$$p(q^f | q^s) = \left(\frac{\sigma(q^s)}{\pi} \right)^{1/2} e^{-\sigma(q^s)(q^f - q^f_0(q^s))^2},$$

where the center manifold is got as a power series in q^s: $q^f = q^f_0(q^s)$. Recall that the center manifold theorem has been used in [Hak93] for providing with a proof for the stability of the synchronized states. The enslaved stable modes are the fast variables which follow the dynamics of the center (critical) modes. We rewrite $f_i(q^1, q^2, \ldots, q^n, \mu_i)$ as

$$f_i(q^1, q^2, \ldots, q^n, \mu_i) = -\frac{\delta F(q^1, q^2, \ldots, q^n, \mu_i)}{\delta q^i}$$

The Fokker–Planck equation for the full system is:

$$\frac{dP(q^i,t)}{dt} = \frac{dP(q^s,q^f,t)}{dt} = \frac{\partial}{\partial q^i}(P(q^i,t)\frac{\partial F}{\partial q^i}) + \frac{\partial^2 P(q^i,t)}{\partial q^{i2}} \tag{5.72}$$

Using (5.71), we can rewrite this as

$$\frac{dP(q^s,q^f,t)}{dt} = \frac{d}{dt}(p(q^f|q^s)P(q^s,t)) = \frac{dp(q^f|q^s)}{dt}P(q^s,t) + p(q^f|q^s)\frac{dP(q^s,t)}{dt}$$

$$= -(H_{FP1}(q^f,q^s) + H_{FP2}(q^s,t))P(q^s,q^f,t) \tag{5.73}$$

Hence in the close proximity of the bifurcation, the operator H_{FP} can be written in a separable form, the part $H_{FP}(q^s,t)$ independent of the fast variables. Here

$$- H_{FP2}(q^s,t))P(q^s,q^f,t) = \frac{\partial}{\partial q^s}(P(q^i,t)\frac{\partial F}{\partial q^s}) + \frac{\partial^2 P(q^i,t)}{\partial q^{s2}} \tag{5.74}$$

We find it convenient to analyze the coupled dynamics in a path integral framework. We follow the procedure of Gozzi [Goz83] to recast the system (5.69), (5.73) as a path integral, and define:

$$\Psi = P(q^f,q^s,t)e^{F(q^i)/2}, \tag{5.75}$$

so that (5.72),(5.74) can be rewritten as

$$\frac{d\Psi}{dt} = -2H_{FP}\Psi, \qquad \text{where} \tag{5.76}$$

$$H_{FP} = -\frac{1}{2}\frac{\partial^2}{\partial q^{i2}} + \frac{1}{8}(\frac{\partial F}{\partial q^i})^2 - \frac{1}{4}\frac{\partial^2 F}{\partial q^{i2}} \tag{5.77}$$

To enable computation of correlation functions within the path integral formalism, we introduce n external sources J_i to probe the full coupled system so that the partition function $Z[J]$ for the system can be written as the time ordered path integral

$$Z[J] = \mathcal{N}\mathcal{T}\prod_i \int \mathcal{D}q_f \mathcal{D}q_s \mathcal{D}\xi^i \times \tag{5.78}$$

$$e^{-\frac{1}{Q}\int J_i(t')q^i(t')dt'}P(q^s,t)p(q^f|q^s)\delta(q^i - q^i{}_\xi)e^{-\int \frac{\xi^i}{4Q}dt'},$$

where $q^i{}_\xi$ denote the solution of the system of Langevin equations (5.69), $\mathcal{T}$ denotes time ordering and $\mathcal{N}$ is the normalization constant. From (21) one can write

$$\delta(q^i - q^i{}_\xi) = \delta(\dot{q}^i - f_i(q^1,q^2,\dots,q^n) - \xi^i)\left\|\frac{\delta\xi^i}{\delta q^i}\right\|.$$

We can rewrite the Jacobian $\left\|\frac{\delta\xi^i}{\delta q^i}\right\|$ of the transformation $\xi^i \to q^i$ as

$$\left\| \frac{\delta \xi^i}{\delta q^i} \right\| = \det \left[\left(\delta_{ij} \partial_t - \frac{\partial f_i(q^1, q^2, \dots, q^n)}{\partial q^j(t')} \right) \delta(t - t') \right]$$

$$= \exp\{ \mathrm{Tr} \ln \partial_t (\delta_{ij} \delta(t - t') - \partial_{t'}{}^{-1} \frac{\partial f_i}{\partial q^j(t')}) \}. \qquad (5.79)$$

The operator $\partial_{t'}{}^{-1}$ satisfies the relation

$$\partial_t G(t - t') = \delta(t - t').$$

Then we can rewrite (5.79) in terms of the Green function in (35) as:

$$\left\| \frac{\delta \xi^i}{\delta q^j} \right\| = \exp\left\{ \mathrm{Tr} \left[\ln \partial_t + \ln \left(\delta(t - t') + G_{ij}(t - t') \frac{\partial f_i}{\partial q^j(t')} \right) \right] \right\}. \qquad (5.80)$$

The system evolves forward in time. Hence

$$G(t - t') = \theta(t - t').$$

Using this and expanding the logarithm in the argument of the exponential, we can simplify (36) to

$$\left\| \frac{\delta \xi^i}{\delta q^j} \right\| = e^{\mathrm{Tr} \ln \partial_t} e^{\int_0^t dt' \theta(0) \frac{\partial f_i}{\partial q^j(t')}}.$$

Substituting this back into (5.78) and using the mid-point prescription $\theta(0) = 1/2$ of Stratonovich, we have

$$Z[J] = \mathcal{N} \mathcal{T} \prod_i \int \mathcal{D} q_f \mathcal{D} q_s \mathcal{D} \xi^i \times \qquad (5.81)$$

$$e^{-\frac{1}{Q} \int_0^t J_i(t') q^i(t') dt'} e^{\frac{1}{2} \int_0^t \frac{\partial f_i}{\partial q^j(t')}} e^{-\frac{1}{4Q} \int_0^t dt' [\dot{q}^i - f_i(q^1, q^2, \dots, q^n)]^2}.$$

(5.81) can be reduced to

$$Z[J] = \mathcal{N} \mathcal{T} \prod_i \int \mathcal{D} q_f \mathcal{D} q_s e^{-\frac{1}{Q} \int_0^t J_i(t') q^i(t') dt'} \times$$

$$e^{-\int_0^t dt' [\frac{1}{2} \frac{\partial^2 F}{\partial q^i \partial q^i} + \frac{1}{4Q} (\dot{q}^i)^2 + \frac{1}{4Q} (\frac{\partial F}{\partial q^i})^2]} e^{-\frac{1}{2Q}(F(t) - F(0))}$$

$$= \mathcal{N} \mathcal{T} \prod_i \int \mathcal{D} q_f \mathcal{D} q_s e^{-\int_0^t dt' [\mathcal{L}^{FP} + \frac{1}{Q} J_i(t') q^i(t')]} e^{-\frac{1}{2Q}(F(t) - F(0))}, \qquad (5.82)$$

where we have defined a *Fokker–Planck Lagrangian*

$$\mathcal{L}^{FP}(q^i, \dot{q}^i, t) = \frac{1}{4Q}(\dot{q}^i)^2 + \frac{1}{4Q}(\frac{\delta F}{\delta q^i})^2 + \frac{1}{2} \frac{\delta^2 F}{\delta q^i \delta q^j} \qquad (5.83)$$

$$= \frac{1}{4Q}(\dot{q}^i)^2 + f_i^2 - \frac{1}{2} \frac{\delta f_j}{\delta q^i \delta q^i},$$

which is related to the *Fokker–Planck Hamiltonian* H_{FP} defined in (5.76),(5.77) through a *Legendre transformation*:

$$H_{FP}(\pi_i, q^i, t) = \pi_i \dot{q}^i - \mathcal{L}^{FP}(q^i, \dot{q}^i, t).$$

Here π_i are the momenta canonically conjugate to the variables q^i:

$$\frac{\delta \mathcal{L}^{FP}}{\delta \dot{q}^i} = \pi_i = \frac{1}{Q}\dot{q}^i, \qquad \text{so that}$$

$$H_{FP}(\pi_i, q^i, t) = Q\pi_i^2 + \frac{1}{4Q}\left(\frac{\delta F}{\delta q^i}\right)^2 + \frac{1}{2}\frac{\delta^2 F}{\delta q^i \delta q^j}.$$

We use these relations in (5.82) to write the partition function as

$$Z[J] = \mathcal{N}\mathcal{T}\prod_i \int \mathcal{D}\pi_i \mathcal{D}q_i e^{-\int_0^t dt' [H_{FP}(\pi_i, q^i, t) + \frac{1}{Q}J_i(t')q^i(t')]}. \tag{5.84}$$

From (5.73),(5.75),(5.76) and (5.77), we see that in the close proximity of the bifurcation H_{FP} can be written in a separable form as

$$H_{FP}(\pi_i, q^i, t) = H_{FP}(\pi_f, q^f; \pi_s, q^s, t) = H_{FP1}(\pi_f, q^f; q^s, t) + H_{FP2}(\pi_s, q^s, t)$$

Thus the corresponding $\mathcal{L}^{FP}$ can also be split up as
$\mathcal{L}^{FP}(q^i, \dot{q}^i, t) = \mathcal{L}_1^{FP}(q^f, \dot{q}_f, t) + \mathcal{L}_2^{FP}(q^s, \dot{q}_s, t)$. Then we can write

$$Z[J] = \mathcal{N}\mathcal{T}\int \mathcal{D}q_f \mathcal{D}q_s e^{-\int_0^t dt' [\mathcal{L}_1^{FP}(q^f, \dot{q}_f) + \mathcal{L}_2^{FP}(q^s, \dot{q}_s, t) + \frac{1}{Q}J_i(t')q^i(t')]}.$$

Averaging over the fast degrees of freedom enables the partition function to be written in terms of an effective Lagrangian as a function of only the slow degrees of freedom. This can be done by first rewriting the fast degrees of freedom in action-angle variables (θ, I). The emergence of a non Abelian gauge structure can then be seen arising from the evolution of the slow dynamics but induced by the fast variables. After tracing the origin of the induced gauge potential to the slow dynamics, we get the conditions necessary to be satisfied in order for the coupled elements to be synchronized in phase.

To begin with, we introduce a generating function $S^{(\alpha)}(q^f, I; q^s)$ which effects the transformation $(q^f, \pi_f) \rightarrow (\theta, I)$ to the action angle variables:

$$\frac{\partial S^{(\alpha)}(q^f, I; q^s)}{\partial q^i} = \pi_i, \qquad \frac{\partial S^{(\alpha)}(q^f, I; q^s)}{\partial I_i} = \theta_i, \tag{5.85}$$

$S^{(\alpha)}(q^f, I; q^s)$ is many-valued and time dependent since the slow variables change with time.

The phase–space structure associated with adiabatic holonomy in classical systems was studied by Gozzi and Thacker [GT87] through Hamiltonian dynamics. We find it useful to employ their methods for our study of coupled oscillatory systems in a fluctuating environment. Using the canonical transformation law, $\mathcal{H}_1(I, q^s, t) = H_{FP1}(q^f(\theta, I, q^s), \pi_f(\theta, I, q^s), t)$ can be expressed in terms of the action-angle variables as:

$$\bar{H}_1(\theta, I, q^s(t)) = \mathcal{H}_1(I, q^s, t) + \dot{q}^s{}_l \frac{\partial S^{(\alpha)}(q^f, I; q^s)}{\partial q^s{}_l}.$$

Using the methods of [GT87] and [Arn89, Ber84, Ber85], we determine the dynamics of the M critical slow variables q^s of the system by averaging out the N fast variables which influence them:

$$\langle\langle \bar{H}_1 \rangle\rangle = \frac{1}{(2\pi)^N} \int d^N\theta\, \bar{H}_1(\theta, I, q^s(t)) \qquad (5.86)$$

$$= \frac{1}{(2\pi)^N} \int d^N\theta \left(\mathcal{H}_1(I, q^s) + \dot{q}^s{}_l \frac{\partial S^{(\alpha)}(q^f, I; q^s)}{\partial q^s{}_l} \right),$$

where the double angular brackets denote the averaging over all θ
$\langle\langle f \rangle\rangle = \frac{1}{(2\pi)^N} \int d^N\theta f$. Since $S^{(\alpha)}(q^f, I; q^s)$ is multi-valued, the single-valued function

$$\zeta(\theta, I, q^s) = S^{(\alpha)}(q^f(\theta, I, q^s), \pi_s(\theta, I, q^s), q^s), \qquad (0 \leq \theta \leq 2\pi).$$

is introduced [Lan86]. We have

$$\frac{\partial \zeta}{\partial q^s{}_l} = \frac{\partial S^\alpha}{\partial q^s{}_l} + \pi_{f\,i} \frac{\partial q^f{}_i}{\partial q^s{}_l}.$$

Hence this can be substituted into (5.86) to get

$$\langle\langle \bar{H}_1 \rangle\rangle = \mathcal{H}_1(I, q^s) + \dot{q}^s{}_l \langle\langle \frac{\partial \zeta}{\partial q^s{}_l} - \pi_{f\,i} \frac{\partial q^f{}_i}{\partial q^s{}_l} \rangle\rangle.$$

The total Hamiltonian of the system is given, after performing the angle averages by:

$$H_{av}(I, \pi_s, q^s) = \langle\langle H_1(q^f, \pi_f; q^s) + H_2(q^s, \pi_s) \rangle\rangle$$
$$= \langle\langle \bar{H}_1(\theta, I; q^s) + H_2(q^s, \pi_s) \rangle\rangle$$
$$= \bar{H}(I, \pi_s, q^s) + \dot{q}^s{}_l \langle\langle \frac{\partial \zeta}{\partial q^s{}_l} - \pi_{f\,i} \frac{\partial q^f{}_i}{\partial q^s{}_l} \rangle\rangle,$$

where we have let

$$\bar{H}(I, \pi_s, q^s) = \mathcal{H}_1(I, q^s) + H_2(q^s, \pi_s).$$

The *Gibbs partition function* in (5.84) can be rewritten in terms of the fast and slow variables as

$$Z[J] = \mathcal{N}\mathcal{T} \int \mathcal{D}\pi_f \mathcal{D}\pi_s \mathcal{D}q_f \mathcal{D}q_s \times$$

$$e^{- \int_0^t dt'[H_{FP1}(q^f,\pi_f;q^s)+H_{FP2}(q^s,\pi_s)+\frac{1}{Q}(J_s(t')q^s(t')+J_f(t')\sigma^f(t'))]}$$

$$= \mathcal{N}\mathcal{T} \int \mathcal{D}\pi_s \mathcal{D}q_s \mathcal{D}I \mathcal{D}\theta \times$$

$$e^{- \int_0^t dt'[\mathcal{H}_1(I,q^s)+H_2(q^s,\pi_s)-\dot{q}^s{}_l\pi_{f_i}\frac{\partial q^f{}_i}{\partial q^s{}_l}+\dot{q}^s{}_l\frac{\partial \zeta}{\partial q^s{}_l}+\frac{1}{Q}(J_s(t')q^s(t')+J_f(t')\theta(t'))]}$$

$$= \mathcal{N}\mathcal{T} \int \mathcal{D}\pi_s \mathcal{D}q_s \mathcal{D}I \mathcal{D}\theta \times$$

$$e^{- \int_0^t dt'[\bar{H}(I,\pi_s,q^s)-\dot{q}^s{}_l\pi_{f_i}\frac{\partial q^f{}_i}{\partial q^s{}_l}+\frac{1}{Q}(J_s(t')q^s(t')+J_f(t')\theta(t'))]}.$$

Performing the θ integration and simplifying the resulting expression, we get

$$Z[J] \approx \mathcal{N}\mathcal{T} \int \mathcal{D}\pi_s \mathcal{D}q_s \mathcal{D}I e^{- \int_0^t dt'(H_{av}(I,\pi_s,q^s)+\frac{1}{Q}J_s q^s)}. \tag{5.87}$$

We use the *Magnus expansion* [Mag54] for expanding the time–ordered integral, which gives the final state properties in terms of integrals over the initial state ones, to rewrite (5.87) as:

$$Z[J] \approx \mathcal{N} \int \mathcal{D}\pi_s \mathcal{D}q_s \mathcal{D}I \, \exp\{-\int_0^t dt'(H_{av} + \frac{1}{Q} J_s q^s$$

$$+ \frac{1}{2}[H_{av}(t'), \int_0^{t'} H_{av}(t_1)dt_1] + \frac{1}{4}[H_{av}(t'),$$

$$\int_0^{t'} [H_{av}(t_2), \int_0^{t_2} H_{av}(t_1)dt_1] \, dt_2] + \frac{1}{12}[\, [H_{av}(t')$$

$$\int_0^{t'} H_{av}(t_2)dt_2], \int_0^{t'} H_{av}(t_1)dt_1] + \dots)\}.$$

The variation of parameters μ_i in time of the system brings about change in its stability. The commutator terms in the Magnus expansion hence arise on account of this parametric time dependence of the Hamiltonian: $[H_{av}(t'), H_{av}(t_1)] = [H_{av}(\mu_i(t')), H_{av}(\mu_j(t_1))]$.

Retaining terms only upto the first commutator in the expansion and substituting for H_{av} from (54), we get after some simplifications:

$$Z[J] \approx \mathcal{N} \int \mathcal{D}\pi_s \mathcal{D}q_s \mathcal{D}I \exp\{-\int_0^t dt'(\bar{H}(I,\pi_s,q^s) + \dot{q}_{s_l}\langle\langle\frac{\partial\zeta}{\partial q^s_l}\rangle\rangle$$

$$-\dot{q}_{s_l}\langle\langle\pi_{f_i}\frac{\partial q^f_i}{\partial q^s_l}\rangle\rangle + \frac{1}{Q}J_s q^s - \frac{1}{2}[\bar{H}(I,\pi_s,q^s,t'),$$

$$\int_0^{t'} \bar{H}(I,\pi_s,q^s,t_1)dt_1] + \frac{1}{2}[\bar{H}(I,\pi_s,q^s,t'),\int_0^{t'} dt_1 \dot{q}_{s_m}\langle\langle\pi_{f_k}\frac{\partial q^f_k}{\partial q^s_m}\rangle\rangle]$$

$$+\frac{1}{2}[\dot{q}_{s_m}\langle\langle\pi_{f_k}\frac{\partial q^f_k}{\partial q^s_m}\rangle\rangle,\int_0^{t'} \bar{H}(I,\pi_s,q^s,t_1)dt_1] - \frac{1}{2}[\dot{q}_{s_l}\langle\langle\pi_{f_i}\frac{\partial q^f_i}{\partial q^s_l}\rangle\rangle,$$

$$\int_0^{t'} dt_1 \dot{q}_{s_m}\langle\langle\pi_{f_k}\frac{\partial q^f_k}{\partial q^s_m}\rangle\rangle]\}$$

From here we can define an effective Hamiltonian H_{eff}:

$$H_{eff} = \bar{H}(I,\pi_s,q^s) + \dot{q}_{s_l}\langle\langle\frac{\partial\zeta}{\partial q^s_l}\rangle\rangle - \dot{q}_{s_l}\langle\langle\pi_{f_i}\frac{\partial q^f_i}{\partial q^s_l}\rangle\rangle$$

$$-\frac{1}{2}[\dot{q}_{s_l}\langle\langle\pi_{f_i}\frac{\partial q^f_i}{\partial q^s_l}\rangle\rangle,\int_0^{t'} dt_1 \dot{q}_{s_m}\langle\langle\pi_{f_k}\frac{\partial q^f_k}{\partial q^s_m}\rangle\rangle].$$

From a Hamiltonian variational principle, it was shown in [GT87] from simple arguments that the averaged fast motion induces an effective gauge field which acts on the slow variables. We follow these arguments closely for the coupled system subject to fluctuations near the instability. The variational principle gives:

$$\delta S_{eff} = \delta \int_0^T dt[\pi_{sl}\dot{q}^s_l - H_{eff}(I,\pi_s,q^s)] = 0$$

$$= \delta \int_0^T dt\{\pi_{sl}\dot{q}^s_l - \bar{H}(I,\pi_s,q^s) - \dot{q}^s_l\langle\langle\frac{\partial\zeta}{\partial q^s_l} - \pi_{f_i}\frac{\partial q^f_i}{\partial q^s_l}\rangle\rangle$$

$$-\frac{1}{2}[\dot{q}_{s_l}\langle\langle\pi_{f_i}\frac{\partial q^f_i}{\partial q^s_l}\rangle\rangle,\int_0^{t'} dt_1 \dot{q}_{s_m}\langle\langle\pi_{f_k}\frac{\partial q^f_k}{\partial q^s_m}\rangle\rangle]\} = 0$$

$$= \delta \int_0^T dt\{[\pi_{sl} + \langle\langle\pi_{f_i}\frac{\partial q^f_i}{\partial q^s_l}\rangle\rangle]\dot{q}^s_l - \bar{H}(I,\pi_s,q^s) + \frac{1}{2}[\dot{q}_{s_l}\langle\langle\pi_{f_i}\frac{\partial q^f_i}{\partial q^s_l}\rangle\rangle,$$

$$\int_0^{t'} dt_1 \dot{q}_{s_m}\langle\langle\pi_{f_k}\frac{\partial q^f_k}{\partial q^s_m}\rangle\rangle]\}.$$

The term having the single-valued function ζ vanishes since it is a total time derivative. Varying S_{eff} with respect to π_s and q^s, keeping the end-points fixed gives:

$$\delta S_{eff} = \int_0^T dt\{\delta\pi_{sl}\left(\dot{q}^s{}_l - \frac{\partial\bar{H}}{\partial\pi_{sl}}\right) +$$

$$\delta q^s{}_l[\left(\frac{\partial}{\partial q^s{}_l}\langle\langle\pi_{fi}\frac{\partial q^f{}_i}{\partial q^s{}_m}\rangle\rangle - \frac{\partial}{\partial q^s{}_m}\langle\langle\pi_{fi}\frac{\partial q^f{}_i}{\partial q^s{}_l}\rangle\rangle + \frac{1}{2}[\langle\langle\pi_{fi}\frac{\partial q^f{}_i}{\partial q^s{}_m}\rangle\rangle, \langle\langle\pi_{fk}\frac{\partial q^f{}_k}{\partial q^s{}_l}\rangle\rangle]\right)\dot{q}^s{}_m$$

$$-\frac{\partial\bar{H}}{\partial q^s{}_l} - \dot{\pi}_{sl}]\}.$$

We define as in [GT87], the quantity in angular brackets as

$$\langle\langle\pi_{fi}\frac{\partial q^f{}_i}{\partial q^s{}_l}\rangle\rangle = A_l. \tag{5.88}$$

Then $\delta S_{eff} = 0$ leads to

$$\dot{q}^s{}_l = \frac{\partial\bar{H}}{\partial\pi_{sl}}, \qquad \dot{\pi}_{sl} = -\frac{\partial\bar{H}}{\partial q^s{}_l} + \left(\frac{\partial A_m}{\partial q^s{}_l} - \frac{\partial A_l}{\partial q^s{}_m} + \frac{1}{2}[A_l, A_m]\right)\dot{q}^s{}_m \tag{5.89}$$

Following [GT87], we can identify A_l with a gauge potential, and a curvature tensor $\mathcal{F}_{lm}$ can be defined as

$$\mathcal{F}_{lm} = \frac{\partial A_m}{\partial q^s{}_l} - \frac{\partial A_l}{\partial q^s{}_m} + \frac{1}{2}[A_l, A_m], \tag{5.90}$$

so that the momenta in (5.89) can be rewritten as

$$\dot{\pi}_{sl} = -\frac{\partial\bar{H}}{\partial q^s{}_l} + \mathcal{F}_{lm}\frac{\partial\bar{H}}{\partial\pi_{sl}}.$$

The commutator terms in the momenta and curvature tensor were absent in [GT87] since the Magnus expansion for the time ordered integral was not used there.

Equation (5.89) shows that the curvature tensor $\mathcal{F}_{lm}$ exerts a velocity dependent force on the slow variables. In order to write canonical equations of motion, one has to therefore introduce modified Poisson bracket relations in the slow-variable space:

$$\{f(q^s, \pi_s), g(q^s, \pi_s)\} = \{\frac{\partial f}{\partial\pi_{sl}}\frac{\partial g}{\partial q^s{}_l} - \frac{\partial f}{\partial q^s{}_l}\frac{\partial g}{\partial\pi_{sl}}\} - \mathcal{F}_{lm}\frac{\partial f}{\partial\pi_{sl}}\frac{\partial g}{\partial\pi_{sm}}.$$

Thus the gauge potential coupled to the slow variables is induced by the fast degrees of freedom as is evident from (5.88), the spontaneous appearance of the gauge symmetry being associated with the phase degrees of freedom of the center modes. The emergence of a gauge structure for the system follows from the crucial property of separability of the variables as slow and fast ones evolving at different time scales, which results from the slaving principle for the stable modes near the bifurcation in a noisy system. This leads to the motion and deformation of the closed orbits in the configuration space.

Having made the correspondence of the gauge potential A_l and the curvature tensor $\mathcal{F}_{lm}$ with the dynamics of the actual coupled system through the fast and slow variables, we proceed to examine under what conditions phase locked behavior and full synchronization would occur in a coupled system.

5.4.3 Synchronization Condition

At any instant of time, the phase difference between two oscillators q^1 and q^2 located at two different points x and y in the configuration space can be found from their inner product:

$$\cos\theta_y = \frac{(q^2(y), q^1(y))}{|(q^2(y), q^1(y))|} = \int d^d y d^d x \, \text{Tr}\left(P(e^{\int_x^y A_\mu^\alpha(s)t^\alpha ds_\mu})\right) \frac{(q^2(y), q^1(x))}{|(q^2(y), q^1(y))|},$$

where θ_y denotes the angle between the oscillators $q^1(x)$ and $q^2(y)$ in configuration space, measured at the coordinate y. The path–ordered Wilson line integral appears in the equation above since $q^1(x)$ must be transported in parallel to the coordinate point y in order to compare it with q^2 located at y. P denotes the path–ordering. Since the q^is are related to each other through a gauge transformation in the nD configuration space, this can be rewritten using (5.68) as

$$\cos\theta_y = \int d^d y d^d x \, \text{Tr}\left(e^{-it^\beta \Lambda^\beta} P(e^{\int_x^y A_\mu^\alpha(s)t^\alpha ds_\mu})\right) \frac{(q^1(y), q^1(x))}{|(q^2(y), q^1(y))|} \quad (5.91)$$

$$= \int d^d y d^d x \, \text{Tr}\left(e^{-it^\beta \Lambda^\beta} P(e^{\int_x^y A_\mu^\alpha(s)t^\alpha ds_\mu}) P(e^{\int_y^x A_\lambda^\gamma(p)t^\gamma dp_\lambda})\right) \frac{(q^1(y), q^1(y))}{|(q^2(y), q^1(y))|}$$

$$= \int d^d y d^d x \, \text{Tr}\left(e^{-it^\beta \Lambda^\beta} P(e^{\int_x^y A_\mu^\alpha(s)t^\alpha ds_\mu}) P(e^{\int_x^y A_\lambda^\gamma(p)t^\gamma dp_\lambda})\right) \frac{1}{|(q^2(y), q^1(y))|},$$

since $(q^1(y), q^1(y)) = 1$. If the angle between q^1 and q^2 remains constant for all times, then the oscillators q^1 and q^2 would be phase–locked; if the angle between them is vanishing for all time, the oscillators would be fully synchronized in phase with each other.

We would like to determine the conditions under which the phases of any two oscillators in a coupled nonlinear system would be locked and fully synchronized. Since each q^i is an oscillator, each undergoes periodic dynamics in the configuration space. Let $q^1(y)$ after being transported in parallel from coordinate x, now return to the point x during the course of its temporal evolution. We denote the state of this oscillator after completing one orbit and returning to x by $q^{1\prime}(x)$. By the time q^1 completes this orbit, q^2 would have evolved to another point z. Hence we would now like to calculate the angle between $q^2(z)$ and $q^{1\prime}(x)$. We have:

$$q^{1\prime}(x) = P\left(e^{\oint A_\mu^\alpha(s)t^\alpha ds_\mu}\right) q^1(x).$$

$q^2(z)$ can be transported in parallel to x to compare it with $q^{1\prime}(x)$:

$$q^2(x) = P\left(e^{\int_z^x A_\mu^\alpha(s)t^\alpha ds_\mu}\right) q^2(z).$$

Then the angle between q^1 and q^2 at x can be calculated:

$$\cos\theta_x = \frac{(q^2(x), q^{1\prime}(x))}{|(q^2(x), q^{1\prime}(x))|} \tag{5.92}$$

$$= \int d^d z d^d x \; \text{Tr} \left(P(e^{\int_z^x A^\alpha_\mu(s)t^\alpha ds_\mu}) \right)^T P \left(e^{\oint A^\beta_\nu(p)t^\beta dp_\nu} \right) \frac{(q^2(z), q^1(x))}{|(q^2(x), q^{1\prime}(x))|}$$

$$= \int d^d z d^d x \; \frac{(q^1(z), q^1(x))}{|(q^2(x), q^{1\prime}(x))|} \times$$

$$\text{Tr} \left[\left(P(e^{\int_z^x A^\alpha_\mu(s)t^\alpha ds_\mu}) e^{it^\beta \Lambda^\beta} \right)^T P \left(e^{\oint A^\beta_\nu(p)t^\beta dp_\nu} \right) \right]$$

$$= \int d^d z d^d x \; \frac{(q^1(x), q^1(x))}{|(q^2(x), q^{1\prime}(x))|} \times \tag{5.93}$$

$$\text{Tr} \left\{ \left[P \left(e^{\int_z^x A^\alpha_\mu(s)t^\alpha ds_\mu} \right) e^{it^\beta \Lambda^\beta} P \left(e^{\int_z^x A^\alpha_\kappa(l)t^\alpha dl_\kappa} \right) \right]^T P \left(e^{\oint A^\beta_\nu(p)t^\beta dp_\nu} \right) \right\}.$$

The change in the angle between q^1 and q^2 during a time interval t can be found using (5.91) and (5.93) and by simplifying the resulting expression, to lowest order in Λ to be

$$\cos\theta_y - \cos\theta_x = -\frac{C(p)}{2} \int d^d x d^d y \{ -i\Lambda^a F^a_{ij} + F^a_{ij} \int_y^x A^a_\mu(s) ds_\mu$$

$$+ (\delta^{\alpha a} + \frac{\epsilon^{\beta\alpha a}}{2} \Lambda^\beta) F^a_{ij} \int_y^x A^\alpha_\mu(s) ds_\mu + \dots \}.$$

In arriving at this expression we have used the relation for the Wilson loop integral

$$P(e^{\oint A^\beta_\mu(s)t^\alpha ds_\mu}) = e^{F_{\mu\nu}},$$

in which the t^a (introduced earlier in (5.68)) are generators of the Lie algebra

$$[t^a, t^b] = i\epsilon^{abc} t^c, \qquad \text{and}$$
$$t^a F^a_{\mu\nu} = \partial_\mu t^a A^a_\nu - \partial_\nu t^a A^a_\mu - i[t^a A^a_\mu, t^b A^b_\nu]$$

is the curvature tensor of the complex abstract vector space. Also we have used the matrix identity

$$e^A e^B = e^{A+B+\frac{1}{2}[A,B]+\dots}$$

and the trace relations:

$$\text{Tr}(t^a_p) = 0, \qquad \text{Tr}(t^a_p t^b_p) = C(p)\delta^{ab},$$

where $C(p)$ is a constant for the representation p. As the system we are considering is subject to fluctuations and is not deterministic, the quantity which is actually of interest to us is the noise average $\langle \cos\theta_y - \cos\theta_x \rangle$ of the phase difference between q^1 and q^2:

$$\langle \cos\theta_y - \cos\theta_x \rangle = -\frac{C(p)}{2}\langle \int d^dx d^dy \{-i\Lambda^a F^a_{ij} \tag{5.94}$$

$$+ F^a_{ij}\int_y^x A^a_\mu(s)ds_\mu + (\delta^{\alpha a} + \frac{\epsilon^{\beta\alpha a}}{2}\Lambda^\beta)F^a_{ij}\int_y^x A^\alpha_\mu(s)ds_\mu + \dots \}\rangle.$$

Using the *Gauss–Bonnet theorem*, we see that the first integral on the right hand side of this equation gives a topological invariant, the *Euler characteristic* χ_E of the surface S over which the integration is performed: $\int_S F_{ij} = \chi_E$. For the situation in which there is perfect phase synchronization between any two oscillators in the system, this constant term on the right–hand side of (5.94) should vanish and the other terms in the equation must also vanish. The 2–torus T^2 is a well known example of a topological space with vanishing Euler characteristic. The limit cycles of the coupled system are therefore constrained to remain on T^2 as they synchronize in phase. For the oscillators q^1 and q^2 to exhibit phase–locked behavior, we observe that we must have, to lowest order in Λ,

$$\frac{C(p)}{2}\langle \int d^dx d^dy \{F^a_{ij}(2\delta^{\alpha a} + \frac{\epsilon^{\beta\alpha a}}{2}\Lambda^\beta)\int_y^x A^\alpha_\mu(s)ds_\mu \}\rangle = \text{constant}. \tag{5.95}$$

While the explicit value of the left hand side of (5.95) would vary from one set of coupled systems to another, it is interesting to note that in all cases external noise seems to play a role in bringing about the phase synchronization. This can be seen as follows. The noise averages of A_μ and $F_{\mu\nu}$ are calculated by solving the coupled Langevin equations in (5.69) using (5.88) and (5.90). From (5.70), (5.85) and (5.88) we see that the lowest order terms of $\langle A_\mu \rangle$ and $\langle F_{\mu\nu} \rangle$ such as $\langle \langle\langle \pi_{f_i} \rangle\rangle \rangle$ would not contribute so that at least to this order, the left hand side of (5.95) is brought very close to zero taking the system towards synchrony, whereas in the case when no fluctuations are present these terms would be non–zero.

The role of noise in bringing about phase synchrony can also be understood in the following way. Consider the analysis by Ermentrout [Erm81] of two weakly coupled oscillators:

$$\frac{1}{\omega_i}\frac{dZ_i}{dt} = F_i(Z_i) + \kappa G_i(Z_i, Z_j), \qquad (i, j = 1, 2, \ i \neq j).$$

$Z_i \in \mathbb{R}^{N_i}$ and F_i are continuous and differentiable, G_i are continuous, and each un–coupled system $\frac{dZ_i}{dt} = F_i(Z_i)$ admits a unique, globally stable periodic solution. It was shown in [Erm81] that the coupled state admits a parameter regime in which $n : m$ phase–locking occurs between the two oscillators, after n cycles of oscillator 1 and m cycles of oscillator 2. Using a multiple-scale perturbation technique, introducing slow τ and fast s time variables: $\frac{d}{dt} = \omega_2\frac{d}{ds} + \kappa\frac{d}{d\tau}$, this $(N_1 + N_2)$D system was reduced to a 1D evolution equation on a slow time–scale for the phase difference Φ between the two oscillators. This was shown to have the form:

$$\frac{d\Phi}{d\tau} = H(\Phi), \tag{5.96}$$

where $H(\Phi) = H(\Phi + 2\pi)$, and the phase shifts Φs vary slowly in the direction of the flow of the limit cycle which is formed due to the coupling of the oscillators. Further it was shown that the phase-locked solution to the coupled system corresponds to the fixed–points of (5.96), H being identified with the *Poincaré map* for the flow of the full system.

For a nonzero value of the noise strength, the sharp transition to the critical point is replaced by a bifurcation region, and hence the time spent by the unstable modes near the bifurcation (critical slowing down of the deterministic system at the bifurcation) is much longer in the stochastic case. It is known that the slower the system moves along any part of the limit cycle, the larger is its statistical weight in that part of the limit cycle. Hence, from (5.96) and the result of [Erm81] mentioned above, the phase-locked solutions of the coupled oscillators are statistically favored. Moreover, since in the stochastic system these (unstable) slowly varying phase differences show an increase in the relaxation time as the instability is approached, as compared to the deterministic case, phase–locking and synchronous solutions have a larger statistical weight in the presence of the weak noise.

Recent experimental observations by Fujii *et al* [FHI00] of two chemical oscillators separated by some distance in the light–sensitive *Belousov–Zhabotinsky reaction* show *self–synchronization* of phase and frequency by application of noise. They observed spontaneous synchronization for small separation distances in the absence of noise and demonstrated the existence of an optimum noise intensity for the self synchronization phenomenon. Phase synchronization in coupled non–identical *FitzHugh–Nagumo neurons* subject to independent external noise was also demonstrated through numerical simulations in [HZ00]. Noise–induced phase and frequency synchronization was also demonstrated recently in stochastic oscillatory systems both analytically and with numerical simulations [TT04].

5.5 Complex Networks and Chaotic Transients

The dynamics of complex networks [BS02] is a challenging research topic in physics, technology and the life sciences. Paradigmatic models of units interacting on networks are pulse– and phase–coupled oscillators [Str]. Often attractors of the network dynamics in such systems are states of collective synchrony (see, e.g., [EPG95, Win01]). Motivated by synchronization phenomena observed in biological systems, such as the heart [Pes75] or the brain [GKE89], many studies have investigated how simple pulse-coupled model units can synchronize their activity. Here a key question is whether and how rapid synchronization can be achieved in large networks. It has been shown that fully connected networks as well as arbitrary networks of non-leaky integrators can synchronize very rapidly [MS90]. Biological networks however, are typically composed of dissipative elements and exhibit a complicated connectivity.

In [ZTG04], the authors investigated the influence of diluted network connectivity and dissipation on the collective dynamics of pulse-coupled oscillators. They found that the dynamics was completely different from that of globally coupled networks or networks of non–leaky units, even for moderate dissipation and dilution: long chaotic transients dominated the network dynamics for a wide range of connectivities, rendering the limit–cycle type *attractors* irrelevant. Whereas the transient length was shortest for very high and very low connectivity, it became very large for networks of intermediate connectivity. Their transient dynamics exhibited a robust form of synchrony that differed strongly from the synchronous dynamics on the limit cycle attractors. They quantified the chaotic nature of the transient dynamics by analytically calculating an approximation to the *largest Lyapunov exponent on the transient*.

Consider a system of N oscillators [MS90, EPG95] that interact on a directed graph by sending and receiving pulses. For concreteness, consider asymmetric random networks in which every oscillator i is connected to an other oscillator $j \neq i$ by a directed link with probability p. A phase variable $0 \leq \phi_j(t) \leq 1$ specifies the state of each oscillator j at time t. In the absence of interactions the dynamics of an oscillator j is given by

$$\dot{\phi}_j(t) = 1. \tag{5.97}$$

When an oscillator j reaches the threshold, $\phi_j(t) = 1$, its phase is reset to zero, $\phi_j(t^+) = 0$, and the oscillator emits a pulse that is sent to all oscillators i possessing an in–link from j. After a delay time τ this pulse induces a phase jump in the receiving oscillator i according to [ZTG04]

$$\phi_i((t+\tau)^+) = \min\{U^{-1}(U(\phi_i(t+\tau)) + \varepsilon_{ij}), 1\}, \tag{5.98}$$

which depends on its instantaneous phase $\phi_i(t+\tau)$, the excitatory coupling strength $\varepsilon_{ij} \geq 0$, and on whether the input is sub– or supra–threshold. The phase dependence is determined by a twice continuously differentiable function $U(\phi)$ that is assumed to be strictly increasing, $U'(\phi) > 0$, concave (down), $U''(\phi) < 0$, and normalized such that $U(0) = 0$ and $U(1) = 1$ [MS90, TWG03].

This model, originally introduced by [MS90], is equivalent to different well known models of interacting threshold elements if $U(\phi)$ is chosen appropriately (see [TWG03]). The results presented in [ZTG04] were obtained for $U_b(\phi) = b^{-1}\ln(1 + (e^b - 1)\phi)$, where $b > 0$ parameterizes the curvature of U, that determines the strength of the dissipation of individual oscillators. The function U approaches the linear, non-leaky case in the limit $\lim_{b \to 0} U_b(\phi) = \phi$. Other nonlinear choices of $U \neq U_b$ give results similar to those reported below. The considered graphs are strongly connected, i.e., there exists a directed path from any node to any other node. We normalize the total input to each node $\sum_{j=1}^{N} \varepsilon_{ij} = \varepsilon$ such that the fully synchronous state (ϕ_i $(t) \equiv \phi_0(t)$ for all i) exists [TWG02]. Furthermore for any node i all its k_i incoming links have the same strength $\varepsilon_{ij} = \varepsilon/k_i$.

Numerical investigations of all–to–all coupled networks ($p = 1$) show rapid convergence from arbitrary initial conditions to periodic orbit attractors (see [MS90, TWG02]), in which several synchronized groups of oscillators (clusters) coexist [EPG95, TWG02]. In general, the transient length T, i.e., the time the system needs to reach an attractor, is short for all-to-all coupled networks and depends only weakly on network size, for instance $T \approx 10^2$ for $N = 100$ oscillators.

In contrast to fully connected networks, diluted networks exhibit largely increased transient times: eliminating just 3% of the links ($p = 0.97$) leads to an increase in T of one order of magnitude $T \approx 10^3$. The system finally settles on an attractor which is similar to the one found in the fully connected network, i.e., a periodic orbit with several synchronized clusters. Further dilution of the network causes the transient length to grow extremely large, $T > 10^5$. The lifetime of an individual transient typically depends strongly on the initial condition. We observed the dynamics started from many randomly chosen initial phase vectors distributed uniformly in $[0, 1]^N$ and typically find a wide range of transient times.

The average transient length, a dynamical feature of the network, depends non-monotonically on the network connectivity p, whereas many structural properties of random graphs such as the average path length between two vertices are monotonic in p [Bol01]. The average transient length is short for low and high connectivity values p, but becomes very large for intermediate connectivities, even for only weakly diluted networks [ZTG04]. Moreover, the mean lifetime $\langle T \rangle$ grows exponentially with the network size N for diluted networks whereas it is almost independent of network size for fully–connected networks. This renders long transients the dominant form of dynamics for all but strongly diluted or fully connected networks. The transient length defines a new, collective time scale that is much larger than the natural period, 1, of an individual oscillator and the delay time, τ, of the interactions. This separation of time scales makes it possible to statistically characterize the dynamics on the transient (see [Tel90]).

What are the main features of the transient dynamics? We determined the statistical distribution of phases of the oscillators during the transient. Interestingly the oscillators exhibit a novel kind of synchrony during the transient: most of the units form a single, roughly synchronized cluster. This cluster is robust in the sense that it always contains approximately the same number of oscillators with the same spread of relative phases although it continuously absorbs and emits oscillators. On the attractor, however, the oscillators are organized in several precisely synchronized clusters, such that the transient dynamics is completely different from the dynamics on the attractor. The nearby trajectories diverge exponentially with time, which is an indication of chaos. To further quantify the chaotic nature of the transient dynamics, the speed of divergence of two nearby trajectories was determined in [ZTG04],

both by numerical measurements and by analytically estimating the largest
Lyapunov exponent on the transient.

Let $\{t_n: n \in \mathbb{N}_0\}$ be a set of firing times of an arbitrary, fixed reference
oscillator and $\phi_i^1(t_n), \phi_i^2(t_n)$, $i \in \{1, \ldots, N\}$, the set of phases of all oscillators
i of two transient trajectories 1 and 2. Define the distance between these two
trajectories at time t_n as

$$D_n = \sum_{i=1}^{N} |\phi_i^1(t_n) - \phi_i^2(t_n)|_c, \tag{5.99}$$

where $|\cdot|_c$ denotes the distance of two points on a circle with circumference
1. For a small initial separation $D_0 \ll 1$ at time t_0 the distance D_n at later
time t_n scales like

$$D_n \approx D_0 e^{\Lambda n}, \tag{5.100}$$

quantifying the speed of divergence by the largest Lyapunov exponent Λ. To
analytically estimate Λ, let us determine the phase advance of oscillator i
evoked by a single pulse received from oscillator j [ZTG04],

$$\phi_i(t^+) = a_{ij}\phi_i(t) + c_{ij}, \tag{5.101}$$

where we have used the definition of $U_b(\phi)$ in (5.98) and considered only
sub–threshold input. We get

$$a_{ij} = e^{b\varepsilon_{ij}} \geq 1, \tag{5.102}$$

independent of the delay τ and the constants c_{ij} which are independent of
$\phi_j(t)$. The magnitude by which a single pulse increases the difference

$$|\phi_i^1(t^-) - \phi_i^2(t^+)|_c = a_{ij}|\phi_i^1(t) - \phi_i^2(t)|_c$$

is thus only determined by a_{ij}. If an oscillator i receives exactly one pulse
from all its upstream (i.e., presynaptic) oscillators between t_n and t_{n+1}, the
total increase due to all these pulses is determined by

$$A_i = \prod_{j=1}^{N} a_{ij} = \exp\left(b \sum_{j=1}^{N} \varepsilon_{ij}\right). \tag{5.103}$$

The normalization $\sum_{j=1}^{N} \varepsilon_{ij} = \varepsilon$ implies $A_i = \exp(b\varepsilon) = A$ for all oscillators i. Noting that the network is in an asynchronous state during the transient
and assuming that between t_n and t_{n+1} each oscillator fires exactly once we
find $D_{n+1} \approx A D_n$. The largest Lyapunov exponent is then approximated by
[ZTG04]

$$\Lambda \approx \ln A = b \cdot \varepsilon. \tag{5.104}$$

Numerical results of [ZTG04] were in good agreement with (5.104) for
intermediate values of b. Note that there is no free parameter in (5.104).

This calculation shows that the curvature b of U_b and the coupling strength ε strongly influence the dynamics during the transient by determining the largest Lyapunov exponent. To further characterize the dynamics we investigated return maps of inter spike intervals (i.e. the time between two consecutive resets of individual oscillators). This analysis revealed broken tori in phase–space (not shown). Moreover we find that the correlation dimension [GP83a] of the transient orbit is low (e.g., $D_{corr.} \approx 5$ for $N = 100$, $p = 0.96$ and $b = 3$, $\varepsilon = 0.1$, $\tau = 0.1$). This indicates that the chaotic motion during the transient takes place in the vicinity of quasi–periodic motion on a low dimensional toroidal manyfold in phase–space. Interestingly quasi–periodic motion was also found in fully-connected networks of similar integrate-and-fire oscillators [Vre96].

6

Josephson Junctions and Quantum Engineering

This Chapter addresses modern electronic devices called *Josephson junctions*, which promise to be a basic building blocks of the future quantum computers. Apparently, they can exhibit chaotic behavior, both as single junctions (which have macroscopic dynamics analogous to those of the forced nonlinear oscillators), and as arrays (or ladders) of junctions, which can show high–dimensional chaos.

A *Josephson junction* is a type of electronic circuit capable of switching at very high speeds, i.e., frequency of typically $10^{10} - 10^{11}$ Hz, when operated at temperatures approaching absolute zero. It is an insulating barrier separating two superconducting materials and producing the *Josephson effect*. The terms are named eponymously after British physicist Brian David Josephson, who predicted the existence of the Josephson effect in 1962 [Jos74]. Josephson junction exploits the phenomenon of *superconductivity*, the ability of certain materials to conduct electric current with practically zero resistance. Josephson junctions have important applications in quantum–mechanical circuits. They have great technological promises as amplifiers, voltage standards, detectors, mixers, and fast switching devices for digital circuits. They are used in certain specialized instruments such as highly–sensitive microwave detectors, magnetometers, and QUIDs. Finally, Josephson junctions allow the realisation of *qubits*, the key elements of *quantum computers*.

Josephson junctions have been particularly useful for experimental studies of nonlinear dynamics as the equation governing a single junction dynamics is the same as that for a pendulum [Str94]. Their dynamics can be analyzed both in a simple overdamped limit and in the more complex underdamped one, either for single junctions and for arrays of large numbers of coupled junctions.

A Josephson junction is made up of two superconductors, separated by a weak coupling non–superconducting layer, so thin that electrons can cross through the insulating barrier. It can be conceptually represented as:

$$Superconductor\ 1\ :\ \psi_1 e^{i\phi_1}$$
$$Weak\ Coupling\ \Updownarrow$$
$$Superconductor\ 2\ :\ \psi_2 e^{i\phi_2}$$

where the two superconducting regions are characterized by simple *quantum–mechanical wave functions*, $\psi_1 e^{i\phi_1}$ and $\psi_2 e^{i\phi_2}$, respectively. Normally, a much more complicated description would be necessary, as there are $\sim 10^{23}$ electrons to deal with, but in the superconducting ground state, these electrons form the so–called *Cooper pairs* that can be described by a single macroscopic wave function $\psi e^{i\phi}$. The flow of current between the superconductors in the

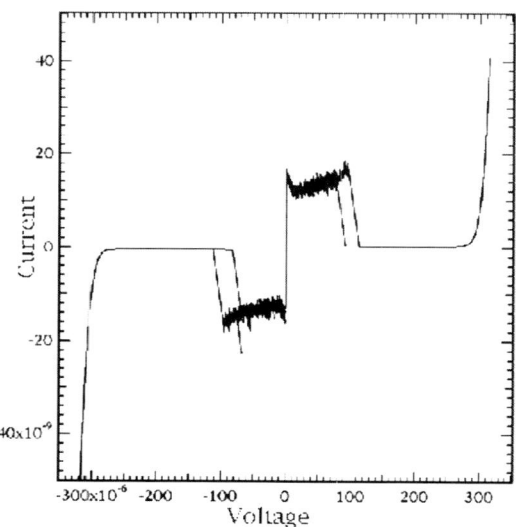

Fig. 6.1. *Josephson junction:* the current–voltage curve obtained at low temperature. The vertical portions (zero voltage) of the curve represent Cooper pair tunnelling. There is a small magnetic field applied, so that the maximum Josephson current is severely reduced. Hysteresis is clearly visible around 100 microvolts. The portion of the curve between 100 and 300 microvolts is current independent, and is the regime where the device can be used as a detector.

absence of an applied voltage is called a *Josephson current*, and the movement of electrons across the barrier is known as *Josephson tunnelling* (see Figure 6.1). Two or more junctions joined by superconducting paths form what is called a *Josephson interferometer*.

One of the characteristics of a Josephson junction is that as the temperature is lowered, superconducting current flows through it even in the absence of voltage between the electrodes, part of the *Josephson effect*. The Josephson effect in particular results from two superconductors acting to preserve their long–range order across an insulating barrier. With a thin enough barrier, the phase of the electron wave–function in one superconductor maintains a fixed relationship with the phase of the wave–function in another superconductor. This linking up of phase is called phase coherence. It occurs throughout a single superconductor, and it occurs between the superconductors in a Josephson junction. The *phase coherence*, or *long–range order*, is the essence of the Josephson effect.

While researching superconductivity, B.D. Josephson studied the properties of a junction between two superconductors. Following up on earlier work by L. Esaki and I. Giaever, he demonstrated that in a situation when there is electron flow between two superconductors through an insulating layer (in the absence of an applied voltage), and a voltage is applied, the current stops flowing and oscillates at a high frequency. The Josephson effect is influenced by magnetic fields in the vicinity, a capacity that enables the Josephson junction to be used in devices that measure extremely weak magnetic fields, such as superconducting quantum interference devices (SQUIDs). For their efforts, Josephson, Esaki, and Giaever shared the Nobel Prize for Physics in 1973.

The *Josephson–junction quantum computer* was demonstrated in April 1999 by Nakamura, Pashkin and Tsai of NEC Fundamental Research Laboratories in Tsukuba, Japan [NPT99]. In the same month, only about one week earlier, Ioffe, Geshkenbein, Feigel'man, Fauchère and Blatter, independently, described just such a computer in Nature [IGF99].

Nakamura, Pashkin and Tsai's computer is built around a *Cooper pair box*, which is a small superconducting island electrode weakly coupled to a bulk superconductor. Weak coupling between the superconductors creates a Josephson junction between them. Like most other junctions, the Josephson junction is also a capacitor, which is charged by the current that flows through it. A gate voltage is applied between the two superconducting electrodes. If the Cooper box is sufficiently small, e.g., as small as a quantum dot, the charging current breaks into discrete transfer of individual Cooper pairs, so that ultimately it is possible to just transfer a single Cooper pair across the junction. The effectiveness of the Cooper pair transfer depends on the energy difference between the box and the bulk and a maximum is reached when a voltage is applied, which equalizes this energy difference. This leads to *resonance* and observable *coherent quantum oscillations* [Ave99].

This contraption, like the Loss–Vincenzo quantum dot computer [LD98], has the advantage that it is controlled electrically. Unlike Loss–Vincenzo computer, this one actually exists in the laboratory. Nakamura, Pashkin and Tsai did not perform any computations with it though. At this stage it was enough of an art to observe the coherence for about 6 cycles of the Cooper pair oscillations, while the chip was cooled to about and carefully shielded from external electromagnetic radiation.

There are two general types of Josephson junctions: overdamped and underdamped. In overdamped junctions, the barrier is conducting (i.e., it is a normal metal or superconductor bridge). The effects of the junction's internal electrical resistance will be large compared to its small capacitance. An over-damped junction will quickly reach a unique equilibrium state for any given set of conditions.

The barrier of an underdamped junction is an insulator. The effects of the junction's internal resistance will be minimal. Underdamped junctions do not have unique equilibrium states, but are hysteretic.

A Josephson junction can be transformed into the so–called *Giaever tunnelling junction* by the application of a small, well defined magnetic field. In such a situation, the new device is called a superconducting tunnelling junction (STJ)[3] and is used as a very sensitive photon detector throughout a wide range of the spectrum, from infrared to hard x-ray. Each photon breaks up a number of Cooper pairs. This number depends on the ratio of the photon energy to approximately twice the value of the gap parameter of the material of the junction. The detector can be operated as a photon-counting spectrometer, with a spectral resolution limited by the statistical fluctuations in the number of released charges. The detector has to be cooled to extremely low temperature, typically below 1 kelvin, to distinguish the signals generated by the detector from the thermal noise. Small arrays of STJs have demonstrated their potential as spectro–photometers and could further be used in astronomy [ESA05]. They are also used to perform energy dispersive X–ray spectroscopy and in principle they could be used as elements in infrared imaging devices as well [Ens05].

6.0.1 Josephson Effect

The basic equations governing the dynamics of the Josephson effect are (see, e.g., [BP82]):

$$U(t) = \frac{\hbar}{2e} \frac{\partial \phi}{\partial t}, \qquad I(t) = I_c \sin \phi(t),$$

where $U(t)$ and $I(t)$ are the voltage and current across the Josephson junction, $\phi(t)$ is the phase difference between the wave functions in the two superconductors comprising the junction, and I_c is a constant, called the *critical current* of the junction. The critical current is an important phenomenological parameter of the device that can be affected by temperature as well as by an applied magnetic field. The physical constant $\hbar/2e$ is the magnetic flux quantum, the inverse of which is the *Josephson constant*.

The three main effects predicted by Josephson follow from these relations:

1.   The DC Josephson effect. This refers to the phenomenon of a direct current crossing the insulator in the absence of any external electromagnetic field, owing to *Josephson tunnelling*. This DC Josephson current is proportional to the sine of the phase difference across the insulator, and may take values between $-I_c$ and I_c.

2.   The AC Josephson effect. With a fixed voltage U_{DC} across the junctions, the phase will vary linearly with time and the current will be an AC

current with amplitude I_c and frequency $2e/\hbar\, U_{DC}$. This means a Josephson junction can act as a perfect *voltage–to–frequency converter*.

3.　　The inverse AC Josephson effect. If the phase takes the form

$$\phi(t) = \phi_0 + n\omega t + a\sin(\omega t),$$

the voltage and current will be

$$U(t) = \frac{\hbar}{2e}\omega[n + a\cos(\omega t)], \qquad I(t) = I_c \sum_{m=-\infty}^{\infty} J_n(a)\sin[\phi_0 + (n + m)\omega t].$$

The DC components will then be

$$U_{DC} = n\frac{\hbar}{2e}\omega, \qquad I(t) = I_c J_{-n}(a)\sin\phi_0.$$

Hence, for distinct DC voltages, the junction may carry a DC current and the junction acts like a perfect *frequency–to–voltage converter*.

6.0.2 Pendulum Analog

To show a driven pendulum analog of a microscopic description of a single Josephson junction, we start with:

1. the *Josephson current–phase relation*

$$I = I_c \sin\phi,$$

where I_c is the *critical current*, I is the bias current, and $\phi = \phi_2 - \phi_1$ is the constant *phase difference* between the phases of the two superconductors that are weakly coupled; and

2. the *Josephson voltage–phase relation*

$$V = \frac{\hbar}{2e}\dot\phi,$$

where $V = V(t)$ is the instantaneous voltage across the junction, $\hbar$ is the Planck constant (divided by 2π), and e is the charge on the electron.

Now, if we apply Kirhoff's voltage and current laws for the parallel RC–circuit with resistence R and capacitance C, we come to the first–order ODE

$$C\dot V + \frac{V}{R} + I_c\sin\phi = I,$$

which can be recast solely in terms of the phase difference ϕ as the second–order pendulum–like ODE,

$$\text{Josephson junction}: \quad \frac{\hbar C}{2e}\ddot\phi + \frac{\hbar}{2eR}\dot\phi + I_c\sin\phi = I, \qquad (6.1)$$

$$\text{Pendulum}: \quad ml^2\ddot\theta + b\dot\theta + mgl\sin\theta = \tau.$$

This mechanical analog has often proved useful in visualizing the dynamics of Josephson Junctions [Str94]. If we divide (6.1) by I_c and define a dimensionless time

$$\tau = \frac{2eI_cR}{\hbar}t,$$

we get the dimensionless oscillator equation for Josephson junction,

$$\beta\phi'' + \phi' + \sin\phi = \frac{I}{I_c}, \qquad (6.2)$$

where $\phi' = d\phi/d\tau$. The dimensionless group β, defined by

$$\beta = \frac{2eI_cR^2C}{\hbar},$$

is called the McCumber parameter and represents a dimensionless capacitance.

In a simple *overdamped limit* $\beta << 1$ with *resistive loading*, the 'inertial term' $\beta\phi''$ may be neglected (as if oscillating in a highly–viscous medium), and so (6.2) reduces to a non–uniform oscillator

$$\phi' = \frac{I}{I_c} - \sin\phi, \qquad (6.3)$$

with solutions approaching a stable fixed–point for $I < I_c$, and periodically varying for $I < I_c$. To find the current–voltage curve in the overdamped limit, we take the average voltage $\langle V \rangle$ as a function of the constant applied current I, assuming that all transients have decayed and the system has reached steady–state, and get

$$\langle V \rangle = I_cR\langle\phi'\rangle.$$

An overdamped *array of N Josephson Junctions* (6.3), parallel with a resistive load R, can be described by the system of first–order dimensionless ODEs [Str94]

$$\phi'_k = \Omega + a\sin\phi_k + \frac{1}{N}\sum_{j=1}^{N}\sin\phi_j, \qquad k = 1,...,N,$$

where

$$\Omega = I_bR_0/I_cr, \qquad a = -(R_0+r)/r, \qquad R_0 = R/N,$$

$$I_b = I_c\sin\dot{\phi}_k + \frac{\hbar}{2eR}\dot{\phi}_k + \frac{\hbar}{2eR}\sum_{j=1}^{N}\dot{\phi}_j.$$

6.1 Dissipative Josephson Junction

The past decade has seen a considerable interest and remarkable activity in an area which presently is often referred to as macroscopic quantum mechanics. Specifically, one has been interested in quantum phenomena of macroscopic objects [Leg86].

In particular, macroscopic quantum tunnelling [CL81] (quantum decay of a meta–stable state), and quantum coherence [LCD87] have been studied. Soon, it became clear that dissipation has a profound influence on these quantum phenomena. Phenomenologically, dissipation is the consequence of an interaction of the object with an environment which can be thought of as consisting of infinitely many degrees of freedom. Specifically, the environmental degrees of freedom may be chosen to be harmonic oscillators such that we may consider the dissipation as a process where excitations, that are phonons, are emitted and absorbed. This, Caldeira–Leggett model has been used in [CL81] where the influence of dissipation on tunnelling has been explored.

As far as quantum coherence is concerned, the most simple system is an object with two different quantum states: it is thought to represent the limiting case of an object in a double-well potential where only the lowest energy states in each of the two wells is relevant and where the tunnelling through the separating barrier allows for transitions that probe the coherence. Since a 2–state system is equivalent to a spin–one–half problem, this standard system is often referred to by this name. In particular, with the standard coupling to a dissipative environment made of harmonic oscillators, it is called the spin–boson problem which has been studied repeatedly in the past [LCD87, SW90].

Level quantization and resonant tunnelling have been observed recently [Vaa95] in a double–well quantum–dot system. However, the influence of dissipation was not considered in this experiment. On the other hand, it seems that Josephson junctions are also suitable systems for obtaining experimental evidence pertaining to macroscopic quantum effects. In this context, evidence for level quantization and for quantum decay have been obtained [MDC85].

Recall that a Josephson junction may be characterized by a *current–phase relation*

$$I(\phi) = I_J \sin \phi, \tag{6.4}$$

where the phase ϕ is related to the voltage difference U by

$$\hbar\dot{\phi} = 2eU . \tag{6.5}$$

Therefore, the phase of a Josephson junction shunted by a capacitance C and biased by an external current I_x obeys a classical type of equation of motion

$$M\ddot{\phi} = -\frac{\partial V(\phi)}{\partial \phi}, \qquad \text{with the mass} \tag{6.6}$$

$$M = \left(\frac{\hbar}{2e}\right)^2 C, \qquad \text{and the potential energy} \tag{6.7}$$

$$V(\phi) = -\frac{\hbar}{2e} \left[I_J \cos \phi + I_x \phi \right] . \tag{6.8}$$

A widely discussed model of a dissipative object is the one where the Josephson junction is also shunted by an Ohmic resistor R. In this case, the classical equation of motion (6.6) has to be replaced by

$$M\ddot{\phi} = -\frac{\partial V(\phi)}{\partial \phi} - \eta\dot{\phi}, \qquad \eta = \left(\frac{\hbar}{2e} \right)^2 \frac{1}{R} . \tag{6.9}$$

The model of a dissipative environment according to the above specification has been discussed by [CL81].

The potential energy $V(\phi)$ of (6.8) displays wells at $\phi \simeq 2n\pi$ with depth shifted by an amount $\Delta \simeq (2\pi\hbar/2e)I_x$. If the wells are sufficiently deep, one needs to concentrate only on transitions between pairs of adjacent wells. Thus, one arrives at the double well problem mentioned above.

The analysis in this paper goes beyond the limiting situation where only the lowest level in each of the two wells is of importance. Roughly, this is realized when the level separation $\hbar(2E_J/M)^{1/2} \simeq (2e\hbar I_J/C)^{1/2}$ is smaller than or comparable with Δ. In particular, we will concentrate on resonance phenomena which are expected to show up whenever two levels in the adjacent wells happen to cross when the bias current I_x, that is Δ, is varied.

For such values of the bias current, there appear sharp asymmetric peaks in the current-voltage characteristic of the Josephson junction. This phenomenon has been studied by [LOS88] within the standard model in the one–phonon approximation. For bias currents that correspond to crossings of the next and next nearest levels (e.g., ground state in the left well and the first or second excited state at the right side), it is possible to neglect processes in the reverse direction provided that the temperature is sufficiently low. Thus, the restriction to a double well system receives additional support.

The transfer of the object from the left to the right potential well is accompanied by the emission of an infinite number of phonons. Therefore, in a paper [OS94] the fact is taken into account that in the resonance region, the contribution of phonons of small energy is important as well as the contribution of resonance phonons with energy equal to the distance between levels in the wells.

6.1.1 Junction Hamiltonian and its Eigenstates

The model of [MS95] consists of a particle, called 'object' (coordinate R_1), which is coupled (in the sense of [CL81]) to a 'bath' of harmonic oscillators (coordinates R_j). We shall use the conventions $j \in \{2, \dots, N\}$ for the bath oscillators and $k \in \{1, \dots, N\}$ for the indices of all coordinates in the model. The double–well potential is approximated by two parabolas about the minima of the two wells.

The phase ϕ of the Josephson contact then corresponds to the object coordinate R_1 of the model, and the voltage U is related to the tunnelling rate J by $2eU = \dot{\phi} = 2\pi J$. As it has already been remarked, the current I_x is proportional to the bias Δ of the two wells. Thus, calculating the transition rate for different values of the bias Δ is equivalent to the determination of the I–V characteristics.

Specifically, following [MS95], we want to write the Hamiltonian of the model in the form

$$\hat{H} = \frac{1}{2m}\sum_k \hat{p}_k^2 + \hat{v}(\hat{R}_1) + \frac{m}{2}\sum_j \omega_j^2(\hat{R}_j - \hat{R}_1)^2,$$

$$\hat{v}(\hat{R}_1) \approx \frac{m}{2}\sum_\pm \Omega^2(\hat{R}_1 \pm a)^2 \pm \frac{\Delta}{2}. \tag{6.10}$$

The states for the two situations 'object in the left well' and 'object in the right well' will be denoted by $|\Lambda_L, L\rangle$ and $|\Lambda_R, R\rangle$, respectively. If one projects onto the eigenstates $|n\rangle$ of the 1D harmonic oscillator and takes into account the shift of the wells, one arrives at the following decomposition $(\phi_n(R) = \langle R|n\rangle)$:

$$\langle n_L, \{R_j\}|\Lambda_L, L\rangle = \int dR_1\, \phi_{n_L}(R_1 + a)\, \phi_{\Lambda_L}^L(\{R_k\}), \tag{6.11}$$

$$\langle n_R, \{R_j\}|\Lambda_R, R\rangle = \int dR_1\, \phi_{n_R}(R_1 - a)\, \phi_{\Lambda_R}^R(\{R_k\}).$$

The situations 'object on the left' and 'object on the right' differ only by the shift and the bias of the wells. Therefore, one can find a unified representation by noting that $\phi_\Lambda^L(\{R_k\}) = \Phi_\Lambda(\{R_k + a\})$ and $\phi_\Lambda^R(\{R_k\}) = \Phi_\Lambda(\{R_k - a\})$. The eigenstates Φ_Λ are defined by the relations

$$\Phi_\Lambda(\{R_k\}) = \langle\{R_k\}|\Lambda\rangle, \qquad \hat{H}_0|\Lambda\rangle = E_\Lambda|\Lambda\rangle,$$

$$\hat{H}_0 = \frac{1}{2m}\sum_k \hat{p}_k^2 + \frac{m}{2}\sum_j \omega_j^2(\hat{R}_j - \hat{R}_1)^2 + \frac{m}{2}\Omega^2\hat{R}_1^2.$$

Thus, it follows from (6.11) that

$$\langle n_L, \{R_j\}|\Lambda_L, L\rangle = \langle n_L, \{R_j\}|\exp\left(ia\sum_j \hat{p}_j\right)|\Lambda_L\rangle, \tag{6.12}$$

$$\langle n_R, \{R_j\}|\Lambda_R, R\rangle = \langle n_R, \{R_j\}|\exp\left(-ia\sum_j \hat{p}_j\right)|\Lambda_R\rangle,$$

where we have used the shift property of the momentum operator $\hat{p}$.

The coupling of the two wells is taken into account by means of a tunnelling Hamiltonian $\hat{H}_T$ which we represent in the form

$$\langle \Lambda_L, L|\hat{H}_T|\Lambda_R, R\rangle = \int d\{R_j\}\sum_{n_L n_R} T_{n_L n_R} \langle \Lambda_L, L|n_L, \{R_j\}\rangle\langle n_R, \{R_j\}|\Lambda_R, R\rangle.$$

Using again the momentum operator, one can write

$$|x\rangle\langle x'| = e^{i\hat{p}(x'-x)}\,|x'\rangle\langle x'| = e^{i\hat{p}(x'-x)}\,\delta(x'-\hat{x})\,.$$

From this, we conclude that

$$\langle \Lambda_L, L|\hat{H}_T|\Lambda_R, R\rangle = \sum_{n_L n_R} T_{n_L n_R} \int dR_1 dR'_1 \frac{dQ}{2\pi}\,\phi^*_{n_R}(R'_1)\,\phi_{n_L}(R_1)$$

$$\times\,\langle \Lambda_L, L|e^{i\hat{p}_1(R'_1-R_1)}e^{iQ(R'_1-\hat{R}_1)}|\Lambda_R, R\rangle\,.$$

6.1.2 Transition Rate

The net transition rate from the left well to the right one is then in second order perturbation theory given by [MS95]

$$J = 2\pi Z_0^{-1} \sum_{\Lambda_L, \Lambda_R} |\langle \Lambda_L, L|\hat{H}_T|\Lambda_R, R\rangle|^2\,\delta(E_{\Lambda_L} - E_{\Lambda_R} + \Delta)$$

$$\times\,[e^{-\beta E_{\Lambda_L}} - e^{\beta E_{\Lambda_R}}],$$

where $Z_0 = \text{Tr}\exp(-\beta H_0)$. The δ−function may be written in Fourier representation, and the fact that the E_Λ are eigen–energies of $\hat{H}_0$ serves us to incorporate the energy conservation into Heisenberg time–dependent operators $\hat{A}(t) = \exp(i\hat{H}_0 t)\hat{A}\exp(-i\hat{H}_0 t)$, i.e.,

$$\langle \Lambda_L, L|\hat{H}_T|\Lambda_R, R\rangle\,\delta(E_{\Lambda_L} - E_{\Lambda_R} + \Delta) =$$

$$= \int dt\,e^{i\Delta t}\,\langle \Lambda_L|e^{-ia\sum_k \hat{p}_k(t)}\,\hat{H}_T(t)\,e^{-ia\sum_k \hat{p}_k(t)}|\Lambda_R\rangle\,.$$

Then, collecting our results from above we arrive at the expression

$$J = Z_0^{-1}(1 - e^{-\beta\Delta})\int dt\,e^{i\Delta t}\sum_{n_L, n_R}\sum_{\overline{n}_L, \overline{n}_R} T_{n_L, n_R} T^*_{\overline{n}_L, \overline{n}_R}$$

$$\times \int \frac{dQ d\overline{Q}}{(2\pi)^2} \int dR_1 dR'_1 d\overline{R}_1 d\overline{R}'_1\,\phi_{n_L}(R_1)\,\phi^*_{n_R}(R'_1)\,\phi^*_{\overline{n}_L}(\overline{R}'_1)\,\phi_{\overline{n}_R}(\overline{R}_1)$$

$$\times \text{Tr}\{e^{-\beta\hat{H}_0}\,e^{-2ia\sum_k \hat{p}_k(t)}\,e^{i\hat{p}_1(t)(R'_1-R_1+2a)}\,e^{-iQ(\hat{R}_1(t)-R'_1)}$$

$$\times e^{2ia\sum_k \hat{p}_k}\,e^{i\hat{p}_1(\overline{R}'_1-\overline{R}_1-2a)}\,e^{-i\overline{Q}(\hat{R}_1-\overline{R}'_1)}\}\,.$$

Let us now use the relation

$$e^{-i(\hat{H}_0+\hat{W})t} = e^{-i\hat{H}_0 t}\,\hat{T}\,e^{-i\int_0^t dt'\,\hat{W}(t')}$$

which holds for $t > 0$ when $\hat{T}$ is the *time–ordering operator* and for $t < 0$ when the anti time–ordering is used. If we define $\langle\hat{A}\rangle = \text{Tr}\exp(-\beta\hat{H}_0)\hat{A}/Z_0$, we can write the following result for the transition rate:

$$J = (1 - e^{-\beta\Delta}) \int dt \, e^{i(\Delta - 2m\Omega^2 a^2)t} \sum_{n_L, n_R} \sum_{\overline{n}_L, \overline{n}_R} T_{n_L, n_R} T^*_{\overline{n}_L, \overline{n}_R}$$

$$\times \int \frac{dQ d\overline{Q}}{(2\pi)^2} \int dR dR' d\overline{R} d\overline{R}' \, e^{iQ\frac{R+R'}{2} + i\overline{Q}\frac{\overline{R}+\overline{R}'}{2}}$$

$$\times \phi_{n_L}(R) \, \phi^*_{n_R}(R' - 2a) \, \phi^*_{\overline{n}_L}(\overline{R}') \, \phi_{\overline{n}_R}(\overline{R} - 2a)$$

$$\times < \hat{T} \exp\left[-iQ\hat{R}_1(t) + i\hat{p}_1(t)(R' - R) + 2im\Omega^2 a \int_0^t dt' \, \hat{R}_1(t')\right.$$

$$\left. - i\overline{Q}\hat{R}_1(0) + i\hat{p}_1(0)(\overline{R}' - \overline{R})\right] > . \tag{6.13}$$

We are now in the position to make use of the fact that the Hamiltonian is quadratic in all coordinates so that we can evaluate exactly

$$\langle \hat{T} e^{i\int dt' \, \eta(t')\hat{R}_1(t')} \rangle = e^{-\frac{i}{2}\int\int dt' dt'' \eta(t')D(t',t'')\eta(t'')}, \tag{6.14}$$

$$D(t', t'') = -i\langle \hat{T}\hat{R}_1(t')\hat{R}_1(t'') \rangle .$$

By comparison with the last two lines in eq. (6.13), the function $\eta(t')$ is given by

$$\eta(t') = -Q\delta(t' - t) - \overline{Q}\delta(t') + 2m\Omega^2 a[\Theta(t') - \Theta(t' - t)]$$

$$+ m(R - R')\delta'(t' - t) + m(\overline{R} - \overline{R}')\delta'(t') .$$

$\Theta(t)$ is meant to represent the step function. The derivatives of the $\delta-$function arise from a partial integration of terms containing $\hat{p}(t) = md\hat{x}(t)/dt$. Note, that these act only on the coordinates but not on the step functions which arise due to the time ordering.

Moreover, the degrees of freedom of the bath can be integrated out in the usual way [CL81] leading to a dissipative influence on the object. One is then lead to the following form of the Fourier transform of $D(t, t') \equiv D(t - t')$:

$$D(\omega) = \frac{D^R(\omega)}{1 - \exp(-\hbar\omega/k_B T)} + \frac{D^R(-\omega)}{1 - \exp(\hbar\omega/k_B T)}, \tag{6.15}$$

$$(D^R)^{-1}(\omega) = m[(\omega + i0)^2 - \Omega^2] + i\eta\omega,$$

where we will use a spectral density $J(\omega) = \eta\omega$, $0 \leq \omega \leq \omega_c$ for the bath oscillators.

From (6.13) and (6.14) one can conclude that the integrations with respect to $Q, \overline{Q}, R, R', \overline{R}, \overline{R}'$ can be done exactly as only Gaussian integrals are involved (note, that the eigenstates of the harmonic oscillator are Gaussian functions and derivatives of these, respectively). Therefore, for given $n_L, n_R, \overline{n}_L, \overline{n}_R$, one has to perform a 6D Gaussian integral [MS95].

6.2 Josephson Junction Ladder (JJL)

2D arrays of Josephson junctions have attracted much recent theoretical and experimental attention. Interesting physics arises as a result of competing vortex-vortex and vortex–lattice interactions. It is also considered to be a convenient experimental realization of the so–called *frustrated XY models*. Following [DT95], we discuss the simplest such system, namely the *Josephson junction ladder* (JJL, see Figure 6.2) [Kar84, Gra90].

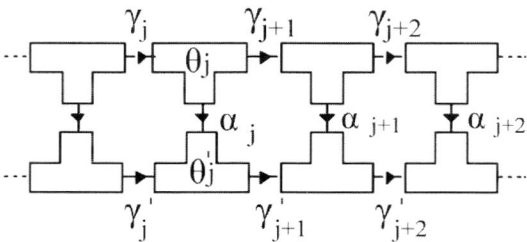

Fig. 6.2. Josephson junction Ladder (JJL).

To construct the system, superconducting elements are placed at the ladder sites. Below the bulk *superconducting–normal transition* temperature, the state of each element is described by its charge and the phase of the superconducting wave function [And64]. In this section we neglect charging effects, which corresponds to the condition that $4e^2/C \ll J$, with C being the capacitance of the element and J the Josephson coupling. Let θ_j (θ'_j) denote the phase on the upper (lower) branch of the ladder at the jth rung. The Hamiltonian for the array [Tin75] can be written in terms the gauge invariant phase differences [DT95],

$$\gamma_j = \theta_j - \theta_{j-1} - (2\pi/\phi_0) \int_{j-1}^{j} A_x dx, \gamma'_j = \theta'_j - \theta'_{j-1} - (2\pi/\phi_0) \int_{j'-1}^{j'} A_x dx,$$

$$\text{and} \quad \alpha_j = \theta'_j - \theta_j - (2\pi/\phi_0) \int_{j}^{j'} A_y dx \quad \text{as}$$

$$H = -\sum_j (J_x \cos\gamma_j + J_x \cos\gamma'_j + J_y \cos\alpha_j), \qquad (6.16)$$

where A_x and A_y are the components of the magnetic vector potential along and transverse to the ladder, respectively, and ϕ_0 the flux quantum. The sum of the phase differences around a plaquette is constrained by

$$\gamma_j - \gamma'_j + \alpha_j - \alpha_{j-1} = 2\pi(f - n_j),$$

where $n_j = 0, \pm1, \pm2, \ldots$ is the vortex occupancy number and $f = \phi/\phi_0$ with ϕ being the magnetic flux through a plaquette. With this constraint, it is convenient to write (6.16) in the form

$$H = -J \sum_j \{2\cos\eta_j \cos[(\alpha_{j-1} - \alpha_j)/2 + \pi(f - n_j)] + J_t \cos\alpha_j\},$$

$$\text{where} \quad \eta_j = (\gamma_j + \gamma'_j)/2, J = J_x \quad \text{and} \quad J_t = J_y/J_x \qquad (6.17)$$

The Hamiltonian is symmetric under $f \to f+1$ with $n_j \to n_j-1$, and $f \to -f$ with $n_j \to -n_j$, thus it is sufficient to study only the region $0 \le f \le 0.5$. Since in one dimension ordered phases occur only at zero temperature, the main interest is in the ground states of the ladder and the low temperature excitations. Note that in (6.17) η_j decouples from α_j and n_j, so that all the ground states have $\eta_j = 0$ to minimize H. The ground states will be among the solutions to the current conservation equations: $\partial_{\alpha_j} H = 0$, i.e., [DT95]

$$J_t \sin \alpha_j = \sin[(\alpha_{j-1} - \alpha_j)/2 + \pi(f - n_j)] - \sin[(\alpha_j - \alpha_{j+1})/2 + \pi(f - n_{j+1})].$$
$$(6.18)$$

For any given f there are a host of solutions to (6.18). The solution that minimizes the energy must be selected to get the ground state.

If one expands the inter–plaquette cosine coupling term in (6.17) about it's maximum, the discrete sine–Gordon model is obtained. A vortex ($n_j = 1$) in the JJL corresponds to a kink in the sine–Gordon model. This analogy was used by [Kar84] as an argument that this system should show similar behavior to the discrete sine–Gordon model which has been studied by several authors [AA80, CS83, PTB86]. This analogy is only valid for J_t very small so that the inter–plaquette term dominates the behavior of the system making the expansion about its maximum a reasonable assumption. However, much of the interesting behavior of the discrete sine–Gordon model occurs in regions of large J_t ($J_t \sim 1$). Furthermore, much of the work by Aubry [AA80] on the sine–Gordon model relies on the convexity of the coupling potential which we do not have in the JJL.

Following [DT95], here we formulate the problem in terms of a transfer matrix obtained from the full partition function of the ladder. The eigenvalues and eigenfunctions of the transfer matrix are found numerically to determine the phases of the ladder as functions of f and J_t. We study the properties of various ground states and the low temperature excitations. As J_t is varied, all incommensurate ground states undergo a superconducting–normal transition at certain J_t which depends on f. One such transition will be analyzed. Finally we discuss the critical current.

The partition function for the ladder, with periodic boundary conditions and $K = J/k_B T$, is

$$Z = \prod_i^N \int_{-\pi}^{\pi} \sum_{\{n_i\}} d\alpha_i d\eta_i \exp\left\{K(2\cos\eta_i \cos[(\alpha_{i-1} - \alpha_i)/2 + \pi(f - n_i)] + J_t \cos\alpha_i)\right\}.$$

The η_i can be integrated out resulting in a simple transfer matrix formalism for the partition function involving only the transverse phase differences:

$$Z = \prod_i^N \int_{-\pi}^{\pi} d\alpha_i P(\alpha_{i-1}, \alpha_i) = Tr\,\hat{P}^N.$$

The transfer matrix elements $P(\alpha, \alpha')$ are

$$P(\alpha, \alpha') = 4\pi \exp[K J_t(\cos\alpha + \cos\alpha')/2] \, I_0(2K \cos[(\alpha - \alpha')/2 + \pi f]), \quad (6.19)$$

where I_0 is the zeroth order modified Bessel function. Note that the elements of $\hat{P}$ are real and positive, so that its largest eigenvalue λ_0 is real, positive and nondegenerate. However, since $\hat{P}$ is not symmetric (except for $f = 0$ and $f = 1/2$) other eigenvalues can form complex conjugate pairs. As we will see from the correlation function, these complex eigenvalues determine the spatial periodicity of the ground states.

The two point correlation function of α_j's is [DT95]

$$\langle e^{i(\alpha_0 - \alpha_l)} \rangle = \lim_{N \to \infty} \frac{\left(\prod_i^N \int_{-\pi}^{\pi} d\alpha_i P(\alpha_{i-1}, \alpha_i) \right) e^{i(\alpha_0 - \alpha_l)}}{Z} = \sum_n c_n \left(\frac{\lambda_n}{\lambda_0} \right)^l,$$

$$(6.20)$$

where we have made use of the completeness of the left and right eigenfunctions. (Note that since $\hat{P}$ is not symmetric both right ψ_n^R and left ψ_n^L eigenfunctions are need for the evaluation of correlation functions.) The λ_n in (6.20) are the eigenvalues ($|\lambda_n| \geq |\lambda_{n+1}|$ and $n = 0, 1, 2, ...$), and the constants

$$c_n = \int_{-\pi}^{\pi} d\alpha' \psi_0^L(\alpha') e^{i\alpha'} \psi_n^R(\alpha') \int_{-\pi}^{\pi} d\alpha \psi_n^L(\alpha) e^{-i\alpha} \psi_0^R(\alpha).$$

In the case where λ_1 is real and $|\lambda_1| > |\lambda_2|$, (6.20) simplifies for large l to

$$\langle e^{i(\alpha_0 - \alpha_l)} \rangle = c_0 + c_1 \left(\frac{\lambda_1}{\lambda_0} \right)^l, \quad |\lambda_1| > |\lambda_2|.$$

In the case where $\lambda_1 = \lambda_2^* = |\lambda_1| e^{i2\pi\Xi}$, (6.20) for large l is [1]

$$\langle e^{i(\alpha_0 - \alpha_l)} \rangle = c_0 + \left(c_1 e^{i2\pi\Xi l} + c_2 e^{-i2\pi\Xi l} \right) \left| \frac{\lambda_1}{\lambda_0} \right|^l, \quad \lambda_1 = \lambda_2^*.$$

There is no phase coherence between upper and lower branches of the ladder and hence no superconductivity in the transverse direction. In this case, we say that the α's are unpinned. If there exist finite intervals of α on which $\rho(\alpha) = 0$, there will be phase coherence between the upper and lower branches and we say that the α's are pinned. In term of the transfer matrix, the phase density is the product of the left and right eigenfunctions of λ_0 [GM79],

$$\rho(\alpha) = \psi_0^L(\alpha) \psi_0^R(\alpha).$$

---

[1] While the correlation length is given by $\xi = [\ln|\lambda_0/\lambda_1|]^{-1}$ the quantity $\Xi = Arg(\lambda_1)/2\pi$ determines the spatial periodicity of the state. By numerical calculation of λ_n, it is found that for f smaller than a critical value f_{c1} which depends on J_t, both λ_1 and λ_2 are real. These two eigenvalues become degenerate at f_{c1}, and then bifurcate into a complex conjugate pair [DT95].

We first discuss the case where $f < f_{c1}$. These are the *Meissner–states* in the sense that there are no vortices ($n_i = 0$) in the ladder. The ground state is simply $\alpha_i = 0$, $\gamma_j = \pi f$ and $\gamma'_j = -\pi f$, so that there is a global screening current $\pm J_x \sin \pi f$ in the upper and lower branches of the ladder [Kar84]. The phase density $\rho(\alpha) = \delta(\alpha)$. The properties of the Meissner state can be studied by expanding (6.17) around $\alpha_i = 0$,

$$H_M = (J/4) \sum_j [\cos(\pi f)(\alpha_{j-1} - \alpha_j)^2 + 2J_t \alpha_i^2].$$

The current conservation (6.18) becomes

$$\alpha_{j+1} = 2\,(1 + J_t/\cos \pi f)\,\alpha_j - \alpha_{j-1}. \tag{6.21}$$

Besides the ground state $\alpha_j = 0$, there are other two linearly independent solutions $\alpha_j = e^{\pm j/\xi_M}$ of (6.21) which describe collective fluctuations about the ground state, where

$$\frac{1}{\xi_M} = \ln \left[1 + \frac{J_t}{\cos \pi f} + \sqrt{\frac{2J_t}{\cos \pi f} + \left(\frac{J_t}{\cos \pi f}\right)^2} \right]. \tag{6.22}$$

ξ_M is the low temperature correlation length for the Meissner state.[2] As f increases, the Meissner state becomes unstable to the formation of vortices. A vortex is constructed by patching the two solutions of (6.21) together using a matching condition. The energy ϵ_v of a single vortex is found to be [DT95]

$$\epsilon_v \approx [2 + (\pi^2/8)\tanh(1/2\xi_M)]\cos \pi f - (\pi + 1)\sin \pi f + 2J_t, \tag{6.23}$$

for J_t close to one. The zero of ϵ_v determines f_{c1} which is in good agreement with the numerical result from the transfer matrix. For $f > f_{c1}$, ϵ_v is negative and vortices are spontaneously created. When vortices are far apart their interaction is caused only by the exponentially small overlap. The corresponding repulsion energy is of the order $J \exp(-l/\xi_M)$, where l is the distance between vortices. This leads to a free energy per plaquette of $F = \epsilon_v/l + J\exp(-l/\xi_M)/l$ [PTB86]. Minimizing this free energy as a function of l gives the vortex density for $f > f_{c1}$: $\langle n_j \rangle = l^{-1} = [\xi_M \ln |f_{c1} - f|]^{-1}$ where a linear approximation is used for f close to f_{c1}.

We now discuss the commensurate vortex states, taking the one with $\Xi = 1/2$ as an example. This state has many similarities to the Meissner state but some important differences. The ground state is

$$\alpha_0 = \arctan \left[\frac{2}{J_t}\sin(\pi f) \right], \qquad \alpha_1 = -\alpha_0, \ \alpha_{i\pm2} = \alpha_i;$$

$$n_0 = 0, \ n_1 = 1, \ n_{i\pm2} = n_i, \tag{6.24}$$

---

[2] Here, $\xi_M < 1$ for $J_t \sim 1$ making a continuum approximation invalid.

so that there is a global screening current in the upper and lower branches of the ladder of $\pm 2\pi J(f-1/2)/\sqrt{4+J_t^2}$. The existence of the global screening, which is absent in an infinite 2D array, is the key reason for the existence of the steps at $\Xi = p/q$. It is easy to see that the symmetry of this vortex state is that of the (antiferromagnetic) Ising model. The ground state is two-fold degenerate. The low temperature excitations are domain boundaries between the two degenerate ground states. The energy of the domain boundary $J\epsilon_b$ can be estimated using similar methods to those used to derive (6.23) for the Meissner state. We found that $\epsilon_b = \epsilon_b^0 - (\pi^2/\sqrt{4+J_t^2})|f-1/2|$, where ϵ_b^0 depends only on

$$J_t = c\arctan^2(2/J_t)J_t^2\coth(1/\xi_b)/\sqrt{4+J_t^2},$$

with c being a constant of order one and

$$\xi_b^{-1} = \ln(1+J_t^2/2+J_t\sqrt{1+J_t^2/4}).$$

Thus the correlation length diverges with temperature as $\xi \sim \exp(2J\epsilon_b/k_BT)$. The transition from the $\Xi = 1/2$ state to nearby vortex states happens when f is such that $\epsilon_b = 0$; it is similar to the transition from the Meissner state to its nearby vortex states. All other steps $\Xi = p/q$ can be analyzed similarly. For comparison, we have evaluated ξ for various values of f and T from the transfer matrix and found that ξ fits $\xi \sim \exp(2J\epsilon_b/k_BT)$ (typically over several decades) at low temperature.

We now discuss the superconducting–normal transition in the transverse direction. For $J_t = 0$, the ground state has $\gamma_i = \gamma_i' = 0$ and

$$\alpha_j = 2\pi f j + \alpha_0 - 2\pi \sum_{i=0}^{i=j} n_i. \tag{6.25}$$

The average vortex density $\langle n_j \rangle$ is f; there is no screening of the magnetic field. α_0 in (6.25) is arbitrary; the α's are unpinned for all f. The system is simply two un–coupled 1D XY chains, so that the correlation length $\xi = 1/k_BT$. The system is superconducting at zero temperature along the ladder, but not in the transverse direction. As J_t rises above zero we observe a distinct difference between the system at rational and irrational values of f. For f rational, the α's become pinned for $J_t > 0$ ($\rho(\alpha)$ is a finite sum of delta functions) and the ladder is superconducting in *both* the longitudinal and transverse directions at zero temperature. The behavior for irrational f is illustrated in the following for the state with $\Xi = a_g$, where $a_g \approx 0.381966\cdots$ is one minus the inverse of the golden mean.

Finally, we consider critical currents along the ladder. One can get an estimate for the critical current by performing a perturbation expansion around the ground state (i.e., $\{n_j\}$ remain fixed) and imposing the current constraint of $\sin\gamma_j + \sin\gamma_j' = I$. Let $\delta\gamma_j$, $\delta\gamma_j'$ and $\delta\alpha_j$ be the change of γ_j, γ_j' and α_j

in the current carrying state. One finds that stability of the ground state requires that $\delta\alpha_j = 0$, and consequently $\delta\gamma_j = \delta\gamma_j' = I/2\cos\gamma_j$. The critical current can be estimated by the requirement that the γ_j do not pass through $\pi/2$, which gives $I_c = 2(\pi/2 - \gamma_{max})\cos\gamma_{max}$, where $\gamma_{max} = \max_j(\gamma_j)$. In all ground states we examined, commensurate and incommensurate, we found that $\gamma_{max} < \pi/2$, implying a finite critical current for all f. See [DT95] for more details.

6.2.1 Underdamped JJL

Recall that the *discrete sine–Gordon equation* has been used by several groups to model so-called hybrid Josephson ladder arrays [UMM93, WSZ95]. Such an array consists of a ladder of parallel Josephson junctions which are inductively coupled together (e.g., by superconducting wires). The sine-Gordon equation then describes the phase differences across the junctions. In an applied magnetic field, this equation predicts remarkably complex behavior, including flux flow resistance below a certain critical current, and a field–independent resistance above that current arising from so–called *whirling modes* [WSZ95]. In the flux flow regime, the fluxons in this ladder propagate as localized solitons, and the IV characteristics exhibit voltage plateaus arising from the locking of solitons to linear *spin–wave modes*. At sufficiently large values of the anisotropy parameter η_J defined later, the solitons may propagate 'ballistically' on the plateaus, i.e., may travel a considerable distance even after the driving current is turned off.

Following [RYD96], here we show that this behavior is all found in a model in which the ladder is treated as a network of coupled small junctions arranged along both the edges and the rungs of the ladder. This model is often used to treat two-dimensional Josephson networks, and includes *no* inductive coupling between junctions, other than that produced by the other junctions. To confirm our numerical results, we derive a discrete sine–Gordon equation from our coupled–network model. Thus, these seemingly quite different models produce nearly identical behavior for ladders. By extension, they suggest that some properties of 2D arrays might conceivably be treated by a similar simplification. In simulations [Bob92, GLW93, SIT95], underdamped arrays of this type show some similarities to ladder arrays, exhibiting the analogs of both voltage steps and whirling modes.

We consider a ladder consisting of coupled superconducting grains, the ith of which has order parameter

$$\Phi_i = \Phi_0 e^{i\theta_i}.$$

Grains i and j are coupled by *resistively–shunted Josephson junctions* (RSJ's) with current I_{ij}, shunt resistance R_{ij} and shunt capacitance C_{ij}, with periodic boundary conditions.

The phases θ_i evolve according to the coupled RSJ equations

$$\hbar\dot{\theta}_i/(2e) = V_i,$$
$$M_{ij}\dot{V}_j = I_i^{ext}/I_c - (R/R_{ij})(V_i - V_j) - (I_{ij}/I_c)\sin(\theta_{ij} - A_{ij}).$$

Here the time unit is $t_0 = \hbar/(2eRI_c)$, where R and I_c are the shunt resistance and critical current across a junction in the $x-$direction; I_i^{ext} is the external current fed into the ith node; the spatial distances are given in units of the lattice spacing a, and the voltage V_i in units of I_cR.

$$M_{ij} = -4\pi eCI_cR^2/h \qquad \text{for} \quad i \neq j,$$
$$\text{and} \qquad M_{ii} = -\sum_{j\neq i} M_{ij},$$

where C is the intergrain capacitance. Finally,

$$A_{ij} = (2\pi/\Phi_0)\int_i^j A \cdot dl,$$

where A is the vector potential. Following [RYD96], we assume N plaquettes in the array, and postulate a current I uniformly injected into each node on the outer edge and extracted from each node on the inner edge of the ring. We also assume a uniform transverse magnetic field $B \equiv f\phi_0/a^2$, and use the *Landau gauge* $A = -Bx\,\hat{y}$.

We now show that this model reduces approximately to a discrete sine–Gordon equation for the *phase differences*. Label each grain by (x, y) where $x/a = 0, \ldots, N-1$ and $y/a = 0, 1$. Subtracting the equation of motion for $\theta(x, 1)$ from that for $\theta(x, 2)$, and defining

$$\Psi(x) = \frac{1}{2}[\theta(x, 1) + \theta(x, 2)], \chi(x) = [\theta(x, 2) - \theta(x, 1)],$$

we get a differential equation for $\chi(x)$ which is second-order in time. This equation may be further simplified using the facts that $A_{x,y;x\pm1,y} = 0$ in the Landau gauge, and that $A_{x,1;x,2} = -A_{x,2;x,1}$, and by defining the *discrete Laplacian*

$$\chi(x+1) - 2\chi(x) + \chi(x-1) = \nabla^2\chi(x).$$

Finally, using the boundary conditions,

$$I^{ext}(x, 2) = -I^{ext}(x, 1) \equiv I,$$

and introducing $\phi(x) = \chi(x) - A_{x,2;x,1}$, we get

$$[1 - \eta_c^2\nabla^2]\beta\ddot{\phi} = i - [1 - \eta_r^2\nabla^2]\dot{\phi} - \sin(\phi) + 2\eta_J^2 \qquad (6.26)$$
$$\times \sum_{i=\pm1} \cos\{\Psi(x) - \Psi(x+i)\}\sin\{[\phi(x) - \phi(x+i)]/2\},$$

where we have defined a dimensionless current $i = I/I_{cy}$, and anisotropy factors

$$2\eta_r^2 = R_y/R_x, \qquad 2\eta_c^2 = C_x/C_y, \qquad 2\eta_J^2 = I_{cx}/I_{cy}.$$

We now neglect all combined space and time derivatives of order three or higher. Similarly, we set the cosine factor equal to unity(this is also checked numerically to be valid *a posteriori*) and linearize the sine factor in the last term, so that the final summation can be expressed simply as $\nabla^2\phi$. With these approximations, (6.26) reduces to *discrete driven sine-Gordon equation with dissipation*:

$$\beta\ddot{\phi} + \dot{\phi} + \sin(\phi) - \eta_J^2\nabla^2\phi = i, \qquad \text{where} \quad \beta = 4\pi eI_{cy}R_y^2C_y/h. \qquad (6.27)$$

Soliton Behavior

In the absence of damping and driving, the continuum version of (6.27) has, among other solutions, the sine–Gordon soliton [Raj82], given by

$$\phi_s(x,t) \sim 4\tan^{-1}\left[\exp\left\{(x - v_v t)/\sqrt{\eta_J^2 - \beta v_v^2,}\right\}\right]$$

where v_v is the velocity. The phase in this soliton rises from ~ 0 to $\sim 2\pi$ in a width $d_k \sim \sqrt{\eta_J^2 - \beta v_v^2}$.

The transition to the resistive state occurs at $n_{min} = 4, 2, 2, 1$ for $\eta_J^2 = 0.5, 1.25, 2.5, 5$. This can also be understood from the *kink–phason resonance* picture. To a phason mode, the passage of a kink of width d_k will appear like the switching on of a step–like driving current over a time of order d_k/v_v. The kink will couple to the phasons only if $d_k/v_v \geq \pi/\omega_1$, the half–period of the phason, or equivalently

$$\frac{1}{\sqrt{\beta}v_v} \geq \frac{\sqrt{1 + \pi^2}}{\eta_J} = \frac{3.3}{\eta_J}.$$

This condition agrees very well with our numerical observations, even though it was obtained by considering soliton solutions from the continuum sine–Gordon equation.

The fact that the voltage in regime I is approximately linear in f can be qualitatively understood from the following argument. Suppose that ϕ for Nf fluxons can be approximated as a sum of well–separated solitons, each moving with the same velocity and described by

$$\phi(x,t) = \sum_{j=1}^{Nf}\phi_j, \qquad \text{where} \qquad \phi_j = \phi_s(x - x_j, t).$$

Since the solitons are well separated, we can use following properties:

$$\sin[\sum_j \phi_j] = \sum_j \sin\phi_j \qquad \text{and} \qquad \int \dot{\phi}_j\dot{\phi}_i dx \propto \delta_{ij}.$$

By demanding that the energy dissipated by the damping of the moving soliton be balanced by that the driving current provides ($\propto \int dx i\dot\phi(x)$), one can show that the Nf fluxons should move with the same velocity v as that for a single fluxon driven by the same current. In the *whirling regime*, the $f-independence$ of the voltage can be understood from a somewhat different argument. Here, we assume a periodic solution of the form

$$\phi = \sum_{j}^{Nf} \phi_w(x - \tilde{v}t - j/f),$$

moving with an unknown velocity $\tilde{v}$ where $\phi_w(\xi)$ describes a whirling solution containing one fluxon. Then using the property $\phi(x + m/f) = \phi(x) + 2\pi m$, one can show that [RYD96]

$$\sin[\sum_{j}^{Nf} \phi_w(x - \tilde{v}t - j/f)] = \sin[Nf\phi_w(x - \tilde{v}t)].$$

Finally, using the approximate property $\phi_w(\xi) \sim \xi$ of the whirling state, one finds $\tilde{v} = v/(Nf)$, leading to an $f-$independent voltage.

Ballistic Soliton Motion & Soliton Mass

A common feature of massive particles is their 'ballistic motion', defined as inertial propagation after the driving force has been turned off. Such propagation has been reported experimentally but as yet has not been observed numerically in either square or triangular lattices [GLW93]. In the so–called *flux–flow regime* at $\eta_J = 0.71$, we also find no ballistic propagation, presumably because of the large pinning energies produced by the periodic lattice.

We can define the fluxon mass in our ladder by equating the *charging energy* $E_c = C/2 \sum_{ij} V_{ij}^2$ to the kinetic energy of a soliton of mass M_v: $E_{kin} = \frac{1}{2} M_v v_v^2$ [GLW93]. Since E_c can be directly calculated in our simulation, while v_v can be calculated from $\langle V \rangle$, this gives an unambiguous way to determine M_v. For $\eta_J^2 = 0.5$, we find $E_c/C \sim 110(\langle V \rangle /I_c R)^2$, in the flux–flow regime. This gives $M_v^I \sim 3.4 C\phi_0^2/a^2$, more than six times the usual estimate for the vortex mass in a 2D square lattice. Similarly, the vortex friction coefficient γ can be estimated by equating the rate of energy dissipation,

$$E_{dis} = 1/2 \sum_{ij} V_{ij}^2/R_{ij}, \qquad \text{to} \qquad \frac{1}{2}\gamma v_v^2.$$

This estimate yields $\gamma^I \sim 3.4\phi_0^2/(Ra^2)$, once again more than six times the value predicted for 2D arrays [GLW93]. This large dissipation explains the absence of ballistic motion for this anisotropy [GLW93]. At larger values $\eta_J^2 = 5$ and 2.5, a similar calculation gives $M_v^I \sim 0.28$ and $0.34\phi_0^2/(Ra^2)$, $\gamma^I \sim 0.28$ and $0.34\phi_0^2/(Ra^2)$. These lower values of γ^I, but especially the low pinning energies, may explain why ballistic motion is possible at these values of η_J. See [RYD96] for more details.

6.3 Synchronization in Arrays of Josephson Junctions

The *synchronization of coupled nonlinear oscillators* has been a fertile area of research for decades [PRK01]. In particular, *Winfree–type phase models* [Win67] have been extensively studied. In 1D, a generic version of this model for N oscillators reads

$$\dot{\theta}_j = \Omega_j + \sum_{k=1}^{N} \sigma_{j,k} \Gamma \left(\theta_k - \theta_j\right), \tag{6.28}$$

where θ_j is the phase of oscillator j, which can be envisioned as a point moving around the unit circle with angular velocity $\dot{\theta}_j = d\theta_j/dt$. In the absence of coupling, this overdamped oscillator has an angular velocity Ω_j. $\Gamma(\theta_k - \theta_j)$ is the coupling function, and $\sigma_{j,k}$ describes the range and nature (e.g., attractive or repulsive) of the coupling. The special case

$$\Gamma(\theta_k - \theta_j) = \sin(\theta_k - \theta_j), \qquad \sigma_{j,k} = \alpha/N, \qquad \alpha = \text{const},$$

corresponds to the uniform, sinusoidal coupling of each oscillator to the remaining $N-1$ oscillators. This mean–field system is usually called the *globally–coupled Kuramoto model* (GKM). Kuramoto was the first to show that for this particular form of coupling and in the $N \to \infty$ limit, there is a continuous dynamical phase transition at a critical value of the coupling strength α_c and that for $\alpha > \alpha_c$ both phase and frequency synchronization appear in the system [Kur84, Str00]. If $\sigma_{j,k} = \alpha \delta_{j,k\pm 1}$ while the coupling function retains the form $\Gamma(\theta_j - \theta_k) = \sin(\theta_k - \theta_j)$, then we have the so–called *locally–coupled Kuramoto model* (LKM), in which each oscillator is coupled only to its nearest neighbors. Studies of synchronization in the LKM [SSK87], including extensions to more than one spatial dimension, have shown that α_c grows without bound in the $N \to \infty$ limit [Sm88].

Watts and Strogatz introduced a simple model for tuning collections of coupled dynamical systems between the two extremes of random and regular networks [WS98]. In this model, connections between nodes in a regular array are randomly rewired with a probability p, such that $p = 0$ means the network is regularly connected, while $p = 1$ results in a random connection of nodes. For a range of intermediate values of p between these two extremes, the network retains a property of regular networks (a large clustering coefficient) and also acquires a property of random networks (a short characteristic path length between nodes). Networks in this intermediate configuration are termed *small–world networks*. Many examples of such small worlds, both natural and human–made, have been discussed [Str]. Not surprisingly, there has been much interest in the synchronization of dynamical systems connected in a small–world geometry [BP02, NML03]. Generically, such studies have shown that the presence of small–world connections make it easier for a network to synchronize, an effect generally attributed to the reduced path length between the linked systems. This has also been found to be true for the special case in which the dynamics of each oscillator is described by a Kuramoto model [HCK02a, HCK02b].

As an example of *physically–controllable systems of nonlinear oscillators*, which can be studied both theoretically and experimentally, Josephson junction (JJ) arrays are almost without peer. Through modern fabrication techniques and careful experimental methods one can attain a high degree of control over the dynamics of a JJ array, and many detailed aspects of array behavior have been studied [NLG00]. Among the many different geometries of JJ arrays, *ladder* arrays deserve special attention. For example, they have been observed to support stable time–dependent, spatially–localized states known as discrete breathers [TMO00]. In addition, the ladder geometry is more complex than that of better understood serial arrays but less so than fully two–dimensional (2D) arrays. In fact, a ladder can be considered as a special kind of 2D array, and so the study of ladders could throw some light on the behavior of such 2D arrays. Also, linearly–stable synchronization of the horizontal, or rung, junctions in a ladder is observed in the absence of a load over a wide range of dc bias currents and junction parameters (such as junction capacitance), so that synchronization in this geometry appears to be robust [TSS05].

In the mid 1990's it was shown that a serial array of zero-capacitance, i.e., overdamped, junctions coupled to a load could be mapped onto the GKM [WCS96, WCS98]. The load in this case was essential in providing an all–to–all coupling among the junctions. The result was based on an averaging process, in which (at least) two distinct time scales were identified: the 'short' time scale set by the rapid voltage oscillations of the junctions (the array was current biased above its critical current) and 'long' time scale over which the junctions synchronize their voltages. If the *resistively–shunted junction* (RSJ) equations describing the dynamics of the junctions are integrated over one cycle of the 'short' time scale, what remains is the 'slow' dynamics, describing the synchronization of the array. This mapping is useful because it allows knowledge about the GKM to be applied to understanding the dynamics of the serial JJ array. For example, the authors of [WCS96] were able, based on the GKM, to predict the level of critical current disorder the array could tolerate before frequency synchronization would be lost. Frequency synchronization, also described as entrainment, refers to the state of the array in which all junctions not in the zero–voltage state have equal (to within some numerical precision) time–averaged voltages: $(\hbar/2e)\langle\dot\theta_j\rangle_t$, where θ_j is the gauge–invariant phase difference across junction j. More recently, the 'slow' synchronization dynamics of finite–capacitance serial arrays of JJ's has also been studied [CS95, WS97]. Perhaps surprisingly, however, no experimental work on JJ arrays has verified the accuracy of this GKM mapping. Instead, the first detailed experimental verification of Kuramoto's theory was recently performed on systems of coupled electrochemical oscillators [KZH02].

Recently, [DDT03] showed, with an eye toward a better understanding of synchronization in 2D JJ arrays, that a ladder array of *overdamped* junctions could be mapped onto the LKM. This work was based on an averaging process, as in [WCS96], and was valid in the limits of weak critical current disorder (less than about 10%) and large dc bias currents, I_B, along the rung junctions ($I_B/\langle I_c \rangle \gtrsim 3$, where $\langle I_c \rangle$ is the arithmetic average of the critical currents of the rung junctions. The result demonstrated, for both open and periodic boundary conditions, that synchronization of the current–biased rung junctions in the ladder is well described by (6.28).

In this section, following [TSS05], we demonstrate that a ladder array of *underdamped junctions* can be mapped onto a second–order Winfree–type oscillator model of the form

$$a\ddot{\theta}_j + \dot{\theta}_j = \Omega_j + \sum_{k=1}^{N} \sigma_{j,k} \Gamma(\theta_k - \theta_j), \tag{6.29}$$

where a is a constant related to the average capacitance of the rung junctions. This result is based on the resistively and *capacitively–shunted junction* (RCSJ) model and a multiple time scale analysis of the classical equations for the array. Secondly, we study the effects of *small world* (SW) connections on the synchronization of both overdamped and underdamped ladder arrays. It appears that SW connections make it easier for the ladder to synchronize, and that a Kuramoto or Winfree type model (6.28) and (6.29), suitably generalized to include the new connections, accurately describes the synchronization of this ladder.

6.3.1 Phase Model for Underdamped JJL

Following [TSS05] we analyze synchronization in disordered Josephson junction arrays. The ladder geometry used consists of an array with $N = 8$ plaquettes, periodic boundary conditions, and uniform dc bias currents, I_B, along the rung junctions (see Figure 6.3). The *gauge–invariant phase difference* across rung junction j is γ_j, while the phase difference across the off–rung junctions along the outer(inner) edge of plaquette j is $\psi_{1,j}(\psi_{2,j})$. The critical current, resistance, and capacitance of rung junction j are denoted I_{cj}, R_j, and C_j, respectively. For simplicity, we assume all off–rung junctions are identical, with critical current I_{co}, resistance R_o, and capacitance C_o. We also assume that the product of the junction critical current and resistance is the same for all junctions in the array [Ben95], with a similar assumption about the ratio of each junction's critical current with its capacitance:

$$I_{cj}R_j = I_{co}R_o = \frac{\langle I_c \rangle}{\langle R^{-1} \rangle} \tag{6.30}$$

$$\frac{I_{cj}}{C_j} = \frac{I_{co}}{C_o} = \frac{\langle I_c \rangle}{\langle C \rangle}, \tag{6.31}$$

where for any generic quantity X, the angular brackets with no subscript denote an arithmetic average over the set of rung junctions,

$$\langle X \rangle \equiv (1/N) \sum_{j=1}^{N} X_j.$$

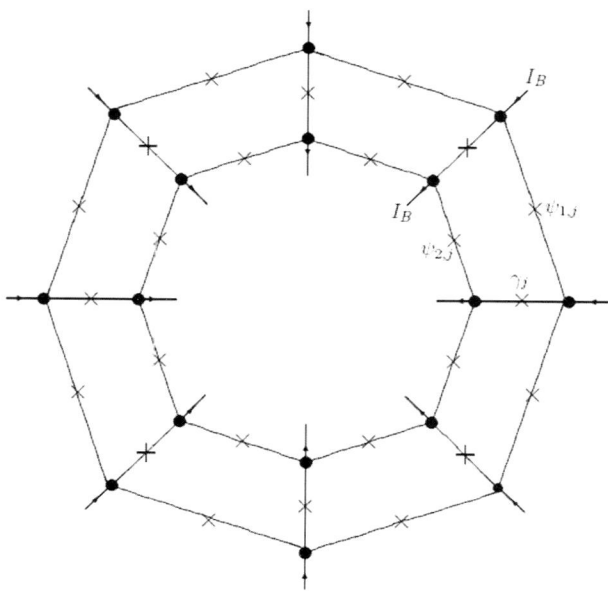

Fig. 6.3. A ladder array of Josephson junctions with periodic boundary conditions and $N = 8$ plaquettes. A uniform, dc bias current I_B is inserted into and extracted from each rung as shown. The gauge–invariant phase difference across the rung junctions is denoted by γ_j where $1 \leq j \leq N$, while the corresponding quantities for the off–rung junctions along the outer(inner) edge are $\psi_{1,j}(\psi_{2,j})$ (adapted and modified from [TSS05]).

For convenience, we work with dimensionless quantities. Our dimensionless time variable is

$$\tau \equiv \frac{t}{t_c} = \frac{2e\langle I_c \rangle t}{\hbar \langle R^{-1} \rangle}, \tag{6.32}$$

where t is the ordinary time. In the following, derivatives with respect to τ will be denoted by *prime* (e.g., $\psi' = d\psi/d\tau$). The dimensionless bias current is

$$i_B \equiv \frac{I_B}{\langle I_c \rangle}, \tag{6.33}$$

while the dimensionless critical current of rung junction j is $i_{cj} \equiv I_{cj}/\langle I_c \rangle$. The McCumber parameter in this case is

$$\beta_c \equiv \frac{2e\langle I_c \rangle \langle C \rangle}{\hbar \langle R^{-1} \rangle^2}. \tag{6.34}$$

Note that β_c is proportional to the mean capacitance of the rung junctions. An important dimensionless parameter is

$$\alpha \equiv \frac{I_{co}}{\langle I_c \rangle}, \tag{6.35}$$

which will effectively tune the nearest–neighbor interaction strength in our phase model for the ladder.

Conservation of charge applied to the superconducting islands on the outer and inner edge, respectively, of rung junction j yields the following equations in dimensionless variables [TSS05]:

$$i_B - i_{cj}\sin\gamma_j - i_{cj}\gamma_j' - i_{cj}\beta_c\gamma_j'' - \alpha\sin\psi_{1,j} - \alpha\psi_{1,j}'$$
$$-\alpha\beta_c\psi_{1,j}'' + \alpha\sin\psi_{1,j-1} + \alpha\psi_{1,j-1}' + \alpha\beta_c\psi_{1,j-1}'' = 0, \tag{6.36}$$
$$-i_B + i_{cj}\sin\gamma_j + i_{cj}\gamma_j' + i_{cj}\beta_c\gamma_j'' - \alpha\sin\psi_{2,j} - \alpha\psi_{2,j}'$$
$$-\alpha\beta_c\psi_{2,j}'' + \alpha\sin\psi_{2,j-1} + \alpha\psi_{2,j-1}' + \alpha\beta_c\psi_{2,j-1}'' = 0, \tag{6.37}$$

where $1 \leq j \leq N$. The result is a set of $2N$ equations in $3N$ unknowns: γ_j, $\psi_{1,j}$, and $\psi_{2,j}$. We supplement (6.37) by the constraint of fluxoid quantization in the absence of external or induced magnetic flux. For plaquette j this constraint yields the relationship

$$\gamma_j + \psi_{2,j} - \gamma_{j+1} - \psi_{1,j} = 0. \tag{6.38}$$

Equations (6.37) and (6.38) can be solved numerically for the $3N$ phases γ_j, $\psi_{1,j}$ and $\psi_{2,j}$ [TSS05].

We assign the rung junction critical currents in one of two ways, randomly or nonrandomly. We generate random critical currents according to a parabolic *probability distribution function* (PDF) of the form

$$P(i_c) = \frac{3}{4\Delta^3}\left[\Delta^2 - (i_c - 1)^2\right], \tag{6.39}$$

where $i_c = I_c/\langle I_c \rangle$ represents a scaled critical current, and Δ determines the spread of the critical currents. Equation (6.39) results in critical currents in the range $1 - \Delta \leq i_c \leq 1 + \Delta$. Note that this choice for the PDF (also used in [WCS96]) avoids extreme critical currents (relative to a mean value of unity) that are occasionally generated by PDF's with tails. The nonrandom method of assigning rung junction critical currents was based on the expression

$$i_{cj} = 1 + \Delta - \frac{2\Delta}{(N-1)^2}\left[4j^2 - 4(N+1)j + (N+1)^2\right], \qquad 1 \leq j \leq N, \tag{6.40}$$

which results in the i_{cj} values varying quadratically as a function of position along the ladder and falling within the range $1 - \Delta \leq i_{cj} \leq 1 + \Delta$. We usually use $\Delta = 0.05$.

Multiple time scale analysis

Now, our goal is to derive a Kuramoto–like model for the phase differences across the rung junctions, γ_j, starting with (6.37). We begin with two reasonable assumptions. First, we assume there is a simple phase relationship between the two off–rung junctions in the same plaquette [TSS05]:

$$\psi_{2,j} = -\psi_{1,j}, \tag{6.41}$$

the validity of which has been discussed in detail elsewhere [DDT03, FW95]. As a result, (6.38) reduces to

$$\psi_{1,j} = \frac{\gamma_j - \gamma_{j+1}}{2}, \tag{6.42}$$

which implies that (6.36) can be written as

$$
\begin{aligned}
i_{cj}\beta_c\gamma_j'' + i_{cj}\gamma_j' &+ \frac{\alpha\beta_c}{2}\left[\gamma_{j+1}'' - 2\gamma_j'' + \gamma_{j-1}''\right] + \frac{\alpha}{2}\left[\gamma_{j+1}' - 2\gamma_j' + \gamma_{j-1}'\right] \\
&= i_B - i_{cj}\sin\gamma_j + \alpha\sum_{\delta=\pm 1}\sin\left(\frac{\gamma_{j+\delta} - \gamma_j}{2}\right).
\end{aligned}
\tag{6.43}
$$

Our second assumption is that we can neglect the discrete Laplacian terms in (6.43), namely

$$\nabla^2\gamma_j' \equiv \gamma_{j+1}' - 2\gamma_j' + \gamma_{j-1}' \qquad \text{and} \qquad \nabla^2\gamma_j'' \equiv \gamma_{j+1}'' - 2\gamma_j'' + \gamma_{j-1}''.$$

We find numerically, over a wide range of bias currents i_B, *McCumber parameters* β_c, and coupling strengths α that $\nabla^2\gamma_j'$ and $\nabla^2\gamma_j''$ oscillate with a time–averaged value of approximately zero. Since the multiple time scale method is similar to averaging over a fast time scale, it seems reasonable to drop these terms. In light of this assumption, (6.43) becomes

$$i_{cj}\beta_c\gamma_j'' + i_{cj}\gamma_j' = i_B - i_{cj}\sin\gamma_j + \alpha\sum_{\delta=\pm 1}\sin\left(\frac{\gamma_{j+\delta} - \gamma_j}{2}\right). \tag{6.44}$$

We can use (6.44) as the starting point for a multiple time scale analysis. Following [CS95] and [WS97], we divide (6.44) by i_B and define the following quantities:

$$\tilde{\tau} \equiv i_B\tau, \qquad \tilde{\beta}_c \equiv i_B\beta_c, \qquad \epsilon = 1/i_B. \tag{6.45}$$

In terms of these scaled quantities, (6.44) can be written as

$$i_{cj}\tilde{\beta}_c\frac{d^2\gamma_j}{d\tilde{\tau}^2} + i_{cj}\frac{d\gamma_j}{d\tilde{\tau}} + \epsilon i_{cj}\sin\gamma_j - \epsilon\alpha\sum_{\delta}\sin\left(\frac{\gamma_{j+\delta} - \gamma_j}{2}\right) = 1. \tag{6.46}$$

Next, we introduce a series of four (dimensionless) time scales,

$$T_n \equiv \epsilon^n \tilde{\tau}, \qquad (n = 0, 1, 2, 3), \tag{6.47}$$

which are assumed to be independent of each other. Note that $0 < \epsilon < 1$ since $\epsilon = 1/i_B$. We can think of each successive time scale, T_n, as being 'slower' than the scale before it. For example, T_2 describes a slower time scale than T_1. The time derivatives in 6.46 can be written in terms of the new time scales, since we can think of $\tilde{\tau}$ as being a function of the four independent T_n's, $\tilde{\tau} = \tilde{\tau}(T_0, T_1, T_2, T_3)$. Letting $\partial_n \equiv \partial/\partial T_n$, the first and second time derivatives can be written as [TSS05]

$$\frac{d}{d\tilde{\tau}} = \partial_0 + \epsilon \partial_1 + \epsilon^2 \partial_2 + \epsilon^3 \partial_3 \tag{6.48}$$

$$\frac{d^2}{d\tilde{\tau}^2} = \partial_0^2 + 2\epsilon \partial_0 \partial_1 + \epsilon^2 \left(2\partial_0 \partial_2 + \partial_1^2\right) + 2\epsilon^3 \left(\partial_0 \partial_3 + \partial_1 \partial_2\right), \tag{6.49}$$

where in (6.49) we have dropped terms of order ϵ^4 and higher.

Next, we expand the phase differences in an ϵ expansion

$$\gamma_j = \sum_{n=0}^{\infty} \epsilon^n \gamma_{n,j}(T_0, T_1, T_2, T_3). \tag{6.50}$$

Substituting this expansion into (6.46) and collecting all terms of order ϵ^0 results in the expression

$$1 = i_{cj} \tilde{\beta}_c \partial_0^2 \gamma_{0,j} + i_{cj} \partial_0 \gamma_{0,j}, \tag{6.51}$$

for which we find the solution

$$\gamma_{0,j} = \frac{T_0}{i_{cj}} + \phi_j(T_1, T_2, T_3), \tag{6.52}$$

where we have ignored a transient term of the form $e^{-T_0/\tilde{\beta}_c}$, and where $\phi_j(T_i)$, $(i = 1, 2, 3)$ is assumed constant over the fastest time scale T_0. Note that the expression for $\gamma_{0,j}$ consists of a rapid phase rotation described by T_0/i_{cj} and slower–scale temporal variations, described by ϕ_j, on top of that overturning. In essence, the goal of this technique is to solve for the dynamical behavior of the slow phase variable, ϕ_j. The resulting differential equation for the ϕ_j is [TSS05]:

$$\beta_c \phi_j'' + \phi_j' = \Omega_j + K_j \sum_{\delta = \pm 1} \sin\left[\frac{\phi_{j+\delta} - \phi_j}{2}\right] + L_j \sum_{\delta = \pm 1} \sin\left[3\left(\frac{\phi_{j+\delta} - \phi_j}{2}\right)\right]$$

$$+ M_j \sum_{\delta = \pm 1} \left\{\cos\left[\frac{\phi_{j+\delta} - \phi_j}{2}\right] - \cos\left[3\left(\frac{\phi_{j+\delta} - \phi_j}{2}\right)\right]\right\}, \tag{6.53}$$

where Ω_j is given by the expression (letting $x_j \equiv i_{cj}/i_B$ for convenience)

$$\Omega_j = \frac{1}{x_j} \left[1 - \frac{x_j^4}{\left(2\beta_c^2 + x_j^2\right)} \right], \tag{6.54}$$

and the three coupling strengths are

$$K_j = \frac{\alpha}{i_{cj}} \left[1 + \frac{x_j^4 \left(3x_j^2 + 23\beta_c^2\right)}{16 \left(\beta_c^2 + x_j^2\right)^2} \right], \tag{6.55}$$

$$L_j = \frac{\alpha}{i_{cj}} \frac{x_j^4 \left(3\beta_c^2 - x_j^2\right)}{16 \left(\beta_c^2 + x_j^2\right)^2}, \tag{6.56}$$

$$M_j = -\frac{\alpha}{i_{cj}} \frac{x_j^5 \beta_c}{4 \left(\beta_c^2 + x_j^2\right)^2}. \tag{6.57}$$

We emphasize that (6.53) is expressed in terms of the original, unscaled, time variable τ and McCumber parameter β_c.

We will generally consider bias current and junction capacitance values such that $x_j^2 \ll \beta_c^2$. In this limit, (6.55)–(6.57) can be approximated as follows [TSS05]:

$$K_j \rightarrow \frac{\alpha}{i_{cj}} \left[1 + \mathcal{O}\left(\frac{1}{i_B^4}\right) \right], \tag{6.58}$$

$$L_j \rightarrow \frac{\alpha}{i_{cj}} \left(\frac{3x_j^4}{16\beta_c^2}\right) \sim \mathcal{O}\left(\frac{1}{i_B^4}\right), \tag{6.59}$$

$$M_j \rightarrow -\frac{\alpha}{i_{cj}} \left(\frac{x_j^5}{4\beta_c^3}\right) \sim \mathcal{O}\left(\frac{1}{i_B^5}\right). \tag{6.60}$$

For large bias currents, it is reasonable to truncate (6.53) at $\mathcal{O}(1/i_B^3)$, which leaves

$$\beta_c \phi_j'' + \phi_j' = \Omega_j + \frac{\alpha}{i_{cj}} \sum_{\delta=\pm 1} \sin\left[\frac{\phi_{j+\delta} - \phi_j}{2}\right], \tag{6.61}$$

where all the cosine coupling terms and the third harmonic sine term have been dropped as a result of the truncation.

In the absence of any coupling between neighboring rung junctions ($\alpha = 0$) the solution to (6.61) is

$$\phi_j^{(\alpha=0)} = A + B e^{-\tau/\beta_c} + \Omega_j \tau,$$

where A and B are arbitrary constants. Ignoring the transient exponential term, we see that $d\phi_j^{(\alpha=0)}/d\tau = \Omega_j$, so we can think of Ω_j as the voltage across rung junction j in the un–coupled limit. Alternatively, Ω_j can be viewed as the angular velocity of the strongly–driven rotator in the un–coupled limit.

Equation (6.61) is our desired phase model for the rung junctions of the underdamped ladder [TSS05]. The result can be described as a locally–coupled

Kuramoto model with a second-order time derivative (LKM2) and with junction coupling determined by α. In the context of systems of coupled rotators, the second derivative term is due to the non–negligible rotator inertia, whereas in the case of Josephson junctions the second derivative arises because of the junction capacitance. The *globally–coupled* version of the second–order Kuramoto model (GKM2) has been well studied; in this case the oscillator inertia leads to a first–order synchronization phase transition as well as to hysteresis between a weakly and a strongly coherent synchronized state [TLO97, ABS00].

6.3.2 Comparison of LKM2 and RCSJ Models

We now compare the synchronization behavior of the RCSJ ladder array with the LKM2. We consider frequency and phase synchronization separately. For the rung junctions of the ladder, frequency synchronization occurs when the time average voltages, $\langle v_j \rangle_\tau = \langle \phi_j' \rangle_\tau$ are equal for all N junctions, within some specified precision. In the language of coupled rotators, this corresponds to phase points moving around the unit circle with the same average angular velocity. We quantify the degree of frequency synchronization via an 'order parameter' [TSS05]

$$f = 1 - \frac{s_v(\alpha)}{s_v(0)}, \qquad (6.62)$$

where $s_v(\alpha)$ is the standard deviation of the N time–average voltages, $\langle v_j \rangle_\tau$:

$$s_v(\alpha) = \sqrt{\frac{\sum_{j=1}^N \left(\langle v_j \rangle_\tau - \frac{1}{N} \sum_{k=1}^N \langle v_k \rangle_\tau \right)^2}{N-1}} \qquad (6.63)$$

In general, this standard deviation will be a function of the coupling strength α, so $s_v(0)$ is a measure of the spread of the $\langle v_j \rangle_\tau$ values for N independent junctions. Frequency synchronization of all N junctions is signaled by $f = 1$, while $f = 0$ means all N average voltages have their un–coupled values.

Phase synchronization of the rung junctions is measured by the usual *Kuramoto order parameter*

$$r \equiv \frac{1}{N} \sum_{j=1}^N e^{i\phi_j}. \qquad (6.64)$$

Lastly in this subsection, we address the issue of the linear stability of the frequency synchronized states ($\alpha > \alpha_c$) by calculating their *Floquet exponents* numerically for the RCSJ model as well as analytically based on the LKM2, (6.61). The analytic technique used has been described in [TM01], giving as a result for the real part of the Floquet exponents:

$$\mathrm{Re}(\lambda_m t_c) = -\frac{1}{2\beta_c} \left[1 \pm \mathrm{Re}\sqrt{1 - 4\beta_c \left(\bar{K} + 3\bar{L} \right) \omega_m^2} \right], \qquad (6.65)$$

where stable solutions correspond to exponents, λ_m, with a negative real part. One can think of the ω_m as the normal mode frequencies of the ladder. We find that for a ladder with periodic boundary conditions and N plaquettes

$$\omega_m^2 = \frac{4\sin^2\left(\frac{m\pi}{N}\right)}{1 + 2\sin^2\left(\frac{m\pi}{N}\right)}, \qquad 0 \le m \le N - 1. \qquad (6.66)$$

To arrive at (6.65) we have ignored the effects of disorder so that $\bar{K}$ and $\bar{L}$ are obtained from (6.55) and (6.56) with the substitution $i_{cj} \to 1$ throughout. This should be reasonable for the levels of disorder we have considered (5%). Substituting the expressions for $\bar{K}$ and $\bar{L}$ into 6.65 results in [TSS05]

$$\mathrm{Re}(\lambda_m t_c) = -\frac{1}{2\beta_c}[1 \pm \mathrm{Re}\sqrt{1 - 2\beta_c\alpha\left\{1 + \frac{2\beta_c^2}{\left(i_B^2\beta_c^2 + 1\right)^2}\right\}\omega_m^2}]. \qquad (6.67)$$

We are most interested in the Floquet exponent of minimum magnitude, $\mathrm{Re}(\lambda_{\min}t_c)$, which essentially gives the lifetime of the longest–lived perturbations to the synchronized state.

6.3.3 'Small–World' Connections in JJL Arrays

Many properties of small world networks have been studied in the last several years, including not only the effects of network topology but also the dynamics of the node elements comprising the network [New00, Str]. Of particular interest has been the ability of oscillators to synchronize when configured in a small–world manner. Such synchronization studies can be broadly sorted into several categories [TSS05]:

(1) Work on coupled lattice maps has demonstrated that synchronization is made easier by the presence of random, *long–range connections* [GH00, BPV03].

(2) Much attention has been given to the synchronization of continuous time dynamical systems, including the first order *locally–coupled Kuramoto model* (LKM), in the presence of small world connections [HCK02a, HCK02b, Wat99]. For example, Hong and coworkers [HCK02a, HCK02b] have shown that the LKM, which does not exhibit a true dynamical phase transition in the thermodynamic limit ($N \to \infty$) in the *pristine* case, does exhibit such a phase synchronization transition for even a small number of shortcuts. But the assertion [WC02] that any small world network can synchronize for a given coupling strength and large enough number of nodes, even when the pristine network would not synchronize under the same conditions, is not fully accepted [BP02].

(3) More general studies of synchronization in small world and scale–free networks [BP02, NML03] have shown that the small world topology does not guarantee that a network can synchronize. In [BP02] it was shown that one

could calculate the average number of shortcuts per node, s_{sync}, required for a given dynamical system to synchronize. This study found no clear relation between this synchronization threshold and the onset of the small world region, i.e., the value of s such that the average path length between all pairs of nodes in the array is less than some threshold value. [NML03] studied arrays with a power–law distribution of node connectivities (scale–free networks) and found that a broader distribution of connectivities makes a network *less* synchronizable even though the average path length is smaller. It was argued that this behavior was caused by an increased number of connections on the hubs of the scale–free network. Clearly it is dangerous to assume that merely reducing the average path length between nodes of an array will make such an array easier to synchronize.

Now, regarding Josephson–junction arrays, if we have a disordered array biased such that some subset of the junctions are in the voltage state, i.e., undergoing limit cycle oscillations, the question is will the addition of random long–range connections between junctions aid the array in attaining frequency and/or phase synchronization? Can we address this question by using the mapping discussed above between the RCSJ model for the *underdamped ladder array* and the second–order, locally–coupled Kuramoto model (LKM2). Based on the results of [DDT03], we also know that the RSJ model for an *overdamped ladder* can be mapped onto a first–order, locally–coupled Kuramoto model (LKM). Because of this mapping, the ladder array falls into category (2) of the previous paragraph. In other words, we should expect the existence of shortcuts to drastically improve the ability of ladder arrays to synchronize [TSS05].

We add connections between pairs of rung junctions that will result in interactions that are longer than nearest neighbor in range. We do so by adding two, nondisordered, off–rung junctions for each such connection. We argue that the RCSJ equations for the underdamped junctions in the ladder array can be mapped onto a straightforward variation of (6.61), in which the sinusoidal coupling term for rung junction j also includes the longer–range couplings due to the added shortcuts. Imagine a ladder with a shortcut between junctions j and l, where $l \neq j, j \pm 1$. Conservation of charge applied to the two superconducting islands that comprise rung junction j will lead to equations very similar to (6.37). For example, the analog to (6.36) will be

$$i_B - i_{cj}\sin\gamma_j - i_{cj}\gamma'_j - \beta_c i_{cj}\gamma''_j - \alpha\sin\psi_{1,j} - \alpha\psi'_{1,j} - \beta_c\alpha\psi''_{1,j} +$$
$$\alpha\sin\psi_{1,j-1} + \alpha\psi'_{1,j-1} + \beta_c\alpha\psi''_{1,j-1} + \sum_l \left[\alpha\sin\psi_{1;jl} + \alpha\psi'_{1;jl} + \beta_c\alpha\psi''_{1;jl}\right] = 0,$$

with an analogous equation corresponding to the inner superconducting island that can be generalized from (6.37). The sum over the index l accounts for all junctions connected to junction j via an added shortcut. Fluxoid quantization still holds, which means that we can augment 6.38 with

$$\gamma_j + \psi_{2;jl} - \gamma_l - \psi_{1;jl} = 0. \tag{6.68}$$

We also assume the analog of (6.41) holds:

$$\psi_{2;jl} = -\psi_{1;jl}.$$

(6.69)

Equations (6.68) and (6.69) allow us to write the analog to (6.42) for the case of shortcut junctions:

$$\psi_{1;jl} = \frac{\gamma_j - \gamma_l}{2}$$

(6.70)

Equation (6.68), in light of (6.70), can be written as

$$i_B - i_{cj}\sin\gamma_j - i_{cj}\gamma_j' - \beta_c i_{cj}\gamma_j'' + \alpha \sum_{\delta=\pm1}\sin(\frac{\gamma_{j+\delta} - \gamma_j}{2}) + \alpha \sum_l \sin(\frac{\gamma_j - \gamma_l}{2})$$
$$+ \frac{\alpha}{2}\nabla^2\gamma_j' + \frac{\alpha}{2}\nabla^2\gamma_j'' + \frac{\alpha}{2}\sum_l(\gamma_j' - \gamma_l') + \frac{\alpha}{2}\sum_l(\gamma_j'' - \gamma_l'') = 0,$$

where the sums Σ_l are over all rung junctions connected to j via an added shortcut. As we did with the pristine ladder, we will drop the two discrete Laplacians, since they have a very small time average compared to the terms $i_{cj}\gamma_j' + i_{cj}\beta_c\gamma_j''$. The same is also true, however, of the terms $\alpha/2\sum_l(\gamma_j' - \gamma_l')$ and $\alpha/2\sum_l(\gamma_j'' - \gamma_l'')$, as direct numerical solution of the full RCSJ equations in the presence of shortcuts demonstrates. So we shall drop these terms as well. Then we have

$$i_B - i_{cj}\sin\gamma_j - i_{cj}\gamma_j' - \beta_c i_{cj}\gamma_j'' + \frac{\alpha}{2}\sum_{k\in\Lambda_j}\sin\left(\frac{\gamma_k - \gamma_j}{2}\right),$$

(6.71)

where the sum is over all junctions in Λ_j, which is the set of all junctions connected to junction j. From above results we can predict that a multiple time scale analysis of (6.71) results in a phase model of the form

$$\beta_c\frac{d^2\phi_j}{d\tau^2} + \frac{d\phi_j}{d\tau} = \Omega_j + \frac{\alpha}{2}\sum_{k\in\Lambda_j}\sin\left(\frac{\phi_k - \phi_j}{2}\right),$$

(6.72)

where Ω_j is give by (7.34). A similar analysis for the *overdamped ladder* leads to the result

$$\phi_j' = \Omega_j^{(1)} + \frac{\alpha}{2}\sum_{k\in\Lambda_j}\sin\left(\frac{\phi_k - \phi_j}{2}\right),$$

(6.73)

where the time-averaged voltage across each overdamped rung junction in the un–coupled limit is

$$\Omega_j^{(1)} = \sqrt{\left(\frac{i_B}{i_{cj}}\right)^2 - 1}.$$

(6.74)

Although the addition of shortcuts makes it easier for the array to synchronize, we should also consider the effects of such random connections on the stability of the synchronized state. The Floquet exponents for the synchronized state allow us to quantify this stability. Using a general technique discussed in [PC98], we can calculate the Floquet exponents λ_m for the LKM based on the expression

$$\lambda_m t_c = \alpha E_m^G, \qquad (6.75)$$

where E_m^G are the eigenvalues of $\mathbf{G}$, the matrix of coupling coefficients for the array. A specific element, G_{ij}, of this matrix is unity if there is a connection between rung junctions i and j. The diagonal terms, G_{ii}, is merely the negative of the number of junctions connected to junction i. This gives the matrix the property $\sum_j G_{ij} = 0$. In the case of the pristine ladder, the eigenvalues of $\mathbf{G}$ can be calculated analytically, which yields Floquet exponents of the form

$$\lambda_m^{(p=0)} t_c = -4\alpha \sin^2\left(\frac{m\pi}{N}\right). \qquad (6.76)$$

See [TSS05] for more details.

7

Fractals and Fractional Dynamics

This Chapter addresses *fractals* and high–dimensional *fractal dynamics*, as well as modern *non–chaotic strange attractors*.

7.1 Fractals

7.1.1 Mandelbrot Set

Recall from Introduction that the *Mandelbrot set*[1] M is defined as the connectedness locus of the family

$$f_c : \mathbb{C} \longrightarrow \mathbb{C}, \qquad z \mapsto z^2 + c,$$

of complex quadratic polynomials. That is, the Mandelbrot set is the subset of the complex plane consisting of those parameters c for which the *Julia set* of f_c is connected. An equivalent way of defining is as the set of parameters for which the critical point does not tend to infinity. That is, $f_c^n(0) \nrightarrow \infty$, where f_c^n is the $n-$fold composition of f_c with itself. The Mandelbrot set is generally considered to be a fractal. However, only the boundary of it is technically a fractal.

The Mandelbrot set M is a compact set, contained in the closed disk of radius 2 around the origin (see Figure 7.1). More precisely, if c belongs to M, then $|f^n(c)| \leq 2$ for all $n \geq 0$. The intersection of M with the real axis

---

[1] Benoit Mandelbrot studied the parameter space of quadratic polynomials in the 1980 article [Man80a]. The mathematical study of the Mandelbrot set really began with work by the mathematicians A. Douady and J.H. Hubbard [DH85], who established many fundamental properties of M, and named the set in honor of Mandelbrot. The Mandelbrot set has become popular far outside of mathematics both for its aesthetic appeal and its complicated structure, arising from a simple definition. This is largely due to the efforts of Mandelbrot (and others), who worked hard to communicate this area of mathematics to the general public.

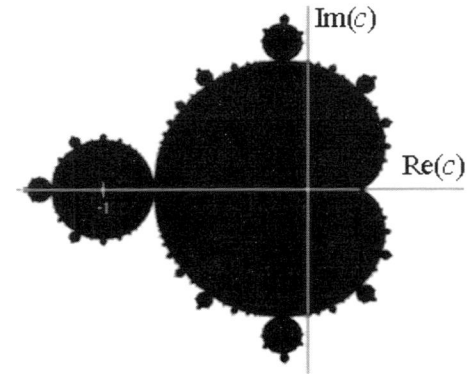

Fig. 7.1. The Mandelbrot set in a complex plane.

is precisely the interval $[-2, 0.25]$. The parameters along this interval can be put in 1–1 correspondence with those of the real *logistic–map family*

$$z \mapsto \lambda z(z - 1), \quad \lambda \in [1, 4].$$

Douady and Hubbard have shown that the Mandelbrot set is connected. In fact, they constructed an explicit conformal isomorphism between the complement of the Mandelbrot set and the complement of the closed unit disk. The dynamical formula for the uniformization of the complement of the Mandelbrot set, arising from Douady and Hubbard's proof of the connectedness of M, gives rise to *external rays* of the Mandelbrot set, which can be used to study the Mandelbrot set in combinatorial terms.

The boundary of the Mandelbrot set is exactly the bifurcation locus of the quadratic family; that is, the set of parameters c for which the dynamics changes abruptly under small changes of c. It can be constructed as the limit set of a sequence of plane algebraic *Mandelbrot curves*, of the general type known as *polynomial lemniscates*. The Mandelbrot curves are defined by setting

$$p_0 = z, \qquad p_n = p_{n-1}^2 + z,$$

and then interpreting the set of points $|p_n(z)| = 1$ in the complex plane as a curve in the real Cartesian plane of degree 2^{n+1} in x and y.

Upon looking at a picture of the Mandelbrot set (Figure 7.1), one immediately notices the large cardioid–shaped region in the center. This *main cardioid* is the region of parameters c for which f_c has an *attracting fixed–point*. It consists of all parameters of the form $c = \frac{1 - (\mu - 1)^2}{4}$, for some μ in the open unit disk.

To the left of the main cardioid, attached to it at the point $c = -3/4$, a circular–shaped bulb is visible. This bulb consists of those parameters c for which f_c has an attracting cycle of period 2. This set of parameters is an actual circle, namely that of radius $1/4$ around -1.

There are many other bulbs attached to the main cardioid: for every rational number p/q, with p and q coprime, there is such a bulb attached at the parameter,

$$c_{\frac{p}{q}} = \frac{1 - \left(e^{2\pi i \frac{p}{q}} - 1\right)^2}{4}.$$

This bulb is called the p/q–bulb of the Mandelbrot set. It consists of parameters which have an attracting cycle of period q and combinatorial rotation number p/q. More precisely, the q–periodic Fatou components containing the attracting cycle all touch at a common α–fixed–point. If we label these components $U_0, \ldots, U_{q-1}$ in counterclockwise orientation, then f_c maps the component U_j to the component $U_{j+p \,(\text{mod } q)}$.

The change of behavior occurring at $c_{\frac{p}{q}}$ is a *bifurcation*: the attracting fixed–point 'collides' with a repelling period q–cycle. As we pass through the bifurcation parameter into the q–bulb, the attracting fixed–point turns into a repelling α–fixed–point, and the period q–cycle becomes attracting.

All the above bulbs were interior components of the Mandelbrot set in which the maps f_c have an attracting periodic cycle. Such components are called *hyperbolic components* (see Figure 7.2). It has been conjectured that these are the only interior regions of M. This problem, known as *density of hyperbolicity*, may be the most important open problem in the field of complex dynamics. Hypothetical non–hyperbolic components of the Mandelbrot set are often referred to as 'queer' components.

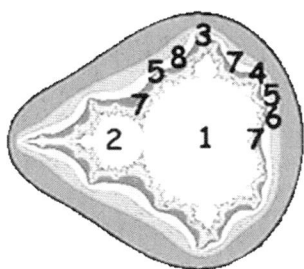

Fig. 7.2. Periods of hyperbolic components of the Mandelbrot set in a complex plane.

The *Hausdorff dimension* of the boundary of the Mandelbrot set equals 2 (see [Shi98]). It is not known whether the boundary of the Mandelbrot set has positive planar *Lebesgue measure*.

Sometimes the connectedness loci of families other than the quadratic family are also referred to as the Mandelbrot sets of these families. The connectedness loci of the unicritical polynomial families $f_c = z^d + c$ for $d > 2$ are often called *Multibrot sets*. For general families of holomorphic functions, the boundary of the Mandelbrot set generalizes to the bifurcation locus, which is a natural object to study even when the connectedness locus is not use–ful. It is also possible to consider similar constructions in the study of non–analytic mappings. Of particular interest is the *tricorn* (also sometimes called the Mandelbar set), the connectedness locus of the anti–holomorphic family: $z \mapsto \bar{z}^2 + c$. The tricorn was encountered by John Milnor in his study of parameter slices of real cubic polynomials. It is not locally connected. This property is inherited by the connectedness locus of real cubic polynomials.

7.2 Robust Strange Non–Chaotic Attractors

Recall that C. Grebogi's phrase *strange non–chaotic attractor* (SNA) [GOP84] refers to an attractor which is non–chaotic in the sense that its orbits are not exponentially sensitive to perturbation (i.e., none of the Lyapunov exponents are positive), but the attractor is strange in the sense that its phase–space structure has nontrivial fractal properties. Past studies indicate that SNA's are typical in nonlinear dynamical systems that are quasi–periodically forced. Here by a typical behavior we mean that the behavior occurs for a positive measure set of parameter values. Alternatively, if parameters are chosen at random from an ensemble with smooth probability density, then the proba–bility of choosing parameters that yield a typical behavior is not zero. The description of a behavior as typical is to be contrasted with the stronger state–ment that a behavior is robust. In particular, we say a behavior of a system is robust if it persists under sufficiently small perturbations; i.e., there exist a positive value δ such that the robust behavior occurs for all systems that can be obtained by perturbation of the original system by an amount less than δ. Thus all robust behaviors are also typical, but not vice versa [KKH03].

With respect to SNA's, examples where they are typical but not robust have been extensively studied [BOA85, DGO89a, RO87, KS97]. Here we give two such examples of quasi–periodically forced maps.

7.2.1 Quasi–Periodically Forced Maps

Forced Logistic Map

Here, following [KLO03], we present the underlying mechanism of the inter–mittency in the *quasi–periodically forced logistic map* M [HH94] which is a representative model for the quasi–periodically forced period-doubling sys–tems:

$$M : \begin{cases} x_{n+1} = (a + \varepsilon \cos 2\pi \theta_n) x_n (1 - x_n), \\ \theta_{n+1} = \theta_n + \omega \pmod{1}, \end{cases} \tag{7.1}$$

where $x \in [0, 1]$, $\theta \in S^1$, a is the nonlinearity parameter of the logistic map, and ω and ε represent the frequency and amplitude of the *quasi–periodic forcing*, respectively. We set the frequency to be the reciprocal of the golden mean, $\omega = (\sqrt{5} - 1)/2$. The intermittent transition is then investigated using the *rational approximations* (RAs) to the quasi–periodic forcing. For the case of the inverse golden mean, its rational approximants are given by the ratios of the Fibonacci numbers, $\omega_k = F_{k-1}/F_k$, where the sequence of $\{F_k\}$ satisfies $F_{k+1} = F_k + F_{k-1}$ with $F_0 = 0$ and $F_1 = 1$. Instead of the quasi–periodically forced system, we study an infinite sequence of periodically forced systems with rational driving frequencies ω_k. We suppose that the properties of the original system M may be obtained by taking the *quasi–periodic limit* $k \to \infty$. Using this technique we observe a new type of invariant unstable set, which will be referred to as the *ring–shaped unstable set* in accordance with its geometry. When a smooth torus (corresponding to an ordinary *quasi–periodic attractor*) collides with this ring–shaped unstable set, a transition to an intermittent SNA is found to occur.

Note that the quasi–periodically forced logistic map M is non–invertible, because its Jacobian determinant becomes zero along the critical curve, $L_0 = \{x = 0.5, \ \theta \in [0,1)\}$. Critical curves of rank k, L_k $(k = 1,2,\dots)$, are then given by the images of L_0, (i.e., $L_k = M^k(L_0)$). Segments of these critical curves can be used to define a bounded trapping region of the phase–space, called an *absorbing area*, inside which, upon entering, trajectories are henceforth confined [MGB96]. It is found that the newly–born intermittent SNA fills the absorbing area. Hence the global structure of the SNA is determined by the critical curves.

Therefore, using RAs we have found the mechanism for the intermittent route to SNAs in the quasi–periodically forced logistic map. When a smooth torus makes a collision with a new type of ring–shaped unstable set, a transition to an intermittent SNA, bounded by segments of critical curves, occurs via a phase-dependent saddle-node bifurcation [KKH03].

Forced Circle Map

Another example of the SNA–type is the *quasi–periodically forced circle map* given by the system [DGO89a],

$$\theta_{n+1} = [\theta_n + \omega] \mod 2\pi, \tag{7.2}$$
$$\varphi_{n+1} = [\varphi_n + \omega_\varphi + \varepsilon \sin \varphi_n + C \cos \theta_n] \mod 2\pi,$$

where $\Omega \equiv \omega/2\pi$ is irrational. Other examples of typical non–robust SNA's involving differential equations have also been studied [BOA85, RO87]. Numerical evidence [DGO89a, RO87] and analysis based on a correspondence [BOA85, KS97] with Anderson localization in a quasi–periodic potential leads to an understanding of the typical but non–robust nature of SNA's in these examples: In particular, it is found that SNA's exist on a positive Lebesgue measure Cantor set in parameter space. In the case of (7.2), for example, consider the rotation number [KKH03]

$$W = \lim_{n \to \infty} (\varphi_n - \varphi_0)/(2\pi n),$$

where for this limit φ_n is *not* computed modulo 2π. For fixed ω, $\varepsilon > 0$, and $C > 0$, a plot of W versus ω_φ yields an incomplete devil's staircase, a nondecreasing graph consisting of intervals of ω_φ where $W(\omega_\varphi)$ is constant and with increase of $W(\omega_\varphi)$ occurring only on a Cantor set of positive measure. For small ε, the values of ω_φ on the Cantor set correspond to orbits that are three frequencies quasi–periodic, but for larger ε they correspond to SNA's. Because an arbitrarily small perturbation of ω_φ from a value in the Cantor set can result in a value of ω_φ outside the Cantor set, these SNA's are not robust. On the other hand, because the Cantor set of ω_φ values has positive Lebesgue measure ('positive length'), these attractors are typical for (7.2).

Other studies suggest that there are situations where SNA's are robust [GOP84, DSS90, HH94, FKP95, NK96, YL96, PBR97, WFP97]. The experiment of [DSS90] on a quasi–periodically forced magneto–elastic ribbon produced evidence of a SNA, and the existence of this SNA appeared to be stable to parameter perturbations. The original paper where the existence of SNA's in quasi–periodically forced systems was first discussed [GOP84] gives numerical evidence of robust SNA's. In addition, the effect of quasi–periodic perturbations on a system undergoing a periodic doubling cascade has been investigated, and evidence has been presented suggesting that, after a finite number of torus doublings, a robust SNA results [HH94, NK96].

Thus there seems to be two types of SNA's: typical, non–robust SNA's, and robust SNA's. Here, following [KKH03], we study a class of models exhibiting robust SNA's. The model class that we study is particularly interesting because it allows the possibility of rigorous analysis. In particular, we are able to prove, under the mild hypothesis that a certain Lyapunov exponent is negative, that the attractor is strange and non–chaotic. Since other cases of SNA's are likely to be accessible only to study by numerical means, it is worthwhile to investigate our, more well–understood models, numerically. By doing this we gain insight into the applicability and limitations of numerical techniques for the study of SNA's.

Here we consider quasi–periodically forced maps which can be motivated by consideration of a system of ordinary differential equations in the form

$$\dot{x} = F(x, \xi, \theta^{(1)}, \theta^{(2)}, \cdots, \theta^{(N)}),$$

where F is 2π periodic in the angles ξ and $\theta^{(i)}$, which are given by

$$\xi = \omega_\xi t + \xi_0, \theta^{(i)} = \omega_{\theta^{(i)}} t + \theta_0^{(i)},$$

and $\omega_\xi, \omega_{\theta^{(1)}}, \cdots, \omega_{\theta^{(N)}}$ are incommensurate. Sampling the state of the system at discrete times t_n given by $\xi = 2n\pi$, we get a mapping of the form

$$
\begin{aligned}
\theta_{n+1}^{(i)} &= [\theta_n^{(i)} + \omega^{(i)}] \mod 2\pi, \\
x_{n+1} &= \tilde{F}(x_n, \theta_n^{(1)}, \theta_n^{(2)}, \cdots, \theta_n^{(N)}),
\end{aligned}
\tag{7.3}
$$

where $x_n = x(t_n)$, $\omega^{(i)} = 2\pi\omega_{\theta^{(i)}}/\omega_\xi$, and there exist no set of integers $(m^{(0)}, m^{(1)}, \cdots, m^{(N)})$ for which $\Sigma_{i=1}^{N} m^{(i)}\omega^{(i)} = 2\pi m^{(0)}$, aside from $(m^{(0)}, m^{(1)}, \cdots, m^{(N)}) = (0, 0, \cdots, 0)$.

For the map (7.3), the simplest possible attractor is an ND torus,

$$x = f(\theta^{(1)}, \theta^{(2)}, \cdots, \theta^{(N)}).$$

In this section, following [KKH03], we consider the case where an attracting $(N + 1)$D torus exists, and the dynamics on the torus is given by

$$\theta_{n+1}^{(i)} = [\theta_n^{(i)} + \omega^{(i)}] \mod 2\pi, \tag{7.4}$$
$$\varphi_{n+1} = [\varphi_n + q^{(1)}\theta_n^{(1)} + q^{(2)}\theta_n^{(2)} + \cdots + q^{(N)}\theta_n^{(N)}$$
$$+ P(\varphi_n, \theta_n^{(1)}, \theta_n^{(2)}, \cdots, \theta_n^{(N)})] \mod 2\pi,$$

where P is periodic in all its variables and $q^{(1)}$, $q^{(2)}$, $\cdots$, $q^{(N)}$ are integers. We are particularly interested in the case that (7.4b) is invertible, so that no chaos is possible, and when at least one $q^{(i)}$ is nonzero, which as we will see prevents the existence of an attracting N−torus.

7.2.2 2D Map on a Torus

Existence of SNA

We investigate the simplest case of (7.3) where $N = 1$ ($\theta^{(i)} \to \theta$) and the state variable $\mathbf{x}$ is 1D. Specifically, we take $\mathbf{x}$ to be an angle variable φ, so that the map operates on a 2D θ–φ torus. Within this class we restrict consideration to maps of the form [KKH03]

$$\theta_{n+1} = [\theta_n + \omega] \mod 2\pi, \tag{7.5}$$
$$\varphi_{n+1} = [\theta_n + \varphi_n + \eta P(\theta_n, \varphi_n)] \mod 2\pi,$$

where $\omega = \pi(\sqrt{5}-1)$, and $P(\theta, \varphi)$ is continuous, differentiable, and 2π periodic in both of its arguments (θ and φ). When η is small enough ($|\eta| < \eta_c$), this map is invertible. That is, the map is solvable for (θ_n, φ_n) when $(\theta_{n+1}, \varphi_{n+1})$ is given. We choose a simple function $P(\theta, \varphi) = \sin\varphi$ for our numerical work. In this case, the system is invertible if $|\eta| < 1$. Furthermore, since the map is invariant under the change of $\eta \to -\eta$ and $\varphi \to \varphi + \pi$, it is sufficient to consider only the case $\eta \geq 0$.

Figure 7.3 illustrates how a curve C on the θ–φ toroidal surface is mapped to a curve C' by the map (7.5). Note that the torus is unrolled in the θ direction to visualize the whole curve C in a 2D plain, but still rolled in the φ direction. The curve C circles around the torus in the θ direction, but does not wrap around the torus in the φ direction. After one iterate of (7.5), the curve C is mapped to a curve C' that wraps once around the torus in the φ direction. This behavior comes about due to the term θ_n on the right–hand

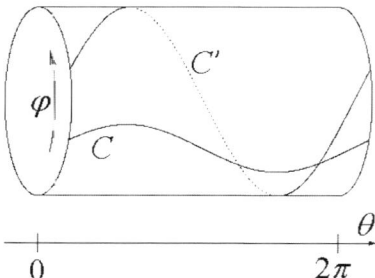

Fig. 7.3. Torus unwrapped in the θ direction. The map (7.5) takes the curve C to the curve C' (adapted and modified from [KKH03]).

side of (7.5b), because $\theta + \varphi + \eta P(\theta, \varphi)$ increases by 2π as θ increases by 2π. Similarly, applying the map to C' produces a curve with two wraps around the torus in the φ direction, and so on [KKH03].

The main results of our numerical experiments and rigorous analysis of (7.5) with $|\eta| < \eta_c$ are as follows:

(i) The map (7.5) has a single attractor.

(ii) For typical $P(\theta, \varphi)$, the attractor has a Lyapunov exponent h_φ that is negative for $\eta \neq 0$.

(iii) The attractor has information dimension one for $\eta \neq 0$.

(iv) The attractor is the entire $\theta - \varphi$ torus and, hence, has box-counting dimension two [DGO89b].

(v) These results are stable to perturbations of the system [BOA85, DGO89a, RBO87].

We first establish (ii) using an approximate formula for h_φ for small η. Our evidence for (ii) is strong but a rigorous mathematical proof is lacking. If we adopt (ii) as a hypothesis, then all the other results rigorously follow.

Lyapunov Exponent

A trajectory of the map (7.5) has two Lyapunov exponents h_θ and h_φ, where $h_\theta = 0$ is associated with (7.5a) and h_φ is associated with (7.5b). The latter exponent is given by the formula,

$$h_\varphi = \int \ln[1 + \eta P_\varphi(\theta, \varphi)]d\mu, \qquad (7.6)$$

where $P_\varphi = \partial P/\partial\varphi$, and μ denotes the measure generated by the orbit from a given initial point (θ_0, φ_0).

If $h_\varphi > 0$ for a particular trajectory, then, since $h_\theta = 0$, the map exponentially expands areas near the trajectory in the limit $n \to \infty$. Since the $\theta - \varphi$ torus has finite area, if the map is invertible, then there cannot be a set of initial points of nonzero area (positive Lebesgue measure) for which $h_\varphi > 0$, and the map thus does not have a chaotic attractor. Thus $h_\varphi \leq 0$ for typical orbits.

Furthermore, we argue that $h_\varphi < 0$ for small nonzero η. We consider first the case $\eta = 0$, for which (7.5b) becomes $\varphi_{n+1} = (\theta_n + \varphi_n) \bmod 2\pi$. If we initialize a uniform distribution of orbit points in the $\theta - \varphi$ torus, then, on one application of the $\eta = 0$ map, the distribution remains uniform. Furthermore, this uniform distribution is generated by the orbit from any initial condition. To verify this, we note that the explicit form of an $\eta = 0$ orbit,

$$\theta_n = (\theta_0 + n\omega) \bmod 2\pi, \qquad \varphi_n = [\varphi_0 + n\theta_0 + \frac{1}{2}(n^2 - n)\omega] \bmod 2\pi,$$

which is shown to generated a uniform density in [Fur61]. We can get an approximation to h_φ for nonzero but small η by expanding $\ln(1+\eta P_\varphi)$ in (7.6) to order η^2 and assuming that, to this order, the deviation of the measure μ from uniformity is not significant $[d\mu \approx d\theta d\varphi/(2\pi)^2]$. Using

$$\ln(1 + \eta P_\varphi) = \eta P_\varphi - (1/2)\eta^2 P_\varphi^2 + O(\eta^3),$$

this gives [KKH03]

$$h_\varphi = -\frac{1}{2}\eta^2 < P_\varphi^2 > +o(\eta^2), \tag{7.7}$$

which is negative for small enough $\eta \neq 0$. Here $< P_\varphi^2 >$ denotes the $\theta - \varphi$ average of P_φ^2, and the order η term is absent by virtue of $\int_0^{2\pi} P_\varphi d\varphi = 0$. Since we cannot show convergence of an expansion in η, our result (7.7) is formal rather than rigorous. However, numerical results strongly support (7.7).

Dimensions of the SNA

For our map the information dimension cannot be less than one due to the quasi–periodic θ dynamics. In addition, the Lyapunov dimension is an upper bound of information dimension [Led81]. Therefore, if we accept (ii), $h_\varphi < 0$, then $h_\theta = 0$ implies (iii).

Results (iii) and (iv) quantify the strangeness of the attractor. In particular, since the information dimension of the attractor is one, orbits spend most of their time on a curve–like set; yet, since the box–counting dimension is two, if one waits long enough, a typical orbit eventually visits any neighborhood on the $\theta - \varphi$ torus.

Topological Transitivity

To establish results (i) and (iv), that the attractor of the map is in the whole $\theta - \varphi$ torus, we prove that the map is topologically transitive: For every pair of open disks A and B, there is a trajectory that starts in A and passes through B. This property is known to imply that a dense set of initial conditions yields trajectories each of which is dense in the torus [KH95]. In particular, any attractor, having an open basin of attraction, must contain a dense orbit, and, hence, must be the entire torus.

In fact, for every pair of line segments $S_a = \{(\theta, \varphi) : \theta \in R_a$ and $\varphi = \varphi_a\}$ and $S_b = \{(\theta, \varphi) : \theta \in R_b$ and $\varphi = \varphi_b\}$, where $R_a = (\theta_a, \theta_a + \delta_a)$ and $R_b = (\theta_b, \theta_b + \delta_b)$, there is a finite trajectory of M that begins on the first segment and ends on the second. In other word, we will show that the nth iterate of S_a intersects S_b for some positive integer n.

The following is a formal definition of g_a. Let M_θ be the map (7.5a). For each θ, let $k(\theta)$ be the smallest nonnegative integer for which $\theta \in M_\theta^k(R_a)$. Let $g_a(\theta)$ be the φ−coordinate of the $k(\theta)$th iterate under M of $(M_\theta^{-k(\theta)}, \varphi) \in S_a$. Then the graph $\varphi = g_a(\theta)$ has a finite number of d_a of discontinuities. Each contiguous piece of this graph is a forward iterate of some piece of S_a.

Now form the curve G_a by taking the graph of g_a and adding line segments in the φ direction at each value of θ where g_a is discontinuous. Thus we make G_a a contiguous curve. Define g_b and G_b similarly to g_a and G_a, but in terms of the backward (not forward) iterates of the S_b. Let d_b be the number of discontinuities of g_b.

Following [KKH03], our goal is to show that for n sufficiently large, the nth iterate of G_a intersects G_b for at least $d_a + d_b + 1$ different values of θ. Then since there are at most $d_a + d_b$ values of θ at which one of these two curves has a connecting segment, there will be at least one intersection point between the nth iterate of the graph of g_a and the graph of g_b. Since the graph of g_a consists of forward iterates of S_a and the graph of g_b consists of backward iterates of S_b, some forward iterated S_a will intersect S_b, as we claimed.

Given a contiguous curve C that, like G_a and G_b, is the graph of a function of θ that is continuous except for a finite number of values at which C has a connecting segment, observe that its image under M_θ is a curve of the same type (in particular, since the map is one–to–one, the heights of the vertical segments remain less than 2π). Furthermore, because of the θ_n term in the φ map (7.5b), the image of C 'wraps around' the torus in the φ direction one more time than C does as one goes around the torus one time in the θ direction.

To formulate what we mean by 'wrapping around', define the winding number of C as follows. As θ increases from 0 to 2π, count the number of times C crosses $\varphi = 0$ in the upward and downward directions. The difference between the number of upward and downward crossings is the winding number of C.

Now if two curves C_1 and C_2, as described above, have different winding numbers w_1 and w_2, then C_1 and C_2 must intersect at least $|w_1 - w_2|$ times. Because of the periodicity of $P(\theta, \varphi)$, the winding number of a curve must increase by 1 each time the map M is applied. Thus for n sufficiently large, the winding number of the nth iterate of G_a differs from the winding number of G_b by at least $d_a + d_b + 1$. Hence the nth iterate of G_a intersects G_b for at least $d_a + d_b + 1$ different values of θ as desired. This establishes claims (i) and (iv).

Notice that the argument above does not depend on the specific form of $P(\theta, \phi)$, only that it is continuous and periodic and that η is sufficiently small ($|\eta| < \eta_c$) that the map (7.5) is one–to–one. This independence of the results from the specific form of $P(\theta, \phi)$ implies that the results are stable to system changes that preserve a quasi–periodic driving component (7.5a).

Discussion

The possible existence of SNA's was originally pointed out in [GOP84], and many numerical explorations of the dynamics on attractors that are apparently strange and non–chaotic have appeared. Recently, there has also been rigorous results on the mathematical properties that SNA's must have if they exist [Sta97]. In spite of these works, a very basic question has remained unanswered: Can it be rigorously established that SNA's generically exist in typical quasi–periodically forced systems? This is an important issue, because, although the numerical evidence for SNA's is very strong, perhaps the attractors observed are non–strange with very fine scale structure (rather than the infinitesimally fine scale structure of a truly strange attractor). Also, there might be the worry that the numerical evidence is somehow an artifact of computational error. Our proof of topological transitivity, combined with the hypothesis that $h_\varphi < 0$, answers the question of the typical existence of SNA's (affirmatively) for the first time ([HO01] contains a preliminary report of our work). The only previous work rigorously establishing the existence of a SNA is that appearing in the original publication on SNA's [GOP84]. These proofs, however, are for a very special class of quasi–periodically forced system such that an arbitrarily small typical change of the system puts it out of the class. Thus this proof does not establish that SNA's exist in typical quasi–periodically forced situations. In order to see that nature of this situation with respect to [GOP84], we recall the example treated in [GOP84]. In that reference the map considered was

$$x_{n+1} = 2\lambda(\tanh x_n)\cos\theta_n \equiv f(x_n, \theta_n),$$

with θ_n evolved as in (7.5a). It was proven in [GOP84] that this map has a SNA for $\lambda > 1$. However, the map has an invariant set, namely, the line $x = 0$, θ in $[0, 2\pi)$, and this fact is essential in the proof of [GOP84]. On the other hand, the existence of this invariant set does not persist under perturbations of the map. Thus, if we perturb $f(x, \theta)$ to $f(x, \theta) + \varepsilon g(x, \theta)$, the invariant set is destroyed, even for small ε, for any typical function $g(x, \theta)$ (in particular, an arbitrarily chosen $g(x, \theta)$ is not expected to satisfy $g(0, \theta) = 0$).

Rational Approximation: Origin of SNA's

Using rational approximations (RA's) to the quasi–periodic forcing, we now investigate the origin for the appearance of SNA's in (7.5) for $P(\theta, \varphi) = \sin \varphi$

and $\omega = \pi(\sqrt{5} - 1)$. For the case of the inverse golden mean $\Omega \equiv \omega/2\pi$, its rational approximants are given by the ratios of the Fibonacci numbers, $\Omega_k = F_{k-1}/F_k$, where the sequence of $\{F_k\}$ satisfies

$$F_{k+1} = F_k + F_{k-1} \qquad \text{with} \qquad F_0 = 0 \qquad \text{and} \qquad F_1 = 1.$$

Instead of the quasi–periodically forced system, we study an infinite sequence of periodically forced systems with rational driving frequencies ω_k. We suppose that the properties of the original system may be obtained by taking the quasi–periodic limit $k \to \infty$ [KKH03].

For each RA of level k, a periodically forced map with the rational driving frequency Ω_k has a periodic or quasi–periodic attractor that depends on the initial phase θ_0 of the external force. Then we take the union of all attractors for different θ_0 to be the kth RA to the attractor in the quasi–periodically forced system. Furthermore, due to the periodicity, it is sufficient to get the RA by changing θ_0 only in an basic interval $\theta_0 \in [0, 1/F_k)$, because the RA to the attractor in the remaining range, $[1/F_k, 1)$, may be obtained through $(F_k - 1)$−times iterations of the result in $[0, 1/F_k)$. For a given k we call the periodic attractors of period F_k the 'main periodic component'.

We first note that for $\eta = 0$ the RA to the regular quasi–periodic attractor consists of only the quasi–periodic component. However, as η becomes positive periodic components appear via phase–dependent (i.e., θ_0−dependent) saddle–node bifurcations.

In what follows we use the RA's to explain the origin of the negative Lyapunov exponent h_φ and the strangeness of the SNA. For a given level k of the RA, let $h_\varphi^{(k)}(\theta)$ denote the Lyapunov exponent of the attractor corresponding to a given θ. Thus $h_\varphi^{(k)}(\theta) = 0$ for θ in the quasi–periodic range and $h_\varphi^{(k)}(\theta) < 0$ for θ in the periodic range (gaps in the gray regions). Since the attractor with irrational ω generates a uniform density in θ, we take the order–k RA to the Lyapunov exponent h_φ to be [KKH03]

$$< h_\varphi^{(k)} >= \frac{1}{2\pi} \int_0^{2\pi} h_\varphi^{(k)}(\theta) d\theta.$$

For $\eta > 0$, due to the existence of periodic components, $< h_\varphi^{(k)} >$ is negative. As η increases for a given level k, the Lebesgue measure in θ for the periodic components increases, and hence $h_\varphi^{(k)}(\theta)$ becomes negative in a wider range in θ. Thus, as η increases $< h_\varphi^{(k)} >$ decreases.

7.2.3 High Dimensional Maps

Radial Perturbations of the Torus Map

We now show that stability to perturbations applies in addition if the system is higher dimensional. In particular, we discuss the case of a 3D system with an attracting invariant torus, and allow perturbations of the toroidal surface. Consider the following map on $\mathbb{R}^3$ [KKH03]:

$$
\begin{aligned}
\theta_{n+1} &= [\theta_n + \omega] \mod 2\pi, \\
\varphi_{n+1} &= [\theta_n + \varphi_n + \eta \bar{P}(\theta_n, \varphi_n, r_n)] \mod 2\pi, \\
r_{n+1} &= \lambda r_n + \rho Q(\theta_n, \varphi_n, r_n).
\end{aligned}
\tag{7.8}
$$

Here θ and φ are coordinates on a torus embedded in $\mathbb{R}^3$ and r is a coordinate transverse to the torus, with $r = 0$ representing the un–perturbed ($\lambda = \rho = 0$) torus. The parameters ω and η, and the dependence of $\bar{P}$ on θ and φ, have the same properties as for map (7.5), and Q is continuously differentiable and 2π periodic in θ and φ. When λ and ρ are small, (7.8) maps a neighborhood of the torus $r = 0$ into itself, and when $\rho = 0$ the torus $r = 0$ is invariant and attracting. It then follows from classical results on the perturbation of invariant manifolds [KH95] that, for λ and ρ sufficiently small, the map (7.8) has a smooth attracting invariant manifold $r = f(\theta, \varphi)$ near the torus $r = 0$. On this attractor, the map (7.8) reduces to a map of the form (7.5), with $P(\theta, \varphi) = \bar{P}[\theta, \varphi, f(\theta, \varphi)]$. Thus statements (i)-(v) above apply also to the attractor of the 3D map (7.8).

The above arguments depend on the existence of a smooth invariant torus on which the attractor is located, and this is guaranteed if λ and ρ are sufficiently small. We now show numerical evidence for the existence of a smooth invariant torus for values of λ and ρ that are appreciable. We consider the example $\bar{P}(\theta, \varphi, r) = \sin \varphi$ and $Q(\theta, \varphi, r) = \sin(r + \varphi)$. We numerically get 2D plots of intersections of the invariant torus with the surfaces $\varphi = \pi$ and $\varphi = \pi$. First consider the case $\varphi = \pi$. Our numerical technique is as follows. We choose an initial value $(\theta_0, \varphi_0 = \pi)$ and get $(\theta_{-n} \; \varphi_{-n})$ by iterating (7.8a) and (7.8b) backward n steps. Since $h_r \sim \ln \lambda < 0$ (when $\rho \ll \lambda < 1$), $r_{-n} \to \pm\infty$ if r_0 is not on the torus. In other words, if $r_{-n} = 0$, then r_0 is on the torus. Thus, we choose $r_{-n} = 0$ and iterate $(r_{-n}, \theta_{-n}, \varphi_{-n})$ forward n steps to $(r_0, \theta_0, \varphi_0 = \pi)$. By varying θ_0, we get the graph, $r_0(\theta_0)$, of the torus intersection with $\varphi = \pi$. Similarly, choosing $(\theta_0 = \pi, \varphi_0)$ and iterating the map (7.8) backward, and then forward, we can get $r_0(\varphi_0)$ of the torus intersection with $\theta = \pi$.

Map on a High–Dimensional Torus

Above we showed that (7.5) is topologically transitive. Here we show how this argument can be modified to higher dimensional maps that include $N > 1$ quasi–periodic driving variables $\theta^{(1)}, \theta^{(2)}, \cdots, \theta^{(N)}$. For exposition we assume $N = 2$, but the argument is virtually identical for all N.

In particular, we consider a map of the form [KKH03]

$$\theta_{n+1}^{(1)} = [\theta_n^{(1)} + \omega^{(1)}] \mod 2\pi, \tag{7.9}$$

$$\theta_{n+1}^{(2)} = [\theta_n^{(2)} + \omega^{(2)}] \mod 2\pi,$$

$$\varphi_{n+1} = [q^{(1)}\theta_n^{(1)} + q^{(2)}\theta_n^{(2)} + \varphi_n + \eta P(\theta_n^{(1)}, \theta_n^{(2)}, \varphi_n)] \mod 2\pi,$$

where $\omega^{(1)}$ and $\omega^{(2)}$ are incommensurate, $(q^{(1)}, q^{(2)})$ is a pair of integers different from $(0, 0)$, and $P(\theta^{(1)}, \theta^{(2)}, \varphi)$ is continuous, differentiable, and 2π periodic in all of its arguments ($\theta^{(1)}$, $\theta^{(2)}$, and φ). We assume without loss of generality that $q^{(1)} \neq 0$.

Let

$$R_a = \{(\theta^{(1)}, \theta^{(2)}) : \theta_a^{(1)} < \theta^{(1)} < (\theta_a^{(1)} + \delta_a) \text{ and } \theta_a^{(2)} < \theta^{(2)} < (\theta_a^{(2)} + \delta_a)\},$$

$$R_b = \{(\theta^{(1)}, \theta^{(2)}) : \theta_b^{(1)} < \theta^{(1)} < (\theta_b^{(1)} + \delta_b) \text{ and } \theta_b^{(2)} < \theta^{(2)} < (\theta_b^{(2)} + \delta_b)\},$$

be two arbitrary squares in the $\theta^{(1)}$-$\theta^{(2)}$ torus, and let

$$S_a = \{(\theta^{(1)}, \theta^{(2)}, \varphi) : (\theta^{(1)}, \theta^{(2)}) \in R_a \text{ and } \varphi = \varphi_a\},$$

$$S_b = \{(\theta^{(1)}, \theta^{(2)}, \varphi) : (\theta^{(1)}, \theta^{(2)}) \in R_b \text{ and } \varphi = \varphi_b\},$$

be a pair of square segments, where φ_a and φ_b are arbitrary. As before, we will show that there is a finite trajectory that begins on S_a and ends on S_b.

In this case, we proceed by iterating R_a forward until the union of its iterates cover all points $(\theta^{(1)}, \pi)$ at least once. The number of iterates needed is finite. Then we select pieces of these iterates that single–cover a thin strip

$$D_a = \{(\theta^{(1)}, \theta^{(2)}) : \pi \leq \theta^{(2)} \leq \pi + \varepsilon_a\},$$

with rectangles of width ε_a. From the corresponding pieces of the corresponding iterates of S_a, we form the graph $\varphi = g_a(\theta^{(1)}, \theta^{(2)})$ of a piecewise continuous function g_a defined on D_a. Similarly we form a graph $\varphi = g_b(\theta^{(1)}, \theta^{(2)})$ on a strip D_b from pieces of backward iterates of $R_b \times \{\varphi_b\}$. As before, we will show that some forward iterate of the graph of g_a must intersect the graph of g_b.

Next, form the strip G_a by taking the graph of g_a and adding 'connecting faces' at each of the d_a values of $\theta^{(1)}$ where g_a is discontinuous, so as to make G_a a contiguous strip. The construction of G_a now has some thickness in the $\theta^{(2)}$ direction (not shown). For each n, the nth iterate of G_a is also a contiguous strip that consists of the graph of a function with d_a discontinuities in the $\theta^{(1)}$ direction, together with d_a connecting faces, over a strip in the $\theta^{(1)}$-$\theta^{(2)}$ torus of width ε_a in the $\theta^{(2)}$ direction. Notice though that the strip moves a distance $\omega^{(2)}$ in the $\theta^{(2)}$ direction with each iteration. Define g_b and G_b similarly to g_a and G_a, but in terms of the backward iterates of S_b, and let d_b be the number of values of $\theta^{(1)}$ at which g_b is discontinuous.

As before, we can define the winding number of strips like G_a and G_b, representing the net number of times the strip wraps in the φ direction as

$\theta^{(1)}$ increases from 0 to 2π. The winding number can be computed for any fixed value of $\theta^{(2)}$ and does not depend on that value. With each iteration of (7.9), the winding number of such a strip changes by $q^{(1)} \neq 0$. Therefore for n sufficiently large, the winding number of the nth iterate of G_a differs from the winding number of G_b by at least $d_a + d_b + 1$. Furthermore, by increasing n if necessary, we can ensure that the domains of these two strips intersect; that is, they have a common value of $\theta^{(2)}$. Then for that value of $\theta^{(2)}$, it follows as before that the nth iterate of the graph of g_a (without the d_a connecting faces of G_a) and the graph of g_b (without the d_b connecting faces of G_b) must intersect as claimed.

These results confirm the existence of SNA's as a generic phenomenon of quasi–periodically forced systems. For details see [KKH03].

7.3 Effective Dynamics in Hamiltonian Systems

A comprehensive understanding of Hamiltonian dynamics is a long outstanding problem in nonlinear and statistical physics, which has important applications in various other areas of physics. Typical Hamiltonian systems are non–hyperbolic as they exhibit mixed phase space with coexisting regular and chaotic regions. Over the past years, a number of ground–breaking works (see [MMG05] and references therein) have increasingly elucidated the asymptotic behavior of such systems and it is now well understood that, because of the stickiness due to Kolmogorov–Arnold–Moser (KAM) tori, the chaotic dynamics of typical Hamiltonian systems is fundamentally different from that of hyperbolic, fully chaotic systems. Here 'asymptotic' means in the limit of large time scales and small length scales. But in realistic situations, the time and length scales are limited. In the case of hyperbolic systems, this is not a constraint because the (statistical) self–similarity of the underlying invariant sets guarantees the fast convergence of the *dynamical invariants* (entropies, Lyapunov exponents, fractal dimensions, escape rates, etc) and the asymptotic dynamics turns out to be a very good approximation of the dynamics at finite scales. In non–hyperbolic systems, however, the *self–similarity* is usually lost because the invariant sets are not statistically invariant under magnifications. As a result, the finite–scale behavior of a Hamiltonian system may be fundamentally different from the asymptotic behavior considered previously, which is in turn hard to come by either numerically or experimentally.

Dynamics of Hamiltonian systems at finite, physically relevant scales has been studied in [MMG05], focusing on *Hamiltonian chaotic scattering*, which is one of the most prevalent manifestations of chaos in open systems, with examples ranging from fluid dynamics to solid–state physics to general relativity. It has been shown that the finite–scale dynamics of a Hamiltonian system is characterized by *effective dynamical invariants* (e.g., effective fractal dimension), which: (1) may be significantly different from the corresponding invariants of the asymptotic dynamics; (2) depend on the resolution but can be regarded as constants over many decades in a given region of the phase space; and (3) may change drastically from one region to another of the *same* dynamically connected (ergodic) component. These features are associated with the slow and nonuniform convergence of the invariant measure due to the *breakdown of self–similarity* in non–hyperbolic systems.

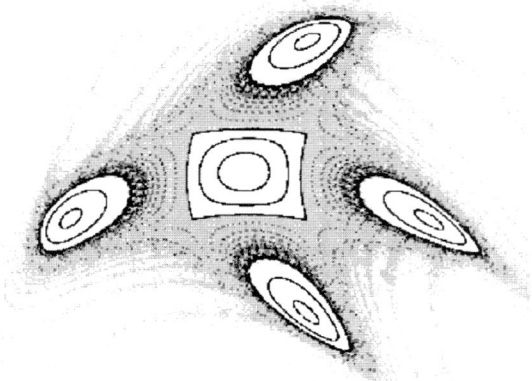

Fig. 7.4. KAM islands (blank) and stable manifold (gray; modified and adapted from [MMG05]).

More specifically, consider a 2D area–preserving map with a major *KAM–island* (see Figure 7.4) surrounded by a *chaotic region*. One such map captures all the main properties of a wide class of Hamiltonian systems with mixed phase–space. When the system is open (scattering), almost all particles initialized in the chaotic region eventually escape to infinity. In the case of *chaotic scattering*, a singularity develops and the invariant measure, given by $\lim_{t\to\infty} \rho(x,t)$, accumulates on the outermost *KAM–torus* of the KAM–island. Physically, this corresponds to the tendency of non–escaping particles to concentrate around the regular regions. Dynamically, the stickiness due to *KAM–tori* underlies two major features of Hamiltonian chaotic scattering:

(i) the algebraic decay of the survival probability of particles in the scattering region; and

(ii) the integer dimension of the chaotic saddle,

and distinguishes this phenomenon from the hyperbolic chaotic scattering characterized by exponential decay and non–integer fractal dimension (see [MMG05] and references therein). However, the convergence of the measure is rather slow and highly nonuniform for typical parameters, which is in sharp contrast with the fast, uniform convergence observed in hyperbolic systems.

Most of works on transport in Hamiltonian systems have used stochastic models, where invariant structures around KAM–islands are smoothened out and the dynamics is given entirely in terms of a diffusion equation or a set of transition probabilities (Markov chains or trees). The stochastic approach is suitable to describe transport properties (as above), but cannot be used to predict the behavior of dynamical invariants such as Lyapunov exponents and fractal dimensions.

Instead, [MMG05] has adopted a deterministic approach using the *Cantori*[2] surrounding the KAM islands to split the non–hyperbolic dynamics of the Hamiltonian system into a chain of hyperbolic dynamical systems. There is a hierarchy of infinitely many Cantori around each island. Let C_1 denote the area of the scattering region outside the outermost *Cantorus*, C_2 denote the annular area in between the first and second Cantorus, and so on. As j is increased, C_j becomes thinner and approaches the corresponding island. For simplicity, consider that there is a single island and that, in each iteration, a particle in C_j may either move to the outer level C_{j-1} or the inner level C_{j+1} or stay in the same level. Let Δ_j^- and Δ_j^+ denote the transition probabilities from level j to $j-1$ and $j+1$, respectively. A particle in C_1 may also leave the scattering region, and in this case we consider that the particle has escaped. The escaping region is denoted by C_0. The chaotic saddle is expected to have points in C_j for all $j \geq 1$. It is natural to assume that the transition probabilities Δ_j^- and Δ_j^+ are constant in time. This means that each individual level can be regarded as a hyperbolic scattering system, with its characteristic exponential decay and non–integer chaotic saddle dimension. Therefore, a non–hyperbolic scattering is in many respects similar to a sequence of hyperbolic scatterings.

7.3.1 Effective Dynamical Invariants

The concept of *effective dynamical invariants* was introduced in [MMG05]. As a specific example, the authors have considered the *effective fractal dimension*, which, for the intersection of a fractal set S with a nD region L, has been defined as

$$D_{eff}(L; \varepsilon) = n - \frac{d \ln f(\varepsilon')}{d \ln \varepsilon'}\bigg|_{\varepsilon'=\varepsilon}, \tag{7.10}$$

where $f(\varepsilon') = N(\varepsilon')/N_0(\varepsilon')$, and $N(\varepsilon')$ and $N_0(\varepsilon')$ are the number of cubes of edge length ε' needed to cover $S \cap L$ and L, respectively. Take L to be a generic segment of line (i.e., $n = 1$ in (7.10)) intersected by S on a fractal set. In the limit $\varepsilon \to 0$, we recover the common *box–counting dimension*

$$D = 1 - \lim_{\varepsilon \to 0} \Delta \ln f(\varepsilon)/\Delta \ln \varepsilon$$

---

[2] Recall that *Cantori* are invariant structures that determine the transversal transport close to the KAM–islands.

of the fractal set $S \cap L$, which is known to be 1 for all our choices of L. However, for any practical purpose, the parameter ε is limited and cannot be made arbitrarily small (e.g., it cannot be smaller than the size of the particles, the resolution of the experiment, and the length scales neglected in modelling the system). At scale ε the system behaves as if the fractal dimension were $D_{eff}(L; \varepsilon)$ (therefore the term 'effective dimension'). In particular, the final state sensitivity of particles launched from L, with the initial conditions known within accuracy ε^*, is determined by $D_{eff}(L; \varepsilon^*)$ rather than D: as ε is varied around ε^*, the fraction of particles whose final state is uncertain scales as $\varepsilon^{1-D_{eff}(L;\varepsilon^*)}$, which is different from the prediction ε^{1-D}. This is important in this context because the value of $D_{eff}(L; \varepsilon)$ may be significantly different from the asymptotic value $D = 1$ even for unrealistically small ε and may also depend on the region of the phase space. Similar considerations apply to many other invariants as well. For more details, see [MMG05].

7.4 Formation of Fractal Structure in Many–Body Systems

Formation of spatial structures is an interesting and important phenomenon in nature. It is seen over a wide range, from protein folding in biological systems [Fer98, Kar00] to large–scale structure in the universe [Pee80]. The theoretical origins of such structures are quite important and will be classified into several classes. One of the most interesting areas within the field of dynamical systems is that some remarkable structure and organization is created dynamically by the mutual interaction among the elements [DHR00, BDR01].

Recently we have discovered that spatial structure with fractal distribution emerges spontaneously from uniformly random initial conditions in a 1D self–gravitating system, that is the sheet model [KK01]. What is noteworthy in this phenomenon is that the spatial structure is not given at the initial condition, but dynamically created from a state without spatial correlation. Succeeding research clarified that the structure is created first in small spatial scale then grows up to large scale through hierarchical clustering [KK02a], and the structure is transient [KK02b]. It is quite interesting that some remarkable spatial structures are emerged instead of monotonous thermal relaxation in Hamiltonian system.

The emergence of fractal structure is a typical example that systems of many degrees of freedom are self–organized by dynamics themselves. Hence to clarify its dynamical mechanism is very important subject toward understanding physics of self–organization of matter.

A way to clarify the dynamical mechanism is to know which class of pair potential can form the fractal structure. Here we note an important fact that fractal structure does not have characteristic spatial scale, nor does the potential of the sheet model, since the pair potential is power of the distance. Hence the scale free property of potential may be a keystone to understand the dynamical mechanism. The question is if the fractal structure can be formed in not only the sheet model, but also other systems with power–law potentials.

In this section, following [KK01, KK02a, KK02b, KK06], we study the possibility that the fractal structure can be formed in more general systems without characteristic spatial scale which is extended from sheet model. Here we adopt the model as the system with *attractive power–law potentials*. At first we examine the formation of the fractal structure by numerical simulation for various values of the power index of the potential. Next we perform linear analysis to consider the numerical results.

7.4.1 A Many–Body Hamiltonian

We consider the model where many particles with an uniform mass interact with purely attractive pair potential of power–law, which is described by the Hamiltonian [MPT98]

$$H = K + U = \sum_{i=1}^{N} \frac{p_i^2}{2} + \sum_{i=1}^{N-1} \sum_{j>i}^{N} |x_i - x_j|^\alpha , \qquad (7.11)$$

where x_i and p_i are the position and momentum a particle, respectively. The first term is kinetic energy and the second term is potential energy. For simplicity, here we consider the system where motion of particles is bounded to 1D direction.

For the special case $\alpha = 1$, the Hamiltonian (7.11) applies to a system of N infinite parallel mass sheets, where each sheet extends over a plane parallel to the yz plane and moves along the x axis under the mutual gravitational attraction of all the other sheets. The Hamiltonian of the sheet model [HF67, Ryb71, LSR84, MR90] is usually written in the form

$$H = \sum_{i=1}^{N} \frac{p_i^2}{2} + \sum_{i=1}^{N-1} \sum_{j>i}^{N} |x_i - x_j| . \qquad (7.12)$$

Previously (see [KK01, KK02a, KK02b]), it has been shown that fractal structure emerges from non–fractal initial conditions in the Hamiltonian (7.12).

7.4.2 Linear Perturbation Analysis

Here we clarify the physical reason analytically why fractal structure can not be constructed for the large value of the power index of the potential. The formation of the fractal structure occurs at the relative early stage in the whole–evolution history [KK02b]. Then it is instructive for understanding the mechanism by which the structure is formed to know the qualitative properties of the short–term behaviors by linear analysis.

Next, we derive the dispersion relation from the *collision–less Boltzmann equation* (CBE), which describes the growing rate of the linear perturbation [BT87].

Collisionless Boltzmann Equation

The CBE is defined by [KK01, KK02a, KK02b, KK06]

$$\left\{ \partial_t + p\partial_x + \int_{-\infty}^{\infty} dx' F(x-x') \left(\int_{-\infty}^{\infty} dp' f(x',p',t) \right) \partial_p \right\} f(x,p,t) = 0,$$
(7.13)

where F is 2–particle force which is related with a pair–potential U by $F(x) = -\partial_x U$, and f is the one–particle distribution function. For simplicity we consider the system is extended in infinite region $-\infty < x < \infty$. It is clear that the state of uniform spatial density with an arbitrary velocity distribution,

$$f(x,p,t) = f_0(p),$$
(7.14)

is a stationary state. We impose the following small perturbation over the stationary state (7.14)

$$f(x,p,t) = f_0(p) + \delta f(x,p,t).$$

The linearized equation for δf is

$$(\partial_t + p\partial_x)\,\delta f(x,p,t) = -\int_{-\infty}^{\infty} dx' F(x-x') \int_{-\infty}^{\infty} dp' \delta f(x',p',t)\partial_p f_0(p).$$
(7.15)

Now we define the *Fourier–Laplace transform* by that x is Fourier transformed and t is Laplace transformed, that is

$$\widetilde{\delta f}(k,p,\omega) \equiv \int_0^{\infty} dt e^{-i\omega t} \int_{-\infty}^{\infty} dx e^{ikx} \delta f(x,p,t).$$

Fourier transform is

$$\widehat{\delta f}(k,p,t) \equiv \int_{-\infty}^{\infty} dx e^{ikx} \delta f(x,p,t), \qquad \text{and} \qquad \hat{F}(k) \equiv \int_{-\infty}^{\infty} dx e^{ikx} F(x).$$

Then the Fourier–Laplace transformed equation of (7.15) is

$$\varepsilon_k(\omega)\widetilde{\delta f_k}(\omega) = \frac{-1}{i(-\omega + kp)} \widehat{\delta f}(k,p,0), \qquad \text{where} \quad \varepsilon_k(\omega) \quad \text{is}$$
(7.16)

$$\varepsilon_k(\omega) \equiv 1 + \int_{-\infty}^{\infty} dp \frac{\hat{F}(k)}{i(-\omega + kp)} \partial_p f_0(p).$$
(7.17)

Dispersion Relation

From the inverse–Laplace transform of (7.16), if follows KK01, KK02a, KK02b, KK06]

$$\delta f_k(t) = \int_{-\infty-i\sigma}^{\infty-i\sigma} e^{i\omega t} \widetilde{\delta f_k}(\omega) \frac{d\omega}{2\pi}$$
$$= \int_{-\infty-i\sigma}^{\infty-i\sigma} e^{i\omega t} \frac{1}{\varepsilon_k(\omega)} \frac{-1}{i(-\omega+kp)} \widehat{\delta f}(k,p,0) \frac{d\omega}{2\pi}.$$

Now we continuously move the integration contour to upper half of complex ω plane while avoiding the singular points. Then the contributions from except of the pole can be neglected, because of the factor $\exp(i\omega t)$ $(Im(\omega) > 0)$.

Then the growth rate of each mode of the fluctuation is obtained by the solution of the *dispersion relation*

$$\varepsilon_k(\omega) = 0. \tag{7.18}$$

If (7.18) has the solution where the inequality $Im(\omega) < 0$ is satisfied, the fluctuation is unstable.

Dynamical Stability

Now we consider the case that the potential is power–law, the pair–potential U is $U(x) = A|x|^\alpha$. Assuming the interaction is attractive, $A > 0$. Also, $\alpha = 1$ for the so–called 'sheet model'. The Fourier–transformed potential (with $(\alpha \neq 0, 2, 4, \cdots, -1, -3, \cdots))$ is [Lig58]

$$\hat{U}(k) \equiv \int_{-\infty}^{\infty} dx e^{ikx} A|x|^\alpha = 2A\left\{-\left(\sin\frac{\alpha\pi}{2}\right)\frac{\Gamma(\alpha+1)}{|k|^{\alpha+1}}\right\}.$$

For simplicity, we choose the stationary state as $f_0 = n_0\delta(p)$, where n_0 is the number density of particles.

The dispersion relation is [KK01, KK02a, KK02b, KK06]

$$\varepsilon_k(\omega) = 1 + 2n_0 A\left\{\left(\sin\frac{\alpha\pi}{2}\right)\frac{\Gamma(\alpha+1)}{|k|^{\alpha-1}}\right\}\frac{1}{\omega^2} = 0,$$

and ω which satisfy the dispersion relation is

$$\omega^2 = -2n_0 A\left\{\left(\sin\frac{\alpha\pi}{2}\right)\frac{\Gamma(\alpha+1)}{|k|^{\alpha-1}}\right\}. \tag{7.19}$$

When $\omega^2 < 0$, the system is unstable. (7.19) can be reduced to

$$\omega = \pm i\sqrt{2n_0 A\left\{\left(\sin\frac{\alpha\pi}{2}\right)\Gamma(\alpha+1)\right\}} \cdot |k|^{(1-\alpha)/2}. \tag{7.20}$$

From (7.20) we can classify the evolution of the perturbation into three types:
    (i) when $0 < \alpha < 1$, the growing rate increase monotonously for $|k|$;
    (ii) when $\alpha = 1$, the fluctuations for all scale grows at same rate;
    (iii) when $1 < \alpha < 2$, the growing rate decrease monotonously for $|k|$.
For more details, see [KK01, KK02a, KK02b, KK06].

7.5 Fractional Calculus and Chaos Control

It is well known that chaos cannot occur in continuous systems of total order less than three. This assertion is based on the usual concepts of order, such as the number of states in a system or the total number of separate differentiations or integrations in the system. The model of system can be rearranged to three single differential equations, where one of the equations contains the non–integer (fractional) order derivative. The total order of system is changed from 3 to $2 + q$, where $0 < q \leq 1$. To put this fact into context, we can consider the fractional–order dynamical model of the system. Hartley *et al.* [HLQ95] consider the fractional–order Chua's system (compare with an ordinary *Chua's circuit* (1.39) described in Introduction) and demonstrated that chaos is possible for systems where the order is less than three. In their work, the limits on the mathematical order of the system to have a chaotic response, as measured from the bifurcation diagrams, are approximately from 2.5 to 3.8. In work [ACF98], chaos was discovered in fractional–order two–cell cellular neural networks and also in work [NE99] chaos was exhibited in a system with total order less than three.

The control of chaos has been studied and observed in experiments (e.g., works [Bai89], [LO97], [PY97], [Ush99]). Especially, the control of the well–known Chua's system [PC89] by sampled data has been studied [YC98]. The main motivation for the control of chaos via sampled data is well–developed digital control techniques.

A sampled–data feedback control of a fractional–order chaotic dynamical system was presented in [Pet99, Pet02], modelled by the state equation $\dot{x} = f(x)$, where $x \in \mathbb{R}^n$ is the state variable, $f : \mathbb{R}^n \to \mathbb{R}^n$ is a nonlinear function and $f(0) = 0$. The approach used was concentrating on the feedback control of the chaotic fractional–order Chua's system, where total order of the system is 2.9.

7.5.1 Fractional Calculus

Fractional Derivatives

Recall that the idea of fractional calculus has been known since the development of the regular calculus, with the first reference probably being associated with Leibniz and L'Hospital in 1695.

Fractional calculus is a generalization of integration and differentiation techniques to a non–integer–order fundamental operator $_aD_t^\alpha$, where a and t are the limits of the operation. This continuous integro–differential operator is defined as [Pet99, Pet02]

$$_aD_t^\alpha = \begin{cases} \frac{d^\alpha}{dt^\alpha} & \text{if } \alpha > 0, \\ 1 & \text{if } \alpha = 0, \\ \int_a^t (d\tau)^{-\alpha} & \text{if } \alpha < 0. \end{cases}$$

The two definitions used for the general fractional integro–differential operator $_aD_t^\alpha$ are the *Grünwald–Letnikov definition* (GL) and the *Riemann–Liouville definition* (RL) (see [OS74, Pod99]). The GL definition is given as

$$_aD_t^\alpha f(t) = \lim_{h \to 0} h^{-\alpha} \sum_{j=0}^{[\frac{t-a}{h}]} (-1)^j \binom{\alpha}{j} f(t - jh), \tag{7.21}$$

where $[x]$ means the integer part of x. The RL definition is given as

$$_aD_t^\alpha f(t) = \frac{1}{\Gamma(n - \alpha)} \frac{d^n}{dt^n} \int_a^t \frac{f(\tau)}{(t - \tau)^{\alpha - n + 1}} d\tau, \tag{7.22}$$

for $(n - 1 < \alpha < n)$ and where $\Gamma(.)$ is the *Gamma* function.

Numerical Calculation of Fractional Derivatives

For numerical calculation of the fractional–order derivations we can use the relation (7.23) derived from the Grünwald–Letnikov definition (7.21). This approach is based on the fact that for a wide class of functions, the above two definitions, (7.21) and (7.22), are equivalent. The relation for the explicit numerical approximation of the α–th derivative at the points kT, $(k = 1, 2, \dots)$ has the following form [Dor94, Pet99, Pod99]

$$_{(k-L/T)}D_{kT}^\alpha f(t) \approx T^{-\alpha} \sum_{j=0}^{k} (-1)^j \binom{\alpha}{j} f_{k-j}, \tag{7.23}$$

where L is the 'memory length', T is the step size of the calculation (sample period), and $(-1)^j \binom{\alpha}{j}$ are binomial coefficients $c_j^{(\alpha)}$, $(j = 0, 1, \dots)$. For its calculation we can use

$$c_0^{(\alpha)} = 1, \qquad c_j^{(\alpha)} = \left(1 - \frac{1 + \alpha}{j}\right) c_{j-1}^{(\alpha)}. \tag{7.24}$$

Some Properties of Fractional Derivatives

Two general properties of the fractional derivative will be used. The first is the composition of fractional derivative with the integer–order derivative, while the second is the property of linearity.

The fractional–order derivative commutes with the integer–order derivation [Pod99],

$$\frac{d^n}{dt^n}\left({}_aD_t^p f(t) \right) = {}_aD_t^p \left(\frac{d^n f(t)}{dt^n} \right) = {}_aD_t^{p+n} f(t), \qquad (7.25)$$

under the condition $t = a$ we have $f^{(k)}(a) = 0$, $(k = 0, 1, 2, \ldots, n-1)$. The relationship (7.25) says that operators $\frac{d^n}{dt^n}$ and ${}_aD_t^p$ commute.

Similar to the integer–order differentiation, the fractional differentiation is a linear operation [Pod99]:

$$_aD_t^p \left(\lambda f(t) + \mu g(t) \right) = \lambda \, _aD_t^p f(t) + \mu \, _aD_t^p g(t). \qquad (7.26)$$

7.5.2 Fractional–Order Chua's Circuit

Recall that classical Chua's circuit (1.39) can be written by

$$\dot{v}_1 = \frac{1}{C_1}\left[G(v_2 - v_1) - f(v_1) \right], \qquad \dot{v}_2 = \frac{1}{C_2}\left[G(v_1 - v_2) + i \right],$$
$$\frac{di}{dt} = \frac{1}{L}\left[-v_2(t) - R_L i \right], \qquad (7.27)$$

where $G = 1/R$ and $f(v_1)$ is the piecewise linear $v - i$ characteristic of non-linear Chua's diode.

Given the above techniques of fractional calculus, there are still a number of ways in which the order of system could be amended. One approach would be to change the order of any or all of three constitutive equations (7.27) so that the total order gives the desired value.

In our case, in the equation one, we replace the first differentiation by the fractional differentiation of order q, $(q \in \mathbb{R})$. The final dimensionless equations of the system for $R_L = 0$ are $(x_1 = v_1, x_2 = v_2, x_3 = i)$ [Pet02]

$$\dot{x}_1(t) = \alpha \, _0D_t^{1-q}\left(x_2(t) - x_1(t) - f(x_1) \right), \qquad (7.28)$$
$$\dot{x}_2(t) = x_1(t) - x_2(t) + x_3(t),$$
$$\dot{x}_3(t) = -\beta x_2(t), \qquad \text{where}$$
$$f(x_1) = bx_1 + \frac{1}{2}(a - b)(|x_1 + 1| - |x_1 - 1|)$$
$$\text{with} \qquad \alpha = C_2/C_1, \qquad \beta = C_2 R^2/L.$$

7.5.3 Feedback Control of Chaos

The structure of the control system with sampled data [YC98] is shown in Figure 7.5. Here, the state variables of the chaotic system are measured and the result is used to construct the output signal $y(t) = Dx(t)$, where D is a constant matrix. The output $y(t)$ is then sampled by sampling block to obtain $y(k) = Dx(k)$ at the discrete moments kT, where $k = 0, 1, 2, \ldots$, and T is the sample period. Then $Dx(k)$ is used by the controller to calculate the control signal $u(k)$, which is fed back into chaotic system.

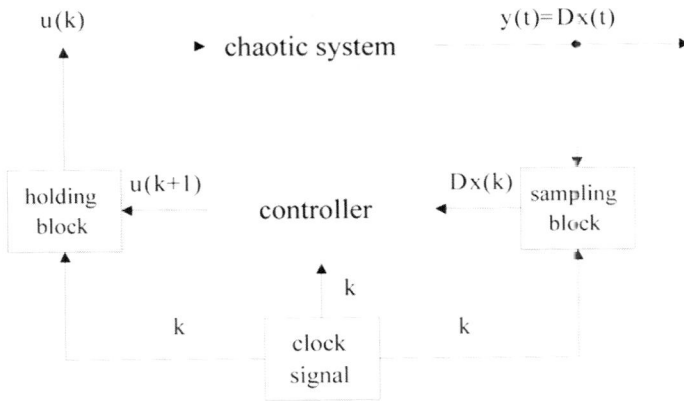

Fig. 7.5. Block–diagram for the chaos–control system (adapted from [YC98]).

In this way controlled chaotic system is defined by relations [YC98]

$$\dot{x}(t) = f(x(t)) + Bu(k), \qquad t \in [kT, (k+1)T) \qquad (7.29)$$
$$u(k+1) = Cu(k) + Dx(k), \qquad (k = 0, 1, 2, \ldots)$$

where $u \in \mathbb{R}^m$, $B \in \mathbb{R}^n \times \mathbb{R}^m$, $C \in \mathbb{R}^m \times \mathbb{R}^m$, $D \in \mathbb{R}^m \times \mathbb{R}^n$ and $t \in R_+$; $x(k)$ is the sampled value of $x(t)$ at $t = kT$. Note that $f(0) = 0$ is an *equilibrium point* of the system (7.29).

The controlled fractional–order Chua's system is now defined by[3]

---

[3] For numerical simulations, the following parameters of the fractional Chua's system (7.28) were chosen in [Pet02]:

$$\alpha = 10, \qquad \beta = \frac{100}{7}, \qquad q = 0.9, \qquad a = -1.27, \qquad b = -0.68,$$

and the following parameters were experimentally found:

$$B = \begin{pmatrix} 1 & 0 & 0 \\ 0 & 0 & 0 \\ 0 & 0 & 0 \end{pmatrix}, \qquad C = \begin{pmatrix} 0.8 & 0 & 0 \\ 0 & 0 & 0 \\ 0 & 0 & 0 \end{pmatrix}, \qquad D = \begin{pmatrix} -3.3 & 0 & 0 \\ 0 & 0 & 0 \\ 0 & 0 & 0 \end{pmatrix}. \qquad (7.30)$$

$$\dot{x}_1(t) = \alpha \ _0D_t^{1-q} \left(x_2(t) - x_1(t) - f(x_1) \right) + u_1(t), \qquad (7.32)$$
$$\dot{x}_2(t) = x_1(t) - x_2(t) + x_3(t) + u_2(t),$$
$$\dot{x}_3(t) = -\beta x_2(t) + u_3(t).$$

For further details, see [Pet02].

7.6 Fractional Gradient and Hamiltonian Dynamics

Recall from the previous section, as well as from, e.g., [SKM93, OS74], that derivatives and integrals of fractional order have found many applications in recent studies in physics. The interest in fractional analysis has been growing continually during the past few years. Fractional analysis has numerous applications: kinetic theories [Zas05, TZ05], statistical mechanics [Tar04a, Tar05a, Tar05b], dynamics in complex media [Tar05c, Tar05d, Tar05e, Tar05f], and many others [Hil00, CM97].

The theory of derivatives of non–integer order goes back to Leibniz, Liouville, Grunwald, Letnikov and Riemann. In the past few decades many authors have pointed out that fractional–order models are more appropriate than integer–order models for various real materials. Fractional derivatives provide an excellent instrument for the description of memory and hereditary properties of various materials and processes. This is the main advantage of fractional derivatives in comparison with classical integer–order models in which such effects are, in fact, neglected. The advantages of fractional derivatives become apparent in modelling mechanical and electrical properties of real materials, as well as in the description of rheological properties of rocks, and in many other fields.

In this section, following [Tar05g], we use a fractional generalization of exterior calculus that had been previously suggested in [CN01]. It allows us to consider the fractional generalization of Hamiltonian and gradient dynamical systems [Gil81, DFN92]. The suggested class of fractional gradient and Hamiltonian systems is wider than the usual class of gradient and Hamiltonian dynamical systems. The gradient and Hamiltonian systems can be considered as a special case of fractional gradient and Hamiltonian systems.

Using the above parameters (7.30) the digital controller in state space form was defined as

$$u_1(k+1) = 0.8u_1(k) - 3.3x_1(k), \qquad (7.31)$$

for $k = 0, 1, 2, \ldots$ The initial conditions for Chua's circuit were $((x_1(0), x_2(0), x_3(0)) = (0.2, -0.1, -0.01)$ and the initial condition for the controller (7.31) was $((u_1(0) = (0))$. The sampling period (frequency) was $T = 100$ Hz. For the computation of the fractional–order derivative in equations (7.32), the relations (7.23), (7.24) and properties (7.25), (7.26) were used. The length of memory was $L = 10$ (1000 coefficients for $T = 100$ Hz).

7.6.1 Gradient Systems

In this subsection, a brief review of gradient systems and exterior calculus [DFN92, II06b] is considered to fix notations and provide a convenient reference.

Gradient systems arise in dynamical systems theory [Gil81, HS74 DFN92]. They are described by the equation

$$\dot{x}^i = -\operatorname{grad} V(x), \qquad \text{where} \quad x \in \mathbb{R}^n.$$

In particular, in Cartesian coordinates, the gradient is given by

$$\operatorname{grad} V = e_i \partial_{x^i} V, \qquad \text{where} \quad x = e_i x^i.$$

More precisely, a dynamical system described by the equations

$$\dot{x}^i = F_i(x), \qquad (i = 1, ..., n) \tag{7.33}$$

is called a *gradient system* in $\mathbb{R}^n$ if the differential 1–form

$$\omega = F_i(x)\, dx^i \tag{7.34}$$

is an *exact form*, i.e., $\omega = -dV$, where $V = V(x)$ is a continuously differentiable function (0–form).

Here d is the exterior derivative [DFN92]. Let $V = V(x)$ be a real, continuously differentiable function on $\mathbb{R}^n$. The exterior derivative of the function V is the 1–form $dV = dx^i \partial_{x^i} V$ written in a coordinate chart $(x^1, ..., x^n)$.

In mathematics [DFN92], the concepts of closed form and exact form are defined for differential forms by the equation $d\omega = 0$ for a given form ω to be a closed form and $\omega = dh$ for an exact form. It is known, that to be exact is a sufficient condition to be closed. In abstract terms, the question of whether this is also a necessary condition is a way of detecting topological information, by differential conditions.

Let us consider the 1–form (7.34). The formula for the exterior derivative d of differential form (7.34) is

$$d\omega = \frac{1}{2}\left(\partial_{x^j} F_i - \partial_{x^i} F_j\right) dx^j \wedge dx^i,$$

where $\wedge$ is the wedge product. Therefore, the condition for ω to be closed is

$$\partial_{x^j} F_i - \partial_{x^i} F_j = 0. \tag{7.35}$$

In this case, if $V(x)$ is a potential function then $dV = dx^i \partial_{x^i} V$. The implication from 'exact' to 'closed' is then a consequence of the symmetry of the second derivatives:

$$\partial^2_{x^i x^j} V = \partial^2_{x^j x^i} V. \tag{7.36}$$

If the function $V = V(x)$ is smooth function, then the second derivative commute, and equation (7.36) holds.

The fundamental topological result here is the Poincaré lemma. It states that for a contractible open subset X of $\mathbb{R}^n$, any smooth p–form β defined on X that is closed, is also exact, for any integer $p > 0$ (this has content only when p is at most n). This is not true for an open annulus in the plane, for some 1–forms ω that fail to extend smoothly to the whole disk, so that some topological condition is necessary. A space X is contractible if the identity map on X is homotopic to a constant map. Every contractible space is simply connected. A space is simply connected if it is path connected and every loop is homotopic to a constant map.

If a smooth vector–field $F = e_i F_i(x)$ of system (7.33) satisfies the relations (7.35) on a contractible open subset X of R^n, then the dynamical system (7.33) is the gradient system such that

$$\dot{x}^i = -\partial_{x^i} V(x). \tag{7.37}$$

This proposition is a corollary of the Poincaré lemma. The Poincaré lemma states that for a contractible open subset X of $\mathbb{R}^n$, any smooth 1–form (7.34) defined on X that is closed, is also exact.

The equations of motion for the gradient system on a contractible open subset X of $\mathbb{R}^n$ can be represented in the form (7.37). Therefore, the gradient systems can be defined by the potential function $V = V(x)$.

If the exact differential 1–form ω is equal to zero ($dV = 0$), then we get the equation

$$V(x) - C = 0, \tag{7.38}$$

which defines the stationary states of the gradient dynamical system (7.37). Here C is a constant.

7.6.2 Fractional Differential Forms

If the partial derivatives in the definition of the exterior derivative $d = dx^i \partial_{x^i}$ are allowed to assume fractional order, a fractional exterior derivative can be defined [CN01] by the equation

$$d^\alpha = (dx^i)^\alpha D_{x^i}^\alpha. \tag{7.39}$$

Here, we use the fractional derivative D_x^α in the *Riemann–Liouville form* [SKM93] that is defined by the equation

$$D_x^\alpha f(x) = \frac{1}{\Gamma(m - \alpha)} \frac{\partial^m}{\partial x^m} \int_0^x \frac{f(y)dy}{(x - y)^{\alpha - m + 1}}, \tag{7.40}$$

where m is the first whole number greater than or equal to α. The initial point of the fractional derivative [SKM93] is set to zero. The derivative of powers k of x is

$$D_x^\alpha x^k = \frac{\Gamma(k+1)}{\Gamma(k+1-\alpha)} x^{k-\alpha}, \tag{7.41}$$

where $k \geq 1$, and $\alpha \geq 0$. The derivative of a constant C need not be zero

$$D_x^\alpha C = \frac{x^{-\alpha}}{\Gamma(1-\alpha)} C. \tag{7.42}$$

For example, the fractional exterior derivative of order α of x_1^k with the initial point taken to be zero and $n = 2$, is given by

$$d^\alpha x_1^k = (dx_1)^\alpha D_{x^1}^\alpha x_1^k + (dx_2)^\alpha D_{x^2}^\alpha x_1^k. \tag{7.43}$$

Using equation (7.41), we get the following relation for the fractional exterior derivative of x_1^k:

$$d^\alpha x_1^k = (dx_1)^\alpha \frac{\Gamma(k+1)x_1^{k-\alpha}}{\Gamma(k+1-\alpha)} + (dx_2)^\alpha \frac{x_1^k x_2^{-\alpha}}{\Gamma(1-\alpha)}.$$

7.6.3 Fractional Gradient Systems

A fractional generalization of exterior calculus was suggested in [CN01], where fractional exterior derivative and the fractional differential forms were defined. It allows us to consider the fractional generalization of gradient systems.

Let us consider a dynamical system that is defined by the equation

$$\dot{x}^i = F_i(x), \qquad \text{on a subset} \quad X \subset \mathbb{R}^n. \tag{7.44}$$

The fractional analog of this definition has the following form. A dynamical system (7.44) is called a fractional gradient system if the fractional differential 1–form

$$\omega_\alpha = F_i(x)(dx^i)^\alpha \tag{7.45}$$

is an exact fractional form $\omega_\alpha = -d^\alpha V$, where $V = V(x)$ is a continuously differentiable function. Using the definition of the fractional exterior derivative, equation (7.45) can be represented as

$$\omega_\alpha = -d^\alpha V = -(dx^i)^\alpha D_{x^i}^\alpha V.$$

Therefore, we have $F_i(x) = -D_{x^i}^\alpha V$.

Note that equation (7.45) is a fractional generalization of equation (7.34). If $\alpha = 1$, then equation (7.45) leads us to equation (7.34). Obviously, a fractional 1–form ω_α can be closed when the 1–form $\omega = \omega_1$ is not closed. So, we have the following proposition. If a smooth vector–field $F = e_i F_i(x)$ on a contractible open subset X of R^n satisfies the relations

$$D_{x^j}^\alpha F_i - D_{x^i}^\alpha F_j = 0, \tag{7.46}$$

then the dynamical system (7.44) is a fractional gradient system such that

$$\dot{x}^i = -D^\alpha_{x^i} V(x), \qquad (7.47)$$

where $V(x)$ is a continuous differentiable function and $D^\alpha_{x^i} V = -F_i$.

This proposition is a corollary of the fractional generalization of *Poincaré lemma*. The Poincaré lemma is shown [CN01] to be true for the exterior fractional derivative. Relations (7.46) are the fractional generalization of relations (7.35). Note that the fractional derivative of a constant need not be zero (7.42). Therefore, we see that constants C in the equation $V(x) = C$ cannot define a stationary state of the gradient system (7.47). It is easy to see that

$$D^\alpha_{x^i} V(x) = D^\alpha_{x^i} C = \frac{(x^i)^{-\alpha}}{\Gamma(1-\alpha)} C \neq 0.$$

In order to define stationary states of fractional gradient systems, we consider the solutions of the system of the equations $D^\alpha_{x^i} V(x) = 0$.

The stationary states of gradient system (7.47) are defined by the equation

$$V(x) - |\prod_{i=1}^{n} x^i|^{\alpha-m} \sum_{k_1=0}^{m-1} \cdots \sum_{k_n=0}^{m-1} C_{k_1...k_n} \left[\prod_{i=1}^{n} (x^i)^{k_i} \right] = 0. \qquad (7.48)$$

The $C_{k_1...k_n}$ are constants and m is the first whole number greater than or equal to α.

In order to define the stationary states of a fractional gradient system, we consider the solution of the equation

$$D^\alpha_{x^i} V(x) = 0. \qquad (7.49)$$

This equation can be solved by using equation (7.40). Let m be the first whole number greater than or equal to α; then we have the solution [SKM93, OS74] of equation (7.49) in the form

$$V(x) = |x^i|^\alpha \sum_{k=0}^{m-1} a_k(x^1, ..., x^{i-1}, x^{i+1}, ..., x^n)(x^i)^k, \qquad (7.50)$$

where a_k are functions of the other coordinates. Using equation (7.50) for $i = 1, ..., n$, we get the solution of the system of equation (7.49) in the form (7.48).

If we consider $n = 2$ such that $x = x^1$ and $y = x^2$, we have the equations of motion for fractional gradient system

$$\dot{x} = -D^\alpha_x V(x, y), \qquad \dot{y} = -D^\alpha_y V(x, y). \qquad (7.51)$$

The stationary states of this system are defined by the equation

$$V(x, y) - |xy|^{\alpha-1} \sum_{k=0}^{m-1} \sum_{l=0}^{m-1} C_{kl} x^k y^l = 0.$$

The C_{kl} are constants and m is the first whole number greater than or equal to α.

Examples

Here we consider a fractional gradient systems that cannot be considered as a gradient system. We prove that the class of fractional gradient systems is wider than the usual class of gradient dynamical systems. The gradient systems can be considered as a special case of fractional gradient systems.

First, let us consider the dynamical system that is defined by the equations

$$\dot{x} = F_x, \qquad \dot{y} = F_y, \tag{7.52}$$

where the right hand sides have the form

$$F_x = acx^{1-k} + bx^{-k}, \qquad F_y = (ax + b)y^{-k}, \tag{7.53}$$

where $a \neq 0$. This system cannot be considered as a gradient dynamical system. Using

$$\partial_y F_x - \partial_x F_y = ay^{-k} \neq 0,$$

we get that $\omega = F_x dx + F_y dy$ is not closed form

$$d\omega = -ay^{-k} dx \wedge dy.$$

Note that the relation (7.46) in the form

$$D_y^{\alpha} F_x - D_x^{\alpha} F_y = 0,$$

is satisfied for the system (7.53), if $\alpha = k$ and the constant c is defined by $c = \Gamma(1-\alpha)/\Gamma(2-\alpha)$. Therefore, this system can be considered as a fractional gradient system with the linear potential function $V(x,y) = \Gamma(1-\alpha)(ax+b)$, where $\alpha = k$.

Second, let us consider the dynamical system that is defined by equation (7.52) with

$$F_x = an(n-1)x^{n-2}+ck(k-1)x^{k-2}y^l, \qquad F_y = bm(m-1)y^{m-2}+cl(l-1)x^k y^{l-2},$$

where $k \neq 1$ and $l \neq 1$. It is easy to derive that

$$\partial_y F_x - \partial_x F_y = ckl\, x^{k-2}y^{l-2}[(k-1)y - (l-1)x] \neq 0,$$

and the differential form $\omega = F_x dx + F_y dy$ is not closed $d\omega \neq 0$. Therefore, this system is not a gradient dynamical system. Using conditions (7.46) in the form

$$D_y^2 F_x - D_x^2 F_y = \partial_{yy}^2 F_x - \partial_{xx}^2 F_y = 0,$$

we get $d^{\alpha}\omega = 0$ for $\alpha = 2$. As the result, we have that this system can be considered as a fractional gradient system with the potential function

$$V(x,y) = ax^n + by^m + cx^k y^l.$$

In the general case, the fractional gradient system cannot be considered as a gradient system. The gradient systems can be considered as a special case of fractional gradient systems such that $\alpha = 1$.

Lorenz System as a Fractional Gradient System

In this section, we prove that dynamical systems that are defined by the well-known Lorenz equations [Lor63, Spa82] are fractional gradient system.

The well–known Lorenz equations [Lor63, Spa82] are defined by

$$\dot{x} = F_x, \qquad \dot{y} = F_y, \qquad \dot{z} = F_z,$$

where the right hand sides F_x, F_y and F_z have the forms

$$F_x = \sigma(y - x), \qquad F_y = (r - z)x - y, \qquad F_z = xy - bz.$$

The parameters σ, r and b can be equal to the following values

$$\sigma = 10, \qquad b = 8/3, \qquad r = 470/19 \simeq 24.74 .$$

The dynamical system which is defined by the Lorenz equations cannot be considered as a gradient dynamical system. It is easy to see that

$$\partial_y F_x - \partial_x F_y = z + \sigma - r, \qquad \partial_z F_x - \partial_x F_z = -y, \qquad \partial_z F_y - \partial_y F_z = -2x.$$

Therefore, $\omega = F_x dx + F_y dy + F_z dz$ is not a closed 1–form and we have

$$d\omega = -(z + \sigma - r)dx \wedge dy + y dx \wedge dz + 2x dy \wedge dz.$$

For the Lorenz equations, conditions (7.46) can be satisfied in the form

$$D_y^2 F_x - D_x^2 F_y = 0, \qquad D_z^2 F_x - D_x^2 F_z = 0, \qquad D_z^2 F_y - D_y^2 F_z = 0.$$

As the result, we get that the Lorenz system can be considered as a fractional gradient dynamical system with potential function

$$V(x, y, z) = \frac{1}{6}\sigma x^3 - \frac{1}{2}\sigma y x^2 + \frac{1}{2}(z - r)xy^2 + \frac{1}{6}y^3 - \frac{1}{2}xyz^2 + \frac{b}{6}z^3. \quad (7.54)$$

The potential (7.54) uniquely defines the Lorenz system. Using equation (7.48), we can get that the stationary states of the Lorenz system are defined by the equation

$$V(x, y, z) + C_{00} + C_x x + C_y y + C_z z + C_{xy}xy + C_{xz}xz + C_{yz}yz = 0, \quad (7.55)$$

where C_{00}, C_x, C_y, C_z C_{xy}, C_{xz}, and C_{yz} are the constants and $\alpha = m = 2$. The plot of stationary states of Lorenz system with the following constants $C_{00} = 1$, $C_x = C_y = C_z = C_{xy} = C_{xz} = C_{yz} = 0$ and parameters $\sigma = 10$, $b = 3$, and $r = 25$ is shown in figure 1 and 2.

Note that the Rossler system [Ros76], which is defined by the equations

$$\dot{x} = -(y + z), \qquad \dot{y} = x + 0.2y, \qquad \dot{z} = 0.2 + (x - c)z,$$

can be considered as a fractional gradient system with the potential function

$$V(x, y, z) = \frac{1}{2}(y + z)x^2 - \frac{1}{2}xy^2 - \frac{1}{30}y^3 - \frac{1}{10}z^2 - \frac{1}{6}(x - c)z^3. \quad (7.56)$$

This potential uniquely defines the Rossler system. The stationary states of the Rossler system are defined by equation (7.55), where the potential function is defined by (7.56).

7.6.4 Hamiltonian Systems

In this subsection, a brief review of Hamiltonian systems is considered to fix notations and provide a convenient reference (for details, see [II06b]).

Let us consider the canonical coordinates $(q^1, ..., q^n, p^1, ..., p^n)$ in the phase–space $\mathbb{R}^{2n}$. We consider a dynamical system that is defined by the equations

$$\dot{q}^i = G^i(q, p), \qquad \dot{p}_i = F_i(q, p). \tag{7.57}$$

The definition of Hamiltonian systems can be realized in the following form [Tar05g].

A dynamical system (7.57) on the phase–space $\mathbb{R}^{2n}$, is called a Hamiltonian system if the differential 1–form

$$\beta = G^i dp_i - F_i dq^i, \tag{7.58}$$

is a closed form $d\beta = 0$, where d is the exterior derivative. A dynamical system is called a non-Hamiltonian system if the differential 1–form β is nonclosed $d\beta \neq 0$.

The exterior derivative for the phase–space is defined as

$$d = dq^i \partial_{q^i} + dp_i \partial_{p_i}. \tag{7.59}$$

Here and later, we mean the sum on the repeated indices i and j from 1 to n.

If the right–hand sides of equations (7.57) satisfy the Helmholtz conditions [Hel986, Tar05g] for the phase–space, which have the following forms:

$$\partial_{p_j} G^i - \partial_{p_i} G^j = 0, \qquad \partial_{q^i} G^j + \partial_{p_j} F_i = 0, \qquad \partial_{q^j} F_i - \partial_q F_j = 0, \tag{7.60}$$

then the dynamical system (7.57) is a Hamiltonian system. In the canonical coordinates (q, p), the vector fields that define the system have the components (G^i, F_i), which are used in equation (7.57). Let us consider the 1–form that is defined by the equation

$$\beta = G^i dp_i - F_i dq^i.$$

The exterior derivative for this form can be written by the relation:

$$d\beta = d(G^i dp_i) - d(F_i dq^i).$$

It now follows that

$$d\beta = \partial_{q^j} G^i dq^j \wedge dp_i + \partial_{p_j} G^i dp_j \wedge dp_i - \partial_{q^j} F_i dq^j \wedge dq^i - \partial_{p_j} F_i dp_j \wedge dq^i.$$

Here, $\wedge$ is the wedge product. This equation can be rewritten in an equivalent form as

$$d\beta = \left(\partial_{q^i} G^j + \partial_{p_j} F_i\right) dq^i \wedge dp_j + \frac{1}{2}\left(\partial_{p_i} G^j - \partial_{p_j} G^i\right) dp_i \wedge dp_j$$

$$+ \frac{1}{2}\left(\partial_{q^j} F_i - \partial_{q^i} F_j\right) dq^i \wedge dq^j.$$

Here, we use the skew-symmetry of $dq^i \wedge dq^j$ and $dp_i \wedge dp_j$ with respect to the index i and j. It is obvious that the conditions (7.60) lead to the equation $d\beta = 0$.

Some of Hamiltonian systems can be defined by the unique function as follows. A dynamical system (7.57) on the phase–space R^{2n} is a Hamiltonian system that is defined by Hamiltonian $H = H(q, p)$ if the differential 1–form

$$\beta = G^i dp_i - F_i dq^i,$$

is an exact form $\beta = dH$, where d is the exterior derivative and $H = H(q, p)$ is a continuous differentiable unique function on the phase–space. Suppose that the differential 1–form β, which is defined by equation (7.58), has the form

$$\beta = dH = \partial_{p_i} H dp_i + \partial_{q^i} H dq^i.$$

In this case, vector fields (G^i, F_i) can be represented in the form

$$G^i(q, p) = \partial_{p_i} H, \qquad F_i(q, p) = -\partial_{q^i} H.$$

If $H = H(q, p)$ is a continuous differentiable function, then condition (7.60) are satisfied and we have a Hamiltonian system. The equations of motion for the Hamiltonian system (7.57) can be written in the form

$$\dot{q}^i = \partial_{p_i} H, \qquad \dot{p}_i = -\partial_{q^i} H, \tag{7.61}$$

which is uniquely defined by the Hamiltonian H.

If the exact differential 1–form β is equal to zero ($dH = 0$), then the equation

$$H(q, p) - C = 0 \tag{7.62}$$

defines the stationary states of the Hamiltonian system (7.57). Here, C is a constant.

7.6.5 Fractional Hamiltonian Systems

Fractional generalization of the differential form (7.58), which is used in definition of the Hamiltonian system, can be defined in the following form:

$$\beta_\alpha = G^i (dp_i)^\alpha - F_i (dq^i)^\alpha.$$

Let us consider the canonical coordinates

$$(x^1, ..., x^n, x^{n+1}, ..., x^{2n}) = (q^1, ..., q^n, p^1, ..., p^n)$$

in the phase–space R^{2n} and a dynamical system that is defined by the equations

$$\dot{q}^i = G^i(q, p), \qquad \dot{p}_i = F_i(q, p). \tag{7.63}$$

The fractional generalization of Hamiltonian systems can be defined by using fractional generalization of differential forms [CN01]. A dynamical system (7.63) on the phase–space R^{2n} is called a fractional Hamiltonian system if the fractional differential 1–form

$$\beta_\alpha = G^i (dp_i)^\alpha - F_i (dq^i)^\alpha,$$

is a closed fractional form

$$d^\alpha \beta_\alpha = 0, \tag{7.64}$$

where d^α is the fractional exterior derivative. A dynamical system is called a fractional non-Hamiltonian system if the fractional differential 1–form β_α is a nonclosed fractional form $d^\alpha \beta_\alpha \neq 0$.

The fractional exterior derivative for the phase–space R^{2n} is defined as

$$d^\alpha = (dq^i)^\alpha D_{q^i}^\alpha + (dp_i)^\alpha D_{p_i}^\alpha. \tag{7.65}$$

For example, the fractional exterior derivative of order α of q^k, with the initial point taken to be zero and $n = 2$, is given by

$$d^\alpha q^k = (dq)^\alpha D_q^\alpha q^k + (dp)^\alpha D_p^\alpha q^k. \tag{7.66}$$

Using equations (7.41) and (7.42), we have the following relation for the fractional exterior derivative (7.65):

$$d^\alpha q^k = (dq)^\alpha \frac{\Gamma(k+1) q^{k-\alpha}}{\Gamma(k+1-\alpha)} + (dp)^\alpha \frac{q^k p^{-\alpha}}{\Gamma(1-\alpha)}.$$

Let us consider a fractional generalization of the Helmholtz conditions.

If the right–hand sides of equations (7.63) satisfy the fractional generalization of the Helmholtz conditions in the following form

$$D_{p_j}^\alpha G^i - D_{p_i}^\alpha G^j = 0, \qquad D_{q^i}^\alpha G^j + D_{p_j}^\alpha F_i = 0, \qquad D_{q_j}^\alpha F_i - D_{q^i}^\alpha F_j = 0, \tag{7.67}$$

then dynamical system (7.63) is a fractional Hamiltonian system.

In the canonical coordinates (q, p), the vector fields that define the system have the components (G^i, F_i), which are used in equation (7.57). The 1–form β_α is defined by the equation

$$\beta_\alpha = G^i (dp_i)^\alpha - F_i (dq^i)^\alpha. \tag{7.68}$$

The exterior derivative for this form can now be given by the relation

$$d^\alpha \beta_\alpha = d^\alpha (G^i (dp_i)^\alpha) - d^\alpha (F_i (dq^i)^\alpha).$$

Using the rule

$$D_x^\alpha (fg) = \sum_{k=0}^\infty \binom{\alpha}{k} (D_x^{\alpha-k} f) \frac{\partial^k g}{\partial x^k},$$

and the relation

$$\frac{\partial^k}{\partial x^k}((dx)^\alpha) = 0, \qquad (k \geq 1),$$

we get that

$$d^\alpha(A^i(dx^i)^\alpha) = \sum_{k=0}^{\infty}(dx_j)^\alpha \wedge \binom{\alpha}{k}(D_{x^j}^{\alpha-k}A^i)\frac{\partial^k}{\partial x^{jk}}(dx^i)^\alpha$$

$$= (dx_j)^\alpha \wedge (dx^i)^\alpha \binom{\alpha}{0}(D_{x^j}^\alpha A^i).$$

Here, we use

$$\binom{\alpha}{k} = \frac{(-1)^{k-1}\alpha\Gamma(k-\alpha)}{\Gamma(1-\alpha)\Gamma(k+1)}.$$

Therefore, we have

$$d^\alpha\beta_\alpha = D_{q_j}^\alpha G^i(dq_j)^\alpha \wedge (dp_i)^\alpha + D_{p_j}^\alpha G^i(dp_j)^\alpha \wedge (dp_i)^\alpha$$
$$- D_{q_j}^\alpha F_i(dq_j)^\alpha \wedge (dq^i)^\alpha - D_{p_j}^\alpha F_i(dp_j)^\alpha \wedge (dq^i)^\alpha.$$

This equation can be rewritten in an equivalent form

$$d^\alpha\beta_\alpha = (D_{q^i}^\alpha G^j + D_{p_j}^\alpha F_i)(dq^i)^\alpha \wedge (dp_j)^\alpha + \frac{1}{2}(D_{p_i}^\alpha G^j - D_{p_j}^\alpha G^i)(dp_i)^\alpha \wedge (dp_j)^\alpha$$

$$+ \frac{1}{2}(D_{q_j}^\alpha F_i - D_{q^i}^\alpha F_j)(dq^i)^\alpha \wedge (dq_j)^\alpha.$$

Here, we use the skew symmetry of $\wedge$. It is obvious that conditions (7.67) lead to the equation $d^\alpha\beta_\alpha = 0$, i.e., β_α is a closed fractional form.

Let us define the Hamiltonian for the fractional Hamiltonian systems. A dynamical system (7.63) on the phase–space $\mathbb{R}^{2n}$ is a fractional Hamiltonian system that is defined by the Hamiltonian $H = H(q,p)$ if the fractional differential 1–form

$$\beta_\alpha = G^i(dp_i)^\alpha - F_i(dq^i)^\alpha,$$

is an exact fractional form

$$\beta_\alpha = d^\alpha H, \qquad (7.69)$$

where d^α is the fractional exterior derivative and $H = H(q,p)$ is a continuous differentiable function on the phase–space. Suppose that the fractional differential 1–form β_α, which is defined by equation (7.68), has the form

$$\beta_\alpha = d^\alpha H = (dp_i)^\alpha D_{p_i}^\alpha H + (dq^i)^\alpha D_{q^i}^\alpha H.$$

In this case, vector fields (G^i, F_i) can be represented in the form

$$G^i(q,p) = D_{p_i}^\alpha H, \qquad F_i(q,p) = -D_{q^i}^\alpha H.$$

Therefore, the equations of motion for fractional Hamiltonian systems can be written in the form

$$\dot{q}^i = D^\alpha_{p_i} H, \qquad \dot{p}_i = -D^\alpha_{q^i} H. \tag{7.70}$$

The fractional differential 1–form β_α for the fractional Hamiltonian system with Hamiltonian H can be written in the form $\beta_\alpha = d^\alpha H$. If the exact fractional differential 1–form β_α is equal to zero ($d^\alpha H = 0$), then we can get the equation that defines the stationary states of the Hamiltonian system.

The stationary states of the fractional Hamiltonian system (7.70) are defined by the equation

$$H(q,p) - |\prod_{i=1}^{n} q^i p_i|^{\alpha-m} \sum_{k_1=0,l_1=0}^{m-1} \cdots \sum_{k_n=0,l_n=0}^{m-1} C_{k_1...k_n l_1...l_n} \prod_{i=1}^{n} (q^i)^{k_i} (p_i)^{l_i} = 0, \tag{7.71}$$

where $C_{k_1...k_n,l_1,...,l_n}$ are constants and m is the first whole number greater than or equal to α.

For example, let us consider a dynamical system in phase–space $\mathbb{R}^2$ ($n = 1$) that is defined by the equation

$$\dot{q} = D^\alpha_p H, \qquad \dot{p} = -D^\alpha_q H, \tag{7.72}$$

where the fractional order $0 < \alpha \le 1$ and the Hamiltonian $H(q,p)$ has the form

$$H(q,p) = ap^2 + bq^2. \tag{7.73}$$

If $\alpha = 1$, then equation (7.72) describes the linear harmonic oscillator.

If the exact fractional differential 1–form

$$\beta_\alpha = d^\alpha H = (dp)^\alpha D^\alpha_p H + (dq)^\alpha D^\alpha_q H$$

is equal to zero ($d^\alpha H = 0$), then the equation

$$H(q,p) - C|qp|^{\alpha-1} = 0$$

defines the stationary states of the system (7.72). Here, C is a constant. If $\alpha = 1$, we get the usual stationary-state equation (7.62).

Using equation (7.73), we get the following equation for stationary states:

$$|qp|^{1-\alpha}(ap^2 + bq^2) = C.$$

If $\alpha = 1$, then we get the equation $ap^2 + bq^2 = C$, which describes the ellipse.

8

Turbulence

This Chapter addresses modern theory of turbulence.

Recall that *chaotic dynamics*, of which *turbulence* is the most extreme form, computationally started in 1963, when Ed Lorenz from MIT took the *Navier–Stokes equations* from viscous fluid dynamics and reduced them into three first–order coupled nonlinear ODEs (1.21), to demonstrate the idea of sensitive dependence upon initial conditions and associated *chaotic behavior*. Starting from the simple Lorenz system, in this Chapter we develop the comprehensive theory of turbulent flows. For start–off, recall from Introduction that D. Ruelle and F. Takens argued in a seminal paper [RT71] that, as a function of an external parameter, the *route to chaos in a fluid flow* is a transition sequence leading from stationary (S) to single periodic (P), double periodic (QP_2), triple periodic (QP_3) and, possibly, quadruply periodic (QP_4) motions, before the flow becomes chaotic (C).

8.1 Parameter–Space Analysis of the Lorenz Attractor

Now recall from Introduction that the *Lorenz dynamical system*

$$\dot{x} = -\sigma x + \sigma y, \qquad \dot{y} = rx - y - xz, \qquad \dot{z} = -bz + xy, \qquad (8.1)$$

derived by Ed Lorenz [Lor63, Spa82] as a discrete model for thermal convection in *Rayleigh–Bénard flow* between two plates perpendicular to the direction of the Earth's gravitational force. The Lorenz system (8.1) has *three control parameters*:

A. The *normalized temperature difference* $r = \Delta T/\Delta T_c = Ra/Ra_c$ between the hotter lower plate and the colder upper one. ΔT_c is the temperature difference at the onset of convection. $Ra = g\alpha_p L^3 \Delta T/\nu\kappa$, the *Rayleigh number*, is the dimensionless temperature difference. g denotes the gravitational acceleration, α_p the isobaric thermal expansion coefficient, L the

529

distance between the plates in upward direction, and ν, κ characterize the kinematic viscosity and thermal diffusivity, respectively.

B. The *Prandtl number* $\sigma = \nu/\kappa$, describing the ratio of the two molecular time scales $\tau_\nu = L^2/\nu$ and $\tau_\kappa = L^2/\kappa$ for momentum and energy transport in the system. σ also measures the ratio of the widths λ_u and λ_T of the kinematic and the thermal boundary layers. Large Prandtl numbers σ indicate a broad kinematic boundary layer containing a small thermal one, fast molecular momentum exchange and small convective Reynolds numbers. Small Prandtl numbers characterize fast molecular thermal exchange, a broad thermal boundary layer containing a thinner kinematic one and larger convective Reynolds numbers. Typical values for σ are 0.7 in air (at room temperature), N_2, O_2, He, and other gases, about 7 in water (room temperature), 600 through 3000 in typical organic liquids as e.g., dipropylene glycol (600), or $\sigma \approx 7250$ in glycerine. Small Prandtl numbers are realized in mercury ($\sigma \approx 0.025$) or liquid sodium with $\sigma \approx 0.005$.

C. The *coupling strength b*, originating from the nonlinearity $(\mathbf{u} \cdot \boldsymbol{\nabla})T$ of the *Boussinesq equation*. It is of order 1 and conventionally chosen as $b = 8/3$; we adopt this value here.

The variables $x(t), y(t)$, and $z(t)$ describe, respectively, the amplitude of the velocity mode (the only one left in the model), the amplitude of the corresponding temperature mode, and the amplitude of the mode measuring the non–dimensional heat flux Nu, the *Nusselt number*. x, y characterize the roll pattern; the heat flux is given by

$$Nu = 1 + 2z/r. \qquad (8.2)$$

For $0 \leq r \leq 1$, there exists one stable fixed–point P_0 ($x_0 = y_0 = z_0 = 0$), with $Nu = 1$, corresponding to pure molecular heat conduction. For r exceeding 1, convective rolls develop, described by two stable fixed–points

$$P_\pm = (x_\pm, y_\pm, z_\pm) = (\pm\sqrt{b(r-1)}, \pm\sqrt{b(r-1)}, r-1), \qquad (8.3)$$

hence $Nu = 3 - 2/r$. The trajectories in phase–space $\Gamma = \{x, y, z\}$ are spirals towards these fixed–points. Stability is lost for

$$r > r_c = \sigma \, \frac{\sigma + b + 3}{\sigma - b - 1}. \qquad (8.4)$$

For the canonical choice $\sigma = 10$ it is $r_c = 24.7368....$ At such large r the three mode approximation for the PDEs describing thermal convection has of course ceased to be physically realistic, but mathematically the model now starts to show its most fascinating properties, because a strange attractor develops for $r > r_c$. With further increasing r, windows of regular behavior emerge out of chaos where stable periodic orbits exist. These stability windows tend to become broader and broader until for sufficiently large r chaos seems to disappear and one type of stable periodic solution takes over.

Of the numerous studies of the Lorenz model we cite as examples the monograph [Spa82] and the more recent survey in Jackson's book [Jac91]. To the best of our knowledge, the parameter dependence of the Lorenz model has mainly been analyzed with respect to its control parameter r, the external forcing, and sometimes with varying b (as in the work [AF84] which is cited in [Jac91]). The Prandtl number σ has usually been fixed to its canonical value $\sigma = 10$, and as a rule, b has been taken as $8/3$. The limit of large r at arbitrary fixed σ was thoroughly studied by Robbins [Rob79] and Sparrow [Spa82].

An important feature of the Lorenz equations (8.1) is their *invariance with respect to the reflection*

$$\Sigma : \quad x, y \mapsto -x, -y \qquad \text{and} \qquad z \mapsto z. \tag{8.5}$$

Evidently the $z-$axis is a $\Sigma-$invariant line. For $r > 1$ it is also the stable manifold of the fixed–point P_0. This characteristic symmetry property with respect to Σ has implications on the nature of the period doubling scenarios which occur over and over again as the parameters r, σ are varied: they do *not* follow the familiar scheme of the paradigmatic logistic iteration $x \mapsto ax(1-x)$ but rather the somewhat more intricate scheme of the cubic iteration $x \mapsto f(x) = (1-c)x + cx^3$. This cubic map possesses the symmetry $f = \tilde{\Sigma}^{-1} f \tilde{\Sigma}$, with $\tilde{\Sigma} : x \mapsto -x$, and may be viewed as one building block for understanding the transition to chaos in the Lorenz system, see Hao et al. [DH90, FH96] for a detailed analysis of this aspect.

The motivation of the recent work of [DSR05] was the simple question why the meteorologist Lorenz would have chosen the Prandtl number $\sigma = 10$, which is adequate for water but less so for air with its order of magnitude $\sigma \approx 1$. The reader will easily find that taking σ this small, he will no longer find chaotic solutions. This is confirmed by a glance on (8.4): with $\sigma = 1$, the critical parameter $r_c = -5/2$ lies in the unphysical range $r < 0$. If one chooses, instead, some $\sigma > b + 1$, implying $r_c > 0$, the window structure in the $r-$dependence is similar to the case $\sigma = 10$ although the details may be nontrivially different. But one hardly sees how these differences in the $r-$behavior for some discrete $\sigma-$values are connected. This motivated us to vary σ in sufficiently small steps and study the 2D $(r, \sigma)-$space of control parameters in a larger domain than has hitherto been done. However, quite apart from such a physical motivation, the mathematical complexity of this innocent looking set of equations warrants an investigation of their asymptotic behavior where it has not yet been explored.

8.1.1 Structure of the Parameter–Space

Following [DSR05], we begin the analysis of the parameter–space structure with a discussion of appropriate scalings of the entire set of Lorenz equations (8.1) when both parameters r and σ become large. Their asymptotic behavior with $r \to \infty$ depends on how σ is assumed to change simultaneously.

Robbins in [Rob79] and also Sparrow [Spa82] considered the case $\sigma = \text{const}$. Here we study the behavior of the Lorenz equations with Prandtl numbers increasing as $\sigma \propto r^s$, with some fixed exponent $s > 0$ to guarantee increasing σ, but $s \le 1$ for σ to stay below the upper stability border $\sigma_u(r) = r - 28/3$.

If r and σ become very large, the motion in x and y is expected to become very fast because d/dt in the equations of motion becomes large. Thus we introduce a rescaled, stretched time τ by defining $d\tau = \varepsilon^{-1}dt$ with a scaling factor $\varepsilon^{-1} \propto r^e$, $e > 0$.

Let us now also rescale the variables x and y which measure, respectively, the amplitudes of the role velocity and temperature. As for the rescaling of z a glance on the second Lorenz equation shows that y is driven by $(r - z)$. This fact is reflected in the amplitudes z of the numerical trajectories, or in the behavior of the fixed–points $P_\pm$, both as functions of r. Thus we define

$$x = \alpha \xi, \qquad y = \beta \eta, \qquad z - r = \gamma \zeta. \tag{8.6}$$

The Lorenz equations then read, with $\xi' = d\xi/d\tau$ etc.,

$$\xi' = -\varepsilon \sigma \xi + \varepsilon \sigma \frac{\beta}{\alpha} \eta, \qquad \eta' = -\varepsilon \eta - \varepsilon \gamma \frac{\alpha}{\beta} \zeta \xi, \qquad \zeta' = -\varepsilon b(\zeta + \frac{r}{\gamma}) + \varepsilon \frac{\alpha \beta}{\gamma} \xi \eta. \tag{8.7}$$

The freedom of choosing scaling parameters may now be used to normalize to unity the coefficients of the driving term in the first equation, and of the two coupling terms $\propto \zeta \xi$ and $\propto \xi \eta$, which leads to

$$\varepsilon \sigma \beta = \alpha, \qquad \varepsilon \gamma \alpha = \beta, \qquad \varepsilon \alpha \beta = \gamma. \tag{8.8}$$

These equations allow us to express α, β, and γ in terms of σ and ε. We find first $\varepsilon^2 \alpha^2 = 1$, thus $\alpha = \pm \varepsilon^{-1}$, then $\beta = \pm \varepsilon^{-2} \sigma^{-1}$, and finally $\gamma = \varepsilon^{-2} \sigma^{-1}$. The $\pm$ signs may be interpreted as corresponding to the symmetry Σ which is clearly preserved under the rescaling. We can therefore restrict ourselves to the plus sign and have

$$\alpha = \varepsilon^{-1} \propto r^e, \qquad \beta = \gamma = \varepsilon^{-2} \sigma^{-1} \propto r^{2e-s}. \tag{8.9}$$

The asymptotic behavior will depend on how the exponent e compares to the other exponent $2e - s$. There appear to be three distinct possibilities. One is that β and γ grow faster than α, i.e., $2e - s > e$ or $e > s$. The second is that all three variables x, y, z scale with the same factor, $2e - s = e$ or $e = s$. The third possibility $e < s$ may be excluded with the argument that then the relaxation term $-\varepsilon\sigma\xi$ in the ξ–equation would diverge.

A representative case of the first possibility is $s = 0$ and $e = \frac{1}{2}$, meaning $\tau = \text{const}$ and $\varepsilon \propto 1/\sqrt{r} \to 0$. This was the choice of Robbins [Rob79] and Sparrow [Spa82]. All dissipative terms vanish in this limit, and the equations reduce to

$$\xi' = \eta, \qquad \eta' = -\zeta \xi, \qquad \zeta' = \xi \eta, \qquad (\sigma = \text{const}, \ \varepsilon = 1/\sqrt{r}). \tag{8.10}$$

This system has the two integrals $A = \frac{1}{2}\xi^2 - \zeta$ and $B^2 = \eta^2 + \zeta^2$; it is equivalent to the equations of a pendulum with length B and energy A. The values of the integrals A and B have to be determined from the ε–correction to (8.10), i. e., by considering their slow dynamics. It was shown in [Rob79] how this can be done in terms of elliptic integrals. In that sense, this case has been solved.

Let us therefore concentrate on the second possibility $e = s$. Then $\varepsilon\sigma$ is independent of r, and we can take $\varepsilon\sigma = 1$. All scaling factors can now be expressed in terms of the Prandtl number σ. The term $-\varepsilon\eta$ in the second Lorenz equation may be neglected as vanishingly small in comparison to with $\zeta\xi$. Hence the rescaled (8.7) become

$$\xi' = -\xi + \eta, \qquad \eta' = -\zeta\xi, \qquad \zeta' = \xi\eta - \frac{b}{\sigma}\zeta - \frac{br}{\sigma^2}, \qquad (e = s \text{ and } \sigma = r^s).$$
$$(8.11)$$

But what about the two last terms in the ζ–equation? Their order of magnitude is br^{-s} and br^{1-2s}, respectively. Again, there are three possibilities:

Case A. With the choice $1 - 2s = 0$, or $s = 1/2$, the second term will tend to a constant and dominate the decreasing first term. We can write $\sigma = w\sqrt{br}$ with some constant w. The set of equations becomes

$$\xi' = -\xi + \eta, \qquad \eta' = -\zeta\xi, \qquad \zeta' = \xi\eta - w^{-2} \qquad (e = s = \frac{1}{2} \text{ and } \sigma = w\sqrt{br}).$$
$$(8.12)$$

When s varies between $1/2$ and 1, the same equations apply because the last term in the ζ–equation of (8.11) dominates the second. Note that no comparison can be made between these two terms and $\xi\eta$.

Case B. When $s = 1$, the two last terms in the ζ–equation are of the same order of magnitude r^{-1}. We write $\sigma = mr$ and get the set of equations

$$\xi' = -\xi + \eta, \qquad \eta' = -\zeta\xi, \qquad \zeta' = \xi\eta - \frac{b}{mr}(\zeta + \frac{1}{m}) \quad (e = s = 1 \text{ and } \sigma = mr).$$
$$(8.13)$$

Case C. When $s > 1$, σ grows faster than r, so we leave the range where $\sigma < \sigma_u(r)$. Formally it is found that both terms besides $\xi\eta$ in the ζ–equation may be neglected, but in contrast to (8.10) there remains a dissipative term in the ξ–equation. The constant of motion $B^2 = \eta^2 + \zeta^2$ constrains the solution to a cylinder. On this cylinder the equation reduces to the damped pendulum, and therefore almost all initial conditions decay towards the stable equilibrium on the respective cylinder.

We thus have to discuss the asymptotic regimes A and B, and do this separately.

We start with the simpler case of the asymptotic Lorenz equations, the set of the rescaled (8.12). Of all the complex dynamics that it contains, we only want to describe periodic orbits of the simplest kind. From our numerical studies we anticipate that these orbits have two clearly distinct phases. There is slow motion down along the z−axis, and a fast return to the original large value of z via a large loop away from the z−axis [DSR05].

The fast motion is readily described when we neglect the constant term in the ζ−equation of (8.12). Then, combining the equations for η and ζ, we get $\eta d\eta = -\zeta d\zeta$ which after integration gives the circles

$$\eta^2 + \zeta^2 = \rho^2, \qquad (8.14)$$

with as yet undetermined radius ρ. A convenient parametrization of these circles is found with a rescaling of time, defining s as an arc length via $ds = \xi(\tau)d\tau$. The equations for $\eta(s)$ and $\zeta(s)$ are thereby linearized, with solutions

$$\eta(s) = \rho \sin s, \qquad \zeta(s) = -\rho \cos s. \qquad (8.15)$$

This explains the observation that ζ oscillates between two equal values of different sign, namely $\pm\rho$.

The equation for ξ,

$$\frac{d\xi}{d\tau} = -(\xi - \eta), \qquad \text{or} \qquad \frac{d\xi}{ds} = -(1 - \frac{\eta(s)}{\xi(s)}), \qquad (8.16)$$

does not appear to be analytically solvable, but as long as ξ is not too small, it describes a relaxation of $\xi(s)$ towards $\eta(s)$. Hence, with a delay in τ of the order of 1, ξ follows η on its excursion to values of the order of ρ.

Consider now the slow motion. It is given by the exact dynamics on the stable manifold $\xi = \eta = 0$ of the fixed–point $(0, 0, 0)$. The solution with initial point $\zeta(0) = \rho$ is $\zeta(\tau) = \rho - \tau/w^2$ and reaches the lower value -ρ at time $T = 2\rho w^2$. Linearizing the set of equations (8.12) around this motion, we get

$$\xi' = -\xi + \eta, \qquad \eta' = -(\rho - \tau/w^2)\xi. \qquad (8.17)$$

The τ−dependent eigenvalues are $-\frac{1}{2} \pm \sqrt{\Delta(\tau)}$ with discriminant $\Delta = \frac{1}{4} - \rho + \tau/w^2$. Three τ−regimes may be identified. Assuming $\rho > 1/4$, the discriminant is initially negative; the motion is oscillatory and decaying according to $e^{-\tau/2}$. This regime ends at the time τ_1 defined by $\Delta(\tau_1) = 0$, or $\tau_1 = w^2(\rho - \frac{1}{4})$. In the second regime, the motion continues to be decaying but is no longer oscillatory; this is the case between τ_1 and $\tau_2 = w^2\rho = T/2$ where $\zeta(\tau_2) = 0$. The third regime, between τ_2 and T, sees ξ and η increasing at a growing rate.

To proceed with the (ξ, η)−equations, we introduce a new variable X by separating the exponentially decaying part from ξ. With $\xi(\tau) = e^{-\tau/2}X(\tau)$, the two linear equations of first order are transformed into

$$X'' - \Delta(\tau)X = 0, \qquad \text{or} \qquad \tilde{X}'' - \frac{\tilde{\tau}}{w^2}\tilde{X} = 0, \qquad (8.18)$$

where $\tilde{\tau} = \tau - \tau_1$, and $\tilde{X}(\tilde{\tau}) = X(\tau - \tau_1)$ lives on the interval -$\tau_1 < \tilde{\tau} < T - \tau_1$. This equation is the well known Airy equation. Its general solution is a superposition of the two Airy functions $\mathrm{Ai}(z)$ and $\mathrm{Bi}(z)$ where $z = w^{-2/3}\tilde{\tau}$. Both solutions oscillate for $z < 0$. For $z > 0$ the solution Ai decays whereas Bi grows exponentially. We are interested in the solution $\tilde{X}(\tilde{\tau}) = \mathrm{Bi}(z)$, or

$$\xi(\tilde{\tau}) = \mathrm{e}^{-\tau_1/2}\mathrm{e}^{-\tilde{\tau}/2}\mathrm{Bi}(w^{-2/3}\tilde{\tau}). \tag{8.19}$$

with $-\tau_1 < \tilde{\tau} < T - \tau_1$. It may be checked that $\eta(\tilde{\tau}) = \mathrm{e}^{-\tau/2}Y(\tau)$ is governed by the same Airy equation and thus, up to a phase shift, has the same solution.

From here it is straightforward to determine the value of ρ, and the periodicity of the pattern of bands.

We know from the analysis of the outer part of the attractor, see (8.15), that the solution starts, for $\tau = 0$ or $\tilde{\tau} = -\tau_1$, at about the same values ξ and η at which it ends for $\tau = T$:

$$\xi(0) = \xi(T). \tag{8.20}$$

Using the asymptotic behavior of Bi for large negative and positive arguments,

$$\mathrm{Bi}(-|z|) \approx \frac{1}{\sqrt{\pi}|z|^{1/4}}\cos\left(\frac{2}{3}|z|^{3/2} + \frac{\pi}{4}\right),$$
$$\mathrm{Bi}(|z|) \approx \frac{1}{\sqrt{\pi}|z|^{1/4}}\exp\{\frac{2}{3}|z|^{3/2}\}, \tag{8.21}$$

and the fact that the exponential terms dominate the algebraic factors by orders of magnitude, we simply equate the exponents on the two sides of (8.20). For $\tilde{\tau} = -\tau_1$ (8.19) gives $\xi(0) = O(1)$ (zero exponent); for $\tilde{\tau} = T - \tau_1$, we add up the exponents in (8.19) and (8.21), with $z = w^{-2/3}(T - \tau_1)$, and get

$$0 = -\frac{\tau_1}{2} - \frac{T - \tau_1}{2} + \frac{2}{3}|w^{-2/3}(T - \tau_1)|^{3/2} = -\frac{1}{2}T + \frac{2}{3}\frac{1}{w}(T - \tau_1)^{3/2}. \tag{8.22}$$

Inserting $T = 2\rho w^2$ and $\tau_1 = w^2\left(\rho - \frac{1}{4}\right)$ we get

$$144\rho^2 = (1 + 4\rho)^3. \tag{8.23}$$

The relevant solution is $\rho = 1.35... = \rho_0$.

To determine the values $w = w_N$ where the orbits have a certain number N of $\xi-$ and $\eta-$zeros at each close encounter with the $z-$axis, we use the asymptotic behavior of the Airy function $\mathrm{Bi}(-|z|)$ as given in (8.21). Consider first the symmetric orbits S_n as they develop from time $\tilde{\tau} = -\tau_1$ to $T - \tau_1$. The coordinates ξ and η come in from large values (positive or negative) and go out to large values of the opposite sign. This means the phase shift between $\xi(\tilde{\tau} = -\tau_1)$ and $\xi(\tilde{\tau} = T - \tau_1)$ is π; the cosine performs an odd number of half oscillations. For the asymmetric orbits A_n, on the other hand, the phase shift is a multiple of 2π. Taken together, this leads to the 'quantum condition' [DSR05]

$$\frac{2}{3}\frac{1}{w}\tau_1^{3/2} + \frac{\pi}{4} = N\pi, \tag{8.24}$$

where $N = 1$ corresponds to the first symmetric and $N = 2$ to the first asymmetric band. With $\rho = \rho_0$ we get

$$w_N = 1.02\sqrt{4N-1}, \qquad N = 1, 2, \dots . \tag{8.25}$$

The results for this regime of large r are the following. The $N-$th band is centered around the line $\sigma_N = w_N\sqrt{br}$; the size of the corresponding attracting periodic orbits in the original variables (x, y, z) is $\rho_0\sigma_N$.

Let us now study the set of equations (8.13)

$$\xi' = -\xi + \eta, \qquad \eta' = -\zeta\xi, \qquad \zeta' = \xi\eta - \frac{b}{mr}(\zeta + \frac{1}{m}),$$

where $m = \sigma/r$ is the constant parameter. Much of the analysis is similar as in the previous subsection, but the explicit occurrence of r in the equations adds complications. The general strategy is the same. For the outer, large ξ and η parts of the solution we neglect the term $\propto r^{-1}$ in the $\zeta-$equation and get identical results as in (8.14)-(8.16).

The slow motion along the $z-$axis is more complicated even though the $\zeta-$equation, neglecting the $\xi\eta-$term, decouples again from the rest: its solution is no longer linear in τ. With the initial condition $\zeta(0) = \rho$ the approximate $\zeta'-$equation is solved by

$$\zeta(\tau) = (\rho + \frac{1}{m})e^{-(b/mr)\tau} - \frac{1}{m}, \tag{8.26}$$

and the time T needed to evolve from ρ to $-\rho$ is

$$T = \frac{mr}{b}\log\frac{1 + m\rho}{1 - m\rho}. \tag{8.27}$$

Using this slow exponential decrease of ζ (with the rate $\propto r^{-1}$) in the equation $\eta' = -\zeta(\tau)\xi$, we have again a time dependent linear system of equations for ξ and η. Its eigenvalues at fixed time are

$$\lambda = -\frac{1}{2}\pm\sqrt{\Delta} \qquad \text{with} \qquad \Delta = \frac{1}{4}-\zeta(\tau) = \frac{1}{4}+\frac{1}{m}-(\rho+\frac{1}{m})e^{-(b/mr)\tau}. \tag{8.28}$$

The discriminant Δ is negative in the $\tau-$range $0 < \tau < \tau_1$ where τ_1 is defined by the condition $\Delta(\tau_1) = 0$:

$$\tau_1 = \frac{mr}{b}\log\frac{1 + m\rho}{1 + m/4}. \tag{8.29}$$

In this $\tau-$range the amplitudes ξ, η oscillate and decay with damping rate $-\frac{1}{2}$. The second regime (further exponential decrease of ξ and η, but now without oscillations) ends at τ_2, defined by the condition that one of the eigenvalues $\lambda(\tau_2)$ be zero, i. e., $\zeta(\tau_2) = 0$:

$$\tau_2 = \frac{mr}{b} \log(1 + m\rho) \,. \tag{8.30}$$

From τ_2 to T one eigenvalue is positive and the variables ξ and η increase at a growing rate.

With the ansatz $\xi(\tau) = e^{-\tau/2} X(\tau)$, the two linear equations of first order are transformed into $X'' - \Delta(\tau)X = 0$ which is, in contrast to the previous case A, not the Airy equation because it is not linear in τ. However, with the substitution $\tau \mapsto u$ via $u = 2r\frac{m}{b} \exp\{-(b/2mr)\tau\}$ it may be reduced to the standard form of the Bessel equation. The ξ, η -solution is then a linear combination of the two Bessel functions of first and second kind of order - $2r\frac{m}{b}\sqrt{1/4 + 1/m}$ and argument u. For fixed, finite τ but large r we have $u \approx 2r\frac{m}{b} - \tau$, linear in τ, but with a large additive constant. Thus in the limit of large r the asymptotics for Bessel functions with large negative order and large arguments is needed. The simplest way to get such asymptotics is to use an approximation by Airy functions from the very beginning when solving the X-equation. This is done by linearizing the discriminant $\Delta(\tau)$ in the original equation at an appropriate time τ^*,

$$\Delta(\tau) = \Delta(\tau^*) + \frac{b}{mr}\left[\zeta(\tau^*) + \frac{1}{m}\right](\tau - \tau^*) + \dots, \tag{8.31}$$

upon which we get the Airy equation for $\tilde{X}(\tilde{\tau}) = X(\tau)$,

$$\tilde{X}'' - \alpha\tilde{\tau}\tilde{X} = 0, \qquad \alpha = \frac{b}{mr}\left[\zeta(\tau^*) + \frac{1}{m}\right], \qquad \text{where} \tag{8.32}$$

$$\tilde{\tau} = \tau - \tau_1^* \qquad \text{with} \qquad \tau_1^* = \tau^* - \Delta(\tau^*)/\alpha \,. \tag{8.33}$$

But what should we take for τ^*? A natural choice seems to be $\tau^* = \tau_1$ where $\Delta = 0$ and $\zeta = \frac{1}{4}$, whence

$$\alpha = \alpha_1 = \frac{b}{m^2 r}(1 + \frac{m}{4}) \qquad \text{and} \qquad \tau_1^* = \tau_1 \qquad (\tau^* = \tau_1). \tag{8.34}$$

However, if we care to treat the decaying and the exploding part of the ξ, η -dynamics with equal weight, then it appears natural to take the minimum of $\xi(\tau) = e^{-\tau/2} X(\tau)$ as the reference point, i.e., $\tau^* = \tau_2$. Here $\zeta(\tau_2) = 0$ and $\Delta(\tau_2) = \frac{1}{4}$, so that from 8.32 and 8.33 we get [DSR05]

$$\alpha = \alpha_2 = \frac{b}{m^2 r} \qquad \text{and} \quad \tau_1^* = \tau_2 - \frac{m^2 r}{4b} = \tau_1 + O\left[(m/4)^2\right], \qquad (\tau^* = \tau_2). \tag{8.35}$$

The last step involves $\tau_2 = \tau_1 + \tau_2 - \tau_1$ and an expansion of $\tau_2 - \tau_1 = \frac{mr}{b}\log(1 + \frac{m}{4})$ to second order in the small quantity $\frac{m}{4}$. The difference $\tau_1^* - \tau_1$

is itself of second order. Finally, when we are mainly interested in the initial phase of the decay, $\tau^* = 0$ ought to be a reasonable choice. Then $\zeta(\tau^*) = \rho$ and $\Delta(\tau^*) = -(\rho - \frac{1}{4})$ which leads to

$$\alpha = \alpha_0 = \frac{b}{m^2 r}(1 + m\rho) \quad \text{and} \quad \tau_1^* = \tau_1 + O\left[(m/4)^2\right], \qquad (\tau^* = 0). \quad (8.36)$$

Again $\tau_1^* - \tau_1$ is of second order in small quantities. Assuming that the m−corrections are not too large, we take $\tau_1^* = \tau_1$ in all three cases and find the solution

$$\xi(\tilde{\tau}) = e^{-\tau_1/2} e^{-\tilde{\tau}/2} \text{Bi}(\alpha^{1/3}\tilde{\tau}), \qquad (8.37)$$

with $-\tau_1 < \tilde{\tau} < T - \tau_1$.

8.1.2 Attractors and Bifurcations

While the periodic orbits discussed in the previous two sections form the backbone of the structure in the (r, σ)−phase diagram on large scales of r and σ, they do not represent the whole picture.

In addition to the periodic attractors of periods 1 and 2, there are small bands of stability for periodic orbits of all periods, and of chaotic motion.

As the structural features of the (r, σ)−phase diagram are different in the regions of low and high σ (separated roughly by $\sigma_s(r) = \sqrt{br}$), we discuss them in turn. To begin with, we present in Figure 8.1 a survey on the bifurcation behavior of attractors along the line $r = 178$ for $4 < \sigma < 65$. This σ−range bridges the two regions of small and large Prandtl numbers, respectively. It starts well below $\sqrt{br}$ ($= 21.8$ for this r) and ends far above. The bifurcation scheme along this line is shown in the left part of Figure 8.1. It records the x−values of the attractors in the Poincaré surface of section $z = r - 1$. Red points were computed with initial conditions $x_0 = y_0 = 0.001, z_0 = r - 1$, green points with $x_0 = y_0 = -0.001, z_0 = r - 1$. When the colors do not mix, they represent asymmetric attractors; symmetric attractors are reached from both initial conditions and appear in color black. Starting with the highest σ, we observe a small black segment of symmetric orbits with period 2, i. e., the attractors are type 2s. The (y, z)−projection for $\sigma = 63.5$ is shown at the right side. Lowering σ to approximately 62, the 2s-orbit splits into two 2a-orbits, one red, the other green, mirror images of each other under Σ. This is illustrated for $\sigma = 57$. Each of these two attractors undergoes a period doubling scenario, 2a $\rightarrow$ 4a $\rightarrow$ 8a $\rightarrow \ldots$, until chaos appears at about $\sigma = 50.5$. When σ is further decreased, the inverse cascade of band merging as described in [GT77] is observed. It ends at $\sigma \approx 48.5$, where the two bands of asymmetric chaos merge into one band of symmetric chaos, thereby restoring the symmetry under Σ.

As σ falls below approximately 41.2, a period 4 window emerges by an intermittency scenario. At first the attractor is symmetric, of type 4s, as shown for $\sigma = 40.7$. Again, this orbit undergoes a pitchfork bifurcation into two

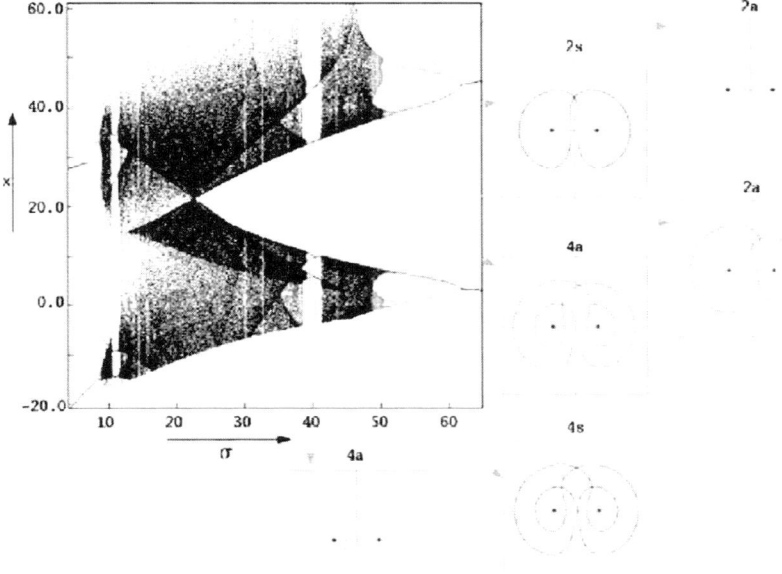

Fig. 8.1. (σ, x)−bifurcation diagram at fixed $r = 178$, $b = 8/3$ for $4 < \sigma < 65$. A few periodic orbits are shown in (y, z)−projection, of periods 2 in the upper right part, and periods 4 in the lower right; arrows indicate the corresponding branches of the bifurcation diagram. The conspicuous merging of the two major chaotic bands, as σ is lowered, seems to occur at or very close to the value $\sigma_\circ(r) = \sqrt{br} = 21.8$ (adapted from [DSR05]).

asymmetric orbits of period 4, shown in red and green for $\sigma = 39.7$, and then the period doubling sequence takes over for both of them (for details, see [DSR05]).

8.2 Periodically–Driven Lorenz Dynamics

In [CFL97], authors discussed how to characterize the behavior of a chaotic dynamical system depending on a parameter that varies periodically in time. In particular, they study the predictability time, the correlations and the mean responses, by defining a local–in–time version of these quantities. In systems where the time scale related to the time periodic variation of the parameter was much larger than the 'internal' time scale, one had that the local quantities strongly depended on the phase of the cycle. In that case, the standard global quantities could give misleading information.

Following this approach, we consider systems in which one can identify two time–scales: the first, that we call T_E, can be thought as due to the coupling with an external time–dependent driver; the second, an 'internal' time–scale T_I, characterizes the system in the limit of a constant external coupling, that we call the 'stationary limit'. In the following, the external time dependence will be assumed periodic with period T, so that $T_E = T$.

If the system, in the stationary limit, is chaotic, one can take as the internal time–scale the Lyapunov time, i.e., the inverse of the maximum Lyapunov exponent, $T_I \sim 1/\lambda$.

In the case $T_E \gg T_I$ one can assume that the system is adiabatically driven through different dynamical regions, so that on observation times short with respect to the long external time, it evolves on a local (in time) attractor. If during its tour across the slowly changing phase–space the system visits regions where the effective Lyapunov exponents are enough different, then the system shows up sensibly different predictability times, that may occur regularly when the driver is time–periodic. Consider, for instance, a slight modification of the *Lorenz system*:

$$\begin{cases} \dot{x} = 10\,(y - x), \\ \dot{y} = -x\,z + r(t)\,x - y, \\ \dot{z} = x\,y - \frac{8}{3}\,z, \end{cases} \qquad (8.38)$$

where the control parameter has a periodic time variation:

$$r(t) = r_o - A\,\cos(2\pi t/T). \qquad (8.39)$$

Since this model describes the convection of a fluid heated from below between two layers whose temperature difference is proportional to the Rayleigh number r, the periodic variations of r roughly mimic the seasonal changing on the solar heat inputs. When r_o is close to the threshold $r_{cr} = 24.74$, where in the standard Lorenz model a transition takes place from stable fixed–points to a chaotic attractor, and the amplitude A of the periodic forcing is such that $r(t)$ oscillates below and above r_{cr}, a good approximation of the solution for very large T, may be given by [CFL97]

$$x(t) = y(t) = \pm\sqrt{\frac{8}{3}(r(t) - 1)}, \qquad z(t) = r(t) - 1, \qquad (8.40)$$

which is got from the fixed–points of the standard Lorenz model by replacing r with $r(t)$. The stability of this solution is a rather complicated issue, which depends on the values of r_o, A , and T. It is natural to expect that if r_o is larger than r_{cr} the solution is unstable. In this case, for A large enough (at least $r_o - A < r_{cr}$) one has a mechanism similar to stochastic resonance in bi–stable systems with random forcing [CFP94]. The value of T is crucial: for large T the systems behaves as follows. If

$$T_n \simeq nT/2 - T/4, \qquad (n = 1, 2, \dots)$$

are the times at which $r(t) = r_{cr}$, one can naively expect that for $0 < t < T_1$ – when $r(t)$ is smaller than r_{cr} – the system is stable and the trajectory is close to one of the two solutions (8.40), while for $T_1 < t < T_2$ – when $r(t) > r_{cr}$ – both solutions (8.40) are unstable and the trajectory relaxes toward a sort of 'adiabatic' chaotic attractor. The chaotic attractor smoothly changes at varying of r above the threshold r_{cr}, but if T is large enough this dependence in first approximation can be neglected. When $r(t)$ becomes again smaller than r_{cr}, the 'adiabatic' attractor disappears and, in general, the system is far from the stable solutions (8.40); but it relaxes toward them, being attractive. If the half–period is much larger than the relaxation time in general, the system follows one of the two regular solutions (8.40) for $T_{2n+1} < t < T_{2n-2}$. However, there is a small but non–zero probability that the system has no enough time to relax to (8.40) and its evolution remains chaotic. It is worth stressing that the system is chaotic. In both cases, in fact the first Lyapunov exponent is positive.

It is rather clear from this example that the Lyapunov exponent is not able to characterize the above behavior, since it just refers to a very long time property of the system, i.e., a property involving times longer than T. A more useful and detailed information can be obtained by computing a 'local' average of the exponential rate of divergence for initially close trajectories. By this we mean an average which explicitly depends on the time t_o, modulus the external period T, to test the behavior of the system in the different states of the external driver. In this way one can make evident different behaviors, if any, of the system.

We therefore define the mean effective Lyapunov exponent, for the time $t_o \in [0, T]$ and for a delay τ, as [CFL97]

$$\langle \gamma(\tau) \rangle_{t_o} = \lim_{N \to \infty} \frac{1}{N} \sum_{k=0}^{N-1} \gamma(t_o + kT, \tau), \qquad (8.41)$$

where

$$\gamma(t, \tau) = \frac{1}{\tau} \ln \frac{|\delta \mathbf{x}(t + \tau)|}{|\delta \mathbf{x}(t)|}, \qquad (8.42)$$

is the local expansion rate, and $\delta \mathbf{x}(t)$ evolves according to the linear equation

$$\delta \dot{x}_i(t) = \frac{\partial f_i(\mathbf{x}(t), t)}{\partial x_j} \delta x_j. \qquad (8.43)$$

From this definition it is clear that $\langle \gamma(\tau) \rangle_{t_o}$ measures the growth of the distance between two trajectories that differ by $|\delta \mathbf{x}(t)|$ when the external driver passes through a fixed value (or state). The maximum Lyapunov exponent of the system gives the global average of $\langle \gamma(\tau) \rangle_{t_o}$:

$$\lambda = \lim_{\tau \to \infty} \langle \gamma(\tau) \rangle_{t_o} = \frac{1}{T} \int_0^T \langle \gamma(\tau) \rangle_{t_o} dt_o. \qquad (8.44)$$

If one is interested on predictability for times much smaller then T, $\langle \gamma(\tau) \rangle_{t_o}$ is a more appropriate quantity than λ, since it distinguishes among different regimes. For example, in the system (8.38) discussed above, for the given values of the parameters, one has $\lambda > 0$, but $\langle \gamma(\tau) \rangle_{t_o} < 0$ when $t_o \in [(n - 1/4)T, (n+1/4)T]$. In the case of weather forecasting, different values of γ for different t_o, correspond to different degree of predictability during the year.

As far as the response properties are concerned, we expect that in a chaotic system the hypothesis of existence of 'adiabatic' attractors implies that a fluctuation/relaxation relation holds also as a time–local property, provided one uses correlation and response functions computed according a local, not a global, average. So, besides the usual correlation function between the variables x_i and x_j,

$$C_{ij}^{(G)}(\tau) = \overline{x_i(t)x_j(t+\tau)} - \overline{x_i}\,\overline{x_j}, \tag{8.45}$$

where $\overline{(\cdot)}$ indicates the global average,

$$\overline{A_i} = \lim_{t \to \infty} \frac{1}{t} \int_o^t A_i(t')dt', \tag{8.46}$$

we introduce their correlation on a delay τ after the time $t_o \in [0, T]$ [CFL97]:

$$C_{ij}(t_o, \tau) = \langle x_i(t_o)x_j(t_o + \tau) \rangle_{t_o} - \langle x_i \rangle_{t_o} \langle x_j \rangle_{t_o}, \tag{8.47}$$

where the local average is defined as in (8.41)

$$\langle A \rangle_{t_o} = \lim_{N \to \infty} \frac{1}{N} \sum_{k=0}^{N-1} A(t_o + kT). \tag{8.48}$$

In a similar way, one can consider two different kinds of mean response function of the variable x_i to a small perturbation δx_j: the global average response,

$$R_{ij}^{(G)}(\tau) = \frac{\overline{\delta x_i(t+\tau)}}{\delta x_j(t)}, \tag{8.49}$$

and the local average response for the time t_o,

$$R_{ij}(t_o, \tau) = \frac{\langle \delta x_i(t_o + \tau) \rangle_{t_o}}{\delta x_j(t_o)}. \tag{8.50}$$

The quantity (8.49) gives the mean response, after a delay τ, to a perturbation occurred at the time t, chosen at random, i.e., with uniform distribution in $[0, T]$. We shall see that $R_{ij}^{(G)}(\tau)$ can be rather different, even at a qualitative level, from $R_{ij}(t_o, \tau)$.

8.2.1 Toy Model Illustration

Consider the *Langevin equation*

$$\dot{q}(t) = -a(t)\,q(t) + \xi(t), \tag{8.51}$$

where $\xi(t)$ is $\delta-$correlated white noise, i.e., $\xi(t)$ is a Gaussian variable with

$$\langle \xi(t) \rangle = 0, \qquad \langle \xi(t)\, \xi(t') \rangle = 2\Gamma\, \delta(t - t'), \tag{8.52}$$

and the coefficient $a(t)$ is a periodic function of period T: $a(t+T) = a(t)$. We require that

$$\int_0^T dt\, a(t) > 0, \tag{8.53}$$

to ensure a well defined asymptotic probability distribution for the stochastic process given by (8.51). Moreover, we assume a slow variation of $a(t)$, i.e.,

$$\min_t a(t) \gg \frac{1}{T}, \tag{8.54}$$

so that, by making an adiabatic approximation, a 'local' probability distribution exists at any time.

Without the noise term, the process described by (8.51) is non–chaotic. Therefore the model (8.51) cannot exhibit the full rich behavior of chaotic systems, nevertheless it catches some of the relevant features. It is easy to see that the characteristic decay time of the local correlation [CFL97]

$$C(t_o, \tau) = \langle q(t_o)\, q(t_o + \tau) \rangle_{t_o} = \lim_{N \to \infty} \frac{1}{N} \sum_{k=0}^{N-1} q(kT + t_o + \tau)\, q(kT + t_o)$$

depends on t_o. This can be easily computed by using the formal solution of (8.51)

$$q(t) = G(t) \left[q(0) + \int_0^t d\tau\, G^{-1}(\tau)\, \xi(\tau) \right], \qquad \text{where} \tag{8.55}$$

$$G(t) = \exp\left[-\int_0^t d\tau\, a(\tau) \right] \tag{8.56}$$

A straightforward calculation leads to

$$C(t_o, \tau) = C(t_o, 0)\, G(t_o, \tau)/G(t_o) \tag{8.57}$$

where the equal–time correlation is

$$C(t_o, 0) = G^2(t_o) \left[\lim_{N \to \infty} \frac{1}{N} \sum_{k=0}^{N-1} q(kT)^2 + 2\Gamma \int_0^{t_o} d\tau\, G^{-2}(\tau) \right] \tag{8.58}$$

In fig. ?? we show $C(t_o, \tau)/C(t_o, 0)$ as a function of τ for

$$a(t) = a + b \cos\left(\frac{2\pi}{T} t \right), \qquad a + b > 0 \tag{8.59}$$

for two different values of t_o, namely $t_o = 0$ and $t_o = T/2$, with $T = 10$, $a = 1$, $b = -0.9$ and $\Gamma = 0.5$. The different behavior is evident. By defining a characteristic time as the time s it takes to have $C(t_o, s) = 0.1$, we get for this case $s_0 \approx 3.475$ and $s_{T/2} \approx 1.275$. When starting from $t_o = T/2$ the decay is almost a factor 3 faster than starting from $t_o = 0$. The usual global average,

$$C^{(G)}(\tau) = \lim_{t \to \infty} \frac{1}{t} \int_0^t dt' \, q(t') \, q(t' + \tau) \tag{8.60}$$

gives an average correlation function, so its characteristic decay time is not able to distinguish different regimes. Moreover, while $C(t_o, \tau)/C(t_o, 0)$ does not depend on the noise strength Γ, $C^{(G)}(\tau)/C^{(G)}(0)$ does.

We consider now how the system responds at time $t_o + \tau$ to a perturbation done at time t_o. This is described by the mean response function $R(t_o, \tau)$, which can be computed as follows. One takes two trajectories differing at time t_o by a quantity ϵ, i.e., $\delta q(kT + t_o) = \epsilon$ for any k, and evolving with the *same* realization of noise. Then the local response function is [CFL97]

$$R(t_o, \tau) = \lim_{N \to \infty} \frac{1}{N} \sum_{k=0}^{N-1} \frac{\delta q(kT + t_o + \tau)}{\delta q(kT + t_o)} = \lim_{N \to \infty} \frac{1}{N} \sum_{k=0}^{N-1} \frac{\delta q(kT + t_o + \tau)}{\epsilon}$$

where $\delta q(kT + t_o + \tau)$ is the difference between the two trajectories at time $t_o + \tau$. Both times t_o and τ run over a cycle, i.e., in the interval $[0, T]$.

By making use of (8.55) it is easy to see that

$$R(t_o, \tau) = \frac{G(t_o, \tau)}{G(t_o)}. \tag{8.61}$$

By combining (8.57) and (8.61) we have the fluctuations/relaxation relation [KT85]

$$C(t_o, \tau) = C(t_o, 0) \, R(t_o, \tau). \tag{8.62}$$

The scenario just described remains basically valid for *nonlinear Langevin equation*. Consider for example the equation

$$\dot{q}(t) = -a(t) \, q^3(t) + \xi(t) \tag{8.63}$$

where $a(t)$ is still given by (8.59).

It is natural to expect that, because of the adiabatic assumption, for (8.63) a probability distribution

$$P_t(q) \propto \exp -[S_t(q)], \tag{8.64}$$

with $S_t(q) = q^4 a(t)/4\Gamma$, exists at any time. Therefore a natural ansatz is to assume that (1.80) becomes:

$$R(t_o, \tau) = \left\langle q(t_o, \tau) \frac{\partial S_{t_o}(q)}{\partial q(t_o)} \right\rangle_{t_o} \tag{8.65}$$

8.3 Lorenzian Diffusion

Chaotic dynamical systems are known to exhibit typical random processes behavior due to their strong sensitivity to initial conditions. Deterministic diffusion arises from the chaotic motion of systems whose dynamics is specified, and it should be distinguished from noise induced diffusion where the evolution is governed by probabilistic laws. Diffusive (standard and anomalous) behaviour have been observed in periodic chaotic maps (see, e.g., [KD99, WMW01] and references therein) and in continuous time dynamical systems [Gas96]. The analysis of deterministic diffusion is relevant for the study of non–equilibrium processes in statistical physics. The major aim is to understand the relationship between the deterministic microscopic dynamics of a system and its stochastic macroscopic description (think, for example, at the connection between *Lyapunov exponents*, *Kolmogorov–Sinai entropy* and macroscopic transport coefficients, firstly highlighted by Gaspard and Nicolis [NP77]).

In this section, mainly following [FMV02a], we present a model of Lorenzian diffusion in a 1D lattice. The steady state chaotic dynamics of the Lorenz system can indeed be re–written as the 1D motion of a classical particle subjected to viscous damping in a past history–dependent potential field (see [FMV01, FMV02b] for an earlier preliminary analysis).

We shortly recall that the (scaled) Lorenz dynamical system is given by

$$\dot{x} = \sigma(y - x), \qquad \dot{y} = -y + x + (r - 1)(1 - z)x, \qquad \dot{z} = b\,(xy - z), \quad (8.66)$$

with $r > 1$. For $1 < r < r_c$, where $r_c = \sigma(\sigma + b + 3)/(\sigma - b - 1)$, three fixed–points exist: $(0, 0, 0)$ (unstable) and $(\pm 1, \pm 1, 1)$ (stable). For $r > r_c$ all fixed–points are unstable, and the Lorenz system can exhibit either periodic or chaotic behavior on a strange attractor set (see, e.g., ref. [Spa82] for a comprehensive exposition on the subject matter).

In the steady state, the system (8.66) can be reduced to a 1D integro–differential equation for the x coordinate [FMV02b]

$$\ddot{x} + \eta\dot{x} + (x^2 - 1)x = -\alpha[x^2 - 1]_\beta x, \qquad (8.67)$$

where $\alpha = (2\sigma/b) - 1$, $\beta = [2b/(r - 1)]^{1/2}$, $\eta = (\sigma + 1)/[(r - 1)b/2]^{1/2}$, and the time is scaled by a factor $[(r - 1)b/2]^{1/2}$ with respect to the time coordinate in 8.66. The square brackets in (8.67) indicate the *exponentially vanishing memory*, which is defined, for any suitable time function $f(t)$, by

$$[f]_k(t) \equiv k \int_0^\infty ds\, e^{-ks} f(t - s)\,.$$

According to (8.67), the Lorenz system chaotic dynamics corresponds to a one dimensional motion in a constant–in–time quartic potential $U(x) = (x^2 - 1)^2/4$. In the presence of friction ($\eta \neq 0$) the motion is sustained by a time dependent memory term, which takes into account the system past evolution.

Although 8.67 has been deduced from the Lorenz system (8.66), it can be generalized to a wider class of equations showing similar dynamical properties [FMV02b]. Indeed, it can be usefully re-casted in the form [FMV02a]

$$\ddot{x} + \eta\dot{x} + \{q(x) + \alpha\,[q(x)]_\beta\}\Phi'(x) = 0 \qquad (8.68)$$

where the prime indicates the derivative with respect to x. Equation (8.67) is obtained for $\Phi(x) = x^2/2$ and $q(x) = x^2 - 1$. The generalized equation (8.68) can be regarded as the description of the motion of a unit mass particle subjected to a viscous force $-\eta\dot{x}$ and interacting with a potential field $\Phi(x)$ through a dynamically varying 'charge' $q_t(x) = q(x) + \alpha[q(x)]_\beta$. This charge depends both on the instantaneous particle position $x(t)$ and on the past history $\{x(t-s)\,|\,0 \le s < \infty\}$. It is just the coupling of $[q(x)]_\beta$ with the fixed potential field $\Phi(x)$ the origin of an endogenous forcing term which can sustain the motion even in the presence of friction: the chaotic behavior can actually arise from the synergy between this term and the viscosity.

Moreover, one can easily verify that (8.68) corresponds to the *generalized Lorenz system* [FMV02a]

$$\dot{x} = \sigma(y - x), \qquad \dot{y} = -y + x + (r-1)(1-z)\,\Phi'(x), \qquad (8.69)$$

$$\dot{z} = -bz + b[\tfrac{1}{2}q'(x)(y-x) + q(x) + 1]\,.$$

The specific Lorenz model can thus be viewed as singled out from a quite general class of dynamical systems which can exhibit chaotic behavior, their common essential property being an exponentially vanishing memory effect together with a viscous damping.

In our previous works [FMV01, FMV02b] the main chaotic dynamical features of the original Lorenz system have been investigated through the analysis of the piecewise linear system corresponding to the choice $\Phi(x) = |x|$ and $q(x) = |x| - 1$. We have thus obtained the piecewise linear Lorenz–like equation

$$\ddot{x} + \eta\dot{x} + \{|x| - 1 + \alpha\,[|x| - 1]_\beta\}\mathrm{sgn}(x) = 0 \qquad (8.70)$$

which has the great advantage of being exactly solvable on each side of $x = 0$.

Now, if in (8.68) one substitutes the quantities $q(x)$ and $\Phi'(x)$ with $x-$periodic functions, the chaotic jumps between the two infinite wells of the original quartic potential $U(x)$ correspond to chaotic jumps among near cells. The result is a deterministic diffusion in an infinite lattice, induced by a Lorenz–like chaotic dynamics. Equation (8.68) will be thus called the *Lorenz diffusion equation*. In order to use as far as possible analytical tools, we shall consider the unit wavelength periodic potential $U(x) = \tfrac{1}{2}(\{x\} - \tfrac{1}{2})^2$ (corresponding to $q(x) = \{x\} - \tfrac{1}{2}$ and $\Phi(x) = \{x\}$), where $\{x\}$ indicates the fractionary part of x.

By simple substitution one easily derives the equation [FMV02a]

$$\ddot{x} + \eta\dot{x} + \{x\} - \frac{1}{2} + \alpha[\{x\} - \frac{1}{2}]_\beta = 0 \qquad (8.71)$$

where η denotes the friction coefficient, α is the memory amplitude and β is related to the inertia whereby the system keeps memory of the past evolution. Note that η and β play a symmetrical role in the dynamics: the solution of (8.71) is indeed left invariant if one changes β with η, while keeping $\beta(1 + \alpha)$ constant.

Inside each potential cell (8.71) can be re–written in a third–order linear differential form. Indeed, by applying the operator $(d/dt + \beta)$ to each side of (8.71), one gets (for $x \neq n$)

$$\dddot{x} + (\beta + \eta)\ddot{x} + (1 + \beta\eta)\dot{x} + \beta(1 + \alpha)(x - n - \frac{1}{2}) = 0. \qquad (8.72)$$

It is worth observing that the nonlinearity of the original model is simply reduced to a change of sign of the forcing term $\beta(1 + \alpha)(\{x\} - \frac{1}{2})$ when x crosses the cell boundaries (in our case the integer values). As we will see, chaotic dynamics essentially results from the unpredictability of the crossing times.

Partial solutions of the third order nonlinear differential equation can be easily calculated inside each open interval $(n, n+1)$. To get a global solution, such partial solutions should be matched at $x = n$ by assuming that the position x, the velocity $\dot{x}$ and the memory $[|x|]_\beta$ are continuous, whereas the acceleration $\ddot{x}$ turns out to be undefined. However, it is easily shown that each pair of acceleration values "immediately" before and after the crossing times are related by

$$\ddot{x}^{(+)} - \ddot{x}^{(-)} = \operatorname{sgn}(\dot{x}).$$

The fixed–points of (8.71) are obviously $x = n + \frac{1}{2}$, $n \in \mathbb{Z}$, and their local stability depends on the roots of the characteristic equation

$$\lambda^3 + (\beta + \eta)\lambda^2 + (1 + \beta\eta)\lambda + \beta(1 + \alpha) = 0. \qquad (8.73)$$

All the fixed–points are unstable when α is larger than the critical value $\alpha_c = \beta^{-1}(1 + \beta\eta)(\beta + \eta) - 1$. In this case (8.73) has one real negative root $(-\lambda_0 < 0)$ and a complex conjugate pair of roots with positive real part $\lambda_\pm = \lambda \pm i\omega$ $(\lambda > 0)$. For $\alpha > \alpha_c$, the partial solution in the generic open interval $(n, n+1)$ can be finally written in the explicit form [FMV02a]

$$x(t) = e^{\lambda t}(C_1 \cos(\omega t) + C_2 \sin(\omega t)) : + : C_3 e^{-\lambda_0 t} + n + 1/2, \qquad (8.74)$$

where the constants C_1, C_2, C_3 are linearly related to the (cell by cell) initial conditions. The motion inside each cell consists of an amplified oscillation around a central point which translates towards the center of the cell. A change of cell yields a discontinuous variation of the acceleration $\ddot{x}$ and consequently of the coefficients C_1, C_2, C_3.

Suppose that, at a given time, say $t = 0$, the particle enters the nth cell at its left boundary with positive velocity (the reverse case can be symmetrically analyzed). In this case $C_3 = -(C_1 + \frac{1}{2})$. The question is now on whether the particle leaves the cell either from the left side (i.e., $x = n$) or from the right side of the cell (i.e., $x = n + 1$). Assigned the model parameters λ_0, λ and ω, the minimum positive time T such that

$$\left| e^{\lambda T} \left(C_1 \cos(\omega T) + C_2 \sin(\omega T) \right) - (C_1 + 1/2) e^{-\lambda_0 T} \right| = 1/2 \qquad (8.75)$$

depends of course on C_1, C_2, and therefore on the initial conditions $\dot{x}_0, \ddot{x}_0$. Unfortunately, the direct problem is transcendent. Moreover, its solution is strongly sensitive to the initial conditions. This fact is a direct consequence of the crossing time definition: T is indeed determined by the intersection of an amplified oscillation and a decreasing exponential. A small change in the initial conditions may thus cause a discontinuous variation of the crossing time. This is the very origin of the system chaotic dynamics which suggests a stochastic treatment of the Lorenz diffusion equation.

Despite the fact we have a 3D space of parameters to investigate the model behaviors, the interesting region is actually a limited portion. This easily follows from the following simple considerations. For η large enough in (8.71), the motion rapidly stops in one of the lattice fixed–points. In order to have non trivial solutions, the viscous coefficient must be smaller than a maximum value which can be explicitly derived from the condition $\alpha > \alpha_c$

$$\eta_{\max} = \frac{1}{2\beta} \left[\sqrt{\left(1 + \beta^2\right)^2 + 4\alpha\beta^2} - 1 - \beta^2 \right].$$

In the opposite limit, if η is too small, the friction term is negligible with respect to the memory term, and the resulting motion is 'ballistic', i.e., $\left\langle [x(t) - x(0)]^2 \right\rangle \sim t^{21}$. Analogous considerations can be repeated for α and β. Inside this region of parameters the Lorenz diffusion equation generates a wide variety of behaviors: as we will see, the observed regimens are strongly sensitive to the control parameters, and furthermore this dependence is often in contrast with the intuitive meaning of α, β, η.

The analysis of the motion in the elementary cell shows that the system rapidly reaches a steady–state. In the diffusive regimes the points corresponding to subsequent cell entering conditions $\dot{x}_0$, $\ddot{x}_0$ are quickly attracted on a particular locus of the plane. In the general situation this attracting set does not define a univocal map between the entering velocity and acceleration. However, it is not difficult to see that, for large T, $\dot{x}_0$ and $\ddot{x}_0$ satisfy the piecewise linear relation [FMV02b]

$$\dddot{x}_0 - 2\lambda \ddot{x}_0 + \frac{1}{2} \left(2 - \lambda^2 - \omega^2\right) \operatorname{sgn}(\dot{x}_0) = 0. \qquad (8.76)$$

---

[1] Due to the deterministic nature of the system, averages should be intended over the initial conditions. In our numerical simulations we have typically chosen the initial conditions to be uniformly distributed over some real interval, and we have naturally assumed that the diffusion coefficient is independent of the choice of the initial ensemble.

If the diffusive regimen admits large enough times of permanence, the couple of straight lines defined by (8.76) therefore belongs to the entering condition attracting set. Moreover, in the limiting case of very large λ_C the attracting set exactly reduces to such lines, and in the steady state (8.76) is satisfied by all entering conditions [FMV02b].

The macroscopic evolution of the system can be statistically analyzed by introducing the diffusion coefficient

$$D \equiv \lim_{t \to \infty} \left\langle [x(t) - x(0)]^2 \right\rangle / (2t).$$

Diffusive regimens are identified by a finite value of D. Different values of the control parameters can lead to qualitatively very different diffusive motions. Despite the somehow strange aspect of some system trajectories, it should be stressed that no anomalies (i.e., super–diffusive transport) have been encountered in the statistical analysis of diffusion, the distribution of long flies having always exponential tails.

8.4 Turbulence

It is well–known that the viscous fluid evolves according to the nonlinear Navier–Stokes PDEs[2]

$$\dot{\mathbf{u}} + \mathbf{u} \cdot \boldsymbol{\nabla} \mathbf{u} + \nabla p / \rho = \nu \Delta \mathbf{u} + \mathbf{f}, \tag{8.77}$$

where $\mathbf{u} = \mathbf{u}(x^i, t)$, $(i = 1, 2, 3)$ is the fluid 3D velocity, $p = p(x^i, t)$ is the pressure field, ρ, ν are the fluid density and viscosity coefficient, while $\mathbf{f} = \mathbf{f}(x^i, t)$ is the nonlinear external energy source. To simplify the problem, we can impose to $\mathbf{f}$ the so–called *Reynolds condition*, $\langle \mathbf{f} \cdot \mathbf{u} \rangle = \varepsilon$, where ε is the average rate of energy injection.

Fluid dynamicists believe that *Navier–Stokes equations* (8.77) *accurately describe turbulence*. A mathematical proof of the global regularity of the solutions to the Navier–Stokes equations is a very challenging problem and yet such a proof or disproof does not solve the problem of turbulence. However, it

---

[2] Recall that the Navier–Stokes equations, named after C.L. Navier and G.G. Stokes, are a set of PDEs that describe the motion of liquids and gases, based on the fact that changes in momentum of the particles of a fluid are the product of changes in pressure and dissipative viscous forces acting inside the fluid. These viscous forces originate in molecular interactions and dictate how viscous a fluid is, so the Navier–Stokes PDEs represent a dynamical statement of the balance of forces acting at any given region of the fluid. They describe the physics of a large number of phenomena of academic and economic interest (they are useful to model weather, ocean currents, water flow in a pipe, motion of stars inside a galaxy, flow around an airfoil (wing); they are also used in the design of aircraft and cars, the study of blood flow, the design of power stations, the analysis of the effects of pollution, etc).

may help understanding turbulence. Turbulence is more of a dynamical system problem. We will see below that studies on chaos in PDEs indicate that turbulence can have *Bernoulli shift dynamics* which results in the wandering of a turbulent solution in a fat domain in the phase–space; thus, turbulence can not be averaged. The hope is that turbulence can be controlled [Li04].

The first demonstration of existence of an unstable recurrent pattern in a 3D turbulent hydrodynamic flow was performed in [KK01], using the full numerical simulation, a 15,422–dimensional discretization of the 3D Plane Couette turbulence at the *Reynold's number* $Re = 400$.[3] The authors found an important unstable spatio–temporally periodic solution, a single unstable recurrent pattern.

8.4.1 Turbulent Flow

Recall that in fluid dynamics, *turbulent flow* is a flow regime characterized by low momentum diffusion, high momentum convection, and rapid variation of pressure and velocity in space and time. Flow that is not turbulent is called *laminar flow*. Also, recall that the *Reynold's number* Re characterizes whether flow conditions lead to laminar or turbulent flow. The structure of turbulent flow was first described by A. Kolmogorov. Consider the flow of water over a simple smooth object, such as a sphere. At very low speeds the flow is laminar, i.e., the flow is locally smooth (though it may involve vortices on a large scale). As the speed increases, at some point the transition is made to turbulent (or, chaotic) flow. In turbulent flow, unsteady vortices[4] appear on many scales and interact with each other. Drag due to boundary layer skin friction increases. The structure and location of boundary layer separation often changes, sometimes resulting in a reduction of overall drag. Because laminar–turbulent transition is governed by Reynold's number, the

---

[3] Recall that the Reynold's number Re is the most important dimensionless number in fluid dynamics and provides a criterion for determining *dynamical similarity*. Where two similar objects in perhaps different fluids with possibly different flow–rates have similar fluid flow around them, they are said to be dynamically similar. Re is the ratio of inertial forces to viscous forces and is used for determining whether a flow will be *laminar* or *turbulent*. Laminar flow occurs at low Reynolds numbers, where viscous forces are dominant, and is characterized by smooth, constant fluid motion, while turbulent flow, on the other hand, occurs at high Res and is dominated by inertial forces, producing random eddies, vortices and other flow fluctuations. The transition between laminar and turbulent flow is often indicated by a critical Reynold's number (Re_{crit}), which depends on the exact flow configuration and must be determined experimentally. Within a certain range around this point there is a region of gradual transition where the flow is neither fully laminar nor fully turbulent, and predictions of fluid behavior can be difficult.

[4] Recall that a *vortex* can be any circular or rotary flow that possesses vorticity. Vortex represents a spiral whirling motion (i.e., a spinning turbulent flow) with closed streamlines. The shape of media or mass rotating rapidly around a center forms a vortex. It is a flow involving rotation about an arbitrary axis.

same transition occurs if the size of the object is gradually increased, or the viscosity of the fluid is decreased, or if the density of the fluid is increased.

Vorticity Dynamics

Vorticity $\omega = \omega(x^i, t)$, $(i = 1, 2, 3)$ is a geometrical concept used in fluid dynamics, which is related to the amount of 'circulation' or 'rotation' in a fluid. More precisely, *vorticity* is the circulation per unit area at a point in the flow field, or formally, $\omega = \nabla \times \mathbf{u}$, where $\mathbf{u} = \mathbf{u}(x^i, t)$ is the fluid velocity. It is a vector quantity, whose direction is (roughly speaking) along the axis of the swirl. The movement of a fluid can be said to be vortical if the fluid moves around in a circle, or in a helix, or if it tends to spin around some axis. Such motion can also be called *solenoidal*. In the atmospheric sciences, vorticity is a property that characterizes large–scale rotation of air masses. Since the atmospheric circulation is nearly horizontal, the 3D vorticity is nearly vertical, and it is common to use the vertical component as a scalar vorticity.

A vortex can be seen in the spiraling motion of air or liquid around a center of rotation. Circular current of water of conflicting tides form vortex shapes. Turbulent flow makes many vortices. A good example of a vortex is the atmospheric phenomenon of a whirlwind or a *tornado*. This whirling air mass mostly takes the form of a helix, column, or spiral. Tornadoes develop from severe thunderstorms, usually spawned from squall lines and *supercell thunderstorms*, though they sometimes happen as a result of a *hurricane*.[5] Another example is a mesovortex on the scale of a few miles (smaller than a hurricane but larger than a tornado). On a much smaller scale, a vortex is usually formed as water goes down a drain, as in a sink or a toilet. This occurs in water as the revolving mass forms a whirlpool.[6] This whirlpool is caused by water flowing out of a small opening in the bottom of a basin or reservoir. This swirling flow structure within a region of fluid flow opens downward from the water surface. In the hydrodynamic interpretation of the behavior of electromagnetic fields, the acceleration of electric fluid in a particular direction creates a positive vortex of magnetic fluid. This in turn creates around itself a corresponding negative vortex of electric fluid.

Dynamical Similarity and Eddies

In order for two flows to be similar they must have the same geometry and equal Reynolds numbers. When comparing fluid behavior at homologous

---

[5] Recall that a hurricane is a much larger, swirling body of clouds produced by evaporating warm ocean water and influenced by the Earth's rotation. In particular, polar vortex is a persistent, large–scale cyclone centered near the Earth's poles, in the middle and upper troposphere and the stratosphere. Similar, but far greater, vortices are also seen on other planets, such as the permanent Great Red Spot on Jupiter and the intermittent Great Dark Spot on Neptune.

[6] Recall that a whirlpool is a swirling body of water produced by ocean tides or by a hole underneath the vortex, where water drains out, as in a bathtub.

points in a model and a full–scale flow, we have $Re^* = Re$, where quantities marked with $*$ concern the flow around the model and the other the real flow. This allows us to perform experiments with reduced models in water channels or wind tunnels, and correlate the data to the real flows. Note that true dynamic similarity may require matching other dimensionless numbers as well, such as the Mach number used in compressible flows, or the Froude number that governs free-surface flows.

In a turbulent flow, there is a range of scales of the fluid motions, sometimes called *eddies*. A single packet of fluid moving with a bulk velocity is called an *eddy*. The size of the largest scales (eddies) are set by the overall geometry of the flow. For instance, in an industrial smoke–stack, the largest scales of fluid motion are as big as the diameter of the stack itself. The size of the smallest scales is set by Re. As Re increases, smaller and smaller scales of the flow are visible. In the smoke–stack, the smoke may appear to have many very small bumps or eddies, in addition to large bulky eddies. In this sense, Re is an indicator of the range of scales in the flow. The higher the Reynold's number, the greater the range of scales.

8.4.2 The Governing Equations of Turbulence

It has been overwhelmingly accepted by fluid dynamicists that the *Navier–Stokes equations* (8.77) are accurate governing equations of turbulence. Their delicate experimental measurements on turbulence have led them to such a conclusion. A simple form of the Navier–Stokes equations, describing viscous incompressible fluids, can be written as [Li05, Li06]

$$u_{i,t} + u_j u_{i,j} = -p_{,i} + \text{Re}^{-1} u_{i,jj} + f_i , \quad u_{i,i} = 0 ; \qquad (8.78)$$

where u_i's are the velocity components, p is the pressure, f_i's are the external force components, and Re is the Reynold's number. There are two ways of deriving the Navier–Stokes equations: (1) The fluid dynamicist's way of using the concept of fluid particle and material derivative, (2) The theoretical physicist's way of starting from *Boltzman equation*. According to either approach, one can replace the viscous term $\text{Re}^{-1} u_{i,jj}$ by for example

$$\text{Re}^{-1} u_{i,jj} + \alpha u_{i,jjkk} + \cdots . \qquad (8.79)$$

Here the only principle one can employ is the Einstein covariance principle which eliminates the possibility of third derivatives for example. According to the fluid dynamicist's way, the viscous term $\text{Re}^{-1} u_{i,jj}$ was derived from a principle proposed by Newton that the stress is proportional to the velocity's derivatives (strain, not velocity). Such fluids are called Newtonian fluids. Of course, there exist non–Newtonian fluids like volcanic lava for which the viscous term is more complicated and can be nonlinear. According to the theoretical physicist's way, the viscous term was obtained from an expansion which has no reason to stop at its leading order term $\text{Re}^{-1} u_{i,jj}$ [Li05, Li06].

8.4.3 Global Well-Posedness of the Navier–Stokes Equations

It is well known that the global well-posedness of the Navier–Stokes equations (8.78) has been selected by the Clay Mathematics Institute as one of its seven one million dollars problems. Specifically, the difficulty lies at the global regularity [Ler34]. More precisely, the fact that

$$\int \int u_{i,j} u_{i,j} \ dx \ dt$$

being bounded only implies

$$\int u_{i,j} u_{i,j} \ dx$$

being bounded for almost all t, is the key of the difficulty. In fact, Leray was able to show that the possible exceptional set of t is actually a compact set of measure zero. There have been a lot of more recent works on describing this exceptional compact set [CKN82]. The claim that this possible exceptional compact set is actually empty, will imply the global regularity and the solution of the problem. The hope for such a claim seems slim.

Even for ordinary differential equations, often one can not prove their global well-posedness, but their solutions on computers look perfectly globally regular and sometimes chaotic. Chaos and global regularity are compatible. The fact that fluid experimentalists quickly discovered shocks in compressible fluids and never found any finite time blow up in incompressible fluids, indicates that there might be no finite time blow up in Navier–Stokes equations (even Euler equations). On the other hand, the solutions of Navier–Stokes equations can definitely be turbulent (chaotic) [Li05, Li06].

Replacing the viscous term $\text{Re}^{-1} \ u_{i,jj}$ by higher order derivatives (8.79), one can prove the global regularity [KP02]. This leaves the global regularity of (8.78) a more challenging and interesting mathematical problem. Assume that the unthinkable event happens, that is, someone proves the existence of a meaningful finite time blow up in (8.78), then fluid experimentalists need to identify such a finite time blow up in the experiments. If they fail, then the choice will be whether or not to replace the viscous term $\text{Re}^{-1} \ u_{i,jj}$ in the Navier–Stokes equations (8.78) by higher order derivatives like (8.79) to better model the fluid motion.

Even after the global regularity of (8.78) is proved or disproved, the problem of turbulence is not solved although the global regularity information will help understanding turbulence. Turbulence is more of a dynamical system problem. Often a dynamical system study does not depend on global well-posedness. Local well-posedness is often enough. In fact, this is the case in my proof on the existence of chaos in partial differential equations [Li04].

8.4.4 Spatio–Temporal Chaos and Turbulence in PDEs

In their first edition of Fluid Mechanics [LL59], Landau and Lifschitz proposed a *route to turbulence* in spatio–temporal fluid systems. Since then, much work, in dynamical systems, experimental fluid dynamics, and many other fields has been done concerning the routes to turbulence. Ever since the discovery of chaos in low–dimensional systems, researchers have been trying to *use the concept of chaos to understand turbulence* (see the seminal paper of Ruelle and Takens [RT71]). recall that there are two types of fluid motions: laminar flows and turbulent flows. Laminar flows look regular, and turbulent flows are non–laminar and look irregular. Chaos is more precise, for example, in terms of the so–called *Bernoulli shift dynamics*. On the other hand, even in low–dimensional systems, there are solutions which look irregular for a while, and then look regular again. Such a dynamics is often called a *transient chaos*.

Low–dimensional chaos is the starting point of a long journey toward understanding turbulence. To have a better connection between chaos and turbulence, one has to study chaos in PDEs [Li04].

Sine–Gordon Equation

Consider the simple perturbed *sine–Gordon equation* [Li04c]

$$u_{tt} = c^2 u_{xx} + \sin u + \epsilon[-a u_t + \cos t \, \sin^3 u], \qquad (8.80)$$

subject to periodic boundary condition

$$u(t, x + 2\pi) = u(t, x) \ ,$$

as well as even or odd constraint,

$$u(t, -x) = u(t, x), \qquad \text{or} \qquad u(t, -x) = -u(t, x) \ ,$$

where u is a real–valued function of two real variables (t, x), c is a real constant, $\epsilon \geq 0$ is a small perturbation parameter, and $a > 0$ is an external parameter. One can view (8.80) as a *flow* (u, u_t) defined in the phase–space manifold $M \equiv H^1 \times L^2$, where H^1 and L^2 are the Sobolev spaces on $[0, 2\pi]$. A point in the phase–space manifold M corresponds to two profiles, $(u(x), u_t(x))$. [Li04c] has proved that there exists a homoclinic orbit $(u, u_t) = h(t, x)$ asymptotic to $(u, u_t) = (0, 0)$. Let us define two orbits segments

$$\eta_0 : \ (u, u_t) = (0, 0) \ , \qquad \text{and} \qquad \eta_1 : \ (u, u_t) = h(t, x) \ , \qquad (t \in [-T, T] \ ).$$

When T is large enough, η_1 is almost the entire homoclinic orbit (chopped off in a small neighborhood of $(u, u_t) = (0, 0)$). To any binary sequence

$$a = \{\cdots a_{-2} a_{-1} a_0, a_1 a_2 \cdots\} \ , \qquad (a_k \in \{0, 1\}), \qquad (8.81)$$

one can associate a pseudo–orbit

$$\eta_a = \{\cdots \eta_{a_{-2}} \eta_{a_{-1}} \eta_{a_0}, \eta_{a_1} \eta_{a_2} \cdots\} \ .$$

The pseudo–orbit η_a is not a true orbit but rather 'almost an orbit'. One can prove that for any such pseudo–orbit η_a, there is a unique true orbit in its neighborhood [Li04c]. Therefore, each binary sequence labels a true orbit. All these true orbits together form a chaos. In order to talk about sensitive dependence on initial data, one can introduce the *product topology* by defining the neighborhood basis of a binary sequence

$$a^* = \{\cdots a^*_{-2} a^*_{-1} a^*_0, a^*_1 a^*_2 \cdots\} \qquad \text{as} \qquad \Omega_N = \{a \ : \quad a_n = a^*_n \ , \quad |n| \leq N\} \ .$$

The Bernoulli shift on the binary sequence (8.81) moves the comma one step to the right. Two binary sequences in the neighborhood Ω_N will be of order Ω_1 away after N iterations of the Bernoulli shift. Since the binary sequences label the orbits, the orbits will exhibit the same feature. In fact, the Bernoulli shift is topologically conjugate to the perturbed sine–Gordon flow.

Replacing a homoclinic orbit by its fattened version – a homoclinic tube, or by a heteroclinic cycle, or by a heteroclinically tubular cycle; one can still get the same Bernoulli shift dynamics. Also, adding diffusive perturbation $\epsilon b u_{txx}$ to (8.80), one can still prove the existence of homoclinics or heteroclinics, but the Bernoulli shift result has not been established [Li04c].

Complex Ginzburg–Landau Equation

Consider the complex–valued *Ginzburg–Landau equation* [Li04a, Li04b],

$$i q_t = q_{xx} + 2[|q|^2 - \omega^2] + i\epsilon[q_{xx} - \alpha q + \beta] \ , \tag{8.82}$$

which is subject to periodic boundary condition and even constraint

$$q(t, x + 2\pi) = q(t, x) \ , \quad q(t, -x) = q(t, x) \ ,$$

where q is a complex–valued function of two real variables (t, x), (ω, α, β) are positive constants, and $\epsilon \geq 0$ is a small perturbation parameter. In this case, one can prove the existence of homoclinic orbits [Li04a]. But the Bernoulli shift dynamics was established under generic assumptions [Li04b].

A real fluid example is the amplitude equation of Faraday water wave, which is also a complex Ginzburg–Landau equation [Li04d],

$$i q_t = q_{xx} + 2[|q|^2 - \omega^2] + i\epsilon[q_{xx} - \alpha q + \beta \bar{q}] \ , \tag{8.83}$$

subject to the same boundary condition as (8.82). For the first time, one can prove the existence of homoclinic orbits for a water wave equation (8.83). The Bernoulli shift dynamics was also established under generic assumptions [Li04d]. That is, one can prove the existence of chaos in water waves under generic assumptions.

The nature of the complex Ginzburg–Landau equation is a parabolic equation which is near a hyperbolic equation. The same is true for the perturbed sine–Gordon equation with the diffusive term $\epsilon b u_{txx}$ added. They contain effects of diffusion, dispersion, and nonlinearity. The Navier–Stokes equations are diffusion–advection equations. The advective term is missing from the perturbed sine–Gordon equation and the complex Ginzburg–Landau equation. However, the modified KdV equation (9.30) does contain an advective term. In principle, perturbed modified KdV equation should have the same feature as the perturbed sine–Gordon equation. Turbulence happens when the diffusion is weak, i.e., in the near hyperbolic regime. One should hope that turbulence should share some of the features of chaos in the perturbed sine–Gordon equation. There is a popular myth that turbulence is fundamentally different from chaos because turbulence contains many unstable modes. In both the perturbed sine–Gordon equation and the complex Ginzburg–Landau equation, one can incorporate as many unstable modes as one likes, the resulting Bernoulli shift dynamics is still the same. On a computer, the solution with more unstable modes may look rougher, but it is still chaos [Li04].

In a word, dynamics of strongly nonlinear classical fields is 'turbulent', not 'laminar'.

On the other hand, field theories such as 4-dimensional QCD or gravity have many dimensions, symmetries, tensorial indices. They are far too complicated for exploratory forays into this forbidding terrain. Instead, we consider a simple spatio–temporally chaotic nonlinear system of physical interest [CCP96].

Kuramoto–Sivashinsky System

One of the simplest and extensively studied spatially extended dynamical systems is the Kuramoto–Sivashinsky (KS) system [Kur76, Siv77]

$$u_t = (u^2)_x - u_{xx} - \nu u_{xxxx}, \qquad (8.84)$$

which arises as an amplitude equation for interfacial instabilities in a variety of contexts. The so–called *flame front* $u(x,t)$ has compact support, with $x \in [0, 2\pi]$ a periodic space coordinate. The u^2 term makes this a nonlinear system, $t \geq 0$ is the time, and ν is a 4–order 'viscosity' damping parameter that irons out any sharp features. Numerical simulations demonstrate that as the viscosity decreases (or the size of the system increases), the *flame front* becomes increasingly unstable and turbulent. The task of the theory is to describe this spatio-temporal turbulence and yield quantitative predictions for its measurable consequences.

For any finite spatial resolution, the KS system (8.84) follows approximately for a finite time a pattern belonging to a finite alphabet of admissible patterns, and the long term dynamics can be thought of as a walk through the space of such patterns, just as chaotic dynamics with a low dimensional attractor can be thought of as a succession of nearly periodic (but unstable) motions. The periodic orbit gives the machinery that converts this intuitive picture into precise calculation scheme that extracts asymptotic time predictions from the short time dynamics. For extended systems the theory gives a description of the asymptotics of partial differential equations in terms of recurrent spatio–temporal patterns.

The KS periodic orbit calculations of Lyapunov exponents and escape rates [CCP96] demonstrate that the *periodic orbit theory* predicts observable averages for deterministic but classically chaotic spatio–temporal systems. The main problem today is not how to compute such averages – periodic orbit theory as well as direct numerical simulations can handle that – but rather that there is no consensus on what the sensible experimental observables worth are predicting [Cvi00].

Burgers Dynamical System

Consider the following *Burgers dynamical system* on a *functional manifold* $M \subset C^k(\mathbb{R};\mathbb{R})$:

$$u_t = uu_x + u_{xx}, \tag{8.85}$$

where $u \in M$, $t \in \mathbb{R}$ is an evolution parameter. The flow of (8.85) on M can be recast into a set of 2–forms $\{\alpha\} \subset \Lambda^2(J(\mathbb{R}^2;\mathbb{R}))$ upon the adjoint jet–manifold $J(\mathbb{R}^2;\mathbb{R})$ as follows [BPS98]:

$$\{\alpha\} = \left\{ du^{(0)} \wedge dt - u^{(1)} dx \wedge dt = \alpha^1, \ du^{(0)} \wedge dx + u^{(0)} du^{(0)} \wedge dt \right.$$
$$\left. + du^{(1)} \wedge dt = \alpha^2 : \ \left(x, t; u^{(0)}, u^{(1)}\right)^T \in M^4 \subset J^1(\mathbb{R}^2;\mathbb{R})\right\}, \tag{8.86}$$

where M^4 is some finite–dimensional submanifold in $J^1(\mathbb{R}^2;\mathbb{R}))$ with coordinates $(x, t, u^{(0)} = u, u^{(1)} = u_x)$. The set of 2–forms (8.86) generates the closed ideal $\mathfrak{J}(\alpha)$, since

$$d\alpha^1 = dx \wedge \alpha^2 - u^{(0)} dx \wedge \alpha^1, \qquad d\alpha^2 = 0, \tag{8.87}$$

the integral submanifold $\bar{M} = \{x, t \in \mathbb{R}\} \subset M^4$ being defined by the condition $\mathfrak{J}(\alpha) = 0$. We now look for a reduced 'curvature' 1–form $\Gamma \in \Lambda^1(M^4) \otimes \mathcal{G}$, belonging to some not yet determined Lie algebra $\mathcal{G}$. This 1–form can be represented using (8.86), as follows:

$$\Gamma = b^{(x)}(u^{(0)}, u^{(1)}) dx + b^{(t)}(u^{(0)}, u^{(1)}) dt, \tag{8.88}$$

where elements $b^{(x)}, b^{(t)} \in \mathcal{G}$ satisfy such determining equations [BPS98]

$$\frac{\partial b^{(x)}}{\partial u^{(0)}} = g_2, \qquad \frac{\partial b^{(x)}}{\partial u^{(1)}} = 0, \qquad \frac{\partial b^{(t)}}{\partial u^{(0)}} = g_1 + g_2 u^{(0)},$$
$$\frac{\partial b^{(t)}}{\partial u^{(1)}} = g_2, \qquad [b^{(x)}, b^{(t)}] = -u^{(1)} g_1. \tag{8.89}$$

The set (8.89) has the following unique solution

$$b^{(x)} = A_0 + A_1 u^{(0)},$$
$$b^{(t)} = u^{(1)} A_1 + \frac{u^{(0)2}}{2} A_1 + [A_1, A_0] u^{(0)} + A_2, \tag{8.90}$$

where $A_j \in \mathcal{G}$, $j = \overline{0,2}$, are some constant elements on M of a Lie algebra $\mathcal{G}$ under search, enjoying the next Lie structure equations:

$$[A_0, A_2] = 0,$$
$$[A_0, [A_1, A_0]] + [A_1, A_2] = 0, \qquad (8.91)$$
$$[A_1, [A_1, A_0]] + \tfrac{1}{2}[A_0, A_1] = 0.$$

From (8.89) one can see that the curvature 2–form $\Omega \in span_{\mathbb{R}}\{A_1, [A_0, A_1] : A_j \in \mathcal{G}, j = 0, 1\}$. Therefore, reducing via the *Ambrose–Singer theorem* the associated principal fibred frame space $P(M; G = GL(n))$ to the principal fibre bundle $P(M; G(h))$. where $G(h) \subset G$ is the corresponding holonomy Lie group of the connection Γ on P, we need to satisfy the following conditions for the set $\mathcal{G}(h) \subset \mathcal{G}$ to be a Lie subalgebra in $\mathcal{G}$: $\nabla_x^m \nabla_t^n \Omega \in \mathcal{G}(h)$ for all $m, n \in \mathbb{Z}_+$.

Let us try now to close the above transfinitive procedure requiring that [BPS98]

$$\mathcal{G}(h) = \mathcal{G}(h)_0 = span_{\mathbb{R}}\{\nabla_x^m \nabla_x^n \Omega \in \mathcal{G} : m + n = 0\} \qquad (8.92)$$

This means that

$$\mathcal{G}(h)_0 = span_{\mathbb{R}}\{A_1, A_3 = [A_0, A_1]\}. \qquad (8.93)$$

To enjoy the set of relations (8.91) we need to use expansions over the basis (8.93) of the external elements $A_0, A_2 \in \mathcal{G}(h)$:

$$A_0 = q_{01}A_1 + q_{13}A_3, \qquad A_2 = q_{21}A_1 + q_{23}A_3. \qquad (8.94)$$

Substituting expansions (8.94) into (8.91), we get that $q_{01} = q_{23} = \lambda$, $q_{21} = -\lambda^2/2$ and $q_{03} = -2$ for some arbitrary real parameter $\lambda \in \mathbb{R}$, that is $\mathcal{G}(h) = span_{\mathbb{R}}\{A_1, A_3\}$, where

$$[A_1, A_3] = A_3/2; \qquad A_0 = \lambda A_1 - 2A_3, \qquad A_2 = -\lambda^2 A_1/2 + \lambda A_3. \quad (8.95)$$

As a result of (8.95) we can state that the holonomy Lie algebra $\mathcal{G}(h)$ is a real 2D one, assuming the following (2×2)−matrix representation [BPS98]:

$$A_1 = \begin{pmatrix} 1/4 & 0 \\ 0 & -1/4 \end{pmatrix}, \qquad A_3 = \begin{pmatrix} 0 & 1 \\ 0 & 0 \end{pmatrix},$$
$$A_0 = \begin{pmatrix} \lambda/4 & -2 \\ 0 & -\lambda/4 \end{pmatrix}, \qquad A_2 = \begin{pmatrix} -\lambda^2/8 & \lambda \\ 0 & \lambda^2/8 \end{pmatrix}. \qquad (8.96)$$

Thereby from (8.88), (8.90) and (8.96) we get the *reduced curvature 1–form* $\Gamma \in \Lambda^1(M) \otimes \mathcal{G}$,

$$\Gamma = (A_0 + uA_1)dx + ((u_x + u^2/2)A_1 - uA_3 + A_2)dt, \qquad (8.97)$$

generating *parallel transport* of vectors from the representation space Y of the holonomy Lie algebra $\mathcal{G}(h)$:

$$dy + \Gamma y = 0 \tag{8.98}$$

upon the integral submanifold $\bar{M} \subset M^4$ of the ideal $\mathcal{I}(\alpha)$, generated by the set of 2–forms (8.86). The result (8.98) means also that the Burgers dynamical system (8.85) is endowed with the standard Lax type representation, having the spectral parameter $\lambda \in \mathbb{R}$ necessary for its integrability in quadratures.

8.4.5 General Fluid Dynamics

In this subsection we will derive the general form of the Navier–Stokes equations (8.77) in nonlinear fluid dynamics.

Continuity Equation

Recall that the most important equation in fluid dynamics, as well as in general continuum mechanics, is the celebrated *equation of continuity*, (we explain the symbols in the following text)

$$\partial_t \rho + \text{div}(\rho \mathbf{u}) = 0. \tag{8.99}$$

As a warm–up for turbulence, we will derive the continuity equation (8.99), starting from the *mass conservation principle*. Let dm denote an infinitesimal mass of a fluid particle. Then, using the absolute time derivative operator $\overline{(\;)} \equiv \frac{D}{dt}$ (see Appendix), the mass conservation principle reads

$$\overline{dm} = 0. \tag{8.100}$$

If we further introduce the fluid density $\rho = dm/dv$, where dv is an infinitesimal volume of a fluid particle, then the mass conservation principle (8.100) can be rewritten as
$$\overline{\rho dv} = 0,$$

which is the absolute derivative of a product, and therefore expands into

$$\dot{\rho} dv + \rho \overline{dv} = 0. \tag{8.101}$$

Now, as the fluid density $\rho = \rho(x^k, t)$ is a function of both time t and spatial coordinates x^k, for $k = 1, 2, 3$, that is, a *scalar–field*, its total time derivative $\dot{\rho}$, figuring in (8.101), is defined by

$$\dot{\rho} = \partial_t \rho + \partial_{x^k} \rho \, \partial_t x^k \equiv \partial_t \rho + \rho_{;k} u^k, \tag{8.102}$$
$$\text{or, in vector form} \quad \dot{\rho} = \partial_t \rho + \text{grad}(\rho) \cdot \mathbf{u},$$

where $u^k = u^k(x^k, t) \equiv \mathbf{u}$ is the velocity vector–field of the fluid.

Regarding $\overline{dv}$, the other term figuring in (8.101), we start by expanding an elementary volume dv along the sides $\{dx^i_{(p)}, dx^j_{(q)}, dx^k_{(r)}\}$ of an elementary parallelepiped, as

$$dv = \frac{1}{3!}\delta^{pqr}_{ijk}dx^i_{(p)}dx^j_{(q)}dx^k_{(r)}, \qquad (i,j,k,p,q,r=1,2,3)$$

so that its absolute derivative becomes

$$\dot{\overline{dv}} = \frac{1}{2!}\delta^{pqr}_{ijk}\dot{\overline{dx^i}}_{(p)}dx^j_{(q)}dx^k_{(r)}$$

$$= \frac{1}{2!}u^i_{;l}\delta^{pqr}_{ijk}dx^l_{(p)}dx^j_{(q)}dx^k_{(r)} \qquad \left(\text{using } \dot{\overline{dx^i}}_{(p)} = u^i_{;l}dx^l_{(p)}\right),$$

which finally simplifies into

$$\dot{\overline{dv}} = u^k_{;k}dv \equiv \operatorname{div}(\mathbf{u})\,dv. \qquad (8.103)$$

Substituting (8.102) and (8.103) into (8.101) gives

$$\dot{\overline{\rho dv}} \equiv \left(\partial_t\rho + \rho_{;k}u^k\right)dv + \rho u^k_{;k}dv = 0. \qquad (8.104)$$

As we are dealing with arbitrary fluid particles, $dv \neq 0$, so from (8.104) follows

$$\partial_t\rho + \rho_{;k}u^k + \rho u^k_{;k} \equiv \partial_t\rho + (\rho u^k)_{;k} = 0. \qquad (8.105)$$

Equation (8.105) is the covariant form of the continuity equation, which in standard vector notation becomes (8.99), i.e., $\partial_t\rho + \operatorname{div}(\rho\mathbf{u}) = 0$.

Forces Acting on a Fluid

A fluid contained in a finite volume is subject to the action of both *volume forces* F^i and *surface forces* S^i, which are respectively defined by

$$F^i = \int_v \rho f^i dv, \qquad \text{and} \qquad S^i = \oint_a \sigma^{ij} da_j. \qquad (8.106)$$

Here, f^i is a force vector acting on an elementary mass dm, so that the elementary volume force is given by

$$dF^i = f^i dm = \rho f^i dv,$$

which is the integrand in the volume integral on l.h.s of (8.106). $\sigma^{ij} = \sigma^{ij}(x^k, t)$ is the stress tensor–field of the fluid, so that the elementary force acting on the closed oriented surface a is given by

$$dS^i = \sigma^{ij} da_j,$$

where da_j is an oriented element of the surface a; this is the integrand in the surface integral on the r.h.s of (8.106).

On the other hand, the elementary momentum dK^i of a fluid particle (with elementary volume dv and elementary mass $dm = \rho dv$) equals the product of dm with the particle's velocity u^i, i.e.,

$$dK^i = u^i dm = \rho u^i dv,$$

so that the total momentum of the finite fluid volume v is given by the volume integral

$$K^i = \int_v \rho u^i dv. \qquad (8.107)$$

Now, the *Newtonian–like force law for the fluid* states that the time derivative of the fluid momentum equals the resulting force acting on it, $\dot{K}^i = \mathcal{F}^i$, where the resulting force $\mathcal{F}^i$ is given by the sum of surface and volume forces,

$$\mathcal{F}^i = S^i + F^i = \oint_a \sigma^{ij} da_j + \int_v \rho f^i dv. \qquad (8.108)$$

From (8.107), taking the time derivative and using $\overline{\rho dv} = 0$, we get

$$\dot{K}^i = \int_v \rho \dot{u}^i dv,$$

where $\dot{u}^i = \dot{u}^i(x^k, t) \equiv \dot{\mathbf{u}}$ is the acceleration vector–field of the fluid, so that (8.108) gives

$$\oint_a \sigma^{ij} da_j + \int_v \rho (f^i - \dot{u}^i) dv = 0. \qquad (8.109)$$

Now, assuming that the stress tensor $\sigma^{ij} = \sigma^{ij}(x^k, t)$ does not have any singular points in the volume v bounded by the closed surface a, we can transform the surface integral in (8.109) in the volume one, i e.,

$$\oint_a \sigma^{ij} da_j = \int_v \sigma^{ij}_{;j} dv, \qquad (8.110)$$

where $\sigma^{ij}_{;j}$ denotes the *divergence of the stress tensor*. The expression (8.110) shows us that the resulting surface force acting on the closed surface a equals the flux of the stress tensor through the surface a. Using this expression, we can rewrite (8.109) in the form

$$\int_v \left(\sigma^{ij}_{;j} + \rho f^i - \rho \dot{u}^i \right) dv = 0.$$

As this equation needs to hold for an arbitrary fluid element $dv \neq 0$, it implies the dynamical equation of motion for the fluid particles, also called the *first Cauchy law of motion*,

$$\sigma^{ij}_{;j} + \rho f^i = \rho \dot{u}^i. \qquad (8.111)$$

Constitutive and Dynamical Equations

Recall that, in case of a homogenous isotropic viscous fluid, the stress tensor σ^{ij} depends on the strain–rate tensor–field $e^{ij} = e^{ij}(x^k, t)$ of the fluid in such a way that

$$\sigma^{ij} = -pg^{ij}, \qquad \text{when} \qquad e^{ij} = 0,$$

where the scalar function $p = p(x^k, t)$ represents the pressure field. Therefore, pressure is independent on the strain–rate tensor e^{ij}. Next, we introduce the viscosity tensor–field $\beta^{ij} = \beta^{ij}(x^k, t)$, as

$$\beta^{ij} = \sigma^{ij} + pg^{ij}, \tag{8.112}$$

which depends exclusively on the strain–rate tensor (i.e., $\beta^{ij} = 0$ whenever $e^{ij} = 0$). A viscous fluid in which the viscosity tensor β^{ij} can be expressed as a function of the strain–rate tensor e^{ij} in the form

$$\beta^{ij} = \alpha_1(e_I, e_{II}, e_{III})g^{ij} + \alpha_2(e_I, e_{II}, e_{III})e^{ij} + \alpha_3(e_I, e_{II}, e_{III})e^i_k e^{kj}, \tag{8.113}$$

where $\alpha_l = \alpha_l(e_I, e_{II}, e_{III})$, $(l = 1, 2, 3)$ are scalar functions of the basic invariants (e_I, e_{II}, e_{III}) of the strain–rate tensor e^{ij}, is called the *Stokes fluid*.

If we take only the linear terms in (8.113), we get the constitutive equation for the *Newtonian fluid*,

$$\beta^{ij} = \alpha_1 e_I g^{ij} + \alpha_2 e^{ij}, \tag{8.114}$$

which is, therefore, a linear approximation of the constitutive equation (8.113) for the Stokes fluid.

If we now put (8.114) into (8.112) we get the dynamical equation for the Newtonian fluid,

$$\sigma^{ij} = -pg^{ij} + \mu e_I g^{ij} + 2\eta e^{ij}, \tag{8.115}$$

If we put $\mu = \eta_V - \frac{2}{3}\eta$, where η_V is called the *volume viscosity* coefficient, while η is called the *shear viscosity* coefficient, we can rewrite (8.115) as

$$\sigma^{ij} = -pg^{ij} + \left(\eta_V - \frac{2}{3}\eta\right) e_I g^{ij} + 2\eta e^{ij}. \tag{8.116}$$

Navier–Stokes Equations

From the constitutive equation of the Newtonian viscous fluid (8.116), by taking the divergence, we get

$$\sigma^{ij}_{;j} = -p_{;j}g^{ij} + \left(\eta_V - \frac{2}{3}\eta\right) e_{I;j}g^{ij} + 2\eta e^{ij}_{;j}.$$

However, as $e_{I;j} = u^k_{;kj}$ as well as

$$e^{ij}_{;j} = \frac{1}{2}(u^{i;j} + u^{j;i})_{;j} = \frac{1}{2}(u^{i;j}_j + u^{j;i}_j) = \frac{1}{2}\Delta u^i + \frac{1}{2}u^{ki}_{;k}$$

we get

$$\sigma^{ij}_{;j} = -p_{;j}g^{ij} + \left(\eta_V - \frac{2}{3}\eta\right) u^k_{;kj}g^{ij} + \eta\Delta u^i + \eta u^k_{;kj}g^{ij},$$

or $\quad \sigma^{ij}_{;j} = -p_{;j}g^{ij} + \left(\eta_V - \frac{1}{3}\eta\right) u^k_{;kj}g^{ij} + \eta\Delta u^i.$

If we now substitute this expression into (8.111) we get

$$\rho\dot{u}^i = \rho f^i - p_{;j}g^{ij} + \left(\eta_V - \frac{1}{3}\eta\right) u^k_{;kj}g^{ij} + \eta\Delta u^i, \qquad (8.117)$$

that is a system of 3 scalar PDEs called the *Navier–Stokes equations*, which in vector form read

$$\rho\dot{\mathbf{u}} = \rho\mathbf{f} - \operatorname{grad} p + \left(\eta_V - \frac{1}{3}\eta\right) \operatorname{grad}(\operatorname{div}\mathbf{u}) + \eta\Delta\mathbf{u}. \qquad (8.118)$$

In particular, for incompressible fluids, $\operatorname{div}\mathbf{u} = 0$, we have

$$\dot{\mathbf{u}} = \mathbf{f} - \frac{1}{\rho}\operatorname{grad} p + \nu\Delta\mathbf{u}, \qquad \text{where} \qquad \nu = \frac{\eta}{\rho} \qquad (8.119)$$

is the coefficient of kinematic viscosity.

8.4.6 Computational Fluid Dynamics

It is possible to numerically solve the Navier–Stokes equations (8.118–8.119) for laminar flow cases and for turbulent flows when all of the relevant length scales can be contained on the grid (a direct numerical simulation).[7] In general however, the range of length scales appropriate to the problem is larger than even today's massively parallel computers can model. In these cases, turbulent flow simulations require the introduction of a turbulence model. The so–called

---

[7] Direct numerical simulation (DNS) captures all of the relevant scales of turbulent motion, so no model is needed for the smallest scales. This approach is extremely expensive, if not intractable, for complex problems on modern computing machines, hence the need for models to represent the smallest scales of fluid motion.

Reynolds–averaged Navier–Stokes equations,[8] and *large eddy simulations*,[9] are two techniques for dealing with these scales.

In many instances, other equations (mostly convective-diffusion equations) are solved simultaneously with the Navier–Stokes equations. These other equations can include those describing species concentration, chemical reactions, heat transfer, etc. More advanced codes allow the simulation of more complex cases involving multi-phase flows (e.g., liquid/gas, solid/gas, liquid/solid) or non–Newtonian fluids, such as blood.

In all of these approaches the same basic procedure is followed: (i) the geometry of the problem is defined, (ii) the volume occupied by the fluid is divided into the mesh of discrete cells, (iii) the physical modelling is defined (e.g., the equations of motions + enthalpy + species conservation), (iv) boundary conditions are defined (this involves specifying the fluid behavior and properties at the boundaries of the problem; for transient problems, the initial conditions are also defined), (v) the equations are solved iteratively as either steady–state or transient, (vi) the resulting numerical solution is visualized and further analyzed using 3D computer graphics methods.

Common discretization methods currently in use are:

(i) *Finite volume method* is the standard approach used most often in commercial software and research codes, in which the governing equations are solved on discrete control volumes. This integral approach yields a method that is inherently conservative (i.e., quantities such as density remain physically meaningful),

---

[8] Reynolds–averaged Navier–Stokes equations (RANS) is the oldest approach to turbulence modelling. In this method, an ensemble version of the governing equations is solved, which introduces new apparent stresses known as Reynolds stress. This adds a second–order tensor of unknown variables for which various models can give different levels of closure. It is a common misconception that the RANS equations do not apply to flows with a time-varying mean flow because these equations are 'time–averaged'. In fact, non–stationary flows can equally be treated. This is sometimes referred to as URANS. There is nothing inherent in Reynolds averaging to preclude this, but the turbulence models used to close the equations are valid only as long as the time over which these changes in the mean occur is large compared to the time scales of the turbulent motion containing most of the energy.

[9] Large eddy simulations (LES) is a technique in which the smaller eddies are filtered and modelled using a sub–grid scale model, while the larger energy carrying eddies are simulated. This method generally requires a more refined mesh than a RANS model, but a far coarser mesh than a DNS solution. The so–called detached eddy simulations (DES) is a modification of RANS, in which the model switches to a subgrid scale formulation in regions fine enough for LES calculations. Regions near solid boundaries and where the turbulent length scale is less than the maximum grid dimension are assigned the RANS mode of solution. As the turbulent length scale exceeds the grid dimension, the regions are solved using the LES mode, so the grid resolution is not as demanding as pure LES.

$$\frac{\partial}{\partial t} \iiint Q dV + \iint F d\mathbf{A} = 0,$$

where Q is the vector of conserved variables, and F is the vector of fluxes.

(ii) *Finite element method* (FEM) is popular for structural analysis of solids, but is also applicable to fluids. The FEM formulation requires, however, special care to ensure a conservative solution.

(iii) *Finite difference method* has historical importance and is simple to program.

(iv) *Boundary element method*, in which the boundary occupied by the fluid is divided into surface mesh.

8.5 Turbulence Kinetics

The application of a filtering procedure to the Navier–Stokes equations (NSE) of hydrodynamics in order to construct a subgrid model is often used for the turbulence modeling [Pop00]. The aim of such models is to take into account the effects of subgrid scales as an extra stress term in the hydrodynamic equations for the resolved scale fields. Further, the subgrid scale terms should be representable in terms of the resolved fields. This procedure, like any other attempt to coarse-grain the NSE, runs into the closure problem due to the nonlinearity of the equation and due to the absence of scale separation. On the other hand, in statistical physics, good schemes to get closure approximations are known for nonlinear evolution equations (with a well–defined separation of scales). Unfortunately, attempts to borrow such schemes fail for the NSE. The fundamental reason for this failure of the coarse–graining procedures on the NSEs is the absence of scale separation. Further, the length over which the equation is coarse-grained (the filter width in the present case) is completely arbitrary (and in practice dictated by the available computational resources), and cannot be justified a priori on physical grounds.

Recently it has been shown that a coarse–grained description of hydrodynamics using the microscopic theories is possible (see [AKO03, AKS04]. Specifically, the authors have applied the standard filtering procedure (isotropic Gaussian filter) not on the NSEs but on the Boltzmann kinetic equation. Recall that the NSEs are a well defined limit of the Boltzmann equation (the hydrodynamic limit), whereas *the filtering operation and going to the hydrodynamic limit are two distinct operations which do not commute*, because kinetic fluctuations generally do not annihilate upon filtering. By doing so we get the following results [AKS04]:

- *Smallness parameter:* The smallness parameter of the present theory is the usual kinetic–theory Knudsen number Kn,

$$\mathrm{Kn} = \frac{\nu}{L c_{\mathrm{s}}} \sim \frac{\mathrm{Ma}}{\mathrm{Re}}, \tag{8.120}$$

where Ma is the Mach number and Re is the Reynold's number, ν is the kinematic viscosity, c_s is the sound speed and L is the characteristic macroscopic length. Smallness of Kn rules emergence of *both*, the usual viscosity terms, and the subgrid contributions, on the viscosity time scale of the filtered Boltzmann equation (that is, in the first–order *Chapman–Enskog solution* to the filtered Boltzmann equation).

- *Scaling:* In the coarse–grained representation obtained by filtering, the filter–width Δ (for the Gaussian filter, Δ^2 is proportional to the covariance) is the smallest length–scale to be resolved. The requirement that contributions from the subgrid scales appear in the kinetic picture at the time scale of molecular relaxation time (viscosity time scale) sets the scaling of Δ with the Knudsen number as follows:

$$\Delta = kL\sqrt{\mathrm{Kn}}, \qquad (8.121)$$

where k is a nonuniversal constant which scales neither with L, nor with Kn. For the sake of simplicity, we set $k = 1$ in all the further computations. Equations (8.120) and (8.121) imply that the filter–width scales with the Reynold's number as follows:

$$\Delta \sim \mathrm{Re}^{-1/2}. \qquad (8.122)$$

While the Kolmogorov length, l_K, scales as $l_\mathrm{K} \sim \mathrm{Re}^{-3/4}$, we have

$$\frac{\Delta}{l_\mathrm{K}} \sim \mathrm{Re}^{1/4}. \qquad (8.123)$$

Thus, the filtering scale is larger than the Kolmogorov scale when Re is large enough.

- *Subgrid model:* With the above smallness parameter (8.120), and the scaling (8.121), we rigorously derive the following subgrid *pressure tensor* $P_{\alpha\beta}^{\mathrm{SG}}$, in addition to the usual (advection and viscosity) terms in the momentum equation:

$$P_{\alpha\beta}^{\mathrm{SG}} = \frac{\mathrm{Kn}L^2\overline{\rho}}{12} \left[\overline{S}_{\alpha\gamma} - \overline{\Omega}_{\alpha\gamma}\right]\left[\overline{S}_{\gamma\beta} + \overline{\Omega}_{\gamma\beta}\right] \qquad (8.124)$$

$$= \frac{\nu\overline{\rho}\,c_s\,L}{12}\left[\overline{S}_{\alpha\gamma} - \overline{\Omega}_{\alpha\gamma}\right]\left[\overline{S}_{\gamma\beta} + \overline{\Omega}_{\gamma\beta}\right].$$

Here $\overline{\rho}$ is the filtered density, and summation convention in spatial components is adopted. For any function X, $\overline{X}$ denotes the filtered value of X. Furthermore, the filtered rate of the strain tensor $\overline{S}_{\alpha\beta}$ and the filtered rate of the rotation tensor $\overline{\Omega}_{\alpha\beta}$ depends only the large scale velocity, $\overline{u_\alpha}$:

$$\overline{\Omega}_{\alpha\beta} = \frac{1}{2}\left\{\partial_\alpha\overline{u_\beta} - \partial_\beta\overline{u_\alpha}\right\}, \qquad \overline{S}_{\alpha\beta} = \frac{1}{2}\left\{\partial_\alpha\overline{u_\beta} + \partial_\beta\overline{u_\alpha}\right\}.$$

The derived subgrid model belongs to the class of Smagorinsky models [Sma63], and the tensorial structure of the subgrid pressure tensor (8.124) corresponds to the so–called tensor–diffusivity subgrid model (TDSG) introduced by [Leo74], and which became popular after the work of [CFR79]. Here, it is interesting to recall that in the class of existing Smagorinsky models the TDSG is one of only a few models in which the sub–grid scale stress tensor remains frame–invariant under arbitrary time–dependent rotations of reference frame [Shi99]. Furthermore, the TDSG model belongs to a subclass of Smagorinsky models which take into account the back–scattering of energy from the small scale to the large scales [Pop00]. Beginning with the seminal work of Kraichnan [Kra76], importance of the back–scattering of energy in turbulence modeling is commonly recognized.

- *Uniqueness:* The result (8.124) requires only isotropy of the filter but otherwise is independent of the particular functional form of the filter. There are no other subgrid models different from (8.124) which can be derived from kinetic theory by the one–step filtering procedure. In other words, higher-order spatial derivatives are not neglected in an uncontrolled fashion, rather, they are of the order Kn^2, and thus do not show up on the viscosity time scale.

- *Nonarbitrary filter–width:* Unlike the phenomenological TDSGM where the prefactor in (8.124) remains an unspecified 'Δ^2', kinetic theory suggests that the filter–width cannot be set at will, rather, it must respect the physical parameterization (specific values of Re, Kn etc) of a given setup when the subgrid model is used for numerical simulation. Recent findings that simulations of the TDSG model become unstable for large Δ [KS03] is in qualitative agreement with the present result that the filter–width Δ cannot be made arbitrary large.

8.5.1 Kinetic Theory

Following [AKO03, AKS04], here we will get a explicit expression for the coarse-grained hydrodynamics equation, where subgrid scales of motions are represented as a function of resolved scales of motion.

For the present discussion, the particular choice of the kinetic model is unimportant as long as the hydrodynamic limit of the kinetic theory is the usual NSE at least up to the order $O(\text{Ma}^3)$. We demonstrate the whole procedure in detail for a recently introduced minimal discrete–velocity kinetic model [KF99, AKO03]. As the final result is just the same in two and three dimensional cases, for the sake of simplicity we chose to demonstrate the whole procedure using the 2D model ($D = 2$). The kinetic equation is

$$\partial_t f_i + C_{i\alpha} \partial_\alpha f_i = -\tau^{-1} \left(f_i - f_i^{\text{eq}} \right), \qquad (8.125)$$

where $f_i(\mathbf{x}, t)$, $i = 1, \ldots 9$ are populations of discrete velocities $\mathbf{C}_i$:

$$C_x = \{0, 1, 0, -1, 0, 1, -1, -1, 1\}, \qquad C_y = \{0, 0, 1, 0, -1, 1, 1, -1, -1\}.$$

The local equilibrium f_i^{eq} is the conditional minimizer of the the entropy function H:

$$H = f_i \ln\left(\frac{f_i}{W_i}\right), \tag{8.126}$$

under the constraint of local conservation laws:

$$f_i^{\text{eq}}\{1, \mathbf{C}_i\}. = \{\rho, \rho\mathbf{u}\}.$$

The weights W_i in the equation (8.126) are:

$$W = \frac{1}{36}\{16, 4, 4, 4, 4, 1, 1, 1, 1\}.$$

The explicit expression for f_i^{eq} reads:

$$f_i^{\text{eq}} = \rho W_i \prod_{\alpha=1}^{D}\left(2 - \sqrt{1 + 3u_\alpha{}^2}\right)\left(\frac{2u_\alpha + \sqrt{1 + 3u_\alpha{}^2}}{1 - u_\alpha}\right)^{C_{i\alpha}},$$

where j is the index for spatial directions.

Below, it will prove convenient to work in the moment representation rather than in the population representation.

We shall move to the moment representation, where the operation of coarse-graining or filtering has the effect of changing the relaxation time for the individual moments. Let us choose the following orthogonal set of basis vectors in the 9D phase–space of the kinetic equation (8.125):

$$\psi_1 = \{1, 1, 1, 1, 1, 1, 1, 1, 1\}, \qquad \psi_2 = \{0, 1, 0, -1, 0, 1, -1, -1, 1\},$$
$$\psi_3 = \{0, 0, 1, 0, -1, 1, 1, -1, -1\}, \qquad \psi_4 = \{0, 0, 0, 0, 0, 1, -1, 1, -1\},$$
$$\psi_5 = \{0, 1, -1, 1, -1, 0, 0, 0, 0\}, \qquad \psi_6 = \{0, -2, 0, 2, 0, 1, -1, -1, 1\},$$
$$\psi_7 = \{0, 0, -2, 0, 2, 1, 1, -1, -1\}, \qquad \psi_8 = \{4, -5, -5, -5, -5, 4, 4, 4, 4\},$$
$$\psi_9 = \{4, 0, 0, 0, 0, -1, -1, -1, -1\}.$$

The orthogonality of the chosen basis is in the sense of the usual Euclidean scalar product, i.e.,

$$\psi_{ik}\psi_{kj} = d_i\delta_{ij},$$

where d_i are some constants needed for the normalization (the basis vectors are orthogonal but not orthonormal). We define new variables M_i, $i = 1, \ldots, 9$ as:

$$M_i = \psi_{ij}f_j,$$

where ψ_{ij} denotes jth component of the 9D vector ψ_i. Basic hydrodynamic fields are $M_1 = \rho$, $M_2 = \rho u_x$, and $M_3 = \rho u_y$. The remaining six moments are related to higher order moments of the distribution (the pressure tensor $P_{\alpha\beta} = f_i C_{i\alpha}C_{i\beta}$ and the third order moment $Q_{\alpha\beta\gamma} = f_i C_{i\alpha}C_{i\beta}C_{i\gamma}$ and so on), as

$$M_4 = P_{xy}, \quad M_5 = P_{xx} - P_{yy}, \quad M_6 = 3f_i C_{iy}^2 C_{ix} - 2M_2,$$
$$M_7 = 3f_i C_{ix}^2 C_{iy} - 2M_3.$$

The explicit form of the stress tensor in term of the new set of variables is:

$$P_{xy} = M_4, \qquad P_{xx} = \frac{2}{3}M_1 + \frac{1}{2}M_5 + \frac{1}{30}M_8 - \frac{1}{5}M_9,$$

$$P_{yy} = \frac{2}{3}M_1 - \frac{1}{2}M_5 + \frac{1}{30}M_8 - \frac{1}{5}M_9.$$

The time evolution equations for the set of moments are:

$$\partial_t M_1 + \partial_x M_2 + \partial_y M_3 = 0, \qquad (8.127)$$

$$\partial_t M_2 + \partial_x \left(\frac{2}{3}M_1 + \frac{1}{2}M_5 + \frac{1}{30}M_8 - \frac{1}{5}M_9 \right) + \partial_y M_4 = 0,$$

$$\partial_t M_3 + \partial_x M_4 + \partial_y \left(\frac{2}{3}M_1 - \frac{1}{2}M_5 + \frac{1}{30}M_8 - \frac{1}{5}M_9 \right) = 0$$

$$\partial_t M_4 + \frac{1}{3}\partial_x (2M_3 + M_7) + \frac{1}{3}\partial_y (2M_2 + M_6) = \frac{1}{\tau} \left(M_4^{\mathrm{eq}}(M_1, M_2, M_3) - M_4 \right),$$

$$\partial_t M_5 + \frac{1}{3}\partial_x (M_2 - M_6) + \frac{1}{3}\partial_y (M_7 - M_3) = \frac{1}{\tau} \left(M_5^{\mathrm{eq}}(M_1, M_2, M_3) - M_5 \right),$$

$$\partial_t M_6 - \frac{1}{5}\partial_x (5M_5 - M_8 + M_9) + \partial_y M_4 = \frac{1}{\tau} \left(M_6^{\mathrm{eq}}(M_1, M_2, M_3) - M_6 \right),$$

$$\partial_t M_7 + \partial_x M_4 + \frac{1}{5}\partial_y (5M_5 + M_8 - M_9) = \frac{1}{\tau} \left(M_7^{\mathrm{eq}}(M_1, M_2, M_3) - M_7 \right),$$

$$\partial_t M_8 + \partial_x (M_2 + 3M_6) + \partial_y (M_3 + 3M_7) = \frac{1}{\tau} \left(M_8^{\mathrm{eq}}(M_1, M_2, M_3) - M_8 \right),$$

$$\partial_t M_9 - \frac{1}{3}\partial_x (2M_2 + M_6) - \frac{1}{3}\partial_y (2M_3 + M_7) = \frac{1}{\tau} \left(M_9^{\mathrm{eq}}(M_1, M_2, M_3) - M_9 \right).$$

The expression for the local equilibrium moments M_i^{eq}, $i = 4, \ldots, 9$ in terms of the basic variables M_1, M_2, and M_3 to the order u^2 is:

$$M_4^{\mathrm{eq}}(M_1, M_2, M_3) = \frac{M_2 M_3}{M_1},$$

$$M_5^{\mathrm{eq}}(M_1, M_2, M_3) = \frac{M_2^2 - M_3^2}{M_1},$$

$$M_6^{\mathrm{eq}}(M_1, M_2, M_3) = -M_2,$$

$$M_7^{\mathrm{eq}}(M_1, M_2, M_3) = -M_3, \qquad (8.128)$$

$$M_8^{\mathrm{eq}}(M_1, M_2, M_3) = -3\frac{M_2^2 + M_3^2}{M_1},$$

$$M_9^{\mathrm{eq}}(M_1, M_2, M_3) = \frac{5}{3}M_1 - \frac{3\left(M_2^2 + M_3^2\right)}{M_1}.$$

The incompressible NSEs are the hydrodynamic limit of the system (8.127)–(8.128).

In the next section, we shall remove small scales through a filtering procedure on the moment system (8.127)–(8.128). A precise definition of the small scales is postponed until later sections. For the time being, let us assume that there exist a length-scale Δ, and we wish to look at the hydrodynamics at length-scale larger than Δ only.

8.5.2 Filtered Kinetic Theory

Coarse-grained versions of the Boltzmann equations have been discussed in the recent literature [CSO99, SKC00, SFC02]. However, a systematic treatment is still lacking. In this section, we shall fill this gap.

For any function X, the filtered function $\overline{X}$ is defined as [AKO03, AKS04]

$$\overline{X}(\mathbf{x}) = \int_{R^D} G(\mathbf{r}) X(\mathbf{x} - \mathbf{r}) d\mathbf{r}. \tag{8.129}$$

Function G is called the filter. In the sequel, we apply the filtering operation (8.129) on the moment system (8.127). We will need two relations. First, for any function X, we have

$$\overline{\partial_\alpha X} = \partial_\alpha \overline{X}. \tag{8.130}$$

This relation is sufficient to filter the propagation terms in the equation (8.127) due to linearity of propagation in the kinetic picture. The latter is a useful property which is *not* shared by the hydrodynamic NSEs, where the nonlinearity and non–locality both come into the same $(\boldsymbol{u}\nabla\boldsymbol{u})$ term. Any isotropic filter, which satisfies the condition of commuting of the derivatives under the application of the filter (8.130), will suffice for the present purpose. We choose a standard Gaussian filter [Pop00] which has the property (8.130):

$$G(\mathbf{r}, \Delta) = \left(\frac{6}{\pi\Delta^2}\right)^D \exp\left(-\frac{6\mathbf{r}^2}{\Delta^2}\right).$$

Let us recall the isotropy properties of a Gaussian filter

$$\int_{R^D} G(\mathbf{r}, \Delta) d\mathbf{r} = 1, \qquad \int_{R^D} G(\mathbf{r}, \Delta)\mathbf{r} d\mathbf{r} = 0,$$

$$\int_{R^D} G(\mathbf{r}, \Delta) r_\alpha r_\beta d\mathbf{r} = \frac{\Delta^2}{12}\delta_{\alpha\beta}.$$

Second, in order to filter the nonlinear terms (8.128) in the right hand side of moment equations (8.127), we will also need the following relation for three arbitrary functions X, Y, Z which follow immediately from the isotropy property by second–order Taylor expansion:

$$\overline{\left(\frac{XY}{Z}\right)} = \left(\frac{\overline{X}\,\overline{Y}}{\overline{Z}}\right) + \frac{\Delta^2}{12\overline{Z}}\{(\partial_\alpha\overline{X})(\partial_\alpha\overline{Y})$$
$$- \frac{2}{\overline{Z}}(\partial_\alpha\overline{Z})(\overline{X}\partial_\alpha\overline{Y} + \overline{Y}\partial_\alpha\overline{X} + \frac{2\overline{XY}}{\overline{Z}}\partial_\alpha\overline{Z})\} + O(\Delta^4). \tag{8.131}$$

The effect of a Gaussian filter need not be truncated to any order at the present step. The higher–order terms lumped under $O(\Delta^4)$ in equation (8.131) can be computed from elementary Gaussian integrals. As we shall see it soon, higher than second order terms disappear in the hydrodynamic limit once the scaling of the filter–width versus Knudsen number is appropriately chosen.

Now, applying the filter (8.129) to the moment system (8.127)–(8.128), using (8.130) and (8.131), and keeping terms up to the order u^2, we get the following filtered moment system:

$$\partial_t \overline{M}_1 + \partial_x \overline{M}_2 - \partial_y \overline{M}_3 = 0,$$

$$\partial_t \overline{M}_2 + \partial_x \left(\frac{2}{3}\overline{M}_1 + \frac{1}{2}\overline{M}_5 + \frac{1}{30}\overline{M}_8 - \frac{1}{5}\overline{M}_9 \right) + \partial_y \overline{M}_4 = 0,$$

$$\partial_t \overline{M}_3 + \partial_x \overline{M}_4 + \partial_y \left(\frac{2}{3}\overline{M}_1 - \frac{1}{2}\overline{M}_5 + \frac{1}{30}\overline{M}_8 - \frac{1}{5}\overline{M}_9 \right) = 0,$$

$$\partial_t \overline{M}_4 + \frac{1}{3}\partial_x \left(2\overline{M}_3 + \overline{M}_7 \right) + \frac{1}{3}\partial_y \left(2\overline{M}_2 + \overline{M}_6 \right) = \qquad (8.132)$$

$$\frac{1}{\tau} \left(M_4^{\mathrm{eq}}(\overline{M}_1, \overline{M}_2, \overline{M}_3) - \overline{M}_4 \right) + \frac{\Delta^2}{12\,\tau\,\overline{M}_1}(\partial_\alpha \overline{M}_2)(\partial_\alpha \overline{M}_3) + O\left(\frac{\Delta^4}{\tau} \right),$$

$$\partial_t \overline{M}_5 + \frac{1}{3}\partial_x \left(\overline{M}_2 - \overline{M}_6 \right) + \frac{1}{3}\partial_y \left(\overline{M}_7 - \overline{M}_3 \right) = \frac{1}{\tau} \left(M_5^{\mathrm{eq}}(\overline{M}_1, \overline{M}_2, \overline{M}_3) - \overline{M}_5 \right)$$

$$+ \frac{\Delta^2}{12\,\tau\,\overline{M}_1}\left\{ (\partial_\alpha \overline{M}_2)(\partial_\alpha \overline{M}_2) - (\partial_\alpha \overline{M}_3)(\partial_\alpha \overline{M}_3) \right\} + O\left(\frac{\Delta^4}{\tau} \right),$$

$$\partial_t \overline{M}_6 - \frac{1}{5}\partial_x \left(5\overline{M}_5 - \overline{M}_8 + \overline{M}_9 \right) + \partial_y \overline{M}_4 = \frac{1}{\tau} \left(M_6^{\mathrm{eq}}(\overline{M}_1, \overline{M}_2, \overline{M}_3) - \overline{M}_6 \right)$$

$$\partial_t \overline{M}_7 + \partial_x \overline{M}_4 + \frac{1}{5}\partial_y \left(5\overline{M}_5 + \overline{M}_8 - \overline{M}_9 \right) = \frac{1}{\tau} \left(M_7^{\mathrm{eq}}(\overline{M}_1, \overline{M}_2, \overline{M}_3) - \overline{M}_7 \right),$$

$$\partial_t \overline{M}_8 + \partial_x \left(\overline{M}_2 + 3\overline{M}_6 \right) + \partial_y \left(\overline{M}_3 + 3\overline{M}_7 \right) = \frac{1}{\tau} \left(M_8^{\mathrm{eq}}(\overline{M}_1, \overline{M}_2, \overline{M}_3) - \overline{M}_8 \right)$$

$$- \frac{3\Delta^2}{12\,\tau\,\overline{M}_1}\left\{ (\partial_\alpha \overline{M}_2)(\partial_\alpha \overline{M}_2) + (\partial_\alpha \overline{M}_3)(\partial_\alpha \overline{M}_3) \right\} + O\left(\frac{\Delta^4}{\tau} \right),$$

$$\partial_t \overline{M}_9 - \frac{1}{3}\partial_x \left(2\overline{M}_2 + \overline{M}_6 \right) - \frac{1}{3}\partial_y \left(2\overline{M}_3 + \overline{M}_7 \right) = \frac{1}{\tau} \left(M_9^{\mathrm{eq}}(\overline{M}_1, \overline{M}_2, \overline{M}_3) - \overline{M}_9 \right)$$

$$- \frac{3\Delta^2}{12\,\tau\,\overline{M}_1}\left\{ (\partial_\alpha \overline{M}_2)(\partial_\alpha \overline{M}_2) + (\partial_\alpha \overline{M}_3)(\partial_\alpha \overline{M}_3) \right\} + O\left(\frac{\Delta^4}{\tau} \right).$$

Thus, we are set up to derive the hydrodynamic equations as the appropriate limit of the filtered kinetic system (8.132). In passing, we note that different moments relax with different effective relaxation time scales, because the subgrid terms are not the same for all kinetic moments.

8.5.3 Hydrodynamic Limit

In the kinetic equation we have a natural length scale set by Knudsen number
Kn (8.120). The Navier–Stokes dynamics is obtained in the limit Kn $\ll$ 1.
By filtering the kinetic equation we have introduced a new length scale as
the size of the filter Δ. The hydrodynamic equations produced by the filtered
kinetic equation will depend on how Δ scales with the Knudsen number. In
order to understand this issue, let us look at the filtered equation for one
of the moments (8.132) in the non-dimensional form. In order to do this,
let us introduce scaled time and space variables $\mathbf{x}' = \frac{\mathbf{x}}{L}$, $t' = \frac{tc_s}{L}$, where
$c_s = 1/\sqrt{3}$ for the present model. Let us also specify Knudsen number in terms
of the relaxation time τ, $\mathrm{Kn} = \frac{\nu}{Lc_s} \equiv \frac{\tau c_s}{L}$, where, ν is the kinematic viscosity,
$\nu = \tau c_s^2$ in the present model. Then, for example, the filtered equation for the
non-dimensional moment M_4 reads [AKO03, AKS04]

$$\partial_{t'}\overline{M}_4 + \frac{1}{3\,c_s}\partial_{x'}\left(2\overline{M}_3 + \overline{M}_7\right) + \frac{1}{3\,c_s}\partial_{y'}\left(2\overline{M}_2 + \overline{M}_6\right)$$

$$= \frac{1}{\mathrm{Kn}}\left(M_4^{\mathrm{eq}}(\overline{M}_1, \overline{M}_2, \overline{M}_3) - \overline{M}_4\right) \qquad (8.133)$$

$$+ \frac{\Delta^2}{12\,\mathrm{Kn}L^2\overline{M}_1}\left\{(\partial_{x'}\overline{M}_2)(\partial_{x'}\overline{M}_3) + (\partial_{y'}\overline{M}_2)(\partial_{y'}\overline{M}_3)\right\} + O\left(\frac{\Delta^4}{L^4\mathrm{Kn}}\right).$$

We see that the in absence of the filter ($\Delta = 0$), the usual situation of
a singularly perturbed kinetic equation is recovered (and this results in the
NSEs in the first–order Chapman–Enskog expansion). Let us consider the
following three possibilities of dependence of Δ on Kn:

- If $\Delta/L \sim \mathrm{Kn}^0$, then we do not have a singularly perturbed equation in
  (8.133) anymore. That is, the filter is too wide, and it affects the advection
  terms in the hydrodynamic equations.
- If $\Delta/L \sim \mathrm{Kn}$, then we do have a singularly perturbed system. However,
  the subgrid terms are of order Kn^2, and they do not show up in the order
  Kn hydrodynamic equation. In other words, the filter is too narrow so that
  it does not affect hydrodynamic equations at the viscous time scale.
- Finally, there is only one possibility to set the scaling of filter–width with
  Kn so that the system is singularly perturbed, and the subgrid terms of the
  order Δ^2 contribute just at the viscous time scale. This situation happens
  if

$$\frac{\Delta}{L} \sim \sqrt{\mathrm{Kn}}. \qquad (8.134)$$

Note that, with the scaling (8.134), all the higher-order terms (of the order
Δ^4 and higher) become of the order Kn and higher, so that they *do not
contribute at the viscous time scale.*

Once the scaling of the filter–width (8.134) is introduced into the filtered moment equations (8.132), the application of the *Chapman–Enskog method* [CC70] becomes a routine. We write:

$$\partial_t = \partial_t^{(0)} + \mathrm{Kn}\,\partial_t^{(1)} + O(\mathrm{Kn}^2), \qquad (8.135)$$

and for $i = 4, \ldots, 9$:

$$\overline{M}_i = M_i^{\mathrm{eq}}(\overline{M}_1, \overline{M}_2, \overline{M}_3) + \mathrm{Kn}\overline{M}_i^{(1)} + O(\mathrm{Kn}^2). \qquad (8.136)$$

Thus, the moment equations at the order $(1/\epsilon^2)$ are: $\overline{M}_i^{(0)} = \overline{M}_i^{\mathrm{eq}}$.

The hydrodynamics equations at the the order $O(1)$ are the Euler equations:

$$\partial_t^{(0)}\overline{M}_1 = -\partial_x \overline{M}_2 - \partial_y \overline{M}_3,$$

$$\partial_t^{(0)}\overline{M}_2 = -\partial_x \left(\overline{M}_1 c_s^2 + \frac{\overline{M}_2\,\overline{M}_2}{\overline{M}_1} \right) - \partial_y \left(\frac{\overline{M}_2\,\overline{M}_3}{\overline{M}_1} \right), \qquad (8.137)$$

$$\partial_t^{(0)}\overline{M}_3 = -\partial_x \left(\frac{\overline{M}_2\,\overline{M}_3}{\overline{M}_1} \right) - \partial_y \left(\overline{M}_1 c_s^2 + \frac{\overline{M}_3\,\overline{M}_3}{\overline{M}_1} \right).$$

Note that no subgrid terms appear at this time scale in the hydrodynamic equations (8.137). This means that large scale motion, even after filtering, is dictated just by the conservation laws. Zero–order time–derivatives of the non–conserved moments are evaluated using the chain rule:

$$\partial_t^{(0)} M_i^{\mathrm{eq}}(\overline{M}_1, \overline{M}_2, \overline{M}_3) = \frac{\partial M_i^{\mathrm{eq}}}{\partial \overline{M}_1}\partial_t^{(0)}\overline{M}_1 + \frac{\partial M_i^{\mathrm{eq}}}{\partial \overline{M}_2}\partial_t^{(0)}\overline{M}_2 + \frac{\partial M_i^{\mathrm{eq}}}{\partial \overline{M}_3}\partial_t^{(0)}\overline{M}_3. \qquad (8.138)$$

In particular, to the order u^2, we have [AKO03, AKS04]

$$\partial_t^{(0)} M_4^{\mathrm{eq}}(\overline{M}_1, \overline{M}_2, \overline{M}_3) = 0, \qquad \partial_t^{(0)} M_5^{\mathrm{eq}}(\overline{M}_1, \overline{M}_2, \overline{M}_3) = 0,$$

$$\partial_t^{(0)} M_6^{\mathrm{eq}}(\overline{M}_1, \overline{M}_2, \overline{M}_3) = \partial_x \left(\overline{M}_1 c_s^2 + \frac{\overline{M}_2\,\overline{M}_2}{\overline{M}_1} \right) + \partial_y \left(\frac{\overline{M}_2\,\overline{M}_3}{\overline{M}_1} \right),$$

$$\partial_t^{(0)} M_7^{\mathrm{eq}}(\overline{M}_1, \overline{M}_2, \overline{M}_3) = \partial_x \left(\frac{\overline{M}_2\,\overline{M}_3}{\overline{M}_1} \right) + \partial_y \left(\overline{M}_1 c_s^2 + \frac{\overline{M}_3\,\overline{M}_3}{\overline{M}_1} \right), \qquad (8.139)$$

$$\partial_t^{(0)} M_8^{\mathrm{eq}}(\overline{M}_1, \overline{M}_2, \overline{M}_3) = 0,$$

$$\partial_t^{(0)} M_9^{\mathrm{eq}}(\overline{M}_1, \overline{M}_2, \overline{M}_3) = -\frac{5}{3}\left[\partial_x \overline{M}_2 + \partial_y \overline{M}_3 \right].$$

At the next order $O(\mathrm{Kn})$, correction to locally conserved moments is equal to zero,

$$\overline{M}_1^{(1)} = \overline{M}_2^{(1)} = \overline{M}_3^{(1)} = 0,$$

whereas corrections to the non–conserved moments, $\overline{M}_i^{(1)}$, $i = 4, \ldots, 9$, are obtained by substituting (8.135), and (8.136) in (8.132) and eliminating the zeroth order time derivatives using (8.139):

$$\overline{M}_4^{(1)} = -Lc_s\left[\partial_x\overline{M}_3 + \partial_y\overline{M}_2\right] + \frac{L^2}{12\,\overline{M}_1}\{(\partial_x\overline{M}_2)\,(\partial_x\overline{M}_3) + (\partial_y\overline{M}_2)(\partial_y\overline{M}_3)\},$$

$$\overline{M}_5^{(1)} = -2\,Lc_s\left[\partial_x\overline{M}_2 - \partial_y\overline{M}_3\right] + \tag{8.140}$$

$$\frac{L^2}{12\,\overline{M}_1}\left\{\left(\partial_x\overline{M}_2\right)^2 - \left(\partial_x\overline{M}_3\right)^2 + \left(\partial_y\overline{M}_2\right)^2 - \left(\partial_y\overline{M}_3\right)^2\right\},$$

$$\overline{M}_6^{(1)} = -Lc_s\left[\partial_x\left\{\frac{3(\overline{M}_3^2)}{\overline{M}_1}\right\} + \partial_y\frac{6\,\overline{M}_2\,\overline{M}_3}{\overline{M}_1}\right],$$

$$\overline{M}_7^{(1)} = -Lc_s\left[\partial_x\frac{6\,\overline{M}_2\,\overline{M}_3}{\overline{M}_1} + \partial_y\left\{\frac{3(\overline{M}_2^2)}{\overline{M}_1}\right\}\right],$$

$$\overline{M}_8^{(1)} = 6\,Lc_s\left[\partial_x\overline{M}_2 + \partial_y\overline{M}_3\right] -$$

$$\frac{3L^2}{12\,\overline{M}_1}\left\{\left(\partial_x\overline{M}_2\right)^2 + \left(\partial_y\overline{M}_2\right)^2 + \left(\partial_x\overline{M}_3\right)^2 + \left(\partial_y\overline{M}_3\right)^2\right\},$$

$$\overline{M}_9^{(1)} = 6\,Lc_s\left[\partial_x\overline{M}_2 + \partial_y\overline{M}_3\right] -$$

$$\frac{3L^2}{12\,\overline{M}_1}\left\{\left(\partial_x\overline{M}_2\right)^2 + \left(\partial_y\overline{M}_2\right)^2 + \left(\partial_x\overline{M}_3\right)^2 + \left(\partial_y\overline{M}_3\right)^2\right\}.$$

and the first–order time derivative of the conserved moments are:

$$\partial_t^{(1)}\overline{M}_1 = 0, \tag{8.141}$$

$$\partial_t^{(1)}\overline{M}_2 = -\partial_x\left(\frac{1}{2}\overline{M}_5^{(1)} + \frac{1}{30}\overline{M}_8^{(1)} - \frac{1}{5}\overline{M}_9^{(1)}\right) - \partial_y M_4^{(1)},$$

$$\partial_t^{(1)}\overline{M}_3 = -\partial_x M_4^{(1)} - \partial_y\left(-\frac{1}{2}\overline{M}_5^{(1)} + \frac{1}{30}\overline{M}_8^{(1)} - \frac{1}{5}\overline{M}_9^{(1)}\right).$$

These equations shows that the viscous term and the subgrid term *both* appear as the $O(Kn)$ contribution. We remind that no assumption was made about relative magnitude of the subgrid term as compared with the viscous terms. The only requirement that is set on the subgrid scale term is that they appear at the viscous time scale rather than the time scale of the advection. We can write the complete hydrodynamics equation by using (8.135), (8.136), (8.137), (8.140), and (8.141) to get the hydrodynamics equations correct up to the order $O(\mathrm{Kn}^2)$ (at this stage one recovers the NSEs using the unfiltered kinetic equation). In the next section, we shall see how the subgrid scale terms affect the NSE description.

8.5.4 Hydrodynamic Equations

The final set of hydrodynamics equation, valid up to the order $O(\mathrm{Kn}^2)$ is [AKS04]

$$\partial_t \overline{\rho} + \partial_x \overline{\rho u_x} + \partial_y \overline{\rho u_y} = 0$$
$$\partial_t \overline{\rho u_x} + \partial_x \overline{P}_{xx} + \partial_y \overline{P}_{xy} = 0$$
$$\partial_t \overline{\rho u_y} + \partial_x \overline{P}_{xy} + \partial_y \overline{P}_{yy} = 0, \qquad \text{where}$$

$$\overline{P}_{xx} = \left[\overline{P} + \frac{\overline{\rho u_x}^2}{\overline{\rho}} - 2\nu\overline{\rho}\overline{S}_{xx} \right] + P_{xx}^{\mathrm{SG}},$$

$$\overline{P}_{xy} = \left[\frac{\overline{\rho u_x}\,\overline{\rho u_y}}{\overline{\rho}} - 2\nu\overline{\rho}\overline{S}_{xy} \right] + P_{xy}^{\mathrm{SG}},$$

$$\overline{P}_{yy} = \left[\overline{P} + \frac{\overline{\rho u_y}^2}{\overline{\rho}} - 2\nu\overline{\rho}\overline{S}_{yy} \right] + P_{yy}^{\mathrm{SG}},$$

with $\overline{P} = \overline{\rho}c_s^2$, as the thermodynamic pressure and

$$P_{\alpha\beta}^{\mathrm{SG}} = \frac{\mathrm{Kn} L^2 \overline{\rho}}{12} \left[\overline{S}_{\alpha\gamma} - \overline{\Omega}_{\alpha\gamma} \right] \left[\overline{S}_{\gamma\beta} + \overline{\Omega}_{\gamma\beta} \right]$$
$$= \frac{\nu\,\overline{\rho}\,L}{12\,c_s} \left[\overline{S}_{\alpha\gamma} - \overline{\Omega}_{\alpha\gamma} \right] \left[\overline{S}_{\gamma\beta} + \overline{\Omega}_{\gamma\beta} \right].$$

Thus, we have obtained a closed set of hydrodynamics equations, for appropriate choice of filtering width. This set of equations, written in the nondimensional form up to the order, u^2 is

$$\partial_\alpha \overline{u}_\alpha = 0, \qquad\qquad (8.142)$$

$$\partial_t \left(\overline{u}_\alpha \right) + \partial_\beta \left(\overline{u}_\alpha\,\overline{u}_\beta \right) = - \partial_\alpha P + 2\,\mathrm{Kn}\,\partial_\beta \left(\overline{S}_{\alpha\beta} \right)$$
$$- \frac{\mathrm{Kn}}{12} \partial_\beta \left\{ \left(\overline{S}_{\alpha\gamma} - \overline{\Omega}_{\alpha\gamma} \right) \left(\overline{S}_{\gamma\beta} + \overline{\Omega}_{\gamma\beta} \right) \right\}.$$

Note that the pressure appearing in the momentum equation is not the thermodynamic pressure anymore, but needs to be computed from the incompressibility condition (8.142) (see [Maj84]).

8.6 Lie Symmetries in the Models of Turbulence

Recall that with the current performance of computers, a direct simulation of a turbulent flow remains difficult, even impossible in many cases, due to the high computational cost that it requires. To reduce this computational cost, use of turbulence models is necessary. At the present time, many turbulence models exist (see [Sag04]). However, derivation of a very large majority of them does not take into account the symmetry group of the basic equations, the Navier–Stokes equations.

In turbulence, symmetries play a fundamental role in the description of the physics of the flow [RH06]. They reflect existence of conservation laws, via Noether's theorem. Notice that, even if Navier–Stokes equations are not directly derived from a Lagrangian, Noether's theorem can be applied and conservation laws can be deduced. Indeed, there exists a Lagrangian (which will be called 'bi-Lagrangian' here) from which Navier–Stokes equations, associated to their 'adjoint' equations, can be derived. A way in which this bi-Lagrangian can be calculated is described by Atherton and Homsy in [AH75].

The importance of symmetries in turbulence is not limited to the derivation of conservation laws. Ünal also used a symmetry approach to show that Navier–Stokes equations may have solutions which have the Kolmogorov form of the energy spectrum [Una94]. Next, symmetries enabled Oberlack to derive some scaling laws for the velocity and the two point correlations [Obe01]. Some of these scaling laws was reused by Lindgren *et al* in [LOJ04] and are proved to be in good agreement with experimental data. Next, symmetries allowed Fushchych and Popowych to get analytical solutions of Navier–Stokes equations [FP94]. The study of self-similar solutions gives also an information on the behaviour of the flow at a large time [CK05]. Lastly, we mention that use of discretisation schemes which are compatible with the symmetries of an equation reduces the numerical errors [Olv01, KO04].

Introduction of a turbulence model in Navier–Stokes equations may destroy symmetry properties of the equations. In this case, physical properties (conservation laws, scaling laws, spectral properties, large-time behaviour, . . .) may be lost. In order to represent the flow correctly, turbulence models should then preserve the symmetries of Navier–Stokes equations. In [RH06] it is shown that most of the commonly used subgrid turbulence models do not have this property. This paper presents a new way of deriving models which are compatible with the symmetries of Navier–Stokes equations and which, unlike many existing models, conform to the second law of thermodynamics.

8.6.1 Lie Symmetries and Prolongations on Manifolds

In this subsection we continue our expose on Lie groups of symmetry, as a link to modern jet machinery, developed below.

Lie Symmetry Groups

Exponentiation of Vector Fields on M

Let $x = (x^1, ..., x^r)$ be local coordinates at a point m on a smooth $n-$manifold M. Recall that the flow generated by the vector–field

$$v = \xi^i(x) \, \partial_{x^i} \in M,$$

is a solution of the system of ODEs

$$\frac{dx^i}{d\varepsilon} = \xi^i(x^1, ..., x^m), \qquad (i = 1, ..., r).$$

The computation of the flow, or one–parameter group of diffeomorphisms, generated by a given vector–field v (i.e., solving the system of ODEs) is often referred to as *exponentiation of a vector–field*, denoted by $\exp(\varepsilon v)\, x$ (see [Olv86]).

If $v, w \in M$ are two vectors defined by

$$v = \xi^i(x)\, \partial_{x^i} \qquad \text{and} \qquad w = \eta^i(x)\, \partial_{x^i},$$

then

$$\exp(\varepsilon v)\, \exp(\theta w)\, x = \exp(\theta w)\, \exp(\varepsilon v)\, x,$$

for all $\epsilon, \theta \in \mathbb{R}, x \in M$, such that both sides are defined, iff they commute, i.e., $[v, w] = 0$ everywhere [Olv86].

A system of vector–fields $\{v_1, ..., v_r\}$ on a smooth manifold M is in *involution* if there exist smooth real–valued functions $h_{ij}^k(x)$, $x \in M$, $i, j, k = 1, ..., r$, such that for each i, j,

$$[v_i, v_j] = h_{ij}^k \cdot v_k.$$

Let $v \neq 0$ be a right–invariant vector–field on a Lie group G. Then the flow generated by v through the identity e, namely

$$g_\varepsilon = \exp(\varepsilon v)\, e \equiv \exp(\varepsilon v),$$

is defined for all $\varepsilon \in \mathbb{R}$ and forms a one–parameter subgroup of G, with

$$g_{\varepsilon + \delta} = g_\varepsilon \cdot g_\delta, \qquad g_0 = e, \qquad g_\varepsilon^{-1} = g_{-\varepsilon},$$

isomorphic to either $\mathbb{R}$ itself or the circle group $SO(2)$. Conversely, any connected 1D subgroup of G is generated by such a right–invariant vector–field in the above manner [Olv86].

For example, let $G = GL(n)$ with Lie algebra $\mathfrak{gl}(n)$, the space of all $n \times n$ matrices with commutator as the Lie bracket. If $A \in \mathfrak{gl}(n)$, then the corresponding right–invariant vector–field v_A on $GL(n)$ has the expression [Olv86]

$$v_A = a_k^i x_j^k\, \partial_{x_j^i}.$$

The one–parameter subgroup $\exp(\varepsilon v_A)\, e$ is found by integrating the system of n^2 ordinary differential equations

$$\frac{dx_j^i}{d\varepsilon} = a_k^i x_j^k, \qquad x_j^i(0) = \delta_j^i, \qquad (i, j = 1, ..., n),$$

involving matrix entries of A. The solution is just the matrix exponential $X(\varepsilon) = e^{\varepsilon A}$, which is the one–parameter subgroup of $GL(n)$ generated by a matrix A in $\mathfrak{gl}(n)$.

Recall that the *exponential map* $\exp : \mathfrak{g} \to G$ is get by setting $\varepsilon = 1$ in the one–parameter subgroup generated by vector–field v :

$$\exp(v) \equiv \exp(v)\, e.$$

Its differential at 0,

$$d\exp : T\mathfrak{g}|_0 \simeq \mathfrak{g} \to TG|_e \simeq \mathfrak{g}$$

is the identity map.

Lie Symmetry Groups and General Differential Equations

Consider a system $\mathcal{S}$ of general differential equations (DEs, to be distinguished from ODEs) involving p independent variables $x = (x^1, ..., x^p)$, and q dependent variables $u = (u^1, ..., u^q)$. The solution of the system will be of the form $u = f(x)$, or, in components, $u^\alpha = f^\alpha(x^1, ..., x^p)$, $\alpha = 1, ..., q$ (so that Latin indices refer to independent variables while Greek indices refer to dependent variables). Let $X = \mathbb{R}^p$, with coordinates $x = (x^1, ..., x^p)$, be the space representing the independent variables, and let $U = \mathbb{R}^q$, with coordinates $u = (u^1, ..., u^q)$, represent dependent variables. A Lie symmetry group G of the system $\mathcal{S}$ will be a local group of transformations acting on some open subset $M \subset X \times U$ in such way that G transforms solutions of $\mathcal{S}$ to other solutions of $\mathcal{S}$ [Olv86].

More precisely, we need to explain exactly how a given transformation $g \in G$, where G is a Lie group, transforms a function $u = f(x)$. We firstly identify the function $u = f(x)$ with its graph

$$\Gamma_f \equiv \{(x, f(x)) : x \in \operatorname{dom} f \equiv \Omega\} \subset X \times U,$$

where Γ_f is a submanifold of $X \times U$. If $\Gamma_f \subset M_g \equiv \operatorname{dom} g$, then the transform of Γ_f by g is defined as

$$g \cdot \Gamma_f = \{(\tilde{x}, \tilde{u}) = g \cdot (x, u) : (x, u) \in \Gamma_f\}.$$

We write $\tilde{f} = g \cdot f$ and call the function $\tilde{f}$ the *transform* of f by g.

For example, let $p = 1$ and $q = 1$, so $X = \mathbb{R}$ with a single independent variable x, and $U = \mathbb{R}$ with a single dependent variable u, so we have a single ODE involving a single function $u = f(x)$. Let $G = SO(2)$ be the rotation group acting on $X \times U \simeq \mathbb{R}^2$. The transformations in G are given by

$$(\tilde{x}, \tilde{u}) = \theta \cdot (x, u) = (x \cos \theta - u \sin \theta, \; x \sin \theta + u \cos \theta).$$

Let $u = f(x)$ be a function whose graph is a subset $\Gamma_f \subset X \times U$. The group $SO(2)$ acts on f by rotating its graph.

In general, the procedure for finding the transformed function $\tilde{f} = g \cdot f$ is given by [Olv86]:

$$g \cdot f = [\Phi_g \circ (1 \times f)] \circ [\Xi_g \circ (1 \times f)]^{-1}, \tag{8.143}$$

where $\Xi_g = \Xi_g(x, u)$, $\Phi_g = \Phi_g(x, u)$ are smooth functions such that

$$(\tilde{x}, \tilde{u}) = g \cdot (x, u) = (\Xi_g(x, u), \; \Phi_g(x, u)),$$

while 1 denotes the identity function of X, so $1(x) = x$. Formula (8.143) holds whenever the second factor is invertible.

Let $\mathcal{S}$ be a system of DEs. A *symmetry group* of the system $\mathcal{S}$ is a local Lie group of transformations G acting on an open subset $M \subset X \times U$ of the space $X \times U$ of independent and dependent variables of the system with the property that whenever $u = f(x)$ is a solution of $\mathcal{S}$, and whenever $g \cdot f$ is defined for $g \in G$, then $u = g \cdot f(x)$ is also a solution of the system.

For example, in the case of the ODE $u_{xx} = 0$, the rotation group $SO(2)$ is obviously a symmetry group, since the solutions are all linear functions and $SO(2)$ takes any linear function to another linear function. Another easy example is given by the classical *heat equation* $u_t = u_{xx}$. Here the group of translations

$$(x, t, u) \mapsto (x + \varepsilon a, \, t + \varepsilon b, \, u), \qquad \varepsilon \in \mathbb{R},$$

is a symmetry group since $u = f(x - \varepsilon a, \, t - \varepsilon b)$ is a solution to the heat equation whenever $u = f(x, t)$ is.

Prolongations

Prolongations of Functions

Given a smooth real–valued function $u = f(x) = f(x^1, ..., x^p)$ of p independent variables, there is an induced function $u^{(n)} = \mathbf{pr}^{(n)} f(x)$, called the nth *prolongation* of f [Olv86], which is defined by the equations

$$u_J = \partial_J f(x) = \frac{\partial^k f(x)}{\partial x^{j_1} \partial x^{j_2} ... \partial x^{j_k}},$$

where the *multi–index* $J = (j_1, ..., j_k)$ is an unordered k–tuple of integers, with entries $1 \leq j_k \leq p$ indicating which derivatives are being taken. More generally, if $f : X \to U$ is a smooth function from $X \simeq \mathbb{R}^p$ to $U \simeq \mathbb{R}^q$, so $u = f(x) = f(f^1(x), ..., f^q(x))$, there are $q \cdot p_k$ numbers

$$u_J^\alpha = \partial_J f^\alpha(x) = \frac{\partial^k f^\alpha(x)}{\partial x^{j_1} \partial x^{j_2} ... \partial x^{j_k}},$$

needed to represent all the different kth order derivatives of the components of f at a point x. Thus $\mathbf{pr}^{(n)} f : X \to U^{(n)}$ is a function from X to the space $U^{(n)}$, and for each $x \in X$, $\mathbf{pr}^{(n)} f(x)$ is a vector whose $q \cdot p^{(n)}$ entries represent the values of f and al its derivatives up to order n at the point x.

For example, in the case $p = 2$, $q = 1$ we have $X \simeq \mathbb{R}^2$ with coordinates $(x^1, x^2) = (x, y)$, and $U \simeq \mathbb{R}$ with the single coordinate $u = f(x, y)$. The second prolongation $u^{(2)} = \mathbf{pr}^{(2)} f(x, y)$ is given by [Olv86]

$$(u; u_x, u_y; u_{xx}, u_{xy}, u_{yy}) = \left(f; \frac{\partial f}{\partial x}, \frac{\partial f}{\partial y}; \frac{\partial^2 f}{\partial x^2}, \frac{\partial^2 f}{\partial x \partial y}, \frac{\partial^2 f}{\partial y^2} \right), \qquad (8.144)$$

all evaluated at (x, y).

The nth prolongation $\mathbf{pr}^{(n)} f(x)$ is also known as the n–jet of f. In other words, the nth prolongation $\mathbf{pr}^{(n)} f(x)$ represents the Taylor polynomial of degree n for f at the point x, since the derivatives of order $\leq n$ determine the Taylor polynomial and vice versa.

Prolongations of Differential Equations

A system $\mathcal{S}$ of nth order DEs in p independent and q dependent variables is given as a system of equations [Olv86]

$$\Delta_r(x, u^{(n)}) = 0, \qquad (r = 1, ..., l), \tag{8.145}$$

involving $x = (x^1, ..., x^p)$, $u = (u^1, ..., u^q)$ and the derivatives of u with respect to x up to order n. The functions $\Delta(x, u^{(n)}) = (\Delta_1(x, u^{(n)}), ..., \Delta_l(x, u^{(n)}))$ are assumed to be smooth in their arguments, so $\Delta : X \times U^{(n)} \to \mathbb{R}^l$ represents a smooth map from the *jet space* $X \times U^{(n)}$ to some lD Euclidean space. The DEs themselves tell where the given map Δ vanishes on the jet space $X \times U^{(n)}$, and thus determine a submanifold

$$\mathcal{S}_\Delta = \left\{ (x, u^{(n)}) : \Delta(x, u^{(n)}) = 0 \right\} \subset X \times U^{(n)} \tag{8.146}$$

of the total the jet space $X \times U^{(n)}$.

We can identify the system of DEs (8.145) with its corresponding submanifold $\mathcal{S}_\Delta$ (8.146). From this point of view, a smooth *solution* of the given system of DEs is a smooth function $u = f(x)$ such that [Olv86]

$$\Delta_r(x, \mathbf{pr}^{(n)} f(x)) = 0, \qquad (r = 1, ..., l),$$

whenever x lies in the domain of f. This is just a restatement of the fact that the derivatives $\partial_J f^\alpha(x)$ of f must satisfy the algebraic constraints imposed by the system of DEs. This condition is equivalent to the statement that the graph of the prolongation $\mathbf{pr}^{(n)} f(x)$ must lie entirely within the submanifold $\mathcal{S}_\Delta$ determined by the system:

$$\Gamma_f^{(n)} \equiv \left\{ (x, \mathbf{pr}^{(n)} f(x)) \right\} \subset \mathcal{S}_\Delta = \left\{ \Delta(x, u^{(n)}) = 0 \right\}.$$

We can thus take an nth order system of DEs to be a submanifold $\mathcal{S}_\Delta$ in the n−jet space $X \times U^{(n)}$ and a solution to be a function $u = f(x)$ such that the graph of the nth prolongation $\mathbf{pr}^{(n)} f(x)$ is contained in the submanifold $\mathcal{S}_\Delta$.

For example, consider the case of *Laplace equation* in the plane

$$u_{xx} + u_{yy} = 0 \qquad (\text{remember}, \quad u_x \equiv \partial_x u).$$

Here $p = 2$ since there are two independent variables x and y, and $q = 1$ since there is one dependent variable u. Also $n = 2$ since the equation is second–order, so $\mathcal{S}_\Delta \subset X \times U^{(2)}$ is given by (8.144). A solution $u = f(x, y)$ must satisfy

$$\frac{\partial^2 f}{\partial x^2} + \frac{\partial^2 f}{\partial y^2} = 0$$

for all (x, y). This is the same as requiring that the graph of the second prolongation $\mathbf{pr}^{(2)} f$ lie in $\mathcal{S}_\Delta$.

Prolongations of Group Actions

Let G be a local group of transformations acting on an open subset $M \subset X \times U$ of the space of independent and dependent variables. There is an induced local action of G on the $n-$jet space $M^{(n)}$, called the nth prolongation $\mathbf{pr}^{(n)}G$ of the action of G on M. This prolongation is defined so that it transforms the derivatives of functions $u = f(x)$ into the corresponding derivatives of the transformed function $\tilde{u} = \tilde{f}(\tilde{x})$ [Olv86].

More precisely, suppose $(x_0, u_0^{(n)})$ is a given point in $M^{(n)}$. Choose any smooth function $u = f(x)$ defined in a neighborhood of x_0, whose graph Γ_f lies in M, and has the given derivatives at x_0:

$$u_0^{(n)} = \mathbf{pr}^{(n)} f(x_0), \qquad \text{i.e.,} \qquad u_{J0}^{\alpha} = \partial_J f^{\alpha}(x_0).$$

If g is an element of G sufficiently near the identity, the transformed function $g \cdot f$ as given by (8.143) is defined in a neighborhood of the corresponding point $(\tilde{x}_0, \tilde{u}_0) = g \cdot (x_0, u_0)$, with $u_0 = f(x_0)$ being the zeroth order components of $u_0^{(n)}$. We then determine the action of the prolonged group of transformations $\mathbf{pr}^{(n)}g$ on the point $(x_0, u_0^{(n)})$ by evaluating the derivatives of the transformed function $g \cdot f$ at $\tilde{x}_0$; explicitly [Olv86]

$$\mathbf{pr}^{(n)}g \cdot (x_0, u_0^{(n)}) = (\tilde{x}_0, \tilde{u}_0^{(n)}),$$

where

$$\tilde{u}_0^{(n)} \equiv \mathbf{pr}^{(n)}(g \cdot f)(\tilde{x}_0).$$

For example, let $p = q = 1$, so $X \times U \simeq \mathbb{R}^2$, and consider the action of the rotation group $SO(2)$. To calculate its first prolongation $\mathbf{pr}^{(1)}SO(2)$, first note that $X \times U^{(1)} \simeq \mathbb{R}^3$, with coordinates (x, u, u_x). given a function $u = f(x)$, the first prolongation is [Olv86]

$$\mathbf{pr}^{(1)} f(x) = (f(x), f'(x)).$$

Now, given a point $(x^0, u^0, u_x^0) \in X \times U^{(1)}$, and a rotation in $SO(2)$ characterized by the angle θ as given above, the corresponding transformed point

$$\mathbf{pr}^{(1)}\theta \cdot (x^0, u^0, u_x^0) = (\tilde{x}^0, \tilde{u}^0, \tilde{u}_x^0)$$

(provided it exists). As for the first–order derivative, we find

$$\tilde{u}_x^0 = \frac{\sin\theta + u_x \cos\theta}{\cos\theta - u_x \sin\theta}.$$

Now, applying the group transformations given above, and dropping the $0-$indices, we find that the prolonged action $\mathbf{pr}^{(1)}SO(2)$ on $X \times U^{(1)}$ is given by

$$\mathbf{pr}^{(1)}\theta \cdot (x, u, u_x) = \left(x\cos\theta - u\sin\theta,\ x\sin\theta + u\cos\theta,\ \frac{\sin\theta + u_x \cos\theta}{\cos\theta - u_x \sin\theta} \right),$$

which is defined for $|\theta| < |\operatorname{arccot} u_x|$. Note that even though $SO(2)$ is a linear, globally defined group of transformations, its first prolongation $\mathbf{pr}^{(1)}SO(2)$ is both nonlinear and only locally defined. This fact demonstrates the complexity of the operation of prolonging a group of transformations.

In general, for any Lie group G, the first prolongation $\mathbf{pr}^{(1)}G$ acts on the original variables (x, u) exactly the same way that G itself does; only the action on the derivative u_x gives an new information. Therefore, $\mathbf{pr}^{(0)}G$ agrees with G itself, acting on $M^{(0)} = M$.

Prolongations of Vector Fields

Prolongation of the infinitesimal generators of the group action turn out to be the *infinitesimal generators* of the *prolonged group action* [Olv86]. Let $M \subset X \times U$ be open and suppose v is a vector–field on M, with corresponding local one–parameter group $\exp(\varepsilon v)$. The nth prolongation of v, denoted $\mathbf{pr}^{(n)}v$, will be a vector–field on the n–jet space $M^{(n)}$, and is defined to be the infinitesimal generator of the corresponding prolonged on–parameter group $\mathbf{pr}^{(n)}[\exp(\varepsilon v)]$. In other words,

$$\mathbf{pr}^{(n)}v|_{(x,u^{(n)})} = \frac{d}{d\varepsilon}\bigg|_{\varepsilon=0} \mathbf{pr}^{(n)}[\exp(\varepsilon v)](x, u^{(n)}) \qquad (8.147)$$

for any $(x, u^{(n)}) \in M^{(n)}$.

For a vector–field v on M, given by

$$v = \xi^i(x, u)\frac{\partial}{\partial x^i} + \phi^\alpha(x, u)\frac{\partial}{\partial u^\alpha}, \qquad (i = 1, ..., p, \ \alpha = 1, ..., q),$$

the nth prolongation $\mathbf{pr}^{(n)}v$ is given by [Olv86]

$$\mathbf{pr}^{(n)}v = \xi^i(x, u)\frac{\partial}{\partial x^i} + \phi^\alpha_J(x, u^{(n)})\frac{\partial}{\partial u^\alpha_J},$$

with $\phi^\alpha_0 = \phi^\alpha$, and J a multiindex defined above.

For example, in the case of $SO(2)$ group, the corresponding infinitesimal generator is

$$v = -u\frac{\partial}{\partial x} + x\frac{\partial}{\partial u},$$

with

$$\exp(\varepsilon v)(x, u) = (x\cos\varepsilon - u\sin\varepsilon, \ x\sin\varepsilon + u\cos\varepsilon),$$

being the rotation through angle ε. The first prolongation takes the form

$$\mathbf{pr}^{(1)}[\exp(\varepsilon v)](x, u, u_x) = \left(x\cos\varepsilon - u\sin\varepsilon, \ x\sin\varepsilon + u\cos\varepsilon, \ \frac{\sin\varepsilon + u_x\cos\varepsilon}{\cos\varepsilon - u_x\sin\varepsilon}\right).$$

According to (8.147), the first prolongation of v is get by differentiating these expressions with respect to ε and setting $\varepsilon = 0$, which gives

$$\mathbf{pr}^{(1)}v = -u\frac{\partial}{\partial x} + x\frac{\partial}{\partial u} + (1 + u_x^2)\frac{\partial}{\partial u_x}.$$

General Prolongation Formula

Let

$$v = \xi^i(x, u)\frac{\partial}{\partial x^i} + \phi^\alpha(x, u)\frac{\partial}{\partial u^\alpha}, \qquad (i = 1, ..., p, \ \alpha = 1, ..., q), \qquad (8.148)$$

be a vector–field defined on an open subset $M \subset X \times U$. The nth prolongation of v is the vector–field [Olv86]

$$\mathbf{pr}^{(n)}v = v + \phi_J^\alpha(x, u^{(n)})\frac{\partial}{\partial u_J^\alpha}, \qquad (8.149)$$

defined on the corresponding jet space $M^{(n)} \subset X \times U^{(n)}$. The coefficient functions ϕ_J^α are given by the following formula:

$$\phi_J^\alpha = D_J\left(\phi^\alpha - \xi^i u_i^\alpha\right) + \xi^i u_{J,i}^\alpha, \qquad (8.150)$$

where $u_i^\alpha = \partial u^\alpha/\partial x^i$, and $u_{J,i}^\alpha = \partial u_J^\alpha/\partial x^i$. D_J is the *total derivative* with respect to the multiindex J, i.e.,

$$D_J = D_{j_1}D_{j_2}...D_{j_k},$$

while the total derivative with respect to the ordinary index, D_i, is defined as follows. Let $P(x, u^{(n)})$ be a smooth function of x, u and derivatives of u up to order n, defined on an open subset $M^{(n)} \subset X \times U^{(n)}$. the total derivative of P with respect to x^i is the unique smooth function $D_i P(x, u^{(n)})$ defined on $M^{(n+1)}$ and depending on derivatives of u up to order $n+1$, with the *recursive property* that if $u = f(x)$ is any smooth function then

$$D_i P(x, \mathbf{pr}^{(n+1)}f(x)) = \partial_{x^i}\{P(x, \mathbf{pr}^{(n)}f(x))\}.$$

For example, in the case of $SO(2)$ group, with the infinitesimal generator

$$v = -u\frac{\partial}{\partial x} + x\frac{\partial}{\partial u},$$

the first prolongation is (as calculated above)

$$\mathbf{pr}^{(1)}v = -u\frac{\partial}{\partial x} + x\frac{\partial}{\partial u} + \phi^x\frac{\partial}{\partial u_x},$$

where

$$\phi^x = D_x(\phi - \xi u_x) + \xi u_{xx} = 1 + u_x^2.$$

Also,

$$\phi^{xx} = D_x\phi^x - u_{xx}D_x\xi = 3u_x u_{xx},$$

thus the infinitesimal generator of the second prolongation $\mathbf{pr}^{(2)}SO(2)$ acting on $X \times U^{(2)}$ is

$$\mathbf{pr}^{(2)}v = -u\frac{\partial}{\partial x} + x\frac{\partial}{\partial u} + (1 + u_x^2)\frac{\partial}{\partial u_x} + 3u_x u_{xx}\frac{\partial}{\partial u_{xx}}.$$

Let v and w be two smooth vector–fields on $M \subset X \times U$. Then their nth prolongations, $\mathbf{pr}^{(n)}v$ and $\mathbf{pr}^{(n)}w$ respectively, have the *linearity property*

$$\mathbf{pr}^{(n)}(c_1 v + c_2 w) = c_1 \mathbf{pr}^{(n)} v + c_2 \mathbf{pr}^{(n)} w, \qquad (c_1, c_2 - \text{constant}),$$

and the *Lie bracket property*

$$\mathbf{pr}^{(n)}[v, w] = [\mathbf{pr}^{(n)}v, \mathbf{pr}^{(n)}w].$$

Generalized Lie Symmetries

Consider a vector–field (8.148) defined on an open subset $M \subset X \times U$. Provided the coefficient functions ξ^i and ϕ^α depend only on x and u, v will generate a (local) one–parameter group of transformations $\exp(\varepsilon v)$ acting pointwise on the underlying space M. A significant generalization of the notion of symmetry group is get by relaxing this geometrical assumption, and allowing the coefficient functions ξ^i and ϕ^α to also depend on derivatives of u [Olv86].

A *generalized vector–field* is a (formal) expression

$$v = \xi^i[u]\frac{\partial}{\partial x^i} + \phi^\alpha[u]\frac{\partial}{\partial u^\alpha}, \qquad (i = 1, ..., p, \ \alpha = 1, ..., q), \tag{8.151}$$

in which ξ^i and ϕ^α are smooth functions. For example,

$$v = xu_x\frac{\partial}{\partial x} + u_{xx}\frac{\partial}{\partial u}$$

is a generalized vector in the case $p = q = 1$.

According to the general prolongation formula (8.149), we can define the *prolonged generalized vector–field*

$$\mathbf{pr}^{(n)}v = v + \phi_J^\alpha[u]\frac{\partial}{\partial u_J^\alpha},$$

whose coefficients are as before determined by the formula (8.150). Thus, in our previous example [Olv86],

$$\mathbf{pr}^{(n)}v = xu_x\frac{\partial}{\partial x} + u_{xx}\frac{\partial}{\partial u} + [u_{xxx} - (xu_{xx} + u_x)u_x]\frac{\partial}{\partial u_x}.$$

Given a generalized vector–field v, its *infinite prolongation* (including all the derivatives) is the formal expression

$$\mathbf{pr}\, v = \xi^i\frac{\partial}{\partial x^i} + \phi_J^\alpha\frac{\partial}{\partial u_J^\alpha}.$$

Now, a generalized vector–field v is a *generalized infinitesimal symmetry* of a system $\mathcal{S}$ of differential equations

$$\Delta_r[u] = \Delta_r(x, u^{(n)}) = 0, \qquad (r = 1, ..., l),$$

iff

$$\mathbf{pr}\, v[\Delta_r] = 0$$

for every smooth solution m $u = f(x)$ [Olv86].

For example, consider the heat equation

$$\Delta[u] = u_t - u_{xx} = 0.$$

The generalized vector–field $v = u_x \frac{\partial}{\partial u}$ has prolongation

$$\mathbf{pr}\, v = u_x \frac{\partial}{\partial u} + u_{xx} \frac{\partial}{\partial u_x} + u_{xt} \frac{\partial}{\partial u_t} + u_{xxx} \frac{\partial}{\partial u_{xx}} + ...$$

Thus

$$\mathbf{pr}\, v(\Delta) = u_{xt} - u_{xxx} = D_x(u_t - u_{xx}) = D_x \Delta.$$

and hence v is a generalized symmetry of the heat equation.

Noether Symmetries

Here we present some results about *Noether symmetries*, in particular for the first–order Lagrangians $L(q, \dot{q})$ (see [BGG89, PSS96]). We start with a *Noether–Lagrangian symmetry*,

$$\delta L = \dot{F},$$

and we will investigate the conversion of this symmetry to the Hamiltonian formalism. Defining

$$G = (\partial L/\partial \dot{q}^i)\, \delta q^i - F,$$

we can write

$$\delta_i L\, \delta q^i + \dot{G} = 0, \tag{8.152}$$

where $\delta_i L$ is the *Euler–Lagrangian functional derivative* of L,

$$\delta_i L = \alpha_i - W_{ik}\, \ddot{q}^k,$$

where

$$W_{ik} \equiv \frac{\partial^2 L}{\partial \dot{q}^i \partial \dot{q}^k} \qquad \text{and} \qquad \alpha_i \equiv -\frac{\partial^2 L}{\partial \dot{q}^i \partial q^k}\, \dot{q}^k + \frac{\partial L}{\partial q^i}.$$

We consider the general case where the mass matrix, or *Hessian* (W_{ij}), may be a singular matrix. In this case there exists a kernel for the pull–back $\mathbb{F}L^*$ of the *Legendre map*, i.e., *fibre–derivative* $\mathbb{F}L$, from the *velocity phase–space manifold* TM (tangent bundle of the biomechanical manifold M) to the *momentum phase–space manifold* T^*M (cotangent bundle of M). This kernel is spanned by the vector–fields

$$\Gamma_\mu = \gamma^i_\mu \frac{\partial}{\partial \dot{q}^i},$$

where γ^i_μ are a basis for the null vectors of W_{ij}. The Lagrangian time–evolution differential operator can therefore be expressed as:

$$X = \partial_t + \dot{q}^k \frac{\partial}{\partial q^k} + a^k(q, \dot{q}) \frac{\partial}{\partial \dot{q}^k} + \lambda^\mu \Gamma_\mu \equiv X_o + \lambda^\mu \Gamma_\mu,$$

where a^k are functions which are determined by the formalism, and λ^μ are arbitrary functions. It is not necessary to use the Hamiltonian technique to find the Γ_μ, but it does facilitate the calculation:

$$\gamma^i_\mu = \mathbb{F}L^* \left(\frac{\partial \phi_\mu}{\partial p_i} \right), \tag{8.153}$$

where the ϕ_μ are the Hamiltonian primary first class constraints.

Notice that the highest derivative in (8.152), $\ddot{q}^i$, appears linearly. Because δL is a symmetry, (8.152) is identically satisfied, and therefore the coefficient of $\ddot{q}^i$ vanishes:

$$W_{ik}\delta q^k - \frac{\partial G}{\partial \dot{q}^i} = 0. \tag{8.154}$$

We contract with a null vector γ^i_μ to find that

$$\Gamma_\mu G = 0.$$

It follows that G is projectable to a function G_{H} in T^*Q; that is, it is the pull–back of a function (not necessarily unique) in T^*Q:

$$G = \mathbb{F}L^*(G_{\mathrm{H}}).$$

This important property is valid for any conserved quantity associated with a Noether symmetry. Observe that G_{H} is determined up to the addition of linear combinations of the primary constraints. Substitution of this result in (8.154) gives

$$W_{ik} \left[\delta q^k - \mathbb{F}L^* \left(\frac{\partial G_{\mathrm{H}}}{\partial p_k} \right) \right] = 0,$$

and so the brackets enclose a null vector of W_{ik}:

$$\delta q^i - \mathbb{F}L^* \left(\frac{\partial G_{\mathrm{H}}}{\partial p_i} \right) = r^\mu \gamma^i_\mu, \tag{8.155}$$

for some $r^\mu(t, q, \dot{q})$.

We shall investigate the projectability of variations generated by diffeomorphisms in the following section. Assume that an infinitesimal transformation δq^i is projectable:

$$\Gamma_\mu \delta q^i = 0.$$

If δq^i is projectable, so must be r^μ, so that $r^\mu = \mathbb{F}L^*(r_H^\mu)$. Then, using (8.153) and (8.155), we see that

$$\delta q^i = \mathbb{F}L^* \left(\frac{\partial(G_H + r_H^\mu \phi_\mu)}{\partial p_i} \right).$$

We now redefine G_H to absorb the piece $r_H^\mu \phi_\mu$, and from now on we will have

$$\delta q^i = \mathbb{F}L^* \left(\frac{\partial G_H}{\partial p_i} \right).$$

Define

$$\hat{p}_i = \frac{\partial L}{\partial \dot{q}^i};$$

after eliminating (8.154) times $\ddot{q}^i$ from (8.152), we get

$$\left(\frac{\partial L}{\partial q^i} - \dot{q}^k \frac{\partial \hat{p}_i}{\partial q^k} \right) \mathbb{F}L^*(\frac{\partial G_H}{\partial p_i}) + \dot{q}^i \frac{\partial}{\partial q^i} \mathbb{F}L^*(G_H) + \mathbb{F}L^* \partial_t G_H = 0,$$

which simplifies to

$$\frac{\partial L}{\partial q^i} \mathbb{F}L^*(\frac{\partial G_H}{\partial p_i}) + \dot{q}^i \mathbb{F}L^*(\frac{\partial G_H}{\partial q^i}) + \mathbb{F}L^* \partial_t G_H = 0 \qquad (8.156)$$

Now let us invoke two identities [BGG89] that are at the core of the connection between the Lagrangian and the Hamiltonian equations of motion. They are

$$\dot{q}^i = \mathbb{F}L^*(\frac{\partial H}{\partial p_i}) + v^\mu(q, \dot{q}) \mathbb{F}L^*(\frac{\partial \phi_\mu}{\partial p_i}),$$

and

$$\frac{\partial L}{\partial q^i} = -\mathbb{F}L^*(\frac{\partial H}{\partial q^i}) - v^\mu(q, \dot{q}) \mathbb{F}L^*(\frac{\partial \phi_\mu}{\partial q^i});$$

where H is any canonical Hamiltonian, so that $\mathbb{F}L^*(H) = \dot{q}^i(\partial L/\partial \dot{q}^i) - L = \hat{E}$, the Lagrangian energy, and the functions v^μ are determined so as to render the first relation an identity. Notice the important relation

$$\Gamma_\mu v^\nu = \delta_\mu^\nu,$$

which stems from applying Γ_μ to the first identity and taking into account that

$$\Gamma_\mu \circ \mathbb{F}L^* = 0.$$

Substitution of these two identities into (8.156) induces (where $\{,\}$ denotes the *Poisson bracket*)

$$\mathbb{F}L^*\{G_\text{H}, H\} + v^\mu \mathbb{F}L^*\{G_\text{H}, \phi_\mu\} + \mathbb{F}L^* \partial_t G_\text{H} = 0.$$

This result can be split through the action of Γ_μ into

$$\mathbb{F}L^*\{G_\text{H}, H\} + \mathbb{F}L^* \partial_t G_\text{H} = 0,$$

and

$$\mathbb{F}L^*\{G_\text{H}, \phi_\mu\} = 0;$$

or equivalently,

$$\{G_\text{H}, H\} + \partial_t G_\text{H} = pc,$$

and

$$\{G_\text{H}, \phi_\mu\} = pc,$$

where pc stands for any linear combination of primary constraints. In this way, we have arrived at a neat characterization for a generator G_H of Noether transformations in the canonical formalism.

Lie Symmetries in Biophysics

In this subsection we consider two most important equations for biophysics:

A. The heat equation, which has been analyzed in muscular mechanics since the early works of A.V. Hill ([Hil38]); and
B. The Korteveg–de Vries equation, the basic equation for solitary models of muscular excitation–contraction dynamics.

Suppose

$$\mathcal{S} : \Delta_r(x, u^{(n)}) = 0, \qquad (r = 1, ..., l),$$

is a system of DEs of maximal rank defined over $M \subset X \times U$. If G is a local group of transformations acting on M, and

$$\mathbf{pr}^{(n)} v[\Delta_r(x, u^{(n)})] = 0, \qquad \text{whenever} \qquad \Delta(x, u^{(n)}) = 0, \qquad (8.157)$$

(with $r = 1, ..., l$) for every infinitesimal generator v of G, then G is a symmetry group of the system $\mathcal{S}$ [Olv86].

The Heat Equation

Recall that the (1+1)D *heat equation* (with the thermal diffusivity normalized to unity)

$$u_t = u_{xx} \tag{8.158}$$

has two independent variables x and t, and one dependent variable u, so $p = 2$ and $q = 1$. Equation (8.158) has the second-order, $n = 2$, and can be identified

with the linear submanifold $M^{(2)} \subset X \times U^{(2)}$ determined by the vanishing of
$\Delta(x, t, u^{(2)}) = u_t - u_{xx}$.

Let

$$v = \xi(x, t, u) \frac{\partial}{\partial x} + \tau(x, t, u) \frac{\partial}{\partial t} + \phi(x, t, u) \frac{\partial}{\partial u}$$

be a vector–field on $X \times U$. According to (8.157) we need to now the second
prolongation

$$\mathbf{pr}^{(2)} v = v + \phi^x \frac{\partial}{\partial u_x} + \phi^t \frac{\partial}{\partial u_t} + \phi^{xx} \frac{\partial}{\partial u_{xx}} + \phi^{xt} \frac{\partial}{\partial u_{xt}} + \phi^{tt} \frac{\partial}{\partial u_{tt}}$$

of v. Applying $\mathbf{pr}^{(2)} v$ to (8.158) we find the infinitesimal criterion (8.157) to
be

$$\phi^t = \phi^{xx},$$

which must be satisfied whenever $u_t = u_{xx}$.

The Korteveg–De Vries Equation

Recall that the *Korteveg–de Vries equation*

$$u_t + u_{xxx} + u u_x = 0 \qquad (8.159)$$

arises in physical systems in which both nonlinear and dispersive effects are
relevant. A vector–field

$$v = \xi(x, t, u) \frac{\partial}{\partial x} + \tau(x, t, u) \frac{\partial}{\partial t} + \phi(x, t, u) \frac{\partial}{\partial u}$$

generates a one–parameter symmetry group iff

$$\phi^t + \phi^{xxx} + u \phi^x + u_x \phi = 0,$$

whenever u satisfies (9.30), etc.

8.6.2 Noether Theorem and Navier–Stokes Equations

Recall that *Noether theorem* can be applied to evolution equations which can
be derived from a Lagrangian, i.e., evolution equations which can be expressed
in an *Euler–Lagrange form* [RH06]:

$$\partial_r L - \operatorname{div} \partial_{\dot{r}} L = 0, \qquad (8.160)$$

where r is the dependent variable, $r = r(y)$, $y = (y^i)$ is the independent
variable, $L = L(r)$ is the Lagrangian and div denotes the divergence vector
operator, defined via the total derivative $\mathrm{d}/\mathrm{d}y^i$ as

$$f \mapsto \operatorname{div} f = \sum_i \frac{\mathrm{d}f}{\mathrm{d}y^i},$$

From the infinitesimal generators of (8.160), conservation laws are deduced.

Navier–Stokes equations cannot be directly written in the form (8.160). However, thanks to an approach of Atherton and Homsy [AH75], see also [IK04], which consists in extending the Lagrangian notion, it will be shown in this appendix that Noether's theorem can be applied to Navier–Stokes equations.

We will say that an evolution equation

$$F(y, r) = 0 \qquad (8.161)$$

is derived from a 'bi-Lagrangian' if there exists an (non necessarily unique) application

$$L \ : \ (r, s) \ \mapsto \ L(r, s) \in \mathbb{R}$$

such that (8.161) is equivalent to

$$\partial_s L(r, s) - \operatorname{div} \partial_{\dot{s}} L(r, s) = 0.$$

s is called the adjoint variable and the equation

$$\partial_r L(r, s) - \operatorname{div} \partial_{\dot{r}} L(r, s) = 0.$$

is called the adjoint equation of (8.161). The Noether theorem can then be applied since the evolution equation, associated to his adjoint, can be written in an Euler–Lagrangian form

$$\partial_w L(w) - \operatorname{div} \partial_{\dot{w}} L(w) = 0,$$

where $w = (r, s)$.

Navier–Stokes equations are derived from a bi-Lagrangian

$$L((u, p), (v, q)) = (\dot{u} \cdot v - u \cdot \dot{v})\,/2 + (q - u \cdot v/2) - p \operatorname{div} v + \nu \operatorname{Tr}\left(^t\nabla u \cdot \nabla v\right).$$

where ^{t} denotes matrix transposition, $v = (v^i)$ and q are the adjoint variables. The corresponding adjoint equations are

$$-\dot{v} + (v \cdot \nabla u^T - u \cdot \nabla v) = \nabla q + \nu \Delta v, \qquad \operatorname{div} v = 0.$$

Noether's theorem can then be applied. The *infinitesimal generators* of the Navier–Stokes equations and their adjoint equations are:

$$X_0 = \partial_t, \qquad Y_0 = \zeta(t)\partial_p,$$
$$X_{ij} = x^j \partial_{x^i} - x^i \partial_{x^j} + u^j \partial_{u^i} - u^i \partial_{u^j} + v^j \partial_{v^i} - v^i \partial_{v^j}, \qquad (i = 1, 2; j > i),$$
$$X_i = \alpha_i(t)\partial_{x^i} + \alpha_i'(t)\partial_{u^i} - x^i\,\alpha_i''(t)\partial_p, \qquad (i = 1, 2, 3),$$
$$Y_1 = 2t\partial_t + x^k \partial_{x^k} - u^k \partial_{u^k} - 2p\partial_p - q\partial_q, \qquad Y_0' = \eta(t)\partial_q,$$
$$X_{ij}' = (x^j u_i - x^i u_j)\partial_q + x^j \partial_{v^i} - x^i \partial_{v^j}, \qquad (i = 1, 2; j > i),$$
$$X_i' = [x^i \sigma'(t) - u^i \sigma(t)]\partial_q - \sigma(t)\partial_{v^i}, \qquad (i = 1, 2, 3),$$
$$Y_1' = v^k \partial_{v^k} + q\partial_q,$$

where ζ, the α_i's, η and σ are arbitrary scalar functions. Conservation laws for Navier–Stokes equations can be deduced from these infinitesimal generators.

8.6.3 Large–Eddy Simulation

Consider a 3D *incompressible Newtonian fluid*, with density ρ and kinematic viscosity ν. The motion of this fluid is governed by Navier–Stokes equations, here expressed as

$$\dot{u} + \mathrm{div}(u \otimes u) + \nabla p/\rho = \mathrm{div}\tau, \qquad \mathrm{div}\, u = 0, \qquad (8.162)$$

where $u = (u^i)_{i=1,2,3}$ and p are respectively velocity and pressure fields and t the time variable. τ is a tensor such that $\rho\tau$ is the viscous constraint tensor. τ can be linked to the strain rate tensor $\sigma = (\nabla u +^t \nabla u)/2$ according to the relation: $\tau = \partial_\sigma \psi$, where ψ is a positive and convex 'potential' defined by: $\psi = \nu \,\mathrm{Tr}\sigma^2$.

Since a direct numerical simulation of a realistic fluid flow requires very significant computational cost, (8.162) is not directly resolved. To circumvent the problem, some methods exist. The most promising one is the large–eddy simulation. It consists in representing only the large scales of the flow. Small scales are dropped from the simulation; however, their effects on the large scales are taken into account. This enables to take a much coarser grid [RH06].

Mathematically, dropping small scales means applying a low–pass filter. The large or resolved scales $\overline{\phi}$ of a quantity ϕ are defined by the convolution: $\overline{\phi} = G_{\bar\delta} * \phi$, where $G_{\bar\delta}$ is the filter kernel with a width $\bar\delta$, and the small scales ϕ' are defined by: $\phi' = \phi - \overline{\phi}$. It is required that the integral of $G_{\bar\delta}$ over $\mathbb{R}^3$ is equal to 1, such that a constant remains unchanged when the filter is applied.

In practice, $(\overline{u}, \overline{p})$ is directly used as an approximation of (u, p). To get $(\overline{u}, \overline{p})$, the filter is applied to (8.162). If the filter is assumed to commute with the derivative operators (that is not always the case in a bounded domain), this leads to:

$$\dot{\overline{u}} + \mathrm{div}(\overline{u} \otimes \overline{u}) + \frac{1}{\rho}\nabla\overline{p} = \mathrm{div}(\overline{\tau} + \tau_s), \qquad \mathrm{div}\overline{u} = 0, \qquad (8.163)$$

where τ_s is the *subgrid stress tensor* defined by: $\tau_s = \overline{u}\otimes\overline{u} - \overline{u \otimes u}$, which must be modelled (expressed by a function of the resolved quantities) to close the equations. Currently, an important number of models exists. Some of the most common ones will be reminded here. They will be classified in four categories: turbulent viscosity, gradient–type, similarity–type and Lund–Novikov–type models [RH06].

Turbulent Viscosity Models

Turbulent viscosity models are models which can be written in the form,

$$\tau_s^d = \nu_s\overline{\sigma}, \qquad (8.164)$$

where ν_s is the turbulent viscosity. The superscript $(^d)$ represents the deviatoric part of a tensor; e.g., for the tensor τ we have:

$$\tau \mapsto \tau^d = \tau - \frac{1}{3}(\mathrm{Tr}\tau)I_d,$$

where I_d is the identity operator. The deviatoric part has been introduced in order to have the equality of the traces in (8.164). In what follows, some examples of turbulent viscosity models are presented as follows [RH06].

Smagorinsky model (see [Sag04]) is one of the most widely used models. It uses the local equilibrium hypothesis for the calculation of the turbulent viscosity. It has the expression,

$$\tau_s^d = (C_S\bar{\delta})^2 |\bar{\sigma}|\bar{\sigma},$$

where $C_S \simeq 0.148$ is the Smagorinsky constant, $\bar{\delta}$ the filter width and $|\bar{\sigma}| = \sqrt{2\mathrm{Tr}(\bar{\sigma}^2)}$.

In order to reduce the modelling error of Smagorinsky model, Lilly [Lil92] proposes a dynamic evaluation of the constant C_S by a least–square approach. This leads to the so–called *dynamic model* defined by:

$$\tau_s^d = C_d\bar{\delta}^2 |\bar{\sigma}|\bar{\sigma}, \qquad \text{with} \quad C_d = \frac{\mathrm{Tr}(LM)}{\mathrm{Tr}M^2}. \tag{8.165}$$

In these terms,

$$L = \widetilde{\bar{u}} \otimes \widetilde{\bar{u}} - \widetilde{\bar{u} \otimes \bar{u}}, \qquad M = \bar{\delta}^2 \widetilde{|\bar{\sigma}|\bar{\sigma}} - \widetilde{\bar{\delta}}^2 |\widetilde{\bar{\sigma}}|\widetilde{\bar{\sigma}},$$

the tilde represents a test filter whose width is $\widetilde{\bar{\delta}}$, with $\widetilde{\bar{\delta}} > \bar{\delta}$.

The last turbulent viscosity model which will be considered is the so–called *structure–function model*. Metais and Lesieur [ML92] make the hypothesis that the turbulent viscosity depends on the energy at the cutoff. Knowing its relation with the energy density in Fourier space, they use the second order structure function and propose the structure–function model:

$$\tau_s^d = C_{SF}\bar{\delta}\sqrt{\bar{F}_2(\bar{\delta})}\,\bar{\sigma}, \tag{8.166}$$

where $\bar{F}_2$ is the spatial average of the filtered structure function:

$$r \mapsto \bar{F}_2(r) = \iint_{||z||=r} ||\bar{u}(x) - \bar{u}(x+z)||^2 \, \mathrm{d}z \, \mathrm{d}x.$$

The next category of models, which will be reminded, consists of the gradient-type models.

Gradient–Type Models

To establish the gradient–type models, the subgrid stress tensor τ_s is decomposed as [RH06]

$$\tau_s = \overline{u} \otimes \overline{u} - (\ \overline{\overline{u} \otimes \overline{u}} + \overline{\overline{u} \otimes u'} + \overline{u' \otimes \overline{u}} + \overline{u' \otimes u'}\ ).$$

Next, each term between the brackets are written in Fourier space. Then, the Fourier transform of the filter, which is assumed to be Gaussian, is approximated by an appropriate function. Finally, the inverse Fourier transform is computed. The models in this category differ by the way in which the Fourier transform of the filter is approximated.

- If a second order Taylor series expansions according to the filter width $\overline{\delta}$ is used in the approximation, one has:

$$\tau_s = -\delta^2 \nabla \overline{u}\ ^t \nabla \overline{u}/12. \tag{8.167}$$

- The gradient model is not dissipative enough and not numerically stable [WWV01, IJL02]. Thus, it is generally combined to Smagorinsky model. This gives *Taylor model*:

$$\tau_s = -\delta^2 \nabla \overline{u}\ ^t \nabla \overline{u}/12 + C\overline{\delta}^2 |\overline{\sigma}|\overline{\sigma}.$$

- The Taylor approximation of the Fourier transform of the filter tends to accentuate the small frequencies rather than attenuating them. Instead, a rational approximation can be used [IJL03, BG04]. This gives the following expression of the model:

$$\tau_s = -\delta^2 \left(I_d - \overline{\delta}^2 \nabla^2/24 \right)^{-1} [\nabla \overline{u}\ ^t \nabla \overline{u}]/12 + C\overline{\delta}^2 |\overline{\sigma}|\overline{\sigma}.$$

To avoid the inversion of the operator $(I_d - \overline{\delta}^2 \nabla^2/24)$, τ_s is approximated by:

$$\tau_s = -\overline{\delta}^2\ G_{\overline{\delta}} * [\nabla \overline{u}\ ^t \nabla \overline{u}]/12 + C\overline{\delta}^2 |\overline{\sigma}|\overline{\sigma}. \tag{8.168}$$

$G_{\overline{\delta}}$ is the kernel of the Gaussian filter. The convolution is done numerically. The model (8.168) is called the *rational model*.

Similarity–Type Models

Models of this category are based on the hypothesis that the statistic structure of the small scales are similar to the statistic structure of the smallest resolved scales. Separation of the resolved scales is done using a test filter (symbolized by $\widetilde{(\ )}$). The largest resolved scales are then represented by $\widetilde{\overline{u}}$ and the smallest ones by $\overline{u} - \widetilde{\overline{u}}$. From this hypothesis, we deduce the *similarity model* [RH06]

$$\tau_s = \widetilde{\overline{u}} \otimes \widetilde{\overline{u}} - \widetilde{\overline{u} \otimes \overline{u}}. \tag{8.169}$$

From this expression, many other models can be obtained by multiplying by a coefficient, by filtering again the whole expression or by mixing with a Smagorinsky–type model.

The last models that we will consider are Lund–Novikov–type models.

Lund–Novikov–Type Models

Lund and Novikov include the filtered vorticity tensor $\overline{\omega} = (\nabla \overline{u} - {}^t \nabla \overline{u})$ in the expression of the subgrid model. Cayley–Hamilton theorem gives then the *Lund–Novikov model* (see [Sag04]):

$$-\tau_s^d = C_1 \overline{\delta}^2 |\overline{\sigma}| \overline{\sigma} + C_2 \overline{\delta}^2 (\overline{\sigma}^2)^d + C_3 \overline{\delta}^2 (\overline{\omega}^2)^d$$
$$+ \ C_4 \overline{\delta}^2 (\overline{\sigma}\,\overline{\omega} - \overline{\omega}\,\overline{\sigma}) + C_5 \overline{\delta}^2 (\overline{\sigma}^2 \overline{\omega} - \overline{\sigma}\,\overline{\omega}^2) / |\overline{\sigma}|, \qquad (8.170)$$

where the coefficients C_i depend on the invariants obtained from $\overline{\sigma}$ and $\overline{\omega}$. The expression of these coefficients are so complex that they are considered as constants and evaluated with statistic techniques. Derivation of these models was done using different hypothesis but did not take into consideration the symmetries of Navier–Stokes equations which may then be destroyed [RH06].

8.6.4 Model Analysis

The (classical) symmetry groups of Navier–Stokes equations have been investigated for some decades (see for example [Dan67, Byt72]). They are generated by the following transformations [RH06]:

- The *time translations*: $(t, x, u, p) \mapsto (t + a, x, u, p)$,
- the *pressure translations*: $(t, x, u, p) \mapsto (t, x, u, p + \zeta(t))$,
- the *rotations*: $(t, x, u, p) \mapsto (t, Rx, Ru, p)$,
- the *generalized Galilean transformations*:
  $(t, x, u, p) \mapsto (t, x + \alpha(t), u + \dot{\alpha}(t), p - \rho\,x \cdot \ddot{\alpha}(t))$,
- and the *first scaling transformations*: $(t, x, u, p) \mapsto (e^{2a}t, e^a x, e^{-a}u, e^{-2a}\,p)$.

In these expressions, a is a scalar, ζ (respectively α) a scalar (resp. vectorial) arbitrary function of t and R a rotation matrix, i.e., $R^t R = I_d$ and $\det R = 1$. The central dot $(\cdot)$ stands for $\mathbb{R}^3$ scalar product.

If it is considered that ν can change during the transformation (which is then an equivalence transformation [IU94]), one has the *second scaling transformations*:

$$(t, x, u, p, \nu) \mapsto (t, e^a x, e^a u, e^{2a} p, e^{2a} \nu),$$

where a is the parameter.

Navier–Stokes equations admit other known symmetries which do not constitute a 1–parameter symmetry group. They are:

- the *reflections*: $(t, x, u, p) \mapsto (t, \Lambda x, \Lambda u, p)$,
  which are discrete symmetries, Λ being a diagonal matrix $\Lambda = \mathrm{diag}(\iota_1, \iota_2, \iota_3)$ with $\iota_i = \pm 1$, $i = 1, 2, 3$,
- and the *material indifference*: $(t, x, u, p) \mapsto (t, \hat{x}, \hat{u}, \hat{p})$,
  in the limit of a 2D flow in a simply connected domain [Can78], with

$$\hat{x} = R(t)\,x, \qquad \hat{u} = R(t)\,u + \dot{R}(t)\,x, \qquad \hat{p} = p - 3a\varphi + \frac{1}{2}a^2 ||x||^2,$$

where $R(t)$ is a 2D rotation matrix with angle at, with a an arbitrary real constant, φ is the common 2D stream function defined by: $u = \mathrm{curl}(\varphi e_3)$, where e_3 is the unit vector perpendicular to the plane of the flow, and $\|.\|$ is the *Euclidean norm*.

Following [RH06], we wish to analyse which of the models cited above is compatible with these symmetries. The set of solutions (u, p) of Navier–Stokes equations (8.162) is preserved by each of the symmetries. We then require that the set of solutions $(\overline{u}, \overline{p})$ of the filtered equations (8.163) is also preserved by all of these transformations, since $(\overline{u}, \overline{p})$ is expected to be a good approximation of (u, p). More clearly, if a transformation

$$T \;:\; (t, x, u, p) \mapsto (\hat{t}, \hat{x}, \hat{u}, \hat{p})$$

is a symmetry of (8.162), we require that the model is such that the same transformation, applied to the filtered quantities:

$$T : (t, x, \overline{u}, \overline{p}) \mapsto (\hat{t}, \hat{x}, \hat{\overline{u}}, \hat{\overline{p}}),$$

is a symmetry of the filtered equations (8.163). When this condition holds, the model will be said *invariant* under the relevant symmetry.

The filtered equations (8.163) may have other symmetries but with the above requirement, we may expect to preserve certain properties of Navier–Stokes equations (conservation laws, wall laws, exact solutions, spectra properties, ...) when approximating (u, p) by $(\overline{u}, \overline{p})$.

We will use the hypothesis that test filters do not destroy symmetry properties, i.e., $\hat{\tilde{\phi}} = \tilde{\hat{\phi}}$ for any quantity ϕ.

For the analysis, the symmetries of (8.162) will be grouped into four categories: (i) translations, containing time translations, pressure translations and the generalized Galilean transformations; (ii) rotations and reflections; (iii) scaling transformations; and (iv) material indifference. The aim is to search which models are invariant under the symmetries within the considered category.

Translational Invariance

Since almost all existing models are autonomous in time and pressure, the filtered equations (8.163) remain unchanged when a time or pressure translation is applied. Almost all models are then invariant under the time and the pressure translations.

The generalized Galilean transformations, applied to the filtered variables, have the following form [RH06]

$$(t, x, \overline{u}, \overline{p}) \;\mapsto\; (\hat{t}, \hat{x}, \hat{\overline{u}}, \hat{\overline{p}}) = (t, x + \alpha(t), \overline{u} + \dot{\alpha}(t), \overline{p} - \rho\, x\ddot{\alpha}(t)).$$

All models in Section 2, in which x and $\overline{u}$ are present only through $\nabla \overline{u}$ are invariant since

$$\hat{\nabla}\hat{\bar{u}} = \nabla\hat{\bar{u}} = \nabla\bar{u}, \qquad (\hat{\nabla} = (\partial_{\hat{x}_1}, \partial_{\hat{x}_2}, \partial_{\hat{x}_3}).)$$

The remaining models, i.e., the dynamic and the similarity models are also invariant because

$$\widetilde{\bar{u} \otimes \bar{u}} - \widetilde{\tilde{\bar{u}}} \otimes \widetilde{\tilde{\bar{u}}} = \widetilde{(\bar{u} + \dot{\alpha}) \otimes (\bar{u} + \dot{\alpha})} - \widetilde{(\bar{u} + \dot{\alpha})} \otimes \widetilde{(\bar{u} + \dot{\alpha})} = \widetilde{\bar{u} \otimes \bar{u}} - \widetilde{\tilde{\bar{u}}} \otimes \widetilde{\tilde{\bar{u}}}.$$

Rotational and Reflective Invariance

The rotations and the reflections can be put together in a transformation [RH06]:

$$(t, x, u, p) \mapsto (t, \Upsilon x, \Upsilon u, p)$$

where Υ is a constant rotation or reflection matrix. This transformation, when applied to the filtered variables, is a symmetry of (8.163) iff

$$\hat{\tau}_s = \Upsilon \tau_s^t \Upsilon. \tag{8.171}$$

Let us check if the models respect this condition.

For Smagorinsky model, we have:

$$\hat{\nabla}\hat{\bar{u}} = [\nabla(\hat{\bar{u}})]^t \Upsilon = [\nabla(\Upsilon\bar{u})]^t \Upsilon = \Upsilon[\nabla\bar{u}]^t \Upsilon. \tag{8.172}$$

This leads to the objectivity of $\bar{\sigma}$:

$$\hat{\bar{\sigma}} = \Upsilon\bar{\sigma}^t \Upsilon.$$

And since $|\hat{\bar{\sigma}}| = |\bar{\sigma}|$, (8.171) is verified. Smagorinsky model is then invariant.

For similarity model (8.169), one has:

$$\hat{\bar{u}} \otimes \hat{\bar{u}} = (\Upsilon\bar{u}) \otimes (\Upsilon\bar{u}) = \Upsilon(\bar{u} \otimes \bar{u})^t \Upsilon.$$

By means of these relations, invariance can easily been deduced.

The same relations are sufficient to prove invariance of the dynamic model since the trace remains invariant under a change of orthonormal basis.

The structure function model (8.166) is invariant because the function $\overline{F}_2$ is not altered under a rotation or a reflection.

Relations (8.172) can be used again to prove invariance of each of the gradient-type models.

Finally, since

$$\hat{\bar{\omega}} = \Upsilon\bar{\omega}^t \Upsilon,$$

Lund–Novikov–type models are also invariant.

Scaling Invariance

The two scaling transformations can be gathered in a 2–parameter transformation which, when applied to the filtered variables, have the following expression [RH06]

$$(t, x, \overline{u}, \overline{p}, \nu) \;\longmapsto\; (e^{2a}t, e^{ab}x, e^{b-a}\overline{u}, e^{2b-2a}\overline{p}, e^{2b}\nu).$$

where a and b are the parameters. The first scaling transformations corresponds to the case $b = 0$ and the second ones to the case $a = 0$.

It can be checked that the filtered equations (8.163) are invariant under the two scaling transformations iff: $\hat{T}_s = e^{2b-2a}T_s$. Since $\hat{\overline{\sigma}} = e^{-2a}\overline{\sigma}$, this condition is equivalent to: $\hat{\nu}_s = e^{2b}\nu_s$ for a turbulent viscosity model.

For Smagorinsky model, we have:

$$\hat{\nu}_s = C_S \overline{\delta}^2 |\hat{\overline{\sigma}}| = e^{-2a} C_S \overline{\delta}^2 |\overline{\sigma}| = e^{-2a}\nu_s.$$

Here, condition $\hat{\nu}_s = e^{2b}\nu_s$ is violated. The model is invariant neither under the first nor under the second scaling transformations. Note that the filter width $\overline{\delta}$ does not vary since it is an external scale length and has no functional dependence on the variables of the flow.

The dynamic procedure used in (8.165) restores the scaling invariance. Indeed, it can be shown that: $\hat{C}_d = e^{2b+2a}C_d$, that implies:

$$\hat{\nu}_s = \hat{C}_d \overline{\delta}^2 |\hat{\overline{\sigma}}| = e^{2b} C_d \overline{\delta}^2 |\overline{\sigma}| = e^{2b}\nu_s.$$

The dynamic model is then invariant under the two scaling transformation.

For the structure function model, we have: $\hat{\overline{F}}_2 = e^{b-a}\overline{F}_2$ and then $\hat{\nu}_s = e^{b-a}\nu_{sm}$, that proves that the model is not invariant.

Since $\hat{\overline{\nabla u}} = e^{2a}\overline{\nabla u}$, the gradient model (8.167) violates $\hat{T}_s = e^{2b-2a}T_s$, if τ_s is varying in the following way: $\hat{\tau}_s = e^{4a}\tau_s$. This also implies that none of the gradient-type models is invariant.

It is straight forward to prove that the similarity model (8.169) is invariant.

At last, Lund–Novikov–type models are not invariant because they comprise a term similar to Smagorinsky model.

In fact, none of the models where the external length scale $\overline{\delta}$ appears explicitly is invariant under the scaling transformations. Note that the dynamic model, which is invariant under these transformations, can be written in the following form:

$$\tau_s^d = \frac{\mathrm{Tr}(LN)}{\mathrm{Tr}(N^2)} \,|\overline{\sigma}|\overline{\sigma}, \qquad \text{where} \qquad N = \widetilde{|\overline{\sigma}|\overline{\sigma}} - (\widetilde{\overline{\delta}}/\overline{\delta})^2 |\widetilde{\overline{\sigma}}|\widetilde{\overline{\sigma}}.$$

It is then the ratio $\hat{\overline{\delta}}/\overline{\delta}$ which is present in the model but neither $\overline{\delta}$ alone nor $\hat{\overline{\delta}}$ alone.

In summary, the dynamic and the similarity models are the only invariant models under the scaling transformations. Though, scaling transformations have a particular importance because it is with these symmetries that Ober-lack [Obe99] derived scaling laws and that Ünal [Una94] proved the existence of solutions of Navier–Stokes equations having Kolmogorov spectrum.

The last symmetry property of Navier–Stokes equations is the material indifference, in the limit of 2D flow, in a simply connected domain.

Material Indifference

The material indifference corresponds to a time-dependent plane rotation, with a compensation in the pressure term as floowos [RH06]. The objectivity of $\overline{\sigma}$ directly leads to invariance of Smagorinsky model.

For similarity model (8.169), we have:

$$\hat{\tau}_s = R\tau_s^t R + R(\widetilde{\overline{u} \otimes x} - \widetilde{\overline{u}} \otimes \widetilde{x})^t R$$
$$+ \dot{R}(\widetilde{x \otimes \overline{u}} - \widetilde{x} \otimes \widetilde{\overline{u}})^t R + \dot{R}(\widetilde{x \otimes x} - \widetilde{x} \otimes \widetilde{x})^t \dot{R}.$$

Consequently, if the test filter is such that

$$(\widetilde{\overline{u} \otimes x} - \widetilde{\overline{u}} \otimes \widetilde{x}) = 0, \qquad (\widetilde{x \otimes \overline{u}} - \widetilde{x} \otimes \widetilde{\overline{u}}) = 0, \qquad (\widetilde{x \otimes x} - \widetilde{x} \otimes \widetilde{x}) = 0, \quad (8.173)$$

then the similarity model is invariant under the material indifference. All filters do not have this property.

Under the same conditions (8.173) on the test filter, the dynamic model is also invariant.

The structure function model is invariant iff

$$\hat{\overline{F}}_2 = \overline{\hat{F}}_2. \tag{8.174}$$

Let us calculate $\hat{\overline{F}}_2$. Let $\overline{u}_z$ be the function $x \mapsto \overline{u}(x + z)$. Then

$$\hat{\overline{F}}_2 = \int_{||z||=\overline{\delta}} ||(R\overline{u} + \dot{R}x) - (Ru_z + \dot{R}x + \dot{R}z)||^2 \, \mathrm{d}z$$
$$= \int_{||z||=\overline{\delta}} ||\overline{u} - \overline{u}_z - {}^t R\dot{R}z|| \, \mathrm{d}z.$$

Knowing that ${}^t R\dot{R}z = ae_3 \times z$, we get:

$$\hat{\overline{F}}_2 = \overline{F}_2 + 2\pi a^2 \overline{\delta}^3 - 2a \int_{||z||=\overline{\delta}} (\overline{u} - \overline{u}_z) \cdot (e_3 \times z) \, \mathrm{d}z.$$

Condition (8.174) is violated. So, the structure function model is not invariant under the material indifference.

For the gradient model, we have:

$$\hat{\nabla}\overline{u} = R\,\nabla\overline{u}\,{}^t R + \dot{R}\,{}^t R, \qquad {}^t\hat{\nabla}\overline{u} = R\,{}^t\nabla\overline{u}\,{}^t R + R\,{}^t\dot{R}.$$

Let J be the matrix such that $\dot{R}^t R = -aJ = -R^t \dot{R}$ or, in a component form:

$$J = \begin{pmatrix} 0 & 1 \\ -1 & 0 \end{pmatrix}. \qquad \text{Then,}$$

$$(\nabla u \, \hat{}\, {}^t \nabla u) = R(\nabla u \, {}^t \nabla u)^t R + aR\nabla \overline{u} \, {}^t RJ - aJR^t \nabla \overline{u} \, {}^t R + a^2 I_d.$$

The commutativity between J and R finally leads to:

$$\hat{\tau}_s = R\tau_s \, {}^t R + aR(\nabla \overline{u} J - J^t \nabla \overline{u}) + a^2 I_d.$$

This proves that the gradient model is not invariant.

The other gradient–type models inherit the lack of invariance of the gradient model.

It remains to review the Lund–Novikov–type models (8.170). For the filtered vorticity tensor ω we have [RH06]

$$\hat{\omega}^2 = R\,\omega^2 \, {}^t R - aR(J\omega + \omega J)^t R - a^2 I_d.$$

Since ω is anti-symmetric and the flow is 2D, ω is in the form:

$$\omega = \begin{pmatrix} 0 & \overline{a} \\ -\overline{a} & 0 \end{pmatrix}.$$

A direct calculation leads then to

$$J\omega = \omega J = -\overline{a} I_d \qquad \text{and} \qquad \hat{\omega}^2 = R\,\omega^2 \, {}^t R - (2\overline{a} - a)aI_d.$$

Let us see now how each term of the model (8.170) containing ω varies under the transformation [RH06]. From the last equation, we deduce the objectivity of $(\omega^2)^d$:

$$(\hat{\omega}^2)^d = R\,(\omega^2)^d \, {}^t R.$$

For the fourth term of (8.170), we have

$$\hat{\overline{\sigma}}\hat{\omega} - \hat{\omega}\hat{\overline{\sigma}} = R(\overline{\sigma}\omega - \omega\sigma)^t R - 2aR\overline{\sigma}J^t R.$$

And for the last term,

$$\hat{\overline{\sigma}}^2 \hat{\omega} - \hat{\overline{\sigma}}\hat{\omega}^2 = R(\overline{\sigma}^2\omega - \overline{\sigma}\,\omega^2)^t R - aR\overline{\sigma}^2 \, {}^t RJ - (2\overline{a} - a)aR\overline{\sigma} \, {}^t R.$$

Putting these results together, we have:

$$\hat{\tau}_s^d = R\tau_s^d \, {}^t R - a\overline{\delta}^2 R \left[2C_4 \overline{\sigma} J - C_5 \frac{1}{|\overline{\sigma}|} \left(\overline{\sigma}^2 J - (2\overline{a} - a)a\overline{\sigma} \right) \right]^t R.$$

We conclude that Lund–Novikov model is not invariant the material indifference.

The dynamic and the similarity models have an inconvenience that they necessitate use of a test filter. Rather constraining conditions, (8.173), are then needed for these models to preserve the material indifference. In addition, the dynamic model does not conform to the second law of thermodynamics since it may induce a negative dissipation. Indeed, $\nu + \nu_s$ can take a negative value. To avoid it, an *a posteriori* forcing is generally done. It consists of assigning to ν_s a value slightly higher than $-\nu$: $\nu_s = -\nu\,(1 - \varepsilon)$, where ε is a positive real number, small against 1.

8.6.5 Thermodynamic Consistence

Following [RH06], we will build a class of models which possess the symmetries of Navier–Stokes equations and next refine this class such that the models also satisfy the thermodynamics requirement.

Invariance under the Symmetries

Suppose that $\overline{\sigma} \neq 0$. Let τ_s be an analytic function of $\overline{\sigma}$. By this way, invariance under the time, pressure and generalised Galilean translations and under the reflections is guaranteed. It means that the *Cayley–Hamilton theorem* and invariance under the rotations lead to:

$$\tau_s^d = A(\chi, \zeta)\,\overline{\sigma} + B(\chi, \zeta)\mathrm{adj}^d\overline{\sigma}, \tag{8.175}$$

where $\chi = \mathrm{Tr}\overline{\sigma}^2$ and $\zeta = \det \overline{\sigma}$ are the invariants of $\overline{\sigma}$ (the third invariant, $\mathrm{Tr}\overline{\sigma}$, vanishes), adj stands for the operator defined by: $(\mathrm{adj}\overline{\sigma})\overline{\sigma} = (\det \overline{\sigma})I_d$, $(\mathrm{adj}\overline{\sigma}$ is simply the comatrix of $\overline{\sigma}$) and A and B are arbitrary scalar functions.

Now, a necessary and sufficient condition for τ_s defined by (8.175) to be invariant under the second scale transformations is that ν can be factorized [RH06]:

$$\tau_s^d = \nu A_0(\chi, \zeta)\,\overline{S} + \nu B_0(\chi, \zeta)\mathrm{adj}^d\overline{S}.$$

Lastly, τ_s is invariant under the first scaling transformations if: $\hat{\tau}_s = \mathrm{e}^{-2a}\tau_s$. Rewritten for A_0 and B_0, this condition becomes:

$$A_0(\mathrm{e}^{-4a}\chi, \mathrm{e}^{-6a}\zeta) = A_0(\chi, \zeta), \qquad B_0(\mathrm{e}^{-4a}\chi, \mathrm{e}^{-6a}\zeta) = \mathrm{e}^{2a}B_0(\chi, \zeta).$$

After differentiating according to a and taking $a = 0$, it follows:

$$-4\chi\partial_\chi A_0 - 6\zeta\partial_\zeta A_0 = 0, \qquad -4\chi\partial_\chi B_0 - 6\zeta\partial_\zeta B_0 = 2B_0.$$

To satisfy these equalities, one can take

$$A_0(\chi, \zeta) = A_1(\zeta/\chi^{3/2}), \qquad B_0(\chi, \zeta) = B_1(\zeta/\chi^{3/2})/\sqrt{\chi}.$$

Finally, if $v = \zeta/\chi^{3/2}$ then

$$\tau_s^d = \nu A_1(v)\ \overline{S} + \nu\, B_1(v)\mathrm{adj}^d\overline{S}//\sqrt{\chi}. \tag{8.176}$$

A subgrid–scale model of class (8.176) remains then invariant under the symmetry transformations of Navier–Stokes equations.

Several authors were interested in building invariant models for a long time. But because they did not use Lie theory, they did not consider some symmetries such as the scaling transformations which are particularly important. Three of the few authors who considered all the above symmetries in the modeling of turbulence are Ünal [Una97] and Saveliev and Gorokhovski [SG05]. The present manner to build invariant models generalises the Ünal's one in the sense that it introduces ν and the invariants of $\overline{\sigma}$ into the models. In addition, Ünal used the Reynolds averaging approach (RANS) instead of the large-eddy simulation approach (LES) for the turbulence modelling. Saveliev and Gorokhovski in [SG05] used the LES approach but derive their model in a different way than in the present article.

Let us now return to considerations which are more specific to large eddy simulation. We know that τ_s represents the energy exchange between the resolved and the subgrid scales. Then, it generates certain dissipation. To account for the second law of thermodynamics, we must ensure that the total dissipation remains positive that is not always verified by models in the literature. In order to satisfy this condition, we refine class (8.176).

Consequences of the Second Law

At the molecular scale, the viscous constraint is [RH06]: $\tau = \partial_\sigma \psi$. The potential $\psi = \nu \mathrm{Tr}\sigma^2$ is convex and positive that ensures that the molecular dissipation is positive: $\Phi = \mathrm{Tr}(\tau\sigma) \geq 0$. The tensor τ_s can be considered as a subgrid constraint, generating a dissipation $\Phi_s = \mathrm{Tr}(\tau_s\overline{\sigma})$. To preserve compatibility with the Navier–Stokes equations, we assume that τ_s has the same form as τ: $\tau_s = \partial_{\overline{\sigma}}\psi_s$, where ψ_s is a potential depending on the invariants χ and ζ of $\overline{\sigma}$. This hypothesis refines class (8.176) in the following way.

Since $\mathrm{Tr}\overline{\sigma} = 0$, it follows that [RH06]

$$\tau_s^d = 2\partial_\chi \psi_s \overline{S} + \partial_\zeta \psi_s \mathrm{adj}^d\overline{S}.$$

Comparing it with (8.176), one gets:

$$\nu A_1(v)/2 = \partial_\chi \psi_s, \qquad \nu B_1(v)/\sqrt{\chi} = \partial_\zeta \psi_s.$$

This leads to:

$$\partial_\zeta[A_1(v)/2] = \partial_\chi[B_1(v)\sqrt{\chi}].$$

If g is a primitive of B_1, a solution of this equation is

$$A_1(v) = 2g(v) - 3vg'(v) \qquad \text{and} \qquad B_1(v) = g'(v). \tag{8.177}$$

Then, the hypothesis $\tau_s = \partial_{\overline{\sigma}}\psi_s$ involves existence of a function g such that:

$$\tau_s^d = \nu[2g(v) - 3vg'(v)]\overline{S} + \nu\frac{1}{\sqrt{\chi}}\,g'(v)\mathrm{adj}^d\overline{S}.$$

Now, let Φ_T be the total dissipation. We have: $\Phi_T = \mathrm{Tr}[(\overline{\tau} + \tau_s)\overline{\sigma}]$. Using (8.176) and (8.177), one can show that

$$\Phi_T \geq 0 \quad \Longleftrightarrow \quad 1 + A_1(v) + 3vB_1(v) \geq 0 \quad \Longleftrightarrow \quad 1 + g(v) \geq 0.$$

8.6.6 Stability of Turbulence Models

After an eventual change of variables such that u vanishes along the boundary Γ of the domain Ω, the filtered equations can be written in the following form:

$$\partial_t \overline{u} + \mathrm{div}(\overline{u} \otimes \overline{u}) + \nabla\overline{p}/\rho - \mathrm{div}(\overline{\tau} - \tau_s) = F, \qquad \mathrm{div}\overline{u} = 0, \qquad (8.178)$$

associated to the conditions

$$\overline{u} = 0, \qquad \text{sur} \;\; \Gamma,$$
$$\overline{u}(0, x) = \gamma(x), \qquad \text{on} \;\; \Omega,$$
$$\int_\Omega \overline{p}(t, x)\,\mathrm{d}x = 0, \qquad \text{for all } t \in [0, t_f].$$

F is an appropriate function of t and x and t_f is the final observation time.

Let $(\overline{u}, \overline{p})$ be a regular solution of (8.178) where τ_s is symmetric and verifies the condition [RH06]:

$$\mathrm{Tr}[(\overline{\tau} - \tau_s)\overline{\sigma}] \geq 0; \qquad \text{then we have}$$

$$||\overline{u}(t, x)||_{L^2(\Omega)} \leq ||\gamma(x)||_{L^2(\Omega)} + \int_0^{t_f} ||F(\tau, x)||_{L^2(\Omega)}\,\mathrm{d}\tau.$$

This proposition ensures a finite energy when the model conforms to the second law of thermodynamics.

To prove it, let $(\,,)$ denote the scalar product of $L^2(\Omega)$ and $(\overline{u}, \overline{p})$ a regular solution of (8.178). From the first equation of (8.178) and the boundary condition, we have:

$$(\partial_t \overline{u}, \overline{u}) + b(\overline{u}, \overline{u}, \overline{u}) - (p, \mathrm{div}\overline{u})/\rho + (\overline{\tau} - \tau_s, \nabla\overline{u}) = (F, \overline{u}),$$

where b is defined by the trilinear form

$$b(u^1, u^2, u^3) = (\mathrm{div}(u^1 \otimes u^2), u^3).$$

From integrals by parts, the boundary condition and the divergence free condition, it can be shown that $b(\overline{u}, \overline{u}, \overline{u}) = 0$ and $(p, \mathrm{div}\overline{u}) = 0$. Since $(\overline{\tau} - \tau_s)$ is symmetric, it follows that

$$(\partial_t \overline{u}, \overline{u}) + (\overline{\tau} - \tau_s, \overline{\sigma}) = (F, \overline{u}). \qquad \text{Consequently,}$$
$$\frac{1}{2}\frac{\mathrm{d}}{\mathrm{d}t}(\overline{u}, \overline{u}) + (\overline{\tau} - \tau_s, \overline{\sigma}) = (F, \overline{u}).$$

Now, using the main hypothesis, we have:

$$(\overline{\tau} - \tau_s, \overline{\sigma}) = \int_\Omega \mathrm{Tr}[(\overline{\tau} - \tau_s)\overline{\sigma}] \, \mathrm{d}x \geq 0, \qquad \text{so that}$$

$$\frac{1}{2} \frac{\mathrm{d}}{\mathrm{d}t} (\overline{u}, \overline{u}) \leq (F, \overline{u}). \qquad \text{Consequently,}$$

$$||\overline{u}||_{L^2(\Omega)} \frac{1}{2} \frac{\mathrm{d}}{\mathrm{d}t} ||\overline{u}||_{L^2(\Omega)} \leq (F, \overline{u}) \leq ||F||_{L^2(\Omega)} ||\overline{u}||_{L^2(\Omega)}.$$

After simplifying by the L^2–norm of $\overline{u}$ and integrating over the time, it follows:

$$||\overline{u}(t, x)||_{L^2(\Omega)} \leq ||\gamma(x)||_{L^2(\Omega)} + \int_0^t ||F(\tau, x)||_{L^2(\Omega)} \, \mathrm{d}\tau$$

$$\leq ||\gamma(x)||_{L^2(\Omega)} + \int_0^{t_f} ||F(\tau, x)||_{L^2(\Omega)} \, \mathrm{d}\tau.$$

8.7 Advection of Vector–Fields by Chaotic Flows

8.7.1 Advective Fluid Flow

When the particles of a fluid are endowed with some scalar density S, the density evolves in time according to [BCI93]

$$\partial_t S + u \cdot \nabla S \equiv \frac{DS}{Dt} = \text{thermal noise.} \qquad (8.179)$$

The right–hand side of (8.179) represents the microscopic spreading of S on the molecular level, and can be thought of as noise added onto the fluid velocity vector u. It is normally described by a term like $\kappa \nabla^2 S$ where κ is a diffusivity. The study of (8.179), especially for chaotic flows and turbulent flows, has been extensively carried on for many decades [Ott39, Tay38].

Fluid motions also move vector–fields around. An evolving vector–field V is governed by an equation of the form

$$\partial_t V + u \cdot \nabla V - V \cdot \nabla u \equiv \frac{DV}{Dt} = \text{thermal noise,} \qquad (8.180)$$

where the *advective derivative* in (8.179) is replaced by a *Lie derivative DV/Dt* in the *evolution operator* [II06b]. The extra term enters because the evolution of a vector–field involves not only being carried about by a fluid particle but also being turned and stretched by the motion of neighboring particles. The thermal noise on the right side is usually a simple diffusion term, at least for the case of constant density. Density variations bring in some unappetizing complications that we shall ignore.

If the right sides of (8.179) and (8.180) are zero (perfect fluid motion) then S and V are frozen–in properties, and the fluid motions can distort any V in a complex fashion. In particular, when the dynamical system

$$\dot{x} = u \qquad (8.181)$$

produces chaotic motion, the effect on the distribution of the advective fields can be rich and surprising, giving rise to intense local concentrations and lacunae of fine structure.

In real fluid settings the situation may be more complicated than even this description suggests. If either S or V can feed back on the dynamics of u itself, the equations lose even their superficially linear appearances. For ordinary fluid motions, we have $\rho V = \omega$, where ρ is the fluid density and the problem is no longer kinematic. Rather, u satisfies [BCI93]

$$\rho(\partial_t u + \omega \times u) = -\nabla p - \frac{1}{2}\rho\nabla u^2 + F + \text{thermal viscous effects}, \qquad (8.182)$$

where F is an *external force density*. So when ρV is vorticity, we have coupled equations for u and V. No wonder that the turbulence problem governed by this pair of equations has proved so difficult.

G.I. Taylor, who made early contributions to the study of (8.179) for turbulent u, observed that vorticity is concentrated by turbulence [Tay38]. To learn what properties of the motion favor this effect, we may begin with the study of (8.180) without worrying, at first, whether these motions correspond to solutions of (8.182). That leads to the search for a field u that may produce local vorticity enhancement when introduced into (8.180). We call this the *kinematic turbulence problem*, after the usage of dynamo theory, where (8.180) applies when ρV is the magnetic field B.

Some authors have sought analogies between the vorticity and magnetic field, since they are both controlled by equation (8.180). However, this viewpoint has been belittled because $\omega = \nabla \times u$, while no analogous relation exists between B and u. Yet this relation is implied by (8.180) and (8.182), and it need not as a result be considered explicitly. When B is the field in question, we couple it to (8.182) through a nonlinear term representing the *Lorentz force*,

$$\rho(\partial_t u + u \cdot \nabla u) = -\nabla p - \frac{1}{4\pi}B \times \nabla \times B + F + \text{thermal viscous effects}. \quad (8.183)$$

Typically the kinematic aspects of the dynamo problem are stressed over those of the analogous vorticity problem. But if the analogy is good, it is not clear whether this should raise interest in the kinematic turbulence problem or throw the kinematic dynamo problem in a bad light. In any case, there does seem to be considerable interest in studying the effect of fluid motion on an immersed vector–field through (8.180), for prescribed u.

These general thoughts lead into our discussion of the effect of a chaotic motion on a vector–field satisfying (8.180). We describe in general terms the

procedures that we have been following, omit the most strenuous portions of
the calculations and refer to the related work on the mixing of a passive scalar
by a continuous flow [CE91], and to the study by [AG39] of fast dynamos in
discrete maps.

Now, to formally solve this problem, let us consider velocity fields u that
are steady. In this autonomous case, the problem can be simplified by writing
$V(x,t) = V_0(x)\exp(\lambda t)$. Then we get, for the diffusion–less case [BCI93],

$$DV/Dt \equiv (u \cdot \nabla V_0 - V_0 \cdot \nabla u)e^{\lambda t} = -\lambda V_0 e^{\lambda t}. \qquad (8.184)$$

We may look for a solution of the form $V_0 = q(x)u(x)$, where

$$u \cdot \nabla q = -\lambda q. \qquad (8.185)$$

This shows how the kinematic problem for vector–fields may be related to the
more extensively studied problem of passive scalar transport [CE91]; if we set

$$S(x,t) = q(x)e^{\lambda t}$$

in (8.179), we get (8.185). Moreover, if q is constant on streamlines, then we
must have $\lambda = 0$. These special solutions arise in the very restricted conditions of steady flow without dispersive effects, and they illustrate the kind of
degeneracy that we encounter when the conditions are too simple.

More generally, we can write the solution of (8.180) formally, as shown by
A. Cauchy. Let $x(t,a)$ be the position of the fluid particle that was at the
point a when $t = 0$. Then the field evolves according to [BC93]

$$V(x,t) = J(a,t)V(a,0), \qquad (8.186)$$

where $J(a,t) = \partial(x)/\partial(a)$ is the Jacobian matrix of the transformation that
moves the fluid into itself with time with $x = x(a,t)$. This result of Cauchy
simply says that V is a contravariant vector.

We can write $x = \phi_t a$, where ϕ_t is the flow that maps the initial positions
of the fluid particles into their positions at time t. Its inverse, $a = \phi_t^{-1}x$ maps
particles at time t and position x back to their initial positions. Then we can
write (8.186) in the seemingly complicated, but quite useful form,

$$V(x,t) = \int \delta(a - \phi_t^{-1}x)J(a,t)\,V(a,0)\,d^3a, \qquad (8.187)$$

where the integral operator introduced here is the analogue of the *Perron–
Frobenius operator* for the case of scalar advection [Rue89]. Having turned the
differential equation (8.180) into an integral equation, we may use analogues
of Fredholm's methods [Ch53] to solve it.

If we were to include the effects of diffusion by using a term of the form
$\eta\nabla^2 V$ on the right hand side of (8.180), with η constant, we would need to
replace D/Dt by $D/Dt - \eta\nabla^2$ in (8.184) and modify (8.187).

8.7.2 Chaotic Flows

To describe the evolution of a frozen–in magnetic field, we need a suitable means to characterize the flow. For the chaotic flows on which we concentrate here, the Lagrangian orbits determined by (8.181) densely contain unstable periodic orbits. Their union forms what we may call a chaotic set. If object is attracting, the term *strange attractor* is appropriate while in the opposite case it becomes a *strange repeller*, from which representative points depart as time proceeds. Even though all periodic orbits within such an ensemble are unstable, trajectories of the flow often spend extended periods tracking these paths. The strange set provides a delicate, skeletal structure embedded within the flow that can be used to systematically approximate such properties as its local stability and geometry [CE91]. This is the basic idea underlying *cycle expansions* [AAC90], the method that we apply here to the problem of vector advection.

To be explicit, we consider a trajectory $x(t)$ generated by the third–order ('jerky') ODE [BCI93]

$$\dddot{x} + \dot{x} - cx + x^3 = 0, \qquad (8.188)$$

with parameter c. This is a special case of equations arising in multiply diffusive convection [ACS85]. The absence of a second derivative in (8.188) ensures that the flow is solenoidal, in line with the kinds of flow most commonly studied in dynamo theory, in spite of the fact that every putative dynamo is compressible. For certain values of c, homoclinic orbits exist and we let c be close to such a value and, in particular, one such that in the neighborhood of the origin the flow satisfies *Shil'nikov criterion* for the existence of an infinity of unstable periodic orbits [Shi65]. Then, an infinite sequence of intertwined saddle–node and period–doubling bifurcations creates a dense chaotic set, and the motion of points in the flow is chaotic.

Without going into details, we mention only that $x(t)$, under the conditions mentioned, is a sequence of pulses. Moreover, the separation of the kth and $(k-1)$st pulses, Δ_k, may be expressed in terms of the two previous spacings. This provides us with a *timing map*, which is a 2D map resembling in form the *Hénon map*. Thus, to good approximation,

$$\Delta_k = \bar{T} + \alpha\tau_k,$$

where $\bar{T}$ is a mean period, α is a small parameter, and τ_k an irregular timing fluctuation satisfying,

$$\tau_{k+1} = 1 - a\tau_k^2 - \tau_{k-1}, \qquad (8.189)$$

which is the orientation and area preserving form of the Hénon map.

The form of any particular pulse is quite close to that of the *homoclinic orbit*, which can be computed at the outset where the map (8.189) determines a sequence of pulse positions. Once these are known, we can generate a complete and reasonably accurate solution for the velocity field. Thus we can reconstruct the entire *flow* from this simple map. Moreover, this map also contains the invariant information about the periodic orbits of the chaotic set (cycle topology and stability eigenvalues) and serves as a powerful tool in the construction of cycle expansions [BCI93].

8.8 Brownian Motion and Diffusion

The simplest model of diffusion is the Brownian motion, the erratic movement of a grains suspended in liquid observed by the botanist Robert Brown as early as in 1827. After the fundamental work of Einstein [Ein05] and Langevin [Lan08], Brownian motion become the prototypical example of stochastic process.

8.8.1 Random Walk Model

In order to study more in detail the properties of diffusion, let us introduce the simplest model of Brownian motion, i.e., the 1D random walk. The walker moves on a line making discrete jumps $v_i = \pm 1$ at discrete times. The position of the walker, started at the origin at $t = 0$, will be [BLV01]

$$R(t) = \sum_{i=1}^{t} v_i. \tag{8.190}$$

Assuming equiprobability for left and right jumps (no mean motion), the probability that at time t the walker is in position x will be

$$p_t(R = x) = prob \left[\frac{\frac{t-x}{2} steps - 1}{\frac{t+x}{2} steps + 1} \right] = \frac{1}{2^t} \left(\begin{array}{c} t \\ \frac{t+x}{2} \end{array} \right). \tag{8.191}$$

For large t and x (i.e., after many microscopic steps) we can use Stirling approximation and get

$$p_t(x) = \sqrt{\frac{t}{2\pi(t^2 - x^2)}} \exp \left[-\frac{t+x}{2} \ln \frac{t+x}{2} - \frac{t-x}{2} \ln \frac{t-x}{2} \right]. \tag{8.192}$$

The core of the distribution recovers the well known Gaussian form, i.e., for $x \ll t$ from (8.192) we get

$$p_t(x) = \sqrt{\frac{1}{2\pi t}} \exp \left(-\frac{x^2}{2t} \right). \tag{8.193}$$

From (8.193) one gets that the variance of the displacement follows diffusive behavior, i.e.,

$$\langle x^2(t) \rangle = 2t. \tag{8.194}$$

We stress that diffusion is geted only asymptotically (i.e., for $t \to \infty$). This is a consequence of central limit theorem which assures Gaussian distributions and diffusive behavior in the limit of many independent jumps. The necessary, and sufficient, condition for observing diffusive regime is the existence of a finite correlation time (here represented by discrete time between jumps) for the microscopic dynamics. Let us stress that this is the important ingredient for diffusion, and *not* a stochastic microscopic dynamics. We will see below that diffusion can arise even in completely deterministic systems.

Another important remark is that Gaussian distribution (8.193) is intrinsic of diffusion process, independent on the distribution of microscopic jumps: indeed only the first two moments of v_i enter into expression (8.193). This is, of course, the essence of the central limit theorem. Following the above derivation, it is clear that this is true only in the core of the distribution. The far tails keep memory of the microscopic process and are, in general, not Gaussian. As an example, in

The Gaussian distribution (8.193) can be geted as the solution of the diffusion equation which governs the evolution of the probability in time. This is the Fokker–Planck equation for the particular stochastic process. A direct way to relate the 1D random walk to the diffusion equation is geted by introducing the master equation, i.e., the time evolution of the probability [Gar94]:

$$p_{t+1}(x) = \frac{1}{2} p_t(x-1) + \frac{1}{2} p_t(x+1). \tag{8.195}$$

In order to get a continuous limit, we introduce explicitly the steps Δx and Δt and write

$$\frac{p_{t+\Delta t}(x) - p_t(x)}{\Delta t} = \frac{(\Delta x)^2}{2\Delta t} \frac{p_t(x+\Delta x) + p_t(x-\Delta x) - 2p_t(x)}{(\Delta x)^2}. \tag{8.196}$$

Now, taking the limit $\Delta x, \Delta t \to 0$ in such a way that $(\Delta x)^2/\Delta t \to 2D$ (the factor 2 is purely conventional) we get the diffusion equation

$$\frac{\partial p(x,t)}{\partial t} = D \frac{\partial^2 p(x,t)}{\partial x^2}. \tag{8.197}$$

The way the limit $\Delta x, \Delta t \to 0$ is taken reflects the scaling invariance property of diffusion equation. The solution to (8.197) is readily geted as

$$p(x,t) = \frac{1}{\sqrt{4\pi D t}} \exp\left(-\frac{x^2}{4Dt}\right). \tag{8.198}$$

Diffusion equation (8.197) is here written for the probability $p(x,t)$ of observing a marked particle (a tracer) in position x at time t. The same equation can have another interpretation, in which $p(x,t) = \theta(x,t)$ represents the concentration of a scalar quantity (marked fluid, temperature, pollutant) as function of time. The only difference is, of course, in the normalization.

8.8.2 More Complicated Transport Processes

As already stated time de–correlation is the key ingredient for diffusion. In the random walker model it is a consequence of randomness: the steps v_i are random uncorrelated variables and this assures the applicability of central limit theorem. But we can have a finite time correlation and thus diffusion also without randomness. To be more specific, let us consider the following deterministic model (standard map [Chi79]):

$$\begin{cases} J(t+1) = J(t) + K \sin\theta(t), \\ \theta(t+1) = \theta(t) + J(t+1). \end{cases} \qquad (8.199)$$

The map is known to display large-scale chaotic behavior for $K > K_c \simeq 0.9716$ and, as a consequence of deterministic chaos, $J(t)$ has diffusive behavior. For large times, $J(t)$ is large and thus the angle $\theta(t)$ rotates rapidly. In this limit, we can assume that at each step $\theta(t)$ decorrelates and thus write

$$J(t)^2 = K^2 \left(\sum_{t'=1}^{t} \sin\theta(t')\right)^2 \simeq K^2 \langle \sin^2\theta \rangle t = 2Dt. \qquad (8.200)$$

The diffusion coefficient D, in the random phase approximation, i.e., assuming that $sin\theta(t)$ is not correlated with $sin\theta(t')$ for $t \neq t'$, is geted by the above expression as $D_{RPA} = K^2/4$.

The two examples discussed above are in completely different classes: stochastic for the random walk (8.190) and deterministic for the standard map (8.199). Despite this difference in the microscopic dynamics, both lead to a macroscopic diffusion equation and Gaussian distribution. This demonstrates how diffusion equation is of general applicability.

8.8.3 Advection–Diffusion

Let us now consider the more complex situation of dispersion in a non-steady fluid with velocity field $\mathbf{v}(\mathbf{x}, t)$ [BLV01]. For simplicity will we consider incompressible flow (i.e., for which $\nabla \cdot \mathbf{v} = 0$) which can be laminar or turbulent, solution of Navier–Stokes equation or synthetically generated according to a given algorithm. In presence of $\mathbf{v}(\mathbf{x}, t)$, the diffusion equation (8.197) becomes the *advection–diffusion* equation for the concentration $\theta(\mathbf{x}, t)$ (1.117). This equation is linear in θ but nevertheless it can display very interesting and non trivial properties even in presence of simple velocity fields, as a consequence of Lagrangian chaos. In the following we will consider a very simple example of diffusion in presence of an array of vortices. The example will illustrate in a nice way the basic mechanisms and effects of interaction between deterministic ($\mathbf{v}$) and stochastic (D) components.

Let us remark that we will not consider here the problem of transport in turbulent velocity field. This is a very classical problem, with obvious and important applications, which has recently attracted a renewal theoretical interest as a model for understanding the basic properties of turbulence [SS00].

Before going into the example, let us make some general consideration. We have seen that in physical systems the molecular diffusivity is typically very small. Thus in (1.117) the advection term dominates over diffusion. This is quantified by the Peclet number, which is the ratio of the typical value of the advection term to the diffusive term

$$Pe = \frac{v_0 l_0}{D}, \tag{8.201}$$

where v_0 is the typical velocity at the typical scale of the flow l_0. With $\tau_0 \simeq l_0/v_0$ we will denote the typical correlation time of the velocity.

The central point in the following discussion is the concept of *eddy diffusivity*. The idea is rather simple and dates back to the classical work of Taylor [Tay21]. To illustrate this concept, let us consider a Lagrangian description of dispersion in which the trajectory of a tracer $\mathbf{x}(t)$ is given by (1.118). Being interested in the limit $Pe \to \infty$, in the following we will neglect, just for simplicity, the molecular diffusivity D, which is generally much lesser that the effective dynamical diffusion coefficient.

Starting from the origin, $\mathbf{x}(0) = 0$, and assuming $\langle \mathbf{v} \rangle = 0$ we have $\langle \mathbf{x}(t) \rangle = 0$ for ever. The square displacement, on the other hand, grows according to

$$\frac{d}{dt}\langle \frac{1}{2}\mathbf{x}(t)^2 \rangle = \langle \mathbf{x}(t) \cdot \mathbf{v}_L(t) \rangle = \int_0^t \langle \mathbf{v}_L(s) \cdot \mathbf{v}_L(t) \rangle ns, \tag{8.202}$$

where we have introduced, for simplicity of notation, the Lagrangian velocity $\mathbf{v}_L(t) = \mathbf{v}(\mathbf{x}(t), t)$. Define the Lagrangian correlation time τ_L from

$$\int_0^\infty \langle \mathbf{v}_L(s) \cdot \mathbf{v}_L(0) \rangle ns = \langle \mathbf{v}_L(0)^2 \rangle \tau_L, \tag{8.203}$$

and assume that the integral converge so that τ_L is finite. From (8.202), for $t \gg \tau_L$ we get

$$\langle \mathbf{x}(t)^2 \rangle = 2\tau_L \langle \mathbf{v}_L^2 \rangle t, \tag{8.204}$$

i.e., diffusive behavior with diffusion coefficient (eddy diffusivity) $D^E = \tau_L \langle \mathbf{v}_L^2 \rangle$.

This simple derivation shows, once more, that diffusion has to be expected in general in presence of a finite correlation time τ_L. Coming back to the advection-diffusion equation (1.117), the above argument means that for $t \gg \tau_L$ we expect that the evolution of the concentration, for scales larger than l_0, can be described by an effective *diffusion equation*, i.e.,

$$\frac{\partial \langle \theta \rangle}{\partial t} = D_{ij}^E \frac{\partial^2 \langle \theta \rangle}{\partial x_i \partial x_j}. \tag{8.205}$$

The computation of the eddy diffusivity for a given Eulerian flow is not an easy task. It can be done explicitly only in the case of simple flows, for example by means of homogenization theory [BCV95]. In the general case it is relatively simple to give some bounds, the simplest one being $D^E \geq D$, i.e., the presence of a (incompressible) velocity field enhances large–scale transport. To be more specific, let us now consider the example of transport in a 1D array of vortices (cellular flow) sketched in Figure 8.2 This simple 2D flow is useful for illustrating the transport across barrier. Moreover, it naturally arises in several fluid dynamics contexts, such as, for example, convective patterns [SG88].

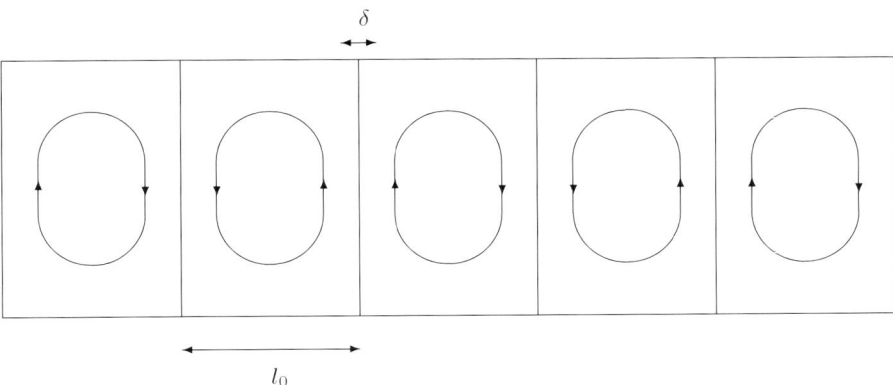

Fig. 8.2. Cellular flow model. l_0 is the size of vortices, δ is the thickness of the boundary layer (adapted from [BLV01]).

Let us denote by v_0 the typical velocity inside the cell of size l_0 and let D the molecular diffusivity. Because of the cellular structure, particles inside a vortex can exit only as a consequence of molecular diffusion. In a characteristic vortex time $\tau_0 \sim l_0/v_0$, only the particles in the boundary layer of thickness δ can cross the separatrix where

$$\delta^2 \sim D\tau_0 \sim D\frac{l_0}{v_0} \,. \tag{8.206}$$

These particles are ballistically advected by the velocity field across the vortex so they see a 'diffusion coefficient' l_0^2/τ_0. Taking into account that this fraction of particles is δ/l_0 we get an estimation for the effective diffusion coefficient as

$$D^E \sim \frac{\delta}{l_0}\frac{l_0^2}{\tau_0} \sim \sqrt{Dl_0v_0} \sim DPe^{1/2}. \tag{8.207}$$

The above result, which can be made more rigorous, was confirmed by nice experiments made by Solomon and Gollub [SG88]. Because, as already stressed above, typically $Pe \gg 1$, one has from (8.207) that $D^E \gg D$. On the other hand, this result do not mean that molecular diffusion D plays no role in the dispersion process. Indeed, if $D = 0$ there is not mechanism for the particles to exit from vortices.

Diffusion equation (8.205) is the typical long–time behavior in generic flow. There exist also the possibility of the so–called anomalous diffusion, i.e., when the spreading of particle do not grow linearly in time, but with a power law

$$\langle x^2(t) \rangle \sim t^{2\nu} \tag{8.208}$$

with $\nu \neq 1/2$. The case $\nu > 1/2$ (formally $D^E = \infty$) is called super–diffusion; sub–diffusion, i.e., $\nu < 1/2$ (formally $D^E = 0$), is possible only for compressible velocity fields.

Super–diffusion arises when the Taylor argument for deriving (8.204) fails and formally $D^E \to \infty$. This can be due to one of the following mechanisms:
a) the divergence of $\langle \mathbf{v}_L^2 \rangle$ (which is the case of *Lévy flights*), or
b) the lack of decorrelation and thus $T_L \to \infty$ (*Lévy walks*). The second case is more physical and it is related to the existence of strong correlations in the dynamics, even at large times and scales.

One of the simplest examples of Lévy walks is the dispersion in a quenched random shear flow [BG90, Isi92]. The flow, sketched in Figure 8.3, is a superposition of strips of size δ of constant velocity v_0 with random directions.

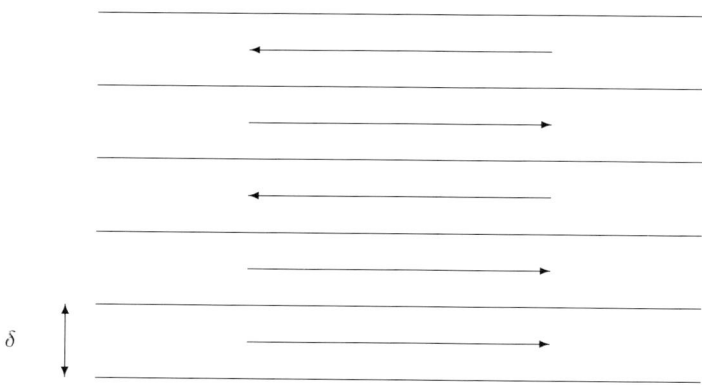

Fig. 8.3. Random shear of $\pm v_0$ velocity in strips of size δ (adapted from [BLV01])

Let us now consider a particle which moves according to the flow of Figure 8.3. Because the velocity field is in the x direction only, in a time t the typical displacement in the y direction is due to molecular diffusion only

$$\delta y \sim \sqrt{Dt} \tag{8.209}$$

and thus in this time the walker visits $N = \delta y/\delta$ strips. Because of the random distribution of the velocity in the strips, the mean velocity in the N strips is zero, but we may expect about $\sqrt{N}$ unbalanced strips (say in the right direction). The fraction of time t spent in the unbalanced strips is $t\sqrt{N}/N$ and thus we expect a displacement

$$\delta x \sim v_0 \frac{t}{\sqrt{N}} . \qquad (8.210)$$

From (8.209) we have $N \sim \sqrt{Dt}/\delta$ and finally

$$\langle \delta x^2 \rangle \sim \frac{v_0^2 \delta}{\sqrt{D}} t^{3/2}, \qquad (8.211)$$

i.e., a super–diffusive behavior with exponent $\nu = 3/4$.

The origin of the anomalous behavior in the above example is in the quenched nature of the shear and in the presence of large stripes with positive (or negative) velocity in the x direction. This leads to an infinite de–correlation time for Lagrangian tracers and thus to a singularity in (8.204). We conclude this example by observing that for $D \to 0$ (8.211) gives $\langle \delta x^2 \rangle \to \infty$. This is not a surprise because in this case the motion is ballistic and the correct exponent becomes $\nu = 1$.

As it was in the case of standard diffusion, also in the case of anomalous diffusion the key ingredient is not randomness. Again, the standard map model (8.199) is known to show anomalous behavior for particular values of K [Ver98].

The qualitative mechanism for anomalous dispersion in the standard map can be easily understood: a trajectory of (8.199) for which $K \sin \theta^* = 2\pi m$ with m integer, corresponds to a fixed–point for θ (because the angle is defined modulo 2π) and linear growth for $J(t)$ (ballistic behavior). It can be shown that the stability region of these trajectories in phase–space decreases as $1/K$ [Ver98, IKH87] and, for intermediate value of K, they play a important role in transport: particles close to these trajectories feel very long correlation times and perform very long jumps. The contribution of these trajectory, as a whole, gives the observed anomalous behavior.

Now, let us consider the cellular flow of Figure 8.2 as an example of sub-diffusive transport. We have seen that asymptotically (i.e., for $t \gg l_0^2/D$) the transport is diffusive with effective diffusion coefficient which scales according to (8.207). For intermediate times $l_0/v_0 \ll t \ll l_0^2/D$, when the boundary layer structure has set in, one expects anomalous sub-diffusive behavior as a consequence of fraction of particles which are trapped inside vortices. A simple model for this problem is the comb model [Isi92, HB87]: a random walk on a lattice with comb-like geometry. The base of the comb represents the boundary layer of size δ around vortices and the teeth, of length l_0, represent the inner area of the convective cells. For the analogy with the flow of Figure 8.2 the teeth are placed at the distance $\delta \sim \sqrt{Dl_0/v_0}$ (8.206).

A spot of random walker (dye) is placed, at time $t = 0$, at the base of the comb. In their walk on the x direction, the walkers can be trapped into the

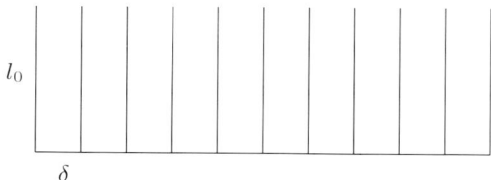

l_0

δ

Fig. 8.4. The comb model geometry (adapted from [BLV01]).

teeth (vortices) of dimension l_0. For times $l_0/v_0 \ll t \ll l_0^2/D$, the dye invades a distance of order $(Dt)^{1/2}$ along the teeth. The fraction $F(t)$ of active dye on the base (i.e., on the separatrix) decreases with time as

$$F(t) \sim \frac{\delta}{(Dt)^{1/2}}, \tag{8.212}$$

and thus the effective dispersion along the base coordinate b is

$$\langle b^2(t) \rangle \sim F(t)Dt \sim \delta (Dt)^{1/2}. \tag{8.213}$$

In the physical space the corresponding displacement will be

$$\langle x^2(t) \rangle \sim \langle b^2(t) \rangle \frac{l_0}{\delta} \sim l_0 (PeDt)^{1/2}, \tag{8.214}$$

i.e., we get a sub–diffusive behavior with $\nu = 1/4$.

The above argument is correct only for the case of free–slip boundary conditions. In the physical case of no–slip boundaries, one gets a different exponent $\nu = 1/3$. The latter behavior has been indeed observed in experiments [CT88].

8.8.4 Beyond the Diffusion Coefficient

From the above discussion it is now evident that diffusion, being an asymptotic behavior, needs large scale separation in order to be observed [BLV01]. In other words, diffusion arises only if the Lagrangian correlation time τ_L (8.203) is finite *and* the observation time is $t \gg \tau_L$ or, according to (8.205), if the dispersion is evaluated on scales much larger than l_0.

On the other hand, there are many physical and engineering applications in which such a scale separation is not achievable. A typical example is the presence of boundaries which limit the scale of motion on scales $L \sim l_0$. In these cases, it is necessary to introduce non–asymptotic quantities in order to correctly describe dispersion.

Before discussing the non-asymptotic statistics let us show, with an example, how it can be dangerous to apply the standard analysis in non–asymptotic situation. We consider the motion of tracers advected by the 2D flow generated by 4 point vortices in a disk. The evolution equation is given by (4.30) and (4.32) but now in (4.32), instead of $\ln r_{ij}$, one has to consider the *Green function* $G(r_{ij})$ on the disk.

In the following we will consider *relative dispersion*, i.e., the mean size of a cluster of particles

$$R^2(t) = \langle |\mathbf{x}(t) - \langle \mathbf{x}(t) \rangle |^2 \rangle, \qquad (8.215)$$

Of course, for separation larger than the typical scale of the flow, l_0, the particles move independently and thus we expect again the asymptotic behavior

$$R^2(t) \simeq 2Dt, \qquad \text{if} \quad R^2(t)^{1/2} \gg l_0. \qquad (8.216)$$

For very small separation we expect, assuming that the Lagrangian motion is chaotic,

$$R^2(t) \simeq R^2(0)e^{2\lambda t}, \qquad \text{if} \quad R^2(t)^{1/2} \ll l_0, \qquad (8.217)$$

where λ is the *Lagrangian Lyapunov exponent* [CFP91].

For intermediate times a power–law behavior with an anomalous exponent $\nu = 1.8$ is clearly observable. Of course the anomalous behavior is spurious: after the discussion of the previous section, we do not see any reason for observing super-diffusion in the point vortex system. The apparent anomaly is simply due to the lack of scale separation and thus to the crossover from the exponential regime (8.217) to the saturation value.

To partially avoid this kind of problem, it has been recently introduced a new indicator based on *fixed scale* analysis [ABC97]. The idea is very simple and it is based on exit time statistics. Given a set of thresholds $\delta_n = \delta_0 r^n$, one measures the exit time $T_i(\delta_n)$ it takes for the separation $R_i(t)$ to grow from δ_n to δ_{n+1}. The factor r may be any value > 1, but it should be not too large in order to have a good separation between the scales of motion.

Performing the exit time experiment over N particle pairs, from the average doubling time $\langle T(\delta) \rangle = 1/N \sum_i T_i(\delta)$, one defines the Finite Size Lyapunov Exponent (FSLE) as

$$\lambda(\delta) = \frac{\ln r}{\langle T(\delta) \rangle}, \qquad (8.218)$$

which recovers the standard Lagrangian Lyapunov exponent in the limit of very small separations $\lambda = \lim_{\delta \to 0} \lambda(\delta)$.

The finite size diffusion coefficient $D(\delta)$ is defined, within this framework, as

$$D(\delta) = \delta^2 \lambda(\delta). \qquad (8.219)$$

For standard diffusion $D(\delta)$ approaches the diffusion coefficient D (see (8.216)) in the limit of very large separations ($\delta \gg l_0$). This result stems from the scaling of the doubling times $\langle T(\delta) \rangle \sim \delta^2$ for normal diffusion.

Thus, according to (8.216)–(8.217), the asymptotic behaviors of the FSLE are

$$\lambda(\delta) \sim \begin{cases} \lambda, & \text{if} \quad \delta \ll l_0, \\ D/\delta^2, & \text{if} \quad \delta \gg l_0. \end{cases} \tag{8.220}$$

In presence of boundary at scales $L \sim l_0$, the second regime is not observable. For separation very close to to the saturation value $\delta_{max} \simeq L$ one expects the following behavior to hold for a broad class of systems [ABC97]:

$$\lambda(\delta) \propto \frac{\delta_{max} - \delta}{\delta}. \tag{8.221}$$

The finite scale method can be easily applied to the analysis of experimental data [BCE00]. An example is the study of Lagrangian dispersion in a experimental convective cell. The cell is a rectangular tank filled with water and heated by a linear heat source placed on the bottom. The heater generates a vertical plume which induces a general 2D circulation of two counter-rotating vortices. For high values of the Rayleigh number (i.e., heater temperature) the flow is not stationary and the plume oscillates periodically. In these conditions, Lagrangian tracers can jump from one side to the other of the plume as a consequence of chaotic advection.

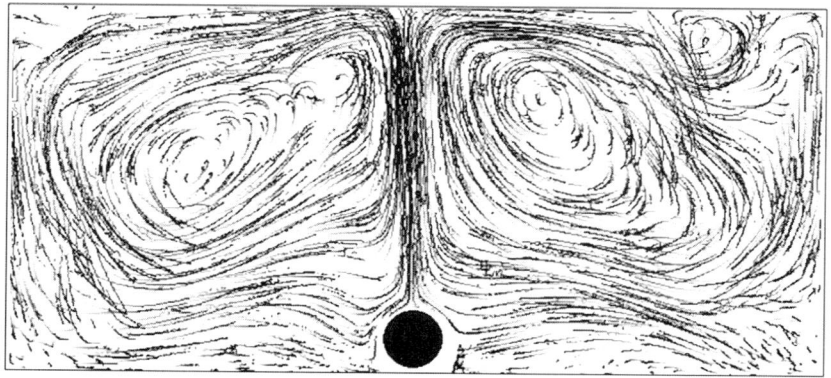

Fig. 8.5. An example of trajectories obtained by PTV technique in the convective cell at $Ra = 2.39 \times 10^8$. The vertical thermal plume is clearly observable. The dark circle on the bottom represents the heat source.

The finite scale tool has been successfully applied to many other numerical and experimental situations, from the dispersion in fully developed turbulence, to the analysis of tracer motion in ocean and atmosphere [LAV01, BLR01, JL02]. to engineering laboratory experiments. It will be probably became a standard tool in the analysis of Lagrangian dispersion [BLV01].

9

Geometry, Solitons and Chaos Field Theory

In this last Chapter we have collected some remaining chaos–related techniques (which did not fit within our previous Chapters) from modern high–dimensional nonlinear dynamics: chaos geometry, gauges, solitons, as well as the chaos field theory.

9.1 Chaotic Dynamics and Riemannian Geometry

Let us now recall very briefly the main points about the geometric theory of Hamiltonian chaos (for details, see [II06b] as well as [Pet93, CP93, CLP95, CCP96]). Despite the fact that this theory is still at its beginning, it has already proved useful not only in its original context, but also in connection with the problem of the relationship between dynamics and statistical mechanics (in particular phase transitions) [CCC98].

Hamiltonian dynamics can be rephrased in geometrical terms owing to the fact that the trajectories of a dynamical system with quadratic kinetic energy can be seen as geodesics of a suitable Riemannian manifold. There are several choices for the ambient manifold as well as for the metric tensor. As previously discussed in [CP93, CLP95, CCP96] a particularly useful ambient space is the enlarged configuration space–time $M \times \mathbb{R}^2$, i.e., the configuration space $\{q^1, \ldots, q^i, \ldots, q^N\}$ with two additional real coordinates q^0 and q^{N+1}. In the following q^0 will be identified with the time t. For standard Hamiltonians $\mathcal{H} = T + V(q)$ where $T = \frac{1}{2} a_{ij} \dot{q}^i \dot{q}^j$, this manifold, equipped with Eisenhart's metric g_E, has a semi–Riemannian (Lorentzian) structure ($\det g_E = -1$). The arc–length is given by

$$ds^2 = a_{ij} dq^i dq^j - 2V(q)(dq^0)^2 + 2dq^0 dq^{N+1},$$

where both i and j run between 1 and N. Let us restrict to geodesics whose arc–length parametrization is affine, i.e., $ds^2 = c_1^2 dt^2$; simple algebra shows that the geodesic equations

$$\ddot{q}^{\mu} + \Gamma^{\mu}_{\nu\lambda}\dot{q}^{\nu}\dot{q}^{\lambda} = 0, \qquad (\mu, \nu, \lambda = 0 \ldots N+1),$$

become Newton equations (without loss of generality $a_{ij} = \delta_{ij}$ is considered)

$$\ddot{q}^i = -\partial_{q^i} V, \qquad (\text{for } i = 1 \ldots N), \tag{9.1}$$

together with two extra equations for q^0 and q^{N+1} which can be integrated to yield

$$q^0 = t, \qquad q^{N+1} = \frac{c_1}{2} t + c_2 - \int_0^t L(q, \dot{q})\, dt, \tag{9.2}$$

where $L(q, \dot{q})$ is the Lagrangian, and c_1, c_2 are real constants. In the following we set $c_1 = 1$ in order that $ds^2 = dt^2$ on the physical geodesics. As stated by Eisenhart theorem [Eis29], the dynamical trajectories in configuration space are projections on M of the geodesics of $(M \times \mathbb{R}^2, g_E)$.

In the geometrical framework, the stability of the trajectories is mapped on the stability of the geodesics, hence it can be studied by the *Jacobi equation for geodesic deviation*

$$\frac{D^2 J}{ds^2} + R(\dot{\gamma}, J)\dot{\gamma} = 0, \tag{9.3}$$

where R is the *Riemann curvature tensor*, $\dot{\gamma}$ is the velocity vector along the reference geodesic $\gamma(s)$, D/ds is the covariant derivative and J, which measures the deviation between nearby geodesics, is referred to as the Jacobi field. The stability – or instability – of the dynamics, and thus deterministic chaos, originates from the curvature properties of the ambient manifold. In local coordinates, (9.3) is written as

$$\frac{D^2 J^{\mu}}{ds^2} + R^{\mu}_{\nu\rho\sigma}\frac{dq^{\nu}}{ds} J^{\rho} \frac{dq^{\sigma}}{ds} = 0,$$

and in the case of *Eisenhart metric* it simplifies to

$$\ddot{J}^i + \frac{\partial^2 V}{\partial q_i \partial q^j} J^j = 0, \tag{9.4}$$

which is nothing but the usual tangent dynamics equation for standard Hamiltonians. The Lyapunov exponents are usually computed evaluating the rate of exponential growth of J by means of a numerical integration of (9.4).

In the particular case of *constant curvature* manifolds, (9.3) becomes very simple [Car92]

$$\frac{D^2 J^{\mu}}{ds^2} + K J^{\mu} = 0, \tag{9.5}$$

and has bounded oscillating solutions $J \approx \cos(\sqrt{K}\, s)$ or exponentially unstable solutions $J \approx \exp(\sqrt{-K}\, s)$ according to the sign of the constant sectional curvature K, which is given by

$$K = \frac{K_R}{N-1} = \frac{\mathcal{R}}{N(N-1)}, \tag{9.6}$$

where $K_R = R_{\mu\nu}\dot{q}^\mu\dot{q}^\nu$ is the Ricci curvature and $\mathcal{R} = R^\mu_\mu$ is the scalar curvature; $R_{\mu\nu}$ is the Ricci tensor. Manifolds with $K < 0$ are considered in abstract ergodic theory (see, e.g., [Sin89]). Krylov [Kry79] originally proposed that the presence of some negative curvature could be the mechanism actually at work to make chaos in physical systems, but in realistic cases the curvatures are neither found constant nor everywhere negative, and the straightforward approach based on (9.5) does not apply. This is the main reason why Krylov's ideas remained confined to abstract ergodic theory with few exceptions.

In spite of these major problems, some approximations on (9.3) are possible even in the general case. The key point is that negative curvatures are not strictly necessary to make chaos, and that a subtler mechanism related to the *bumpiness* of the ambient manifold is actually at work. Upon an assumption of quasi–isotropy of the ambient manifold, i.e., that the manifold can be obtained as a small deformation of a constant-curvature space (for details, see [CCP96]), (9.3) can be approximated by an effective scalar equation which reads

$$\ddot{\psi} + K(t)\,\psi = 0, \tag{9.7}$$

where ψ is a generic component of the vector J (in this approximation all the components are considered equivalent), and $K(t)$ is a stochastic process which models the curvature along the geodesic curve. Such a stochastic model is defined by

$$K(t) = \langle k_R \rangle + \langle \delta^2 k_R \rangle^{1/2}\eta(t), \tag{9.8}$$

where $k_R = K_R/N$, $\langle \cdot \rangle$ stands for an average taken along a geodesic, which, for systems in thermal equilibrium, can be substituted with a statistical average taken with respect to a suitable probability measure (e.g., the micro-canonical or the canonical measure); $\eta(t)$ is a stationary $\delta-$correlated Gaussian stochastic process with zero mean and variance equal to one. Using Eisenhart metric, and for standard Hamiltonians, the non–vanishing components of the Riemann tensor are $R_{0i0j} = \partial_{q_i}\partial_{q_j}V$, hence the Ricci curvature has the remarkably simple form

$$k_R = \frac{1}{N}\nabla^2 V, \tag{9.9}$$

where ∇^2 is the Euclidean Laplacian operator. Equation (9.7) becomes a stochastic differential equation, i.e., the evolution equation of a random oscillator [Kam92]. It is worth noticing that (9.7) is no longer dependent on the dynamics, since the random process depends only on statistical averages. The estimate of the *Lyapunov exponent* λ is then obtained through the evolution of the second moments of the solution of (9.7) as

$$\lambda = \lim_{t\to\infty}\frac{1}{2t}\log\frac{\psi^2(t) + \dot{\psi}^2(t)}{\psi^2(0) + \dot{\psi}^2(0)}.$$

As shown in [CLP95, CCP96], this yields the following expression for λ:

$$\lambda(k, \sigma_k, \tau) = \frac{1}{2}\left(\Lambda - \frac{4k}{3\Lambda}\right), \qquad \text{where} \tag{9.10}$$

$$\Lambda = \left(\sigma_k^2 \tau + \sqrt{\frac{64k^3}{27} + \sigma_k^4 \tau^2}\right)^{1/3}, \qquad \tau = \frac{\pi\sqrt{k}}{2\sqrt{k(k+\sigma_k)} + \pi\sigma_k}.$$

In the above expressions k is the average *Ricci curvature* $k = \langle k_R\rangle$ and σ_k stands for the mean–square fluctuation of the Ricci curvature, $\sigma_k = \langle\delta^2 k_R\rangle^{1/2}$.

The advantages in using the geometric approach to Hamiltonian chaos are thus evident. In fact, it is possible to give reliable estimates of the Lyapunov exponent without actually computing the time evolution of the system: the estimate (9.10) of λ depends only on statistical averages which can be either computed analytically in some cases (for instance in the case of the FPU model [CLP95]) or, in general, extracted from a Monte Carlo or a dynamical simulation, as it is the case of the model studied in the present work.

The behavior of the average geometric observables as the control parameter (e.g., the energy density or the temperature) is varied conveys an information which goes beyond the possibility of computing the Lyapunov exponent. In fact, one can look at the random oscillator equation (9.7) as an effective Jacobi equation for a geodesic flow on a surface M whose Gaussian curvature is given by the random process $K(t)$. As long as nonlinear coupled oscillators are considered, the average Ricci curvature is positive, hence M can be regarded as a sphere with a fluctuating radius. In the limit of vanishing fluctuations, one recovers the bounded evolution of the Jacobi field associated with integrable dynamics. Chaos suddenly appears as curvature fluctuations are turned on, nevertheless it it will be 'weak' as long as $\sigma_k \ll k$, i.e., as long as M can be considered as a weakly perturbed sphere. On the contrary as the size of curvature fluctuations becomes of the same order of the average curvature, $\sigma_k \simeq k$, M can no longer resemble a sphere, and the dynamics will no longer 'feel' the integrable limit. Hence we expect the dynamics to be strongly chaotic. This is by no means a deep explanation of the existence of weakly and strongly chaotic regimes in Hamiltonian dynamics. Nevertheless it shows how the simple geometric concepts which enter the Riemannian description of Hamiltonian chaos, besides providing effective computational tools, are also useful in helping one's physical intuition with images and analogies which would be difficult to find elsewhere [CGP99].

9.2 Chaos in Physical Gauge Fields

Classical dynamical aspects of lattice gauge theories have recently attracted some interest [BMM95, GS88]. The classical limit of a lattice gauge theory is interesting both from the point of view of classical dynamical system theory and in the perspective of its quantum counterpart. As far as the latter aspect is concerned, the interest mainly resides in the fact that very few nonperturbative tools are available to study quantum gauge field theories, while in the classical limit it is in principle possible to exactly simulate the real time evolution of the system at any energy. From the point of view of the theory of classical dynamical systems, lattice gauge theories are highly non–trivial many DOF Hamiltonian systems which exhibit a rich and interesting phenomenology. The classical Hamiltonian dynamics of such systems is known to be chaotic [BMM95]; however, a precise characterization of the different chaotic regimes which may occur in these systems is still lacking.

Many particular aspects of this general problem have been considered in the literature, concerning the properties of pure gauge theories and of theories coupled with matter (mainly Higgs fields), e.g., the systematic study of the Lyapunov spectra [BMM95], the study of thermalization processes [HHL97], and the relation between Lyapunov exponents and observable quantities like the *plasmon damping rate* [BMM95].

A particular problem which is still open is the dependence of the largest Lyapunov exponent λ of the pure Yang-Mills lattice theory on the energy density ε, energy per plaquette, or energy per degree of freedom, particularly at low ε. First, the ε scaling of λ seems different according to the fact that the theory is Abelian $U(1)$, or non–Abelian $SU(2)$ and $SU(3)$; in the latter case two different scalings have been measured, namely $\lambda \propto \varepsilon^{1/4}$ and $\lambda \propto \varepsilon$ while in the former case a rapid decrease of λ at low ε was observed [BMM95]. As we shall see in the following, our results suggest that in the $U(1)$ case the power law $\lambda \propto \varepsilon^2$ holds. As pointed out by [MT92], such a problem is interesting because it is tightly related with the problem of the relevance of the chaotic lattice dynamics for the continuum limit of the gauge theory. In fact, intrinsic dimensional arguments can be used to show that the lattice spacing a, which can be set equal to one in all the numerical simulations after a convenient choice of units, enters the relation between λ and ε as $a\lambda(a) = f(a\varepsilon(a))$, hence if one observes numerically a power law $\lambda \propto \varepsilon^k$, the latter can be read as $\lambda(a) \propto a^{k-1}\varepsilon(a)$ (see [CGP99]). This means that considering a continuum limit $a \to 0$ in which $\varepsilon(a = 0)$ is finite, corresponding to a finite temperature of the resulting field, then the Lyapunov exponent is finite in the limit $a \to 0$ only for the particular exponent $k = 1$. Larger exponents $k > 1$ would imply that $\lim_{a \to 0} \lambda(a) = 0$, thus leading to a regular continuum theory, while exponents $k < 1$ would mean that in the continuum theory the Lyapunov exponent diverges. The linear behavior plays then a very special role. As regards non–Abelian theories, some evidence has been reported supporting the fact that the correct scaling is the linear one, $\lambda \propto \varepsilon$, the other one ($\lambda \propto \varepsilon^{1/4}$) being a spurious result due to finite–time effects. According to these results the continuum limit, for small but finite energy densities, of the Lyapunov exponent of non-Abelian lattice gauge theories is finite. In fact, extracting reliable informations from numerical simulations in the low–energy regime is very difficult, mainly because of finite–time effects, which become very important at small energy densities as the characteristic instability time scales grow, and of finite-size effects which are known to be rather large in typical lattice gauge theory simulations [CGP99].

A previously proposed formalism based on a Riemannian geometrization of Hamiltonian dynamics [Pet93, CP93, CLP95, CCP96] was applied in [CGP99] to the classical dynamics of lattice gauge theories. Such a formalism allows one to relate chaotic dynamics and curvature properties of suitable manifolds, and to get an analytic formula for the Lyapunov exponent in terms of average curvature fluctuations [CLP95, CCP96]. The quantities entering this formula are statistical averages which can be computed regardless of the knowledge of the dynamics, either by Monte–Carlo or molecular dynamics simulations, or analytically in some cases [CCP96]. As a first step, we apply this formalism to the Abelian – $U(1)$ – lattice gauge theory, which is the simplest one, leaving the non–Abelian case to future work. In the case of a $U(1)$ gauge theory we perform a precise numerical measurement of the Lyapunov exponent by simultaneous integration of the Hamilton's equations and of the tangent dynamics equations using a precise symplectic algorithm [Cas95] for several lattice sizes. In these simulations we measure also the relevant geometric observables that allow for a characterization of the dynamics and that enter the above-mentioned theoretical expression for the Lyapunov exponent. We find that the analytic estimate compares very well with the outcomes of the numerical simulations for the Lyapunov exponents. Moreover, we find that the theoretical estimate is almost free from finite–size effects; for small energies we find the law $\lambda \propto \varepsilon^2$ already when inserting in the formula the values of the geometric observables obtained with small ($4 \times 4 \times 4$) lattices, while the numerical values of the Lyapunov exponents are affected by large finite–size effects.

The dynamical system that we are now considering is the classical lattice gauge theory based on the Abelian gauge group $U(1)$ [CGP99]. The Hamiltonian of such a system can be derived in two ways, either as the dual Wegner model of a lattice planar spin system [Kog79], or starting from the Kogut-Susskind Hamiltonian for a generic gauge group G [Kog79, KS75] and then specializing to the group $U(1)$. In the latter case we get the same Hamiltonian as in the former case by choosing the $SO(2)$ real matrix representation of the group $U(1)$.

The lattice Lagrangian, obtained from the Wilson action by fixing the temporal gauge (all the temporal components of the gauge fields are set to zero) and then by taking the continuum limit in the temporal direction, is

$$\mathcal{L} = \frac{ag^2}{2} \sum_{\text{links}} \langle \dot{U}_{x,\mu}, \dot{U}_{x,\mu} \rangle - \frac{1}{ag^2} \sum_{\text{plaquettes}} \left(1 - \frac{1}{2} \text{Tr } U_{\mu\nu} \right), \qquad (9.11)$$

where $U_{x,\mu} \in G$ is a group element defined on a link of a d-dimensional cubic lattice, labeled by the site index x and the oriented lattice direction μ, g^2 is a coupling constant, a is the lattice spacing, $\langle \cdot, \cdot \rangle$ stands for the scalar product between group elements, defined as [CGP99]

$$\langle A, B \rangle = \frac{1}{2} \text{Tr } (AB^\dagger),$$

and $U_{\mu\nu}$ is a shorthand notation for the plaquette operator,

$$U_{\mu\nu} = U_{x,\mu}U_{x+\mu,\nu}U_{x+\mu+\nu,-\mu}U_{x+\nu,-\nu}.$$

We can pass to a standard Hamiltonian formulation by putting

$$P = \frac{\partial \mathcal{L}}{\partial \dot{U}} = ag^2\dot{U}, \qquad \text{which implies}$$

$$ag^2\mathcal{H} = \frac{1}{2}\sum_{\text{links}}\langle P_{x,\mu}, P_{x,\mu}\rangle + \sum_{\text{plaquettes}}\left(1 - \frac{1}{2}\text{Tr }U_{\mu\nu}\right). \qquad (9.12)$$

The parameters a and g^2 can be scaled out, so we set $g = a = 1$.

The Hamiltonian (9.12) is the classical Hamiltonian for a lattice gauge theory with a generic gauge group G. Let us now specialize to the Abelian group $G = U(1)$. Choosing the representation [CGP99]

$$U = \begin{pmatrix} \cos\varphi & \sin\varphi \\ -\sin\varphi & \cos\varphi \end{pmatrix}, \qquad \text{we have}$$

$$P = \dot{U} = \dot{\varphi}\begin{pmatrix} -\sin\varphi & \cos\varphi \\ -\cos\varphi & -\sin\varphi \end{pmatrix}, \qquad \text{so that we get}$$

$$\frac{1}{2}\langle P, P\rangle = \frac{1}{2}\dot{\varphi}^2.$$

To write the plaquette operator, we use the fact that the group $U(1) \simeq SO(2)$ is Abelian. Then the product of the four rotations is a rotation of the sum of the oriented angles along the plaquette (the discrete curl of the field φ)

$$\varphi_{x,\mu\nu} = \varphi_{x,\mu} + \varphi_{x+\mu,\nu} - \varphi_{x+\nu,\mu} - \varphi_{x,\nu},$$

and the magnetic energy, i.e., the potential energy of the dynamical system can be written as

$$V = \sum_{\text{plaquettes}}\left(1 - \frac{1}{2}\text{Tr }U_{\mu\nu}\right) = \sum_{\text{plaquettes}}\left(1 - \cos\varphi_{x,\mu\nu}\right). \qquad (9.13)$$

Our canonical variables are then the angles $\varphi_{x,\mu}$ and the angular momenta $\pi_{x,\mu} = \dot{\varphi}_{x,\mu}$, and the Hamiltonian (9.12) becomes

$$\mathcal{H} = \frac{1}{2}\sum_{\text{links}}\pi_{x,\mu}^2 + \sum_{\text{plaquettes}}\left(1 - \cos\varphi_{x,\mu\nu}\right). \qquad (9.14)$$

Constant–energy (micro–canonical) simulations have been performed on 3D lattices – with lattice sizes ranging from 4^3 to 15^3 – by integrating the canonical equations of motion

$$\dot{\varphi}_{x,\mu} = \pi_{x,\mu}, \qquad \dot{\pi}_{x,\mu} = -\frac{\partial V}{\partial \varphi_{x,\mu}}, \qquad (9.15)$$

where x runs over the L^3 lattice sites and $\mu = 1, 2, 3$, by using a precise third–order bilateral symplectic algorithm [Cas95]. We remind that symplectic algorithms are integration schemes which exactly preserve the canonical form of the equations of motion. The exact Hamilton equations are replaced, in the time discretization procedure, by a map that is symplectic, hence the discrete-time flow that approximates the true Hamiltonian flow generated by Hamilton's equations is still Hamiltonian. All the geometric constraints on the phase–space trajectories which are enforced by the canonical form of the equations of motion are thus exactly preserved during the numerical integration procedure and the Hamiltonian flow is faithfully represented. As a by–product of this features, symplectic algorithms conserve very well the total energy of the system: in our simulations relative energy fluctuations were of the order of $10^{-7} \div 10^{-8}$. In (9.2) V is given by (9.13), whose explicit expression on a 3D lattice is [CGP99]

$$V = \sum_x \sum_{(\mu\nu)} \left(1 - \cos \varphi_{x,\mu\nu}\right),$$

where $(\mu\nu) = 12, 13, 23$. The forces (rhs of 9.15) are given by

$$-\frac{\partial V}{\partial \varphi_{x,\mu}} = \sum_{\delta=1,2} \sin \varphi_{x-\mu-\delta,\mu\mu+\delta} - \sin \varphi_{x,\mu\mu+\delta}. \tag{9.16}$$

In order to compute the largest Lyapunov exponent λ by the standard method [BGS76], the tangent dynamics equations (9.4), which now reads as

$$\ddot{J}_{x,\mu} + \sum_y \sum_\nu \frac{\partial^2 V}{\partial \varphi_{x,\mu} \partial \varphi_{y,\nu}} J_{y,\nu} = 0, \tag{9.17}$$

have been integrated simultaneously with the Hamilton's equations (9.2) by means of the same algorithm. The largest Lyapunov exponent has then been computed according to the definition

$$\lambda = \lim_{t\to\infty} \frac{1}{t} \log \frac{\left[|\dot{J}|^2(t) + |J|^2(t)\right]^{1/2}}{\left[|\dot{J}|^2(0) + |J|^2(0)\right]^{1/2}},$$

where $|J|^2 = \sum_x \sum_\mu J_{x,\mu}^2$ is the squared Euclidean norm of the tangent vector J.

According to the above geometrical discussion, the relevant geometric observable which is able to characterize the chaotic dynamics of the model is the Ricci curvature k_R, computed with the Eisenhart metric, defined by (9.9), which now can be rewritten as [CGP99]

$$k_R = \frac{1}{L^3} \sum_x \sum_\mu \sum_{\nu\neq\mu} [\cos \varphi_{x,\mu\nu} + \cos\varphi_{x-\nu,\mu\nu}]. \tag{9.18}$$

9.3 Solitions

Recall that synergetics teaches us that *order parameters* (and their spatio–temporal evolution) are *patterns, emerging from chaos*. In our opinion, the most important of these order parameters, both natural and man made, are *solitons*, because of their self–organizing quality to create order out of chaos. From this perspective, *nonlinearity* – the essential characteristic of nature – is the *cause* of both *chaos* and *order*. Recall that the solitary particle–waves, also called the 'light bullets', are localized space–time excitations $\Psi(x,t)$, propagating through a certain medium Ω with constant velocities v_j. They describe a variety of nonlinear wave phenomena in one dimension and playing important roles in optical fibers, many branches of physics, chemistry and biology.

9.3.1 History of Solitons in Brief

In this subsection we will give a brief of soliton history, mainly following [Sco99, Sco04].

It all started in 1834, when young Scottish engineer named John Scott Russell was conducting experiments on the Union Canal (near Edinburgh) to measure the relationship between the speed of a boat and its propelling force, with the aim of finding design parameters for conversion from horse power to steam. One August day, a rope parted in his measurement apparatus and [Rus844] "The boat suddenly stopped – not so the mass of water in the channel which it had put in motion; it accumulated round the prow of the vessel in a state of violent agitation, then suddenly leaving it behind, rolled forward with great velocity, assuming the form of a large solitary elevation, a rounded, smooth and well defined heap of water, which continued its course along the channel without change of form or diminution of speed." Russell did not ignore this unexpected phenomenon, but "followed it on horseback, and overtook it still rolling on at a rate of some eight or nine miles an hour, preserving its original figure some thirty feet long and a foot to a foot and a half in height" until the wave became lost in the windings of the channel. He continued to study the solitary wave in tanks and canals over the following decade, finding it to be an independent dynamic entity moving with constant shape and speed.

Using a wave tank he demonstrated four facts [Rus844]:

(i) Solitary waves have the shape $h \operatorname{sech}^2[k(x - vt)]$;

(ii) A sufficiently large initial mass of water produces two or more independent solitary waves;

(iii) Solitary waves cross each other "without change of any kind";

(iv) A wave of height h and travelling in a channel of depth d has a velocity given by the expression

$$v = \sqrt{g(d + h)}, \tag{9.19}$$

(where g is the acceleration of gravity), implying that a large amplitude solitary wave travels faster than one of low amplitude.

Although confirmed by observations on the Canal de Bourgogne, near Dijon, most subsequent discussions of the hydrodynamic solitary wave missed the physical significance of Russell's observations. Evidence that to the end of his life Russell maintained a much broader and deeper appreciation of the importance of his discovery is provided by a posthumous work, where he correctly estimated the height of the earth's atmosphere from (9.19) and the fact that "the sound of a cannon travels faster than the command to fire it" [Rus885].

In 1895, D. Korteweg and H. de Vries published a theory of shallow water waves that reduced Russell's problem to its essential features. One of their results was the nonlinear PDE

$$u_t + c\,u_x + \varepsilon u_{xxx} + \gamma u\,u_x = 0, \tag{9.20}$$

which would play a key role in soliton theory [KdV95]. In this equation, $u(x,t)$ is the wave amplitude, $c = \sqrt{gd}$ is the speed of small amplitude waves, $\varepsilon \equiv c(d^2/6 - T/2\rho g)$ is a dispersive parameter, $\gamma \equiv 3c/2d$ is a nonlinear parameter, and T and ρ are respectively the surface tension and the density of water. The authors showed that (9.20) has a family of exact travelling wave solutions of the form $u(x,t) = \tilde{u}(x - vt)$, where $\tilde{u}(\cdot)$ is Russell's "rounded, smooth and well defined heap" and v is the wave speed. If the dispersive term (ε) and the nonlinear term (γ) in (9.20) are both zero, then the *Korteweg–de Vries equation* (KdV) becomes linear,

$$u_t + c\,u_x = 0,$$

with a travelling wave solution for any pulse shape at the fixed speed $v = c = \sqrt{gd}$. In general the KdV equation (9.20) is nonlinear with exact travelling wave solutions

$$u(x,t) = h\,\mathrm{sech}^2[k(x - vt)], \tag{9.21}$$

where $k = \sqrt{\gamma h/12\varepsilon}$, implying that higher amplitude waves are more narrow. With this shape, the effects of dispersion balance those of nonlinearity at an adjustable value of the pulse speed. Thus the solitary wave is recognized as an independent dynamic entity, maintaining a dynamic balance between these two influences. Interestingly, solitary wave velocities are related to amplitudes by

$$v = c + \gamma h/3 = \sqrt{gd}\,(1 + h/2d), \tag{9.22}$$

in accord with Russell's empirical results, given in (9.19), to $O(h)$.

Although unrecognized at the time, such an energy conserving solitary wave is related to the existence of a *Bäcklund transform* (BT), which was proposed by J.O. Bäcklund in 1885 [Lam76]. In such a transform, a known solution generates a new solution through a single integration, after which the new solution can be used to generate another new solution, and so on. It is straightforward to find a BT for any linear PDE, which introduces a new eigenfunction into the total solution with each application of the transform.

Only special nonlinear PDEs are found to have BTs, but late nineteenth mathematicians knew that these include

$$u_{\xi\tau} = \sin u, \tag{9.23}$$

which arose in research on the geometry of curved surfaces [Ste36].

In 1939, Y. Frenkel and T. Kontorova introduced a seemingly unrelated problem arising in solid state physics to model the relationship between dislocation dynamics and plastic deformation of a crystal [FK39]. From this study, a PDE describing dislocation motion is

$$u_{xx} - u_{tt} = \sin u, \tag{9.24}$$

where $u(x, t)$ is atomic displacement in the x-direction and the 'sin' function represents periodicity of the crystal lattice. A *travelling wave solution* of (9.24), corresponding to the propagation of a dislocation, is

$$u(x,t) = 4 \arctan \left[\exp \left(\frac{x - vt}{\sqrt{1 - v^2}} \right) \right], \tag{9.25}$$

with velocity v in the range $(-1, +1)$. Since (9.24) is identical to (9.23) after an independent variable transformation $[\xi = (x-t)/2$ and $\tau = (x+t)/2]$, exact solutions of (9.24) involving arbitrary numbers of dislocation components as in (9.25) can be generated through a succession of Bäcklund transforms, but this was not known to Frenkel and Kontorova.

In the late 1940s, E. Fermi, J. Pasta and S. Ulam (the famous FPU–trio) suggested one of the first scientific problems to be assigned to the Los Alamos MANIAC computing machine: the dynamics of energy equipartition in a slightly nonlinear crystal lattice, which is related to thermal conductivity. The system they chose was a chain of 64 equal mass particles connected by slightly nonlinear springs, so from a linear perspective there were 64 normal modes of oscillation in the system. It was expected that if all the initial energy were put into a single vibrational mode, the small nonlinearity would cause a gradual progress toward equal distribution of the energy among all modes (thermalization), but the numerical results were surprising. If all the energy is originally in the mode of lowest frequency, it returns almost entirely to that mode after a period of interaction among a few other low frequency modes. In the course of several numerical refinements, no thermalization was observed [FPU55].

Pursuit of an explanation for this "FPU recurrence" led Zabusky and Kruskal to approximate the nonlinear spring-mass system by the KdV equation. In 1965, they reported numerical observations that KdV solitary waves pass through each other with no change in shape or speed, and coined the term "soliton" to suggest this property [ZK65].

N.J. Zabusky and M.D. Kruskal were not the first to observe nondestructive interactions of energy conserving solitary waves. Apart from Russell's tank measurements [Rus844], Perring and Skyrme, had studied solutions of Equation (9.24) comprising two solutions as in Equation (9.25) undergoing a collision. In 1962, they published numerical results showing perfect recovery of shapes and speeds after a collision and went on to discover an exact analytical description of this phenomenon [PS62].

This result would not have surprised nineteenth century mathematicians; it is merely the second member of the hierarchy of solutions generated by a Bäcklund transform. Nor would it have been unexpected by Seeger and his colleagues, who had noted in 1953 the connections between the nineteenth century work [Ste36] and the studies of Frenkel and Kontorova [SDK53]. Since Perring and Skyrme were interested in Equation (9.24) as a nonlinear model for elementary particles of matter, however, the complete absence of scattering may have been disappointing.

Throughout the 1960s, (9.24) arose in a variety of problems including the propagation of ferromagnetic domain walls, self–induced transparency in non-linear optics, and the propagation of magnetic flux quanta in long Josephson transmission lines. Eventually (9.24) became known as the *sine–Gordon (SG) equation*, which is a nonlinear version of the linear *Klein–Gordon equation*:

$$u_{xx} - u_{tt} = u.$$

Perhaps the most important contribution made by Zabusky and Kruskal in their 1965 paper was to recognize the relation between nondestructive soliton collisions and the riddle of FPU recurrence. Viewing KdV solitons as independent and localized dynamic entities, they explained the FPU observations as follows. The initial condition generates a family of solitons with different speeds, moving apart in the $x - t$ plane. Since the system studied was of finite length with perfect reflections at both ends, the solitons could not move infinitely far apart; instead they eventually reassembled in the $x - t$ plane, approximately recreating the initial condition after a surprisingly short *recurrence time*.

By 1967, this insight had led Kruskal and his colleagues to devise a non-linear generalization of the Fourier transform method for constructing solutions of the KdV equation emerging from arbitrary initial conditions [GGK67]. Called the inverse scattering (or inverse spectral) method (ISM), this approach proceeds in three steps.

A. The nonlinear KdV dynamics are mapped onto an associated linear problem, where each eigenvalue of the linear problem corresponds to the speed of a particular KdV soliton.
B. Since the associated problem is linear, the time evolution of its solution is readily computed.
C. An inverse calculation then determines the time evolved KdV dynamics from the evolved solution of the linear associated problem. Thus the solution of a nonlinear problem is found from a series of linear computations.

Another development of the 1960s was M. Toda's discovery of exact two–soliton interactions on a nonlinear spring–mass system, called *Toda lattice* [Tod67]. As in the FPU system, equal masses were assumed to be interconnected with nonlinear springs, but Toda chose the potential

$$\left(\frac{a}{b}\right)[e^{-bu_j} - 1] + au_j, \qquad (9.26)$$

where $u_j(t)$ is the longitudinal extension of the jth spring from its equilibrium value and both a and b are adjustable parameters. (In the limit $a \to \infty$ and $b \to 0$ with ab finite, this reduces to the quadratic potential of a linear spring. In the limit $a \to 0$ and $b \to \infty$ with ab finite, it describes the interaction between hard spheres.) Thus by the late 1960s, it was established – although not widely known – that solitons were not limited to PDEs (KdV and SG). Local solutions of difference–differential equations could also exhibit the unexpected properties of unchanging shapes and speeds after collisions.

These events are only the salient features of a growing panorama of nonlinear wave activities that became gradually less parochial during the 1960s. Solid state physicists began to see relationships between their solitary waves (magnetic domain walls, self–shaping pulses of light, quanta of magnetic flux, polarons, etc.), and those from classical hydrodynamics and oceanography, while applied mathematicians began to suspect that the ISM might be used for a broader class of nonlinear wave equations. It was amid this intellectual ferment that A.C. Newell and his colleagues organized the first soliton research workshop during the summer of 1972 [New74]. Interestingly, one of the most significant contributions to this conference came by post. From the Soviet Union arrived a paper by V.E. Zakharov and A.B. Shabat, formulating Kruskal's ISM for the nonlinear PDE [ZS72]

$$iu_t + u_{xx} + 2|u|^2 u = 0 \,. \qquad (9.27)$$

In contrast to KdV, SG and the Toda lattice, the dependent variable in this equation is complex rather than real, so the evolutions of two quantities (magnitude and phase of u) are governed by the equation. This reflects the fact that (9.27) is a nonlinear generalization of a linear PDE

$$iu_t + u_{xx} + u = 0,$$

solutions of which comprise both an envelope and a carrier wave. Since this linear equation is a *Schrödinger equation* for the quantum mechanical probability amplitude of a particle (like an electron) moving through a region of uniform potential, it is natural to call (9.27) the *nonlinear Schrödinger (NLS) equation*. When the NLS equation is used to model wave packets in such fields as hydrodynamics, nonlinear optics, nonlinear acoustics, plasma waves and biomolecular dynamics, however, its solutions are devoid of quantum character.

Upon appreciating the Zakharov and Shabat paper, many left the 1972 conference convinced that four nonlinear equations (KdV, SG, NLS, and the

Toda lattice) display solitary wave behavior with the special properties that led Zabusky and Kruskal to coin the term soliton [New74]. Within two years, ISM formulations had been constructed for the SG equation and also for the Toda lattice.

Since the mid–1970s, the soliton concept has become established in several areas of applied science, and dozens of dynamic systems are now known to be integrable through the ISM. Even if a system is not exactly integrable, additionally, it may be close to an integrable system, allowing analytic insight to be gleaned from perturbation theory. Thus one is no longer surprised to find stable spatially localized regions of energy, balancing the opposing effects of nonlinearity and dispersion and displaying the essential properties of objects. In the last two decades, soliton studies have been focused on working out the details of such object–like behavior in a wide range of research areas [Sco99].

9.3.2 The Fermi–Pasta–Ulam Experiments

Perhaps the single most important event leading up to the explosive growth of soliton mathematics in the last decades was a seemingly innocuous computer computation, carried out by Enrico Fermi, John Pasta, and Stanislaw Ulam in 1954–55, on the Los Alamos MANIAC computer (originally published as Los Alamos Report LA1940 (1955) and reprinted in [FPU55]).

The following quotation is taken from Stanislaw Ulam's autobiography, *Adventures of a Mathematician* [Ula91].

> Computers were brand new; in fact the Los Alamos Maniac was barely finished. . . . As soon as the machines were finished, Fermi, with his great common sense and intuition, recognized immediately their importance for the study of problems in theoretical physics, astrophysics, and classical physics. We discussed this at length and decided to formulate a problem simple to state, but such that a solution would require a lengthy computation which could not be done with pencil and paper or with existing mechanical computers. . . [W]e found a typical one. . . the consideration of an elastic string with two fixed ends, subject not only to the usual elastic force of stress proportional to strain, but having, in addition, a physically correct nonlinear term. . . . The question was to find out how. . . the entire motion would eventually thermalize. . .
>
> John Pasta, a recently arrived physicist, assisted us in the task of flow diagramming, programming, and running the problem on the Maniac. . .
>
> The problem turned out to be felicitously chosen. The results were entirely different qualitatively from what even Fermi, with his great knowledge of wave motion, had expected.

What Fermi, Pasta, and Ulam (FPU) were trying to do was to verify numerically a basic article of faith of statistical mechanics; namely the belief that if a mechanical system has many degrees of freedom and is close to a stable equilibrium, then a generic nonlinear interaction will "thermalize" the energy of the system, i.e., cause the energy to become equidistributed among the normal modes of the corresponding linearized system. In fact, Fermi believed he had demonstrated this fact in [Fer23]. Equipartition of energy among the normal modes is known to be closely related to the ergodic properties of such a system, and in fact FPU state their goal as follows: "The ergodic behavior of such systems was studied with the primary aim of establishing, experimentally, the rate of approach to the equipartition of energy among the various degrees of freedom of the system."

FPU make it clear that the problem that they want to simulate is the vibrations of a "1D continuum" or "string" with fixed end–points and non-linear elastic restoring forces, but that "for the purposes of numerical work this continuum is replaced by a finite number of points ... so that the PDE describing the motion of the string is replaced by a finite number of ODE." To rephrase this in the current jargon, FPU study a 1D lattice of N oscillators with nearest neighbor interactions and zero boundary conditions (for their computations, FPU take $N = 64$) (see [Pal97]).

We imagine the original string to be stretched along the x−axis from 0 to its length ℓ. The N oscillators have equilibrium positions

$$p_i = ih, \qquad (i = 0, \ldots, N - 1), \qquad \text{where} \qquad h = \ell/(N - 1)$$

is the lattice spacing, so their positions at time t are $X_i(t) = p_i + x_i(t)$, (where the x_i represent the displacements of the oscillators from equilibrium). The force attracting any oscillator to one of its neighbors is taken as $k(\delta + \alpha\delta^2)$, δ denoting the "strain", i.e., the deviation of the distance separating these two oscillators from their equilibrium separation h. (Note that when $\alpha = 0$ this is just a linear Hooke's law force with spring constant k.) The force acting on the ith oscillator due to its right neighbor is

$$F(x)_i^+ = k[(x_{i+1} - x_i) + \alpha((x_{i+1} - x_i)^2],$$

while the force acting on it due to its left neighbor is

$$F(x)_i^- = k[(x_{i-1} - x_i) - \alpha((x_{i-1} - x_i)^2].$$

Thus the total force acting on the ith oscillator will be the sum of these two forces, namely:

$$F(x)_i = k(x_{i+1} + x_{i-1} - 2x_i)[1 + \alpha(x_{i+1} - x_{i-1})],$$

and assuming that all of the oscillators have the same mass, m, Newton's equations of motion read:

$$m\ddot{x}_i = k(x_{i+1} + x_{i-1} - 2x_i)[1 + \alpha(x_{i+1} - x_{i-1})],$$

with the boundary conditions

$$x_0(t) = x_{N-1}(t) = 0.$$

In addition, FPU looked at motions of the lattice that start from rest, i.e., they assumed that $\dot{x}_i(0) = 0$, so the motion of the lattice is completely specified by giving the $N - 2$ initial displacements $x_i(0)$, $i = 1, \ldots, N - 2$. We shall call this the FPU initial value problem (with initial condition $x_i(0)$).

It will be convenient to rewrite Newton's equations in terms of parameters that refer more directly to the original string that we are trying to model. Namely, if ρ denotes the density of the string, then $m = \rho h$, while if κ denotes the Young's modulus for the string (i.e., the spring constant for a piece of unit length), then $k = \kappa/h$ will be the spring constant for a piece of length h. Defining $c = \sqrt{\kappa/\rho}$ we can now rewrite Newton's equations as [Pal97]

$$\ddot{x}_i = c^2 \left(\frac{x_{i+1} + x_{i-1} - 2x_i}{h^2} \right) [1 + \alpha(x_{i+1} - x_{i-1})],$$

and in this form we shall refer to them as the FPU Lattice Equations. We can now 'pass to the continuum limit'; i.e., by letting N tend to infinity (so h tends to zero) we can attempt to derive a PDE for the function $u(x,t)$ that measures the displacement at time t of the particle of string with equilibrium position x. We shall leave the nonlinear case for later, and here restrict our attention to the linear case, $\alpha = 0$. If we take $x = p_i$, then by definition $u(x,t) = x_i(t)$, and since $p_i + h = p_{i+1}$ while $p_i - h = p_{i-1}$, with $\alpha = 0$ the latter form of Newton's equations gives:

$$u_{tt}(x,t) = c^2 \frac{u(x+h,t) + u(x-h,t) - 2u(x,t)}{h^2}.$$

By Taylor's formula:

$$f(x \pm h) = f(x) \pm hf'(x) + \frac{h^2}{2!}f''(x) \pm \frac{h^3}{3!}f'''(x) + \frac{h^4}{4!}f''''(x) + O(h^5),$$

and taking $f(x) = u(x,t)$ gives:

$$\frac{u(x+h,t) + u(x-h,t) - 2u(x,t)}{h^2} = u_{xx}(x,t) + \left(\frac{h^2}{12}\right)u_{xxxx}(x,t) + O(h^4);$$

so letting $h \to 0$, we find $u_{tt} = c^2 u_{xx}$, i.e., u satisfies the linear wave equation, with propagation speed c (and of course the boundary conditions $u(0,t) = u(\ell,t) = 0$, and initial conditions $u_t(x,0) = 0$, $u(x,0) = u_0(x)$).

This is surely one of the most famous initial value problems of mathematical physics, and nearly every mathematician sees a derivation of both the d'Alembert and Fourier version of its solution early in their careers. For each positive integer k there is a normal mode or 'standing wave' solution [Pal97]

$$u_k(x,t) = \cos\left(\frac{k\pi ct}{\ell}\right) \sin\left(\frac{k\pi x}{\ell}\right),$$

and the solution to the initial value problem is

$$u(x,t) = \sum_{k=1}^{\infty} a_k u_k(x,t),$$

where the a_k are the Fourier coefficients of u_0:

$$a_k = \frac{2}{l} \int_0^{\ell} u_0(x) \sin\left(\frac{k\pi x}{\ell}\right) dx.$$

Replacing x by $p_j = jh$ in $u_k(x,t)$ (and using $\ell = (N-1)h$) we get functions

$$\xi_j^{(k)}(t) = \cos\left(\frac{k\pi ct}{(N-1)h}\right) \sin\left(\frac{kj\pi}{N-1}\right),$$

and it is natural to conjecture that these will be the normal modes for the FPU initial value problem (with $\alpha = 0$ of course). This is easily checked using the addition formula for the sine function. It follows that, in the linearized case, the solution to the FPU initial value problem with initial conditions $x_i(0)$ is given explicitly by

$$x_j(t) = \sum_{k=1}^{N-2} a_k \xi_j^{(k)}(t),$$

where the Fourier coefficients a_k are determined from the formula:

$$a_k = \sum_{j=1}^{N-2} x_j(0) \sin\left(\frac{kj\pi}{N-1}\right).$$

Clearly, when α is zero and the interactions are linear, we are in effect dealing with $N-2$ un–coupled harmonic oscillators (the above normal modes) and there is no thermalization. On the contrary, the sum of the kinetic and potential energy of each of the normal modes is a constant of the motion.

But if α is small but non–zero, FPU expected (on the basis of then generally accepted statistical mechanics arguments) that the energy would gradually shift between modes so as to eventually roughly equalize the total of potential and kinetic energy in each of the $N-2$ normal modes $\xi^{(k)}$. To test this they started the lattice in the fundamental mode $\xi^{(1)}$, with various values of α, and integrated Newton's equations numerically for a long time interval, interrupting the evolution from time to time to compute the total of kinetic plus potential energy in each mode. What did they find? Here is a quotation from their report:

> Let us say here that the results of our computations show features which were, from the beginning, surprising to us. Instead of a gradual, continuous flow of energy from the first mode to the higher modes,

all of the problems showed an entirely different behavior. Starting in one problem with a quadratic force and a pure sine wave as the initial position of the string, we did indeed observe initially a gradual increase of energy in the higher modes as predicted (e.g., by Rayleigh in an infinitesimal analysis). Mode 2 starts increasing first, followed by mode 3, and so on. Later on, however, this gradual sharing of energy among the successive modes ceases. Instead, it is one or the other mode that predominates. For example, mode 2 decides, as it were, to increase rather rapidly at the cost of the others. At one time it has more energy than all the others put together. Then mode 3 undertakes this rôle. It is only the first few modes which exchange energy among themselves, and they do this in a rather regular fashion. Finally, at a later time, mode 1 comes back to within one percent of its initial value, so that the system seems to be almost periodic.

There is no question that Fermi, Pasta, and Ulam realized they had stumbled onto something big. In his autobiography [Ula91], Ulam devotes several pages to a discussion of this collaboration. Here is a little of what he says:

> I know that Fermi considered this to be, as he said, "a minor discovery." And when he was invited a year later to give the Gibbs Lecture (a great honorary event at the annual American Mathematical Society meeting), he intended to talk about it. He became ill before the meeting, and his lecture never took place....
> The results were truly amazing. There were many attempts to find the reasons for this periodic and regular behavior, which was to be the starting point of what is now a large literature on nonlinear vibrations. Martin Kruskal, a physicist in Princeton, and Norman Zabusky, a mathematician at Bell Labs, wrote papers about it. Later, Peter Lax contributed signally to the theory.

Unfortunately, Fermi died in 1955, even before the paper cited above was published. It was to have been the first in a series of papers, but with Fermi's passing it fell to others to follow up on the striking results of the Fermi–Pasta–Ulam experiments.

The MANIAC computer, on which FPU carried out their remarkable research, was designed to carry out some computations needed for the design of the first hydrogen bombs, and of course it was a marvel for its day. But it is worth noting that it was very weak by today's standards–not just when compared with current supercomputers, but even when compared with modest desktop machines. At a conference held in 1977 Pasta recalled, "The program was of course punched on cards. A DO loop was executed by the operator feeding in the deck of cards over and over again until the loop was completed!"

9.3.3 The Kruskal–Zabusky Experiments

Following the FPU experiments, there were many attempts to explain the surprising quasi-periodicity of solutions of the FPU Lattice Equations. However it was not until ten years later that Martin Kruskal and Norman Zabusky took the crucial steps that led to an eventual understanding of this behavior [ZK65].

In fact, they made two significant advances. First they demonstrated that, in a continuum limit, certain solutions of the FPU Lattice Equations could be described in terms of solutions of the so–called Korteweg–de Vries (or KdV) equation. And second, by investigating the initial value problem for the KdV equation numerically on a computer, they discovered that its solutions had remarkable behavior that was related to, but if anything even more surprising and unexpected than the anomalous behavior of the FPU lattice that they had set out to understand.

Finding a good continuum limit for the nonlinear FPU lattice is a lot more sophisticated than one might at first expect after the easy time we had with the linear case. In fact the approach to the limit has to be handled with considerable skill to avoid inconsistent results, and it involves several non–obvious steps.

Let us return to the FPU Lattice Equations

$$\ddot{x}_i = c^2 \left(\frac{x_{i+1} + x_{i-1} - 2x_i}{h^2} \right) [1 + \alpha(x_{i+1} - x_{i-1})], \qquad (9.28)$$

and as before let $u(x,t)$ denote the function measuring the displacement at time t of the particle of string with equilibrium position x, so if $x = p_i$ then, by definition, $x_i(t) = u(x,t)$, $x_{i+1}(t) = u(x+h,t)$, and $x_{i-1}(t) = u(x-h,t)$. Clearly, $\ddot{x}_i = u_{tt}(x,t)$, and Taylor's Theorem with remainder gives

$$\frac{x_{i+1} + x_{i-1} - 2x_i}{h^2} = \frac{u(x+h,t) + u(x-h,t) - 2u(x,t)}{h^2}$$

$$= u_{xx}(x,t) + (\frac{h^2}{12})u_{xxxx}(x,t) + O(h^4).$$

By a similar computation

$$\alpha(x_{i+1} - x_{i-1}) = (2\alpha h)u_x(x,t) + (\frac{\alpha h^3}{3})u_{xxx}(x,t) - O(h^5),$$

so substitution in (FPU) gives

$$(\frac{1}{c^2})u_{tt} - u_{xx} = (2\alpha h)u_x u_{xx} + (\frac{h^2}{12})u_{xxxx} + O(h^4).$$

As a first attempt to derive a continuum description for the FPU lattice in the nonlinear case, it is tempting to just let h approach zero and assume that $2\alpha h$ converges to a limit ϵ. This would give the PDE [Pal97]

$$u_{tt} = c^2(1 + \epsilon u_x)u_{xx}$$

as our continuum limit for the FPU Lattice equations and the nonlinear generalization of the wave equation. But this leads to a serious problem. This equation is familiar in applied mathematics, it was studied by Rayleigh in the last century, and it is easy to see from examples that its solutions develop discontinuities (shocks) after a time on the order of $(\epsilon c)^{-1}$, which is considerably shorter than the time scale of the almost periods observed in the Fermi–Pasta–Ulam experiments. It was Zabusky who realized that the correct approach was to retain the term of order h^2 and study the equation

$$(\frac{1}{c^2})u_{tt} - u_{xx} = (2\alpha h)u_x u_{xx} + (\frac{h^2}{12})u_{xxxx}. \tag{9.29}$$

If we differentiate this equation with respect to x and make the substitution $v = u_x$, we see that it reduces to the more familiar *Boussinesq equation*

$$(\frac{1}{c^2})v_{tt} = v_{xx} + \alpha h \frac{\partial(v^2)}{\partial x^2} + (\frac{h^2}{12})v_{xxxx}.$$

(The effect of the fourth order term is to add dispersion to the equation, and this smoothes out incipient shocks before they can develop.)

It is important to realize that, since $h \neq 0$, (ZK) cannot logically be considered a true continuum limit of the FPU lattice. It should rather be regarded as an asymptotic approximation to the lattice model that works for small lattice spacing h (and hence large N). Nevertheless, we shall now see how to pass from (ZK) to a true continuum description of the FPU lattice.

The next step is to notice that, with α and h small, solutions of (ZK) should behave qualitatively like solutions of the linear wave equation

$$u_{tt} = c^2 u_{xx},$$

and increasingly so as α and h tend to zero. Now the general solution of the linear wave equation is obviously

$$u(x, t) = f(x + ct) + g(x - ct),$$

i.e., the sum of an arbitrary left moving travelling wave and an arbitrary right moving travelling wave, both moving with speed c. Recall that it is customary to simplify the analysis in the linear case by treating each kind of wave separately, and we would like to do the same here. That is, we would like to look for solutions $u(x, t)$ that behave more and more like (say) right moving travelling waves of velocity c—and for longer and longer periods of time–as α and h tend to zero.

It is not difficult to make precise sense out of this requirement. Suppose that $y(\xi, \tau)$ is a smooth function of two real variables such that the map $\tau \mapsto y(\cdot, \tau)$ is uniformly continuous from $\mathbb{R}$ into the bounded functions on $\mathbb{R}$ with the sup norm–i.e., given $\epsilon > 0$ there is a positive δ such that

$$|\tau - \tau_0| < \delta \qquad \text{implies} \qquad |y(\xi, \tau) - y(\xi, \tau_0)| < \epsilon.$$

Then for

$$|t - t_0| < T = \delta/(\alpha hc) \qquad \text{we have}$$
$$|\alpha hct - \alpha hct_0| < \delta, \qquad \text{so} \qquad |y(x - ct, \alpha hct) - y(x - ct, \alpha hct_0)| < \epsilon.$$

In other words, the function $u(x, t) = y(x - ct, \alpha hct)$ is uniformly approximated by the travelling wave $u^0(x, t) = y(x - ct, \alpha hct_0)$ on the interval $|t - t_0| < T$ (and of course $T \to \infty$ as α and h tend to zero). To restate this a little more picturesquely, $u(x, t) = y(x - ct, \alpha hct)$ is approximately a travelling wave whose shape gradually changes in time. Notice that if $y(\xi, \tau)$ is periodic or almost periodic in τ, the gradually changing shape of the approximate travelling wave will also be periodic or almost periodic.

To apply this observation, we define new variables $\xi = x - ct$ and $\tau = (\alpha h)ct$. Then by the chain rule, $\partial^k/\partial x^k = \partial^k/\partial \xi^k$, $\partial/\partial t = -c(\partial/\partial \xi - (\alpha h)\partial/\partial \tau)$, and $\partial^2/\partial t^2 = c^2(\partial^2/\partial \xi^2 - (2\alpha h)\partial^2/\partial \xi \partial \tau) + (\alpha h)^2 \partial^2/\partial \tau^2)$. Thus in these new coordinates the wave operator transforms to [Pal97]

$$\frac{1}{c^2}\frac{\partial^2}{\partial t^2} - \frac{\partial^2}{\partial x^2} = -2\alpha h\frac{\partial^2}{\partial \xi \partial \tau} + (\alpha h)^2\frac{\partial^2}{\partial \tau^2},$$

so substituting $u(x, t) = y(\xi, \tau)$ in (ZK) (and dividing by -2αh) gives:

$$y_{\xi\tau} - (\frac{\alpha h}{2})y_{\tau\tau} = -y_\xi y_{\xi\xi} - (\frac{h}{24\alpha})y_{\xi\xi\xi\xi},$$

and, at last, we are prepared to pass to the continuum limit. We assume that α and h tend to zero at the same rate, i.e., that as h tends to zero, the quotient h/α tends to a positive limit, and we define

$$\delta = \lim_{h \to 0} \sqrt{h/(24\alpha)} \qquad \text{which implies} \qquad \alpha h = O(h^2),$$

so letting h approach zero gives

$$y_{\xi\tau} + y_\xi y_{\xi\xi} + \delta^2 y_{\xi\xi\xi\xi} = 0.$$

Finally, making the substitution $v = y_\xi$ we arrive at the KdV equation:

$$v_\tau + vv_\xi + \delta^2 v_{\xi\xi\xi} = 0. \qquad (9.30)$$

Note that if we re–scale the independent variables by $\tau \to \beta\tau$ and $\xi \to \gamma\xi$, then the KdV equation becomes:

$$v_\tau + (\frac{\beta}{\gamma})vv_\xi + (\frac{\beta}{\gamma^3})\delta^2 v_{\xi\xi\xi} = 0,$$

so by appropriate choice of β and γ we can get any equation of the form

$$v_\tau + \lambda v v_\xi + \mu v_{\xi\xi\xi} = 0,$$

and any such equation is referred to as "the KdV equation". A commonly used choice that is convenient for many purposes is

$$v_\tau + 6 v v_\xi + v_{\xi\xi\xi} = 0,$$

although the form

$$v_\tau - 6 v v_\xi + v_{\xi\xi\xi} = 0$$

(obtained by replacing v by $-v$) is equally common. We will use both these forms.

Let us recapitulate the relationship between the FPU Lattice and the KdV equation. Given a solution $x_i(t)$ of the FPU Lattice, we get a function $u(x,t)$ by interpolation, i.e.,

$$u(ih, t) = x_i(t), \qquad (i = 0, \ldots, N).$$

For small lattice spacing h and nonlinearity parameter α there will be solutions $x_i(t)$ so that the corresponding $u(x,t)$ will be an approximate right moving travelling wave with slowly varying shape, i.e., it will be of the form

$$u(x,t) = y(x - ct, \alpha h c t)$$

for some smooth function $y(\xi, \tau)$, and the function $v(\xi, \tau) = y_\xi(\xi, \tau)$ will satisfy the KdV equation

$$v_\tau + v v_\xi + \delta^2 v_{\xi\xi\xi} = 0, \qquad \text{where} \qquad \delta^2 = h/(24\alpha).$$

Having found this relationship between the FPU Lattice and the KdV equation, Kruskal and Zabusky made some numerical experiments, solving the KdV initial value problem for various initial data. Before discussing the remarkable results that came out of these experiments, it will be helpful to recall some of the early history of this equation.

9.3.4 A First Look at the KdV Equation

Korteweg and de Vries derived their equation in 1895 to settle a debate that had been going on since 1844, when the naturalist and naval architect John Scott Russell, in an oft–quoted paper [Rus844], reported an experience a decade earlier in which he followed the bow wave of a barge that had suddenly stopped in a canal. This "solitary wave", some thirty feet long and a foot high, moved along the channel at about eight miles per hour, maintaining its shape and speed for over a mile as Russell raced after it on horseback. Russell became fascinated with this phenomenon and made extensive further experiments with such waves in a wave tank of his own devising, eventually deriving a (correct) formula for their speed as a function of height. The mathematicians Airy and Stokes made calculations which appeared to show that any such wave would be unstable and not persist for as long as Russell claimed. However, later work by Boussinesq (1872), Rayleigh (1876) and finally the Korteweg–de Vries paper in 1895 [KdV95] pointed out errors in the analysis of Airy and Stokes and vindicated Russell's conclusions.

The KdV equation is now accepted as controlling the dynamics of waves moving to the right in a shallow channel. Of course, Korteweg and de Vries did the obvious and looked for travelling–wave solutions for their equation by making the Ansatz [Pal97]

$$v(x,t) = f(x - ct).$$

When this is substituted in the standard form of the KdV equation, it gives

$$-cf' + 6ff' + f''' = 0.$$

If we add the boundary conditions that f should vanish at infinity, then a fairly routine analysis leads to the 1–parameter family of travelling–wave solutions

$$v(x,t) = 2a^2 \operatorname{sech}^2(a(x - 4a^2 t)),$$

now referred to as the one–soliton solutions of KdV (these are the solitary waves of Russell). Note that the amplitude $2a^2$ is exactly half the speed $4a^2$, so that taller waves move faster than their shorter brethren.

Now, back to Zabusky and Kruskal. For numerical reasons, they chose to deal with the case of periodic boundary conditions, in effect studying the KdV equation

$$u_t + uu_x' + \delta^2 u_{xxx} = 0 \tag{9.31}$$

on the circle instead of on the line. For their published report, they chose $\delta = 0.022$ and used the initial condition

$$u(x,0) = \cos(\pi x).$$

Here is an extract from their report (containing the first use of the term "soliton") in which they describe their observations:

(I) Initially the first two terms of (9.31) dominate and the classical overtaking phenomenon occurs; that is u steepens in regions where it has negative slope. (II) Second, after u has steepened sufficiently, the third term becomes important and serves to prevent the formation of a discontinuity. Instead, oscillations of small wavelength (of order δ) develop on the left of the front. The amplitudes of the oscillations grow, and finally *each* oscillation achieves an almost steady amplitude (that increases linearly from left to right) and has the shape of an individual solitary–wave of (9.31). (III) Finally, each "solitary wave pulse" or *soliton* begins to move uniformly at a rate (relative to the background value of u from which the pulse rises) which is linearly proportional to its amplitude. Thus, the solitons spread apart. Because of the periodicity, two or more solitons eventually overlap spatially and interact nonlinearly. Shortly after the interaction they reappear virtually unaffected in size or shape. In other words, solitons "pass through" one another without losing their identity. *Here we have a nonlinear physical process in which interacting localized pulses do not scatter irreversibly.*

Zabusky and Kruskal go on to describe a second interesting observation, a recurrence property of the solitons that goes a long way towards accounting for the surprising recurrence observed in the FPU Lattice. Let us explain again, but in somewhat different terms, the reason why the recurrence in the FPU Lattice is so surprising. The lattice is made up of a great many identical oscillators. Initially the relative phases of these oscillators are highly correlated by the imposed cosine initial condition. If the interactions are linear ($\alpha = 0$), then the oscillators are harmonic and their relative phases remain constant. But, when α is positive, the an–harmonic forces between the oscillators cause their phases to start drifting relative to each other in an apparently uncorrelated manner. The expected time before the phases of all of the oscillators will be simultaneously close to their initial phases is enormous, and increases rapidly with the total number N. But, from the point of view of the KdV solitons, an entirely different picture appears. As mentioned in the above paragraph, if δ is put equal to zero in the KdV equation, it reduces to the so–called *inviscid Burgers' equation*, which exhibits steepening and breaking of a negatively sloped wave front in a finite time T_B. (For the above initial conditions, the breaking time, T_B, can be computed theoretically to be $1/\pi$.) However, when $\delta > 0$, just before breaking would occur, a small number of solitons emerge (eight in the case of the above initial wave shape, $\cos(\pi x)$) *and this number depends only on the initial wave shape, not on the number of oscillators.* The expected time for their respective centers of gravity to all eventually "focus" at approximately the same point of the circle is of course much smaller than the expected time for the much larger number of oscillators to all return approximately to their original phases. In fact, the recurrence time T_R for the solitons turns out to be approximately equal to $30.4T_B$, and at this time the wave shape $u(x, T_R)$ is uniformly very close to the initial wave form $u(x, 0) = \cos(\pi x)$. There is a second (somewhat weaker) focusing at time $t = 2T_R$, etc. (Note that these times are measured in units of the "slow time", τ, at which the shape of the FPU travelling wave evolves, not in the "fast time", t, at which the travelling wave moves.) In effect, the KdV solitons are providing a hidden correlation between the relative phases of the FPU oscillators.

Notice that, as Zabusky and Kruskal emphasize, it is the persistence or shape conservation of the solitons that provides the explanation of recurrence. If the shapes of the solitons were not preserved when they interacted, there would be no way for them to all get back together and approximately reconstitute the initial condition at some later time. Here in their own words is how they bring in solitons to account for the fact that thermalization was not observed in the FPU experiment:

> Furthermore, because the solitons are remarkably stable entities, preserving their identities throughout numerous interactions, one would expect this system to exhibit thermalization (complete energy sharing among the corresponding linear normal modes) only after extremely long times, if ever.

But this explanation, elegant as it may be, only pushes the basic question back a step. A full understanding of FPU recurrence requires that we comprehend the reasons behind the remarkable new phenomenon of solitonic behavior, and in particular *why* solitons preserve their shape. In fact, it was quickly recognized that the soliton was itself a vital new feature of nonlinear dynamics, so that understanding it better and discovering other nonlinear wave equations that had soliton solutions became a primary focus for research in both pure and applied mathematics. The mystery of the FPU Lattice recurrence soon came to be regarded as an important but fortuitous spark that ignited this larger effort.

The next few short sections explain some elementary but important facts about 1D wave equations. If you know about shock development, and how dispersion smooths shocks, you can skip these sections without loss of continuity.

9.3.5 Split–Stepping KdV

In the KdV equation,
$$u_t = -6uu_x - u_{xxx},$$
if we drop the nonlinear term, we have a constant coefficient linear PDE whose initial value problem can be solved explicitly by the Fourier Transform. On the other hand, if we ignore the linear third–order term, then we are left with the inviscid Burgers' equation, whose initial value problem can be solved numerically by a variety of methods. Note that it can also be solved in implicit form analytically, for short times, by the method of characteristics,

$$u = u_o(x - 6ut),$$

but the solution is not conveniently represented on a fixed numerical grid. So, can we somehow combine the methods for solving each of the two parts into an efficient numerical method for solving the full KdV initial value problem?

In fact we can, and indeed there is a very general technique that applies to such situations. In the pure mathematics community it is usually referred to as the Trotter Product Formula, while in the applied mathematics and numerical analysis communities it is called *split–stepping*. Let me state it in the context of ordinary differential equations. Suppose that Y and Z are two smooth vector–fields on $\mathbb{R}^n$, and we know how to solve each of the ODEs

$$\dot{x} = Y(x) \qquad \text{and} \qquad \dot{x} = Z(x),$$

meaning that we know both of the flows ϕ_t and ψ_t on $\mathbb{R}^n$ generated by X and Y respectively. The Trotter Product Formula is a method for constructing the flow θ_t generated by $Y + Z$ out of ϕ and ψ; namely, letting [Pal97]

$$\Delta t = \frac{t}{n}, \qquad \theta_t = \lim_{n \to \infty} (\phi_{\Delta t} \psi_{\Delta t})^n.$$

The intuition behind the formula is simple. Think of approximating the solution of

$$\dot{x} = Y(x) + Z(x)$$

by Euler's Method. If we are currently at a point p_0, to propagate one more time step Δt we go to the point $p_0 + \Delta t (Y(p_0) + Z(p_0))$. Using the split–step approach on the other hand, we first take an Euler step in the $Y(p_0)$ direction, going to $p_1 = p_0 + \Delta t Y(p_0)$, then take a second Euler step, but now from p_1 and in the $Z(p_1)$ direction, going to $p_2 = p_1 + \Delta t Z(p_1)$. If Y and Z are constant vector–fields, then this gives exactly the same final result as the simple full Euler step with $Y + Z$, while for continuous Y and Z and small time step Δt it is a good enough approximation that the above limit is valid.

The situation is more delicate for flows on infinite dimensional manifolds. Nevertheless it was shown by F. Tappert in [Tap74] that the *Cauchy Problem for KdV* can be solved numerically by using split–stepping to combine solution methods for

$$u_t = -6 u u_x \qquad \text{and} \qquad u_t = -u_{xxx}.$$

In addition to providing a perspective on an evolution equation's relation to its component parts, split-stepping allows one to modify a code from solving KdV to the *Kuramoto–Sivashinsky equation*

$$u_t + u u_x = -u_{xx} - u_{xxxx},$$

or study the joint zero–diffusion–dispersion limits *KdV–Burgers' equation*

$$u_t + 6 u u_x = \nu u_{xx} + \epsilon u_{xxxx},$$

by merely changing one line of code in the Fourier module.

Tappert uses an interesting variant, known as Strang splitting, which was first suggested in [Str68] to solve multi–dimensional hyperbolic problems by split–stepping 1D problems. The advantage of splitting comes from the greatly reduced effort required to solve the smaller bandwidth linear systems which arise when implicit schemes are necessary to maintain stability. In addition, Strang demonstrated that second-order accuracy of the component methods need not be compromised by the asymmetry of the splitting, as long as the pattern $\phi_{\frac{\Delta t}{2}} \psi_{\frac{\Delta t}{2}} \psi_{\frac{\Delta t}{2}} \phi_{\frac{\Delta t}{2}}$ is used, to account for possible non-commutativity of Y and Z. (This may be seen by multiplying the respective exponential

series.) No higher–order analogue of Strang splitting is available. Serendipi-
tously, when output is not required, several steps of Strang splitting require
only marginal additional effort:

$$(\phi_{\frac{\Delta t}{2}} \psi_{\frac{\Delta t}{2}} \psi_{\frac{\Delta t}{2}} \phi_{\frac{\Delta t}{2}})^n = (\phi_{\frac{\Delta t}{2}} \psi_{\Delta t}(\phi_{\Delta t} \psi_{\Delta t})^{n-1} \phi_{\frac{\Delta t}{2}}.$$

Now, the FPU Lattice is a classical finite dimensional mechanical system,
and as such it has a natural Hamiltonian formulation. However its relation to
KdV is rather complex, and KdV is a PDE rather than a finite dimensional
system of ODE, so it is not clear that it too can be viewed as a Hamiltonian
system. We shall now see how this can be done in a simple and natural way.
Moreover, when interpreted as the infinite dimensional analogue of a Hamiltonian system, KdV turns out to have a key property one would expect from any
generalization to infinite dimensions of the concept of complete integrability
in the Liouville sense, namely the existence of infinitely many functionally
independent constants of the motion that are in involution.

In 1971, Gardiner and Zakharov independently showed how to interpret
KdV as a Hamiltonian system, starting from a Poisson bracket approach, and
from this beginning Poisson brackets have played a significantly more important rôle in the infinite dimensional theory of Hamiltonian systems than they
did in the more classical finite dimensional theory, and in recent years this has
led to a whole theory of *Poisson manifolds* and *Poisson Lie groups*. However,
we will start with the more classical approach to Hamiltonian systems, defining a symplectic structure for KdV first, and then get the Poisson bracket
structure as a derived concept (see [AMR88], as well as [II06b]). Thus, we can
exhibit a symplectic structure Ω for the phase–space P of the KdV equation
and a Hamiltonian function, $H : P \to \mathbb{R}$, such that the KdV equation takes
the form

$$\dot{u} = (\nabla_s H)_u.$$

9.3.6 Solitons from a Pendulum Chain

Consider the linear chain of N pendula, each coupled to the nearest neighbors
by the elastic bonds. Let us denote the angle of rotation of ith pendulum by
$\varphi_i = \varphi_i(t)$ and the angular velocity by $\omega_i = \omega_i(t)$. Each pendulum experiences
the angular momentum due to the gravity and the angular momenta from the
two elastic bonds, which are proportional to the difference in angles of rotation
of the coupled pendula with the coefficient κ. Under the assumption that all
pendula have the same inertia moment J, the set of equations of motion takes
the form

$$J\dot{\omega}_i = p_i^{el} + p_i^{gr}, \tag{9.32}$$

where p_i^{el} is the angular momentum from the elastic bonds and p_i^{gr} is the
gravity angular momentum. The angular momenta from the left and right
elastic bonds are $\kappa(\varphi_{i-1} - \varphi_i)$ and $\kappa(\varphi_{i+1} - \varphi_i)$, respectively and hence,
$p_i^{el} = \kappa(\varphi_{i+1} - 2\varphi_i + \varphi_{i-1})$. The gravity angular momentum can be expressed

as $p_i^{gr} = -J\,g\sin\varphi_i$, where g is the gravity constant. In view of the expressions for p_i^{el} and p_i^{gr}, the equations of motion (9.32) can be rewritten as

$$J\ddot{\varphi}_i = \kappa(\varphi_{i+1} - 2\varphi_i + \varphi_{i-1}) - J\,g\sin\varphi_i. \tag{9.33}$$

In the *continuous limit*, when $\kappa/J\,g$ tends to 0, the ODE system (9.33) transforms into a single PDE

$$J\,\Phi_{tt} = \kappa\Phi_{xx} - K_G\sin\Phi, \tag{9.34}$$

where $\Phi = \Phi(x,t)$, and $K_G = J\,g$.

In the new spatial and time variables, $X = x\sqrt{\frac{K_G}{\kappa}}$, $T = t\sqrt{\frac{K_G}{J}}$, the PDE (9.34) gets the standard form of the *sine–Gordon equation* for solitary waves,

$$\Psi_{tt} = \Psi_{xx} - \sin\Psi,$$

with $\Psi = \Psi(X,T)$.

In this way the system of coupled pendula represents the discrete analog to the continuous sine–Gordon equation. In the framework of the pendulum–chain the *kink* is the solitary wave of counter–clockwise rotation of pendula through the angle 2π. The *antikink* is the solitary wave of clockwise rotation of pendula through the angle 2π.

9.3.7 1D Crystal Soliton

Crystal is considered as a chain of the un–deformable molecules, which are connected with each other by the elastic hinges with elastic constant f.

The chain is compressed by the force p along its axis. The vertical springs make elastic support for the nodes of the chain. u_n denotes the transversal displacement of nth node.

The Hamiltonian of the crystal may be written in the dimensionless form:

$$H = \frac{1}{2}\sum_n[\dot{u}_n^2 + f(u_{n+1} - 2u_n + u_{n-1})^2 - p(u_n - u_{n-1})^2 + u_n^2 + \frac{1}{2}u_n^4], \tag{9.35}$$

where the first term is the kinetic energy, the second one is the potential energy of the elastic hinge, the third one is the work done by the external force and the last two terms give the energy of the elastic nonlinear support due to the vertical springs.

Equation of motion for nth hinge has the form

$$\ddot{u}_n + f(u_{n+2} - 4u_{n+1} + 6u_n - 4u_{n-1} + u_{n-2}) + P(u_{n+1} - 2u_n + u_{n-1}) + u_n + u_n^3 = 0. \tag{9.36}$$

The role of the elastic hinges is to keep the chain as a straight line while the compression force plays a destructive role, it tends to destroy the horizontal arrangement of bars. The competition between these two factors, f and p, gives rise to modulation instability in the model.

9.3.8 Solitons and Chaotic Systems

In the past four decades, both the solitons [KM89] and the chaos [GC00] have been widely studied and applied in many natural sciences and especially in almost all the physics branches such as the condense matter physics, field theory, fluid dynamics, plasma physics and optics etc. Usually, one considers that the solitons are the basic excitations of the integrable models, and the chaos is the basic behavior of the non–integrable models. Actually, the above consideration may not be complete especially in higher dimensions. When one says a model is integrable, one should emphasize two important facts. The first one is that we should point out the model is integrable under what special meaning(s). For instance, we say a model is Painlevé integrable if the model possesses the Painlevé property, and a model is Lax or IST (inverse scattering transformation) integrable if the model has a Lax pair and then can be solved by the IST approach. An integrable model under some special meanings may not be integrable under other meanings. For instance, some Lax integrable models may not be Painlevé integrable [CE97]. The second fact is that for the general solution of a higher dimensional integrable model, say, a Painlevé integrable model, there exist some lower dimensional arbitrary functions, which means any lower dimensional chaotic solutions can be used to construct exact solutions of higher dimensional integrable models.

In [LTZ01, TLZ02], it was shown that an IST integrable model and/or a Painlevé integrable model might have some lower dimensional reductions with chaotic behaviors and then a lower dimensional chaos system might have some higher dimensional Lax pairs. They used the (2+1)D *Davey–Stewartson (DS) equation* [DS74]

$$iu_t + 2^{-1}(u_{xx} + u_{yy}) + \alpha|u|^2 u - uv = 0, \qquad v_{xx} - v_{yy} - 2\alpha(|u|^2)_{xx} = 0, \quad (9.37)$$

as a concrete example at first. The DS equation is an isotropic Lax integrable extension of the well known $(1+1)$D nonlinear Schrödinger (NLS) equation. The DS system is the shallow water limit of the Benney–Roskes equation [DS74], where u is the amplitude of a surface wave–packet and v characterizes the mean motion generated by this surface wave. The DS system (9.37) can also be derived from the plasma physics and the self–dual Yang–Mills field. The DS system has also been proposed as a $(2+1)$D model for quantum field theory [SAB87]. It is known that the DS equation is integrable under some special meanings, namely, it is IST integrable and Painlevé integrable [BLM88]. Many other interesting properties of the model like a special bilinear form, the Darboux transformation, finite dimensional integrable reductions, infinitely many symmetries and the rich soliton structures [BLM88, LH94] have also been revealed.

To select out some chaotic behaviors of the DS equation, we make the following transformation [LTZ01, TLZ02]

$$v = v_0 - f^{-1}(f_{x'x'} + f_{y'y'} + 2f_{x'y'}) + f^{-2}(f_{x'}^2 + 2f_{y'}f_{x'} + f_{y'}^2), \qquad u = gf^{-1} + u_0, \quad (9.38)$$

with real f and complex g, where $x' = (x+y)/\sqrt{2}$, $y' = (x-y)/\sqrt{2}$, and $\{u_0,\ v_0\}$ is an arbitrary seed solution of the DS equation. Under the transformation (9.37), the DS system (9.38) is transformed to a general bilinear form:

$$(D_{x'x'} + D_{y'y'} + 2iD_t)g \cdot f + u_0(D_{x'x'} + 2D_{x'y'} + D_{y'y'})f \cdot f$$
$$+ 2\alpha u_0 gg^* + 2\alpha u_0^2 g^* f - 2v_0 gf + G_1 fg = 0, \tag{9.39}$$

$$2(D_{x'y'} + \alpha|u_0|^2)f \cdot f + 2\alpha gh + 2\alpha g f u_0^* + 2\alpha u_0 g^* f - G_1 ff = 0, \tag{9.40}$$

where D is the usual bilinear operator [Hir71] defined as

$$D_x^m A \cdot B \equiv (\partial_x - \partial_{x_1})^m A(x)B(x_1)|_{x_1=x},$$

and G_1 is an arbitrary solution of the associated equation [LTZ01, TLZ02]

To discuss further, we fix the seed solution $\{u_0,\ v_0\}$ and G_1 as

$$u_0 = G_1 = 0, \qquad v_0 = p_0(x,t) + q_0(y,t), \tag{9.41}$$

where $p_0 \equiv p_0(x,t)$ and $q_0 \equiv q_0(y,t)$ are some functions of the indicated variables.

To solve the bilinear equations (9.39) and (9.40) with (9.41), we make the ansatz

$$f = C + p + q, \quad g = p_1 q_1 e^{ir+is}, \tag{9.42}$$

where $p \equiv p(x,\ t)$, $q \equiv q(y,\ t)$, $p_1 \equiv p_1(x,\ t)$, $q_1 \equiv q_1(y,\ t)$, $r \equiv r(x,\ t)$, $s \equiv s(y,\ t)$ are all real functions of the indicated variables and C is an arbitrary constant. Substituting (9.42) into (9.39) and (9.40), and separating the real and imaginary parts of the resulting equation, we have

$$2p_x q_y - \alpha p_1^2 q_1^2 = 0. \tag{9.43}$$

$$(q_1 p_{1xx} + p_1 q_{1yy} - p_1 q_1 (2r_t + 2s_t + 2(p_0 + q_0) + s_y^2 + r_x^2))(C + p + q)$$
$$+ q_1(p_1 p_{xx} - 2p_{1x} p_x) + p_1(q_1 q_{yy} - 2q_{1y} q_y) = 0, \tag{9.44}$$

$$(-q_1(2r_x p_{1x} + 2p_{1t} + p_1 r_{xx}) - p_1(2s_y q_{1y} + 2q_{1t} + q_1 s_{yy}))(C + p + q)$$
$$+ 2q_1 p_1(q_t + s_y q_y) + 2q_1 p_1(r_x p_x + p_t) = 0 \tag{9.45}$$

Because the functions p_0, p, p_1 and r are only functions of $\{x,\ t\}$ and the functions q_0, q, q_1 and s are only functions of $\{y,\ t\}$, the equation system (9.43), (9.44) and (9.45) can be solved by the following variable separated equations [LTZ01, TLZ02]:

$$p_1 = \delta_1 \sqrt{2\alpha^{-1} c_1^{-1} p_x}, \quad q_1 = \delta_2 \sqrt{c_1 q_y}, \quad (\delta_1^2 = \delta_2^2 = 1), \tag{9.46}$$
$$p_t = -r_x p_x + c_2, \quad q_t = -s_y q_y - c_2, \tag{9.47}$$

$$4(2r_t + r_x^2 + 2p_0)p_x^2 + p_{xx}^2 - 2p_{xxx}p_x + c_0 p_x^2 = 0, \qquad (9.48)$$
$$4(2s_t + s_y^2 + 2q_0)q_y^2 + q_{yy}^2 - 2q_y q_{yyy} - c_0 q_y^2 = 0. \qquad (9.49)$$

In (9.46)–(9.49), c_1, c_2 and c_0 are all arbitrary functions of t.

Generally, for a given p_0 and q_0 the equation systems $\{(9.46), (9.48)\}$ and $\{(9.47), (9.49)\}$ may not be integrable. However, because of the arbitrariness of p_0 and q_0, we may treat the functions p and q are arbitrary while p_0 and q_0 are determined by (9.48) and (9.49). Because p and q are arbitrary functions, in addition to the stable soliton selections, there may be various chaotic selections. For instance, if we select p and q are solutions of $(\tau_1 \equiv x + \omega_1 t, \ \tau_2 \equiv x + \omega_2 t)$

$$p_{\tau_1\tau_1\tau_1} = \left(p_{\tau_1\tau_1}p_{\tau_1} + (c+1)p_{\tau_1}^2\right)p^{-1} - (p^2 + bc + b)p_{\tau_1} \qquad (9.50)$$
$$-(b+c+1)p_{\tau_1\tau_1} + pc(ba - b - p^2),$$
$$q_{\tau_2\tau_2\tau_2} = \left(q_{\tau_2\tau_2}q_{\tau_2} + (\gamma+1)q_{\tau_2}^2\right)q^{-1} - (q^2 + \beta\gamma - \beta)q_{\tau_2} \qquad (9.51)$$
$$-(\beta+\gamma+1)q_{\tau_2\tau_2} + qc(\beta\alpha - \beta - q^2),$$

where ω_1, ω_2, a, b, c, α, β and γ are all arbitrary constants, then

$$c_0 = c_2 = 0, \qquad r = -\omega_1(x + \omega_1 t/2), \qquad s = -\omega_2(y + \omega_2 t/2), (9.52)$$
$$p_0 = -4^{-1}cp^3 p_{\tau_1}^{-1} + p^2 - bcp_{\tau_1}^{-1}(a-1)p + b(c+1)$$
$$+ \ (b+c+1)p_{\tau_1\tau_1}p_{\tau_1}^{-1} + p_{\tau_1\tau_1}^2 2^{-1}p_{\tau_1}^{-2} - p^{-1}(p_{\tau_1}(c+1) + p_{\tau_1\tau_1}), \ (9.53)$$
$$q_0 = -4^{-1}\gamma q^3 q_{\tau_2}^{-1} + q^2 - \beta\gamma q_{\tau_2}^{-1}(a-1)q + \beta(\gamma+1)$$
$$+ \ (\beta+\gamma+1)q_{\tau_2\tau_2}q_{\tau_2}^{-1} + q_{\tau_2\tau_2}^2 2^{-1}q_{\tau_2}^{-2} - q^{-1}(q_{\tau_2}(\gamma+1) + q_{\tau_2\tau_2}). \ (9.54)$$

Substituting (9.42) with (9.46)–(9.54), we get a general solution of the DS equation

$$u = \delta_1\delta_2\sqrt{2\alpha^{-1}p_{\tau_1}q_{\tau_2}} \exp(-i(\omega_1 x + \omega_2 y + \frac{1}{2}(\omega_1^2 + \omega_2^2)t)(C + p + q)^{-1}(9.55)$$
$$v = p_0 + q_0 - (q_{\tau_2\tau_2} + p_{\tau_1\tau_1})(C + p + q)^{-1} \qquad (9.56)$$
$$+ \ (q_{\tau_2}^2 + 2q_{\tau_2}p_{\tau_1} + p_{\tau_1}^2)(C + p + q)^{-2},$$

where p_0 and q_0 are determined by (9.53) and (9.54), while p and q are given by (9.50) and (9.51).

It is straightforward to prove that (9.50) (and (9.51)) is equivalent to the well known chaos system, the *Lorenz system* [Lor63, Spa82]:

$$p_{\tau_1} = -c(p-g), \qquad g_{\tau_1} = p(a-h) - g, \qquad h_{\tau_1} = pg - bh. \qquad (9.57)$$

Actually, after cancelling the functions g and h in (9.57), one can find (9.50) immediately.

From the above discussions, some interesting things are worth emphasizing:

Firstly, because of the arbitrariness of the functions p and q, any types of other lower dimensional systems may be used to construct the exact solutions of the DS system.

Secondly, the lower dimensional chaotic behaviors may be found in many other higher dimensional soliton systems. For instance, by means of the direct substitution or the similar discussions as for the DS equation, one can find that ($\tau_1 \equiv x + \omega_1 t$)

$$u = 2p_x w_y (p + w)^{-2}, \tag{9.58}$$

$$v = \frac{2p_x^2}{(p+w)^2} + \frac{(c+1)p_x}{3p} - \frac{(5p-w)p_{xx}}{3p(p+w)} - \frac{1}{3}p^2 - \frac{1}{3p_x}[(1+b+c)p_{xx}+cp^3-cb(a-1)p], \tag{9.59}$$

with w being an arbitrary function of y, and p being determined by the (1+1)D extension of the Lorenz system

$$p_t = -p_{xxx} + p^{-1}[p_{xx}p_x + (c+1)p_x^2] - p^2 p_x - (b+c+1)p_{xx} - pc(b-ba+p^2), \tag{9.60}$$

solves the following IST and Painlevé integrable KdV equation which is known as the ANNV model [BLM87]

$$u_t + u_{xxx} - 3(uv)_x = 0, \qquad u_x = v_y. \tag{9.61}$$

It is clear that the Lorenz system (9.50) is just a special reduction of (9.61) with $p = p(x + b(c+1)t) \equiv p(\tau_1)$. Actually, p of equations (9.58) may also be arbitrary function of $\{x,\ t\}$ after changing (9.59) appropriately [BLM87]. In other words, any lower dimensional chaotic behavior can also be used to construct exact solutions of the ANNV system.

The third thing is more interesting. The Lax pair plays a very important and useful role in integrable models. Nevertheless, there is little progress in the study of the possible Lax pairs for chaos systems like the Lorenz system. In [CE97], the Lax pairs for two discrete non–Painlevé integrable models have been proposed.

Now, from the above discussions, we know that both the Lax pairs of the DS equation and those of the ANNV system may be used as the special higher dimensional Lax pairs of arbitrary chaos systems like the Lorenz system (9.50) and/or the generalized Lorenz system (9.60) by selecting the fields appropriately like (9.55)–(9.56) and/or (9.58)–(9.59). For instance, the $(1+1)$D generalized Lorenz system (9.60) has the following Lax pair [LTZ01, TLZ02]

$$\psi_{xy} = u\psi, \qquad \psi_t = -\psi_{xxx} + 3v\psi_x, \tag{9.62}$$

with $\{u,\ v\}$ being given by (9.58) and (9.59). From (9.58)–(9.61), we know that a lower dimensional chaos system can be considered as a consistent condition of a higher dimensional linear system. For example, $\psi_{xyt} = \psi_{txy}$ of (9.62) just gives out the generalized Lorenz system (9.60).

Now a very important question is what the effects of the lower dimensional chaos to the higher dimensional soliton systems are. To answer this question, we use the numerical solutions of the Lorenz system to see the behaviors of the corresponding solution (9.58) of the ANNV equation by taking

$$w = 200 + \tanh(y - y_0) \equiv w_s \qquad p = p(\tau_1) \equiv p'(X),$$

as the numerical solution of the Lorenz system (9.50). Under the selection $w = w_s$, (9.58) is a line soliton solution located at $y = y_0$. Due to the entrance of the function p, the structures of the line soliton become very complicated. For some types of the parameters, the solutions of the Lorenz system have some kinds of periodic behavior, then the line soliton solution (9.58) with $w = w_s$ becomes an x periodic line soliton solution that means the solution is localized in y direction (not identically equal to zero only near $y = y_0$) and periodic in x direction.

In summary, though some $(2+1)$D soliton systems, like the DS equation and the ANNV equation, are Lax and IST integrable, and some special types of soliton solutions can be found by IST and other interesting approaches [BLM88], any types of chaotic behaviors may still be allowed in some special ways. Especially, the famous chaotic Lorenz system and its $(1+1)$D generalization are derived from the DS equation and the ANNV equation. Using the numerical results of the lower dimensional chaotic systems, we may get some types of nonlinear excitations like the periodic and chaotic line solitons for higher dimensional soliton systems. On the other hand, the lower dimensional chaos systems like the generalized Lorenz system may have some particular Lax pairs in higher dimensions (for details, see [LTZ01, TLZ02]).

It is also known that both the ANNV system and the DS systems are related to the Kadomtsev–Petviashvili (KP) equation while the DS and the KP equation are the reductions of the self–dual Yang–Mills (SDYM) equation (see [II06b]). So both the KP and the SDYM equations may possess arbitrary lower dimensional chaotic behaviors induced by arbitrary functions.

9.4 Chaos Field Theory

In [Cvi00], Cvitanovic re–examined the path–integral formulation and the role that the classical solutions play in quantization of strongly nonlinear fields. In the path integral formulation of a field theory the dominant contributions come from saddle–points, the classical solutions of equations of motion. Usually one imagines *one dominant saddle point*, the 'vacuum' (see Figure 9.1, (a)).

The *Feynman diagrams* of quantum electrodynamics (QED) and quantum chromodynamics (QCD), associated to their path integrals, give us a visual and intuitive scheme to calculate the correction terms to this starting semiclassical, *Gaussian saddlepoint approximation*. But there might be other saddles (Figure 9.1, (b)). That field theories might have a rich repertoire of classical solutions became apparent with the discovery of *instantons* [BPS75], analytic solutions of the classical $SU(2)$ *Yang–Mills relation*, and the realization that the associated *instanton vacua* receive contributions from countable ∞'s of saddles. What is not clear is whether these are the important classical saddles. Cvitanovic asks the question: could it be that the strongly nonlinear theories are dominated by altogether different classical solutions?

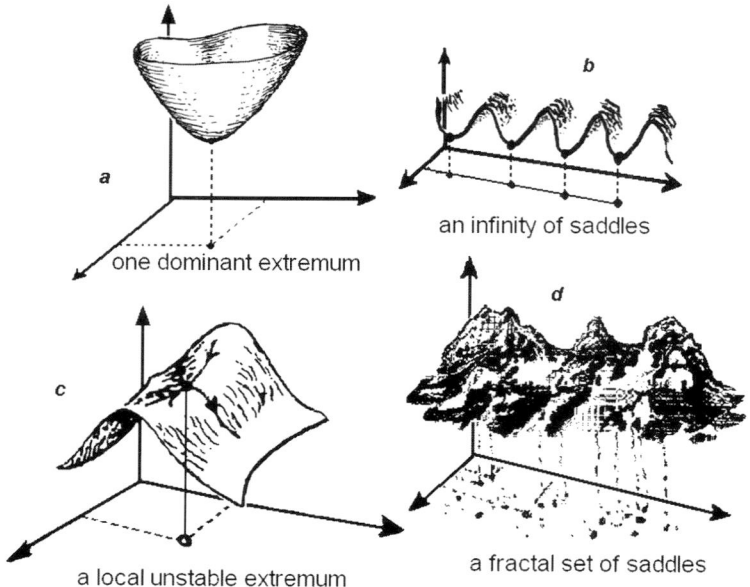

Fig. 9.1. Path integrals and chaos field theory (see text for explanation).

The search for the classical solutions of nonlinear field theories such as the *Yang–Mills* and *gravity* has so far been neither very successful nor very systematic. In modern field theories the main emphasis has been on symmetries (compactly collected in action functionals that define the theories) as guiding principles in writing down the actions. But writing down a differential equation is only the start of the story; even for systems as simple as 3 coupled ordinary differential equations one in general has no clue what the nature of the long time solutions might be.

These are hard problems, and in explorations of modern field theories the dynamics tends to be is neglected, and understandably so, because the wealth of the classical solutions of nonlinear systems can be truly bewildering. If the classical behavior of these theories is anything like that of the field theories that describe the classical world – the hydrodynamics, the magneto–hydrodynamics, the *Burgers dynamical system* (8.85), *Ginzburg–Landau equation* (8.82), or *Kuramoto–Sivashinsky equation* (8.84), there should be very many solutions, with very few of the important ones analytical in form; the strongly nonlinear classical field theories are turbulent, after all. Furthermore, there is not a dimmest hope that such solutions are either beautiful or analytic, and there is not much enthusiasm for grinding out numerical solutions as long as one lacks ideas as what to do with them.

By late 1970's it was generally understood that even the simplest nonlinear systems exhibit chaos. Chaos is the norm also for generic Hamiltonian flows, and for path integrals that implies that instead of a few, or countably few saddles (Figure 9.1, (c)), classical solutions populate fractal sets of saddles (Figure 9.1, (d)). For the path–integral formulation of quantum mechanics such solutions were discovered and accounted for by [Gut90] in late 1960's. In this framework the spectrum of the theory is computed from a set of its unstable classical periodic solutions and quantum corrections. The new aspect is that the individual saddles for classically chaotic systems are nothing like the harmonic oscillator degrees of freedom, the quarks and gluons of QCD – they are all unstable and highly nontrivial, accessible only by numerical techniques.

So, if one is to develop a semiclassical field theory of systems that are *classically chaotic* or *turbulent*, the problem one faces is twofold [Cvi00]

A. Determine, classify, and order by relative importance the classical solutions of nonlinear field theories.
B. Develop methods for calculating perturbative corrections to the corresponding classical saddles.

References

Arn61. Arnold, V.I.: On the stability of positions of equilibrium of a Hamiltonian system of ordinary differential equations in the general elliptic case. Sov. Math. Dokl. **2**, 247, (1961)

Arn65. Arnold, V.I.: Sur une propriété topologique des applications globalement canoniques de la mécanique classique. C. R. Acad. Sci. Paris A, **261**, 17, (1965)

AA68. Arnold, V.I., Avez, A.: Ergodic Problems of Classical Mechanics. Benjamin, New York, (1968)

Arn78. Arnold, V.I.: Ordinary Differential Equations. MIT Press, Cambridge, (1978)

Arn88. Arnold, V.I.: Geometrical Methods in the Theory of Ordinary differential equations. Springer, New York, (1988)

Arn89. Arnold, V.I.: Mathematical Methods of Classical Mechanics (2nd ed). Springer, New York, (1989)

Arn92. Arnold, V.I.: Catastrophe Theory. Springer, Berlin, (1992)

Arn93. Arnold, V.I.: Dynamical systems. Encyclopaedia of Mathematical Sciences, Springer, Berlin, (1993)

Arn99. Arnaud, M.C.: Création de connexions en topologie C^1. Preprint Université de Paris-Sud 1999, to appear in Ergodic Theory and Dynamical Systems. C. R. Acad. Sci. Paris, Série I, **329**, 211–214, (1999)

Aba96. Abarbanel, H.D.I.: Analysis of Observed Chaotic Data. Springer-Verlag, Berlin, (1996)

Ala99. Alaoui, A.M.: Differential Equations with Multispiral Attractors. Int. Journ. of Bifurcation and Chaos, **9**(6), 1009-1039, (1999)

ACF98. Arena, P., Caponetto, R., Fortuna, L., Porto, D.: Bifurcation and chaos in non-integer order cellular neural networks. Int. J. Bif. Chaos **8**, 1527–1539, (1998)

AA80. Aubry, S., André, G.: Colloquium on Computational Methods in Theoretical Physics. In Group Theoretical Methods in Physics, Horowitz and Ne'eman (ed), Ann. Israel Phys. Soc. **3**, 133-164, (1980)

AAC90. Artuso, R., Aurell, E., Cvitanovic, P.: Recycling of strange sets: I & II. Nonlinearity **3**, 325-359 and 361, (1990)

AB86. Aref, H., Balachandar, S.: Chaotic advection in a Stokes flow Phys. Fluids, **29**, 3515-3521, (1986)

ABC96. Aurell, E., Boffetta, G., Crisanti, A., Paladin, G., Vulpiani, A.: Predictability in Systems with Many Characteristic Times: The Case of Turulence. Phys. Rev. E **53**, 2337, (1996)

ABC97. Artale, V., Boffetta, G., Celani, A., Cencini, M., Vulpiani, A.: Dispersion of passive tracers in closed basins: Beyond the diffusion coefficient. Phys. Fluids A **9**, 3162, (1997)

ABS00. J. A. Acebrón, L. L. Bonilla, and R. Spigler: Synchronization in populations of globally coupled oscillators with inertial effects. Phys. Rev. E **62**, 34373454, (2000)

Ach97. Acheson, D.: From Calculus to Chaos. Oxford Univ. Press, Oxford, (1997)

ACS85. Arneodo, A., Coullet, P. and Spiegel, E.A.: The dynamics of triple convection. Geophys. Astrophys. Fluid Dynamics **31**, 148, (1985)

AF84. Alfsen, K.H., Frøyland, J.: Systematics of the Lorenz model at $\sigma = 10$. Technical report, University of Oslo, Inst. of Physics, Norway, (1984)

AG39. Aurell, E., Gilbert, A.: Fast dynamos and determinants of singular integral operators. Geophys. Astrophys. Fluid Dyn. **73**, 532, (1993)

AGI98. Amaral, L.A.N., Goldberger, A.L., Ivanov, P.Ch., Stanley, H.E.: Scale–independent measures and pathologic cardiac dynamics. Phys. Rev. Lett., **81**, 2388–2391, (1998)

AGS85. Amit, D.J., Gutfreund, H., Sompolinsky, H.: Spin-glass models of neural networks. Phys. Rev. A **32**, 10071018, (1985)

AH75. Atherton, R.W., Homsy, G.M.: On the existence and formulation of variational principles for nonlinear differential equations, Stud. Appl. Math., **54**, 31–60, (1975)

AKO03. Ansumali, S., Karlin, I.V., Öttinger, H.C.: Minimal Entropic Kinetic Models for Hydrodynamics, Europhys. Lett. **63**, 798-804, (2003)

AKS04. Ansumali, S., Karlin, I.V., Succi, S.: Kinetic theory of turbulence modeling: smallness parameter, scaling and microscopic derivation of Smagorinsky model. Physica A 338, 379, (2004)

Ale68. Alekseev, V.: Quasi-random dynamical systems I, II, III. Math. USSR Sbor. **5**, 73–128, (1968)

AM78. Abraham, R., Marsden, J.: Foundations of Mechanics. Benjamin, Reading, (1978)

AM88. Amari, S.I., Maginu, K.: Statistical neurodynamics of associative memory. Neu. Net. **1**, 63-73, (1988)

Ama77. Amari, S.I.: Neural theory of association and concept-formation. Biol Cybern. **26**:(3),175185, (1977)

Ama78. Amari, S., Takeuchi, A.: Mathematical theory on formation of category detecting nerve cells. Biol Cybern. **29**:(3),127136, (1978)

AMR88. Abraham, R., Marsden, J., Ratiu, T.: Manifolds, Tensor Analysis and Applications. Springer, New York, (1988)

AN99. Aoyagi, T., Nomura, M.: Oscillator Neural Network Retrieving Sparsely Coded Phase Patterns. Phys. Rev. Lett. **83**, 1062–1065, (1999)

And01. Andrecut, M.: Biomorphs, program for $Mathcad^{TM}$, Mathcad Application Files, Mathsoft, (2001)

And64. Anderson, P.W.: Lectures on the Many Body Problem. E.R. Caianiello (ed), Academic Press, New York, (1964)

AR95. Antoni, M., Ruffo, S.: Clustering and relaxation in long-range Hamiltonian dynamics. Phys. Rev. E, **52**, 2361–2374, (1995)

Arb95. Arbib, M.A.: The Handbook of Brain Theory and Neural Networks, MIT Press, (1995)

Are83. Aref, H.: Integrable, chaotic and turbulent vortex motion in two-dimensional flows. Ann. Rev. Fluid Mech. **15** 345-89, (1983)

Are84. Aref, H.: Stirring by chaotic advection. J. Fluid Mech. **143** 1-21, (1984)

AS72. Abramowitz, M., Stegun, I.A.: Handbook of Mathematical Functions. Dover, New York, (1972)

AS92. Abraham, R., Shaw, C.: Dynamics: the Geometry of Behavior. Addison-Wesley, Reading, (1992)

Att71. Attneave, F.: Multistability in Perception. Scientific American, **225**, 62–71, (1971)

ATT90. Aihara, K., Takabe, T., Toyoda, M.: Chaotic Neural Networks, Phys.Lett. A **144**, 333–340, (1990)

Ave99. Averin, D.V.: Solid-state qubits under control, Nature **398**, 748-749, (1999)

Bae98. Von Baeyer, H.C.: All Shook Up. The Sciences, **38**(1), 12–14, (1998)

Bal06. Balakrishnan, J.: A geometric framework for phase synchronization in coupled noisy nonlinear systems. Phys. Rev. E **73**, 036206, (2006)

BBP94. Babiano, A., Boffetta, G., Provenzale, A., Vulpiani, A.: Chaotic advection in point vortex models and two-dimensional turbulence. Phys. Fluids A **6**:7, 2465-2474, (1994).

BCE00. Boffetta, G., Cencini, M., Espa, S., Querzoli, G.: Chaotic advection and relative dispersion in a convective flow. Phys. Fluids **12**, 3160-3167, (2000)

BC85. Benedicks, M., Carleson, L.: On iterations of $1 - ax^2$ on $(-1, 1)$, Ann. of Math. **122**, 1–25, (1985)

BC91. Benedicks, M., Carleson, L.: Dynamics of the Hénon map, Ann. of Math. **133**, 73–169, (1991)

BBV99. Béguin, F., Bonatti, C., Vieitez, J.L.: Construction de flots de Smale en dimension 3. Ann. Fac. Sci. de Toulouse Math. **6**, 369–410, (1999)

BS70. Bhatia, N.P., Szego, G.P.: Stability theory of dynamical systems. Springer-Verlag, Heidelberg, (1979)

BD99. Bonatti, C., Díaz, L.J.: Connexions heterocliniques et genericité d'une infinité de puits ou de sources. Ann. Sci. École Norm. Sup. **32**, 135–150, (1999)

BDP99. Bonatti, C., Díaz, L.J., Pujals, E.: A C^1-dichotomy for diffeomorphisms: weak forms of hyperbolicity or infinitely many sinks or sources. Preprint, (1999)

BCB92. Borelli, R.L., Coleman, C., Boyce, W.E.: Differential Equations Laboratory Workbook. Wiley, New York, (1992)

BCF02. Boffetta, G., Cencini, M., Falcioni, M., Vulpiani, A.: Predictability: a way to characterize Complexity, Phys. Rep., **356**, 367–474, (2002)

BCI93. Balmfort, N.J., Cvitanovic, P., Ierley, G.R., Spiegel, E.A., Vattay, G.: Advection of vector–fields by chaotic flows. Ann. New York Acad. Sci. **706**, 148, (1993)

BCR82. Borsellino, A., Carlini, F., Riani, M., Tuccio, M.T., Marzo, A.D., Penengo, P., Trabucco, A.: Effects of visual angle on perspective reversal for ambiguous patterns. Perception. **11**, 263–273, (1982)

BCV95. Biferale, L., Crisanti, A., Vergassola, M., Vulpiani, A.: Eddy diffusivities in scalar transport. Phys. Fluids **7**, 2725, (1995)

BDG93. Bielawski, S., Derozier, D., Glorieux, P.: Experimental characterization of unstable periodic orbits by controlling chaos. Phys. Rev. A, **47**, 2492, (1993)

BDR01. Barré, J., Dauxois, T., Ruffo, S.: Clustering in a Hamiltonian with repulsive long range interactions. Physica **A 295**, 254 (2001)

Bel81. Belbruno, E.A.: A new family of periodic orbits for the restricted problem, Celestial Mech., **25**, 397–415, (1981)

Ben84. Benettin, G.: Power law behaviour of Lyapunov exponents in some conservative dynamical systems. Physica D **13**, 211-213, (1984)

Ben95. S. P. Benz: Superconductor-normal-superconductor junctions for programmable voltage standards. Appl. Phys. Lett. 67, 2714-2716, (1995)

Ber84. Berry, M.V.: Quantal phase factor accompanying adiabatic change. Proc. R. Soc. Lond. A **392**, 45-57, (1984)

Ber85. Berry, M.V.: Classical adiabatic angles and quantal adiabatic phase. J. Phys. A **18**, 15-27, (1985)

Ber91. Bernier, O: Stochastic analysis of synchronous neural networks with asymmetric weights. Europhys. Lett. **16**, 531-536, (1991)

BF79. Boldrighini, Franceschini, V.: A Five-Dimensional Truncation of the Plane Incompressible Navier-Stokes Equations. Commun. Math. Phys., **64**, 159-170, (1979)

BFL02. Barbara, P. Filatrella, G., Lobb, C.J., Pederson, N.F.: Studies of High Temperature Superconductors, NOVA Science Publishers **40**, Huntington, (2002)

BFR04. Bagnoli, F., Franci, F., Rechtman, R.: Chaos in a simple cellular automata model of a uniform society. In Lec. Not. Comp. Sci., **3305**, 513–522, Springer, London, (2004)

BG04. Berselli L.C., Grisanti C.R.: On the consistency of the rational large eddy simulation model, Comput. Vis. Sci., 6, N 2–3, 75–82, (2004)

BG79. Barrow-Green, J.: Poincaré and the Three Body Problem. American Mathematical Society, Providence, RI, (1997)

BG90. Bouchaud, J.P., Georges, A.: Anomalous diffusion in disordered media - statistical mechanisms, models and physical applications. Phys. Rep. **195**, 127, (1990)

Bai89. Bai-Iin, H.: Elentary Symbolic Dynamics and Chaos in Dissipative Systems, World Scientific, Singapore, (1989)

BG96. Baker, G.L., Gollub, J.P.: Chaotic Dynamics: An Introduction (2nd ed). Cambridge Univ. Press, Cambridge, (1996)

BGG80. Benettin, G., Giorgilli, A., Galgani, L., Strelcyn, J.M.: Lyapunov exponents for smooth dynamical systems and for Hamiltonian systems; a method for computing all of them. Part 1: theory, and Part 2: numerical applications. Meccanica, **15**, 9-20 and 21-30, (1980)

BGG89. Batlle, C., Gomis, J., Gràcia, X., Pons, J.M.: Noether's theorem and gauge transformations: application to the bosonic string and CP_2^{n-1} model. J. Math. Phys. **30**, 1345, (1989)

BGL00. Boccaletti, S., Grebogi, C., Lai, Y.-C., Mancini, H., Maza, D.: The Control of Chaos: Theory and Applications. Physics Reports **329**, 103–197, (2000)

BGS76. Benettin, G., Galgani, L., Strelcyn, J.M.: Kolmogorov Entropy and Numerical Experiments. Phys. Rev. A **14** 2338, (1976)

BH83. Bhatnagar, K.B., Hallan, P.P.: The effect of perturbation in coriolis and centrifugal forces on the non-linear stability of equilibrium points in the restricted problem of three bodies. Celest. Mech. & Dyn. Astr. Vol. **30**, 97, (1983)

BH95. Bestvina, M., Handel, M.: Train-tracks for surface homeomorphisms. Topology, **34**(1), 109–140, (1995)

Bil65. Billingsley, P.: Ergodic theory and information, Wiley, New York, (1965)

Bir15. Birkhoff, G.D.: The restricted problem of three–bodies, Rend. Circolo Mat. Palermo, **39**, 255–334, (1915)

Bir17. Birkhoff, G.D.: Dynamical Systems with Two Degrees of Freedom. Trans. Amer. Math. Soc. **18**, 199-300, (1917)

Bir27. Birkhoff, G.D.: Dynamical Systems, Amer. Math. Soc., Providence, RI, (1927)

Bla84. Blanchard, P.: Complex analytic dynamics on the Riemann sphere. Bull. AMS **11**, 85–141, (1984)

BLM87. Boiti, M., Leon, J.J., Manna, M., Penpinelli, F.: On a spectral transform of a KdV-like equation related to the Schrodinger operator in the plane. Inverse Problems **3**, 25, (1987)

BLM88. Boiti, M., Leon, J.J., Martina, L., Penpinelli, F.: Scattering of Localized Solitons in the Plane, Phys. Lett. A 132, 432439, (1988)

BLR01. Boffetta, G., Lacorata, G., Redaelli, G., Vulpiani, A.: Barriers to transport: a review of different techniques. Physica D, **159**, 58-70, (2001)

BLV01. Boffetta, G., Lacorata, G., Vulpiani, A.: Introduction to chaos and diffusion. Chaos in geophysical flows, ISSAOS, (2001)

BMA72. Borsellino, A., Marco, A. D., Allazatta, A., Rinsei, S. Bartolini, B.: Reversal time distribution in the perception of visual ambiguous stimuli. Kybernetik, **10**, 139–144, (1972)

BMB82a. Van den Broeck, C., Mansour, M., Baras, F.: Asymptotic Properties of Coupled Non-linear Langevin Equations in the Limit of Weak Noise, I: Cusp Bifurcation. J.Stat.Phys. **28**, 557-575, (1982)

BMB82b. Van den Broeck, C., Mansour, M., Baras, F.: Asymptotic Properties of Coupled Non-linear Langevin Equations in the Limit of Weak Noise, II: Transition to a Limit Cycle. J.Stat.Phys. 28, 577-587, (1982)

BPT94. Van den Broeck, C., Parrondo, J.M.R., Toral, R.: Noise–Induced Non–equilibrium Phase Transition. Phys. Rev. Lett. **73**, 3395–3398, (1994)

BPT97. Van den Broeck, C., Parrondo, J.M.R., Toral, R., Kawai, R.: Non–equilibrium phase transitions induced by multiplicative noise. Phys. Rev. E **55**, 4084–4094, (1997)

BMM95. Biró, T.S., Matinyan, S.G., Müller, B.: Chaos and gauge field theory. World Scientific, Singapore, (1995)

BOA85. Bondeson, A., Ott, E., Antonsen, T.M.: Quasiperiodically Forced Damped Pendula and Schrodinger Equations with Quasi-periodic Potentials: Implications of Their Equivalence. Phys. Rev. Lett. **55**, 2103, (1985)

Bob92. Bobbert, P.A.: Simulation of vortex motion in underdamped two-dimensional arrays of Josephson junctions. Phys. Rev. B **45**, 75407543, (1992)

Bol01. Bollobás, B.: Random Graphs, (2nd ed). Cambridge Univ. Press, Cambridge, (2001)

Bow70. Bowen, R.: Markov partitions for Axiom A diffeomorphisms. Amer. J. Math. **92**, 725-747 (1970)

Bow73. Bowen, R.: Symbolic dynamics for hyperbolic flows. Amer. J. Math. **95**, 429-460, (1973)

Bow75. Bowen, R.: Equilibrium states and the ergodic theory of Anosov diffeomorphisms. Lecture Notes in Math. **470**, Springer-Verlag, Berlin. (1975)

BP02. Barahona, M., Pecora, L.M.: Synchronization in Small-World Systems. Phys. Rev. Lett. **89**, 054101 (2002)

BP82. Barone, A., Paterno, G.: Physics and Applications of the Josephson Effect. New York: John Wiley & Sons, (1982)

BRR95. Botina, J., Rabitz, H., Rahman, N.: Optimal control of chaotic Hamiltonian dynamics. Phys. Rev. A **51**, 923933, (1995)

658   References

BPS75. Belavin, A.A., Polyakov, A.M., Swartz, A.S., Tyupkin, Yu.S.: SU(2) instantpons discovered. Phys. Lett. B **59**, 85, (1975)

BPS98. Blackmore, D.L., Prykarpatsky, Y.A., Samulyak, R.V.: The Integrability of Lie-invariant Geometric Objects Generated by Ideals in the Grassmann Algebra. J. Nonlin. Math. Phys., **5**(1), 54–67, (1998)

BPV03. Batista, A.M., de Pinto, S.E., Viana, R.L., Lopes, S.R.: Mode locking in small-world networks of coupled circle maps. Physica A **322**, 118, (2003)

BR75. Bowen, R., Ruelle, D.: The ergodic theory of Axiom A flows. Invent. Math. **29**, 181-202, (1975)

BRR95. Botina, J., Rabitz, H., Rahman, N.: Optimal control of chaotic Hamiltonian dynamics. Phys. Rev. A **51**, 923–933, (1995)

BS02. Bornholdt, S., Schuster, H.G. (eds): Handbook of Graphs and Networks. Wiley-VCH, Weinheim, (2002)

BS91. Beals, R., Sattinger. D.H.: On the complete integrability of complete integrable systems. Commun. Math. Phys. **138**, 409–436, (1991)

BS93. Beck, C., Schlögl, F.: Thermodynamics of chaotic systems. Cambridge Univ. Press, Cambridge, (1993)

BT00. Bellet, L.R., Thomas, L.E.: Asymptotic Behavior of Thermal Non-Equilibrium Steady States for a Driven Chain of Anharmonic Oscillators. Commun. Math. Phys. **215**, 1–24, (2000)

BT87. Binney, J., Tremaine, S.: Galactic Dynamics. Princeton Univ. Press, Princeton, NJ, (1987).

Buc41. Buchanan, D.: Trojan satellites – limiting case. Trans. Roy. Soc. Canada Sect. III, **35**(3), 9–25, (1941)

Byt72. Bytev, V.O.: Group-theoretical properties of the Navier–Stokes equations, Numerical Methods of Continuum Mechanics, **3**(3), 13–17 (in Russian), (1972)

CMP00. Carballo, C.M., Morales, C.A., Pacifico, M.J.: Homoclinic classes for generic C^1 vector fields. Preprint MAT. **07**, PUC-Rio, (2000)

CA93. Chialvo, D.R., Apkarian, V.: Modulated noisy biological dynamics: Three examples. Journal of Statistical Physics, **70**, 375–391, (1993)

CA97. Chen, L., Aihara, K.: Chaos and asymptotical stability in discrete-time neural networks, Physica D **104**, 286–325, (1997)

CAM05. Cvitanovic, P., Artuso, R., Mainieri, R., Tanner, G., Vattay, G.: Chaos: Classical and Quantum. ChaosBook.org, Niels Bohr Institute, Copenhagen, (2005)

Can78. Cantwell B.J.: Similarity transformations for the two-dimensional, unsteady, stream-function equation. J. Fluid Mech. **85**, 257–271, (1978)

Car92. Do Carmo, M.P.: Riemannian geometry. Birkhäuser, Boston, (1992)

Cas95. Casetti, L.: Efficient symplectic algorithms for numerical simulations of Hamiltonian flows. Phys. Scr., **51**, 29, (1995)

Cat94. Caticha, N.: Consistency and Linearity in Quantum Theory. J. Phys. A: Math. Gen. **27**, 5501, (1994)

CB88. Casson, A., Bleiler, S.: Automorphisms of surfaces, after Nielsen and Thurston, volume 9 of London Mathematical Society Student Texts. Cambridge Univ. Press, (1988)

CBS91. Chan, C.K., Brumer, P., Shapiro, M.: Coherent radiative control of IBr photodissociation via simultaneous (w1,w3) excitation. J. Chem. Phys. **94**, 2688-2696, (1991)

CC70. Chapman, S., Cowling, T.: Mathematical Theory of Non-Uniform Gases. Cambridge Univ. Press, Cambridge, (1970)

CCC97. Caiani, L., Casetti, L., Clementi, C., Pettini, M. Geometry of dynamics, Lyapunov exponents and phase transitions. Phys. Rev. Lett. **79**, 4361, (1997)

CCC98. Caiani, L., Casetti, L., Clementi, C., Pettini, G., Pettini, M., Gatto, R.: Geometry of dynamics and phase transitions in classical lattice φ^4 theories Phys. Rev. E **57**, 3886, (1998)

CCP96. Christiansen, F., Cvitanovic, P., Putkaradze, V.: Hopf's last hope: spatio-temporal chaos in terms of unstable recurrent patterns. Nonlinearity, **10**, 1, (1997)

CCP96. Casetti, L., Clementi, C., Pettini, M.: Riemannian theory of Hamiltonian chaos and Lyapunov exponents. Phys. Rev. E **54**, 59695984. (1996)

CCP99. Casetti, L., Cohen, E.G.D., Pettini, M.: Origin of the Phase Transition in a Mean–Field Model. Phys. Rev. Lett., **82**, 4160, (1999)

CCT87. Chaiken, J., Chu, C.K., Tabor, M., Tan, Q.M.: Lagrangian Turbulence in Stokes Flow. Phys. Fluids, **30**, 687, (1987)

CCV98. Castellanos, A., Coolen, A.C., Viana, L.: Finite-size effects in separable recurrent neural networks. J. Phys. A: Math. Gen. **31**, 6615-6634, (1998)

CD98. Chen, G., Dong, X.: From Chaos to Order. Methodologies, Perspectives and Application. World Scientific, Singapore, (1998)

CE91. Cvitanovic, P., Eckhardt, B.: Periodic orbit expansions for classical smooth flows. J. Phys. A **24**, L237, (1991)

CE97. Chandre, C., Eilbeck, J.C.: Does the existence of a Lax pair imply integrability. Preprint, (1997)

CFL97. Crisanti, A. Falcioni, M., Lacorata, G., Purini, R., Vulpiani, A.: Characterization of a periodically driven chaotic dynamical system. J. Phys. A, Math. Gen. **30**, 371-383, (1997)

CFP90. Crisanti, A. Falcioni, M., Provenzale, A., Vulpiani, A.: Passive advection of particles denser than the surrounding fluid. Phys. Lett. A **150**, 79, (1990)

CFP91. Crisanti, A., Falcioni, M., Paladin, G., Vulpiani, A.: Lagrangian Chaos: Transport, Mixing and Diffusion in Fluids. Riv. Nuovo Cim. **14**, 1, (1991)

CFP94. Crisanti, A., Falcioni, M., Paladin, G., Vulpiani, A.: Stochastic Resonance in Deterministic Chaotic Systems. J. Phys. A **27** L597, (1994)

CFR79. Clark, R.A., Ferziger, J.H., Reynolds, W.C.: Evaluation of Subgrid-Scale Turbulence Models Using an Accurately Simulated Turbulent Flow. J. Fluid. Mech. **91**, 1-16, (1979)

CK97. Cutler, C.D., Kaplan, D.T.: (eds): Nonlinear Dynamics and Time Series. Fields Inst. Comm. **11**, American Mathematical Society, (1997)

CGP88. Cvitanovic, P., Gunaratne, G., Procaccia, I.: Topological and metric properties of Hénon-type strange attractors. Phys. Rev. A **38**, 1503-1520, (1988)

CGP99. Casetti, L., Gatto, R., Pettini, M.: Geometric approach to chaos in the classical dynamics of Abelian lattice gauge theory. J. Phys. A **32**, 3055, (1999)

CH01. de Carvalho, A., Hall, T.: Pruning theory and thurston's classification of surface homeomorphisms. J. European Math. Soc., **3**(4), 287–333, (2001)

CH02. De Carvalho, A., Hall, T.: The forcing relation for horseshoe braind types. Experimental Math., **11**(2), 271–288, (2002)

CH03. De Carvalho, A., Hall, T.: Conjugacies between horseshoe braids. Nonlinearity, **16**, 1329–1338, (2003)

Ch53. Courant, R., Hilbert, D.: Methods of Mathematical Physics, Vol 1, Interscience Pub. (1953)

Cha43. Chandrasekhar, S.: Stochastic Problems in Physics and Astronomy. Rev. Mod. Phys. **15**, 1-89, (1943)

Che70. Chernikov, Yu. A.: The photogravitational restricted three body problem. Sov. Astr. AJ. **14**(1), 176-181, (1970)

Chi79. Chirikov, B.V.: A universal instability of many-dimensional oscillator systems. Phys. Rep. **52**, 264-379, (1979)

CJP93. Crisanti, A, Jensen, M.H., Paladin, G., Vulpiani, A.: Intermittency and Predictability in Turbulence. Phys. Rev. Lett. **70**, 166, (1993)

CK05. Cannone, M., Karch, G.: About the regularized Navier–Stokes equations, J. Math. Fluid Mech., **7**, 1–28, (2005)

CKN82. Caffarelli, L., Kohn, R., Nirenberg, L.: Partial regularity of suitable weak solutions of the Navier-Stokes equations, Comm. Pure Appl. Math. **35**, 771-831, (1982)

CKS05. Coolen, A.C.C., Kuehn, R., Sollich, P.: Theory of Neural Information Processing Systems. Oxford Univ. Press, Oxford, (2005)

CL81. Caldeira, A.O., Leggett, A.J.: Influence of Dissipation on Quantum Tunneling in Macroscopic Systems. Phys. Rev. Lett. **46**, 211, (1981)

CMP98a. Claussen, J.C., Mausbach, T., Piel, A. Schuster, H.G.: Improved difference control of unknown unstable fixed–points: Drifting parameter conditions and delayed measurement. Phys. Rev. E, **58**(6), 7256–7260, (1998)

CMP98b. Claussen, J.C., Mausbach, T., Piel, A. Schuster, H.G.: Memory difference control of unknown unstable fixed–points: Drifting parameter conditions and delayed measurement. Phys. Rev. E **58**(6), 7260–7273, (1998)

CS98. Claussen, J.C., Schuster, H.G.: Stability borders of delayed measurement from time–discrete systems. arXiv nlin. CD/0204031, (1998)

Cla02a. Claussen, J.C.: Generalized Winner Relaxing Kohonen Feature Maps. arXiv cond–mat/0208414, (2002)

Cla02b. Claussen, J.C.: Floquet Stability Analysis of Ott-Grebogi-Yorke and Difference Control. arXiv:nlin.CD/0204060, (2002)

CLP95. Casetti, L., Livi, R., Pettini, M.: Gaussian Model for Chaotic Instability of Hamiltonian Flows. Phys. Rev. Lett. **74**, 375378, (1995)

CLS96. Colombo, G. Lautman, D., Shapiru I.I.: The Earth's dust belt: Fact or fiction? Gravitational focusing and Jacobi capture. J. Geophys. Res. **71**, 5705-5717, (1996)

CM97. Carpinteri, A., Mainardi, F.: Fractals and Fractional Calculus in Continuum Mechanics. Springer Verlag, Wien & New York, (1997)

CN01. Cottrill-Shepherd, K., Naber M.: Fractional Differential Forms. J. Math. Phys. **42**, 2203-2212, (2001)

Col02. Collins, P.: Symbolic dynamics from homoclinic tangles. Intern. J. Bifur. Chaos, **12**(3), 605–617, (2002)

Col05. Collins, P.: Forcing relations for homoclinic orbits of the Smale horseshoe map, Experimental. Math. **14**(1), 75-86, (2005)

Con63. C. C. Conley, Some new long period solutions of the plane restricted body problem of three–bodies, Comm. Pure Appl. Math., **16**, 449–467, (1963)

Coo01. Coolen, A.C.C.: Statistical Mechanics of Recurrent Neural Networks. In F. Moss, S. Gielen (eds.) Handbook of Biological Physics, **4**, Elsevier, (2001)

Coo89. Cook, J.: The mean-field theory of a Q-state neural network model. J. Phys. A **22**, 2057, (1989)

CP93. Casetti, L., Pettini, M.: Analytic computation of the strong stochasticity threshold in Hamiltonian dynamics using Riemannian geometry. Phys. Rev. E **48**, 43204332, (1993)

CPC03. Casetti, L., Pettini, M., Cohen, E.G.D.: Phase transitions and topology changes in configuration space. J. Stat. Phys., **111**, 1091, (2003)

CPS93. Coolen, A.C., Penney, R.W., Sherrington, D.: Coupled dynamics of fast spins and slow interactions: An alternative perspective on replicas. Phys. Rev. B **48**, 1611616118, (1993)

CR88. Cohen, E.G.D., Rondoni, L.: Note on phase–space contraction and entropy production in thermostatted Hamiltonian systems. Chaos **8** 357-365, (1998)

CS83. Coppersmith, S.N., Fisher, D.S.: Pinning transition of the discrete sine-Gordon equation. Phys. Rev. B **28**, 25662581, (1983)

CS94. Coolen, A.C.C., Sherrington, D.: Order-parameter flow in the fully connected Hopfield model near saturation. Phys. Rev. E **49** 1921-934; and Erratum: Order-parameter flow in the fully connected Hopfield model near saturation, 5906, (1994)

CS95. Chernikov, A.A., Schmidt, G.: Conditions for synchronization in Josephson-junction arrays. Phys. Rev. E **52**, 34153419, (1995).

CSO99. Chen, H., Succi, S., Orszag, S.: Analysis of Subgrid Scale Turbulence Using the Boltzmann Bhatnagar–Gross–Krook kinetic equation. Phys. Rev E **59**, R2527-R2530, (1999)

Chu94. Chua, L.O: Chua's Circuit: an overview ten years later. J. Circ. Sys. Comp. **4**(2), 117-159, (1994)

CT88. Cardoso, O., Tabeling, P.: Anomalous diffusion in a linear array of vortices. Europhys. Lett. **7**(3), 225230, (1988)

CC95. Christini, D.J., Collins, J.J.: Controlling Nonchaotic Neuronal Noise Using Chaos Control Techniques. Phys. Rev. Lett. **75**, 2782-2785, (1995)

Cvi00. Cvitanovic, P.: Chaotic field theory: a sketch. Physica A **288**, 61–80, (2000)

Cvi91. Cvitanovic, P.: Periodic orbits as the skeleton of classical and quantum chaos. Physica, D **51**, 138, (1991)

CY89. Crutchfield, J.P., Young, K.: Computation at the onset of chaos. In Complexity, Entropy and the Physics of Information, 223, SFI Studies in the Sciences Complexity, Vol. VIII, W.H. Zurek (Ed.), Addison-Wesley, Reading, MA, (1989)

Dan67. Danilov, Yu.A.: Group properties of the Maxwell and Navier–Stokes equations, Khurchatov Inst. Nucl. Energy, Acad. Sci. USSR, (in Russian) (1967)

DBO01. DeShazer, D.J., Breban, R., Ott, E., Roy, R.: Detecting Phase Synchronization in a Chaotic Laser Array. Phys. Rev. Lett. **87**, 044101, (2001)

DCS98. Düring, A, Coolen, A.C., Sherrington, D.: Phase diagram and storage capacity of sequence processing neural networks. J. Phys. A: Math. Gen. **31**, 8607, (1998)

DD67. Deprit, A., Deprit-Bartholomé, A.: Stability of the triangular Lagrangian points. Astron. J. **72**, 173-179, (1967)

DDT03. Daniels, B.C., Dissanayake, S.T.M., Trees, B.R.: Synchronization of coupled rotators: Josephson junction ladders and the locally coupled Kuramoto model. Phys. Rev. E **67**, 026216 (2003).

Dev89. Devaney, R.: An Introduction to Chaotic Dynamical Systems. Addison Wesley Publ. Co. Reading MA, (1989)

DFM94. Dotsenko, V., Franz, S., Mézard, M.: Memorizing polymers' shapes and permutations. J. Phys. A: Math. Gen. **27**, 2351, (1994)

DFN92. Dubrovin, B.A., Fomenko, A.N., Novikov, S.P.: Modern Geometry - Methods and Applications. Part I. Springer-Verlag, New York, (1992)

DGO89a. Ding, M., Grebogi, C., Ott, E.: Evolution of attractors in quasi–periodically forced systems: From quasiperiodic to strange nonchaotic to chaotic. Phys. Rev. A **39**, 25932598, (1989)

DGO89b. Ding, M., Grebogi, C., Ott, E.: The Dimension of Strange Nonchaotic Attractors. Phys. Lett. A **137**, 167172, (1989)

DGV95. Diekmann, O., van Gils, S.A., Verduyn-Lunel, S.M., Walther, H.-O.: Delay Equations. Appl. Math. Sci. **110**, Springer–Verlag, New York, (1995)

DGY97. Ding, M., Grebogi, C., Yorke, J.A.: Chaotic dynamics. In The Impact of Chaos on Science and Society, Eds. C. Grebogi and J.A. Yorke, 1-17, United Nations Univ. Press, Tokyo, (1997)

DGZ87. Derrida, B., Gardner, E., Zippelius, A.: An exactly solvable asymmetric neural network model. Europhys. Lett. **4**, 167, (1987)

DH68. Deprit, A., Henrard, J.: A manifold of periodic solutions, Advan. Astron. Astr. **6**, 6–12, (1968)

DH85. Douady, A., Hubbard, J.: On the dynamics of polynomial-like mappings. Ann. scient. Éc. Norm. Sup., 4e ser. **18**, 287–343, (1985)

DH90. Ding, M.-Z., Hao, B.-L.: Systematics of the periodic windows in the Lorenz model and its relation with the antisymmetric cubic map. Commun. Theor. Phys. **9**, 375, (1988)

DHR00. Dauxois, T., Holdsworth, P., Ruffo, S.: Violation of ensemble equivalence in the antiferromagnetic mean-field XY model. Eur. Phys. J. **B 16**, 659, (2000)

DHS91. Domany, E., van Hemmen, J.L., Schulten, K. (eds): Models of neural Networks I, Springer, Berlin, (1991)

DN79. Devaney, R., Nitecki, Z.: Shift automorphisms in the Hénon mapping. Comm. math. Phys. **67**, 137–48, (1979)

Dri77. Driver, R.D.: Ordinary and delay differential equations. Applied Mathematical Sciences 20, Springer Verlag, New York, (1977)

DS74. Davey, A., Stewartson, K.: On Three-Dimensional Packets of Surface Waves. Proc. R. Soc. A **338**, 101-110, (1974)

DSR05. Dullin, H.R., Schmidt, S., Richter, P.H., Grossmann, S.K.: Extended Phase Diagram of the Lorenz Model. Chaos (to appear) (2005)

DSS90. Ditto, W.L., Spano, M.L., Savage, H.T., Rauseo, S.N., Heagy, J., Ott, E.: Experimental observation of a strange nonchaotic attractor. Phys. Rev. Lett. **65**, 533-536, (1990)

DT89. Ditzinger, T., Haken, H.: Oscillations in the perception of ambiguous patterns: A model based on synergetics. Biological Cybernetics, **61**, 279–287, (1989)

DT90. Ditzinger, T., Haken, H.: The impact of fluctuations on the recognition of ambiguous patterns. Biological Cybernetics, **63**, 453–456, (1990)

DT95. Denniston, C., Tang, C.: Phases of Josephson Junction Ladders. Phys. Rev. Lett. **75**, 3930, (1995)

DTP02. Dauxois, T., Theodorakopoulos, N., Peyrard, M.: Thermodynamic instabilities in one dimension: correlations, scaling and solitons. J. Stat. Phys. **107**, 869, (2002)

Dor94. Dorcak, L.: Numerical Models for Simulation the Fractional-Order Control Systems, UEF SAV, The Academy of Sciences, Inst. of Exper. Phys., Kosice, Slovak Republic, (1994)

DW95. Dormayer, P., Lani–Wayda, B.: Floquet multipliers and secondary bifurcations in functional differential equations: Numerical and analytical results, Z. Angew. Math. Phys. **46**, 823–858, (1995)

Duf18. Duffing, G.: Erzwungene Schwingungen bei vernderlicher Eigenfrequenz. Vieweg Braunschweig, (1918)

Ein05. Einstein, A.: Die von der Molekularkinetischen Theorie der Warme Gefordete Bewegung von in ruhenden Flussigkeiten Suspendierten Teilchen. Ann. Phys. **17**, 549, (1905)

Eis29. Eisenhart, L.P.: Dynamical trajectories and geodesics. Ann. of Math. **30**, 591-606, (1929)

EM90. Evans, D.J., Morriss, G.P.: Statistical mechanics of norequilibrium fluids. Academic Press, New York, (1990)

EMN99. Ellis, J., Mavromatos, N., Nanopoulos, D.V.: A microscopic Liouville arrow of time. Chaos, Solit. Fract., **10**(2–3), 345–363, (1999)

Ens05. Enss, C. (ed): Cryogenic Particle Detection. Topics in Applied Physics **99**, Springer, New York, (2005)

EP98. Ershov, S.V., Potapov, A.B.: On the concept of Stationary Lyapunov basis. Physica D, **118**, 167, (1998)

EPG95. Ernst, U., Pawelzik, K., Geisel, T.: Synchronization induced by temporal delays in pulse-coupled oscillators. Phys. Rev. Lett. **74**, 1570, (1995)

EPR99. Eckmann, J.P., Pillet, C.A., Rey-Bellet, L.: Non-equilibrium statistical mechanics of anharmonic chains coupled to two heat baths at different temperatures. Commun. Math. Phys., **201**, 657–697, (1999)

ER85. Eckmann, J.P., Ruelle, D.: Ergodic theory of chaos and strange attractors, Rev. Mod. Phys., **57**, 617–630, (1985)

Erm81. Ermentrout, G.B.: The behavior of rings of coupled oscillators. J. Math. Biol. **12**, 327, (1981)

ESA05. European Space Agency. Payload and Advanced Concepts: Superconducting Tunnel Junction (STJ). Last updated: February 17 (2005)

ESH98. Elson, R.C., Selverston, A.I., Huerta, R. *et al.*: Synchronous Behavior of Two Coupled Biological Neurons. Phys. Rev. Lett. **81**, 5692–5695, (1998)

Fat19. Fatou, P.: Sur les équations fonctionnelles. Bull. Soc. math. France **47**, 161–271, (1919)

Fat22. Fatou, P.: Sur les fonctions méromorphes de deux variables and Sur certaines fonctions uniformes de deux variables. C.R. Acad. Sc. Paris **175**, 862-65 and 1030–33, (1922)

FCS99. Franzosi, R., Casetti, L., Spinelli, L., Pettini, M.: Topological aspects of geometrical signatures of phase transitions. Phys. Rev. E, **60**, 5009–5012, (1999)

FD94. Feldman, D.E., Dotsenko, V.S.: Partially annealed neural networks. J. Phys. A: Math. Gen. **27**, 4401-4411, (1994)

Fed69. Federer, H.: Geometric Measure Theory. Springer, New York, (1969)

Fei78. Feigenbaum, M.J.: Quantitative universality for a class of nonlinear transformations. J. Stat. Phys. **19**, 25-52, (1978)

Fei79. Feigenbaum, M.J.: The universal metric properties of nonlinear transformations. J. Stat. Phys. **21**, 669-706, (1979)

Fer23. Fermi, E.: Beweis dass ein mechanisches Normalsysteme im Allgemeinen quasi-ergodisch ist, Phys. Zeit. **24**, 261–265, (1923)

Fer98. Fersht, A.: Structure and Mechanism in Protein Science. W.H.Freeman, New York, (1998)

Fey72. Feynman, R.P.: Statistical Mechanics, A Set of Lectures. WA Benjamin, Inc., Reading, MA, (1972)

FH96. Fang, H.-P., Hao, B.-L.: Symbolic dynamics of the Lorenz equations. Chaos, Solitons, and Fractals, **7**, 217–246, (1996)

FHI00. Fujii, K., Hayashi, D., Inomoto, O., Kai, S: Noise-Induced Entrainment between Two Coupled Chemical Oscillators in Belouzov-Zhabotinsky Reactions. Forma **15**, 219, (2000)

Fit61. FitzHugh, R.A.: Impulses and physiological states in theoretical models of nerve membrane. Biophys. J., **1**, 445–466, (1961)

FK39. Frenkel, J., Kontorova, T.: On the theory of plastic deformation and twinning. Phys. Z. Sowiet Union, **1**, 137–49, (1939)

FK88. Fontanari, J.F., Köberle, R.: Information Processing in Synchronous Neural Networks. J. Physique **49**, 13, (1988)

FKP95. Feudel, U., Kurths, J., Pikovsky, A.S.: Strange Nonchaotic Attractor in a Quasiperiodically Forced Circle Map. Physica D **88**, 176-186, (1995)

FM93. Franks, J., Misiurewicz, M.: Cycles for disk homeomorphisms and thick trees. In Nielsen Theory and Dynamical Systems, **152** in Contemporary Mathematics, 69–139, (1993)

FMV01. Festa, R., Mazzino, A., Vincenzi, D.: An analytical approach to chaos in Lorenz-like systems. A class of dynamical equations. Europhys. Lett. **56**, 47-53, (2001)

FMV02a. Festa, R., Mazzino, A., Vincenzi, D.: Lorenz deterministic diffusion. Europhys. Lett., **60**(6), 820-826, (2002)

FMV02b. Festa, R., Mazzino, A., Vincenzi, D.: Lorenz-like systems and classical dynamical equations with memory forcing: An alternate point of view for singling out the origin of chaos. Phys. Rev. E **65**, 046205, (2002)

FP04. Franzosi, R., Pettini, M.: Theorem on the origin of phase transitions. Phys. Rev. Lett., **92**(6), 60601, (2004)

FP94. Fushchych, W.I., Popowych, R.O.: Symmetry reduction and exact solutions of the Navier–Stokes equations. J. Nonlinear Math. Phys. **1**, 75–113, 156–188, (1994)

FPS00. Franzosi, R., Pettini, M. Spinelli, L.: Topology and phase transitions: a paradigmatic evidence. Phys. Rev. Lett. **84**(13), 2774–2777, (2000)

FPU55. Fermi, E., Pasta, J.R., Ulam, S.M.: Studies of nonlinear problems, Los Alamos Scientific Laboratory Report No. LA–1940, (1955)

FPV88. Falcioni, M., Paladin, G., Vulpiani, A.: Regular and chaotic motion of fluid particles in a two-dimensional fluid. J. Phys. A: Math. Gen., **21**, 3451-3462, (1988)

FS92. Freeman, J.A., Skapura, D.M.: Neural Networks: Algorithms, Applications, and Programming Techniques. Addison-Wesley, Reading, (1992)

FT87. Faddeev, L.D., Takhtajan, L.A.: Hamiltonian Methods in the Theory of Solitons, Springer-Verlag, Berling, (1987)

Fur61. Furstenberg, H.: Strict ergodicity and transformation of the torus. Am. J. Math. **83**, 573-601, (1961)

fW80. Franks, J., Williams, R.F.: Anomalous Anosov flows. 158-174 in Lecture Notes in Math. **819**, Springer, Berlin, (1980)

FW95. Filatrella, G., Wiesenfeld, K.: Magnetic-field effect in a two-dimensional array of short Josephson junctions. J. Appl. Phys. **78**, 1878-1883, (1995)

FY83. Fujisaka, H., Yamada, T.: Amplitude Equation of Higher-Dimensional Nikolaevskii Turbulence. Prog. Theor. Phys. **69**, 32 (1983)

Gar94. Gardiner, C.W.: Handbook of Stochastic Methods. Springer, Berlin, (1994)

Gas96. Gaspard, P.: Hydrodynamic modes as singular eigenstates of the Liouvillian dynamics: Deterministic diffusion. Phys. Rev. E **53**, 43794401, (1996)

GC00. Gollub, J.P., Cross, M.C.: Many systems in nature are chaotic in both space and in time. Nature, **404**, 710-711, (2000)

GC95a. Gallavotti, G., Cohen, E.G.D.: Dynamical ensembles in nonequilibrium statistical mechanics. Phys. Rev. Letters **74**, 2694-2697, (1995)

GC95b. Gallavotti, G., Cohen, E.G.D.: Dynamical ensembles in stationary states. J. Stat. Phys. **80**, 931-970, (1995)

Geo88. Georgii, H.O.: Gibbs Measures and Phase Transitions. Walter de Gruyter, Berlin, (1988)

GGK67. Gardner, C.S., Greene, J.M., Kruskal, M.D., Miura, R.M.: Method for solving the Korteweg-de Vries equation. Phys. Rev. Let., **19**, 1095–97, (1967)

GSD92. Garfinkel, A., Spano, M.L., Ditto, W.L., Weiss, J.N.: Controlling cardiac chaos. Science **257**, 1230-1235, (1992)

GH00. Gade, P.M., Hu, C.K.: Synchronous chaos in coupled map lattices with small-world interactions. Phys. Rev. E **62**, 64096413, (2000)

Gil81. Gilmor, R.: Catastrophe Theory for Scientists and Engineers. Wiley, New York, (1981)

GZ94. Glass, L., Zeng, W.: Bifurcations in flat-topped maps and the control of cardiac chaos. Int. J. Bif. Chaos **4**, 1061-1067, (1994)

GKE89. Gray, C.M., König, P., Engel, A.K., Singer, W.: Oscillatory responses in cat visual cortex exhibit intercolumnar synchronization which reflects global stimulus properties. Nature **338**, 334, (1989)

Gle87. Gleick, J.: Chaos: Making a New Science. Penguin–Viking, New York, (1987)

GLW93. Geigenmüller, U., Lobb, C.J., Whan, C.B.: Friction and inertia of a vortex in an underdamped Josephson array. Phys. Rev. B **47**, 348358, (1993)

GM04. Grinza, P., Mossa, A.: Topological origin of the phase transition in a model of DNA denaturation. Phys. Rev. Lett. **92**(15), 158102, (2004)

GM79. Guyer, R.A., Miller, M.D.: Commensurability in One Dimension at $T \neq 0$. Phys. Rev. Lett. **42**, 718722, (1979)

GM84. De Groot, S.R., Mazur, P.: Nonequilibrium thermodynamics. Dover, New York, (1984)

GMS95. Golubitsky, M., Marsden, J., Stewart, I., Dellnitz, M : The constrained Lyapunov–Schmidt procedure and periodic orbits. Fields Institute Communications, **4**, 81–127, (1995)

GN90. Gaspard, P., Nicolis, G.: Transport properties, Lyapunov exponents, and entropy per unit time. Phys. Rev. Lett. **65**, 1693-1696, (1990)

GNR91. Gross, P., Neuhauser, D., Rabitz, H.: Optimal Control of Unimolecular Reactions in the Collisional Regime. J. Chem. Phys. **94**, 1158, (1991)

GNR92. Gross, P., Neuhauser, D., Rabitz, H.: Optimal control of curve-crossing systems. J. Chem. Phys. **96**, 2834-2845, (1992)

GNR93. Gross, P., Neuhauser, D., Rabitz, H.: Teaching lasers to control molecules in the presence of laboratory field uncertainty and measurement imprecision. J. Chem. Phys. **98**, 4557-4566, (1993)

Gol99. Goldberger, A.L.: Nonlinear Dynamics, Fractals, and Chaos Theory: Implications for Neuroautonomic Heart Rate Control in Health and Disease. In: Bolis CL, Licinio J, eds. The Autonomic Nervous System. World Health Organization, Geneva, (1999)

GOP84. Grebogi, C., Ott, E., Pelikan, S., Yorke, J.A.: Strange attractors that are not chaotic. Physica D **3**, 261-268, (1984)

Got96. Gottlieb, H.P.W.: Question #38. What is the simplest jerk function that gives chaos? Am. J. Phys., **64**(5), 525, (1996)

GOY87. Grebogi, C., Ott, E., Yorke, J.A.: Chaos, strange attractors, and fractal basin boundaries in nonlinear dynamics. Science, **238**, 632–637, (1987)

Goz83. Gozzi, E.: Functional-integral approach to Parisi-Wu stochastic quantization: Scalar theory. Phys. Rev. D **28**, 1922, (1983)

GP83. Grassberger, P., Procaccia, I.: Measuring the strangeness of strange attractors, Physica D **9**, 189–208, (1983)

GP83a. Grassberger, P., Procaccia, I.: Measuring the Strangeness of Strange Attractors. Phys. D **9**, 189-208, (1983)

GP83b. Grassberger, P., Procaccia, I.: Characterization o f Strange Attractors. Phys. Rev. Lett. **50**, 346-349, (1983)

Gra90. Granato, E.: Phase transitions in Josephson-junction ladders in a magnetic field. Phys. Rev. B **42**, 47974799, (1990)

Gre51. Green, M.S.: Brownian motion in a gas of non-interacting molecules. J. Chem. Phys. **19**, 1036-1046, (1951)

Gre79. Greene, J.M.: A method for determining a stochastic transition. J. Math. Phys. **20**, 1183-1201, (1979)

GS88. Giansanti, A., Simic, P.D.: Onset of dynamical chaos in topologically massive gauge theories. Phys. Rev. D **38**, 13521355, (1988)

GT77. Grossmann, S., Thomae, S.: Invariant distributions and stationary correlation functions of one-dimensional discrete processes. Z. Naturforsch. A **32**, 1353–1363, (1977)

GT87. Gozzi, E., Thacker, W.D.: Classical adiabatic holonomy and its canonical structure. Phys. Rev. D **35**, 2398, (1987)

GGS81. Guevara, M.R., Glass, L., Shrier, A.: Phase locking, period-doubling bifurcations, and irregular dynamics in periodically stimulated cardiac cells. Science **214**, 1350-53, (1981)

GH83. Guckenheimer, J., Holmes, P.: Nonlinear Oscillations, Dynamical Systems, and Bifurcations of Vector Fields. Springer-Verlag, Berlin, (1983)

Gut90. Gutzwiller, M.C.: Chaos in Classical and Quantum Mechanics. Springer, New York, (1990)

Had75. Hadjidemetriou, J.D.: The continuation of periodic orbits from the restricted to the general three–body problem, Celestial Mech., **12**, 155–174, (1975)

Hak83. Haken, H.: Synergetics: An Introduction (3rd ed). Springer, Berlin, (1983)

Hak91. Haken, H.: Synergetic Computers and Cognition. Springer-Verlag, Berlin, (1991)

Hak93. Haken, H.: Advanced Synergetics: Instability Hierarchies of Self-Organizing Systems and Devices (3nd ed.). Springer, Berlin, (1993)

Hak96. Haken, H.: Principles of Brain Functioning: A Synergetic Approach to Brain Activity, Behavior and Cognition, Springer, Berlin, (1996)

Hak00. Haken, H.: Information and Self-Organization: A Macroscopic Approach to Complex Systems. Springer, Berlin, (2000)

Hak02. Haken, H.: Brain Dynamics, Synchronization and Activity Patterns in Pulse-Codupled Neural Nets with Delays and Noise, Springer, New York, (2002)

Hal94. Hall, T.: The creation of horseshoes. Nonlinearity, **7**(3),861–924, (1994)

HCT97. Hall, K., Christini, D.J., Tremblay, M., Collins, J.J., Glass, L., Billette, J.: Dynamic Control of Cardiac Alternans. Phys. Rev. Lett. **78**, 4518–4521, (1997)

Han99. Handel, M.: A fixed-point theorem for planar homeomorphisms. Topology, **38**(2), 235–264, (1999)

Har92. Harris-Warrick, R.M. (ed): The Stomatogastric Nervous System. MIT Press, Cambridge, MA, (1992)

HLQ95. Hartley, T.T., Lorenzo, C.F., Qammer, H.K.: Chaos on a Fractional Chua's System, IEEE Trans. Circ. Sys. Th. App. **42**, 485–490, (1995)

HB87. Havlin, S., Ben-Avraham, D.: Diffusion in disordered media. Adv. Phys. **36**, 695-798, (1987)

Hay97. Hayashi, S.: Connecting invariant manifolds and the solution of the C^- stability and Ω-stability conjectures for flows. Annals of Math., **145**, 81–137, (1997)

Hay98. Hayashi, S.: Hyperbolicity, stability, and the creation of homoclinic points. In Documenta Mathematica, Extra Volume ICM, Vol. II, (1998)

Hay00. Hayashi, S.: A C^1 make or break lemma. Bull. Braz. Math. Soc., **31**, 337–350, (2000)

Heb49. Hebb, D.O.: The Organization of Behaviour. Wiley, New York, (1949)

Hel986. Helmholtz, H.: Uber der physikalische Bedeutung des Princips der kleinsten. J. Reine Angew. Math. **100**, 137-166, (1886)

Hen66. Hénon, M.: Sur la topologie des lignes de courant dans un cas particulier. C. R. Acad. Sci. Paris A, **262**, 312-314, (1966).

Hen69. Hénon, M.: Numerical study of quadratic area preserving mappings. Q. Appl. Math. **27**, (1969)

Hen76. Hénon, M.: A two-dimensional mapping with a strange attractor. Com. Math. Phys. **50**, 6977, (1976)

HF67. Hohl, F., Feix, M.R.: Numerical Experiments with a One-Dimensional Model for a Self-Gravitating Star System. Astrophys. J. **147**, 1164-1180, (1967)

HH94. Heagy, J.F., Hammel, S.M.: The birth of strange nonchaotic attractors. Physica D **70**, 140-153, (1994)

HHL97. Heinz, U., Hu, C., Leupold, S., Matinyan, S. Müller, B.: Thermalization and Lyapunov exponents in Yang-Mills-Higgs theory. Phys. Rev. D **55**, 2464, (1997)

Hil94. Hilborn, R.C.: Chaos and Nonlinear Dynamics: An Introduction for Scientists and Engineers. Oxford Univ. Press, Oxford, (1994)

HI97. Hoppensteadt, F.C., Izhikevich, E.M.: Weakly Connected Neural Networks. Springer, New York, (1997)

Hil00. Hilfer, R. (ed): Applications of Fractional Calculus in Physics. World Scientific, Singapore, (2000)

Hil38. Hill, A.V.: The heat of shortening and the dynamic constants of muscle, Proc. R. Soc. B **76**, 136–195, (1938)

Hir71. Hirota, R.: Exact Solution of the Kortewegde Vries Equation for Multiple Collisions of Solitons. Phys. Rev. Lett. **27**, 11921194, (1971)

HPS70. Hirsch, M., Pugh, C., Shub, M.: Stable manifolds and hyperbolic sets. Proc. of Symposia in Pure Mathematics-Global Analysis, **14**, 133–163, (1970)

HS74. Hirsch, M.W., Smale, S.: Differential Equations, Dynamical Systems and Linear Algebra. Academic Press, New York, (1974)

Hir76. Hirsch, M.W.: Differential Topology. Springer, New York, (1976)

HL84. Horsthemke, W., Lefever, R.: Noise–Induced Transitions. Springer, Berlin, (1984)

HL86a. Hale, J.K., Lin, X.B.: Symbolic dynamics and nonlinear semiflows, Ann. Mat. Pur. Appl. **144**(4), 229–259, (1986)

HL86b. Hale, J.K., Lin, X.B.: Examples of transverse homoclinic orbits in delay equations, Nonlinear Analysis **10**, 693–709, (1986)

HL93. Hale, J.K., Lunel, S.M.V.: Introduction to Functional Differential Equations. Springer, New York. (1993)

HCK02a. Hong, H., Choi, M.Y., Kim, B.J.: Synchronization on small-world networks. Phys. Rev. E **65**, 26139, (2002)

HCK02b. Hong, H., Choi, M.Y., Kim, B.J.: Phase ordering on small-world networks with nearest-neighbor edges. Phys. Rev. E **65**, 047104, (2002)

HO90. Henkel, R.D., Opper, M.: Distribution of internal fields and dynamics of neural networks. Europhys. Lett. **11**, 403-408, (1990)

Hoo86. Hoover, W.G.: Molecular dynamics. Lecture Notes in Physics **258**, Springer, Heidelberg, (1986)

Hoo95. Hoover, W.G.: Remark on Some Simple Chaotic Flows. Phys. Rev. E, **51**(1), 759–760, (1995)

Hop82. Hopfield, J.J.: Neural networks and physical systems with emergent collective computational abilities. Proc. Natl. Acad. Sci. USA **79**, 2554, (1982)

HS88. Hale, J.K., Sternberg, N.: Onset of Chaos in Differential Delay Equations, J. Comp. Phys. **77**(1), 221–239, (1988)

HV93. Hale, J.K., Verduyn-Lunel, S.M.: Introduction to Functional Differential Equations, Applied Math. Sci. **99**, Springer Verlag, New York, (1993)

HW83. Van der Heiden, U., Walther, H.-O.: Existence of chaos in control systems with delayed feedback, J. Diff. Equs. **47**, 273–295, (1983)

HZ00. Hu, B., Zhou, C.: Phase synchronization in coupled nonidentical excitable systems and array-enhanced coherence resonance. Phys. Rev. E **61**, R1001, (2000)

HO01. Hunt, B.R., Ott, E.: Fractal Properties of Robust Strange Nonchaotic Attractors. Phys. Rev. Lett. **87**, 254101, (2001)

IAG99. Ivanov, P. Ch., Amaral, L. A. N., Goldberger, A. L., Havlin, S., Rosenblum, M. B., Struzik, Z. & Stanley, H.E.: Multifractality in healthy heartbeat dynamics. Nature, **399**, 461–465, (1999)

IGF99. Ioffe, L.B., Geshkenbein, V.B., Feigel'man, M.V., Fauchère, A.L., Blatter, G.: Environmentally decoupled sds-wave Josephson junctions for quantum computing, Nature **398**, 679-681, (1999)

II05. Ivancevic, V., Ivancevic, T.: Human–Like Biomechanics. Springer, Series: Microprocessor-Based and Intelligent Systems Engineering, Vol. 28, (2005)

II06a. Ivancevic, V., Ivancevic, T.: Natural Biodynamics. World Scientific, Series: Mathematical Biology, (2006)

II06b. Ivancevic, V., Ivancevic, T.: Geometrical Dynamics of Complex Systems. Springer, Series: Microprocessor-Based and Intelligent Systems Engineering, Vol. 31, (2006)

IJL02. Iliescu T., John V., Layton W.: Convergence of finite element approximations of large eddy motion. Numer. Methods Partial Differential Equations, **18**, 689–710, (2002)

IJL03. Iliescu T., John V., Layton W.J., Matthies G., Tobiska L.: A numerical study of a class of LES models, Int. J. Comput. Fluid Dyn. **17**, 75–85, (2003)

IK04. Ibragimov, N.H., Kolsrud, T.: Lagrangian approach to evolution equations: symmetries and conservation laws, Nonlinear Dyn. **36**, 29–40, (2004)

IKH87. Ichikawa, Y.H., Kamimura, T., Hatori, T.: Physica D **29**, 247, (1987)

ILW92. Ivanov, A., Lani–Wayda, B., Walther, H.-O.: Unstable hyperbolic periodic solutions of differential delay equations, Recent Trends in Differential Equations, ed. R.P. Agarwal, 301–316, World Scientific, Singapore, (1992)

IN96. Inoue, M., Nishi, Y.: Dynamical Behavior of Chaos Neural Network of an Associative Schema Model. Prog. Theoret. Phys. **95**, 837–850, (1996)

Ish97. Ishwar, B.: Nonlinear stability in the generalised restricted three body problem. Celest. Mech. Dyn. Astr. **65**, 253-289, (1997)

Isi92. Isichenko, M.B.: Percolation, statistical topography, and transport in random media. Rev. Mod. Phys. **64**, 9611043, (1992)

IU94. Ibragimov, N.H., Ünal G.: Equivalence transformations of Navier–Stokes equations, Istanbul Tek. Üniv. Bül., **47**, 203–207, (1994)

JAB00. Jongen, G., Anemüller, J., Bollé, D., Coolen A.C., Pérez-Vicente, C.J.: J. Phys. A: Math. Gen. **34**, 3957, (2000)

Jac91. Jackson, E.A.: Perspectives of Nonlinear Dynamics **2**. Cambridge Univ. Press, Cambridge, (1991)

JBC98. Jongen, G., Bollé, D., Coolen, A.C.: The XY spin-glass with slow dynamic couplings. J. Phys. A **31**, L737-L742, (1998)

JBO97. Just, W., Bernard, T., Ostheimer, M., Reibold, E., Benner, H.: Mechanism of time–delayed feedback control. Phys. Rev. Lett., **78**, 203–206, (1997)

JL02. Joseph, B., Legras, B.: Relation between kinematic bouncaries, stirring and barriers for the Antarctic polar vortex. J. Atmos. Sci. **59**, 1198-1212, (2002)

JLR90. Judson, R., Lehmann, K., Rabitz, H., Warren, W.S.: Optimal Design of External Fields for Controlling Molecular Motion – Application to Rotation J. Mol. Struct. **223**, 425-446, (1990)

Jos74. Josephson, B.D.: The discovery of tunnelling supercurrents. Rev. Mod. Phys. **46**(2), 251-254, (1974)

JP98. Jakšić, V., Pillet, C.-A.: Ergodic properties of classical dissipative systems I. Acta mathematica **181**, 245–282, (1998)

Jul18. Julia, G.: Mémoires sur l'itération des fonctions rationelles. J. Math. **8**, 47–245, (1918)

Jun93. Jung, P.: Periodically driven stochastic systems. Phys. Reports **234**, 175, (1993)

KA85. Kawamoto, A.H., Anderson, J.A.: A Neural Network Model of Multistable Perception. Acta Psychol. **59**, 35–65, (1985)

Kal60. Kalman, R.E.: A new approach to linear filtering and prediction problems. Transactions of the ASME, Ser. D, J. Bas. Eng., **82**, 34–45, (1960)

Kam92. Van Kampen, N.G.: Stochastic Processes in Physics and Chemistry. North-Holland, Amsterdam, (1992)

Kar00. M. Karplus: Aspects of Protein Reaction Dynamics: Deviations from Simple Behavior. J. Phys. Chem. B **104**, 11-27, (2000)

Kar84. (a) M. Kardar: Free energies for the discrete chain in a periodic potential and the dual Coulomb gas. Phys. Rev. B **30**, 63686378, (1984)

Kat75. Kato, T.: Quasi-linear equations of evolution, with applications to partial differential equations, Lecture Notes in Math. **448**, Springer-Verlag, Berlin, 25–70, (1975)

Kat83. Kato, T.: On the Cauchy Problem for the (Generalized) Korteweg-de Vries Equation, Studies in Applied Math., Adv. in Math. Stud. **8**, 93–128, (1983)

KD99. Klages, R., Dorfman, J.R.: Simple deterministic dynamical systems with fractal diffusion coefficients. Phys. Rev. E **59**, 53615383, (1999)

KdV95. Korteweg, D.J., de Vries, H.: On the change of form of long waves advancing in a rectangular canal, and on a new type of long stationary waves. Philosophical Magazine, **39**, 422–43, (1895)

Kai97. Kaiser, D.: Preheating in an expanding universe: Analytic results for the massless case. Phys. Rev. D 56, 706–716, (1997)

Kai99. Kaiser, D.: Larger domains from resonant decay of disoriented chiral condensates. Phys. Rev. D **59**, 117901, (1999)

KF99. Karlin, I.V., Ferrante, A., Öttinger, H.C.: Perfect Entropy Functions of the Lattice Boltzmann Method. Europhys. Lett. **47**, 182-188, (1999)

KFA69. Kalman, R.E., Falb, P., Arbib, M.A.: Topics in Mathematical System Theory. McGraw Hill, New York, (1969)

KFP91. Kaplan, D.T., Furman, M.I., Pincus, S.M., Ryan, S.M., Lipsitz, L.A., Goldberger, A.L.: Aging and the complexity of cardiovascular dynamics,' Biophys. J., **59**, 945–949, (1991)

KH95. Katok, A., Hasselblatt, B.: Introduction to the Modern Theory of Dynamical Systems. Cambridge Univ. Press., Cambridge, (1995)

Kha92. Khalil, H.K.: Nonlinear Systems. New York: MacMillan, (1992)

Khi57. Khinchin, A.I.: Mathematical foundations of Information theory. Dover, (1957)

KI04. Kushvah, B.S., Ishwar, B.: Triangular equilibrium points in the generalised photogravitational restricted three body problem with Poynting-Robertson drag. Rev. Bull. Cal. Math. Soc. **12**, 109-114, (2004)

KK00. Kye, W.-H., Kim, C.-M.: Characteristic relations of type-I intermittency in the presence of noise. Phys. Rev. E **62**, 6304–6307, (2000)

KK01. Koyama, H., Konishi, T.: Emergence of power-law correlation in 1-dimensional self-gravitating system. Phys. Lett. **A 279**, 226230, (2001)

KK02a. Koyama, H., Konishi, T.: Hierarchical clustering and formation of power-law correlation in 1-dimensional self-gravitating system. Euro. Phys. Lett. **58**, 356, (2002)

KK02b. Koyama, H., Konishi, T.: Long-time behavior and relaxation of power-law correlation in one-dimensional self-gravitating system. Phys. Lett. **A 295**, 109, (2002)

KK06. Koyama, H., Konishi, T.: Formation of fractal structure in many-body systems with attractive power-law potentials. Phys. Rev. E **73**, 016213, (2006)

KK01. Kawahara, G., Kida, S.: Periodic motion embedded in plane Couette turbulence: regeneration cycle and burst. J. Fluid Mech. **449**, 291–300, (2001)

KKH03. Kim, J.-W., Kim, S.-Y. Hunt, B., Ott, E.: Fractal properties of robust strange nonchaotic attractors in maps of two or more dimensions. Phys. Rev. E **67**, 036211, (2003)

KLO03. Kim, S.-Y., Lim, W., Ott, E.: Mechanism for the intermittent route to strange nonchaotic attractors. Phys. Rev. E **67**, 056203, (2003)

KYR98. Kim, C.M., Yim, G.S., Ryu, J.W., Park, Y.J.: Characteristic Relations of Type-III Intermittency in an Electronic Circuit. Phys. Rev. Lett. **80**, 5317–5320, (1998)

KO04. Kim, P., Olver, P.J.: Geometric integration via multi-space, Regul. Chaotic Dyn. **9**(3), 213–226, (2004)

KLR03. Kye, W.-H., Lee, D.-S., Rim, S., Kim, C.-M., Park, Y.-J.: Periodic Phase Synchronization in coupled chaotic oscillators. Phys. Rev. E **68**, 025201(R), (2003)

KM89. Kivshar, Y.S., Malomend, B.A.: Dynamics of solitons in nearly integrable systems. Rev. Mod. Phys. **61**, 763915, (1989)

KN00. Kotz, S., Nadarajah, S.: Extreme Value Distributions. Imperial College Press, London, (2000)

KP05. Kobes, R., Peles, S.: A Relationship Between Parametric Resonance and Chaos. arXiv:nlin/0005005, (2005)

KLS94. Kofman, L., Linde, A., Starobinsky, A.: Reheating after Inflation. Phys. Rev. Lett. **73**, 3195–3198, (1994)

Kog79. Kogut, J.: An introduction to lattice gauge theory and spin systems. Rev. Mod. Phys. **51**, 659713, (1979)

KP02. Katz, N., Pavlovic, N.: A cheap Caffarelli-Kohn-Nirenberg inequality for the Navier-Stokes equation with hyper-dissipation. Geom. Funct. Anal. **12**(2), 355-379, (2002)

KP95. Kocarev, L., Parlitz, U.: General Approach for Chaotic Synchronization with Applications to Communication. Phys. Rev. Lett. **74**, 5028–5031, (1995)

KP96. Kocarev, L., Parlitz, U.: Generalized Synchronization, Predictability, and Equivalence of Unidirectionally Coupled Dynamical Systems. Phys. Rev. Lett. **76**, 1816–1819, (1996)

Kra76. Kraichnan, R.H.: Eddy Viscosity in Two and Three dimensions J. Atmos. Sci. **33**, 1521-1536, (1976)

KRG89. Kosloff, R., Rice, S.A., Gaspard, P., Tersigni, S., Tannor, D.J.: Wavepacket Dancing: Achieving Chemical Selectivity by Shaping Light Pulses. Chem. Phys. **139**, 201-220, (1989)

Kry79. Krylov, N.S.: Works on the foundations of statistical physics. Princeton Univ. Press, Princeton, (1979)

KS03. Kobayashi, H., Shimomura, Y.: Inapplicability of the Dynamic Clark Model to the Large Eddy Simulation of Incompressible Turbulent Channel Flows. Phys. Fluids **15**, L29-L32, (2003)

KS75. Kogut, J., Susskind, L.: Hamiltonian formulation of Wilson's lattice gauge theories. Phys. Rev. D **11**, 395408, (1975)

KS90. Kirchgraber, U., Stoffer, D.: Chaotic behavior in simple dynamical systems. SIAM Review **32**(3), 424–452, (1990)

KS97. Ketoja, J.A., Satija, I.I.: Harper equation, the dissipative standard map and strange non-chaotic attractors: Relationship between an eigenvalue problem and iterated maps. Physica D **109**, 70-80, (1997)

Kur68. Kuratowski, K.: Topology II. Academic Press- PWN-Polish Sci. Publishers Warszawa, (1968)

KSI06. Kushvah, B.S., Sharma, J.P., Ishwar, B.: Second Order Normalization in the Generalized Photogravitational Restricted Three Body Problem with Poynting-Robertson Drag. arXiv:math.DS/0602528, (2006)

KT01. Kye, W.-H., Topaj, D.: Attractor bifurcation and on-off intermittency. Phys. Rev. **E 63**, 045202(R), (2001)

KT85. Kubo, R., Toda, M., Hashitsume, N.: Statistical Physics. Springer-Verlag, Berlin, (1985)

Kub57. Kubo, R.: Statistical-mechanical theory of irreversible processes. I. J. Phys. Soc. (Japan) **12**, 570-586, (1957)

Kur76. Kuramoto, Y., Tsuzuki, T.: Persistent propagation of concentration waves in dissipative media far from thermal equilibrium. Progr. Theor. Physics **55**, 365, (1976)

Kur84. Kuramoto, Y.: Chemical Oscillations. Waves and Turbulence. Springer, New York, (1984)

KW83. Knobloch, E., Weisenfeld, K.A.: Bifurcations in fluctuating systems: The center manifold approach. J.Stat.Phys. 33, 611, (1983)

KY75. Kaplan, J.L., Yorke, J.A.: On the stability of a periodic solution of a differential delay equation, SIAM J. Math. Ana. **6**, 268–282, (1975)

KY79. Kaplan, J.L., Yorke, J.A.: Numerical Solution of a Generalized Eigenvalue Problem for Even Mapping. Peitgen, H.O., Walther, H.O. (Eds): Functional Differential Equations and Approximations of Fixed Points, Lecture Notes in Mathematics, **730**, 228-256, Springer, Berlin, (1979)

KY79. Kaplan, J.L., Yorke, J.A.: Preturbulence: a regime observed in a fluid flow of Lorenz. Commun. Math. Phys. **67**, 93-108, (1979)

Kap00. Kaplan, D.T.: Applying Blind Chaos Control to Find Periodic Orbits. arXiv:nlin.CD/0001058, (2000)

KZ91. Kree, R., Zippelius, A.: In Models of Neural Networks, I, Domany, R., van Hemmen, J.L., Schulten, K. (Eds), 193,, Springer, Berlin, (1991)

KZH02. Kiss, I.Z., Zhai, Y., Hudson, J.L.: Emerging coherence in a population of chemical oscillators. Science **296**, 1676-1678, (2002)

Lai94. Lai, Y.-C.: Controlling chaos. Comput. Phys., **8**, 62–67, (1994)

Lak97. Lakshmanan, M.: Bifurcations, Chaos, Controlling and Synchronization of Certain Nonlinear Oscillators. In Lecture Notes in Physics, **495**, 206, Y. Kosmann-Schwarzbach, B. Grammaticos, K.M. Tamizhmani (ed), Springer-Verlag, Berlin, (1997)

Lak03. Lakshmanan, M., Rajasekar, S: Nonlinear Dynamics: Integrability, Chaos and Patterns, Springer-Verlag, New York, (2003)

Lam45. Lamb, H.: Hydrodynamics. New York Dover Publ. New York,(1945)

Lam76. Lamb, G.L. Jr.: Bäcklund transforms at the turn of the century. In Bäcklund Transforms, (ed) by R.M. Miura, Springer, Berlin, (1976)

Lan00. Landsberg, A.S.: Disorder-induced desynchronization in a 2×2 circular Josephson junction array. Phys. Rev. B **61**, 36413648, (2000)

Lan08. Langevin, P.: Comptes. Rendue **146**, 530, (1908)

Lan75. Lanford, O.E.: Time evolution of large classical systems. 1-111 in Lecture Notes in Physics **38**, Spinger-Verlag, Berlin, (1975)

Lan86. Lanczos, C.: The variational principles of mechanics. Dover Publ. New York, (1986)

Lan95a. Lani–Wayda, B: Persistence of Poincaré maps in functional differential equations (with application to structural stability of complicated behavior). J. Dyn. Diff. Equs. **7**(1), 1–71, (1995)

Lan95b. Lani–Wayda, B: Hyperbolic Sets, Shadowing and Persistence for Noninvertible Mappings in Banach spaces, Research Notes in Mathematics **334**, Longman Group Ltd., Harlow, Essex, (1995)

LW95. Lani–Wayda, B, Walther, H.-O.: Chaotic motion generated by delayed negative feedback, Part I: A transversality criterion. Diff. Int. Equs. **8**(6), 1407–52, (1995)

LW96. Lani–Wayda, B, Walther, H.-O.: Chaotic motion generated by delayed negative feedback, Part II: Construction of nonlinearities. Math. Nachr. **180**, 141–211, (1996)

Lan99. Lani–Wayda, B: Erratic solutions of simple d elay equations. Trans. Amer. Math. Soc. **351**, 901-945, (1999)

Lia83. Liao, S.T.: On hyperbolicity properties of nonwandering sets of certain 3-dimensional differential systems. Acta Math. Sc., **3**, 361–368, (1983)

Las77. Lasota, A.: Ergodic problems in biology. Asterisque **50**, 239–250, (1977)

LO97. Lenz, H., Obradovic, D.: Robust control of the chaotic Lorenz system. Int. J. Bif. Chaos **7**, 2847–2854, (1997)

LAT83. Lobb, C.J., Abraham, D.W., Tinkham, M.: Theoretical. interpretation of resistive transitions in arrays of superconducting. weak links. Phys. Rev. **B27**, 150, (1983)

LAV01. Lacorata, G., Aurell, E., Vulpiani, A.: Drifter dispersion in the Adriatic Sea: Lagrangian data and chaotic model. Ann. Geophys. **19**, 1-9, (2001)

Laz94. Lazutkin, V.A.: Positive Entropy for the Standard Map I, 94-47, Université de Paris-Sud, Mathématiques, Bâtiment **425**, 91405 Orsay, France, (1994)

LCD87. Leggett, A.J., Chakravarty, S., Dorsey, A.T., Fisher, M.P.A., Chang, A., Zwerger, W.: Dynamics of the dissipative two-state system. Rev. Mod. Phys. **59**, 1, (1987)

LD98. Loss, D., DiVincenzo, D.P.: Quantum computation with quantum dots. Physical Review A **57**(1), 120-126, (1998)

Leb93. Lebowitz, J.L.: Boltzmann's entropy and time's arrow. Physics Today **46**(9), 32-38, (1993)

Led81. Ledrappier, F.: Some Relations Between Dimension and Lyapunov Exponents. Commun. Math. Phys. **81**, 229-238, (1981)

Leg86. Leggett, A.J.: In The Lesson of Quantum Theory, Niels Bohr Centenary Symposium 1985; J. de Boer, E. Dal, and O. Ulfbeck (ed). North Holland, Amsterdam, (1986)

Lei75. Leith, C.A.: Climate response and fluctuation dissipation. J. Atmos. Sci. **32**, 2022, (1975)

Lei78. Leith, C.A.: Predictability of climate. Nature **276**, 352, (1978)

Leo62. Leontovic, An.M.: On the stability of the Lagrange periodic solutions for the reduced problem of three body. Sov. Math. Dokl. **3**, 425-429, (1962)

Leo74. Leonard, A.: Energy Cascade in Large-Eddy Simulations of Turbulent Fluid Flows. Adv. Geophys. **18**, 237-248, (1974)

Ler34. Leray, J.: Sur le mouvement d'un liquide visquex emplissant l'espace, Acta Math. **63**, 193-248, (1934)

LH94. Lou, S., Hu, X.: Symmetries and algebras of the integrable dispersive long wave equations in (2+1)-dimensional spaces. J. Phys. A: Math. Gen. **27**, L207, (1994)

Li04. Li, Y.: Chaos in Partial Differential Equations. Int. Press Sommerville, MA, (2004)

Li04a. Li, Y.: Persistent homoclinic orbits for nonlinear Schrödinger equation under singular perturbation. Dyn. PDE, **1**(1), 87–123, (2004)

Li04b. Li, Y.: Existence of chaos for nonlinear Schrödinger equation under singular perturbation. Dyn. PDE. **1**(2), 225-237, (2004)

Li04c. Li, Y.: Homoclinic tubes and chaos in perturbed sine-Gordon equation. Cha. Sol. Fra., **20**(4), 791–798, 2004)

Li04d. Li, Y.: Chaos in Miles' equations. Cha. Sol. Fra. **22**(4), 965–974, (2004)

Li05. Li, Y.: Invariant manifolds and their zero-viscosity limits for Navier-Stokes equations. Dynamics of PDE. **2, no.2**, 159-186, (2005)

Li06. Li, Y.: On the True Nature of Turbulence. arXiv:math.AP/0507254, (2005)

Lig58. Lighthill, M.J.: An Introduction to Fourier Analysis and Generalised Functions. Campridge Univ. Press, (1958)

Lil92. Lilly D.: A proposed modification of the Germano subgrid-scale closure method, Phys. Fluids 4, 633–635, (1992)

Lin97. Linz, S.J.: Nonlinear Dynamical Models and Jerky Motion. Am. J. Phys., **65**(6), 523–526, (1997)

LL59. Landau, L.D., Lifshitz, E.M.: Fluid Mechanics. Pergamon Press, (1959)

LL78. Landau, L.D., Lifshitz, E.M.: Statitsical Physics. Pergamon Press, Oxford, (1978)

LOJ04. Lindgren B., Österlund J., Johansson A.: Evaluation of scaling laws derived from Lie group symmetry methods in zero-pressure-gradient turbulent boundary layers. J. Fluid Mech. **502**, 127–152, (2004)

Lor63. Lorenz, E.N.: Deterministic Nonperiodic Flow. J. Atmos. Sci., **20**, 130–141, (1963)

LOS88. Larkin, A.I., Ovchinnikov, Yu.N., Schmid, A.: Physica B **152**, 266, (1988)

LS82. Ledrappier, F., Strelcyn, J.-M.: A proof of the estimation from below in Pesin's entropy formula. Ergod. Th. and Dynam. Syst. **2**, 203-219, (1982)

LSR84. Luwel, M., Severne, G., Rousseeuw, P.J.: Numerical Study of the Relaxation of One-Dimensional Gravitational Systems. Astrophys. Space Sci. **100**, 261277, (1984)

LTZ01. Lou, S., Tang, X., Zhang, Y.: Chaos in soliton systems and special Lax pairs for chaos systems. arXiv:nlin/0107029, (2001)

LW76. Lasota, A., Wazewska–Czyzewska, M.: Matematyczne problemy dynamiki ukladu krwinek czerwonych, Mat. Stosowana **6**, 23–40, (1976)

LY77. Li, T.Y., Yorke, J.A.: Period three implies chaos. Am. Math. Monthly **82**, 985–992, (1977)

LY85. Ledrappier, F., Young, L.S.: The metric entropy of diffeomorphisms: I. Characterization of measures satisfying Pesin's formula, II. Relations between entropy, exponents and dimension. Ann. of Math. **122**, 509-539, 540-574, (1985)

Lya47. Lyapunov, A.: Problème générale de la stabilité du mouvement. Ann. Math. Studies 17, Princeton Univ. Press, Princeton, NJ, (1947)

Lya56. Lyapunov, A.M.: A general problem of stability of motion. Acad. Sc. USSR, (1956)

LZJ95. Liou, J.C., Zook, H.A., Jackson, A.A.: Radiation pressure, Poynting-Robertson drag and solar wind drag in the restricted three body problem. Icarus **116**, 186-201, (1995)

Mag54. Magnus, W.: On the exponential solution of differential equations for a linear operator. Commun. Pure Appl. Math. **7**, 649-673, (1954)

Maj84. Majda, A.: Compressible Fluid Flow and System of Conservation Laws in Several Space Variables. Springer-Verlog, Berlin, (1984)

Mal00. Malasoma, J.M.: What is the Simplest Dissipative Chaotic Jerk Equation which is Parity Invariant? Phys. Lett. A, **264**(5), 383–389, (2000)

Mal798. Malthus, T.R.: An essay on the Principle of Population. Originally published in 1798. Penguin, (1970)

Mal88. Mallet–Paret, J.: Morse decompositions for delay differential equations, J. Diff. Equs. **72**, 270–315, (1988)

Man80a. Mandelbrot, B.: Fractal aspects of the iteration of $z \mapsto \lambda z(1-z)$ for complex λ, z, Annals NY Acad. Sci. **357**, 249–259, (1980)

Man80b. Mandelbrot, B.: The Fractal Geometry of Nature. WH Freeman and Co., New York, (1980)

Mar99. Marsden, J.E.: Elementary Theory of Dynamical Systems. Lecture notes. CDS, Caltech, (1999)

May73. May, R.M. (ed.): Stability and Complexity in Model Ecosystems. Princeton Univ. Press, Princeton, NJ, (1973)

May76. May, R.: Simple Mathematical Models with Very Complicated Dynamics. Nature, **261**(5560), 459-467, (1976)

May76. May, R.M. (ed.): Theoretical Ecology: Principles and Applications. Blackwell Sci. Publ. (1976)

MDC85. Martinis, J.M., Devoret, M.H., Clarke, J.: Energy-Level Quantization in the Zero-Voltage State of a Current-Biased Josephson Junction. Phys. Rev. Lett. **55**, 15431546, (1985)

MDT00. Mangioni, S.E., Deza, R.R., Toral, R., Wio, H.S.: Non–equilibrium phase ransitions induced by multiplicative noise: effects of self–correlation. Phys. Rev. E, **61**, 223–231, (2000)

Man82. Mañé, R.: An ergodic closing lemma. Annals of Math., **116**, 503–540, (1982)

MP92. De Melo, W., Palis, J.: Geometric Theory of Dynamical Systems-An Introduction. Springer Verlag, Berlin, (1982)

Men98. Mendes, R.V.: Conditional exponents, entropies and a measure of dynamical self-organization. Phys. Let. A **248**, 167–1973, (1998)

Mey70. Meyer, K.R.: Generic bifurcation of periodic points. Trans. Amer. Math. Soc. **149**, 95–107, (1970)

Mey71. Meyer, K.R.: Generic stability properties of periodic points, Trans. Amer. Math. Soc. **154**, 273–277, (1971)

Mey73. Meyer, K.R.: Symmetries and integrals in mechanics. In Dynamical Systems (M. Peixoto, Ed.), 259–272, Academic Press, New York, (1973)

Mey81a. Meyer, K.R.: Periodic solutions of the N-body problem. J. Dif. Eqs. **39**(1), 2–38, (1981)

Mey81b. Meyer, K.R.: Periodic orbits near infinity in the restricted N–body problem, Celest. Mech. **23**, 69–81, (1981)

Mey94. Meyer, K.R.: Comet like periodic orbits in the N–body problem. J. Comp. and Appl. Math. **52**, 337–351, (1994)

MG77. Mackey, M.C., Glass, L.: Oscillation and chaos in physiological control systems. Science **197**, 287–295, (1977)

MGB96. Mira, C., Gardini, L., Barugola, A., Cathala, J.-C.: Chaotic Dynamics in Two-Dimensional Noninvertible Maps. World Scientific, Singapore, (1996)

MH38. Morse, M., Hedlund, G.: Symbolic Dynamics. Am. J. Math. **60**, 815–866, (1938)

MF53. Morse, M., Feshbach, H.: Methods of Theoretical Physics. Part I, McGraw-Hill, New York, (1953)

MH92. Meyer, K.R., Hall, G.R.: Introduction to Hamiltonian Dynamical Systems and the N–body Problem. Springer, New York, (1992)

MI95. Mishra, P., Ishwar, B: Second-Order Normalization in the Generalized Restricted Problem of Three Bodies, Smaller Primary Being an Oblate Spheroid. Astr. J. **110**(4), 1901-1904, (1995)

MK05. Moon, S.J., Kevrekidis, I.G.: An equation-free approach to coupled oscillator dynamics: the Kuramoto model example. Submitted to Int. J. Bifur. Chaos, (2005)

ML92. Méais, O., Lesieur, M.: Spectral large-eddy simulation of isotropic and stably stratified turbulence. J. Fluid Mech. **256**, 157–194, (1992)

MM75. Marsden, J.E., McCracken, M.: The Hopf bifurcation and its applications. MIT Press, Cambridge, MA, (1975)

MR99. Marsden, J.E., Ratiu, T.S.: Introduction to Mechanics and Symmetry: A Basic Exposition of Classical Mechanical Systems. (2nd ed), Springer, New York, (1999)

MM95. Matsui, N., Mori, T.: The efficiency of the chaotic visual behavior in modeling the human perception-alternation by artificial neural network. In Proc. IEEE ICNN'95. **4**, 1991–1994, (1995)

Mof83. Moffat, H.K.: Transport effects associated with turbulence, with particular attention to the influence of helicity. Rep. Prog. Phys. **46**, 621-664, (1983)

MP01. Morales, C.A., Pacifico, M.J.: Mixing attractors for 3-flows. Nonlinearity, **14**, 359–378, (2001)

Mor21. Morse, M.: A one-to-one representation of geodesics on a surface of negative curvature, Am. J. Math. **43**, 33–51, (1921)

Mos62. Moser, J.: On invariant curves of area - preserving mappings of an annulus., Nach. Akad. Wiss., Gottingen, Math. Phys. Kl. **2**, 1, (1962)

Mos73. Moser, J.: Stable and Random Motions in Dynamical Systems. Princeton Univ. Press, Princeton, (1973)

Mos96. Mosekilde, E.: Topics in Nonlinear Dynamics: Application to Physics, Biology and Economics. World Scientific, Singapore, (1996)

MMG05. Motter, A.E., de Moura, A.P.S., Grebogi, C., Kantz, H.: Effective dynamics in Hamiltonian systems with mixed phase space. Phys. Rev. E **71**, 036215, (2005)

Mou06. Moulton, F.R.: A class of periodic solutions of the problem of three–bodies with applications to lunar theory. Trans. Amer. Math. Soc. **7**, 537–577, (1906)

Mou12. Moulton, F.R.: A class of periodic orbits of the superior planets, Trans. Amer. Math. Soc. **13**, 96–108, (1912)

MP70. Meyer, K.R.: Palmore, J.I.: A new class of periodic solutions in the restricted three–body problem, J. Diff. Eqs. **44**, 263–272, (1970)

MPT98. Milanović, Lj., Posch, H.A., Thirring, W.: Statistical mechanics and computer simulation of systems with attractive positive power-law potentials. Phys. Rev. E **57**, 27632775, (1998)

MPV87. Mézard, M., Parisi, G., Virasoro, M.A.: Spin Glass Theory and Beyond. World Scientific, Singapore, (1987)

MR90. Miller, B.N., Reidl, C.J. Jr.: Gravity in one dimension - Persistence of correlation. Astrophys. J. **348**, 203-211, (1990)

MS00. Meyer, K.R., Schmidt, D.S.: From the restricted to the full three-body problem. Trans. Amer. Math. Soc., **352**, 2283–2299, (2000)

MS71. Meyer, K.R., Schmidt, D.S.: Periodic orbits near L_4 for mass ratios near the critical mass ratio of Routh, Celest. Mech. **4**, 1971, 99–109, (1971)

MS90. Mirollo, R.E., Strogatz, S.H.: Synchronization of pulse-coupled biological oscillators. SIAM J. Appl. Math. **50**, 1645, (1990)

MS90. Mallet–Paret, J., Smith, H.L.: The Poincaré–Bendixson theorem for monotone cyclic feedback systems, J. Dyn. Diff. Eqns. **2**, 367–421, (1990)

MS96. Mallet–Paret, J., Sell,G.: The Poincaré–Bendixson theorem for monotone cyclic feedback systems with delay, J. Diff. Eqns. **125**, 441–489, (1996)

MS95. Müllers, J., Schmid, A.: Resonances in the current-voltage characteristics of a dissipative Josephson junction. cond-mat/9508035, (1995)

MT92. Müller, B., Trayanov, A.: Deterministic chaos in non-Abelian lattice gauge theory. Phys. Rev. Lett. **68**, 3387 (1992)

Mun75. Munkres, J.R.: Topology a First Course. Prentice-Hall, New Jersey, (1975)

Mur02. Murray, J.D.: Mathematical Biology, Vol. I: An Introduction (3rd ed.), Springer, New York, (2002)

Mur94. Murray, C.D.: Dynamical effect of drag in the circular restricted three body problem 1. Location and stability of the Lagrangian equilibrium points. Icarus **112**, 465-484, (1994)

MW74. Marsden, J., Weinstein, A.: Reduction of symplectic manifolds with symmetries, Rep. Math. Phy. **5**, 121–130, (1974)

MW89. Marcus, C.M., Westervelt, R.M.: Dynamics of iterated-map neural networks. Phys. Rev. A **40**(1), 501–504, (1989)

MW94. Mallet–Paret, J., Walther, H.-O.: Rapid oscillations are rare in scalar systems governed by monotone negative feedback with a time lag. Math. Inst. Univ. Giessen, (1994)

MWH01. Michel, A.N., Wang, K., Hu, B.: Qualitative Theory of Dynamical Systems (2nd ed). Dekker, New York, (2001)

NAY60. Nagumo, J., Arimoto, S., Yoshizawa, S.: An active pulse transmission line simulating 1214-nerve axons, Proc. IRL **50**, 2061–2070, (1960)

Nay73. Nayfeh, A.H.: Perturbation Methods. Wiley, New York. (1973)

New00. Newman, M.E.J.: Models of the small world. J. Stat. Phys. **101**, 819, (2000)

New74. Newell, A.C. (ed): Nonlinear Wave Motion. Am. Mat. Soc., Providence, R.I., (1974)

New80. Newhouse, S.: Lectures on dynamical systems. In Dynamical Systems, C.I.M.E. Lectures, 1–114, Birkhauser, Boston, MA, (1980)

Nie94. Niedzielska, Z.: Nonlinear stability of the libration points in the photogravitational restricted three body problem. Celest. Mech. Dyn. Astron., **58**(3), 203-213, (1994)

NK96. Nishikawa, T., Kaneko, K.: Fractalization of a torus as a strange nonchaotic attractor. Phys. Rev. E **54**, 61146124, (1996)

NKF97. Nishimura, H., Katada, N., Fujita, Y.: Dynamic Learning and Retrieving Scheme Based on Chaotic Neuron Model, In: R. Nakamura et al. (eds): Complexity and Diversity, Springer-Verlag, New York, 64–66, (1997)

NLG00. Newrock, R.S., Lobb, C.J., Geigenmüller, U., Octavio, M.: Solid State Physics. Academic Press, San Diego, Vol. 54, (2000)

NE99. Nimmo, S., Evans, A.K.: The Effects of Continuously Varying the Fractional Differential Order of Chaotic Nonlinear Systems. Chaos, Solitons & Fractals **10**. 1111–1118, (1999)

NML03. Nishikawa, T., Motter, A.E., Lai, Y.C., Hoppensteadt, F.C.: Heterogeneity in Oscillator Networks: Are Smaller Worlds Easier to Synchronize? Phys. Rev. Lett. **91**, 014101, (2003)

NNM00. Nagao, N., Nishimura, H., Matsui, N.: A Neural Chaos Model of Multistable Perception. Neural Processing Letters **12**(3): 267-276, (2000)

NNM97. Nishimura, H., Nagao, N., Matsui, N.: A Perception Model of Ambiguous Figures based on the Neural Chaos, In: N. Kasabov et al. (eds): Progress in Connectionist-Based Information Systems, **1**, 89–92, Springer-Verlag, New York, (1997)

NP77. Nicolis, G., Prigogine, I.: Self–Organization in Nonequilibrium Systems: From Dissipative Structures to Order through Fluctuations. Wiley Europe, (1977)

NPT99. Nakamura, Y., Pashkin, Yu.A., Tsai, J.S.: Coherent control of macroscopic quantum states in a single-Cooper-pair box, Nature, **398**, 786-788, (1999)

Obe00. Oberste-Vorth, R.: Horseshoes among Hénon mappings. In Mastorakis, N. (ed.) Recent Advances in Applied and Theoretical Mathematics, WSES Press, 116–121, (2000)

Obe01. Oberlack, M.: A unified approach for symmetries in plane parallel turbulent shear flows. J. Fluid Mech. **427**, 299–328, (2001)

Obe87. Oberste-Vorth, R.: Complex Horseshoes and the Dynamics of Mappings of Two Complex Variables. PhD thesis, Cornell Univ., (1987)

Obe99. Oberlack M.: Symmetries, invariance and scaling-laws in inhomogeneous turbulent shear flows. Flow, Turbulence and Combustion **62**, 111–135, (1999)

OGY90. Ott, E., Grebogi, C., Yorke, J.A.: Controlling chaos. Phys. Rev. Lett., **64**, 1196–1199, (1990)

Oka95. Okada, M.: A hierarchy of macrodynamical equations for associative memory. Neural Networks **8**, 833-838, (1995)

OS74. Oldham, K.B., Spanier, J.: The Fractional Calculus. Academic Press, New York, (1974)

Olv01. Olver, P.J.: Geometric foundations of numerical algorithms and symmetry. Appl. Algebra Engrg. Comm. Comput. **11**, 417–436, (2001)

Olv86. Olver, P.J.: Applications of Lie Groups to Differential Equations (2nd ed). Graduate Texts in Mathematics, **107**, Springer, New York, (1986)

Ons35. Onsager, L.: Reciprocal relations in irreversible processes. II. Phys. Rev. **38**, 2265-2279, (1931)

OS74. Oldham, K.B, Spanier. J.: The Fractional Calculus. Acad. Press, New York, (1974)

OS94. Ovchinnikov, Yu.N., Schmid, A.: Resonance phenomena in the current-voltage characteristic of a Josephson junction. Phys. Rev. B **50**, 63326339, (1994)

Ose68. Oseledets, V.I.: A Multiplicative Ergodic Theorem: Characteristic Lyapunov Exponents of Dynamical Systems. Trans. Moscow Math. Soc., **19**, 197–231, (1968)

Ose68. Oseledets, V.I.: A multiplicative ergodic theorem. Lyapunov characteristic numbers for dynamical systems. Trans. Moscow Math. Soc. **19**, 197-221, (1968)

Ott89. Ottino, J.M.: The kinematics of mixing: stretching, chaos and transport. Cambridge Univ. Press, Cambridge, (1989)

Ott93. Ott, E.: Chaos in dynamical systems. Cambridge University Press, Cambridge, (1993)

PT93. Palis, J., Takens, F.: Hyperbolicity and sensitive-chaotic dynamics at homoclinic bifurcations. Cambridge University Press, (1993)

Pal67. Palmore, J.I.: Bridges and Natural Centers in the Restricted Three–Body Problem, University of Minnesota Report, Minneapolis, MN, (1969)

Pal88. Palmer, K.J.: Exponential dichotomies, the shadowing lemma and transversal homoclinic points, U. Kirchgraber and H.–O. Walther (eds), 265–306, Dynamics Reported, vol. I, Teubner-Wiley, Stuttgart/Chichester, (1988)

Pal97. Palais, R.S.: The symmetries of solitons. Bull. Amer. Math. Soc. **34**, 339-403, (1997)

PY97. Pan, S., Yin, F.: Optimal control of chaos with synchronization. Int. J. Bif. Chaos **7**, 2855–2860, (1997)

PC89. Parker, T.S., Chua, L.O.: Practical Numerical Algorithm for Chaotic Systems, Springer-Verlag, New York, (1989)

PB89. Peyrard, M., Bishop, A.R.: Statistical mechanics of a nonlinear model for DNA denaturation. Phys. Rev. Lett. **62**, 2755, (1989)

PB94. Plischke, M. Bergersen, B.: Equilibrium Statistical Mechanics. World Scientific, Singapore, (1994)

Pod99. Podlubny, I.: Fractional Differential Equations. Academic Press, San Diego, (1999)

PBR97. Prasad, A., Mebra, V., Ramaswamy, R.: Intermittency Route to Strange Nonchaotic Attractors. Phys. Rev. Lett. **79**, 41274130, (1997)

PC90. Pecora, L.M., Carroll, T.L.: Synchronization in chaotic systems. Phys. Rev. Lett. **64**, 821–824, (1990)

PC91. Pecora, L.M., Carroll, T.L.: Driving systems with chaotic signals. Phys. Rev. A **44**, 2374-2383, (1991)

PC98. Pecora, L.M., Carroll, T.L.: Master stability functions for synchronized coupled systems. Phys. Rev. Lett. **80**, 2109-2112, (1998)

PCS93. Penney, R.W., Coolen, A.C., Sherrington D.: Coupled dynamics of fast and slow interactions in neural networks and spin systems. J. Phys. A: Math. Gen. **26**, 3681, (1993)

Pee80. Peebles, P.J.E.: The Large Scale Structure of the Universe. Princeton Univ. Press, (1980)

Pen89. Penrose, R.: The Emperor's New Mind. Oxford Univ. Press, Oxford, (1989)

Per37. Perron, O.: Neue periodische Lösungen des ebenen Drei und Mehrkörperproblem, Math. Z., **42**, 593–624, (1937)

Per84. Peretto, P.: Collective properties of neural networks: a statistical physics approach. Biol. Cybern. **50**, 51, (1984)

Per92. Peretto, P.: An Introduction to the Theory of Neural Computation. Cambridge Univ. Press, Cambridge, (1992)

Pes75. Peskin, C.: Mathematical Aspects of Heart Physiology. Courant Inst. Math. Sci., New York Univ, (1975)

Pes76. Pesin, Ya.B.: Invariant manifold families which correspond to non-vanishing characteristic exponents. Izv. Akad. Nauk SSSR Ser. Mat. **40**(6), 1332-1379, (1976)

Pes77. Pesin, Ya.B.: Lyapunov Characteristic Exponents and Smooth Ergodic Theory. Russ. Math. Surveys, **32**(4), 55–114, (1977)

Pet93. Peterson, I.: Newton's Clock: Chaos in the Solar System. W.H. Freeman, San Francisco, (1993)

Pet99. Petras, I.: The Fractional-order controllers: Methods for their synthesis and application. J. El. Eng. **9-10**, 284–288, (1999)

Pet02. Petras, I.: Control of Fractional-Order Chua's System. J. El. Eng. **53**(7-8), 219-222, (2002)

Pet96. Petrov, V., Showalter, K.: Nonlinear Control from Time-Series. Phys. Rev. Lett. **76**, 3312, (1996)

Pet93. Pettini, M.: Geometrical hints for a nonperturbative approach to Hamiltonian dynamics. Phys. Rev. E **47**, 828850, (1993)

PGY06. Politi, A., Ginelli, F., Yanchuk, S., Maistrenko, Y.: From synchronization to Lyapunov exponents and back. arXiv:nlin.CD/0605012, (2006)

PHV86. Posch, H.A., Hoover, W.G., Vesely, F.J.: Canonical Dynamics of the Nosé Oscillator: Stability, Order, and Chaos. Phys. Rev. A, **33**(6), 4253–4265, (1986)

Pic86. Pickover, C.A.: Computer Displays of Biological Forms Generated From Mathematical Feedback Loops. Computer Graphics Forum, **5**, 313, (1986)

Pic87. Pickover, C.A.: Mathematics and Beauty: Time–Discrete Phase Planes Associated with the Cyclic System. Computer Graphics Forum, **11**, 217, (1987)

PK97. Pikovsky, A., Kurth, J.: Coherence Resonance in a Noise-Driven Excitable Systems. Phys. Rev. Lett. **78**, 775-778, (1997)

Poi899. Poincaré, H.: Les méthodes nouvelles de la mécanique céleste. Gauther–Villars, Paris, (1899)

Pop00. Pope, S.B.: Turbulent Flows. Cambridge Univ. Press, Cambridge, (2000)

POR97. Pikovsky, A., Osipov, G., Rosenblum, M., Zaks, M., Kurths, J.: Attractor–repeller collision and eyelet intermittency at the transition to phase synchronization. Phys. Rev. Lett. **79**, 47-50, (1997)

Poy03. Poynting, J.H.: Radiation in the solar system: i ts effect on temperature and its pressure on small bodies. Mon. Not. Roy. Ast. Soc. **64**, 1-5, (1903)

Pri62. Prigogine, I.: Introduction to thermodynamics of irreversible processes. John Wiley, New York, (1962)

PRK01. Pikovsky, A., Rosenblum, M., Kurths, J.: Synchronization: A Universal Concept in Nonlinear Sciences (Cambridge University Press, Cambridge, UK (2001)

PS62. Perring, J.K., Skyrme, T.R.H.: A model unified field equation. Nuc. Phys., **31**, 550–55, (1962)

PS89. Pugh, C.C., Shub, M.: Ergodic attractors. Trans. Amer. Math. Soc. **312**, 1-54, (1989)

PS94. Penney, R.W., Sherrington, D.: Slow interaction dynamics in spin-glass models. J. Phys. A: Math. Gen. **27**, 4027-4041, (1994)

PSS96. Pons, J.M., Salisbury, D.C., Shepley, L.C.: Gauge transformations in the Lagrangian and Hamiltonian formalisms of generally covariant theories. arXiv:gr-qc/9612037, (1996)

PSV87. Paladin, G., Serva, M., Vulpiani, A.: Complexity in Dynamical Systems with Noise. Phys. Rev. Lett. **74**, 6669, (1995)

PTB86. Pokrovsky, V.L., Talapov, A.L., Bak, P.: In Solitons, 71-127, edited by Trullinger, Zakharov, and Pokrovsky. Elsevier Science, (1986)

Put93. Puta, M.: Hamiltonian Mechanical Systems and Geometric Quantization, Kluwer, Dordrecht, (1993)

Pyr92. Pyragas, K.: Continuous control of chaos, by self-controlling feedback. Phys. Lett. A, **170**, 421–428, (1992)

Pyr95. Pyragas, K.: Control of chaos via extended delay feedback. Phys. Lett. A, **206**, 323–330, (1995)

PZR97. Pikovsky, A., Zaks, M., Rosenblum, M., Osipov, G., Kurths, J.: Phase synchronization of chaotic oscillations in terms of periodic orbits. Chaos **7**, 680, (1997)

Raj82. Rajaraman, R.: Solitons and Instantons. North-Holland, Amsterdam, (1982)

Rat69. Ratner, M.: Markov partitions for Anosov flows on 3-dimensional manifolds. Mat. Zam. **6**, 693-704, (1969)

RBO87. Romeiras, J., Bodeson, A., Ott, E., Antonsen, T.M. Jr., Grebogi, C.: Quasiperiodically forced dynamical systems with strange nonchaotic attractors. Physica D **26**, 277-294, (1987)

RGL99. Rodriguez, E., George, N., Lachaux, J., Martinerie, J., Renault, B., Varela, F.: Long-distance synchronization of human brain activity. Nature, **397**, 430, (1999)

RGO92. Romeiras, F.J., Grebogi, C., Ott, E., Dayawansa, W.P.: Using small perturbations to control chaos. Physica D **58**, 165-192, (1992)

RH06. Razafindralandy, D., Hamdouni, A.: Consequences of symmetries on the analysis and construction of turbulence models. Sym. Int. Geo. Met. Appl. **2**, 052, (2006)

Ric01. Richter, P.H.: Chaos in Cosmos. Rev. Mod. Ast. **14**, 53-92, (2001)

Ric93. Ricca, R.L.: Torus knots and polynomial invariants for a class of soliton equations, Chaos, **3**(1), 83–91, (1993)

RKH88. Riedel, U., Kühn, R., van Hemmen, J.L.: Temporal sequences and chaos in neural nets. Phys. Rev. A **38**, 1105, (1988)

RM86. Rumelhart, D. E., McClelland, J. L. and the PDP Research Group: Parallel Distributed Processing. vol. 1, MIT Press, (1986)

RMS90. Riani, M., Masulli, F., Simonotto, E.: Stochastic dynamics and input dimensionality in a two-layer neural network for modeling multistable perception. In: Proc. IJCNN. 1019–1022, (1990)

RO87. Romeiras, F.J., Ott, E.: Strange nonchaotic attractors of the damped pendulum with quasiperiodic forcing. Phys. Rev. A **35**, 44044413, (1987)

Rob37. Robertson, H.P.: Dynamical effects of radiation in the solar system, Mon. Not. Roy. Ast. Soc. **97**, 423-437, (1937)

Rob79. Robbins, K.: Periodic solutions and bifurcation structure at high r in the Lorenz system. SIAM J. Appl. Math. **36**, 457–472, (1979)

Rob95. Robinson, C.: Dynamical Systems. CRC Press. Boca Raton, FL, (1995)

ROH98. Rosa, E., Ott, E., Hess, M.H.: Transition to Phase Synchronization of Chaos. Phys. Rev. Lett. **80**, 1642–1645, (1998)

Ros76. Rössler, O.E.: An Equation for Continuous Chaos. Phys. Lett. A, **57**(5), 397–398, (1976)

RPK96. Rosenblum, M., Pikovsky, A., Kurths, J.: Phase synchronization of chaotic oscillators. Phys. Rev. Lett. **76**, 1804, (1996)

RPK97. Rosenblum, M., Pikovsky, A., Kurths, J.: From Phase to Lag Synchronization in Coupled Chaotic Oscillators. Phys. Rev. Lett. **78**, 4193–4196, (1997)

RSW90. Richter, P.H., Scholz, H.-J., Wittek, A.: A breathing chaos. Nonlinearity **3**, 45-67, (1990)

RT71. Ruelle, D., Takens, F.: On the nature of turbulence. Comm. Math. Phys., **20**, 167–192, (1971)

Rue76. Ruelle, D.: A measure associated with Axiom A attractors. Am. J. Math. **98**, 619-654, (1976)

Rue78. Ruelle, D.: An inequality for the entropy of differentiable maps. Bol. Soc. Bras. Mat. **9**, 83-87, (1978)

Rue78. Ruelle, D.: Thermodynamic formalism. Addison-Wesley, Reading, MA, (1978)

Rue78. Ruelle, D.: Thermodynamic formalism. Encyclopaedia of Mathematics and its Applications. Addison–Wesley, Reading, (1978)

Rue79. Ruelle, D.: Ergodic theory of differentiable dynamical systems. Publ. Math. IHES **50**, 27-58, (1979)

Rue89. Ruelle, D.: The thermodynamical formalism for expanding maps. Commun. Math. Phys. **125**, 239-262, (1989)

Rue98. Ruelle, D.: Nonequillibrium statistical mechanics near equilibrium: computing higher order terms. Nonlinearity **11**, 5-18, (1998)

Rue99. Ruelle, D.: Smooth Dynamics and New Theoretical Ideas in Non–equilibrium Statistical Mechanics. J. Stat. Phys. **95**, 393–458, (1999)

Rul01. Rulkov, N.F.: Regularization of Synchronized Chaotic Bursts. Phys. Rev. Lett. **86**, 183–186, (2001)

Rus844. Russell, J.S.: Report on Waves, 14th meeting of the British Association for the Advancement of Science, BAAS, London, (1844)

Rus885. Russell, J.S.: The Wave of Translation in the Oceans of Water, Air and Ether. Trübner, London, (1885)

Ryb71. Rybicki, G.B.: Exact statistical mechanics of a one-dimensional self-gravitating system. Astrophys. Space Sci. **14**, 5672, (1971)

RYD96. Ryu, S., Yu, W., Stroud, D.: Dynamics of an underdamped Josephson-junction ladder. Phys. Rev. E **53**, 2190–2195, (1996)

SAB87. Schultz, C.L., Ablowitz, M.J., BarYaacov, D.: Davey-Stewartson I System: A Quantum (2+1)-Dimensional Integrable System. Phys. Rev. Lett. **59**, 2825282S, (1987)

Sag04. Sagaut, P.: Large eddy simulation for incompressible flows. An introduction. Scientific Computation, Springer, (2004)

SB89. Shapiro, M., Brumer, P.: Coherent Radiative Control of Unimolecular Reactions: Selective Bond Breaking with Picosecond Pulses. J. Chem. Phys. **90**, 6179, (1989)

SC00. Skantzos, N.S., Coolen, A.C.C.: $(1 + \infty)$-Dimensional Attractor Neural Networks. J. Phys. A **33**, 5785-5807, (2000)

SC01. Skantzos, N.S., Coolen, A.C.C.: Attractor Modulation and Proliferation in $(1 + \infty)$-Dimensional Neural Networks. J. Phys. A **34**, 929-942, (2001)

Sch74. Schmidt, D.S.: Periodic solutions near a resonant equilibrium of a Hamiltonian system, Celest. Mech. **9**, 81–103, (1974)

Sch78. Schot, S.H.: Jerk: The Time Rate of Change of Acceleration. Am. J. Phys. **46**(11), 1090–1094, (1978)

Sch80. Schuerman, D.W.: The restricted three body problem including radiation pressure Astrophys. J. **238**(1), 337-342, (1980)

Sch90. Schmidt, D.S.: Transformation to versal normal form. In Computer Aided Proofs in Analysis, 235-240, (Ed. K. R. Meyer and D. S. Schmidt), IMA Series **28**, Springer–Verlag, (1990)

Sch94. Schmidt, D.S.: Versal normal form of the Hamiltonian function of the restricted problem of three–bodies near L_4, J. Com. Appl. Math. **52**, 155–176, (1994)

Sch91. Schroeder, M.: Fractals, Chaos, Power Laws. Freeman, New York, (1991)

Sch88. Schuster, H.G.(ed): Handbook of Chaos Control. Wiley-VCH, (1999)

Sco04. Scott, A. (ed): Encyclopedia of Nonlinear Science. Routledge, New York, (2004)

Sco99. Scott, A.C.: Nonlinear Science: Emergence and Dynamics of Coherent Structures, Oxford Univ. Press, Oxford, (1999)

SDG92. Shinbrot, T., Ditto, W., Grebogi, C., Ott, E., Spano, M., Yorke, J.A.: Using the sensitive dependence of chaos (the butterfly effect) to direct trajectories in an experimental chaotic system. Phys. Rev. Lett. **68**, 28632866, (1992)

SDK53. Seeger, A., Donth, H., Kochendörfer, A.: Theorie der Versetzungen in eindimensionalen Atomreihen. Zeitschrift für Physik, **134**, 173–93 (1953)

SFC02. Succi, S., Filippova, O., Chen, H., Orszag, S.: Towards a Renormalized Lattice Boltzmann Equation for Fluid Turbulence. J. Stat. Phys. **107**, 261-278, (2002)

SG05. Saveliev V., Gorokhovski M., Group-theoretical model of developed turbulence and renormalization of the Navier–Stokes equation. Phys. Rev. E **72**, 016302, (2005)

SG88. Solomon, T.H., Gollub, J.P.: Passive transport in steady Rayleigh-Benard convection. Phys. Fluids **31**, 1372, (1988)

Shi65. Shilnikov, L.P.: On the generation of a periodic motion from a trajectory which leaves and re-enters a saddle-saddle state of equilibrium. Soc. Math. Dokl **6**, 163166, (in Russian), (1965)

Shi87. Shinomoto, S.: Memory maintenance in neural networks. J. Phys. A Math. Gen. **20**, L1305-L1309, (1987)

Shi98. Shishikura, M.: The Hausdorff dimension of the boundary of the Mandelbrot set and Julia sets. Ann. of Math. **147**, 225-267, (1998)

Shi99. Shimomura, Y.: A Family of Dynamic Subgrid-Scale Models Consistent with Asymptotic Material Frame Indifference, J. Phys. Soc. Jap. **68**, 2483-2486, (1999)

Sie50. Siegel, C.L.: Über eine periodische Lösung im Dreikörperproblem, Math. Nachr. **4**, 28–64, (1950)

Sin68a. Sinai, Ya.G.: Markov partitions and C-diffeomorphisms. Funkts. Analiz i Ego Pril. **2**(1), 64-89, (1968)

Sin68b. Sinai, Ya.G.: Constuction of Markov partitions. Funkts. Analiz i Ego Pril. **2**(3), 70-80, (1968)

Sin72. Sinai, Ya.G.: Gibbsian measures in ergodic theory. Uspehi Mat. Nauk **27**(4), 21-64, (1972)

Sin89. Sinai, Ya.G.: Dynamical Systems II, Encyclopaedia of Mathematical Sciences, **2**, Springer-Verlag, Berlin, (1989)

Sha06. Sharma, S.: An Exploratory Study of Chaos in Human-Machine System Dynamics, IEEE Trans. SMC B. 36(2), 319-326, (2006)

SIT95. Shea, H.R., Itzler, M.A., Tinkham, M.: Inductance effects and dimensionality crossover in hybrid superconducting arrays Phys. Rev. B **51**, 1269012697, (1995)

Siv77. Sivashinsky, G.I.: Nonlinear analysis of hydrodynamical instability in laminar flames – I. Derivation of basic equations. Acta Astr. **4**, 1177, (1977)

SJD94. Schiff, S.J., Jerger, K., Duong, D.H., Chang, T., Spano, M.L., Ditto, W.L : Controlling chaos in the brain. Nature, **370**, 615–620, (1994)

SK75. Sherrington, D. and Kirkpatrick S.: Solvable Model of a Spin-Glass. Phys. Rev. Lett. **35**, 17921796, (1975)

SK86. Sompolinsky, H., Kanter, I.: Temporal Association in Asymmetric Neural Networks. Phys. Rev. Lett. **57**, 2861, (1986)

SKC00. Succi, S., Karlin, I.V., Chen, H., Orszag, S.: Resummation Techniques in the Kinetic-Theoretical Approach to Subgrid Turbulence Modeling. Physica A **280**, 92-98, (2000)

SKM93. Samko, S.G., Kilbas, A.A., Marichev, O.I.: Fractional Integrals and Derivatives Theory and Applications. Gordon & Breach, New York, (1993)

SKW95. Sakai, K., Katayama, T., Wada, S., Oiwa, K.: Chaos causes perspective reversals for ambiguous patterns. In: Advances in Intelligent Computing IPMU'94, 463–472, (1995)

SL00. Sprott, J.C., Linz, S.J.: Algebraically Simple Chaotic Flows. Int. J. Chaos Theory and Appl., **5**(2), 3–22, (2000)

SM71. Siegel, C.L., Moser, J.K.: Lectures on Celestial Mechanics. Springer–Verlag, New York, (1971)

Sm88. Strogatz, S.H., Mirollo, R.E.: Phase-locking and critical phenomena in lattices of coupled nonlinear oscillators with random intrinsic frequencies. Physica D **31**, 143, (1988)

Sma63. Smagorinsky, J.: General Circulation Experiments with the Primitive Equations: I. The Basic Equations. Mon. Weath. Rev. **91**, 99-164, (1963)

Sma67. Smale, S.: Differentiable Dynamical Systems. Bull. AMS **73**, 747–817, (1967)

SOG90. Shinbrot, T., Ott, E., Grebogi, C., Yorke, J.A.: Using chaos to direct trajectories to targets. Phys. Rev. Lett. **65**, 32153218, (1990)

Sok78. Sokol'skii, A.: On the stability of an autonomous Hamiltonian system. J. Appl. Math. Mech. **38**, 741–49, (1978)

Spa82. Sparrow, C.: The Lorenz Equations: Bifurcations, Chaos, and Strange Attractors. Springer, New York, (1982)

Spr93a. Sprott, J.C.: Automatic Generation of Strange Attractors. Comput. & Graphics, **17**(3), 325–332, (1993)

Spr93b. Sprott, J.C.: Strange Attractors: Creating Patterns in Chaos. M&T Books, New York, (1993)

Spr94. Sprott, J.C.: Some Simple Chaotic Flows. Phys. Rev. E, **50**(2), R647–R650, (1994)

Spr97. Sprott, J.C.: Some Simple Chaotic Jerk Functions. Am. J. Phys., **65**(6), 537–543, (1997)

SR90. Shi, S., Rabitz, H.: Quantum mechanical optimal control of physical observables in microsystems. J. Chem. Phys. **92**(1), 364-376, (1990)

SR91. Schwieters, C.D., Rabitz, H.: Optimal control of nonlinear classical systems with application to unimolecular dissociation reactions and chaotic potentials. Phys. Rev. A **44**, 52245238, (1991)

SR91. Shi, S., Rabitz, H.: Optimal Control of Bond Selectivity in Unimolecular Reactions. Comp. Phys. Com. **63**, 71, (1991)

SRK98. C. Schäfer, M.G. Rosenblum, J. Kurths, and H.-H Abel: Heartbeat Synchronized with Ventilation. Nature **392**, 239-240 (1998)

SS00. Shraiman, B.I., Siggia, E.D.: Scalar Turbulence. Nature **405**, 639-46, (2000)

SSK87. Sakaguchi, H., Shinomoto, S., Kuramoto, Y.: Local and global self-entrainments in oscillator-lattices. Prog. Theor. Phys. **77**, 1005-1010, (1987)

SSR93. Shen, L., Shi, S., Rabitz, H., Lin, C., Littman, M., Heritage, J.P., Weiner, A.M.: Optimal control of the electric susceptibility of a molecular gas by designed nonresonant laser pulses of limited amplitude. J. Chem. Phys. **98**, 7792-7803, (1993)

Sta97. Stark. J.: Invariant Graphs for Forced Systems. Physica D **109**, 163-179, (1997)

Ste36. Steuerwald, R.: Über Enneper'sche Flächen und Bäcklund'sche Transformation. Abhandlungen der Bayerischen Akademie der Wissenschaften München, 1–105, (1936)

Str. Strogatz, S.H.: Exploring complex networks. Nature, **410**, 268, (2001)

Str00. Strogatz, S.H.: From Kuramoto to Crawford: exploring the onset of synchronization in populations of coupled oscillators. Physica D, **143**, 1–20, (2000)

Str68. Strang, G.: On the Construction and Comparison of Difference Schemes. SIAM J. Num. Anal. **5**, 506–517, (1968)

Str94. Strogatz, S.: Nonlinear Dynamics and Chaos. Addison-Wesley, Reading, MA, (1994)

SW87. Shapere, A. Wilczek, F.: Two Applications of Axion Electrodynamics. Phys. Rev. Lett. **58**, 2051, (1987)

SW89a. Shapere, A. Wilczek, F.: Geometry of self-propulsion at low Reynold's number. J. Fluid. Mech. **198**, 557, (1989)

SW89b. Shapere, A. Wilczek, F.: Geometric Phases in Physics. World Scientific, Singapore, (1989)

SN79. Shimada, I., Nagashima, T.: A Numerical Approach to Ergodic Problem of Dissipative Dynamical Systems. Prog. The. Phys. **61**(6), 1605-16, (1979)

SW90. Steinlein, H., Walther, H.-O.: Hyperbolic Sets, Transversal Homoclinic Trajectories, and Symbolic Dynamics for C^1-maps in Banach Spaces. J. Dyn. Diff. Equs. **2**, 325–365, (1990)

SW90. Sassetti, M., Weiss, U.: Universality in the dissipative two-state system. Phys. Rev. Lett. **65**, 22622265, (1990)

Tap74. Tappert, F.: Numerical Solutions of the Korteweg-de Vries Equations and its Generalizations by the Split-Step Fourier Method. In Nonlinear Wave Motion, 215–216, Lectures in Applied Math. **15**, Amer. Math. Soc., (1974)

Tar04a. Tarasov, V.E.: Fractional Generalization of Liouville Equation. Chaos **14**, 123-127, (2004)

Tar04b. Tarasov, V.E.: Path integral for quantum operations. J. Phys. A **37**, 3241-3257, (2004)

Tar05a. Tarasov, V.E.: Fractional Systems and Fractional Bogoliubov Hierarchy Equations. Phys. Rev. E **71**, 011102, (2005)

Tar05b. Tarasov, V.E.: Fractional Liouville and BBGKI equations. J. Phys.: Conf. Ser. **7**, 17-33, (2005)

Tar05c. Tarasov, V.E.: Continuous Medium Model for Fractal Media. Phys. Lett. A **336**, 167-174, (2005)

Tar05d. Tarasov, V.E.: Possible Experimental Test of Continuous Medium Model for Fractal Media. Phys. Let. A **341**, 467-472, (2005)

Tar05e. Tarasov, V.E.: Fractional Hydrodynamic Equations for Fractal Media. Ann. Phys. **318**, 286-307, (2005)

Tar05f. Tarasov, V.E.: Fractional Fokker-Planck Equation for Fractal Media. Chaos **15**, 023102, (2005)

Tar05g. Tarasov, V.E.: Fractional Generalization of Gradient and Hamiltonian Systems. J. Phys. A **38**(26), 5929–5943, (2005)

Tay21. Taylor, G.I.: Diffusion by continuous movements. Proc. London Math. Soc. **20**, 196, (1921)

Tay38. Taylor, G.I.: Frozen-in Turbulence Hypothesis. Proc. R. Soc. London A **164**, 476, (1938)

Tel90. Tél, T.: Transient Chaos. In Directions in Chaos, H. Bai-Lin (ed.), World Scientific, Singapore, (1990)

Tho75. Thom, R.: Structural Stability and Morphogenesis. Addison–Wesley, Reading, (1975)

Tho79. Thorpe, J.A.: Elementary Topics in Differential Geometry. Springer, New York, (1979)

Tin75. Tinkham, M.: Introduction to Superconductivity. McGraw-Hill, New York, (1975)

TLO97. Tanaka, H.A., Lichtenberg, A.J., Oishi, S.: First Order Phase Transition Resulting from Finite Inertia in Coupled Oscillator Systems. Phys. Rev. Lett. **78**, 21042107, (1997)

TLO97. Tanaka, H.A., Lichtenberg, A.J., Oishi, S.: Self-synchronization of coupled oscillators. with hysteretic response. Physica D **100**, 279, (1997)

TLZ02. Tang, X., Lou, S., Zhang, Y.: Chaos in soliton systems and special Lax pairs for chaos systems. Localized excitations in (2+ 1)-dimensional systems. Phys. Rev. E **66**, 046601, (2002)

TM01. Trees, B.R., Murgescu, R.A.: Phase locking in Josephson ladders and the discrete sine-Gordon equation: The effects of boundary conditions and current-induced magnetic fields. Phys. Rev. E **64**, 046205, (2001)

TMO00. Tras, E., Mazo, J.J., Orlando, T.P.: Discrete Breathers in Nonlinear Lattices: Experimental Detection in a Josephson Array. Phys. Rev. Lett. **84**, 741, (2000)

Tod67. Toda, M.: Vibration of a chain with nonlinear interactions. J. Phys. Soc. Jap. **22**, 431–36; also, Wave propagation in anharmonic lattices. J. Phys. Soc. Jap. **23**, 501–06, (1967)

TR85. Tannor, D.J., Rice, S.A.: Control of selectivity of chemical reaction via control of wave packet evolution. J. Chem. Phys. **83**, 5013-5018, (1985)

Tri88. Tritton, D.J.: Physical fluid dynamics. Oxford Science Publ., Oxford, (1988)

TRW98. Tass, P., Rosenblum, M.G., Weule, J., Kurths, J. *et al.*: Detection of $n : m$ Phase Locking from Noisy Data: Application to Magnetoencephalography. Phys. Rev. Lett. **81**, 3291–3294, (1998)

TS01. Thompson, J.M.T., Stewart, H.B.: Nonlinear Dynamics and Chaos: Geometrical Methods for Engineers and Scientists. Wiley, New York, (2001)

TS99. Trees, B.R., Stroud, D.: Two-dimensional arrays of Josephson junctions in a magnetic field: A stability analysis of synchronized states. Phys. Rev. B **59**, 71087115, (1999)

TSS05. Trees, B.R., Saranathan, V., Stroud, D.: Synchronization in disordered Josephson junction arrays: Small-world connections and the Kuramoto model. Phys. Rev. E **71**, 016215, (2005)

TT04. Teramae, J.N., Tanaka, D.: Robustness of the noise-induced phase synchronization in a general class of limit cycle oscillators. Phys.Rev.Lett. **93**, 204103, (2004)

TWG02. Timme, M., Wolf, F., Geisel, T.: Prevalence of unstable attractors in networks of pulse-coupled oscillators. Phys. Rev. Lett. **89**, 154105, (2002)

TWG03. Timme, M., Wolf, F., Geisel, T.: Unstable attractors induce perpetual synchronization and desynchronization. Chaos, **13**, 377, (2003)

TZ05. Tarasov, V.E., Zaslavsky, G.M.: Fractional Ginzburg-Landau equation for fractal media. Physica A **354**, 249-261, (2005)

UC02. Uezu, T., Coolen, A.C.C.: Hierarchical Self-Programming in Recurrent Neural Networks. J. Phys. A **35**, 2761-2809, (2002)

Ula91. Ulam, S.M.: Adventures of a Mathematician. Univ. Calif. Press, (1991)

Ush99. Ushio, T.: Synthesis of Synchronized Chaotic Systems Based on Observers. Int. J. Bif. Chaos **9**, 541–546, (1999)

UMM93. Ustinov, A.V., Cirillo, M., Malomed, B.A.: Fluxon dynamics in one-dimensional Josephson-junction arrays. Phys. Rev. B **47**, 83578360, (1993)

Una94. Ünal, G.: Application of equivalence transformations to inertial subrange of turbulence, Lie Groups Appl. **1**, 232–240, (1994)

Una97. Ünal, G.: Constitutive equation of turbulence and the Lie symmetries of Navier–Stokes equations. In Modern Group Analysis VII, Editors N.H. Ibragimov, K. Razi Naqvi and E. Straume, Trondheim, Mars Publishers, 317–323, (1997)

Vaa95. Van der Vaart, N.C. *et al.*: Resonant Tunneling Through Two Discrete Energy States. Phys. Rev. Lett. **74**, 47024705, (1995)

Ver838. Verhulst, P. F.: Notice sur la loi que la population pursuit dans son accroissement. Corresp. Math. Phys. **10**, 113-121, (1838)

Ver845. Verhulst, P.F.: Recherches Mathematiques sur La Loi D'Accroissement de la Population (Mathematical Researches into the Law of Population Growth Increase). Nouveaux Memoires de l'Academie Royale des Sciences et Belles-Lettres de Bruxelles, **18**(1), 1-45, (1845)

Ver98. Vergassola, M.: In Analysis and Modelling of Discrete Dynamical Systems, eds. D. Benest & C. Froeschlé, 229, Gordon & Breach, (1998)

Via97. Viana, M.: Multidimensional non–hyperbolic attractors. Publ. Math. IHES **85**, 63-96, (1997)

Vre96. Van Vreeswijk, C.: Partial synchronization in populations of pulse-coupled oscillators. Phys. Rev. E **54**, 5522, (1996)

Wal81. Walther, H.-O.: Homoclinic solution and chaos in $\dot{x}(t) = f(x(t-1))$. Nonl. Ana. **5**, 775–788, (1981)

Wal89. Walther, H.-O.: Hyperbolic periodic solutions, heteroclinic connections and transversal homoclinic points in autonomous differential delay equations. Memoirs AMS **402**, (1989)

Wal95. Walther, H.-O.: The 2–dimensional attractor of $x'(t) = -\mu x(t) + f(x(t-1))$, Memoirs AMS **544**, (1995)

Wat99. Watts, D.J.: Small Worlds. Princeton University Press, Princeton, (1999)

WBB94. Wiesenfeld, K., Benz, S.P., Booi, P.A.: Phase-locked oscillator optimization for arrays of Josephson junctions. J. Appl. Phys. **76**, 3835-3846, (1994)

WC02. Wang, X.F., Chen, G.: Synchronization in small-world dynamical networks. Int. J. Bifur. Chaos **12**, 187, (2002)

WCL96. Whan, C.B., Cawthorne, A.B., Lobb, C.J.: Synchronization and phase locking in two-dimensional arrays of Josephson junctions. Phys. Rev. B **53**, 12340, (1996)

WCS96. Wiesenfeld, K., Colet, P., Strogatz, S.H.: Synchronization Transitions in a Disordered Josephson Series Array. Phys. Rev. Lett. **76**, 404407, (1996)

WCS98. Wiesenfeld, K., Colet, P., Strogatz, S.H.: Frequency locking in Josephson arrays: Connection with the Kuramoto model. Phys. Rev. E **57**, 15631569, (1998)

Wei05. Weisstein, E.W.: MathWorld–A Wolfram Research Web Resource. http://mathworld.wolfram.com, (2005)

WFP97. Witt, A., Feudel, U., Pikovsky, A.S.: Birth of Strange Nonchaotic Attractors due to Interior Crisis. Physica D **109**, 180-190, (1997)

Whi27. Whittaker, E.T.: A Treatise on the Analytical Dynamics of Particles and Rigid Bodies, Cambridge Univ. Press, Cambridge, (1927)

Wig90. Wiggins, S.: Introduction to Applied Dynamical Systems and Chaos. Springer, New York, (1990)

Wik05. Wikipedia, the free encyclopedia. http://wikipedia.org, (2005)

Wil00. Wilson, D.: Nonlinear Control, Advanced Control Course (Student Version), Karlstad Univ., (2000)

Win01. Winfree, A.T.: The Geometry of Biological Time. Springer, New York, (2001)

Win67. Winfree, A.T.: Biological rhythms and the behavior of populations of coupled oscillators. J. Theor. Biol. **16**, 15, (1967)

WMW01. Weibert, K., Main, J., Wunner, G.: Periodic orbit quantization of chaotic maps by harmonic inversion. Phys. Lett. A **292**, 120, (2001)

WS91. Watkin, T.L.H., Sherrington, D.: The parallel dynamics of a dilute symmetric Hebb-rule network. J. Phys. A: Math. Gen. **24**, 5427-5433, (1991)

WS97. Watanabe, S., Swift, J.W.: Stability of periodic solutions in series arrays of Josephson junctions with internal capacitance. J. Nonlinear Sci. **7**, 503, (1997)

WS98. Watts, D.J., Strogatz, S.H.: Collective dynamics of small-world networks. Nature **393**, 440, (1998)

WSS85. Wolf, A., Swift, J.B., Swinney, H.L., Vastano, J.A.: Determining Lyapunov Exponents from a Time Series. Physica D, **16**(3), 285-317, (1985)

WSZ95. Watanabe, S., Strogatz, S.H., van der Zant, H.S.J., Orlando, T.P.: Whirling Modes and Parametric Instabilities in the Discrete Sine-Gordon Equation: Experimental Tests in Josephson Rings. Phys. Rev. Lett. **74**, 379382, (1995)

WWV01. Winckelmans, G.S., Wray, A., Vasilyev, O.V., Jeanmart, H.: Explicit filtering large-eddy simulation using the tensor-diffusivity model supplemented by a dynamic Smagorinsky term. Phys. Fluids, **13**, 1385–1403, (2001)

WZ98. Waelbroeck, H., Zertuche, F.: Discrete Chaos. J. Phys. A **32**, 175, (1998)

YAS96. Yorke, J.A., Alligood, K., Sauer, T.: Chaos: An Introduction to Dynamical Systems. Springer, New York, (1996)

Yeo92. Yeomans, J.M.: Statistical Mechanics of Phase Transitions. Oxford Univ. Press, Oxford, (1992)

YL52. Yang, C.N., Lee, T.D.: Statistical theory of equation of state and phase transitions I: Theory of condensation. Phys. Rev. **87**, 404–409, (1952)

YL96. Yalcinkaya, T., Lai, Y.-C.: Blowout bifurcation route to strange nonchaotic attractors. Phys. Rev. Lett. **77**, 50395042, (1996)

YL97. Yalcinkaya, T., Lai, Y.-C.: Phase Characterization of Chaos. Phys. Rev. Lett. **79**, 3885–3888, (1997)

YML00. Yanchuk, S., Maistrenko, Yu., Lading, B., Mosekilde, E.: Effects of a parameter mismatch on the synchronizatoin of two coupled chaotic oscillators. Int. J. Bifurcation and Chaos **10**, 2629-2648, (2000)

YC98. Yang, T., Chua, L.O.: Control of chaos using sampled-data feedback control. Int. J. Bif. Chaos **8**, 2433–2438, (1998)

Zas05. Zaslavsky, G.M.: Hamiltonian Chaos and Fractional Dynamics. Oxford Univ. Press, Oxford, (2005)

ZF71. Zakharov, V.E., Faddeev, L.D.: Korteweg-de Vries equation is a fully integrable Hamiltonian system, Funkts. Anal. Pril. **5**, 18–27, (1971)

Zin93. Zinn-Justin, J.: Quantum Field Theory and Critical Phenomena. Oxford Univ. Press, Oxford, (1993)

ZK65. Zabusky, N.J., Kruskal, M.D.: Interactions of solitons in a collisionless plasma and the recurrence of initial states. Phys. Rev. Let., **15**, 240–43, (1965)

ZS72. Zakharov, V.E., Shabat, A.B.: Exact theory of two-dimensional self-focusing and one-dimensional self-modulation of waves in nonlinear media. Sowiet Physics, JETP **34**, 62–69, (1972)

ZTG04. Zumdieck, A., Timme, M., Geisel, T.: Long Chaotic Transients in Complex Networks. Phys. Rev. Lett. **93**, 244103, (2004)

Index

International Series on
INTELLIGENT SYSTEMS, CONTROL AND AUTOMATION: SCIENCE AND ENGINEERING

Editor: Professor S. G. Tzafestas, *National Technical University, Athens, Greece*

International Series on
INTELLIGENT SYSTEMS, CONTROL AND AUTOMATION: SCIENCE AND ENGINEERING